THE SELECTED PAPERS OF
WILLIAM N. LIPSCOMB, JR.
A Legacy in Structure–Function Relationships

ICP Selected Papers

ISSN: 2045-0990

Published

Vol. 1 The Selected Papers of Sir Alan Fersht: Development of Protein Engineering
edited by Alan Fersht and Qinghua Wang

Vol. 2 The Selected Papers of Denis Noble CBE FRS: A Journey in Physiology Towards Enlightenment
edited by Denis Noble, Zhu Chen, Charles Auffray and Eric Werner

Vol. 3 The Selected Papers of George E. Andrews (With Commentary)
by George E. Andrews and edited by Andrew V. Sills

Vol. 4 The Selected Papers of William N. Lipscomb, Jr.: A Legacy in Structure–Function Relationships
edited by Jianpeng Ma

ICP Selected Papers — Vol. 4

THE SELECTED PAPERS OF WILLIAM N. LIPSCOMB, JR.

A Legacy in Structure–Function Relationships

Editor

Jianpeng Ma

Baylor College of Medicine and Rice University, USA

ICP Imperial College Press

Published by

Imperial College Press
57 Shelton Street
Covent Garden
London WC2H 9HE

Distributed by

World Scientific Publishing Co. Pte. Ltd.
5 Toh Tuck Link, Singapore 596224
USA office: 27 Warren Street, Suite 401-402, Hackensack, NJ 07601
UK office: 57 Shelton Street, Covent Garden, London WC2H 9HE

Library of Congress Cataloging-in-Publication Data
Lipscomb, William N.
The selected papers of William N. Lipscomb, Jr. : a legacy in structure-function relationships / editor, Jianpeng Ma.
p. ; cm. -- (ICP selected papers ; v. 4)
Includes bibliographical references.
ISBN 978-1-84816-840-4 (hardcover : alk. paper)
I. Ma, Jianpeng. II. Title.
[DNLM: 1. Lipscomb, William N. 2. Chemistry--Anecdotes. 3. Chemistry--Collected Works. 4. Chemistry--Festschrift. 5. Faculty--Anecdotes. 6. Faculty--Collected Works. 7. Faculty--Festschrift. 8. Biology--Anecdotes. 9. Biology--Collected Works. 10. Biology--Festschrift. 11. Structure-Activity Relationship--Anecdotes. 12. Structure-Activity Relationship--Collected Works. 13. Structure-Activity Relationship--Festschrift. WZ 305]

540.92--dc23

2012051674

British Library Cataloguing-in-Publication Data
A catalogue record for this book is available from the British Library.

In-house Editor: Veronica Low

Printed by Fulsland Offset Printing (S) Pte Ltd Singapore.

Preface

This book was originally planned as a volume dedicated to the celebration of the 90th birthday of Professor William N. Lipscomb, also known as the Colonel. We did celebrate his 90th birthday in Cambridge, Massachusetts, near the end of 2009. It is very sad that he passed away on April 14, 2011 at the age of 91, while the manuscript of the book was being finalized. The completion of this book involved enormous efforts from all the contributors, from getting permission to reproduce the original scientific articles to writing their personal commentaries. One of the most important contributions is from Jim Lipscomb, the son of Bill Lipscomb. He contributed the first chapter from a family point of view. In that chapter, one can see many aspects of Lipscomb's life as a scientist, son, and father. It also has many anecdotal stories of Lipscomb from the early days of his life and career. In my personal opinion, these are the types of story that are important for developing a more in-depth and complete view of Lipscomb — what made him the person that we all knew.

One of the aspects of Lipscomb that is not reflected in depth in this book is his musical achievement. He was a very accomplished musician and often said that his scientific accomplishments were related to his practice of music. He was a classical clarinetist in chamber groups and had been a principal clarinetist in several orchestras as well. Hopefully, there will be more publications on this issue in the future.

The order of chapters in this book correlates roughly to the order of the contributors' association with Lipscomb. Thus, Jim Lipscomb's chapter naturally comes first. The book also contains photos collected from various sources that the editor had access to. They are certainly not a complete record, but they do reflect some of the key moments of Lipscomb's life.

For many years, Professor Lipscomb was the most revered living icon in chemistry and biophysics after the two-time Nobel Laureate Linus C. Pauling, of whom Lipscomb was a protégé. He himself won a Nobel Prize in 1976 for his research on the structures of boron molecules and on chemical bonding. There are several Nobel Laureates among his scientific descendants, and more may come in the future. This multi-generational lineage of Nobel Laureates is very rare, if not completely unique, in the history of science. It truly reflects the importance of mentorship, originality, and creativity in scientific research. One of the reasons that Lipscomb's research accomplishments cover such a wide range of topics is that he had no barrier whatsoever in switching fields. He did not have any prejudice regarding techniques, and would use, for instance, both computational and experimental techniques as necessary. He always said, "I pick an important problem, then try to use whatever method I can put my hands on to solve it." Unfortunately, this is not a trait shared by most scientists nowadays.

In daily life, Lipscomb was a very kind person with a great and wry sense of humor.

The death of Professor Lipscomb is a great loss to his former associates and to the scientific community in general. We hope that this book, which contains anecdotal comments from many of Lipscomb's former associates and reproduces a number of original publications, can serve as a memorial volume for this larger-than-life figure in our personal and professional lives. We, all the contributors to this book and many other associates of Lipscomb not included here, believe that Lipscomb's influence will project far into the future and continue to inspire generations of scientists.

Jianpeng Ma, Ph.D.
Lodwick T. Bolin Professor in Biochemistry
Department of Biochemistry and Molecular Biology
Baylor College of Medicine and
Department of Bioengineering
Rice University
Houston, Texas

Contents

Papers Reproduced

Chapter 1

James S. Lipscomb

Biography

I am Bill Lipscomb's son.

Physics was my first love fueled by lectures at home from Bill, and I even got a physics B.S. degree. But computer science was my next and lasting love. Bill Lipscomb mentioned just once that the IBM 1620 computer in the Harvard Chemistry Department lobby was available. A year later I said yes, I would like to use it. Some of our conversations traded sentences on this time scale. A graduate student showed me in about ten minutes how to compile and run a program, and I was on my own. I was last priority, having to leave if anyone approached. The hours between 3 a.m. and 6 a.m. were for me, as the tough people who could stay up late or get up early were not around. Bill and I shared a car, which I had to return in time for work. My first publications were computer art done in college using the 1620.

Bill gave me other opportunities. One summer I translated 2,000 lines of 7094 assembly language to Fortran for Bill, and the program worked too, after graduate student George Reeke corrected my oversight that a revised subroutine call needed the rows and columns interchanged in a triangular matrix. Another summer I programmed molecular dipole moment calculations with Bill's former student Don Boyd at Eli Lilly & Co.

I flunked out of the Ph.D. track temporarily. My advisor was the sink-or-swim type. For years I treaded water, upping my master's thesis to a dissertation in my eyes if in no one else's. Signed by a doctoral committee as a dissertation, it was put on a shelf while I took the required courses to "prepare" me and passed the written and oral doctoral exams.

IBM is where I have worked since. Some of my projects there have been:

1. A recognizer to understand strokes made by a pen on a digitizing tablet, much as on a pocket organizer or on a cell phone nowadays.
2. An early virtual reality system, one of the first in which people could interact.
3. A stereoscopic and head-mounted display for a computer assisted design system.

4. A panoramic display as a plug-in or as an Active X control in a web browser.
5. Rich media (panoramas, animations, video, etc.) in a web browser using Java.
6. TV set-top box purchase dialogs.
7. A video storage and search system used by CNN.
8. Applied virtual worlds. See http://qp.research.ibm.com/LotusQuickr/sldemos/Main.nsf or search for Federal Demo Virtual Worlds in Second Life, whenever Google indexes it.

With this book chapter I have 72 publications, consisting of 40 papers, 25 patents, 5 computer art publications, and 2 records on which I was a back-up musician playing the French horn.

In conclusion: "Do the children of Nobelists in fact ever amount to anything? Many, of course, do not." A published answer: "But there are at least five sets of laureate parents and offspring on the Nobel roster." No other measure of success is mentioned.[1]

A Family Perspective

1.1. Scientific Aggression as a Way of Life

1.1.1. *Bill's ultimate scientific aggression*

Not in the usual way was Bill aggressive. A more usual aggression goes against the person, scientific aggression goes against the facts. More on that difference later. First, to Bill.

Bill's scientific aggression was the force of science, applied even in everyday life where in other hands it may not be effective. Bill was not a part-timer. By applying scientific attitudes at all times he strengthened his commitment to science at work, expanded the scope of problems that he could address with science, and improved his skill to do so. It was a conscious decision by Bill, a design to make himself both a better scientist and a better person.

Eight things make up Bill Lipscomb's ultimate scientific aggression in my view:

1) Learn from the masters — a basis
2) Be tough, resourceful, independent
3) Work more hours
4) Step around the person to address issues technically
5) Do not learn alternatives to scientific argument
6) Be correct rather than modest or immodest
7) Collect fields of study
8) Set intermediate goals — the most important

There is little wisdom in the list itself. We already do most of this and do it well, we think. Bill's wisdom was in how to push these to an ultimate extreme at all times to make his ultimate scientific aggression his exceptional life.

1) Learn from the masters — a basis

Some people must have help and guidance to succeed. Others do better without help, because help may make them weak and dependent. Bill became stronger both ways, when he had help and when he did not.

In high school Bill visited the famous mathematician Fritz John, who lived two houses down the street, for help in figuring out what calculus was.

In college Professor MacIntosh drove Bill to meetings of the American Chemical Society. Also,

> ... Bertram Ramsey used to help me read some elementary books in quantum mechanics, and I bought Douchman's book, ... called *Quantum Mechanics*, and I worked on it as a separate project, and Bertram Ramsey helped a bit.

Bill entered the California Institute of Technology as a physics graduate student but soon switched to chemistry under the influence of Linus Pauling.

> [From] Linus Pauling, I learned a lot of things. ... I thought the worst thing that could happen to me would be if I published something wrong. That is not the worst thing. [Pauling told me that] the worst thing that can happen to you is [that] you publish something that is uninteresting. And Pauling was sometimes wrong about things. Then he went on to something else. And he says to think of lots of ideas and throw away the bad ones. And he didn't always succeed at throwing away the bad ones. He ... published them occasionally. But he had so many good ideas that he really changed chemistry.
>
> Also, the scientific method is not what people think. It's not just the systematic collection of data, and then you reason it out and figure out what the principle is, and then you test it and all that sort of stuff. That's wrong. What you do is you learn all you can around the problem and then make a guess. [Pauling] calls it the "stochastic method" to make it sound good. And then see if the guess is right by testing. ...
>
> Another thing I learned [from Pauling] was that if you want to find something interesting to do don't work on the same field that other people are working on, work on the borderline between that field and something else, or work on something that other people are not working on. ... So, that means ... we often would extend beyond the structure to theoretical methods, to chemical methods, to biochemistry, [and] to molecular biology. ... You could take up a whole new field and bring it to bear on the problem, and most people don't do that. They stick [to] the same field.

Bill told these and more stories about Pauling, and also the wonderful story of how Bill got a grade of C in economics (in the video, not in the transcript), in "Reflections."[2] Jack Dunitz tells a similar story about Pauling's attitude towards mistakes.[3]

Linus Pauling won a Nobel Prize in Chemistry for his work on the nature of the chemical bond, but there was more to learn about it. Bill followed in Pauling's footsteps in a way, winning a Nobel Prize in Chemistry himself for his work on the nature of the chemical bond.

There are echoes from these experiences and quotes throughout the remaining points. Without this step there might be little else.

2) Be tough, resourceful, independent

Bill in his youth was tough, resourceful, and independent, firstly in aggressively seeking out mentors, but also in that even while having a mentor, some of Bill's projects were largely solo ventures.

Bill did chemistry projects in high school without a mentor.

> Mr. Jones [his high school chemistry teacher] realized that I knew all the chemistry in the course, essentially. He said to me, 'I'll give you my college textbooks in first year ... organic chemistry, and all I want you to do is to show up for the final examination.' Well, that was the greatest thing in the world for me because I sat in the back of the room while they were having class and produced an original piece of research (at least it was original to me) bridging organic and inorganic chemistry, making hydrogen from sodium formate and sodium hydroxide. ... I analyzed the hydrogen and found it was pure and wrote it up. I never published it, but [Mr. Jones] certainly made a good decision.

Bill told me that he got a C in high school chemistry. His grade was based on just the final exam. The teacher required students to memorize the first ten elements of the periodic table in order, hardly difficult, but, "I thought that was silly. I could just look it up. So I didn't do it." Bill did some of his own thinking.

Bill's college chemistry professors did not teach quantum mechanics because they did not believe in it, so Bill taught quantum mechanics first to himself and then to other chemistry students.

In graduate school, Bill found that Linus Pauling was not the hand-holding type, as Bill describes in "Reflections."

As a professor, the issue of independence had to do with Bill's students. At the University of Minnesota his research students needed so much help that Bill had time for only a small number. His research needed more students, so he put out the word that he wanted to move. Harvard was the first to call. Students there were on average tougher, more resourceful, and more independent. At any given time as an active Harvard professor he had about 25 research students and postdocs.

3) Work more hours

It frustrated Bill to hear other scientists talk of what they do after work, to get away from work: "You can do that, but you will never be good."

He told me to examine what one does for play and make that one's work. Then, with the mind on work at all times one may become really good.

As I grew up Bill worked about 80 hours a week: twelve hours a day, six days a week, plus nine hours on Sundays. The Lipscomb house was a quiet place. His children became tough, resourceful, and independent. Vacations and holidays were unknown to Bill, except when he attended summer music camp.

Bill noticed that another Harvard professor's car was in its space before he arrived at his usual time of 9:30 a.m. and was still there at night when he went home each night at 9:30 p.m. In frustration Bill said to his first wife, Marydell, "How can I compete with that?"

For a time Bill was required to be the chairman of the Chemistry Department at Harvard. None of the other Harvard professors wanted the job. They had their own forms of ultimate scientific aggression, which chairmanship duties would dilute, so, as Bill told me, they passed the chairmanship around like a hot potato. He took out his frustration in his usual way, by advancing science: "If they are going to make me be chairman, I'll just have to do something with it."

So, Bill sought construction funds for the Conant Chemistry Building, or maybe for the Science Center. Neither of us could remember. Harvard has friends who fund grand adventures destined to bring great change. They are followers. First, one leader must establish that destiny with a down payment. Bill scheduled lunch with Henry Cabot Lodge. Before they sat down, before Bill could speak, Henry Cabot Lodge looked at him and said, "Bill, if I were you, I would not worry about that new building." When Bill told me this he smiled, and said, "When Henry Cabot Lodge tells me not to worry, I don't worry." Henry Cabot Lodge shows that one can have both ultimate aggression in one's domain and the manner of a gentleman.

Bill managed his limited time in part according to an inflexible priority list:

1) Family
2) Students
3) Everyone else

One could be the most important person, but if a student walked in the office, we shall talk later. When I called or visited Bill he was never busy. Except only once, when I telephoned his office and he said that the faculty and a student were in the middle of a Ph.D. exam, and even then he did not tell me to call back, he asked.

Once, when Bill had the flu, he headed for bed with some molecular models, the kind with the plastic straws for the bonds and flexible rubber connectors for the atoms. A few days later he emerged to write his landmark paper on the diamond-square-diamond transformation for boron cage polyhedral rearrangement.[4] What did we accomplish when we were sick with the flu?

When Bill was 87 years old, he was still working all day every day. He told me that he was going to a retirement party for a chemist who had just turned 65. He spread his hands in disbelief, "He's giving up."

4) Step around the person to address issues technically

Bill appeared to suppress all personal aggression to channel that energy into scientific aggression. Dealing with personal issues takes time and energy that does not get chemistry done. Avoid personal problems and forget them, perhaps by speaking of them scientifically, squelching discussion by dropping the bomb of an unarguable fact to avoid argument.

Bill used rhetoric to avoid personal conflict. Bill's rhetoric was quiet, concise, and lacking personal confrontation except through bare facts. Consider when you heard Bill give you his heart, and then reconsider that perhaps really he gave you a fact from which you inferred his heart. Example: "He's giving up," just above. Then you may have found that it is hard to argue with a fact or with your own conclusions from it.

Even better for Bill was to avoid personal issues to start with. Bill's first wife Marydell was a gifted conversationalist who freely spoke her heart to all. Bill told her, "You must never do that. They will get a handle on you."

And which of us has not revealed too much, a weakness, and been skewered on it? But as Marydell spoke her heart, all who spoke with her loved her. Bill's admonition was not meaningful to her. Bill divorced Marydell for reasons that he never shared.

Years ago when a newspaper asked for a comment on the Russians stealing the secrets of the atomic bomb, Bill told them something like, "I have been trying for years to give away secrets in my chemistry classes and have met with only indifferent success." He expected his implications, what he was really saying here, to be understood.

A great educator can bridge the complex and the simple. Bill used this ability to turn what might be normal conversations into technical discussions. Three examples:

It seemed that there might be big money in high-energy boron zip fuels for jet engines. A major American company went around to groups working with boron to destroy them. They tried to steal patent rights to some boron compounds from the University of Minnesota. They also got Bill to commit to a larger research group at the University of Minnesota with the promise of additional funding, which they canceled at the last minute. Bill came home and said to his young son, "I'm going to teach you about business."

He easily lifted the boy up on the kitchen counter and held out his arms. "Jump," he said, "and I'll catch you." His son had never jumped from this high onto something so hard. He looked to his father. "Jump," Bill said, "and I'll catch you," waving his outstretched arms. Bill stepped back as young Jimmy crashed to the floor. Jimmy looked up with big, saucer eyes and heard, "Now, you know about business."

Postscript: The University of Minnesota, in appreciation for Bill's help on patent rights, found money for his extra incoming students, and years later the company wrote Bill a letter of apology, saying that they were sorry for what they did and that they did not do business that way anymore.

"What is a molecule?" Bill was visiting the home of a professor, who had told his young child that Professor Lipscomb would have an answer to this question. For children so young, talk of atoms and molecular bonds would lose them immediately. Bill was so proud of his answer that when he returned home he told me what he had said:

> Take a salt crystal, cut it in half to get two small salt crystals, and throw one of those halves away. Keep doing this until you have a salt crystal so small that if you cut it in half again, the two halves will not be salt. That is a molecule. (Approximate quote)

When I was young, a noted mathematician was to visit. Bill had an idea for a trick, so he coached me on my lines. I asked the mathematician, "Is it five plus seven is eleven or five plus seven are eleven?" The mathematician replied, "Five plus seven is eleven." I smiled, "Five plus seven is twelve."

Bill's technical, aggressive reach extended to the arts. He played the Mozart Clarinet Quintet with the Asam Quartet in 1980 in Lindau, the poster for which appears in Figure 1.1. The following review of that performance is all the more impressive if one has experience with music reviewers:

> The slim American Nobel Prize winner, William N. Lipscomb, . . . was as relaxed as can be as he played Mozart's clarinet quintet with the Asam Quartet from Munich. Like a professional! He filled the role exactly that Mozart had assigned to the clarinet: not as a solo part, but an inner partnership, an integrated clarinet string combination. Striking was the beautiful tone formation of the clarinetist, the evenness of the rise and fall of the musical phrases, and the extended breath control, as if this were the easiest thing in the world! Also the triplet runs were no problem for Lipscomb. Consequently, one could listen with ease to the full excitement of the precious sweetness of the music.[5]

Once, at a Benny Goodman jazz concert in downtown Boston, Bill occasionally burst out laughing. I noticed that when he laughed another instrumentalist was playing the lead, and Goodman turned and pointed his clarinet at the lead player. Later Bill explained that the lead player was mischievously inching up the key changes to bring in more sharps or flats and that when Goodman stared him down the lead player backed off to a simpler key. Nobody else had laughed. It was over their heads.

Figure 1.1 Poster for Bill's concert with the Asam Quartet, 1980. Poster scanned by James S. Lipscomb. Creative Commons License above applies.

When there is a game to be played, tennis, softball, ping pong, chess, or double solitaire, Bill always played to win. Bill also played to win his personal fights, but not by fighting.

A policeman pulled Bill over while he drove down Massachusetts Avenue in Cambridge, MA. Bill had not slowed down for an intersection. The law is that one must slow down at every intersection. This street was one-way, exiting Massachusetts Ave. It was the end of the month. Bill told me that he would pay the ticket. He said to think of it as making a contribution to local government. He was trying to convince himself to put this behind him. His rationalization was a kind of white lie, the most that Bill ever lied. One giveaway is that he spoke wistfully. We shall hear him speaking wistfully again.

This might be considered passive-aggressive behavior. Bill handled attacks in this way when he can. By doing nothing his victory was to give science an extra day of his work.

In England the salesman at the tailor shop measured Bill for a new suit. It became clear that Bill would pay in dollars. The salesman became concerned that delay in converting dollars, which were in decline, to pounds might reduce his commission. The salesman pushed the American out the door, literally. A normal person might have argued that this is hardly the way to treat an honorary don at the town's university. Bill, rather, made the facts known around the high tables. Later, at home near Boston, a package arrived from the tailor shop with the completed suit and a request for payment at Bill's convenience. (I do not recall if this story is from Bill's visit to Oxford University or to Churchill College, Cambridge, UK, where Bill was an overseas fellow.)

At the Hermitage Museum in Russia Bill asked if he could see a section that he had heard about, but which was not usually open to the public. The official looked Bill's slim, athletic figure up and down before saying no. A normal person might have argued. Bill was not normal and does not engage in non-scientific argument. Bill told me that he believed that the head-to-toe inspection was a way of communicating that he was not fat enough. He explained that being fat was a status symbol in Russia, reminding me of news footage of Politburo members. I mentally filled in the blanks, as one must with Bill, that fat meant rich and powerful enough to buy too much food and not do manual labor. Bill may not argue, but he may later tell a story, complete with an educational insight.

But what of ultimate confrontation, when a man is pushed so hard that he must respond? Will not the quiet man be made to shout back, and give the lie to stories of his non-confrontational character? I have been there. Bill throws up his hands, says, "Oh, for heaven's sake!" and leaves.

The United States federal government kicked Bill out of boron chemistry. Senator Mike Mansfield got an amendment passed that stopped the military from funding research with no acknowledged military application. Other government agencies would not fund Bill's boron research. He spoke to me of losing an entire career of research into the nature of the chemical bond by way of boron chemistry. I shall not summarize. Here is everything Bill ever said over the years on this subject: "Nowadays, research must be — *relevant*. The nature of the chemical bond is not — *relevant*."

Once the amendment passed, the Office of Naval Research wrote to Bill suggesting that he withdraw his request for funding, because the inevitable refusal would reflect badly on him. Bill wrote back that no, he would not, and that their funding refusal would reflect badly not on him but on them, which is both logical and true. Somewhat to the government's credit, after Bill won the Nobel Prize in Chemistry for his boron work, a year or so of additional funding was made available to him.

5) Do not learn alternatives to scientific argument

A good way to expand scientific discourse to solve people problems is to have little alternative.

My sister Dorothy asked for Bill's help with decisions that were too hard for her make. The difficulty was that the alternatives were equal. Bill said something like, "I spend my life solving problems, but I cannot solve Dorothy problems. They have no solution."

Bill was evidently unaware that this is a well-known management principle that in organizations hard choices, where the better alternative is not clear to lower-level employees, are likewise kicked upstairs to higher management, who are less qualified to decide than those close to the problems, but alternatives being nearly equal, it matters little which way the decision goes. Bill was unaware, I believe deliberately, that he chose not to study books on management. How else to explain this odd gap in his otherwise wide knowledge of art, music, etc.? This would be consistent with Bill honing and strengthening his own scientific approach to human problems.

Bill visited another lab where a student asked about a problematical result. Bill replied that yes, he knew why. In this case the samples were impure. But when the student replied, "Well, I did not get that out of the paper," Bill did not know how to respond. The student's statement did not logically follow. The explanation not being in the paper was the reason that the student asked. The student had an idealized and false view of the refereed scientific paper as the only trusted source of scientific knowledge. It is not hard in such a case to employ the rhetorical technique of agreeing with your adversary, and then to logically follow through the consequences of that agreement to show their position to be untenable. Again, Bill did without such rhetorical skills, and I believe his choice to be deliberate.

6) Be correct rather than modest or immodest

I have heard that Bill was considered by many to be immodest. But I conclude that he sought only to be correct, which usually sounds immodest, though sometimes not. He did not care.

In "Reflections," Bill could not tell the story of his big baseball accomplishment without a technical description of the exact variant of baseball being played, so that the accomplishment is not oversold or undersold. Later in that presentation, Bill gives his views of who discovered the structure of the protein alpha helix without regard for whose toes he might be stepping on, focused only on attributing credit correctly.

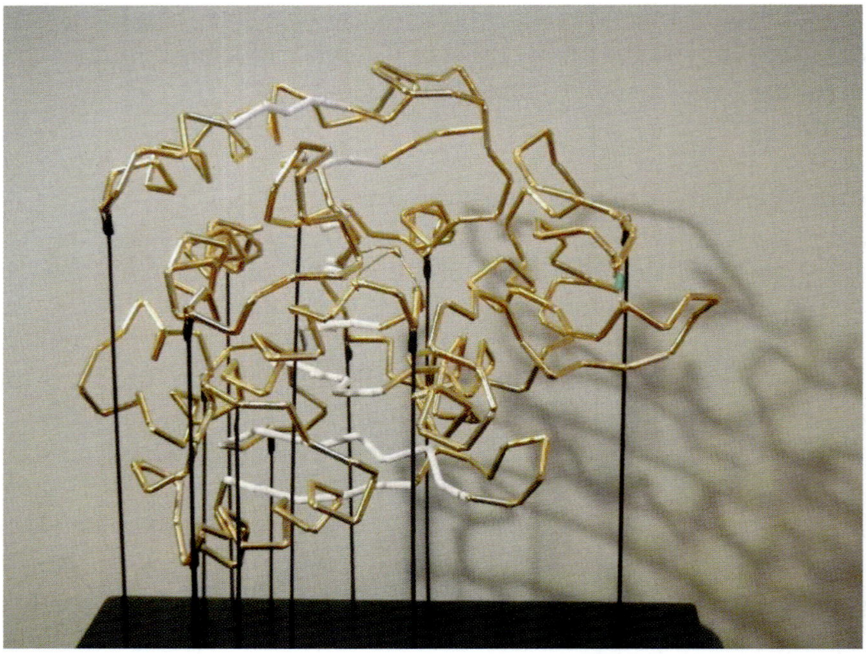

Figure 1.2 Carboxypeptidase A. Photo by James S. Lipscomb. Creative Commons License applies.

Scientific accuracy can sound immodest.

Bill's sense of humor may not have helped with this perception of immodesty. Bill's group determined the structure of carboxypeptidase A, the largest molecule whose structure was determined at the time. Bill had a desktop brass model made (by Kathleen Seyfarth) of the overall shape, see Figure 1.2.

When Bill showed it, he was always sure to mention, "It's gold plated."

If no question is asked of this, a little prodding: "You're supposed to ask why it's gold plated."

"OK, why is it gold plated?"

"Because solid gold would be too expensive."

This may be the joking side of a serious, scientific observation by Bill. Bill sought to say or hear correct statements that are outrageous, because, he told me, they are most often the most important or insightful.

One occasion for the outrageous but true was teaching. He told me that if by the end of class he did not have his students shouting at him, that he would not consider the class to have been entirely successful.

"The stars are too hot to burn." When Bill read this in an astronomy book[6] in secondary school, it helped to turn the course of his life towards physics. The seemingly outrageous nature of the statement caught his attention. It could be understood only by physics then little known, and which Bill did not know at all.

But seeking the outrageous was the occasional stretch. I have heard Bill quick to stop a conversation that seemed to give him too much credit, so that he could carefully list the contributions of other researchers. Careful correctness was Bill's custom.

So here are modest-sounding statements from Bill to challenge the reader. In the shadow of the Nobel Prize one might expect modesty, especially if one frames Bill as a normal person. But from my experience these statements are no different from Bill's everyday rhetoric. The challenge is to read these statements as ordinary and accurate.

Chemistry Nobel Prize winner Harold Urey had written the letter rejecting Bill's application to graduate school at Columbia University. Bill recalled thinking as he threw away this letter, before he had won the Nobel Prize himself, that there would never be an occasion when he could show the letter to Harold Urey in a way that they could both have a laugh over it.

Two years before winning the prize,

> My sister, Helen, asked me ... if I would win the Nobel Prize and I said, 'Well, I'm sure the work is of that quality but I'm not sure the committee wants to recognize this particular branch of chemistry." ... But I didn't try. A lot of people try to win the prizes and they sort of are not very nice ... I wasn't particularly interested in ... getting into what might be thought of as the politics.

When Bill's students told him that he had won, his first questions were,

> What for? Whom did I share it with? ... because it could have been the early protein work ... I would share it with David Phillips ..., or Wally Higgins or [M. Frederick] Hawthorne if it were the boron chemistry, or [Russell M.] Pitzer if it were the ethane barrier.

And to me, "I am honored to be a reporter for what my students have done."

I told Bill that I thought that his boron work, though good, was not sufficient to win the Nobel Prize in Chemistry and that the Swedish Academy was really honoring him for his three outstanding contributions to chemistry together: nuclear magnetic resonance, boron chemistry, and large molecule crystallography. He replied that he had independently come to the same conclusion, but that the terms of the Nobel Prize in Chemistry are that it must be awarded for a single thing, a single most important chemical discovery or improvement.

I heard of some people asking the same question in a different way, wondering if Bill's nuclear magnetic resonance or large molecule work might win him another Nobel Prize. I asked Bill about this, and his reply was on the surface nothing more than to list many deserving chemists who had not won a Nobel Prize and what they had done, and then to note how few years were available in which these many people could win.

7) Collect fields of study

"Other people move from one field to another. I collect them." Bill was referring to how he did not leave behind his early interests in mathematics and physics but combined them with chemistry to become one of the early physical chemists. It is his way of applying Pauling's advice to work on the borderline between two scientific fields.

It may be hard in the current context of interdisciplinary research at colleges and universities to remember back to the 1950s and 1960s when chemists were chemists and physicists were physicists. As Bill put it,

> A physical chemist is someone to whom physicists give all the honor and respect that they would a chemist, and to whom chemists give all the honor and respect that they would a physicist. (Approximate quote)

He was only half joking.

Bill collected people too. After high school, Bill and mathematician Fritz John continued to correspond, and Bill says that Fritz John was important in Bill's applying mathematics to chemistry.

Bill did not just move from mathematics to theoretical chemistry to structure determination to organic chemistry, etc.; he collected these fields for cross-discipline studies in his research, applying mathematics to all of them, and crossing inorganic and organic chemistry by analyzing hints of partial bonding in organic structures.

On the subject in which Bill was awarded the Nobel Prize in Chemistry, Bill's research group came to better understand partial bonding, where covalent bonds have a deficit or an excess of electrons, including with Crawford and Eberhardt the covalent version of the three-center bond, also explained by Longuet-Higgins.[7] In secondary-school chemistry covalent bonds have two centers, the atom at each end of the bond. In a three-center bond a single bond spans three atoms with one in the middle. Christopher Longuet-Higgins employed a three-center bond to propose the correct structure of diborane, as seen in Figure 1.3.

The two hydrogen atoms in the middle of the structure are midway along two three-center bonds stretching from one boron atom to another, as seen in Figure 1.4.

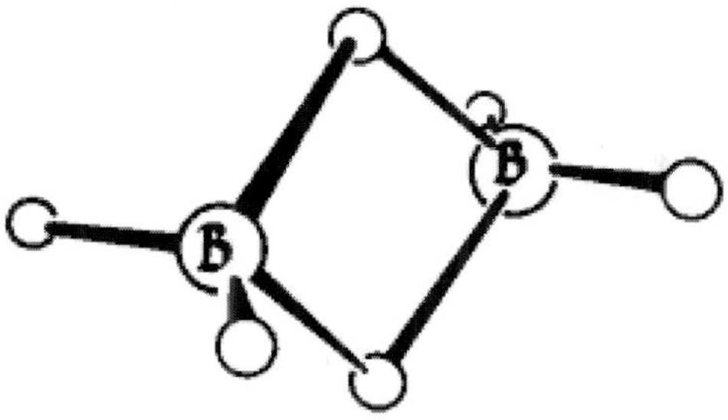

Figure 1.3 Diborane. From William N. Lipscomb, Jr.'s collection, drawn for him as work under a government contract and therefore in the public domain.

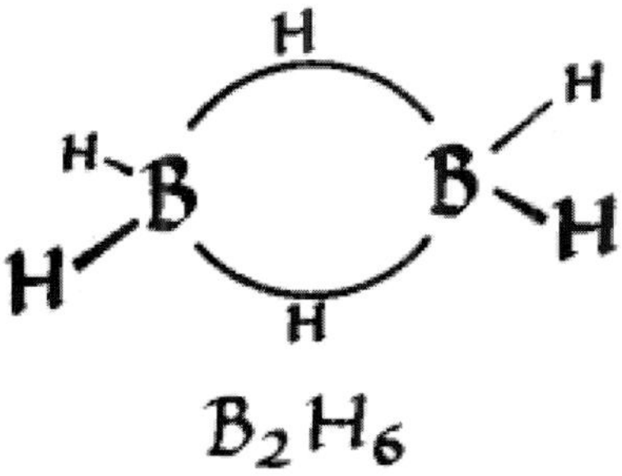

Figure 1.4 Curving lines designate three-center bonds in diborane. From William N. Lipscomb, Jr.'s collection, drawn for him as work under a government contract and therefore in the public domain.

Bill elucidated how this works through electron orbital calculations using formulas by Edmiston and Ruedenberg,[8] and by Boys,[9] each of which, according to Bill, "... when you plug [the situation] into it, gives you the three-center bond, whether you are looking for it or not."

Collaboration between other experts could have discovered this, but Bill was himself an expert in the two fields necessary and had the inclination. A single person knowing two fields has no barrier to collaboration between them. Indeed, Bill advised me that to become the world's expert in something it is sometimes sufficient only to be the only person who is an ordinary expert in just the right two or three things that complement each other. Not just any combination will do, as seen from an Internet search for quotations from Bill: "If your position is everywhere, your momentum is zero."

8) Set intermediate goals — the most important

This is what Bill told me most separates him from other high-performers.

Marydell once asked him, "Bill, what happened to all those other Caltech students? We never hear about them, and some of them are smarter than you."

Bill replied that some of the Caltech students had succeeded at everything they had done, choosing easy problems, so nobody cared. Others had tried problems too hard and had failed. What he meant was that they were wrong in thinking that the ultimate aggression is to seek the most success or challenge. Indeed, thinking of any specific level of success or challenge is the key fault.

Bill later crystallized the point: "I have failed at everything [big] I've tried, but I set intermediate goals." By this he meant that he picked those grand challenges that might well be impossible but that can be broken down into smaller parts, and each part completed would guarantee the biggest, most significant accomplishment.

The following are examples from his research as he them explained to me.

With nuclear magnetic resonance (NMR), his first field, Bill's goal was to determine molecular structures solely from their NMR spectra, which would be much faster than by X-ray diffraction. Bill tells the story in his Nobel lecture:

> I thought then that progress in structure determination, for new polyborane species and for substituted boranes and carboranes, would be greatly accelerated if the 11B nuclear magnetic resonance spectra, rather than X-ray diffraction, could be used.[10]

Bill's dream even came true for some small structures, but bigger molecules were problematic. With NMR structure determination mined out, Bill had to move on.

The intermediate goals he achieved include:

- Developing calculation methods, both empirical[11] and from quantum mechanical theory.[12]
- Calculating by these methods accurate self-consistent field molecular orbitals, which were used to study boranes, carboranes, and most notably to carry out an *ab initio* calculation that produced the first accurate value of the barrier to internal rotation in ethane.[13]
- Laying out a general theory of the chemical shift.[14]

Of his original intention when examining the nature of the chemical bond through studies of boron, Bill wrote:

> My original intention in the late 1940's was to spend a few years understanding the boranes, and then to discover a systematic valence description of the vast numbers of "electron deficient" intermetallic compounds. I have made little progress toward this latter objective. Instead, the field of boron chemistry has grown enormously, and a systematic understanding of some of its complexities has now begun.[15]

Examples of these intermetallic compounds are KHg_{13} and Cu_5Zn_7. Bill told me that of perhaps 24,000 such compounds, the structures of only 4,000 are known and that we cannot predict structures for the others because we do not sufficiently understand the nature of the chemical bond, even now after two Nobel Prizes have been awarded for work on understanding bonding. Compounds of boron and hydrogen were to be only a stepping stone. Calculation time for metallic compounds was out of reach in the 1960s. In Bill's opinion: "For the person who figures this out there may be a big prize."

The intermediate goals he achieved include:

- Pioneering X-ray diffraction at low temperatures[16] at about the same time as parallel work in Fankuchen's laboratory[17] at the then Polytechnic Institute of Brooklyn.
- Calculations that led to an understanding of the three-center bond, as discussed above.
- A detailed examination of partial bonding through "theoretical studies of multicentered chemical bonds including both delocalized and localized molecular orbitals."[18]
- Developing extended Hückel theory with Lawrence Lohr and Roald Hoffmann.[19,20] Hoffmann tells the story of the theory's early history.[21] This theory was extended by Hoffmann[22] and in Bill's lab reconciled with self-consistent field theory by Newton[23] and Boer.[24]

Finally, while researching large biochemical molecules, Bill's early achievements included:

- Determining the structure and function of carboxypeptidase A, the largest atomic structure determined at that time.
- Later, determining the structure and function of aspartate carbamoyltransferase, which became the largest atomic structure determined in its time.

Bill also used this stepwise career plan for his students. For step one, whenever possible, he gave them a small problem that would be publishable, interesting, small enough for a training task, and yet big enough for a Ph.D. Bill told me that finding these problems was one of the hardest things he had to do. Then, for step two, he would give them a big problem or make them part of a big project that might fail, so at worst they could fall back on the results of the first problem. I wonder if they understood the career implications.

A minimum intermediate goal that Bill set for his students was to publish. Bill said that it is well known that students who do not publish in graduate school tend not to publish afterwards. To further motivate his published students to publish again, Bill made them lead authors on publications with his own name last. Exceptions were extreme cases where he had the idea and did 90% of the work, or when the students could not agree on whose name should go first.

1.1.2. *Confronting destructive personal aggression*

Bill, in using scientific aggression in everyday life, had to find ways for it to prevail against a more usual sort of aggression in people, call it destructive personal aggression. First, the difference between the two.

C. P. Snow, in his book, *The Two Cultures and the Scientific Revolution*,[25] hypothesized that the sciences are on the rise; that there are two ways to live that are exemplified by the scientist and by the artist; that these constitute cultures; that communication between them matters; and that effective communication between them is difficult if not impossible. I shall simplify C. P. Snow's two cultures to scientific aggression, as exemplified by Bill Lipscomb, and personal aggression. Personal aggression in my view comes in two forms. Both solve problems by going against the person.

Constructive personal aggression solves problems by going to the person in a positive and constructive way. For example, in *The Last Lecture* by Randy Pausch,[26] Randy tells of advancing computer science. This was achieved technically of course, but also by techniques of salesmanship, teamwork, and thinking quickly on his feet in pivotal meetings. Unlike Bill, Randy talked his way out of his traffic ticket. In Randy's personal contests there is no loser, only two winners. Constructive personal aggression is good, but it is not Bill's way.

Destructive personal aggression solves problems by going to the person in a negative and destructive way. Many consider sociopaths to be plentiful among business leaders. A person who is sued may hire for the fight the meanest lawyer available among the many such to choose from. Some people choose their politicians this way too. Other people are attacked, diminished, or used as things. Facts may be dismissed if they do not align with an imagined reality. Destructive personal aggression is not Bill's way either.

Bill's scientific aggression, in contrast to both of these approaches, solves problems by going to the facts.

The percentage of Harvard students who were personally aggressive in a destructive way towards Bill is not high, but it does seem to have been higher than anything I saw from students at my college. Harvard is the proving ground for some of the future leaders of America, if one thinks of leadership in a certain way. Paraphrasing a bumper sticker, "Nice people do not make history." It is bumper sticker wisdom.

I shall present the students in groups. A conventional presentation would give each group a name, a label to aid memory and organize thought. No such luck here. Bill does not label people. Christian teaching is not to call a man a liar. He must be given a way out. He is rather a man who has lied. Bill goes one step further. He will give only the facts, the data, of what the man has done. If the man did lie, it is for you to give it that label. There are those who would drive a wedge between science and religion. Not Bill.

Bill's telephone rang. I was there. It was a student from the *Harvard Crimson*, the student newspaper. Previously, Bill had given them deep background on a story, the part not to write, but to help make correct the part to be written. They wrote it all. On another day they said that happened years ago, that they had a whole new staff with a whole new outlook. Then they did the same thing. Bill began to speak into the telephone. He said that because of these past two events he would have no further contact with them. He listened. Then he said the same thing. Eventually the conversation ended.

Bill put in his outer office, which was open day and night, his extra books for his chemistry students to borrow. His books began to disappear. His first calculus book, gone. For the pre-meds he set up a special library of the books that they would need. Many of these disappeared too, as the pre-meds undercut each other.

One may wonder how an intelligent person would trust a reporter, twice, or would twice put books out to be stolen. Bill could not work with someone other than as an equal, with the same high standard regarding integrity. He would not adjust. With Bill it was all or nothing.

Bill's project to determine the molecular structure of aspartate carbamoyltransferase was going badly. Don Wiley was not getting crystals. Another graduate student decided to take charge. Behind Bill's back this other student went to each of the students and postdocs on the project to tell them to quit, that it would never work. Each student individually refused. Eventually Don got the crystals and the team got the ACTase structure. Thank you, Don. Mr. Take-Charge-Guy left to become a medical doctor. One day I brought him up in conversation, and Bill offered this dry commentary: "If I were sick, I would not exactly go to him exactly."

It was Bill's way of saying that Dr. Take-Charge-Guy was foolish; his career choice, when combined with his tendency to give up easily, may make him dangerous.

I was using the Harvard Chemistry Department's IBM 1620 computer. Up walked a graduate student, who began to yell at me. I was using too much paper. I had used the equivalent of about 15 sheets. As he continued, I tuned out his words and tuned in the meta-level implications. His rage came not from reason, but from a rationale. I asked Bill about him, and a few days later Bill replied, "He was kicked out of one graduate school after another, until the only place that would take him was Harvard." He paused, then added, "To understand that statement you have to understand Harvard Admissions."

A few years later Bill found a short note in a newspaper that appeared to show this former student on trial for his life in his home country, accused of a double murder, and if I recall correctly, one victim was a clergyman. Most such troubles Bill put behind him.

But one trouble stuck. Bill spoke of it several times. His graduate students needed shelf space. While looking to create some, they found many small boxes filled with old papers in a student office adjacent to Bill's. Out to the trash all these boxes went. The students did not ask Bill. He had planned to use these papers to write his memoirs. Once those papers were gone, it was impossible to do it right.

Most of the conflicts above were with students in the sciences. So one hypothesis is that while Harvard of course selects its science students for their scientific aggression, in some cases they are also selecting for high personal aggression too, as perhaps with some non-science students.

These data suggest a disconnect between Harvard students and faculty. The students, even the science students, are to be future leaders of society, so high personal aggression seems to be valued. The science faculty by contrast are expected only to revolutionize their fields. For them high personal aggression seems not to be a consideration.

Bill may have considered such a hypothesis. The commas in the following quotation from him indicate pauses as he carefully constructed this remark on another occasion: "Harvard, has, historically, been a rich boys' school, which may never have fully integrated the sciences."

In these conflicts there was little to no effective communication between Bill with his scientific aggression and others with their personal aggression, in line with C. P. Snow's hypothesis. There were of course winners and losers. In a way, the power of Bill's scientific aggression won many of these conflicts.

1.1.3. *Molding a life*

I was an underachiever. For a way out I looked to my father, Bill Lipscomb, except that I could not be as smart as he was, or work as hard as he did.

First, a question. I asked Bill what good was a Ph.D.? He replied, "With a Ph.D. you will have a better chance of spending the rest of your life doing what you want to do, instead of what someone else wants you to do." I was sold.

I took part in a campaign to make Bill's scientific aggression my own by re-interpreting Bill's eight points into eleven points that might work for me.

1) Do not ask for hand-holding

Bill might help me with a single homework problem once every few years when I asked. Evidently he was giving his children what he had had, solo responsibility that made us ultimately tough, resourceful, and independent, although my performance in secondary school may have suffered along the way.

No matter. My projects outside of class I did alone, with no mentor. I built a computer grading system for my high school and worked independently on my college art, music, physics, and computer projects, and ultimately became independent at work.

2) Equate play and work

Traditionally, students major in what they enjoy working on. Bill's advice was similar but more precise, to do professionally what one does for play.

In my spare time in secondary school I reread the Mr. Tompkins books, particle physics books, and posed and solved physics problems, so that led me to major in physics at college.

As college ended, I looked back to see that in my spare time I played with computers, so in graduate school I focused on computer science. I was unprepared for this change and so took a few summer computer science courses. Moderating one's reach in response to worry and fear was not part of Bill's advice.

As Bill was sparing with words, I listened also to what Bill did not say.

3) Do not waste dreams on the unimportant

Some students and parents dream of particular colleges. I dreamed instead of a Ph.D. and of becoming a scientist. I saw how going to the modest University of Kentucky catapulted Bill to Caltech, and so understood that the college I chose may not matter. It is what one does there.

The only important educational location is within one's own head.

4) Seek freedom from slavery to grades

As the University of Kentucky was under-equipped for the likes of Bill, it gave him time on his own initiative and outside of class to study physics with Professor Ramsey so he could pursue a degree in physics as well as his major in chemistry.

Likewise at my college was easy enough for me to take time on my own initiative and outside of class to win international awards for computer art and to do more precise physics computations using a computer. I also installed the APL programming language on the college IBM 1130 computer and taught the rudiments of the language to mathematics professor William R. Jones, who later wrote the book, *APL with a Mathematical Accent* with Clifford Reiter.[27] These and other undirected projects catapulted me, a little like Bill, to a good graduate school and then on to a good job that I enjoy. Nowadays many colleges have instituted undergraduate science research programs to similar effect.

At a "better" college I might have been consumed by the struggle for grades and come off the end of the assembly line, better educated in a way, but fashioned the same as the other identical units and distinguishable from them only by being the fourth clown from the left.

5) Put in the time by limiting entertainment

Since I could not free up time for projects by working as hard as Bill, in college I extended my hours by staying out of the TV room and later stayed away from computer entertainment. This created time and induced boredom, which gave me the motivation to pursue other projects.

6) When done do not stop

Like Bill, I was never off duty. Other students sometimes failed to solve problems. I treated such problems as not yet solved. In my sleep I solved the day's problem in recursive functions. The shock of it woke me up. Here too, staying away from entertainment was a first step to shaping the idle time of the mind.

7) Instead of receiving intimidation, project curiosity

When I was young, Bill's wife Marydell told him, "You must talk to your son. You are losing your son." And so, sometimes at the end of Bill's 12-hour days he would lecture his young son. He would show me the Schrödinger equation of the hydrogen atom, and how to calculate the Bohr radius of the hydrogen atom from first principles, or he would just blow off some steam explaining how some of the concepts that his thermodynamics students found so hard were just not that difficult.

In college I faced physics and math problems with the curiosity, flexibility, and focus I learnt from Bill's private lectures. I had at any given time only a single problem in front of me. No one can be intimidated by just a single problem. I did not even realize until after college that other students dropping out of the physics program may have been intimidated.

To win the fight against intimidation is to lose, since that gives intimidation a place at the table, even if an empty place. Fighting intimidation does not produce correct solutions to problems.

8) In the snowstorm, instead of freezing, see as far as one's gaze can penetrate

Related to intimidation is fear. From Bill's advanced science books on my grade-school bookshelf and from his private science and math lectures I learned how to learn what I could and to put aside what I did not understand. Reading books above my level became easy and natural. There were no questions, no tests, and no visible interest from Bill in whether I understood anything or not. I learned not to fear.

9) Advanced does not mean hard

Bill's occasional private science and math lectures after work taught me how quantum mechanics is simpler than classical mechanics because with just a few numbers and variables the situation is described completely, and also how partial differential equations are simpler than ordinary differential equations because variables become constants that one can pretty much ignore.

I saw how these courses would be simple, so of course they were. It is a mystery to me only why such simple courses are called advanced. Perhaps the students bailing out to other majors did not have this outlook.

10) Step around the person to address issues technically

In the two biggest professional troubles of my life, when I showed weakness, I won through quiet, steady accomplishments. Argument, even if seemingly effective, confirms weakness.

Scientific aggression was sufficient. They could not get a handle on me.

11) Have a distant gaze that is short

With both my father and grandfather having lost careers, I understood that even my biggest dreams were only possible directions and options. It is about the journey. My eyes looked down. I saw my shoes.

I contracted my vision to one-week plans and focused on one foot in front of the other every day, uphill if possible. I did not count the years.

1.1.3.1. *Summary*

Personal aggression is found between people in a noisy contest to win the day. Scientific aggression is in the collaboration between people in a quiet contest to win for the world something lasting. The decision to work this way is taken alone in the quiet of the night. The practitioner may not have the eye of the tiger but rather a distant gaze, which may extend little further than to one's shoes.

1.1.4. *A Harvard Hypothesis*

One of the tenants of a Harvard education is that in mixing with some of the world's greatest minds, students may become greater than they might have had they gone to another university. It is a Harvard hypothesis, though it is not unique to Harvard, of course. Bill always sought to make the most of the opportunities he had, and to improve his knowledge and skills by applying the lessons he learned from the great minds he encountered.

Recall from above how in high school and in college Bill sought out the greatest minds available: in high school, mathematician Fritz John, and in college, Bertram Ramsey who believed in quantum mechanics. Combined with Bill's conventional academic skills these additional studies catapulted Bill to a top graduate school, Caltech, and he used his knowledge both there and onward in his career in physical chemistry.

Recall how Bill applied what he learned at Caltech from two-time Nobel Prize winner Linus Pauling (1954 and 1962, whose studies of the nature of the chemical bond Bill extended) about not to being uninteresting. This turned Bill's mind to big problems with a plan to set intermediate goals, and to work on the borderline between fields, which Bill implemented though studying the quantum mechanics of chemistry. Pauling's demonstration that Caltech-quality students can thrive without hand-holding showed the way for Bill's plan to move from the University of Minnesota to Harvard to enable him to enlarge his research group. Recall too how Bill's accomplishments may have outstripped those of his Caltech classmates despite being judged not the smartest among them in a way.

Bill's students also achieved great results as a result of the Harvard hypothesis-style education they received from Bill. Look at the accomplishments of Bill's Nobel Prize for Chemistry-winning students, Roald Hoffmann (1981), Tom Steitz (2009), and Ada Yonath (2009) who worked closely with Bill when she was a postdoc at MIT.

As for myself, I may have had something of a Harvard education at home from Bill through the implementation of his eight-point list above, along with my eleven-point interpretation. Professor Fred Brooks, who graduated from Harvard, at the University of North Carolina at Chapel Hill taught me more when I was a graduate student there.

At the University of North Carolina I was one of the four principal builders of GRIP-75,[28,29] a system on which crystallographers claim this milestone: "This is the first instance in which a complete three-dimensional structure [of a large molecule] was solved with the use of computer graphics alone, without the construction of a physical model,"[30] a neurotoxin in the venom of a sea snake. Another large molecular structure was also solved on GRIP-75 at about the same time.[31] At least two dozen large structures were solved (fitted to their electron density maps) on this system.

I have worked for more than 20 years at IBM, as a research staff member at the IBM T. J. Watson Research Center and later as a senior software engineer.

These achievements of Bill, his students, and his son are data. These data may be consistent with this Harvard hypothesis

1.2. Science Humor

"Humor, in my opinion, is an essential part of presentation in science." — W. Lipscomb, Jr.

1.2.1. *Humor Direct from Bill*

1.2.1.1. *Publication zingers*

Bill planted humor in some of his scientific publications. Here is his unpublished monograph about it:

HUMOR IN SCIENCE

William N. Lipscomb

There are two aspects of the Ig Nobel Prize occasions that appeal to me. The first is that they show that science is fun, and the second is that they recognize accomplishment that could not or should not be repeated. Moreover these prizes occasionally expose the misuse of science, for example, those who advocate the teaching of creationism in science courses.

Having wondered from time to time about whether the Ig Nobel Prizes are viewed with concern by the members of Nobel Prize committees, I asked one such member, who replied, "Oh, we just laugh."

Indeed! Humor, in my opinion, is an essential part of presentation in science. I frequently make a joke or (untranslatable) pun (e.g. there is no vice versa) in my lectures to see if anyone is listening. Some examples from my published papers are as follows:

1) "We wish to thank the Computer for several valuable suggestions during the course of this study." J. Reddy and W. N. Lipscomb, *J. Chem. Phys.* **31**, 610 (1959).
2) "We admittedly made this observation with the benefit of hindsight. This science is known as retrospectroscopy." E. A. Laws, R. M. Stevens, and W. N. Lipscomb, *J. Am. Chem. Soc.* **94**, 4473 (1972).
3) From "Crystal Structure of Beryllium Borohydride": "Concerning the probable gas phase structure, we wish to suggest a close relationship to the appropriate fragment of this solid-state structure, but we dare not do so!" D. S. Marynick and W. N, Lipscomb, *Inorg. Chem.* **11**, 823 (1972).

4) In the early 1970s we frequently superimposed a diffraction pattern of aluminum powder on single crystal X-ray photographs (for calibration). The mounted sample, labeled as Al Powder, was thought by one member of my research group to have been labeled by the name of a previous member of the group. Hence the acknowledgement, "We also wish to thank Al Powder for assistance in the calibration of the unit cell parameters." C. H. Schwalbe, and W. N. Lipscomb, *Inorg. Chem.* **10**, 160 (1971).
5) The occurrence of three references to our structure of B_6H_{10} in a review (W. N. Lipscomb, *Adv. Inorg. Chem. Radiochem.* **1**, 119 (1959)) led to the footnote, "What I tell you three times is true." From the Bellman in Lewis Carroll, "The Hunting of the Snark."
6) "We also thank Alice, the cow, who while nevertheless still living, supplied the pancreatic juice from which our crystals of the enzyme carboxypeptidase were obtained." W. N. Lipscomb, J. A. Hartsuck, G. N. Reeke, Jr., F. A. Quiocho, P. H. Bethge, M. L. Ludwig, T. A. Steitz, H. Muirhead, and J. C. Coppola, *Brookhaven Symp. Biol.* **21**, 85 (1968).
7) "Xue and Lipscomb, personal communication" in Y. M. Chook, J. V. Gray, H. Ke and W. N. Lipscomb, *J. Mol. Biol.* **240**, 498 (1994). A communication to myself!
8) Acknowledgement: "We wish to thank R. Dean Dragsforf, who raised a question leading to our discovery of these mislabelings, for reading the N_2O_2 paper."
9) Perhaps this last correction inspired my comment to a reporter following the announcement of the Nobel Prize in Chemistry for 1976: "I knew that I had written a lot of nice papers about boranes, but I never actually knew that anyone read them."
10) The structure of HCN tetramer is not in doubt: "It is diaminomaleonitrile (DAMN)." B. R. Penfold and W. N. Lipscomb, *Acta Cryst.* **14**, 593 (1961). Of course, we were hoping the structure would be different. The acronym DAMN then appears nine more times in the paper.
11) "We have even ventured a few predictions, knowing that if we must join the ranks of boron-hydride predictors later proved wrong, we shall be in the best of company." W. H. Eberhardt, B. Crawford, Jr., and W. N. Lipscomb, *J. Chem. Phys.* **22**, 989 (1954). The best of company refers to, among others, Linus Pauling and Kenneth S. Pitzer. The styx numbers were introduced in this paper.
12) Visual estimates of X-ray diffraction intensities in 1975 were called a "great leap backwards." Y.-Y. H. Chiu and W. N. Lipscomb, *J. Am. Chem. Soc.* **97**, 2526 (1975).
13) Termination of a low temperature experiment by the "Great Blackout" (*New York Times*. November 10–15, 1965). L. B. Friedman and W. N. Lipscomb, *Inorg. Chem.* **5**, 1752 (1966).
14) "The structure of SiF_4 at $-145°C$ occurred as a result of an attempt to determine the structure of HF in a pyrex capillary." M. Atoji and W. N. Lipscomb, *Acta Cryst.* **7**, 597 (1954).
15) "A possible 'new' boron hydride, isolated on a vacuum line in which silicone stopcock grease was used, turned out to be cyclic $(Me_2SiO)_4$." F. P. Boer and W. N. Lipscomb, unpublished.

16) The footnote, "M. V. King and W. N. Lipscomb, to be unpublished," W. N. Lipscomb, *J. Chem. Phys.* **22**, 986 (1954).
17) The title of "A One Dimensional Wave Equation for Hydrogen Bonds," C. E. Nordman and W. N. Lipscomb, *J. Chem. Phys.* **21**, 2077 (1953) was printed in the list of papers on the journal cover as "A One Dimensional Wave Equation for Hydrogen Bombs." It was a nervous time!
18) "We wish to thank the Office of Navel Research." R. Hegstrom and W. N. Lipscomb, *J. Chem. Phys.* **46**, 4539 (1967). I had written "Naval", and when I saw "Navel" in the proofs, I confess that it was left uncorrected! [But Bill had it corrected for the reprints — JL]
19) "Our crystallographic studies of protein structures, which began in 1962 with Martha Ludwig, were greatly accelerated by the termination in 1964 of my grant from National Science Foundation for research in boron chemistry (resumed in 1977)." *Crystallography in North America*, D. McLachlan and J. P. Glusker, eds. ACA 1983, p. 100. Note that my Nobel Prize was received in 1976.
20) "If, instead of closo, nido, arachno and hypho, the names groucho, chico, harpo and zeppo had been used, the new class would be called gummo." *Boron Chemistry-4*, R. W. Parry and G. Kodama, eds. INPAC (Inorganic Chemistry Division), Pergamon Press 1980 p. 9. It has been my experience that Nomeclature Committees have little or no sense of humor.
21) From "Structure of $(OC)_3$ FeC_8H_8 $Fe(CO)_3$": "As usual, the most naive way of counting electrons around Fe is unbelievably satisfactory." B. Dickens and W. N. Lipscomb, *J. Am. Chem. Soc.* **83**, 489 (1961).
22) From "Complexes of Cylo-Octatetraene with Iron Tricarbonyl": "The three classes of investigators who proposed structures, all incorrect, for the geometry of the COT ring in these complexes are those who prepared the compounds, those who could not await our promised X-ray diffraction results, and those (including ourselves) who did not publish." B. Dickens and W. N. Lipscomb, *J. Chem. Phys.* **37**, 2084 (1962).
23) "Practically everyone knows that most closed-shell molecules are diamagnetic." R. A. Hegstrom and W. N. Lipscomb. *Rev. Mod. Phys.* **40**, 354 (1968).
24) At a Meeting in Landau, I took my seat at lunch beside a lady to whom I introduced myself. She replied. "How do you do, I am Mrs. Heisenberg." Immediately I asked "Are you certain?" She was. Indeed. She was the widow of Werner Heisenberg. The reference is *W. N. Lipscomb: An Aesthetic Aspects of Science.* D. W. Curing, editor. Philosophical Library. New York, 1982. page 1.
25) *Nachrichten aus Chemie und Technik* **19**, 122 (1971). Especially note Figure 1, a blank electron density map, with acknowledgment to Lewis Carroll's "The Hunting of the Snark"; in particular the Bellmans' speech: "He gave them a map they could all understand, a perfect and absolute blank."

1.2.1.2. *Bill's Ph.D. dissertation*

Bill planted these in the Propositions section of his Ph.D. dissertation, p. 71.[32]

11(a) "Research and study at the Institute has been unnecessarily hampered by the present policy of not heating the buildings on weekends."

11(b) "Manure should not be used as a fertilizer on ground adjacent to the Campus Coffee Shop."

1.2.1.3. *A new enzyme*

Figure 1.5 shows a draft of Bill's spoof paper (no. 25 in Bill's list above), published as "Ein neues Enzym" in *Nachrichten aus Chemie und Technik,* **19**, no. 7, p. 122 (1971).

A New Enzyme

In the course of studies of proteolysis of small peptides, we have isolated an enzyme which slowly hydrolyzed substrates such as $(L\text{-}Ala)_4$ (hereinafter abbreviated as LLLL) at the penultimate peptide bond. Even slower hydrolysis at the other two peptide bonds was noted, and was attributed to shifted modes of binding, in which either the C-terminal or N-terminal peptide bond of these tetrapeptides was productively bound at the same active site. Further studies in which D-Ala was substituted for L-Ala at various positions revealed a new function: the fairly rapid change in configuration from D to L primarily at the penultimate amino acid side chain. The predominance of this rate over the proteolysis rate thus led us to regard this enzyme as an invertase rather than a proteinase. As the proportion of D-Ala was increased, a/slow [very] step in the over-all reaction became apparent, after which slow rates of hydrolysis of $(D\text{-}Ala)_4$ were noted.

As this chemical study was being carried out, a parallel X-ray diffraction investigation of single crystals was begun. Because of the slow rate of proteolysis of LLLL, it was possible, by cooling the crystals to $4^{\circ}C$, to obtain single crystal patterns from the enzyme-substrate complex. Single crystal studies of the enzyme which slowly hydrolyzed LLLL and its modified form which

2

hydrolyzed DDDD revealed no apparent differences in the diffraction patterns. After preliminary results showed that this modified form also bound DDDD in the crystalline state, we collected three-dimensional data on both the initial form complexed with LLLL, and the modified form with DDDD. The three-dimensional difference electron density is summarized in Fig. 1, where it is clear that the diffraction patterns were identical. Indeed, the low background of this electron density map shows that the X-ray data on these two complexes were quite accurately measured. The conclusion, that the two complexes are identical except for absolute configuration, seems inescapable. Because the enzyme reverses the chirality of not only its substrates, but also itself, at a slower rate, we have named this enzyme autoinvertase.

Theoretical studies have suggested that the intermediate in the transformation is an excited complex (exiplex) of the polymeric structure (polyexiplex), and that the change was triggered by the bacteriocidal agent, cyclooctene-5-one, used in trace amounts as a preservative. This reagent reacts with the phenylalanine side chains in the wax-like hydrophobic regions, so the excited complex can be regarded as a polyexiplex ball of wax. The long-lived intermediate is stabilized by supra-nonhyperconjugative antidestabilization, and, furthermore, the reaction does obey in detail the elegant rules recently deduced by Woodman and Hoffward. Permutation theory, in the form of an ab initio calculation of the electronic structure of the complex, shows that the whole effect may be attributed to the exclusion principle

$$\psi(1,2) = -\psi(2,1) \qquad (1)$$

where ψ is the angle of pseudorotation, and 1 and 2 are numbers. The physical interpretation is simple: the exclusion principle means that all other explanations are excluded. The reason is, of course, that no other effects have been included.

Quite aside from the modern trend in chemistry, as illustrated by the progress of this study, we have had to abandon further studies of autoinvertase, because of mutual agreement with our supporting Agency that our work is so conclusive that no further studies seemed necessary to them, and, not least, because of the unfortunate physiological effects of this enzyme. My coworkers found an unavoidable tendency to drive on the left side of the road. Subsequent studies in England by Gordon Hartley of Oxbridge

4

showed a tendency for drivers to overturn (rather than the expected, symmetrical, result). Only after a world-wide search have we discovered, in Zurich, a scientist whose interests lie in enzymes and organic mechanisms, and who does not drive an automobile. We have sent the world's supply of autoinvertase to him, in the hope that he will someday write an Epilogue to this study.

Contributions (?) from members (nameless but not blameless) of the add hock committee of the Burgenstock Stereochemical Conference (1970) are gratefully acknowledged. Fig. 1 has ~~previously~~ been plagarized from Lewis Carroll, The Hunting of the Snark, the Bellman's Speech.

William N. Lipscomb
Department of Chemistry
Harvard University
Cambridge, Massachusetts 02138 U.S.A.

Figure 1. Composites of three-dimensional electron density difference map (see text).

Figure 1.5 A draft of Bill's spoof paper "Ein neues Enzym." Typewritten manuscript from William N. Lipscomb, Jr.'s collection. Reprinted with permission.

1.2.1.4. *Periodic Table*

What is so good about boron for studying the nature of the chemical bond can be seen in boron's position in the periodic table. Bill's rendition of the periodic table is shown in Figure 1.6.

We see that boron has one more electron than beryllium and one less electron than carbon. Bill sought to learn more about the nature of the chemical bond through calculation.

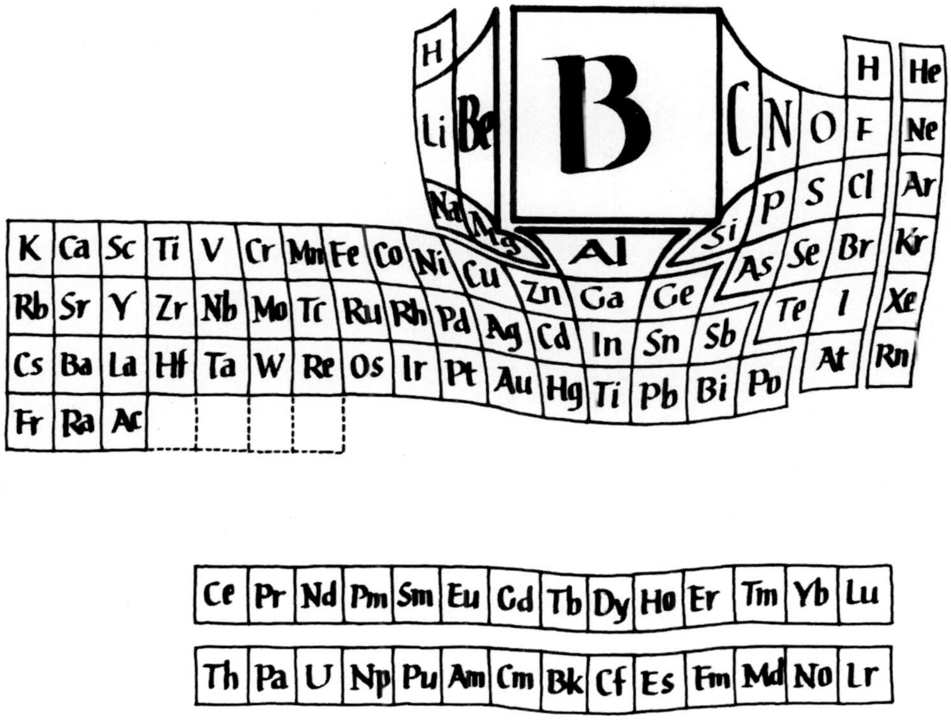

Figure 1.6 Bill's periodic table of the elements. From William N. Lipscomb, Jr.'s collections, drawn for him by his second wife Jean C. Evans as work under a government contract and therefore in the public domain.

But the computing time grows dramatically and beyond the possible very quickly with the number of electrons, placing most carbon compounds out of reach in the 1960s. The fewer electrons the better for calculations, but the chemistry of beryllium and lithium is correspondingly less complex. That leaves boron.

1.2.1.5. *Dow Board*

Bill served in a scientist position on the Dow Chemical Company board of directors from 1982 to 1990. He served on the Environment, Health and Safety Committee and the Public Interest Committee of the Dow board of directors. By 1984 Bill concluded that the group photos looked dull and added a Hawaiian shirt; compare Figures 1.7 and 1.8.

1.2.1.6. *Ig Nobel*

Bill was a frequent and enthusiastic participant in the Ig Nobel Prize ceremonies and other zaniness sponsored by Marc Abrahams of the Annals of Improbable Research.

Issues that feature Bill include:

- Professor Lipscomb makes home-brew tea, *Annals of Improbable Research (AIR)*, Volume 7, Issue 6, Nov/Dec 2001.

Figure 1.7 1984 Dow board with Bill in a Hawaiian shirt. From William N. Lipscomb, Jr.'s photo collection. Reprinted with permission.

Figure 1.8 Bill also brought his regular shirt (back row, third from the left). From William N. Lipscomb, Jr.'s photo collection. Reprinted with permission.

- Professor Lipscomb goes shopping, *Annals of Improbable Research (AIR)*, Volume 6, Issue 3, May/June 2000.
- Nobel Thoughts: William Lipscomb, *Annals of Improbable Research (AIR)*, Volume 5, Issue 4, July/Aug 1999.
- Professor Lipscomb demonstrates how to tie a string tie. Video from *Improbable TV of Improbable Research* Collection #123.

1.2.2. *Humor About Bill*

1.2.2.1. *Songs*

Herbert Beall, songwriter and Bill's Ph.D. student, composed loving and borderline-satirical views of life in Bill's lab at Harvard in the 1960s. Not so evident in the lyrics below, the satirical slant to the songs is much clearer in the way they are sung.

The Clock

The hall clock outside Bill's office was five minutes slow. Bill was tired of being five minutes late to lectures, so he called Harvard Buildings and Grounds to request someone come to fix it. They replied that no, they would not, because it was impossible. The reason, they explained, was that because all clocks were ticked forward once each minute from a central control, if they sent five extra ticks to make his clock correct, then all the other clocks on campus would be five minutes ahead.

No argument from Bill. He avoids personal confrontations, preferring technical replies. He got a ladder and found that by flipping the clock upside down at the half-minute mark he could advance it an extra tick every minute. But that solution was unsatisfying, so he took off the minute hand and went to the vise. Figure 1.9 shows the result of his efforts.

Figure 1.9 The clock, 1976. Photo by James S. Lipscomb. Creative Commons License above applies.

The line "...bent and banged and swore," lampoons the fact that Bill was well known never, under any circumstances, to swear.

There's a clock that hangs high upon the wall and it calls its mind its own,
For it gives wrong time to both great and small and its will is made of stone.
It's a model of life to weaker folk, inspirational is to me.
Though the mightiest man it may provoke, it will never humbled be.

Oh, sing aloud and n'er forget the strong proud clock that won't be set.

Now there came a man with an active mind, who prefers the world his way.
Should he something less than perfection find, he will set it right that day.
He eyed the clock with a visage cold and he stripped it from the wall,
For to set it right he then made bold, 'neath his will that clock must fall.

Oh, sing aloud and n'er forget the strong proud clock that won't be set.

Then the fiercest battle did take place as he bent and banged and swore,
But at last sad tears wet the clock's proud face, and it could resist no more.
Now it's mutilated quiet and sad. It is far too tired to fight.
T'was a fine proud clock that once we had, but now our clock is right.

Oh, sing aloud with sad regret the strong proud clock has now been set.

The Commander

Herb Beall composed this song for, I think, the going-away party for Bill's sabbatical leave in 1966 to Cambridge, UK, to learn large-molecule crystallography of proteins. There is a reference to Bill "working every night," which he did and which may have made the graduate students feel obliged do the same.

When with eye both sad and tearful off the great Commander goes.
Cynics say some men are cheerful now instead of work they'll doze.
Cynics say there'll be much Hell-ing, when the great Commander leaves.
But the truth I'm truly telling, every Trooper truly grieves.
But the truth I'm truly telling, every Trooper truly grieves.

Who will guide us, who will lead us, who will set our clock aright,
Never chide us or impede us, help us on to greater height.
When the Leader says adieu, those not sighing are the few.
Those not sighing, those not sighing, those not sighing are the few.

When with eye both sad and tearful the Commander goes away,
Cynics say that they are fearful he'll not work a single day.
Cynics say that he'll be drinking in the pub by flickering light,
But the truth I'm truly thinking he'll be working every night.
But the truth I'm truly thinking he'll be working every night.

Who will guide us, who will lead us, who will set our clock aright,
Never chide us or impede us, help us on to greater height.
When the Leader says adieu, those not sighing are the few.
Those not sighing, those not sighing, those not sighing are the few.

Oh for Days

Herb, after a few years out of school, looked back at days in Bill's lab at Harvard in this song.

In a golden sea of glory dwells an isle with silver tower.
There stands Gibbs Lab famed in story where we spent our heaped on hour.
Though the king was just and kindly to his goodness we were dim,
And we criticized him blindly, un-appreciating him.

Now with life we have to grips come. Now it's life, which we must face.
Oh, for days with Colonel Lipscomb and his easy going pace.
Oh, for days with Colonel Lipscomb and his easy going pace.

Sometimes he would give suggestion in his unassuming way,
And his wisdom we would question though we rue our words today.
He would wander off dejected to his parapet above,
Wondering how one so neglected can win students lasting love.

From our mouths unkindly quips come. Yes, we have abused him so.
Oh, for days with Colonel Lipscomb and his nature calm and slow.
Oh, for days with Colonel Lipscomb and his nature calm and slow.

Now that island seems so distant. 'Tis a wonder to behold,
How it can be so resistant to a world that's harsh and cold.
Quotas deadlines they're not given. All exists for joy and light.
No one ever sternly driven, no one ever works at night.

To that island hopeful ships come that they may imbibe the cheer.
Oh, for days with Colonel Lipscomb and the parties and the beer.
Oh, for days with Colonel Lipscomb and the parties and the beer.

Dr. Herbert Beall became a professor of chemistry at Worcester Polytechnic Institute and died in 2003.

1.2.2.2. *Hypho-Man*

The Periodic Table of Comic Books is the effort of John P. Selegue and F. James Holler of the Department of Chemistry at the University of Kentucky to create a database of the appearance of chemical elements in comic books.[33] Bill was enlisted to help Hypho-Man in the boron entry. As no real comic book mentioning boron could be found, they made up a fragment of one, complete with a story.

"Hypho" signifies a polyhedral cage that is missing some vertices. This may be the most obscure and nerdy superhero name ever.

1.2.2.3. *Deoxypreposterase*

The *Harvard Lampoon* had fun with Bill's project to determine the atomic structure of carboxypeptidase A by dubbing it deoxypreposterase in their 1968 parody of *Life* magazine.

Amazingly, the students spoofed the molecule correctly having not, as far as we know, talked to anyone in Bill's lab. They got it right that carboxypeptidase A is an enzyme that digests proteins. At least one of Bill's graduate students puzzled over what was up while watching them photograph the cheeseburger in Bill's large Kendrew model of carboxypeptidase A. The *Lampoon* showed Harvard students at their best: aggressive, creative, knowledgeable, and motivated.

Although, had the students spoken to someone in the lab they might have been persuaded to place the cheeseburger higher up in the model at the active site (of digestion), which would have been a cool inside joke for the perhaps two or three people in the world who would care.

1.3. Family Stories

1.3.1. *Bill, the son*

When he talked about his early years, Bill Lipscomb usually told two stories with special pride. One I will discuss a bit later, but one was about a game of a variant of baseball Bill played while in graduate school in Linus Pauling's group at the California Institute of Technology.

> ... with the bases loaded and nobody out I was playing center field and I was playing in and a guy hit a line drive over second base and I came charging in and got it at the shoe-tops, that's one out, and I stepped on second base, that's two out, and tagged the guy that came in from first base ... I didn't realize I had three outs. I threw it to third base and we got the fourth out. But it was an unassisted triple play, and it made the Pasadena newspapers.

When Bill was young, he set up a chemistry lab at home.

> This dates from age eleven when my mother gave me one of those Gilbert chemistry sets that boys were allowed to play with. Most boys would ... play for two weeks, and then quit, but I set up on the desk in my bedroom a home laboratory which I was able to add many things to ... I would sometimes order apparatus, glassware, from a chemical company, but mostly I would buy chemicals from the drug store using my father's privilege of cut rate because he was a physician, and I augmented the laboratory. ... I did many, many experiments mostly beginning with color changes and evil smells, like hydrogen sulfide, to send the girls [Bill's two sisters] away.

Bill made gunpowder for homemade fireworks from chemicals he purchased at a local pharmacy. First, he mixed the ingredients. Then, he began to grind. He stepped back. He does not know why. Up rose a fireball.

Bill formed a musical troop with his sisters. He played the clarinet while they danced.

Then, tragedy. In 1937 Bill's sister Helen, at age 17, caught polio. To help her finish the last two years of high school from her bed, Bill taught her mathematics and a teacher visited to teach other courses. Helen lost her dreams of an ordinary life, a husband, and children, and Bill's father lost not just his job, he lost his career. No one would bring their child to the doctor from the house of polio.

Consequently, Bill had to go to an inexpensive college, the University of Kentucky, on a clarinet music scholarship with help from his summer job distributing seeds to farmers. He donated his home chemistry lab to his high school, more than doubling their equipment.

When his college chemistry professors saw how much Bill could handle, they bumped up the quality of some of the courses to the dismay of the other students, who asked Bill to tell the professors that he could not follow at this greater pace. Bill's answer is unknown, but I have never known Bill to lie.

Bill's father, however, could tell a tall tale. On his 50th wedding anniversary he and his wife dressed up and took their daughter, Helen, to a nice restaurant. The waitress asked, "What's the occasion?" Bill's father indicated his middle-aged, wheelchair-bound daughter and deadpanned, "For her peace of mind we finally decided to get married." I am told that the waitress did not speak to Bill's father for the rest of the evening. Bill's dry sense of humor may have its origins here.

1.3.2. *Bill, the father*

Bill Lipscomb was distinguished long before he was awarded the Nobel Prize in Chemistry in 1976. By way of coping, we children dubbed him Sir Father. Bill took particular delight that when his guests remarked on how polite his children were, calling their father sir, that they did not also notice that we were smirking.

For now, he will be Bill. Formality did not suit him. To his students he was the Colonel, directly addressed as Colonel. "His first doctoral student, Murray Vernon King, pinned the label on him, and it was quickly adopted by other students, who wanted to use an appellation that showed informal respect. ... Lipscomb's Kentucky origins as the rationale for the designation."[34]

Indeed, Bill really was made an honorary Kentucky Colonel (see Figures 1.10 and 1.11), of all his awards the only one that hung in his bedroom. He gleefully explained that the award was in a sense meaningless in that it was nothing more than an honor.

There were two rules in the early Lipscomb household:

1) Don't wake daddy. If he did not get a long night's sleep, then his science work would lose a day, as he could not work on anything hard. From an Internet search for quotations from Bill:

   > For me, the creative process, first of all, requires a good nine hours of sleep a night. Second, it must not be pushed by the need to produce practical applications.

2) As your knife traveled from the butter dish to your dinner roll, the butter on the knife was fair game to a swooping attack, diverting the butter to the roll of another. This was a surprise to the occasional dinner guest. That was when Bill did the honors. The rest of us were too polite.

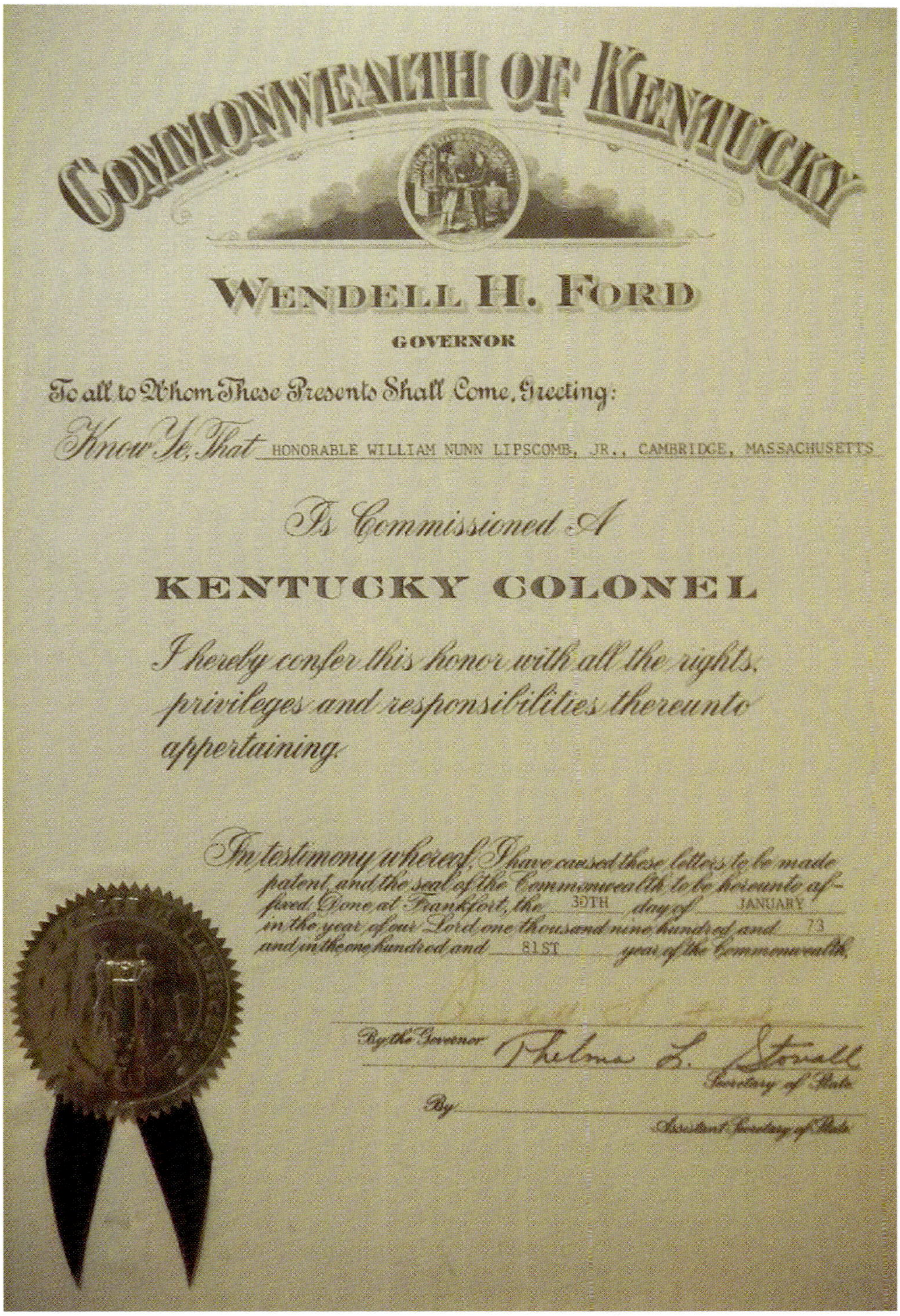

COMMONWEALTH OF KENTUCKY

WENDELL H. FORD
GOVERNOR

To all to Whom These Presents Shall Come, Greeting:

Know Ye, That HONORABLE WILLIAM NUNN LIPSCOMB, JR., CAMBRIDGE, MASSACHUSETTS

Is Commissioned A

KENTUCKY COLONEL

I hereby confer this honor with all the rights, privileges and responsibilities thereunto appertaining.

In testimony whereof, I have caused these letters to be made patent, and the seal of the Commonwealth to be hereunto affixed. Done at Frankfort, the 30TH day of JANUARY in the year of our Lord one thousand nine hundred and 73 and in the one hundred and 81ST year of the Commonwealth.

By the Governor

Secretary of State

By

Assistant Secretary of State

Figure 1.10 Bill is an honorary Kentucky Colonel. Scanned by James S. Lipscomb. Creative Commons License applies.

Figure 1.11 Bill in Japan holding his Kentucky Colonel card by a statue of Colonel Sanders. Photo © 1979 Jean C. Evans. Reprinted with permission.

Bill pushed a bowl towards the dinner guest, "Have a walnut."

The guest picked the one on top, and cracked it open as she continued talking. When she saw just the cracked shell and nothing more in her hand, she stared in disbelief and disappointment. The day before, Bill had gently cracked that walnut into two unbroken halves, scooped out the inside, and glued the shell back together.

For the son of Bill Lipscomb the early years included advanced science books given to a young boy. In grade school, along with Dr. Seuss books, Bill stocked my bookshelf with *The Restless Universe* by Max Born, *One, Two, Three, Infinity* by George Gamow, and the Mr. Tompkins physics books. He also gave me a college textbook, *Astronomy: An Introduction* by Robert Baker, and I studied it. I began to be interested in physics and particularly, in one of these books, in the Minkowski diagram, which really puzzled me. The axes of space and time rotate to calculate the distortions of space-time near the speed of light, but it wasn't quite clear if these axes really rotated in the natural world, or if they rotated only mathematically to explain the phenomenon, and I puzzled over that a while.

One of Bill's World War II military research projects during graduate school had him sometimes carrying vials of pure nitroglycerin. Bill told me that it was safest to hold the vials by the bare fingertips, so that the heat of one's hand might not detonate the solution, and that not wearing gloves reduced the chance of dropping a vial. Later, Bill taught me four

Figure 1.12 Bill's air rifle. Photo by James S. Lipscomb. Creative Commons License applies.

precise steps for how to handle dry ice with my bare hands. When I was in early grade school, Bill showed me that centrifugal force would reduce the chance that we would slip and fall when we left the fenced trail to run behind and half-way up the Minnehaha Falls in Minneapolis, Minnesota, on a semi-circular, wet, narrow ledge that sloped down to the rocks perhaps 20 feet below. I suppose that all fathers give their sons safety tips.

Bill bought a Sears and Roebuck air rifle (see Figure 1.12)[35] when I was in high school so that I could dispose of old experimental solutions, glass ampoules of boranes that exploded on contact with air. There was no company or agency that would take care of it. Bill paid for the air rifle himself, as he did for the many sets of poker (or maybe bingo or roulette) chips he bought to tape to the hanging clear plastic sheets of the carboxypeptidase A electron density map to indicate atomic positions. He did not want to have to explain the purchase if he billed the granting agency.

I encountered my fireball in the air rifle era, when a graduate student heated a tube on a glass vacuum line and pulled off a sealed ampoule. It held 1 cubic centimeter of liquid diborane — which explodes with a sharp crack (and violently; 1 cc is a lot) — mixed with 5 cc of pentaborane, which flames for a second or two. The ampoule was too dangerous to carry up to our roof-top shooting range, because the compounds within explode on contact with air, so the tiniest crack in the glass could set them off. Gently the student put the ampoule in a metal dewar (heavy-duty thermos bottle) and placed it, now a flamethrower pointing up, on the Gibbs Lab fire escape. On the grating a floor above someone laid down a sheet of safety glass. I knelt down on it and shot the air rifle just over the edge diagonally down and to the right, into the upward-pointing mouth of the dewar.

I was too close to the explosion to see it. Behind my safety glasses I flinched from the intense heat as the dewar blasted a fireball upwards. I opened my eyes to see, across the parking lot, visitors queued for a back door of the Peabody Museum all looking towards Gibbs. Graduate students in the room above came out surprised and slightly concerned as the expanding radius of the fireball was sufficient for it to pass their window. Having a chemist for a dad can be very cool.

1.3.3. *Billy finds his profession*

Whenever Bill's father served as a physician at Boy Scout summer camp he took young Bill with him. Bill became a Star Scout (two ranks below Eagle Scout) and an enthusiastic outdoorsman. Indeed, years later there were two summers when he drove his young family 200 miles north of St. Paul, Minnesota, for a few weeks in a log cabin on the shore of Deer Lake, close to Little Deer Lake, living on the fish we caught. Back in his youth, while camping in the open in his sleeping bag, a meteor shower started in which the meteors were so bright and frequent, falling every few seconds, that they woke him up.

The other story Bill told about his early years with particular pride was about an advanced science book given to him as a young boy.

> ...I used to go to the open nights at the observatory run by Professor Downing, who was a great friend of mine. He gave me a college textbook, and I studied it.[36]

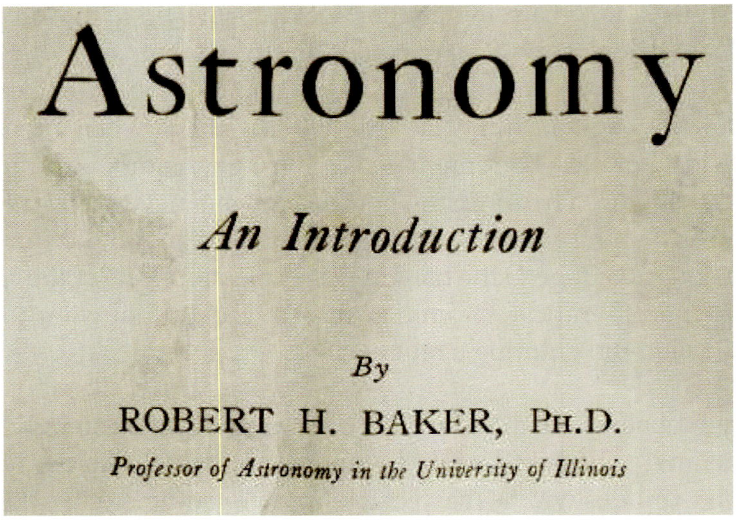
Astronomy

An Introduction

By

ROBERT H. BAKER, PH.D.

Professor of Astronomy in the University of Illinois

Figure 1.13 The title page of the college textbook Professor Downing gave to the young Bill. Presented under the copyright doctrine of Fair Use.

The title page of that book can be seen in Figure 1.13. When we open the front cover we see that Bill's mother, Edna, has written Bill's name, address, and phone number. Bill wrote the date, 1932 (see Figure 1.14). He was about 12 years old, the same year that his mother gave him a chemistry set, which grew into a home chemistry lab.[37]

As we flip the pages, we see that they are wrinkled from a night in the rain in the little tree house that Bill built. As we flip the pages time passes, tastes change.

> But I began to be interested more in the mechanics and particularly in one statement in the astronomy book that really puzzled me, "The stars are too hot to burn." Well, nuclear energy wasn't known then and the only source of energy is the gravitational infall as the Sun radiates, but it wasn't quite clear why if the stars were too hot to burn, they were so luminous, and I puzzled over that a while ...

Bill highlighted the problem in pencil on p. 432 (see Figure 1.15).

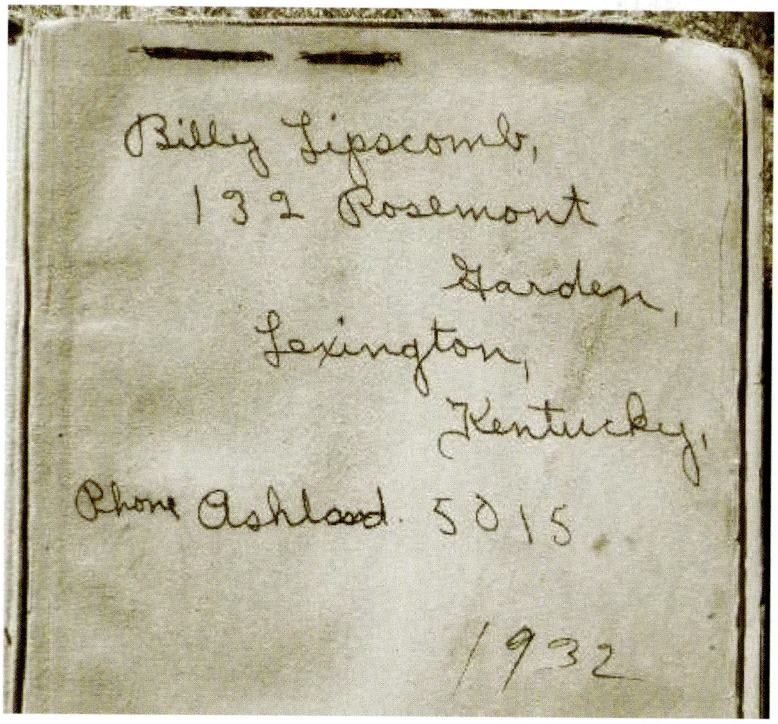

Figure 1.14 Inside the front cover of Bill's copy of *Astronomy: An Introduction*. Scanned by James S. Lipscomb. Creative Commons License above applies.

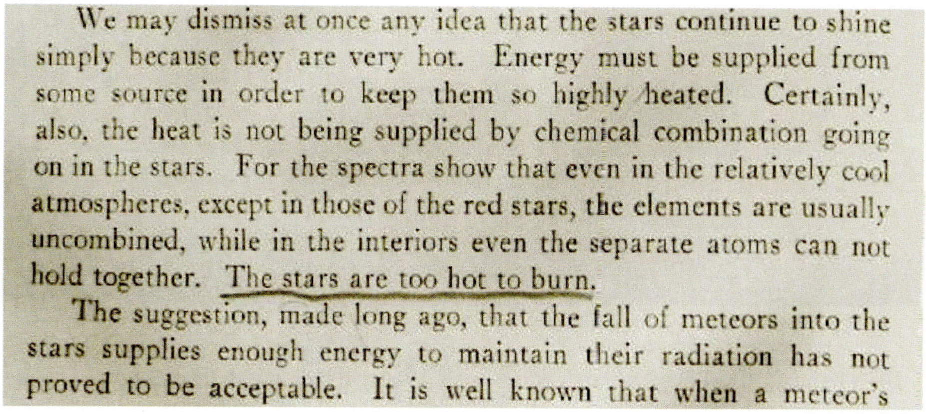

We may dismiss at once any idea that the stars continue to shine simply because they are very hot. Energy must be supplied from some source in order to keep them so highly heated. Certainly, also, the heat is not being supplied by chemical combination going on in the stars. For the spectra show that even in the relatively cool atmospheres, except in those of the red stars, the elements are usually uncombined, while in the interiors even the separate atoms can not hold together. The stars are too hot to burn.

The suggestion, made long ago, that the fall of meteors into the stars supplies enough energy to maintain their radiation has not proved to be acceptable. It is well known that when a meteor's

Figure 1.15 Bill underlined the statement that most struck him in *Astronomy: An Introduction*.

Reaching the inside back cover we see that Bill had started a list that grew with time.

Penciled in a young hand is MOLECULE. Below that, ATOM. Following, is PROTON and next to it a plus sign. After that, ELECTRON, which is marked as negative. Then, POSITRON, DEUTRON, NEUTRON. Then, in an older hand the wavelengths of light. Then, in ink, Cathode Rays, Alpha, and Beta (see Figure 1.16).

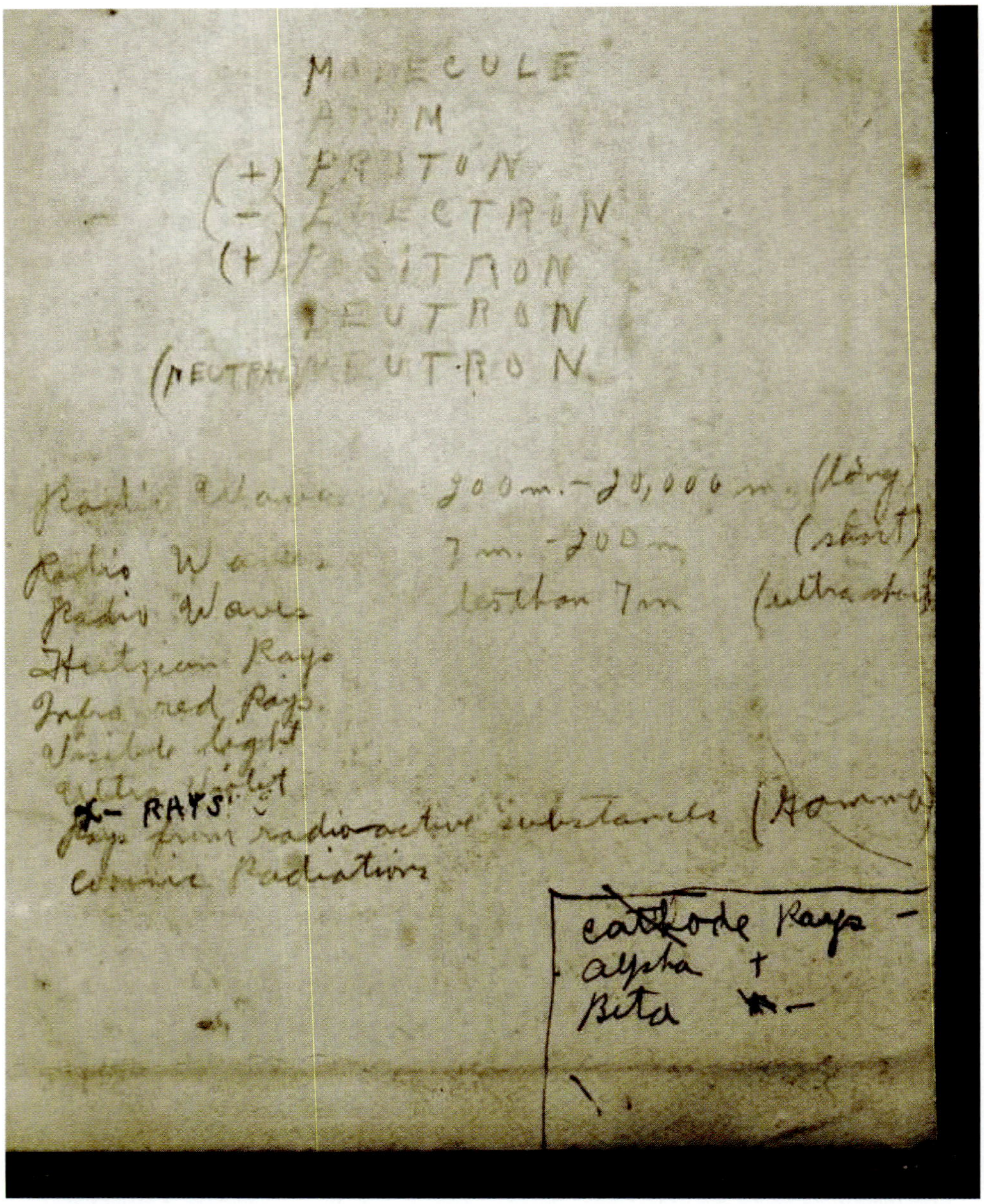

Figure 1.16 Bill's handwritten list in the back of the book. Scanned by James S. Lipscomb. Creative Commons License above applies.

Here, it begins.

1.3.4. *Here it Ends*

Behind every exquisite thing that existed, there was something tragic. — Oscar Wilde

The eulogy I gave for my father on April 15, 2011 was a shortened version of some of the following sections.

1.3.4.1. *Eldest Son*

It may be traditional for the eldest son to speak now. He would lead us to the heart of Bill Lipscomb, so that we might reach out to it one last time.

Bill Lipscomb's eldest son was born October 8, 1945. He rests in Calvary Cemetery in Los Angeles, having lived for two hours. I speak in his name, though he had none. I am James, the younger brother.

1.3.4.2. *Swingle Accident*

When I feel proud of Bill, I am most often thinking of what he did on September 23, 1943.

At the California Institute of Technology, in the Crellin Laboratory's sub-basement, 29-year-old Elizabeth Swingle waited for the elevator. The kilogram bottle of ethyl chlorocarbonate in her hand blew off its cap, spraying liquid over her head and shoulders.[38] Bill continues the story:

> I was down the hall, and I got her and put her in the shower. I was too late. She had already inhaled enough of it. I was the one who got the next largest dose as I stood there holding the shower. I couldn't get away. ...half the molecule is phosgene, so it does the same thing as phosgene. It gives HCl, hydrochloric acid, in your lungs.[39]

Phosgene first dissolves in water to produce hydrochloric acid, which kills lung tissue in the lower airway. Then, phosgene attacks the capillaries causing fluid to leak into the lungs, which, combined with tissue swelling in an inflammatory cascade, can cause drowning.[40]

About eight hours later, Elizabeth Swingle died. Bill's lung function was good enough for him to play tennis and the clarinet, but he would cough terribly from the slightest cigarette smoke or diesel exhaust.

1.3.4.3. *Next Generation*

An old person may pull the child aside to pass along important information. I received four such lectures from Bill. He shunned debate (except when upholding scientific truth), and preferred to speak for himself by advising others and taking a distant, professorial tone, which I have had practice interpreting. First, the lectures:

1) What Bill told me most separates him from the other high-performers is that he picked those grand challenges that might be impossible but that could be broken down into a stepwise path. In this way the highest step that one can climb will give the biggest accomplishment possible. Bill later crystallized the point: "I have failed at everything [big] I've tried, but I set intermediate goals."

2) Even now, in the 21st century, the nature of the chemical bond is in some ways little understood. We do not have "... a systematic valence description of the vast numbers of 'electron deficient' intermetallic compounds"[41] or a full understanding of the chemical bond in, for example, KHg_{13} and Cu_5Zn_7. Bill concluded, "For the person who figures this out there may be a big prize."
3) Bill's research group showed how the three-center two-electron bond works through electron orbital calculations not using something new, but using existing formulas by Edmiston and Ruedenberg,[42] and by Boys.[43] Of these, Bill said, "when you plug [the situation] into it, [each] gives you the three-center bond, whether you are looking for it or not." He also said, "Anyone could have done it."
4) The Bohr radius and the Rydberg constant of the hydrogen atom can be derived from first principles. Bill showed my son and me the derivations on one page.

On today's occasion there may be lists by which we might know Bill, lists of Lipscomb this and Lipscomb that. But Bill's list for his son looks outward.

He gave me:

1) Advice on how to choose and to attack big scientific challenges.
2) A career suggestion of a big chemistry challenge.

And he advised me to:

3) Give high credit to key contributors.
4) Appreciate the beauty of simplicity and elegance in nature.

1.3.4.4. *Friends*

Bill had many quiet friends whom he heard from when there was an occasion. But close friends we shall always hear from — there are never so many of those.

Bill kept an emotional distance. Marydell, Bill's first wife, was a gifted conversationalist who freely spoke her heart to all. Bill told her, "You must never do that. They will get a handle on you." These two spirits would never understand each other.

When Bill could no longer work, his days were spent at home. Among my routine visits I recall on four occasions the phone ringing to bring the voice of a friend.

Sometimes I would drive Bill to the Harvard mailroom. There were many magazines.

1.3.4.5. *Forgiveness*

Forgiveness is not something that one can give to oneself.

To the son

When I was in secondary school, Bill came to my room. It was time to speak to the son with the mediocre grades. Bill spoke of how the Lipscomb male line had produced four generations in a row of doctors of one sort or another. He said that logically it could not go on forever. It must end sometime. Then he left.

Bill did not give up on me. He later sent me to a private high school whose tuition he could scarcely afford. Perhaps by way of reply I was eventually awarded a Ph.D. I wonder if Bill was proud of me in this regard. We never spoke of it.

But for now in his young son's room we did not see this future. Bill had stopped talking. Bill gives the surface facts, leaving the listener to read his heart. To my reading the conclusion was self evident. Bill was letting go of a dream that his son would be the fifth doctor in a row. Bill was giving forgiveness to me in advance.

To the father

Biographers beware of the drive-by interview. Over time, complexity and ambiguity may emerge.

The Lipscomb male line had produced three generations in a row of medical doctors, ending with William Lipscomb, Sr. His son, William Lipscomb, Jr., did not want to become the fourth medical doctor. Instead, he earned a Ph.D. and became a chemistry professor. William Lipscomb, Sr. died three years before his son was awarded the Nobel Prize.

There was a day when Bill Lipscomb spoke wistfully, "I was told jokingly that if I were to be a physician I would be the fourth straight generation." And there were more of these days when Bill recalled that his physician father told Bill that he should do whatever he wanted, that his father was content with Bill's choice to get a Ph.D. rather than an M.D.

But, there was another day when he said, "I have one regret about the Nobel Prize. I regret that it did not happen a few years earlier. If it had, then maybe my father would have forgiven me." These were lingering doubts from the heart. Bill had just done what he had told Marydell one must never do.

After some thought, I spoke, but not as the son. I spoke as the grandson of the medical doctor. I am by some accounting one quarter my grandfather. I told Bill that like the early anatomists who pioneered dissection of the human body to see its structure and then to deduce how the body works, so too Bill's research group was among the first, just after the solutions of myoglobin and lysozyme [1] to find the atomic structure of, to start with, two of the body's large proteins, carboxypeptidase A and apartate carbamoyltransferase (see Figure 1.17) — each the largest structure in its time — and to deduce how they work. They were pioneering 20th century dissections.

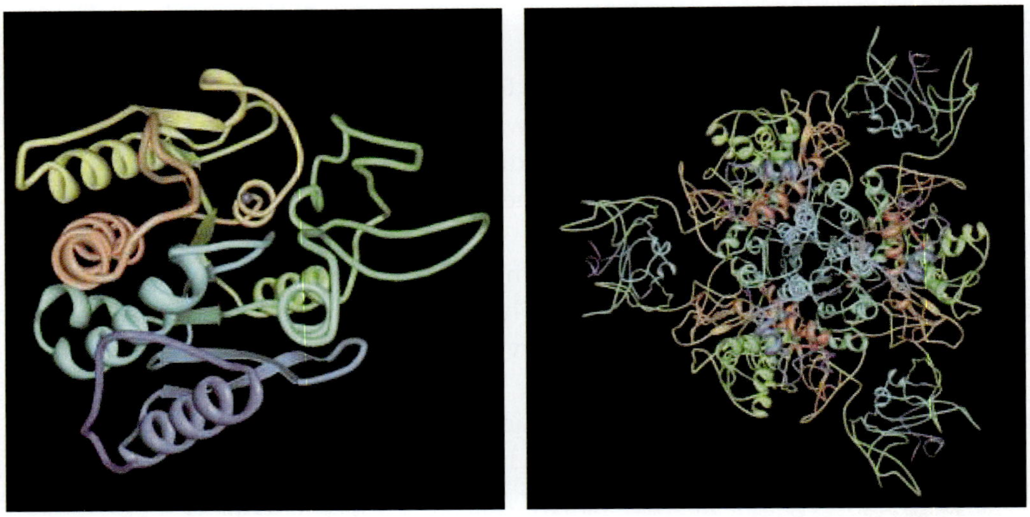

(a) (b)

Figure 1.17 (a) Carboxypeptidase A[44] and (b) Apartate carbamoyltransferase.[45] Images by James S. Lipscomb using the data from Protein Data Bank and rendered using the PDB Simple Viewer. Creative Commons License above applies.

Figure 1.18 Fructose-1,6-bisphosphatase.[46,47] Images by James S. Lipscomb using the data from Protein Data Bank and rendered using the PDB Simple Viewer. Creative Commons License above applies.

I said that the earliest anatomists had no medical degrees; there was no such thing then. But now the medical community counts them as fully their own.

I further spoke of how Bill's recent work on fructose-1,6-bisphosphatase (see Figure 1.18) and its inhibitor MB06322 (CS-917) with Metabasis Therapeutics, Inc. [2] may come close to clinical application for diabetes therapy.

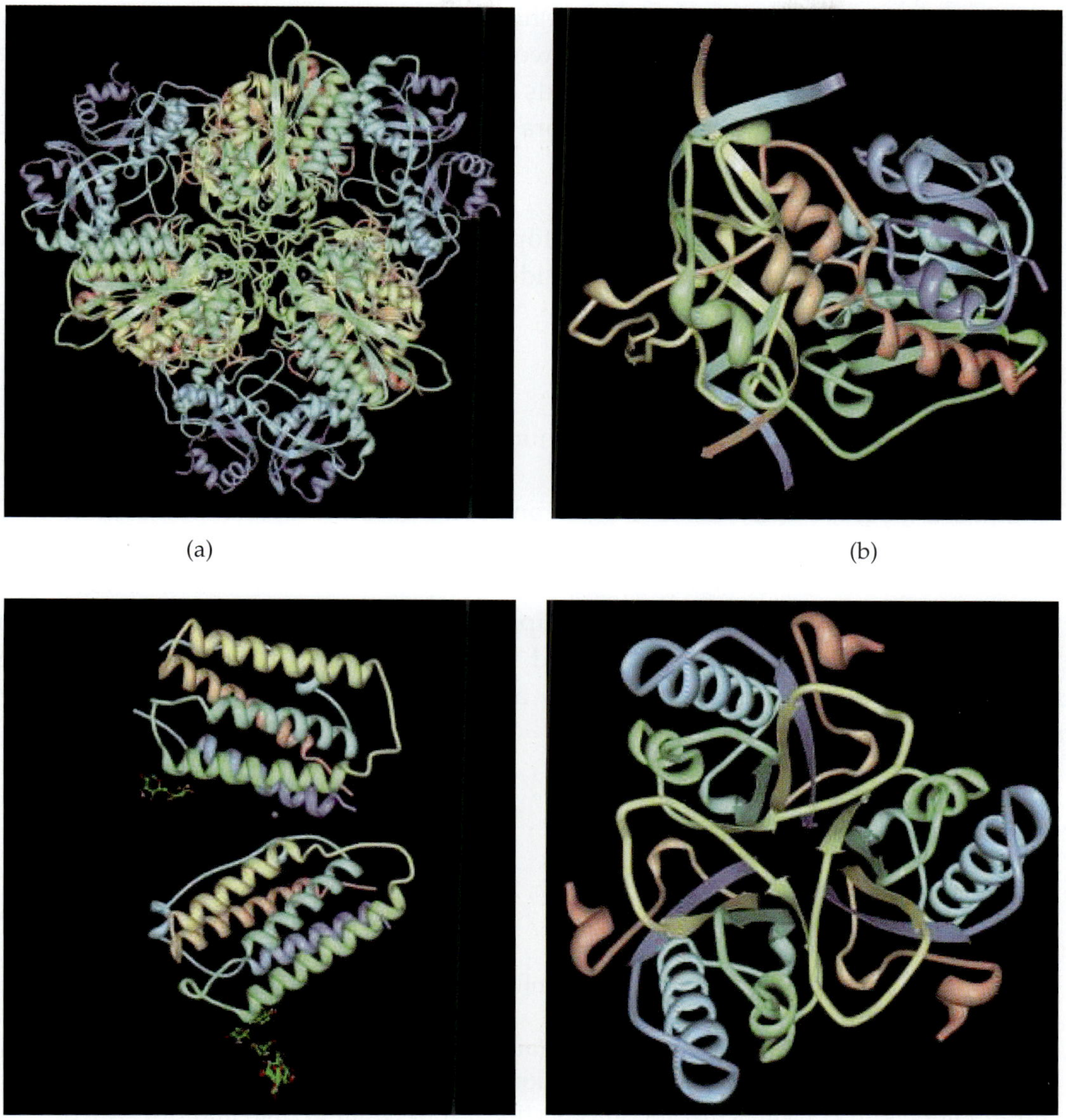

Figure 1.19 (a) Leucine aminopeptidase,[48] (b) HaeIII methyltransferase/DNA,[49] (c) Human interferon beta,[50] and (d) Chorismate mutase.[51] Images by James S. Lipscomb using the data from Protein Data Bank and rendered using the PDB Simple Viewer. Creative Commons License above applies.

I told Bill that considering his research group's determination of the structure and function [3] of four other large proteins as well — leucine aminopeptidase, HaeIII methyltransferase/DNA, human interferon beta, and chorismate mutase (see Figure 1.19) — that I judged this phase of Bill's career to be not merely the equal of that of a distinguished M.D., Ph.D. researcher, but, because of the historical significance of his first two large structures, the career of a pioneering research physician.

Finally, I repeated to Bill his statement that his father had not forgiven him for becoming something other than a medical doctor. Two things now stood opposed. They suggest several conclusions, among which one stands out. Perhaps by way of reply, as I left for my home later that day, Bill shook my hand more firmly and for longer than on any other day that I can recall.

But for now in my father's house I had stopped talking, for in the Lipscomb way one states the facts, and the conclusion is self evident.

Notes

[1] And perhaps I should have added ribonuclease-S to the list.

[2] Metabasis Therapeutics, Inc. is now Ligand Pharmaceuticals after acquisition in 2010. http://www.ligand.com/

[3] I neglected to mention the Moncrief, Lipscomb structural determination of the anti-cancer drug vincristine. Moncrief J. W. and Lipscomb W. N., Structures of leurocristine (vincristine) and vincaleukoblastine. X-ray analysis of leurocristine methiodide. *J. Am. Chem. Soc.*, **87**, 4963–4964 (1965).

References

1. Broad W. J., A bank for Nobel sperm. *Science*, **207**, 1326–1327 (1980).
2. Lipscomb W. N., Reflections (video, 42 min.) in *The Life and Work of Linus Pauling (1901–1994): A Discourse on the Art of Biography* (1995). http://osulibrary.oregonstate.edu/specialcollections/events/1995paulingconference/video-s3-4-lipscomb.html
3. Dunitz J., Don't be afraid to make mistakes (audio clip, 45 min.) in *Lifestory: Linus Pauling*. Produced by the British Broadcasting Corporation (1997). http://library.oregonstate.edu/specialcollections/coll/pauling/dna/audio/1997v.1-mistakes.html
4. Lipscomb W. N., Framework rearrangement in boranes and carboranes. *Science*, **153**, 373–378 (1966).
5. Winfried Wild, July 4, 1980, no. 152 (translated).
6. Baker R. H., *Astronomy: An Introduction*. New York: D. van Nostrand Company, Inc. (1930).
7. Longuet-Higgins H. C. and Roberts M. de V., The electronic structure of an icosahedron of boron atoms. *P. R. Soc. London*, **A230**, 110–119 (1955).
8. Edmiston C. K. and Ruedenberg K., Localized atomic and molecular orbitals. *Rev. Mod. Phys.*, **35**, 457–465 (1963).
9. Boys S. F., in Löwdin P. O. (Ed.), *Quantum Theory of Atoms, Molecules and the Solid State*. New York: Academic Press, pp. 253–262 (1966).
10. Lipscomb W. N., The boranes and their relatives. *Science*, **196**, 1047–1055 (1977).
11. Eaton G. R. and Lipscomb W. N., *Nuclear Magnetic Resonance Studies of Boron Hydrides and Related Compounds*. New York: W. A. Benjamin, Inc. (1969).
12. Stevens R. M., Pitzer R. M., and Lipscomb W. N., Perturbed Hartree–Fock calculations. I. Magnetic susceptibility and shielding in the LiH molecule. *J. Chem. Phys.*, **38**, 550–560 (1963).

13. Pitzer R. M. and Lipscomb W. N., Calculation of the barrier to internal rotation in ethane. *J. Chem. Phys.*, **39**, 1995–2004 (1963).
14. Lipscomb W. N., The chemical shift and other second-order magnetic and electric properties of small molecules. In Waugh J. (Ed.), *Advances in Nuclear Magnetic Resonance*. Vol. 2. New York: Academic Press, pp. 137–176 (1966).
15. Lipscomb W. N., The boranes and their relatives. In *Les Prix Nobel en 1976*. Stockholm: Imprimerie Royale PA Norstedt & Soner, pp. 110–131 (1977). Quote in next to last paragraph omitted from *Science* version of the paper (1977).
16. Abrahams S. C., Collin R. L., Lipscomb W. N., and Reed T. B., Further techniques in single-crystal X-ray diffraction studies at low temperatures. *Rev. Sci. Instrum.*, **21**, 396–397 (1950).
17. Kaufman H. S. and Fankuchen I. A low temperature single crystal X-ray diffraction technique. *Rev. Sci. Instrum.*, **20**, 733–734 (1949).
18. Lipscomb W. N. The boranes and their relatives. *Science*, **196**, 1047–1055 (1977).
19. Hoffmann R. and Lipscomb W. N., Theory of polyhedral molecules. I. Physical factorizations of the secular equation. *J. Chem. Phys.*, **36**, 2179–2189 (1962).
20. Lipscomb W. N., *Boron Hydrides*. New York: W.A. Benjamin Inc., Chapter 3 (1963).
21. Hoffmann R., A simple quantum chemical method. *This Week's Citation Classic*, **19** (1989). http://garfield.library.upenn.edu/classics/classics_h.html
22. Hoffmann R., An extended Hückel theory. I. Hydrocarbons. *J. Chem. Phys.*, **39**, 1397–1412 (1963).
23. Newton M. D., Boer F. P., and Lipscomb W. N., Molecular orbital theory for large molecules. Approximation of the SCF LCAO Hamiltonian matrix. *J. Am. Chem. Soc.*, **88**, 2353–2360 (1966).
24. Boer F. P., Newton M. D., and Lipscomb W. N., Molecular orbitals for boron hydrides parameterized from SCF model calculations. *J. Am. Chem. Soc.*, **88**, 2361–2366 (1966).
25. Snow C. P., *The Two Cultures and The Scientific Revolution. The Rede Lecture*. New York: Cambridge University Press (1959).
26. Pausch R., *The Last Lecture*. New York: Hyperion (2008).
27. Reiter C. A. and Jones W. R., *APL with a Mathematical Accent*. Belmont, CA: Brooks/Cole Pub. Co. (1990).
28. Brooks F. P., Jr., The computer "scientist" as toolsmith: studies in interactive computer graphics. *Proc. IFIP*, 625–634 (1977).
29. Lipscomb J. S., Brooks F. P., Jr., and Pique M. E, The GRIP-75 man–machine interface. *ACM SIGGRAPH Video Review*, **4** (1981).
30. Tsernoglou D., Petsko G. A., and Tu A. T., Protein sequencing by computer graphics. *Biochim. Biophys. ACTA*, **491**, 605–608 (1977).
31. Richardson D. C., Three-dimensional structure of Cu, Zn superoxide dismutase. In Michelson A. M., McCord J. M., and Fridovich I. (Eds), *Superoxide and Superoxide Dismutases*. New York: Academic Press (1977).
32. Lipscomb W. N., Electron diffraction investigations of vanadium tetrachloride, dimethylketene dimer, tetrachloroethylene, and trichloroethylene. Part 2. The crystal structure of methylammonium chloride. (Parts 3 and 4 omitted, confidential.) Dissertation, California Institute of Technology, p. 71 (1946).
33. The Comic Book Periodic Table of the Elements. http://www.uky.edu/Projects/Chemcomics
34. Katz, L. Letter presented to Bill on the occasion of his 80th birthday, collected with others in a Festschrift (party book), May 12–14, 2000.
35. Lipscomb W. N., The boranes and their relatives. *Science*, **196**, 1047–1055 (1977). See "air gun" on p. 227 of http://nobelprize.org/nobel_prizes/chemistry/laureates/1976/lipscomb-lecture.pdf
36. Baker R. H., *Astronomy: An Introduction*. New York: D. van Nostrand Company, Inc. (1930).

37. Lipscomb W. N., Process of discovery (1977); an autobiographical sketch. In Eaton G. R., Wiley D. C., and Jardetzky O. (Eds), *Structures and Mechanisms: From Ashes to Enzymes.* ACS Symposium Series. Washington, DC: American Chemical Society, pp. xv–xxviii (2002). http://pubs.acs.org/isbn/9780841237360
38. Pauling L., Letter to Eastman Kodak Company, October 1, 1943, which gives a technical description of the Swingle accident. http://osulibrary.oregonstate.edu/specialcollections/coll/pauling/war/corr/corr111.1-lp-eastmankodak-19431001-transcript.html
39. Elizabeth Swingle's laboratory accident (audio clip, 1 min. 33 sec. of interview of William Lipscomb, November 3, 1991) by Thomas Hager for use in *Force of Nature: The Life of Linus Pauling.* New York: Simon & Schuster (1995).
40. Phosgene toxicity http://emedicine.medscape.com/article/820649-overview
41. Lipscomb W. N., The boranes and their relatives. In *Les Prix Nobel en 1976.* Stockholm: Imprimerie Royale PA Norstedt & Soner, pp. 110–131 (1977). Quote in the next to last paragraph omitted from the Science magazine version of the paper. See p. 242 of http://nobelprize.org/nobel_prizes/chemistry/laureates/1976/lipscomb-lecture.pdf
42. Edmiston C. K. and Ruedenberg K., Localized atomic and molecular orbitals. *Rev. Mod. Phys.*, **35**, 457–465 (1963).
43. Boys S. F., in Löwdin P. O. (Ed), *Quantum Theory of Atoms, Molecules and the Solid State.* New York: Academic Press, pp. 253–262 (1966).
44. Lipscomb W. N., Hartsuck J. A., Reeke G. N., Jr., Quiocho F. A., Bethge P. H., Ludwig M. L., Steitz T. A., Muirhead H., and Coppola J. C., The structure of carboxypeptidase A. VII. The 2.0-angstrom resolution studies of the enzyme and of its complex with glycyltyrosine, and mechanistic deductions. *Brookhaven Sym. Biol.*, **21**, 24–90 (1968).
45. Honzatko R. B., Crawford J. L., Monaco H. L., Ladner J. E., Edwards B. F. P., Evans D. R., Warren S. G., Wiley D. C., Ladner R. C., and Lipscomb W. N., Crystal and molecular structures of native and CTP-liganded aspartate carbamoyltransferase from Escherichia coli. *J. Mol. Biol.*, **160**, 219–263 (1983).
46. Ke H., Thorpe C. M., Seaton B. A., Lipscomb W. N., and Marcus F., Structure refinement of fructose-1,6-bisphosphatase and its fructose-2,6-bisphosphate complex at 2.8 A resolution. *J. Mol. Biol.*, **212**, 513–539 (1989).
47. Erion M. D., van Poelje P. D., Dang Q., Kasibhatla S. R., Potter S. C., Reddy M. R., Reddy K. R., Jiang T., and Lipscomb W. N., MB06322 (CS-917): a potent and selective inhibitor of fructose 1,6-bisphosphatase for controlling gluconeogenesis in type 2 diabetes. *P. Natl. Acad. Sci. USA*, **102**, 7970–7975 (2005).
48. Burley S. K., David P. R., Sweet R. M., Taylor A., and Lipscomb W. N., Structure determination and refinement of bovine lens leucine aminopeptidase and its complex with bestatin. *J. Mol. Biol.*, **224**, 113–140 (1992).
49. Reinisch K. M., Chen L., Verdine G. L., and Lipscomb W. N., The crystal structure of the Hae III methyltransferase covalently complexed to DNA: an extrahelical cytosine and rearranged base pairing. *Cell*, **82**, 143–153 (1995).
50. Karpusas M., Nolte M., Benton C. B., Meier W., and Lipscomb W. N., The crystal structure of human interferon beta at 2.2-A resolution, *P. Natl. Acad. Sci. USA*, **94**, 11813–11818 (1997).
51. Strater N., Schnappauf G., Braus G., and Lipscomb W. N., Mechanisms of catalysis and allosteric regulation of yeast chorismate mutase from crystal structures, *Structure*, **5**, 1437–1452 (1997).

Acknowledgments

Thanks to my wife Margaret and to my son Thomas for their proofreading and suggestions. Some quotes from Bill Lipscomb are from an interview conducted by Margaret Lipscomb.

Copyright Note: Creative Commons License

Chapter 2

Oleg Jardetzky

Biography

Oleg Jardetzky was born on February 11, 1929 in Belgrade, Yugoslavia, into a family of Russian émigrés. His father was a well-known geophysicist and authority on celestial mechanics, Wenceslas S. Jardetzky, at the time a professor at the University of Belgrade. His mother, Tatiana, née Taranovsky, was a professor of French, Russian, German, and comparative literature.

His early education was in Russian and German schools. After the war, in 1947, he graduated from secondary school in Graz, Austria and subsequently finished the first two years of medical school at the University of Graz. In 1949 he received a displaced person scholarship to Macalester College, St. Paul, Minnesota, from which he graduated in 1950 with majors in chemistry and biology. He obtained his M.D. degree from the University of Minnesota in 1954 and his Ph.D. in 1956 under the auspices of an interdisciplinary committee chaired by W. N. Lipscomb and M. B. Visscher. The year 1956–1957 he spent as a National Research Council Fellow in the laboratory of Linus Pauling at Caltech.

In 1952 he married Christine Drakakis, who received her Ph.D. in biochemistry in Minnesota in 1955 and was a postdoctoral fellow with Linus Pauling in 1956–1957. Their three children are Alexander (born 1957), subsequently a Stanford engineering graduate, who had the good Silicon Valley fortune to retire at the age of 37, Theodore (born 1960), who became a protein crystallographer and professor of structural biology at Stanford, and Paul (born 1963), who obtained his Ph.D. in computer science from Cambridge University and has held a variety of research and administrative positions in Silicon Valley.

In the years 1957–1967 Oleg was on the faculty of the Department of Pharmacology at the Harvard Medical School, where in 1959 he founded the first nuclear magnetic resonance laboratory dedicated to biological research, which was funded by the National Science Foundation. Much of the groundwork for the future use of NMR as a method for protein structure determination was laid there.

Between 1967 and 1969 he served first as director of the Department of Biophysics, then as executive director of the Merck Institute at the Merck, Sharp and Dohme Laboratories in Rahway, NJ. There it became possible to carry out the first selective deuteration — isotopic spectral editing — experiments on a protein (staphylococcal nuclease) and obtain the first partial protein (active site) structures determined by NMR. A by-product of this structural work was the formulation of the catalytic mechanism of pancreatic ribonuclease, which stood the test of time. Since 1969 he has served as professor of pharmacology at the Stanford Medical School (since 1990 molecular pharmacology and since 2006 chemical and systems biology), and 1972–1997 as director of the Stanford Magnetic Resonance Laboratory (SMRL).

A monograph *NMR in Molecular Biology*, coauthored with a former postdoctoral fellow, Gordon Roberts, appeared in 1982. In 1984 he reported the first protein structure determined by NMR — the lac repressor headpiece — and in 1993 his laboratory reported the then largest (36kD) NMR protein structure — the trp repressor — and its complex with operator DNA.

In 1975 he married a friend from his Austrian student days, the Austrian writer and art historian Erika Albensberg, who helped with the upbringing of his sons, and died in 2008.

Since 1991 Oleg has also served as director of the summer school on magnetic resonance and structural biology at the Ettore Majorana Centre for Science and Culture in Erice, Sicily and has edited several volumes of its proceedings, at first jointly with a former postdoctoral fellow, Jean Francois Le Fevre, who died in 1999, and then with his successor as head of SMRL, Joseph D. Puglisi.

Apart from natural science, history — in particular heraldry and genealogy in eastern Europe — has been a lifelong scholarly interest. He published several articles on such subjects in the *Herald*, the journal of the Austrian Society of Genealogy and Heraldry, and two books. The first of these, on the history of the Polish clan Ciolek, published in 1991, received a prize from the International Union of Heraldry and Genealogy in 1998.

He retired and closed his laboratory in 2006, but continued to publish, mostly historical articles, on the early history of NMR and of Russian emigration after the revolution of 1917.

William N. Lipscomb: Lifelong Teacher and Friend

I first met the young assistant professor of chemistry at the University of Minnesota, William N. Lipscomb, in the fall of 1950. I had just enrolled in the university as a graduate student in physiology; my advisor, the distinguished American physiologist Maurice B. Visscher, required that his graduate students have a minor in physical chemistry and advised me to talk to the university's "bright young physical chemist." Having grown up in Europe in an era in which an audience with a professor was a rare and dreaded event in a student's life, I was amazed and delighted to find a professor who was approachable, friendly, and easy to speak with.

We spoke for a long time, beginning with the questions of what a biologist should know about chemistry and why, and ending with the unity of science. I enrolled in Lipscomb's course on quantum mechanics, having previously had a very excellent general course in physical chemistry from Otto Kratky at the University of Graz. That proved to be a very interesting experience. I have never understood quantum mechanics. Lipscomb took a pragmatic point of view — that it was a set of rules, not quite understandable in terms of familiar images, but which allowed predictions to be made. I could not find it satisfying, and each of us, each for different reasons, grudgingly accepted Max Born's statistical interpretation of electron distributions — it provides no clue of the nature of the periodicity implied in the existence of quantum states. Only years later, after conversations with Linus Pauling, Robert Mulliken, Christopher Longuet-Higgins, Paul Dirac, and Werner Heisenberg, did I accept the fact that even its framers did not understand quantum mechanics the way I thought it should be understood.

My return to medical school in 1951 did not interrupt our contact, but it took a different form. Lipscomb spent a couple of lunch hours every week playing clarinet with David Schafer, at the time a fellow graduate student in physiology, accompanying him on the piano, simulating as needed, both the Brahms trio and the Mozart quintet. I was often invited as a listener, but I never had the courage to join in, to my later regret. Still, the memories are unforgettable, not only for the beauty of the music, but also the camaraderie, which Bill Lipscomb's openness and ease created among students and coworkers. Everyone by then called him by his favorite title — the Colonel — except me, because it was the one English word I found impossible to pronounce correctly. Dropping Old World prejudices that this was disrespectful, I got to call him Bill.

When I finished medical school in 1954 and decided to finish a Ph.D. and try a career in science rather than clinical medicine, I consulted Bill on the choice of a thesis topic. He was by then was a full professor and chief of the Division of Physical Chemistry, but still just as accessible and easygoing as he had been in his younger days. Not that I had an interest in boron hydrides or X-ray diffraction at low temperatures, but Bill was one of few people who was willing to spend time seriously thinking about subjects outside his own specialty and was clearly interested in the general problem of understanding nature, including biology, in terms of physics. My interests at the time were in the mechanisms of active transport of the sodium ion across cell membranes, and I was wondering whether a statistical analysis would allow us to gain a better understanding of this phenomenon. Lipscomb thought about it and said, "You know, with statistical analysis you get out what you put in, even if in a more sophisticated form. Why don't you try something really new? There is this new method — nuclear magnetic resonance. John Wertz downstairs is working on it. Why don't you talk to him?" I did. My thesis, defended in 1956, was the first NMR study of the sodium ion, and the rest is history.[1]

There were at the time no interdisciplinary graduate programs at the university, and so a special advisory committee was appointed to allow me to write a thesis in what would normally be my minor — physical chemistry — without changing departments and meeting all of the Chemistry Department's requirements. Bill Lipscomb and Maurice Visscher were co-chairs of the committee, with John Wertz and Victor Lorber as the other members from chemistry and physiology respectively. It is quite interesting that the thinking of those interested in active transport at the time was dominated by the idea that there were carriers. So John Wertz and I examined weak complexes of the sodium ion, in search of a potential carrier.[2] Only a decade later did I come to realize that the secret of active transport lay in allosteric transitions.[3]

My Ph.D. final in July 1956, however, remained a trauma remembered for life. After presenting my thesis and answering a few specific and a few general philosophical questions, which I thought were rather easy, I was sent out of the room and I waited . . . I waited for at least half an hour, which seemed like eternity. Finally Bill opened the door and congratulated me. "What did I do wrong that it took you so long to reach a decision?" I asked. "Oh, no," he answered, "we were done with you in thirty seconds, but we see each other so seldom that we had a lot of other things to discuss!"

In 1955 Bill invited Linus Pauling to give a lecture in Minnesota, which inspired me to spend a postdoctoral year with Pauling at Caltech. There I encountered an amusing restriction. No radios or musical instruments were allowed in the Gates and Crellin Laboratories. Why? Because some years ago a graduate student named Bill Lipscomb used to practice clarinet there, and distracted everyone else.

Our contact remained interrupted for the year I spent at Caltech and my first two years on the Harvard faculty. However, as soon as Bill arrived at Harvard in 1959, we began seeing each other socially as well as at school, since my first wife, Christine, who got her

Ph.D. in biochemistry and also minored in physical chemistry, knew Lipscomb well and offered to help them get settled in Massachusetts. His first complaint was, "You know what they did to me? They made me department chairman!" The implication was clear: he did not come to Harvard to become an administrative flunky. That is how the Great Gods of the Faculty of Arts and Sciences regarded their chairs, in contrast to the medical school where department chairmen were the only Gods. But he discharged his duties with the same seemingly effortless excellence that characterized everything else he did.

The 1960s were an exciting time in the still barely emerging field of molecular biology — better known at the time as the "practice of biochemistry without a license." Practically all of the two or three dozen — mostly young — people engaged in it knew each other personally and encountered each other frequently in one of its three centers — Cambridge, Caltech, and Harvard. Structural biology did not yet exist, although the discovery of the DNA structure by Watson and Crick and the determination of the myoglobin and hemoglobin structures by Perutz and Kendrew made it clear that the use of physical methods would allow an understanding of biological phenomena in terms of chemical structure. It is in this climate that Bill Lipscomb branched out from inorganic chemistry to X-ray diffraction studies of protein structure and function, while I decided to explore the potential of NMR for the same purpose.

The Establishment was not always helpful. Arriving at Harvard I was surprised that there was a serious debate as to whether Jim Watson should be given tenure. For many years there were those who doubted the correctness of the Watson–Crick structure. I was called in by the dean of the medical school and was advised not to apply for a large grant to explore NMR in molecular biology in order to not embarrass the university since I was not likely to get it. Fortunately, I did get it and Bill Lipscomb communicated some of our earliest results on peptides to the Proceedings of the National Academy of Sciences. As it became known that a large part of his program was shifting to protein crystallography, a distinguished member of the Chemistry Department grumbled in my presence, "We hired this guy to be our inorganic chemist. I don't know what he is thinking about, but he is surely not thinking of the best interests of the university!" Lipscomb's response to this was, "That is why there is stagnation in universities."

In the spring of 1966 Lipscomb spent a few weeks in Cambridge, UK, where I was also spending a sabbatical year. Among discussions with Max Perutz and his colleagues at the then still new, now old, MRC Lab, and dinners at our house, one memory stands out: a concert at Pembroke College featuring, among others, a Brahms sonata — William Lipscomb, clarinet, and Christopher Longuet-Higgins, piano.

The first structure determination of a protein in Lipscomb's lab, that of the enzyme carboxypeptidase A, was not nearly as straightforward and painless as it may appear from the subsequent published literature. The precedent set by the determination of the myoglobin structure by Kendrew, and subsequently of the enzyme lyzozyme by David Phillips in London, was to use the known primary covalent sequence of the protein as an additional constraint in fitting a three-dimensional model to the X-ray data. The sequence of carboxypeptidase A had been determined, but not yet published, and Hans Neurath and Bert Vallee, who had it, refused to release it to the Lipscomb group. On more than one occasion during that time the normally warm and hospitable corner office in Gibbs felt as if it was constraining a boiling temper. Lipscomb's ultimate response was remarkably constructive — to obtain data at sufficiently high resolution, so that the sequence could be determined from the X-ray data. So much for non-cooperation.

As I moved to Stanford in 1969, our meetings became less frequent, but still allowed an exchange of views on all major developments in science, as they occurred. After Bill received the Nobel Prize in 1976, we both attended a meeting of the enzyme group in Santa Barbara. I was sitting next to him, and, since this was still in the days before complete prohibition, I started filling my pipe. Suddenly, he grabbed my arm and said, "Don't! Don't smoke, I can't stand it." I was startled. I don't remember ever having smoked in his presence before — I only started after several years on the Harvard faculty — but others, notably Maurice Visscher, who was rarely seen without his pipe, must have. So I was left to ponder the effects of a Nobel Prize on an allergy to smoking. A record of our divergent views on the subject was made at his 65th birthday symposium in Denver — again, with my sitting on his right holding a pipe, and his holding his nose in a gesture of disgust.

As I took on the directorship of the School of Structural Biology and Magnetic Resonance at the Ettore Majorana Centre for Scientific Culture in Erice, Sicily, in 1989, I asked Bill to always give the introductory overview lecture on protein structure and dynamics, which he continued to do well into the 21st century.

Two subjects dominated our many discussions in the 1990s. The first was protein dynamics and some of the thoughts we discussed were reflected in reviews which each of us wrote. But the reality is easier to express uncensored in conversation than it is in print. The grim reality of the subject is that one can imagine whatever one wants, but there are no methods that would allow one to describe any dynamic process precisely and without frightening uncertainty. Crystallography allows only the definition of the initial and final states of any dynamic transition — or the presence of disorder, which may be motional. NMR is very sensitive to dynamics, but reflects time and ensemble averages of several processes simultaneously. One can build models that fit the data, but then one can usually build more than one and have no basis for making a choice. The only theoretical method — molecular dynamics — that in principle could provide a complete description of all dynamic events, in practice suffers from far too many limitations to provide more than a rough guess of the answer — a multitude of simplifying assumptions and the prohibitive cost of carrying computations on the longer time scales on which protein dynamic events really occur. What one can say about such events borders on science fiction. But this does not mean that it is uninteresting and could not be true.

The other subject was peer review. Both of us had applications to the National Institutes of Health (NIH) rejected on the basis of reviewers' comments that ranged from simply illiterate to ludicrous. Letting off steam on the subject privately was useful, since it was not useful to do it publicly. I was particularly impressed by a rejection of Bill's NIH application on the grounds that the proposed program was far too ambitious and should be scaled down. Bill's answer was to complete all the proposed structures while fighting the rejection and present them as a progress report in a re-application. It did not do all that much good, because only a very modest grant was finally funded.

I last saw Bill Lipscomb when he came to give a talk at the symposium marking my retirement from Stanford in October 2006 at the age of 77. I reminded him of Maurice Visscher's saying that you don't feel old until you go to the retirement parties of your students. "Nonsense," he responded, "it is just like another Ph.D. exam."

References

1. Jardetzky O. and Roberts G. C. K., *NMR in Molecular Biology*. New York: Academic Press (1981).
2. Jardetzky O. and Wertz J. E., The complexing of the sodium ion with some common metabolites. *Arch. Biochem. Biophys.*, **65**, 569–572 (1956), a paper first summarily rejected for its irrelevance to biology, but then praised by the editor, Paul Boyer, in the acceptance letter as "the beginning of biological applications of NMR."
3. Jardetzky O., Simple allosteric model for membrane pumps. *Nature*, **211**, 969–270 (1966).

Reprinted from the PROCEEDINGS OF THE NATIONAL ACADEMY OF SCIENCES
Vol. 58, No. 6, pp. 2212–2219. December, 1967.

SYSTEMATIC ANALYSIS OF CHEMICAL SHIFTS IN THE NUCLEAR MAGNETIC RESONANCE SPECTRA OF PEPTIDE CHAINS, I. GLYCINE-CONTAINING DIPEPTIDES*

BY ASAO NAKAMURA† AND OLEG JARDETZKY‡

DEPARTMENT OF PHARMACOLOGY, HARVARD MEDICAL SCHOOL

Communicated by William N. Lipscomb, August 16, 1967

The interpretation of the nuclear magnetic resonance (NMR) spectra of proteins in structural terms is predicated on a quantitative knowledge of the spectral changes which accompany the polymerization of amino acids and their arrangement in specific secondary and tertiary structures. Despite some elegant work,[1–4] the decade which elapsed since the high-resolution NMR spectra of amino acids[5, 6] and proteins[7, 8] were first observed and qualitatively accounted for[9] has left unanswered some of the most important questions, such as (1) the extent to which chemical shifts observed upon the incorporation of an amino acid into a peptide chain reflect the formation of a peptide bond at the N and the C terminal, respectively; (2) the effects of a given residue on the shifts of its nearest and next-nearest neighbors along the peptide chain; and (3) the side-chain dependence of the shifts observed on titration[5, 6, 10] and helix-coil transition.[11–15]

It is the purpose of this and subsequent papers to show that the chemical shifts accompanying the incorporation of a given amino acid into a polypeptide chain can be systematized in terms of a few, relatively simple rules. The findings offer little hope for the use of NMR for the analysis of the primary sequence of large peptides, but underscore its usefulness in the study of secondary and tertiary structure.

Experimental.—Most of the peptides were purchased and, if spectroscopically pure, were used without purification.

All spectra were recorded on a Varian DP-60 high-resolution spectrometer, as previously described.[15] The measurements were made in DCl, D_2O, or NaOD solution at 30°C, unless otherwise specified. Sodium 2,2-dimethyl-2-silopentane-5-sulfonate (DSS) was used as an internal and hexamethyl-disiloxane (HMS) as an external reference. All shifts are reported in counts per second (cps) or parts per million (ppm) from DSS. The concentration of each peptide was 0.05 *M*. The ionic form of each solute was identified by pD measurements using a glass electrode with a Beckman model G pH meter. The empirical correction pD = pH + 0.4 has been applied to all measurements. Solutions of peptide cations showed pD values of 0.9–1.0, those of zwitterions 5.4–6.0, and those of anions 12–13. In the case of peptides containing either more than one acidic or basic group, the pD value of the solution was adjusted so as to correspond to the fully ionized form.

Results and Discussion.—Additivity of shifts: The chemical shifts of glycine-CH_2 protons in peptides containing a glycine residue are summarized in Table 1. The data show several interesting features: (1) All chemical shifts of the CH_2 group of glycine tend to cluster about six mean values, as follows: (*a*) glycyl-amino acid cations, 231.8 cps from DSS; (*b*) glycyl-amino acid zwitterions, 228.8 cps; (*c*) glycyl-amino acid anions, 199.0 cps; (*d*) aminoacyl-glycine cations, 242.5 cps; (*e*) aminoacyl-glycine zwitterions, 227.2 cps; and (*f*) aminoacyl-glycine anions, 225.0 cps. This immediately suggests that the shifts resulting from dif-

TABLE 1

Chemical Shifts of Glycine-CH_2 Protons in Peptides Containing a Glycine Residue

Peptide	Cation	Zwitterion	Anion
Glycine	231.2	212.8	185.6
Glycyl-L-alanine	231.4	228.8	199.3
Glycyl-L-valine	233.7	232.6	201.3
Glycyl-L-leucine	232.3	229.9	199.4
Glycyl-L-isoleucine	233.5	231.7	200.8
Glycyl-L-methionine	232.9	230.7	201.2
Glycyl-L-serine	234.8	233.8	202.1
Glycyl-L-proline	239.8	236.8	206.2
Glycyl-L-aspartic acid	232.5	230.9,[a] 228.9[b]	199.4
Glycyl-L-glutamic acid	233.2	230.7,[c] 230.2[d]	200.1
Glycyl-L-lysine	233.0	233.5[e]	206.9,[f] 200.3
Glycyl-L-histidine	231.9	230.1,[g] 220.4[h]	196.3
Glycyl-L-tryptophan	227.4 / 221.2	—*	189.0
Glycyl-L-phenylalanine	226.5	227.1 / 220.3	192.9
Glycyl-β-alanine	226.9	226.4	196.6
L-alanylglycine	243.8	231.3 / 224.8	226.1
L-valylglycine	244.8	235.2 / 222.4	226.7
L-leucylglycine	244.0	233.5 / 221.5	225.5
L-serylglycine	245.5	235.2 / 226.4	227.4
L-prolylglycine	244.8	230.5 / 225.5	225.8
L-hydroxyprolylglycine	244.8	228.3	225.8
L-lysylglycine	243.8	228.2,[i] 226.3[j]	225.9
L-histidylglycine	245.8 / 240.2	235.1 / 221.7,[k] 226.3[l]	223.1
L-tryptophylglycine	239.2 / 232.6	239.7 / 215.4	216.3
L-phenylalanylglycine	239.0	231.6 / 213.3	223.4 / 217.3
L-tyrosylglycine	239.9	232.0 / 214.2	225.6 / 219.4
β-alanylglycine	241.0	226.0	225.2

Cps from DSS as an internal reference (60 Mc, conc. 0.05 M/l).
(a) pD 3.61, (b) pD 6.94, (c) pD 3.79, (d) pD 6.85, (e) pD 3.59, (f) pD 8.47, (g) pD 4.80, (h) pD 7.60, (i) pD 5.88, (j) pD 9.58, (k) pD 4.11, (l) pD 6.61.
* Solubility is not adequate for measurement.

ferences in the chemical nature of the nearest neighbor (at either the N- or the C-terminal) are an order of magnitude smaller than the changes in the chemical shift resulting from titration or from the formation of a peptide bond. (2) The chemical shifts of glycine in glycyl-amino acid cations and glycyl-amino acid zwitterions are almost identical (within 3 cps), and so are the shifts in aminoacyl-glycine zwitterions and aminoacyl-glycine anions, respectively (within 2.5 cps), indicating that the charge on the glycine-amino group dominates the shift, whereas the charge on the neighboring amino acid has only a small effect. (3) The shift of the CH_2 peak in aminoacyl-glycine anions (226.0 cps) is approximately 40 cps (0.66 ppm) to lower fields as compared to the glycine anion (185.6 cps). Since the state of protonation of glycine is the same in the two cases, this large shift can be attributed directly to the formation of the peptide bond at the N-terminal. In contrast, (4) the shifts of the CH_2 peak in glycyl-amino acid cations (231.8 cps) are almost identical to that in free glycine cation (231.2 cps). From this one may infer that formation of a peptide bond at the C-terminal makes only a small contribution to the over-all chemical shift. Finally, (5) in a number of aminoacyl-glycines and some of the glycyl-amino acids, a chemical shift nonequivalence for the two methylene protons is observed, giving rise to an AB pattern in place of the usual glycine singlet. This has been previously observed by Mandel[16] for the middle residue of leucyl-glycylglycine and has been recently discussed by Morlino and Martin.[17] Our findings are in agreement with theirs, except that because of the better resolution of our instrument we have been able to detect splitting in L-alanylglycine, L-serylglycine, and L-prolylglycine. Our interpretation of the phenomenon differs somewhat from theirs and will be discussed in the last section.

The chemical shift differences apparent in Table 1 can be compared in a variety of ways. The most obvious and most informative comparison is between two species which differ by a single chemical change such as the glycine anion and the X-glycine anion, where the only chemical change is the substitution of the uncharged amino acid residue X- for a proton on the uncharged amino group of glycine. A comparison between the glycine zwitterion and the X-glycine zwitterion, on the other hand, reflects at least two chemical steps, i.e., the removal of the charge from the glycine-amino group, *and* the substitution of a positively charged residue X at the N-terminal.

There are, in all, eight single-step comparisons which can be made on the data in Table 1. These are indicated diagrammatically in Figure 1. A single chemical shift term is assigned to each step. The remaining comparisons in Figure 1 are between species which differ by more than one chemical step. According to the above analysis, these over-all shifts may be expressed as a sum of terms, each reflecting one individual step.

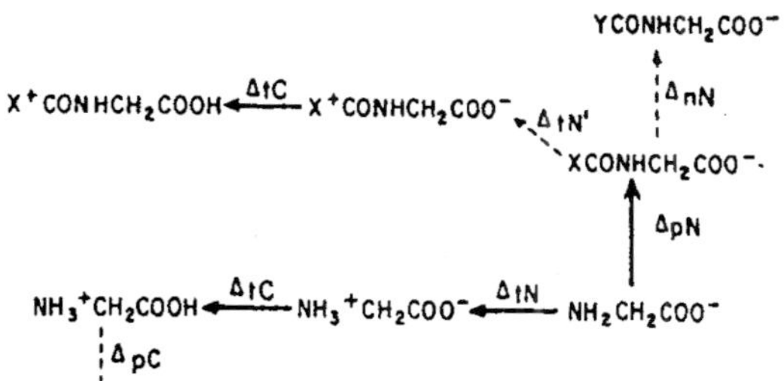

FIG. 1.—Systematics of single-step chemical shifts in glycine peptides.

The analysis of the data in Table 1 and related data reported elsewhere[15, 18] in the foregoing manner leads to two important conclusions: (1) Chemical shifts which accompany the incorporation of a given amino acid residue into a polypeptide with a defined primary and secondary structure fall into two nonoverlapping classes, when grouped according to their magnitude. (2) The shifts are indeed additive to a very high degree of approximation. For convenience we shall refer to all shifts which fall into the range of 0.3–1 ppm as *first-order shifts* and to those which fall into the range of 0.01–0.2 ppm as *second-order shifts*. It is noteworthy that first-order shifts resulting from changes in the primary or secondary structure are observed only for α-C protons, whereas second-order effects may be seen for both the α-C and the side-chain protons.[1–8, 11, 15] Large chemical shifts affecting the side-chain protons may thus be taken as reflecting the tertiary structure of a polypeptide chain. The magnitude of the tertiary structure shifts for α-C protons remains to be established, but because of the geometry of the polypeptide chain it is likely to be large only in exceptional circumstances.

First-order shifts: Three first-order shifts are apparent in Table 1:

(1) The shift due to titration of the N-terminal Δ_{tN} is given by

$$\begin{aligned}\Delta_{tN} &\equiv \delta(A)^{\pm} - \delta(A)^{-}\\ &\cong \delta(AX)^{\pm} - \delta(AX)^{-}\\ &= 0.45 \pm 0.01 \text{ ppm},\end{aligned}$$

where $\delta(A)^{\pm}$ and $\delta(A)^{-}$ are chemical shift values of the α-CH protons of the free amino acids A in its zwitterionic and anionic form, and where $\delta(AX)^{\pm}$ and $\delta(AX)^{-}$ denote the chemical shifts of the α-CH protons of amino acid A which forms the N-terminal residue of a dipeptide. Note that Δ_{tN} is a positive quantity denoting a downfield shift.

(2) The C-terminal titration shift Δ_{tC} is given by

$$\begin{aligned}\Delta_{tC} &\equiv \delta(A)^{+} - \delta(A)^{\pm}\\ &\cong \delta(AX)^{+} - \delta(AX)^{\pm}\\ &= 0.3 \pm 0.01 \text{ ppm},\end{aligned}$$

where $\delta(AX)^{+}$ and $\delta(AX)^{\pm}$ denote the chemical shifts of the α-CH protons of amino acid A which forms the C-terminal residue of a dipeptide in the cationic and zwitterionic form. Note that Δ_{tC} is also positive, since the corresponding shift is also to lower fields. The two titration shifts, Δ_{tN} and Δ_{tC}, have been known since the initial NMR studies of amino acids and have been used by Scheinblatt[19, 20] to identify the N- and C-terminal amino acids in peptides. The only noteworthy feature of the present data lies in the virtual independence of the titration shift from the nature of the neighboring amino acid at either the N- or C-terminal end. This is apparent from the small standard error of the mean for titration shifts in different peptides.

(3) A third type of shift, the N-terminal peptide shift Δ_{pN} is given by

$$\begin{aligned}\Delta_{pN} &\equiv \delta(XA)^{-} - \delta(A)^{-}\\ &\cong 0.66 \pm 0.01 \text{ ppm}.\end{aligned}$$

In arriving at the value for Δ_{pN}, the unusual chemical shifts seen with L-tryptophylglycine, L-phenylalanylglycine, and L-tyrosylglycine were omitted.

For the sake of completeness, a fourth first-order shift should be added to those discussed above:

(4) This is the helix-formation shift Δ_h given by

$$\begin{aligned}\Delta_h &= \delta(A)_{\text{helix}} - \delta(A)_{\text{coil}}\\ &\cong -0.4 \text{ ppm},\end{aligned}$$

where $\delta(A)_{\text{helix}}$ and $\delta(A)_{\text{coil}}$ represent the chemical shifts of the α-CH protons of amino acid A in the α-helical and random coil configuration.[15] Poly-L-glycine homopolymers are not known to form α-helices, and the helix-formation shift has not been measured in glycine-containing helical polypeptides. In view of the regularities observed in other polypeptides,[13–15] however, Δ_h for glycine is not likely to fall outside the range to +0.3 to +0.6 ppm.

The chemical shift of an α-CH peak in the spectrum of a polypeptide δ_{obs} may thus be predicted by the equation

$$\delta_{obs} \cong \delta(A)^{\pm} + \Delta_{tC} - \Delta_{tN} + \Delta_{pN} + \Delta_h. \qquad (1)$$

In the case of glycine CH_2 peaks in nonhelical polypeptides, the uncertain term Δ_h is not required, and chemical shifts may be predicted to within 0.1 ppm.

Second-order shifts: A closer inspection of the data reveals several much smaller, but nevertheless discernible shifts.

(1′) The C-terminal peptide shift

$$\Delta_{pC} \equiv \delta(AX)^{+} - \delta(A)^{+}$$
$$\cong 0.00\text{–}0.02 \text{ ppm.}$$

The nature of the nearest-neighbor peptide may influence the chemical shift of α-CH protons. Depending on the side on which the neighboring residue is attached, two second-order shifts are defined

(2′) The N-terminal neighbor shift

$$\Delta_{nN} \equiv \delta(XA)^{-} - \delta(YA)^{-}$$
$$\cong \pm 0.05\text{–}0.1 \text{ ppm.}$$

(3′) The C-terminal neighbor shift

$$\Delta_{nC} = \delta(AX)^{+} - \delta(AY)^{+}$$
$$\cong 0.0\text{–}0.2 \text{ ppm,}$$

where X and Y represent different amino acids. These two shifts are of particular interest because one might hope to utilize them for the identification of nearest neighbors, and hence for sequence analysis. Shifts of the order of 0.2 ppm, if subject to a systematic interpretation, would be quite adequate for this purpose, since they can now be measured with reasonable accuracy.

The titration of nearest neighbors may also influence the magnitude of α-CH chemical shifts. Two second-order titration shifts are defined by

(4′) The N′-terminal titration shift

$$\Delta_{tN}' \equiv \delta(XA)^{\pm} - \delta(XA)^{-}$$
$$\cong 0.05\text{–}0.1 \text{ ppm.}$$

(5′) The C′-terminal titration shift

$$\Delta_{tC}' \equiv \delta(AX)^{+} - \delta(AX)^{\pm}$$
$$\cong 0.05\text{–}0.1 \text{ ppm.}$$

Thus, a better approximation for the chemical shift of a glycine residue in the interior of a peptide chain is given by

$$\delta_{obs} = \delta(A)^{\pm} + A_{tC} - \Delta_{tN} + \Delta_{pN} + \Delta_h + (\Delta_{pC} + \Delta_{nN} + \Delta_{nC}). \qquad (2)$$

VOL. 58, 1967 CHEMISTRY: NAKAMURA AND JARDETZKY 2217

The validity of this approximation is illustrated in a sample calculation given in Table 2. The data to be presented in the second and third papers of this series as well as still incomplete observations on other amino acids are consistent with this notion of approximate additivity of chemical shifts attributable to different effects.

TABLE 2

COMPARISON OF OBSERVED AND CALCULATED LINE POSITION FOR GLYCINE-CONTAINING DIPEPTIDES

Residue	Calculated*	Observed
Gly (2) in triglycine (±)	4.11	4.07
Gly (2) in tetraglycine (±)	4.11	4.10
Gly (3) in tetraglycine (±)	4.06	3.98
Gly (2) in pentaglycine (±)	4.11	4.12
Gly (3) in pentaglycine (±)	4.06	4.00
Gly (4) in pentaglycine (±)	4.06	3.98

* δ_{calc} from equation (2), using values $\delta(Gly)^{\pm}$ = 3.55 ppm. Δ_{tN} = 0.45, Δ_{tC} = 0.3, Δ_{pN} = 0.66 ppm, Δ_{pC} = 0, $\Delta_{tN'}$ = 0.05.

Chemical shift nonequivalence of the glycine protons: This effect deserves further comment, inasmuch as two different views[16, 17] have been put forth as an explanation—i.e., (*a*) the existence of preferred rotamers and (*b*) incomplete averaging of electric field gradients in the presence of free rotation. The latter effect has been predicted by Pople[21] for all molecules containing an asymmetric carbon atom.

If the observed chemical shift nonequivalence were *solely* the result of the incomplete averaging of the electric field gradients of the positive charge on the asymmetric carbon atom of the substituent in the presence of free rotation, one would expect the following: (*a*) A pronounced nonequivalence of the βCH_2 protons in all amino acids containing an asymmetric α-carbon, since the effect of such gradients should be greatest for the immediate neighbors of the α-carbon. This is generally not found.[6, 10, 16, 22, 23] (*b*) A substantial effect of the positive charge of the neighboring amino acid on the average shift of the glycine protons, i.e., $\Delta_{tN'}$ comparable in magnitude to the shift difference of the A and B protons in the glycine AB quartet. In fact $\Delta_{tN'}$ is very much smaller. (*c*) Complete temperature independence of the relative shift of the A and B protons. The temperature dependence of the relative shifts in L-phe-gly, L-tyr-gly, L-val-gly is shown in Figure 2. It is in good agreement with the findings of Morlino and Martin[17] and indicates that the contributions reflecting a redistribution of rotamer populations are measurable.

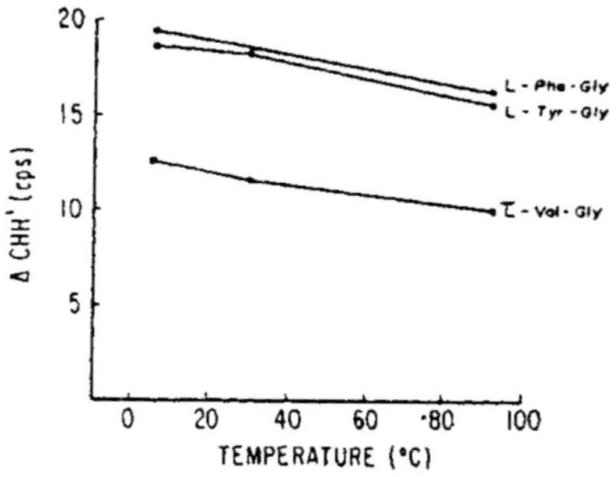

FIG. 2.—The temperature dependence of the relative shift of the glycine protons in L-phe-gly, L-tyr-gly, L-val-gly.

Thus our conclusion is that the observed nonequivalence reflects largely the restriction of rotation about these bonds. This, however, does not necessarily signify the existence of one or a small number of preferred rotamers, but rather a range of probable and a range of less probable rotamers.

The argument is diagrammatically represented in Figures 3 and 4. Figure 3 defines the two angles ϕ and ψ corresponding to rotations about the C_α'—CO and N—C_α bonds. Viewing the molecule along these bonds (which are almost parallel) in a direction from NH_3^+ to CO_2^-, we find that the values of ϕ allowed to the NH_3^+ and X groups, respectively, are not equal. This is indicated by the shading in Figures 4*a* and *b*. The energy barriers between the rotamers in which these two

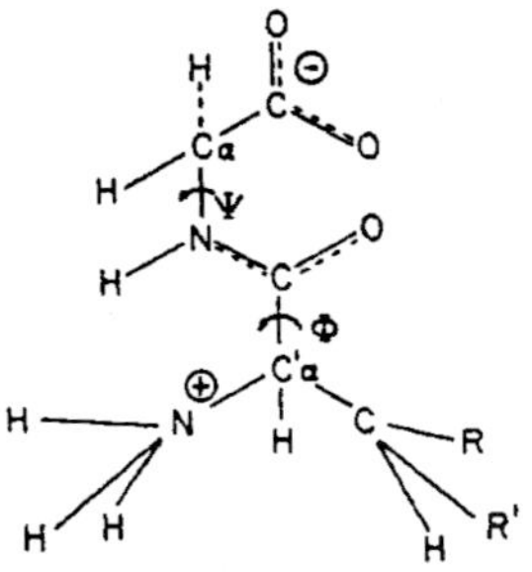

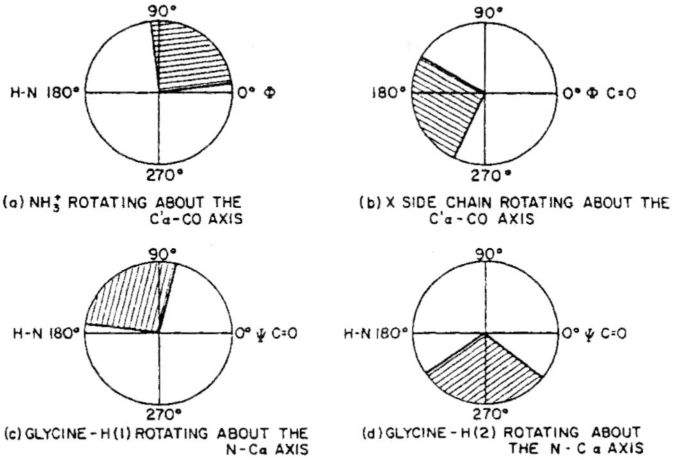

FIGS. 3 and 4.—Diagramatic representation of allowed and forbidden rotamers in aminoacyl glycines.

Forbidden regions are shaded; the view is along the C'α-CO and N-Cα axis in the NH_3^+ to COO^- direction.

groups fall into the unshaded areas are relatively low. On the other hand, barriers for rotation of each group through the shaded areas are relatively high. The angles ψ relatively accessible to each of the two CH_2 protons are similarly defined in Figures 4*c* and *d*. Superposition of (*c*) or (*d*) on (*a*) and (*b*), respectively, shows the range of configurations over which each of the CH_2 protons can be *cis* (with respect to the plane of the peptide bond) with either the amino group or with the side chain of its N-terminal neighbor. It is in these configurations that the direct effect of the neighbor will be largest. The extent of overlap is clearly different for the two protons of an aminoacyl-glycine, but much less so in the case of glycyl-amino acids, because the barriers for the rotation of an N-terminal glycine about the C—CO bond are relatively low. This accounts for the preponderance of chemical shift nonequivalence in the former.

The primary effect of charge in this view is to further restrict the averaging by increasing the lifetime of all rotamers in which the positive and negative charges appear *cis* to each other. Only where there is a large contribution from a magnetically anisotropic side chain are the shifts apparent without additional restrictions.

* This work has been supported by grants from the National Science Foundation (NSF GB-4697) and the U.S. Public Health Service (5-K3-GM-15,379-08).

† Present address: Central Research Laboratories, Ajinomoto Chemical Company, Kawasaki, Japan.

‡ To whom all correspondence should be directed. Present address: Department of Biophysics and Pharmacology, Merck Sharp & Dohme Research Laboratories, Rahway, New Jersey.

[1] Kowalsky, A., *J. Biol. Chem.*, **237**, 1807 (1962).

[2] Kowalsky, A., *Biochemistry*, **4**, 2382 (1965).

[3] McDonald, C. C., and W. D. Phillips, in *Proceedings of the 2nd International Conference on Magnetic Resonance in Biological Systems* (New York: Pergamon Press, in press).

[4] McDonald, C. C., and W. D. Phillips, *J. Am. Chem. Soc.*, in press.

[5] Takeda, M., and O. Jardetzky, *J. Chem. Phys.*, **26**, 1346 (1957).

[6] Jardetzky, O., and C. D. Jardetzky, *J. Biol. Chem.*, **233**, 383 (1958).

[7] Saunders, M., A. Wishnia, and J. Kirkwood, *J. Am. Chem. Soc.*, **79**, 3289 (1957).

[8] Saunders, M., and A. Wishnia, *Ann. N.Y. Acad. Sci.*, **70**, 870 (1958).

[9] Jardetzky, O., and C. D. Jardetzky, *J. Am. Chem. Soc.*, **79**, 5322 (1957).

[10] Bovey, F. A., and G. V. D. Tiers, *J. Am. Chem. Soc.*, **81**, 2870 (1959).
[11] Bovey, F. A., G. V. D. Tiers, and G. Filipovich, *J. Polymer Sci.*, **38**, 73 (1959).
[12] Glick, R. E., W. E. Stewart, and L. Mandelkern, *Biochim. Biophys. Acta*, **120**, 302 (1966).
[13] Stewart, W. E., L. Mandelkern, and R. E. Glick, *Biochemistry*, **6**, 143 (1967).
[14] Marlborough, D. I., K. G. Orell, and H. N. Rydon, *Chem. Commun.* (1965), p. 518.
[15] Markley, J. L., D. H. Meadows, and O. Jardetzky, *J. Mol. Biol.*, **27**, 25 (1967).
[16] Mandel, M., *J. Biol. Chem.*, **240**, 1586 (1965).
[17] Morlino, V. J., and R. B. Martin, *J. Am. Chem. Soc.*, **89**, 3107 (1967).
[18] Nakamura, A., and O. Jardetzky, submitted for publication.
[19] Scheinblatt, M., *J. Am. Chem. Soc.*, **87**, 572 (1965).
[20] Scheinblatt, M., *J. Am. Chem. Soc.*, **88**, 2845 (1966).
[21] Pople, J. A., *J. Mol. Phys.*, **1**, 1 (1958).
[22] Fujiwara, S., and Y. Arata, *Bull. Chem. Soc. Japan*, **37**, 44 (1964).
[23] Nakamura, A., *J. Chem. Soc. Japan*, **86**, 780 (1965).

Pergamon

Prog. Biophys. molec. Biol., Vol. 65, No. 3, pp. 171–219, 1996

0079–6107/96 $32.00 + 0.00

PII: S0079–6107(96)00010–7

PROTEIN DYNAMICS AND CONFORMATIONAL TRANSITIONS IN ALLOSTERIC PROTEINS

OLEG JARDETZKY

Stanford Magnetic Resonance Laboratory, Stanford University, Stanford, CA 94305-5055, U.S.A.

CONTENTS

I. INTRODUCTION

Understanding a protein as a molecular mechanism—which it is—requires detailed and accurate knowledge of its structure and dynamics. Our knowledge of protein structure is by now very extensive and growing almost exponentially. Over 1000 structures modeled at atomic resolution[1] have been determined by X-ray diffraction and to a lesser extent by high resolution nuclear magnetic resonance (NMR). Our knowledge of protein dynamics is by comparison still rather fragmentary. To a large extent this is due to the fact that the most powerful method for structure determination—X-ray diffraction—provides no direct information on motion, as B-factors and missing electron density do not distinguish between static and dynamic disorder. Even for a protein undergoing a structural transition, it presents a *fait accompli*—the detailed structures of the initial and final states. The same type of information can be derived from structural studies by NMR, which almost always assume that the structure is rigid.

Structural studies nevertheless set the standard and define what we need to know to understand the dynamics of a conformational transition—the time-course, the sequence of

[1] The term "atomic resolution" is used here in a broader sense than is customary among crystallographers who refer to the quality of the data and the electron density map. The broadening of the term to denote that atomic coordinates can be defined, whether by the data or by modeling, is desirable, because NMR data are always at atomic resolution.

events leading from the initial to the final state and the forces that drive it. It is the aim of this review to summarize what we do know about protein dynamics in the light of this standard. The focus is on key concepts and findings relating dynamics and function in systems for which dynamic information is available at atomic resolution. No attempt is made to provide an encyclopedic coverage of the literature.

The two atomic resolution methods—crystallography and NMR—differ in an important respect. High resolution NMR allows a number of time-dependent measurements—notably measurements of relaxation parameters, chemical shift averaging and backbone proton exchange rates—that can be made independently of a structure determination and directly reflect the dynamics of a protein structure. These now make it possible to study different types of motions in proteins in considerable detail.

Internal motions in proteins can also be detected, though not described in detail, by a variety of spectroscopic methods monitoring environmental changes at a single site in a complex structure. Fluorescence transfer and depolarization, depolarized light scattering, electron spin resonance (ESR), resonance Raman and Mössbauer spectroscopy and flash photolysis are among those most widely used. From NMR and other spectroscopic studies we know that such motions occur over a wide range of time-scales—from picoseconds to minutes and longer. Theoretical molecular dynamics simulations of protein motions in an empirically calibrated force-field have also added greatly to our understanding of protein dynamics, but remain limited to the study of very fast (pico- to nanosecond) motions for both theoretical and practical reasons.

The fact that proteins move and undergo structural changes has been appreciated since the observation of Felix Haurowitz (1938) that deoxyhemoglobin crystals spontaneously shatter upon oxygenation. The structural representation of this conformational change—the first of its kind—was accomplished with the determination of the crystal structures of deoxy- and oxyhemoglobin by Perutz (1962). The dynamics and kinetics of this structural transition have generated a large literature and are still subject to debate (Kim *et al.*, 1995; Holt and Ackers, 1995). Theoretical arguments supporting the notion of protein mobility have been advanced even earlier (for a review see Klotz and Hunston, 1975). An important early line of evidence for the flexibility of globular proteins in solution came from the deuterium exchange experiments of Linderstrøm-Lang (1955) and Hvidt and Nielsen (1966). Extensive side chain mobility in proteins was also apparent in the early protein NMR spectra (Jardetzky and Jardetzky, 1962), making it possible to differentiate different ligand binding sites by differences in their dynamics (Jardetzky, 1964; Jardetzky and Wade-Jardetzky, 1965). Later it became evident that aromatic rings embedded in a closely packed structure were able to rotate relatively freely (Jardetzky, 1970; Campbell *et al.*, 1976) and that some proteins, such as the tobacco mosaic virus (TMV) contained flexible segments (Jardetzky *et al.*, 1978), or even entire independently moving domains, as the *lac*-repressor headpiece (Wade-Jardetzky *et al.*, 1979).

Three types of mobility identified by the early crystallographic and early NMR experiments respectively remain useful to define a framework for the understanding of motions in proteins and of their functional role. (1) *En-bloc* translocations of entire, relatively rigid domains which result from changes in relatively few angles in the backbone, but are amplified by a lever effect, are typically the movements seen by crystallography. The time-scale of such movements is usually in the micro- to millisecond range. The movements detected by NMR reflect two additional categories of processes. (2) Diffuse concerted motions within the structure—the structure yielding to allow the rotation of normally closely packed side chains, and possibly of larger sub-domains. Typically these are observed as aromatic "ring flipping," but the movements of the surrounding structure which make ring flipping possible are poorly understood. The known ring rotations are on the time-scale of microseconds to seconds, although ring librations on a nanosecond time-scale are also known to occur (Nelson *et al.*, 1976). (3) Flexibility of a longer contiguous segment of the polypeptide backbone. This may be evident on time-scales varying from pico- to milliseconds or longer.

Each of these types of motions has a functional correlate. The functional correlate of domain translocation *en-bloc* is to facilitate or break contacts—either to substrates and products, as in the case of allosteric enzymes—or to other macromolecules as a key step in a variety of control and signal transmission mechanisms. The functional correlate of the concerted movements within a protein domain is to facilitate the binding of substrates and the diffusion of products from an enzyme-active site. The functional correlate of segmental flexibility, seen particularly in DNA-binding proteins, is to optimize local contacts or "induced fit" between a protein and its ligand.

II. DESCRIPTION OF CONFORMATIONAL TRANSITIONS FROM STRUCTURAL STUDIES

2.1. *Allosteric Changes of Quaternary Structure*

Structure determination provides several clues to the motions that are necessary to produce a structural rearrangement—a conformational change—but does not offer any detailed insight into their nature. The difference in the conformational states of deoxy- and oxyhemoglobin, and the two state allosteric model of Monod *et al.* (1965) which it has inspired, defining a "relaxed" (oxygenated) R state and a "tense" (deoxygenated) T state, have become textbook classics. The "concerted" or "MWC" model focuses on the rearrangement of the quaternary structure to define the two conformational states, as does the alternative "sequential" model proposed by Koshland *et al.* (1966), which permits a larger number of intermediate states. This focus on changes in quaternary structure is hardly surprising, since it reflects the most dramatic structural rearrangement seen in crystallographic studies. The applicability of either model, or of their more sophisticated variants, such as the "symmetry-rule mechanism" proposed by Ackers (Ackers *et al.*, 1992; Huang and Ackers, 1995; Holt and Ackers, 1995), is tested by kinetic or thermodynamic experiments in which the protein structure can be represented at a cartoon level and otherwise treated as a black box. It was noted 20 yr ago (Stryer, 1975) that the kinetics of some allosteric systems seem to be adequately described by the concerted MWC model, those of others by the sequential model and those of some by neither model. The important point to consider in the present context is that kinetic models of an allosteric transition beg the question of the dynamic mechanism in its entirety. It is therefore not surprising that, as noted by Franzen *et al.* (1994), "despite a large number of studies of the molecular structure of hemoglobin, the detailed mechanism of the cooperative transition ... is still not fully understood." The same conclusion was reached in a recent review of cooperativity in enzyme function by Neet (1995). Since the focus of the present review is on studies that can elucidate the sequence of events in structural transitions at an atomic level, thermodynamic and kinetic analysis is beyond its scope. The reader interested in the thermodynamic and kinetic description of allostery is referred to the article by Neet and the references cited therein.

The problems of relating kinetic measurements to a mechanical model have recently also been discussed by Tian *et al.* (1996), who used double pulse flash-photolysis experiments on carbon-monoxymyoglobin (MbCO) to define the time-scale of conformational averaging between two states they called "open" and "closed." The rates were found to lie in the range 10^4–10^6/sec and to depend on solvent composition, but the structural changes involved in the observed transition remain undefined.

The first structurally well-defined step in the sequence of events constituting the R→T transition in hemoglobin has been identified by time-resolved resonance Raman spectroscopy following dissociation of carbon monoxide from carbomonoxyhemoglobin (HbCO) by flash photolysis in the cited paper by Franzen *et al.* (1994). The transition could be followed by observing three changes in the vibrational spectra: the disappearance of the iron-ligand stretching vibration at 505/cm upon the breaking of the iron–CO bond, the appearance of the iron-histidine F8 stretching vibration at 212/cm seen only in the deoxy-form, and the appearance of a vibrational mode at 304/cm, also seen only in the deoxy-form and assigned to the "doming" of the heme—the out-of-plane displacement of the heme iron. The three changes are apparent within 350 fsec and are essentially complete within 1 psec, followed by a frequency shift of the iron-histidine F8 line on a microsecond time-scale which can be attributed to larger scale conformational changes involving the rotation of the histidine and possibly of the entire F-helix. These findings directly confirm the suggestion originally made by Perutz (1962, 1990) that the out-of-plane movement of the heme iron serves as a trigger for the R→T transition, which had been corroborated less directly by previous spectroscopic experiments and theoretical simulations (e.g., Henry *et al.*, 1985).

A more detailed picture of subsequent slower conformational changes, on the time-scale of nano- to microseconds, has been proposed by Spiro and coworkers (Jayaraman *et al.*, 1995), who had followed changes in the UV resonance Raman spectra of different tryptophan and tyrosine chromophores scattered throughout the Hb structure giving rise to bands labeled W17, W16, Y9a, W7, W18, W3. On a microsecond time-scale, four time constants (ranging from 0.01 to 0.25 to 17 to 500 at pH 7.4) can be identified and the authors associate them with four intermediates (B, R, S, T′). The interpretation of the spectroscopic changes is however complicated by the fact that the photodissociated CO tends to recombine with the iron (geminate recombination, as high as 55%), with a time constant of 20 nsec, so that one is not dealing with the spectra of pure species appearing in sequence, but with a mixture of states, and further by the fact that the bands are not uniquely assigned to individual tryptophans in the sequence. For example, W3 has components from βTrp37 as well as from αTrp14 and βTrp15. The basic problems of interpreting resonance Raman spectra in terms of specific structural changes have been previously discussed by Cho *et al.* (1994). The method has an excellent time resolution, but no significant spatial resolution. The structure of kinetic intermediates identified by a sequence of Raman spectra cannot be defined and any structural interpretation remains speculative.

With a different approach, a three-state mechanism has been suggested by Srinivasan and Rose (1994), based on the crystallographic finding of an additional liganded state, called Y or R2 (Smith and Simmons, 1994). The three crystal structures (T, R and R2) were analysed in terms of difference distance maps, which suggested that the classic R structure was intermediate between T and R2/Y, raising the possibility that the R2/Y structure corresponds to the true, physiological fully oxygenated state of hemoglobin, with R being a crystallizable intermediate of questionable physiological relevance. The structural comparison also provided important clues to a possible underlying change in tertiary structure: in the transition from T to R the β-turns at the N-termini of helices E and F assume a strained conformation, but in the transition R to R2/Y they return to the conformation found in the T state.

A different three-state mechanism has been proposed by Philo *et al.* (1996), who invoked a "locked" T state to account for their kinetic measurements on hemoglobin valency hybrids and came to the conclusion that the stereochemical mechanism proposed by Perutz (1979) cannot provide a complete (though possibly a partial) explanation for the coupling of heme ligation to conformational changes in the protein.

A promising, though very labor intensive approach to the study of intermediates in the R→T transition may be the crystallographic examination of cross-linked hemoglobins, reported by Schumacher *et al.* (1995). It is too early to say to what extent cross-linking is simply a trap for short-lived naturally occurring intermediates, and to what extent it results in structural changes not found in the natural pathway. In the latter case, sorting out intermediates on the normal pathway from experimental artifacts may present serious problems. To complicate matters, Perutz *et al.* (1994) found by studying the chloride effect on the oxygen affinity of hemoglobin that chloride exerted this effect by neutralizing electrostatic repulsion from an excess of positive charges in the central cavity of the molecule, without binding to any specific "allosteric" site; thus defining a new kind of allosteric mechanism, which would not be expected to have a unique pathway for the transmission of the allosteric signal. Abraham *et al.* (1995) have described that different effectors bound to the same residue can have opposite effects on the RT equilibrium. Allosteric behavior of hemoglobin can be abolished by mutations and restored by compensatory mutations (Kim *et al.*, 1994). A newly constructed recombinant hemoglobin (α96Val→Trp) can undergo a switch in quaternary structure without changing the ligation state of the heme iron (Kim *et al.*, 1995).

In summary, we are led to conclude not only that the RT transition in hemoglobin, the prototype of all allosteric transitions, remains an unsolved problem, but also that the methods by which it can currently be studied do not contain enough information to solve it completely—in the sense of yielding the time-course of the structural change and the forces driving it.

Several additional examples of quaternary structure rearrangements in proteins subject to allosteric control have been described in recent years. The subclass of five proteins for which both the R-state and the T-state crystal structures are known and which show positive cooperativity on ligand titration (sigmoidal titration curve)—including hemoglobin, aspartate transcarbamylase (EC 2.1.3.2, ATCase), glycogen phosphorylase (EC 2.4.1.1, GP), phosphofructokinase (EC 2.7.1.11, PFK) and fructose-1,6-bisphosphatase (EC 3.1.2.11, FBP)—has been reviewed by Lipscomb (1991). The common feature of these quaternary rearrangements are rotations of the subunits with respect to each other, which vary from 7° for PFK to 10° for GP, to 15° for hemoglobin, to 19° in FBP, and are most pronounced in ATCase, where the catalytic trimer rotates by 12°, whereas the regulatory dimer rotates by 15°, leading to an 11 Å separation of the subunit surfaces and the establishment of new contacts. The establishment of a new interface and changes in the affinity at the substrate binding site are the key features of the structural changes. In all instances the secondary and tertiary features of the structures are largely preserved and the large rearrangement of structured domains is accompanied only by small changes in a few Ramachandran angles of the backbone.

In the case of PFK the opening of a turn of helix 6 in the R→T transition has been seen and the loop between helix 6 and β-strand F has been thought to be important in the allosteric transition (Schirmer and Evans, 1990). More recently, residue 161 located in this loop has been directly implicated by mutant studies (Auzat *et al.*, 1995a) in the inhibition of the enzyme by phosphoenolpyruvate (PEP), but not in the activation by GDP. Thus the mutant Q161A was shown to be insensitive to PEP, but not to GDP. A more detailed mechanism for the conversion of tertiary into quaternary conformational changes has been proposed (Auzat *et al.*, 1995b), based on the observation that an R152K mutation led to a two-fold increase in the Hill coefficient, pointing to the importance of electrostatic interactions between E148 and R152 in regulating the movement of the two helices containing the respective residues. The proposal is plausible, but the evidence is very indirect. The understanding of the allosteric changes in PFK may be complicated further by studies of tryptophan fluorescence on ligand binding which suggest the existence of more than two conformational states (Auzat *et al.*, 1995c).

In the case of FBP a comparison of the R-state structures (fructose-2,6-bisphosphate and fructose-6-phosphate complexes) was made with the T-state structures (complex with the allosteric regulator AMP and the quaternary AMP-fructose-2,6-bisphosphate, Zn^{++} complex, Xue *et al.*, 1994, and references therein). FBP is a tetramer of four identical subunits (C1–C4) each of which contains a substrate binding site and an allosteric site. The monomers are paired as two dimers C1-C2 and C3-C4 and the R→T transition involves a 17° rotation of the C1-C2 dimer with respect to the C3-C4 dimer (Y. Zhang *et al.*, 1994). This is shown in Fig. 1. In the T-state helix H4 (residues 123–127) is better defined than in the R-state and interacts with Tyr258 and Arg243 of the adjacent monomer.

During the R→T transition caused by the binding of AMP a relative rotation of 2.3° occurs between the active site domain and the AMP domain (Y. Zhang *et al.*, 1994). This movement is accompanied by conformational changes at most of the metal-binding residues at the active site, and may be transmitted by the eight-stranded β-sheet. Strand β3 (113–118) of this β-sheet connects Lys112 and Tyr113, both of which bind to AMP, to AMP118 and other nearby metal-binding residues. These conformational changes could be effected by movements of the helices H1, H2 and H3 which propagate structural

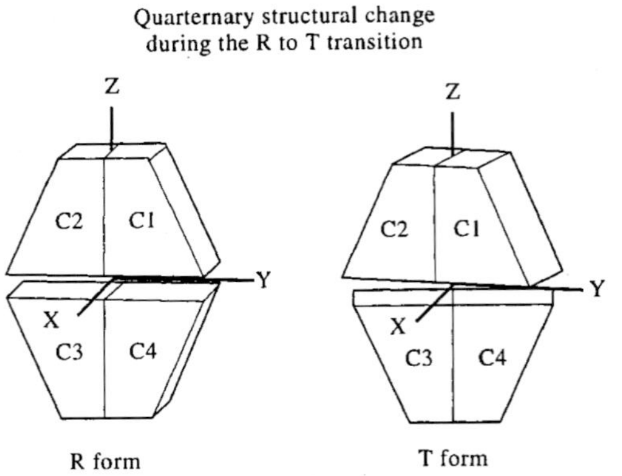

Fig. 1. Schematic drawing showing the coordinate system used in describing the R and T structure of Fru-1,6-Pase. The origin superimposes with the geometric centers of both the R and T structures, and the *x*, *y* and *z* axes correspond to their 3 principal axes. The three two-fold symmetry axes are also coincident with the *x*, *y* and *z* axes. The monomers are labeled counter-clockwise as C1, C2, C3 and C4. The R to T transition involves a 17° dimer rotation around the *Z* axis. (Fig. 1 from Y. Zhang *et al.* © 1994, reprinted by permission of the publisher, Academic Press Limited, London.)

changes from the AMP site to the C1–C4, C1–C3 and C1–C2 interfaces. The quaternary change displaces the less strongly bound catalytic metal ion so that it cannot bind to the 1-phosphate of the substrate, fructose-1,6-bisphosphate. In addition to inhibition by AMP at the allosteric site, FBP is inhibited at the active site by fructose-2,6-bisphosphate. This inhibition is synergistic with that of AMP, and adds another feature to the regulator mechanism which involves the short helix H4 (123–127). In particular, H4 is more ordered when AMP in the presence of Zn^{++} converts the R-state binary complex (enzyme + F-2,6-P_2) to the T-state, and new interactions are established from H4 to residues Tyr258 and Arg243 from an adjacent monomer. The synergistic cooperativity between the two inhibitors AMP and F-2,6-P_2 involves a separate pathway from that for inhibition by AMP of the hydrolysis of the substrate F-1,6-P_2. In particular, this synergism very probably involves the AMP site, the adjacent $\beta 3$ strand, the metal site immediate to the active site, the helix H4 and Tyr258 and Arg243 from the adjacent monomer within the upper (or lower) dimer (Fig. 1). In summary, the AMP-only pathway and the additional pathway through H4 for the synergism may be important components of a more complex set of pathways (Lipscomb, 1991). Moreover, mutations may create entirely new pathways for allosteric effects (Lipscomb, 1994).

The relationship between tertiary and quaternary conformational changes appears to be even more complex in ATCase, and remains to be unraveled (Lipscomb, 1994). In accord with the X-ray diffraction studies and with earlier studies of mutants, Glu50 in the catalytic chain has been implicated in the stabilization of the R-structure (Tuac *et al.*, 1994). The tertiary change in the 240s loop has been monitored by fluorescence spectroscopy, while the quaternary transition was monitored by small angle X-ray scattering, leading to the conclusion that the quaternary change is complete when four of the six catalytic sites are occupied, while the change in the 240s loop reached its maximum value only when all sites were occupied (Fetler *et al.*, 1995a, b). Residues 146–149 of the regulatory chain and residues 242–245 of the catalytic chain were found in mutant studies to be involved in transmitting the effect of ATP (De Staercke *et al.*, 1995), but the picture of the signaling pathway is far from complete. More results emerge from low angle X-ray scattering of the titration of the T→R equilibrium by n-phosphonoacetyl-L-aspartate (PALA). These scattering curves are given to good approximation by a two-state model in the presence of various concentrations of PALA for ATCase-PALA, ATCase-PALA-CTP or ATCase-PALA-ATP. The results indicate that CTP slightly decreases the proportion of molecules in the R-state, whereas ATP has no detectable effect. Thus homotropic cooperativity in the presence of PALA is simple and concerted, and at least part of the CTP effect acts directly on the equilibrium between quaternary structures, whereas ATP employs a totally different mechanism not directly involving the T→R equilibrium (Fetler *et al.*, 1995b). Other circumstances in which these various effects are separable have been reviewed (Lipscomb, 1994).

Allosteric control by phosphorylation has recently been discussed by Johnson (1994). In the case of muscle glycogen phosphorylase, a dimer, phosphorylation of Ser14 is followed by a −120° swing of the 20 N-terminal residues in both units, forming new contacts at the interface between the two subunits. In the process the N-terminal residues, which are poorly ordered in the unphosphorylated state become ordered. The formation of new contacts is accompanied by changes in both the tertiary and quaternary structure of both monomers, reaching from the vicinity of the serine–phosphate site to the catalytic site at a distance of 30°. A similar mechanism is proposed for the allosteric control of yeast phosphorylase, whose structure has been determined by Rath and Fletterick (1994), with phosphorylation of Thr10 playing a similar role. In this case, steric repulsion between the phosphate and the adjacent non-polar groups is thought to be the trigger for a refolding of the N-terminal region (39 residues long), so that the residues blocking the active site are rearranged, freeing the access to the active site—hence the proposed name for this mechanism as "relief of steric and conformational blocking." Though data are as yet lacking, the possibility that relief of conformational blocking by loops carrying the phosphorylated residue is a general mechanism for the allosteric control of other enzymes regulated by phosphorylation (such as twitchin kinase and the cAMP-dependent protein kinase) has been considered.

Characterization of the R- and T-states of the insulin hexamer (Brzovic *et al.*, 1994), of bacterial L-lactate dehydrogenase (Iwata *et al.*, 1994) and of the *E. coli* pyruvate kinase (Mattevi *et al.*, 1995) and the design of a Ni-histidine trigger for the activation of the allosteric transition in glycogen phosphorylase (Browner *et al.*, 1994) add to the rapidly growing list of allosteric proteins whose mechanisms remain to be explained.

A somewhat different type of quaternary structure rearrangement—a domain closure—has already some time ago been observed in citrate synthase (Remington *et al.*, 1982). Yet here again, the mechanism by which this rearrangement is accomplished remains ill-defined. A caliper-like movement of two helix-turn-helix domains activated by Fe or Ni ions has been observed in the metal ion activation of the diphtheria toxin repressor (DtxR, an asymmetric dimer from *C. diphtheriae*), with the metal ion functioning as a pin at the pivot point to rotate the two monomers, each bearing an HTH domain by 2° toward each other (Qiu *et al.*, 1995; Schiering *et al.*, 1995). The definitive analysis of the structure is still outstanding, especially as only the structure of the N-terminal domains could be determined, the C-terminal segments appearing to be largely disordered.

Of very major interest are the conformational transitions occurring in proteins responsible for mobility in cells. With the solution of the crystal structures of myosin (Rayment *et al.*, 1993) and more recently the kinesin (Kull *et al.*, 1996) and the kinesin-related ncd motor domain (Sablin *et al.*, 1996), the foundations for studying the dynamics of motile proteins have been laid. However, at this stage the nature of the different conformational states involved in motion remains to be defined (Goody, 1996) and little can be said about the dynamics of the process.

The same is true of the conformational transition of the modular binding domains (Src homology modules (SH2, SH3); pleckstrin homology, PH; and others) in signal transduction proteins, recently reviewed by Cohen *et al.* (1995). Here examination of the biological evidence in the light of the known structure has led to the development of plausible and imaginative concepts about possible mechanisms, but hard physical evidence to define them is almost entirely lacking.

2.2. *Changes of Tertiary Structure*

There are a number of proteins in which rearrangement of the tertiary, rather than quaternary structure is responsible for the allosteric effect. A prime example is the *trp-*

repressor from *E. coli* (Schevitz *et al.*, 1985; Zhao *et al.*, 1993), where the DNA-binding helix-turn-helix is seen to shift with respect to the intertwined dimeric core upon binding of the corepressor tryptophan. This is illustrated in Fig. 2. Subsequent studies have shown the conformational difference to be a reflection of a general flexibility of the region, as discussed below. A conformational transition within a tertiary structure provides a complete description of the function of a number of the simpler control and signal transduction mechanisms; and may prove generally necessary to account for the more dramatic changes in quaternary structure discussed above. A change in at least one backbone Ramachandran angle is required to produce any change in the tertiary structure. Units of secondary structure, particularly the α-helices, can act as levers to amplify the excursions at the other end. The initial angular changes may be quite small. Indeed the *en-bloc* translocation of the DNA-binding helix-turn-helix in *trp*-repressor observed on ligand binding involves changes of only two angles in the hinges flanking the domain, in excess of 10° as shown in Fig. 3 and Table 1. Numerous small (1–5°) changes in adjacent angles are also seen, but it is questionable whether they lie outside the experimental error. Larger angular changes are seen in the turn of the helix-turn-helix and upon formation of the DNA complex. A similar translocation is observed in the liganded protein between two different crystal forms, but a different set of angles is changed.

A very interesting example is provided by the cell cycle regulatory protein suc1, which has been found to exist in two thermodynamically stable forms; a monomer which forms weak dimers linked by Zn^{++} ions, and an intertwined dimer (Endicott *et al.*, 1995; Bourne *et al.*, 1995). The monomer domain consists of an N-terminal α-helix, followed by two β-strands, two more α-helices and a pair of β-strands forming a hairpin. In the dimer the C-terminal β-strand is extended and interchanged with the same strand from the other subunit. The β-hinge can be identified in the sequence H(88)VPEPH(93) and provides a conformational switch. The trigger for the conformational transition is not yet known, nor is the detailed sequence of movements. The conformational transition significantly changes exposed surface contacts, as would be expected for a mechanism involved in protein signaling and regulation.

Conformational flexibility observed in many non-allosteric enzymes may ultimately prove relevant to the understanding of allosteric mechanisms. A particularly striking example is provided by the Met6→Ile mutant of T4 phage lysozyme which was found to assume five different conformations in the crystalline state (Faber and Matthews, 1990). It

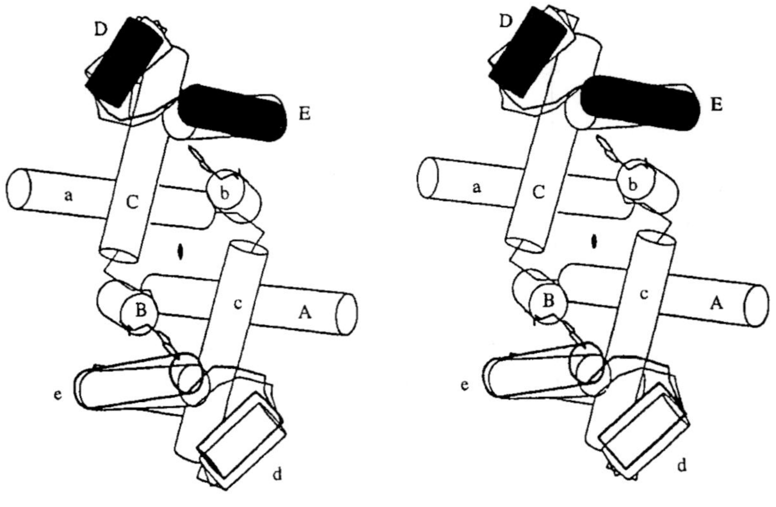

Fig. 2. Tertiary structure change in the *trp*-repressor found upon *trp* binding and in different crystal forms of the repressor. (Fig. 3 from Lawson *et al.*, 1988 reproduced by permission.)

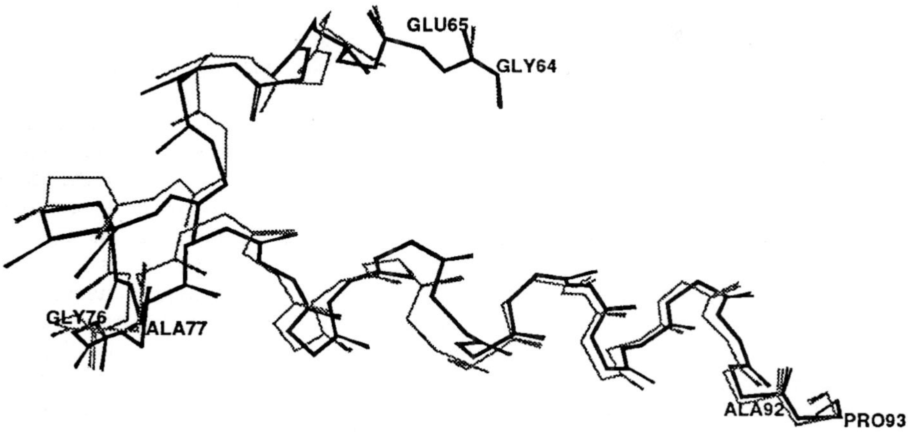

Fig. 3. Backbone angle change in the DNA-binding domain (DE helix-turn-helix) of the *trp*-repressor accompanying the *en-bloc* translocation shown in Fig. 2.

Table 1.

		Difference Apo-Holorepressor (orthorhombic)	Difference Holo (orthorhombic)–DNA complex
Hinge 1			
Asn 32	φ	4.57	−5.02
	ψ	−20.27	17.98
Asp 33	φ	7.00	−15.40
	ψ	4.52	9.07
Leu 34	φ	−3.06	−4.30
	ψ	5.83	−4.06
Hinge 2			
Leu43	φ	−2.58	9.85
	ψ	−3.52	0.27
Thr44	φ	2.92	4.32
	ψ	2.77	10.50
Pro45	φ	−7.68	−7.12
	ψ	−4.46	3.94
Hinge 3			
Gly64	φ	−6.53	19.39
	ψ	−13.46	−24.06
Glu65	φ	3.36	11.59
	ψ	8.41	6.23
Met66	φ	−2.51	7.16
	ψ	4.06	−21.47
Ser67	φ	−4.35	10.10
	ψ	−8.68	−5.47
Hinge 4			
Leu75	φ	−11.67	−2.28
	ψ	−13.95	−3.17
Gly76	φ	−8.97	28.34
	ψ	30.46	−41.56
Ala77	φ	−25.62	30.10
	ψ	−4.39	−16.88
Gly78	φ	15.10	1.78
	ψ	−17.48	16.11
Hinge 5			
Ala91	φ	8.07	−3.01
	ψ	2.82	−13.23
Ala92	φ	4.54	7.11
	ψ	−16.92	19.95
Pro93	φ	−1.80	−4.75
	ψ	−0.28	−12.46
Val94	φ	6.28	3.47
	ψ	2.03	6.82

is worth noting that the mutated residue is located in the N-terminal α-helix which in the three-dimensional structure lies near, but is not part of the hinge connecting the N-terminal and C-terminal domains. This type of flexibility may provide the simplest examples of processes involved in allosteric regulation. Gerstein *et al.* (1994) have surveyed domain motion upon substrate or inhibitor binding in a series of enzymes and proposed to distinguish two classes of domain motion; or rather, domain displacements. In the first class domain closure represents a displacement perpendicular to the principal plane of the cleft, with adenylate kinase, calmodulin, T4 lysozyme, glutamate dehydrogenase being the prime examples. In the second class the observed displacement is parallel to the plane of the cleft, as(it is in citrate synthase, hexokinase, aspartate amino transferase, *trp*-repressor. The first type they call hinge motion; the second, shear motion. The nomenclature helps to develop a vivid imagery, but is unfortunate on two counts. Firstly, it suggests that the mechanisms can be distinguished, which they can not. Changes of backbone angles which are the essence of the hinge motion also occur in the "shearing" motion. Secondly, the terminology refers to motions, whereas the observable is the net displacement, the result of multiple, mostly unknown, motions. A precise terminology that will allow a detailed description of the underlying mechanisms is very much needed, but as discussed below, only its elements can be defined at the present state of our knowledge. Not surprisingly, using a terminology with imprecise category boundaries and no reference to the underlying mechanism, Gerstein *et al.* find that although some proteins fall into one or the other of the two categories, there are others that defy classification—notably immunoglobulins, discussed in terms of a ball-and-socket joint, HIV-1 reverse transcriptase, thermolysin, the TATA-box binding protein and others. Special cases are the serpins, where one finds an additional β-strand inserted into a β-sheet as the essential part of the conformational transition and tobacco mosaic virus (TMV) coat protein, where an ordered secondary structure is formed from a disordered segment upon binding to RNA. The dynamic processes underlying all of these phenomena remain to be described.

Although no attempt has been made to summarize all of the work potentially relevant to the structural transitions in allosteric proteins, some general conclusions are evident from the reviews and examples cited. First and foremost, it is not yet possible to formulate a coherent mechanism, a detailed and clearly demonstrable sequence of movements within the tertiary structure to explain the observed end states of the RT transition for any protein. The sum total of the data reported are not sufficient for this purpose and the difficulty in identifying a pathway may stem from several causes. For one, the accuracy and precision of the existing protein structures may be inadequate to reliably detect subtle changes in average structure that suffice for the transmission of information. In the evaluation of subtle changes it is important to bear in mind that even though the reported precision of the refined X-ray and NMR structures can be very high, their accuracy is generally ill-defined and cannot be said to be any better than that of the unrefined topology given by the electron density or cluster of NMR structures derived from experimental constants only. The uncertainty about the accuracy of protein structures remains much greater than commonly assumed (Liu *et al.*, 1992; Zhao and Jardetzky, 1994a). Furthermore, the transmission of conformational information may not depend on changes in average structure, but on a modulation of protein dynamics, as has been repeatedly suggested (Frauenfelder, 1983; Jardetzky and King, 1983; Bialek and Onuchic, 1988). This includes the possibility that there exists a transient change in the average tertiary structure, which relaxes to its usual configuration after the remote surface contacts have been changed. Time-resolved spectroscopic studies are possible for only a limited number of chromophores and their number is insufficient to define the time-course of structural changes at atomic resolution, even if they can be unequivocally identified, which often they cannot be.

Structural studies of intermediates "trapped" by chemical reactions suffer from the general drawback of all chemical modification studies that artifacts produced by the chemical modifications are difficult, if not impossible, to either define or rule out

(Jardetzky and Roberts, 1981). Studies of mutants, though very popular, are also inherently ambiguous when attempts are made to interpret them in detailed mechanistic terms. If a mutant shows different allosteric behavior than the native protein, does this mean that the residue is directly involved in the allosteric mechanism, or does the mutation indirectly change the structure and properties at a remote site that is directly involved? If a mutation has no effect, does it mean that the affected residue is not involved, or that any amino acid can perform its function, or that there are alternative pathways that can compensate? One can only hope that very extensive, very systematic studies will allow a distinction between the various possibilities. For the time being it is best to list the mechanistic alternatives, as was done in the study of the Lys34→Ala and Lys34→Glu mutants of the chaperonin GroES, where the mutations were found to modulate the allosteric transition of its ligand GroEL (Kovalenko *et al.*, 1994). A new phenomenon is added, when the chemical (or genetic) modification produces a major change in structure or function, as in the case of the ligand independent conformational switch in hemoglobin (Kim *et al.*, 1995), but ambiguities in relating the observed change to the original mechanism remain. In addition, in all solution studies one has to be aware that the observed "state" of a protein may not be a single uniquely defined conformation, but a mixture or average of several states, which requires caution in interpretation (Jardetzky, 1980; Campbell *et al.*, 1985). Given such serious methodological limitations it is not surprising that the literature is replete with plausible hypotheses, based on model building, which go far beyond the available hard data, and that in essence the problem of the mechanism of the allosteric transition defined by its crystallographic endpoints, remains unsolved.

Analysis at the level of angular changes in the backbone will be crucial to the understanding of functionally significant motions, but thus far they are few in number. Changes in backbone angles are clearly required for a change in the tertiary structure. Evidence is accumulating that changes in quaternary structure are predicated on a change in tertiary structure, although it could be argued that they can be achieved by changes in surface contacts between perfectly rigid subunits. In any event, when the change in contacts occurs in response to the binding of a ligand at a remote site within the tertiary structure as in hemoglobin and PFK, transmission of the information through the subunit, either through its secondary or tertiary structure, is required. Defining the signal pathway, and the nature of the trigger, remains a major challenge.

III. THE LESSONS OF MOLECULAR DYNAMICS SIMULATIONS

Theoretical study of protein dynamics has made a significant impact on the evolution of concepts essential for an understanding of protein function. The method is based on a numerical solution of Newton's equations of motion for each atom comprising a given protein structure, assuming empirically defined potentials and an experimental set of atomic coordinates. After a decade of such studies—now a decade ago—Dobson and Karplus (1986) were able to summarize its impact by saying that "the *accepted* view of the structure of a protein is now a time-dependent one with ever present, significant fluctuations about an average structure." Many more theoretical simulations of the dynamics of many different proteins have appeared in the literature since (for a recent review see Karplus and Petsko, 1990). Much has been learned about the role of the solvent in damping internal motions and the prediction of certain macroscopic properties has been substantially improved. The method has also found wide acceptance in the refinement of crystal as well as NMR protein structures, in the form of restrained molecular dynamics (Brünger, 1992).

The beauty of the method is that it provides a clear physical picture of the motions that can occur within a protein and that in principle it can account for all motional phenomena, including the mechanism of allosteric transitions and a correct prediction of the pathway of protein folding. In reality, the situation is very different, and the usefulness of the method in explaining motions that can be detected experimentally remains severely limited. For one, the method is limited to very fast motions (on the pico- to nanosecond time-scale) which can be simulated with existing computer technology. Description of slower motions would require simulations longer by several orders of magnitude. Perhaps even more serious is the limitation resulting from the relative inaccuracy of the potentials and the neglect of coupling. Even though some improvements have been and continue to be made, the basic difficulties of choosing the "right" potential that could describe a structure stable over a long period of time and its conformational transitions remain.

Attempts to describe motions in proteins by a normal mode analysis (Brüschweiler, 1992; Brüschweiler and Case, 1994), though strictly not part of molecular dynamics, share with the latter two significant features—dependence on empirical potentials and a limitation to fast motions on the time-scale of protein function. The terminology used in connection with this approach can be confusing as one finds frequent references to "slow" motions, which are indeed slower than the random picosecond motions, but still on the nanosecond time-scale.

Although molecular dynamics has made an enormous contribution to our ability to visualize internal motions in proteins, it must be said that within its present limits it has contributed very little to our understanding of functionally significant motions. Occasionally a correlation between a molecular dynamics simulation and crystallographic data suggests a plausible mechanism for an observed conformational change. One example is the closure of the active site upon substrate binding seen in crystals of triosephosphate isomerase, studied by Joseph *et al.* (1990). The experimental observation is that an 11-residue loop closes the active site like a lid, by moving seemingly as a rigid body by more than 7 Å. Molecular dynamics simulations "confirm" that the mobility of this loop rests in the two lateral hinges (residues 166–167 and 174–176) by which it is attached to the remainder of the protein; in the sense that the angular excursions found on simulation are greater at the hinges than elsewhere in the loop. However, there is no correlation between the time-scale of the simulation and the time-scale of the actual movement. Generally, molecular dynamics simulations fail to account for slow motions detectable by NMR relaxation, by dielectric relaxation or by fluorescence transfer measurements, for the "breathing" of the protein structure detectable by solvent or ligand access experiments, for the exchange of peptide backbone protons, for coherent signal transmission within the peptide chain that can be inferred from such experiments—in fact for any functionally significant conformational change or for the process of protein folding. Nevertheless it remains a useful guide to the thinking of the experimentalist and harbors the hope of ultimately attaining a theoretical understanding of all aspects of protein dynamics.

IV. EXPERIMENTAL DETECTION OF INTERNAL MOTIONS IN PROTEINS AND THE "GLASS TRANSITION"

The detection of internal motions in proteins by many different methods—fluorescence transfer and quenching, Mössbauer spectroscopy, NMR, ESR, dielectric relaxation, ligand diffusion into the protein interior and backbone hydrogen exchange—have provided extensive (and conclusive) evidence that a variety of internal motions on different time-scales occur in proteins. Larger scale fluctuations, between different local or global conformations are often evident in such "probe" experiments. Recent studies of slow fluctuations permitting exchange of water molecules from the protein interior and penetration of small molecules into cavities in protein structures have been discussed by Petsko (1996). However, the serious limitation of such "probe" methods is that they do not provide information at atomic resolution and do not permit the structural identification of either (1) the different fluctuating conformations which may be present in solution, or of (2) their fluctuations in detail. They are therefore not discussed in this review. Our understanding of the (mostly slower) functionally important fluctuations will ultimately depend on the use of atomic resolution methods—X-ray crystallography and NMR. Nevertheless, the inevitable conclusion from this large body of evidence is that the "accepted picture" of a protein structure as "a time-dependent one, with ever-present significant fluctuations around an average structure" is still a gross oversimplification, even though in its time it represented a significant conceptual advance over the static picture provided by the early crystal structures. Especially for proteins known to undergo large structural changes the more complete picture probably involves a relatively slow interconversion between two or more well-defined conformations superimposed on the ubiquitous rapid "fluctuations about an average structure," and this needs to be constantly borne in mind (Jardetzky and Roberts, 1981 Ch. V; Campbell *et al.*, 1985).

The difference between fluctuation about a single average structure and a conformational equilibrium is of fundamental importance both for the understanding of protein dynamics and for structure determination, especially by NMR, and is therefore illustrated in Fig. 4 (modified from Jardetzky and Roberts, 1981). The distinction is clearly made on the basis of the existence of two (or more) identifiable conformations with a sizeable energy barrier between them, but the possibility of each energy well containing a multitude of substrates which allow the system to gradually "creep" over the barrier with a multitude of intermediate states, must always be borne in mind.

Crystallography normally provides no information on motions and when a part of a protein appears disordered in a crystal it is not possible to establish whether the disorder is of static or dynamic origin. The Debye–Waller or B-factor measures, strictly speaking, the uncertainty of atomic position. A similar uncertainty ellipsoid can be used to measure the uncertainty of an NMR structure (Altman and Jardetzky, 1989). Still, even if the factor does not permit a distinction between static and dynamic disorder, or a definition of any motion, it can be taken as a measure of motional freedom, since both static and dynamic disorder imply freedom to move. Thus, information on motional freedom about an average position can be derived by crystallographic studies of the temperature dependence of the Debye–Waller factor, as has been demonstrated by Petsko and

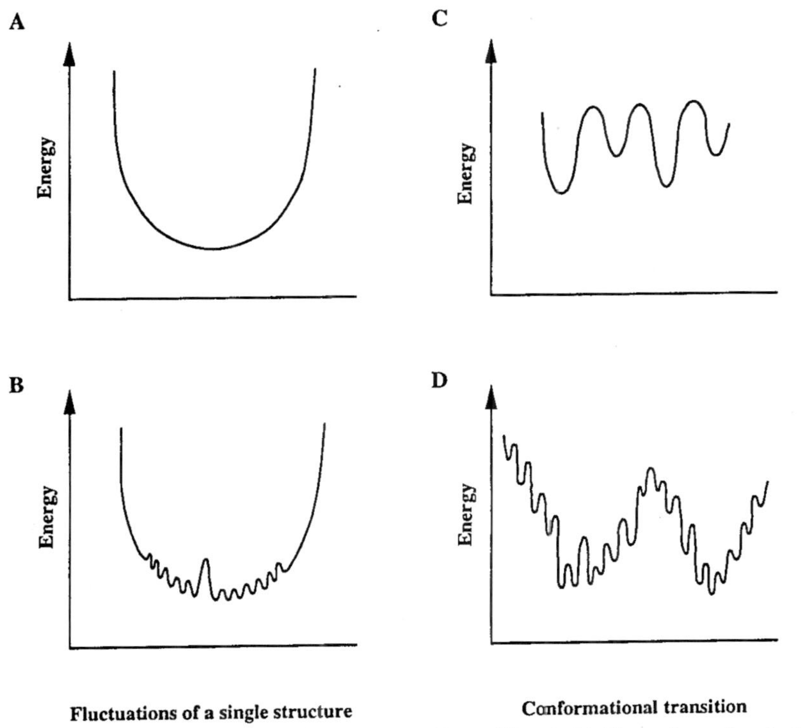

Fig. 4. Fluctuations vs R→T conformation transitions. (A) and (B) Fluctuations of a single structure. (C) and (D) Conformational transitions between two or more stable structures.

coworkers (Alber *et al.*, 1983; Tilton *et al.*, 1992; Ding *et al.*, 1994). The range of motions that can be detected by this method is clearly limited to those that do not disrupt the crystal and this generally excludes motions that produce major conformational changes. In a classic study of this type Tilton *et al.* (1992) have determined the structure of ribonuclease A to 1.5 Å resolution at nine different temperatures, ranging from 98 to 320 K. Two significant findings emerge from this type of study. (1) The radius of gyration of the molecule increases by about 4.7% over this temperature range and so does the probe accessible volume (Connolly, 1985), indicating that the protein expands slightly as the temperature is raised. The expansion is anisotropic and linear over the entire range and can be attributed mostly to increased motion of loops protruding from the surface, with some internal repacking. Similar observations have earlier been made on myoglobin (Frauenfelder *et al.*, 1979, 1987), suggesting that the expansion coefficient may be larger for purely helical proteins. (2) The average atomic thermal coefficients (Debye–Waller or B-factors) increase more than two-fold, from 6.6 $Å^2$ to 15.7 $Å^2$ over the same temperature range. In this case the increase is not linear, but biphasic, with a small positive slope of 1.2 $Å^2$/100 K at temperatures below 200 K and a larger slope of 6.4 $Å^2$/100 K in the range between 200–320 K. Below approximately 212 K, substrate and inhibitor binding are abolished and so is enzyme activity, which is observable only at temperatures above 228 K. The transition monitored by the average Debye–Waller factors and occurring in the temperature range 212–228 K, as shown in Fig. 5, has been termed the "glass transition" (Rasmussen *et al.*, 1992).[2] Molecular dynamics simulations suggest that below 220 K the predominant motion in the protein is harmonic vibration of individual atoms, while at higher temperatures anharmonic collective motions of bonded and non-bonded groups of atoms begin to dominate (Loncharich and Brooks, 1990).

These conclusions are very similar to those reported for hydrated myoglogin by Doster *et al.* (1989), who used inelastic neutron scattering to show that below 180 K the protein behaved as a harmonic solid, but above 80 K non-vibrational modes—torsional jumps—came into play. Whether this is indeed a complete and correct explanation for the glass transition remains to be established, since the conventional isotropic Debye–Waller factors do not contain sufficient information on anisotropic and concerted motions. The basic fact is clear: the amplitude of thermal motions dramatically increases above the glass

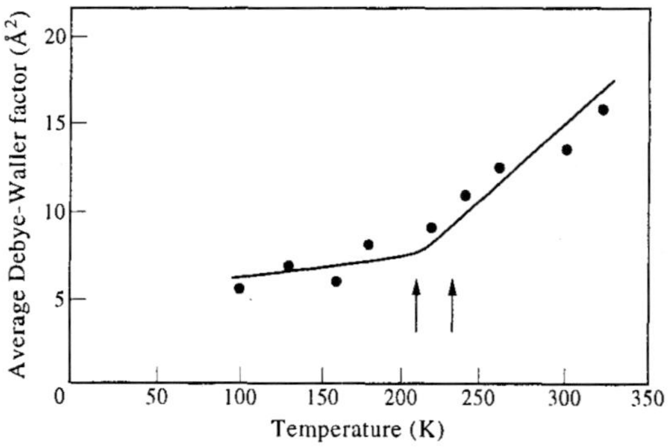

Fig. 5. The "glass transition." Plot of the average Debye–Waller factor for all of the non-hydrogen protein atoms in crystalline bovine pancreatic ribonuclease A as a function of temperature. The plot is biphasic, but the exact transition temperature is hard to estimate. This difficulty in part arises because the transition is broad, but there is also evidence that it depends on solvent composition and on the rate of cooling. The estimate of the midpoint at 220K is approximate. (Fig. 1 from Rasmussen *et al.* © 1992, Macmillan Magazines Limited, reprinted with permission.)

[2] The term was chosen by analogy with spin glasses and was not intended to suggest that proteins have any glass-like properties, which they do not, but only that the transition has characteristics similar to those of glass melting—it is broad, dependent on the rate of cooling and involves a change in the type of atomic motion (Petsko, personal communication).

transition temperature and new motions, "frozen out" at low temperatures, come into play. This justifies the important conclusion of Rasmussen *et al.* (1992), that motion—an overall flexibility—is essential for enzyme activity, even in non-allosteric systems. Whether the more highly localized types of motion involved in allosteric transitions also have a correlate detectable by this method remains to be determined. No similar studies of an allosteric structure have been reported thus far. If the applications of time-resolved X-ray diffraction to proteins prove successful, they may provide the ultimate answers to the description of allosteric transitions in atomic detail (Ringe *et al.*, 1990).

Conformational changes and conformational equilibria also remain undetected in conventional NMR structure determination, because the usual method of structure calculation is predicated on the assumption that all distance data refer to a single rigid protein structure. An exception to this is discussed below, but it requires data not obtainable from NMR to verify the correctness of the analysis. The "disorder" often observed in a family of conventionally calculated solution structures cannot be automatically taken to indicate motional disorder, as it simply reflects the absence of relevant distance data, which may be missing for a variety of reasons. However, a distinction between "apparent" and "true" motional disorder can be made by NMR relaxation and proton exchange measurements, as illustrated for the *trp*-repressor below.

Our current understanding of protein dynamics is therefore limited on two fronts. On one side the methods that provide a detailed visual image of very fast motions and large scale displacements, X-ray diffraction and molecular dynamics, are inadequate to deal with motions on slower time-scales. On the other side the methods that can detect slow motions and define their time-scales do not provide sufficient information to completely characterize and visualize the motions they detect. As discussed below, evidence is accumulating that NMR studies can to some extent bridge this gap.

V. MAPPING OF PROTEIN MOTIONS AT ATOMIC RESOLUTION ON THE BASIS OF NMR DATA

5.1. *General*

The study of individual atomic motions in proteins by NMR has become possible in the last decade, as multidimensional spectroscopy, the availability of high frequency spectrometers and isotopic labeling techniques have allowed a complete assignment of protein NMR spectra. The method has so far been largely limited to small recombinant proteins, but even recombinant hemoglobins are now amenable to study (Kim *et al.*, 1995). It is to be expected that the information derived from such studies will become increasingly important for the understanding of allosteric mechanisms and induced fit.

Three types of NMR experiments on a protein whose spectrum has been completely assigned make it possible to describe its internal motions with atomic resolution: (1) measurements of relaxation parameters; (2) observations of chemical shift averaging, a particularly clear example being the rotation or 180° flipping of aromatic rings; and (3) measurements of proton exchange rates. Relaxation data can reflect motions on time-scales ranging from pico- and nanoseconds to about 1 msec. Ring flipping typically occurs on a micro- to millisecond time-scale. Proton exchange data reflect motions within the secondary structure on time-scales from milliseconds to minutes and hours.

The relaxation parameters measured (for any of the nuclei used) are: (1) the longitudinal relaxation time T_1, sensitive to rotational diffusion of the protein, and motions faster than the rotational diffusion rate; (2) the transverse relaxation time T_2, sensitive to the same motions, but in addition to slow motions and to chemical exchange between environments with non-equivalent chemical shifts, which results in line broadening; and (3) the nuclear Overhauser effect (NOE), which reflects rapid motions, but may also be affected by chemical exchange. Sometimes it is also useful to measure (4) T_1 in the rotating frame ($T_{1\rho}$), which is sensitive to motions on slower time-scales, as is T_2. A satisfactory model for the motions giving rise to relaxation must be able to account for all relaxation parameters simultaneously. At the present state of our understanding, this condition is not always met.

There exist several theoretical methods for the analysis of relaxation data on proteins. A variety of simple models have been used, but mostly found to be inadequate (for discussion and references see Jardetzky and Roberts, 1981, Ch. XII). The rigorous model-free formalism (King and Jardetzky, 1978; Jardetzky, 1981) suffers from the drawback that it requires a massive set of relaxation data at different frequencies to ensure internal consistency of the analysis. Widely used is the simpler "model-free" formalism of Lipari and Szabo (1982), which assumes isotropic overall tumbling of the protein and describes the internal motion by an order parameter S^2, with a range from 0, for completely isotropic motion, to 1 for a nucleus immobilized in a rigid structure, and an "effective internal correlation time," τ. The formalism is mathematically equivalent to the wobble-on-a-cone model of Howarth (1979) and is model-free only in the sense that it gives no specific physical significance to the amplitude of the motion measured by the order parameter. Otherwise it is a very restrictive model allowing only isotropic rigid body tumbling and one degree of internal motional freedom. It has gradually been found not to be adequate for the analysis of relaxation data on macromolecules (Powers *et al.*, 1992; Rischel *et al.*, 1994; Buck *et al.*, 1995) and has required numerous *ad hoc* modifications (Clore *et al.*, 1990; Jardetzky, 1992). Nagayama and coworkers (Ishima *et al.*, 1995; Ishima and Nagayama, 1995a,b) have recently introduced a "Quasi Spectral Density Function (QSDF) analysis" based on the King–Jardetzky formalism in order to allow a larger number of degrees of freedom. Operationally identical, but conceptually distinct, is the method of reduced spectral density mapping of Lefèvre *et al.* (1996), a simplification of the full spectral density mapping technique of Peng and Wagner (1992a) and Peng and Wagner (1992b). Here the mobility of an NH-vector is described by values of the spectral density calculated from the experimental relaxation data without any reference to a specific motional model. The method brings out the common denominator in a set of relaxation data, but basically contains and conveys the same information as the raw data. All of the cited methods of analysis can be useful for comparative assessment of mobility along the polypeptide chain—the identification of "more mobile" and "less mobile" regions—but fall short of providing a clear physical picture of the motions involved.

^{1}H relaxation measurements are of limited use in the study of protein dynamics, because the motional relaxation behavior of individual protons is obscured by rapid magnetization transfer or spin diffusion (Kalk and Berendsen, 1976). They remain very useful for identifying longer freely mobile segments, in which spin diffusion is minimal (Jardetzky *et al.*, 1978; Perham *et al.*, 1981), and also in the study of backbone proton exchange, where the relaxation process is dominated by the exchange process (Krishna *et al.*, 1979; Englander *et al.*, 1980; Spera *et al.*, 1991; Zheng *et al.*, 1995a,b; see below).

5.2. *Side Chain Dynamics*

^{13}C relaxation studies of proteins have proved to be particularly valuable for the study of side chain dynamics. Already the early experiments at natural ^{13}C abundance, conducted by Allerhand and coworkers and ourselves (Allerhand *et al.*, 1971; Komoroski and Allerhand, 1972; Opella *et al.*, 1974; Nelson *et al.*, 1974; Ribeiro *et al.*, 1980) have unequivocally established that generally side chains, especially the side chains of residues located on the protein surface are much more mobile than the polypeptide backbone. The interpretation of ^{13}C relaxation experiments is relatively simple, because at the usual spectrometer frequencies (up to 500 MHz) the dominant relaxation mechanism is dipolar and the dominant term stems from the proton covalently attached to the carbon. At higher frequencies chemical shift anisotropy begins to play a role, and while it potentially provides additional information, the analysis is more complicated (Jardetzky and Roberts, 1981). The motions monitored by this method are largely on the pico- to nanosecond time-scale and are clearly part of the normal thermal motion of protein atoms.

In recent years methods for detailed quantitative ^{13}C relaxation studies on individually assigned ^{13}C-labeled residues have been developed (Brüschweiler *et al.*, 1989; Kay, 1993; Yamazaki *et al.*, 1994). Ernst and Ernst (1994) have introduced a heteronuclear (^{13}C–^{1}H) cross-correlated cross relaxation (SIIS) technique for measuring the rates of rotation of CH_2 and CH_3 groups around the χ_1, χ_2, χ_3, etc. angles in amino acid side chains and have applied it to the cyclic peptide antamanide and several residues of the 8.5 kD human ubiquitin. In both cases fast puckering of proline residues could be detected. In ubiquitin, rotation about χ_1 was found to be either unrestricted and "slow" $\tau_i > 1.37$ nsec, or restricted to $< 120°$ but fast $\tau_i < 1.37$ nsec, whereas the rotation around χ_2, χ_3 and χ_4 in all six of the lysines examined was unrestricted and fast. Distinction between lysine side chains sticking out freely into the solvent and those involved in intramolecular hydrogen bonding should be possible on this basis. In earlier studies of antamanide dynamics by Richard Ernst and his colleagues (Ernst, 1992 and references therein), a 25-μsec lifetime for a hydrogen bond was detected by $T_{1\rho}$ measurements and 1176 interconverting low-energy structures were identified with the aid of energy minimization techniques, yielding a Boltzmann distribution. The approach represents the ultimate in detail for the description of internal motions in peptides, but is unfortunately limited to small structures for practical reasons. Much attention has also been devoted to a detailed analysis of rotamer populations and potentials for internal rotations from NMR spin-coupling and NOE data. The recently proposed Continuous Probability Distribution (CUPID) method (Dzakula *et al.*, 1992a,b; 1996) promises to avoid both extremes of analysis—oversimplification and intractability.

Nicholson *et al.* (1992) have studied methyl group rotation of the leucine residues in staphylococcal nuclease, noting an increase in the order parameter upon Ca^{2+} or inhibitor binding for those residues in the immediate vicinity of the binding site; a "local stiffening" of the structure near the binding site. Rotation of side chains in a binding site, to accommodate different conformations of an inhibitor has been demonstrated in dihydrofolate reductase (Gronenborn *et al.*, 1981; Birdsall *et al.*, 1984), where the ternary complex enzyme, trimethoprim–$NADP^+$, was shown to exist in two slowly interconverting conformational states, defined by two different orientations of the trimethoprim and nicotinamide rings. Thus far relatively few studies of this type have been reported, and the most important question—whether the motions sensed by ^{13}C relaxation in any way reflect structural features responsible for allosteric transitions, or allow the identification of regions involved in functionally significant motions—remains open. In a relaxation study of [2-^{13}C]-labeled tryptophan 24 in different mutants of the chymotrypsin inhibitor 2 (9.2 kD), Matthews *et al.* (1993) were able to establish a correlation between the picosecond dynamics of trp 24 and the thermodynamic stability of the mutant and also found that mutations can influence the dynamic behavior of this residue though it is relatively distant both in the primary sequence and in the three-dimensional structure. If this could be shown to be a direct effect, not accompanied by a change in the intervening structure and transmitted as a modulation of dynamics, it would be highly significant. Yet it could also be an indirect effect resulting from a change in the intervening structure; e.g., the packing density around the monitored residue, which might be expected to change the local dynamics. The absence of a global structural change has been argued, but not clearly demonstrated in this case and the ultimate interpretation of the finding remains uncertain.

5.3. *Relaxation Studies of Backbone Dynamics*

Studies of the dynamics of the polypeptide backbone are potentially the most relevant to an understanding of allosteric transitions since, as already mentioned, changes in the dihedral angles along the backbone are most likely a common denominator of all large conformational changes. Dynamics of the polypeptide backbone can be monitored, (1) by ^{15}N relaxation measurements in a uniformly or selectively ^{15}N-labeled protein, and (2) by hydrogen–deuterium exchange, which can be followed by NMR either, (1) directly by measuring the decrease in line intensity as a function of time after dissolving the protein in $^{2}H_2O$ for protons exchanging on a time-scale of minutes or longer, or (2) indirectly, by observing the relaxation of the proton itself, when exchange rates are of the same order as the intrinsic relaxation rates, i.e. on time-scales from 20 msec to 5 sec. Only upper and lower limits can be given for exchange rates falling on the time-scale of ~5 sec–5 min. Several detailed studies using each of the three methods have been reported in the literature (Zheng *et al.*, 1995b and references therein).

The simplest findings in studies of backbone relaxation are obtained for small globular proteins, such as BPTI (Richarz *et al.*, 1980; Ribeiro *et al.*, 1980), where the ^{13}C data were originally interpreted in terms of an isotropic overall motion of a rigid rotor, with an overall tumbling time of 4–12 nsec and one degree of internal motional freedom on a picosecond time-scale (Jardetzky and Roberts, 1981; Lipari and Szabo, 1982). In ^{15}N studies the simple model-free model accounts for the relaxation behavior of the glucocorticoid receptor DNA-binding domain (GR-DBD, 9.2 kD, Berglund *et al.*, 1992), which can be described as a relatively rigid structure with an overall rotational tumbling time of 6.3 nsec, and a uniform order parameter of $\sim 0.86 \pm 0.07$ for a single internal motion between 11 and 60 psec. It may be noted that the internal correlation time is rather ill-defined in this method of analysis and cannot be specified with great precision. The analysis holds for most residues in this protein, but the possibility of slower motions has not been completely ruled out, as only those values of $T_{1\rho}$ (instead of T_2) were used in the analysis which gave results consistent with the measured T_1 and NOE. Schneider *et al.* (1992) have reported an analysis of ^{15}N relaxation data on human ubiquitin in terms of the simple model, finding an average order parameter of 0.80 for hydrogen-bonded residues and 0.69 for those not involved in hydrogen bonding and concluded that packing interactions and to a lesser degree, hydrogen bonding, provide the most important mechanisms for damping fast motions in proteins. The backbone dynamics of the c-Jun leucine-zipper, a 2-helix strand, can be described by anisotropic diffusion and a single order parameter for internal motion, leading to the conclusion that the structure is rigid throughout (Mackay *et al.*, 1996).

On the other hand, already in the first ^{15}N relaxation study of backbone relaxation using the now standard method of inverse (proton) detection, Kay *et al.* (1989), found the situation in the 18 kD staphylococcal nuclease to be more complicated. Comparison of the values of T_1, which are sensitive only to rapid motions and T_2, which are sensitive to both rapid (picosecond) and slow (nano- to millisecond) motions led to the conclusion that in addition to the overall tumbling (9.1 nsec) and two rapid motions (0.020 and 0.12–0.54 nsec) for most residues, a slow motion was necessary to account for the data for the loop P42–E57. In a study of interleukin-1-β (Clore *et al.*, 1990) evidence was again found for motions on three time-scales. Common to all the residues were very fast motions (< 20–50 psec), showing a highly ordered structure with an order parameter of 0.82, corresponding to a semiangle of 20.7° for free diffusion within a cone. In addition however the data for 32 residues reflected a second fast motion (0.5–4 nsec) and those for 42 residues slow conformational transition between different species on time-scales from 30 nsec to 10 msec; 19 residues were found to undergo a common conformational equilibrium between a major and a minor species. Most of the residues showing the additional slow motions are located in the loops connecting the 12 β-strands of the protein. All of the residues showing the common conformational equilibrium (Val3, Arg4 ... Lys92, Lys93 and Glu96) are located in the same corner of the structure, where the loop connecting strands VII and VIII (residues 86–88) lies near the N-terminus. Although the nature of the conformations could not be identified, it is apparent that this part of the structure possesses greater conformational flexibility—on a relatively slow time-scale, compared to the rest of the molecule. On a nanosecond time-scale the molecule still behaves as a single rigid unit with one or two degrees of internal freedom.

Most ^{15}N relaxation studies of backbone dynamics reported to date reveal at least this degree of complexity in the type of motions that determine the relaxation process; internal motions on one or two rapid time-scales and one or more motions, representing conformational exchange on slower time-scales. Thus in the study of human interleukin-4 by Redfield *et al.* (1992), two classes of backbone nitrogens could be distinguished; those in the helical core, characterized by highly restricted conformational flexibility, as measured by large order parameters (S^2~0.9) and those in the N- and C-termini, and in two of the longer loops connecting helices A and B and C and D, which showed extensive conformational flexibility with order parameters S^2~0.3–0.8, with an overall tumbling time of 7.6 nsec. However, relaxation for only 47 residues could be accounted for by isotropic rotation with one rapid internal motion. For 20 residues a second rapid motion had to be invoked, as in the case of staphylococcal nuclease and interleukin-1-β. For 21 residues conformational exchange terms on still slower time-scales had to be added. Similarly in the four helix bundle human granulocyte colony stimulating factor (G-CSF, 18.8 kD, Zink *et al.*, 1994), with an overall tumbling time of ~13 nsec only one of the loops, connecting helices C and D was found to have significant conformational flexibility. However, again substantial exchange contributions had to be invoked for the majority of residues, including those located within well formed helices, in order to account for the T_2 data.

Powers *et al.* (1992) have analysed in detail the backbone dynamics of the ribonuclease H domain of the HIV reverse transcriptase and found it necessary to invoke six "model-free functions" using different combinations of two order parameters, two internal correlation times and an exchange term. Very significant is that all residues exhibit the fast motions described by the original order parameter, on a time-scale < 20 psec. However the ^{15}N relaxation data for only 14 residues can be described by the simple model. A slower fast motion on a time-scale of 28.8 psec to 3.9 nsec is found for 39 residues, an exchange term for 15 residues and both a nanosecond internal motion and a conformational exchange for an additional 39 residues, with a total of 93 residues displaying multiple modes of motion. The most mobile regions are the C-terminus and loops between strands β1 and β2 and α-helixB and β4. It has been noted in the commentary on this paper (Jardetzky, 1992) that the basic conclusions concerning different multiple motions in different parts of the backbone can be drawn from the raw relaxation data regardless of the specific mode of analysis. This is important, because the analysis proposed by the authors, and similar amendments to the original model-free formalism used by others, involve several *ad hoc* additions to the basic model, whose physical meaning is not defined. In addition the quantitative analysis in some studies is made uncertain by using the approximation of isotropic rotation for structures showing an appreciable degree of anisotropy (e.g., 1.81: 1.49: 1.00, as in Redfield *et al.*, 1992). The danger of such approximations is that the anisotropy will appear as a spurious contribution to internal motional freedom, since the formal description of anisotropy and internal motion are equivalent (Woessner, 1962a,b; King and Jardetzky, 1978). No accurate statements about the physical nature of backbone motions on different time-scales will be possible, until a more complete theory of NMR relaxation in complex structures is developed and tested. In none of the cases cited have the implications of the slow exchange terms for the accuracy of the calculated structures been addressed (see below). However, the detection of motions on different time-scales in different parts of the polypeptide chain by this method is on safe ground.

Similar findings have been reported for interleukin 8 (Grasberger *et al.*, 1993), bacterioopsin (Orekhov *et al.*, 1994), the glucose permease IIA domain from *B. subtilis* (Stone *et al.*, 1992), human profilin (Constantine *et al.*, 1993a), thioredoxin from *E. coli* (Stone *et al.*, 1993), the unliganded FK-506 binding protein (Cheng *et al.*, 1993), the chemotaxis Y protein from *E. coli* (Moy *et al.*, 1994) ribonuclease HI from *E. coli* (Mandel *et al.*, 1995), and hen egg white lysozyme (Buck *et al.*, 1995), among others. Most authors agree that relative lack of Van der Waals contacts is a major determinant of regional main chain as well as side chain mobility in proteins, but a clear proof of the proposition is still outstanding.

Of special interest is the damping of backbone motions by ligand binding. This has been observed for the acyl-coenzyme A binding protein by Rischel *et al.* (1994), affecting residues in the immediate vicinity of the ligand binding site. The simple Lipari–Szabo model has been found not to apply to the analysis of data in either the liganded or unliganded form of the protein and in both forms of the protein several residues, notably Ile27 and His30, were found to require large exchange contributions on the millisecond to second time-scale, to explain the data. No effect on the picosecond motions was seen on ligand binding in the phosphopeptide complexed Src Homology 2 (SH2) domain, but a damping of the "exchange terms" on a micro- to millisecond time-scale was observed (Farrow *et al.*, 1994). A restriction of local side chain mobility in the SH2 domain has also been observed on ligand binding (Kay *et al.*, 1996). A restriction of flexibility upon binding of the FK-506 ligand, as reflected in increased order parameters, was seen in the relatively mobile loop formed by residues 82–87 in the FK-506 binding protein, which surrounds the binding site.

In a rather dramatic example, ^{15}N relaxation studies have been used to establish that the central helix of calmodulin (CaM) found in the crystal structure is in fact disordered in solution (Barbato *et al.*, 1992). CaM (16.7 kD) is a calcium-binding protein widely implicated in cellular signaling. Its crystal structure (Babu *et al.*, 1988) shows it to have a dumb-bell shape, consisting of two globular domains with two helix-turn-helix Ca^{2+} binding units each, connected by a rigid central helix. As small angle X-ray scattering studies had suggested that the helix was flexible in solution, the problem was investigated by NMR. The result indeed showed two well-defined globular lobes comprising the N-terminal segment L4–K77 and the C-terminal segment E82–S147, with slightly different rotational correlation times of 6.3 and 7.1 nsec respectively, indicating that their tumbling was independent. Order parameters in both were of the order of 0.8, as would be expected for highly structured domains. In contrast, the order parameters for the connecting link K77–S81, which is helical in the crystal, were in the range of 0.5–0.6, consistent with a high degree of disorder or flexibility. Also consistent with a high degree of flexibility was the finding that the residues which had a low S^2 value also had fast hydrogen exchange rates (Spera *et al.*, 1991). The findings unequivocally support a "flexible hinge" model for calmodulin in solution. A very similar conclusion was reached in a study of the smaller Cα-binding protein calbindin D_{9k} (Kördel *et al.*, 1992; Akke *et al.*, 1993). The two helix-turn-helix Cα-binding loops showed a high degree of rigidity, with $S^2 \sim 0.83$–0.85, while the linker region had an S^2 of 0.59 as is readily apparent from the data shown in Fig. 6. In the C-terminal binding domain order parameters tended to be somewhat higher in the Ca-bound state than in the ion-free apoprotein. ^{13}C relaxation experiments were used similarly to provide evidence for the flexibility of the central helix in cardiac troponin C (Kleerekoper *et al.*, 1995). Recent determinations of the solution structures of Ca-free and Ca-calmodulin have revealed major conformational differences between the two forms (Finn *et al.*, 1995; Zhang *et al.*, 1995; Kuboniwa *et al.*, 1995), which may in the future be correlated with changes in dynamics upon Ca^{2+} binding.

A somewhat similar finding of two independently tumbling domains was reported for the amino terminal fragment of a plasminogen activator consisting of a kringle domain and a growth factor domain (Hansen *et al.*, 1994). Surprisingly however the linker between the two domains showed a higher degree of rigidity on a nanosecond time-scale than the flexible linker in calmodulin, although it required "exchange" terms to account for the T_2 data, suggesting flexibility on a longer time-scale. It is not possible for two domains to rotate independently on a nanosecond time-scale, if the linker remains rigid on a millisecond time-scale. The conclusion that they did rotate independently was drawn from the finding that the anisotropy of the rotation was different for the two domains. This however would also be the case if there were a population of conformers interconverting on a millisecond time-scale, in which the anisotropy of rotation were different. The anisotropy of rotation averaged independently for each domain would also be different in such a case. This would reconcile the otherwise contradictory findings.

In contrast a high degree of correlation between the motions of the three zinc finger domains of the *Xenopus* transcription factor TFIIIA on a time-scale shorter than 10 nsec was found by analysing the anisotropy of ^{15}N relaxation (Brüschweiler *et al.*, 1995).

The existence of a flexible "hinge" between the Fab and Fc segments of antibodies has been known since the early experiments of Yguerabide *et al.* (1970). More recently,

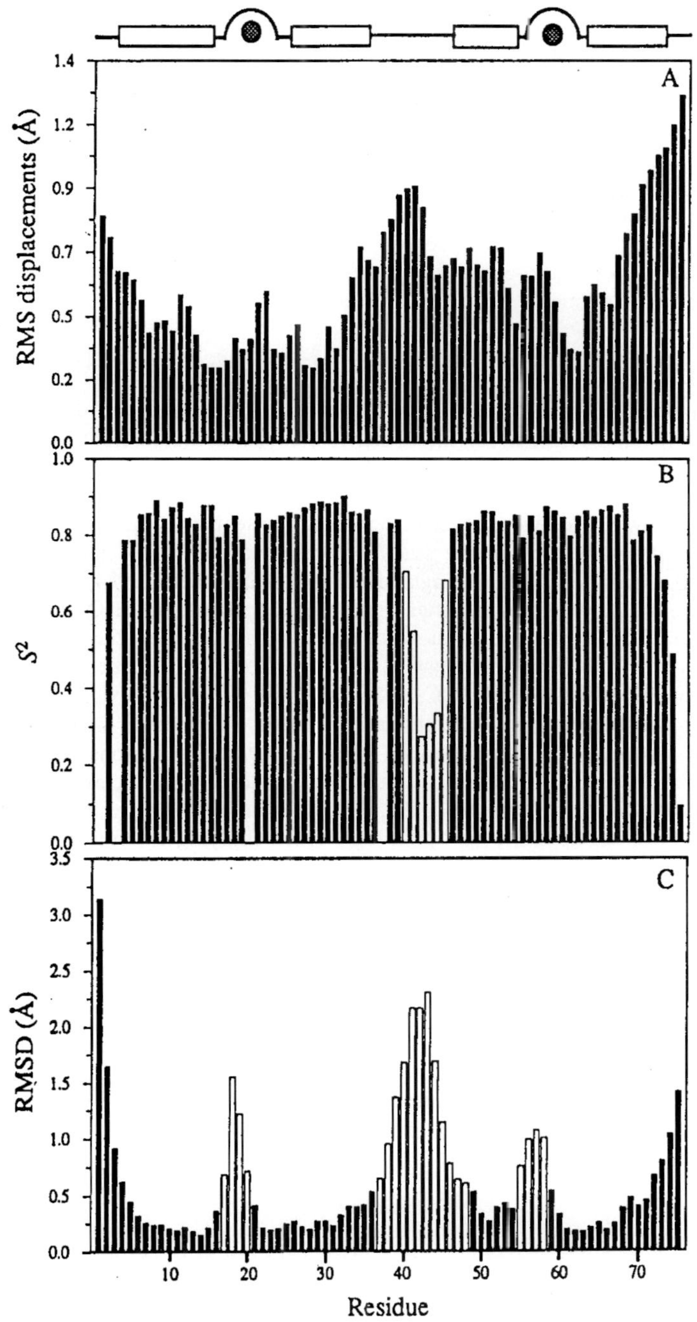

Fig. 6. Comparison of structural and dynamic parameters of the backbone amide nitrogens of calcium-loaded bovine calbindin D_{9k}: (A) crystallographic RMS displacements; (B) order parameters from the ^{15}N relaxation studies; (C) average RMS deviations of the refined solution structure from the mean structure. The elements of regular secondary structure are identified above the diagram with the α-helices indicated by rectangular boxes, the calcium-binding loops by semicircular arcs, and the Ca^{2+} ions by filled circles. Disregarding the termini, order parameters in (B) smaller than 0.75 and RMSD values in (C) larger than 0.6 Å are indicated with open bars. (Fig. 5 from Kördel *et al.* © 1992, American Chemical Society, reprinted with permission.)

detailed maps of the flexible regions in antibody fragments have been made by Arata and coworkers (Kato *et al.*, 1991; Arata, 1996), indicating that the antibody binding sites possess considerable flexibility as well. In the case of the dansyl-lysine binding antibody, stabilization of the heavy chain hypervariable region H3 upon hapten binding has been demonstrated. Comparison of the NMR findings to the finding from the analysis of crystallographic data (Chothia and Lesk, 1987; Chothia *et al.*, 1989) that H3, in contrast to the five other hypervariable regions of the heavy and light chains, has no canonical structures preserved in different antibodies naturally leads to the suggestion that upon crystallization H3 is frozen in different conformations for different antibodies, by virtue of its flexibility (Takahashi *et al.*, 1992). In contrast, a very detailed relaxation analysis of the VL domain of an anti-digoxin antibody led to the conclusion that this domain has low flexibility on a pico- to nanosecond time-scale, flexibility being limited to turns, loops and N- and C-terminal segments (Constantine *et al.*, 1993b).

5.4. *Backbone Proton Exchange as an Indicator of Backbone Dynamics*

The exchange of amide hydrogens in the peptide backbone with deuterium in the solvent has provided evidence for the existence of dynamic fluctuations and of regional differences in exchange rates ever since the initial experiments of Linderstrøm-Lang (1955). As it became possible to resolve the amide peaks in the NMR spectra of proteins, and thus to follow the individual exchange rates in the sequence, the method became one of the most important tools in the study of protein dynamics (Wemmer *et al.*, 1981; Wagner and Wüthrich, 1982; Englander and Kellenbach, 1984). Although the disappearance of individual peaks after dissolving the sample in ^2H_2O can be followed in 1D and 2D COSY spectra, as it was in the earlier references cited, the method of choice now is to carry out the measurements on a uniformly ^{15}N-labeled protein, with heteronuclear 2D detection (HMQC, HSQC). This provides maximal resolution of individual protons, especially needed in larger proteins. The method is largely limited to recombinant proteins, which can easily be labeled *in vitro*. For protons exchanging on a time-scale of milliseconds to seconds, it is also possible to monitor the exchange by NMR relaxation measurements (Zheng *et al.*, 1995a,b). As a pulse-labeling technique the isotope exchange method has found wide application for identifying intermediates in protein folding (Roder *et al.*, 1985), but a discussion of the extensive literature on protein folding is beyond the scope of this review.

The kinetic interpretation of the exchange data is generally in terms of the model proposed by Linderstrøm-Lang, which allows exchange from the "open" or "accessible" state, but not from the "closed" or "inaccessible" state:

$$A_{\text{inaccessible}} B_{\text{accessible}} \text{solvent}$$

Since the exchange rates of amides in random coil peptides ("the intrinsic" rates) are well known and their sequence dependence can be predicted (Molday *et al.*, 1972; Bai *et al.*, 1993), the retardation of exchange in the protein structure is usually measured by the "protection factor" P:

$$P = k_{\text{intrinsic}}/k_{\text{observed}}$$

Empirically, protection factors are found to vary from 2 to 10^8 or more. Although the original picture underlying the Linderstrøm-Lang model equated the protected state with a hydrogen-bonded state and the unprotected with an exposed peptide, it is important to bear in mind that exchange kinetics by themselves contain no information whatsoever on the number of states or on the nature of either state. It has recently been pointed out that a two-step model of the L-L type is not required to account for data when exchange can be described by a single exponential (Finucane and Jardetzky, 1996). Only when biexponential exchange is observed, as it sometimes is when exchange is monitored by NMR relaxation, does a two-step model represent a minimum required to account for the data (Zheng *et al.*, 1995b; Gryk *et al.*, 1995). The definition of a protection factor is of course model-independent and in any case provides an empirical measure of the structural barrier to exchange. However, in biexponential cases separate protection factors have to be defined to describe the barriers at each step (Finucane and Jardetzky, 1995).

Different hypothetical models for the structural interpretation of exchange data have been advanced—the most prominent being the "local unfolding" and the "solvent accessibility" models. There is no unanimity in the literature on the "right" interpretation. The controversial field has recently been reviewed (Miller and Dill, 1995), and we shall limit our discussion to the findings that unequivocally follow from the data: there are significant differences in amide exchange rates along the peptide chain in most of the proteins studied. Even without detailed structural interpretation these can be taken as indicators of structural or dynamic differences. In the case of solvent exposed amides, the structural difference may suffice to account for a more rapid exchange. In the case of amide protons that are no more exposed to the solvent (when the protein is assumed to be rigid) than those exchanging more slowly, the faster exchange rate signals structural instability or fluctuation at the specific site. It is worth underscoring that any attempt to interpret amide exchange data in more concrete structural terms requires information that is not contained in the NMR exchange experiment. Overinterpretation of exchange data in terms of specific models has unfortunately become very common and strict distinction between conclusions that can be drawn from the data and those that only follow from additional *ad hoc* assumptions is rarely made in the literature.

In a careful study of hydrogen exchange in native and denatured states of hen egg-white lysozyme, Radford *et al.* (1992) have established a baseline for the use of the method in protein folding studies. Protection factors for the denatured states were very small (1–5), indicating that there was minimal residual structure; their distribution showed no relationship to the protection pattern in the native structure. In the native enzyme, four classes of amide protons could be distinguished on the basis of their exchange behavior (Pedersen *et al.*, 1991). Those in class I exchanged too rapidly to be measured. Classes II–IV were distinguished by the magnitude of protection factors, which varied from 10^4 to 10^8 at pH 4.2–7.5 and 30°C. The variation of protection factors along the sequence is striking and shows that exchange does not result from a cooperative unfolding which would be reflected in equal protection factors for the unfolding unit, but from local fluctuations of widely differing frequencies. There are large variations in exchange rates within the three longest helices (residues 4–15, 24–36 and 88–99). Similar protection factors and a similar variation is seen in the triple-stranded β-sheet (residues 41–60). The most rapidly exchanging amide protons are found in loops that have no secondary structure. The simple correlation between distance from the surface and exchange rate was poor, and so was the correlation of the exchange rate with the hydrogen bonding pattern, yet the two factors taken together may account for the retardation of exchange in most cases. It is fair to say that the variability of individual amide proton exchange rates, observed in this and other proteins, such as BPTI (Wagner and Wüthrich, 1982), ribonuclease A (Robertson and Baldwin, 1991), and cytochrome c (Elöve *et al.*, 1992), is as yet incompletely understood. A correlation between individual exchange rates and calculated local thermodynamic stability has been proposed for BPTI (Rashin, 1987), but the calculation is based on the uncertain assumption that the exchange involves local unfolding of the structure.

In a study of hydrogen exchange in staphylococcal nuclease by Loh *et al.* (1993) the protection factors were found to be appreciably smaller than in lysozyme (of the order of 10^4–10^5), but with a similar variation along the sequence. A strong correlation with the features of secondary and tertiary structure was observed. Thus for almost all amides in unstructured regions, which are exposed to solvent, notably those in the loop of residues 40–60 and in the turns between units of secondary structure exchange was too fast for a measurement. An exception which showed sizeable protection factors were the four amides in exposed helices (F61, K64, A130 and Q131) and V39 for which no obvious structural explanation could be found. The slowest exchanging protons were located in the three α-helices and five β-strands. Some variation of protection factors within these structures could be attributed to a "fraying" at the ends, but for the most part it is awaiting an explanation. The overall picture that emerged from the study is of a molecule subject to rather larger local fluctuations than the slightly smaller lysozyme. The enzyme is known to undergo multiple conformational exchange attributable to cis-trans isomerization of the K116-P117 and H46-P47 bonds (Markley and Jardetzky, 1970; Fox *et al.*, 1986; Evans *et al.*, 1987). Both dipeptides are part of segments in which proton exchange too fast to measure. A comparison with the ^{15}N relaxation of Kay *et al.* (1989) is of interest. The order parameter describing very fast motions is rather uniform throughout the structure, suggesting that most of it behaves as a rigid unit on a nanosecond, though apparently not on a minute–hour scale. Flexibility of the loop between residues 42–57 seen in relaxation experiments is reflected in the rapid exchange rates. The four residues showing conformational exchange in relaxation experiments (Thr22, Val23, Glu52 and Lys53) behave differently with respect to hydrogen exchange. The rates for the latter two is too fast to measure, but the first two show sizeable protection factors (10^4) in unliganded form. Exchange in the denatured state was found, as in the case of lysozyme, to reflect intrinsic rates, with no indication of residual structure. The most striking finding in the study was a dramatic (up to 100-fold) increase of protection factors throughout the structure upon binding of Ca^{2+} ions and the inhibitor pdTp. As in the case of the relaxation studies cited above, this represents an action of the ligand at a distance. This non-localized effect of ligands on the stability of the structure as measured by hydrogen exchange on a time-scale of hours can be contrasted with only local stabilization of picosecond motions, confined to residues in the immediate vicinity of the binding site. This finding potentially has very significant implications for the possible role of slow motions in the mechanism of information transfer, and will be further discussed below.

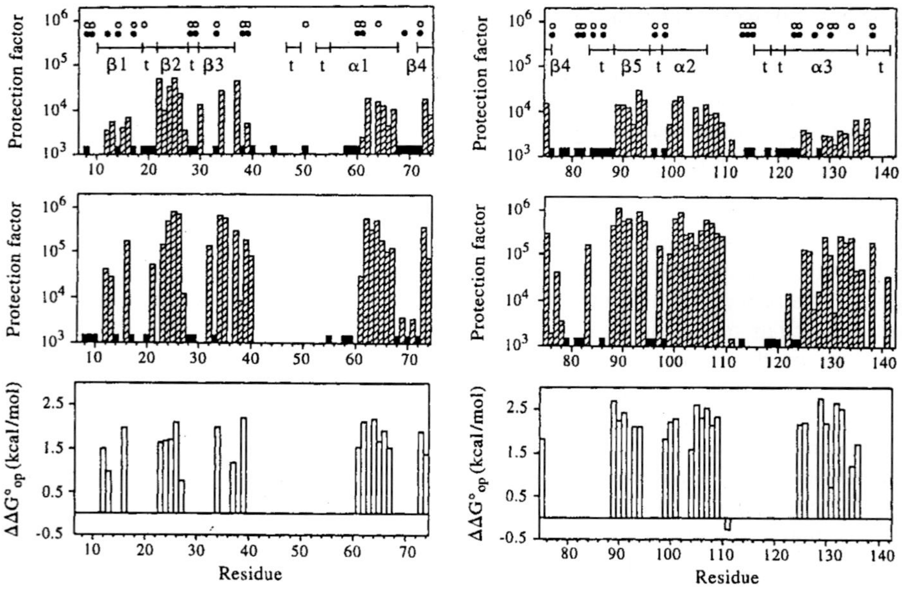

Fig. 7. Experimental protection factors for the backbone amide NHs of unligated (top) and ligated (middle) staphylococcal nuclease H124L displayed by residue number. Black bars indicated that exchange was nearly complete within the experimental dead time of ~20 min. Open and closed circles indicate residues whose amide group (nitrogen or hydrogen) are exposed to solvent in the X-ray crystal structures of unligated (Hynes and Fox, 1991) and ligated (Loll and Lattman, 1989) nuclease wt, respectively. The secondary structure of nuclease wt (Hynes and Fox, 1991) is shown for reference. The ratios of protection factors for the complexed and uncomplexed enzymes, expressed as $\Delta\Delta G^0_{op}$, are shown in the bottom trace. (Fig. 2 from Loh *et al.* © 1993, American Chemical Society, reprinted with permission.)

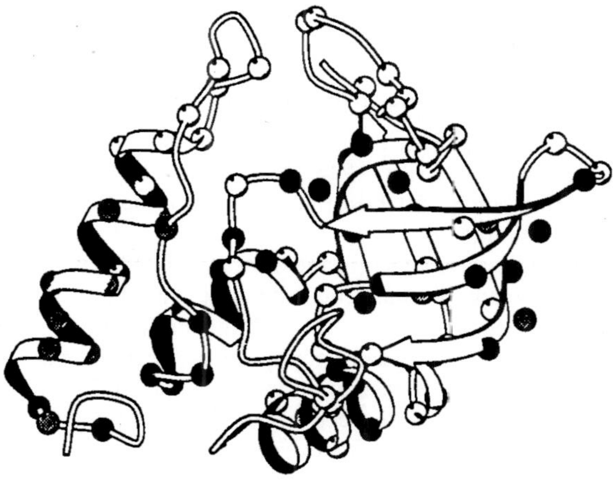

Fig. 8. Ribbon diagram of unligated nuclease wt (Hynes and Fox, 1991) showing positions of slowly exchanging amide hydrogens encoded by level of protection: (white) exchange essentially complete within the dead time of the experiment (20 min.); (grey) protection factors between 1×10^3 and 5×10^3; (black) protection factors in excess of 5×10^3. Helix 2 and helix 3 appear as the central and left-most helices, respectively. (Fig. 3 from Loh *et al.* © 1993, American Chemical Society, reprinted with permission).

The plot of protection factors as a function of sequence and a map of backbone dynamics obtained by proton exchange measurements on staphylococcal nuclease are shown in Figs 7 and 8, respectively. At least within the limits of the unrefined resolution of protein structures no significant structural changes are seen on ligand binding in this enzyme (Arnone *et al.*, 1969; Loll and Lattman, 1989; Hynes and Fox, 1991). More recent studies of point mutations in this enzyme indicate that some mutations, notably PG42, PG47 and PG117 stabilize the nuclease structure against denaturation (Vidugiris *et al.*, 1996) and by altering its dynamics (Truckses *et al.*, 1996).

A global retardation of exchange (by factors of 5–24) on ligand (DNA) binding was also observed for the helix-loop-helix GAL-4-DNA binding domain (Mau *et al.*, 1992). In addition it was found that the native Zn-containing protein exhibited slower exchange (by factors of 3–20) throughout the molecule than the Cd analog, suggesting that Cd substitution globally destabilized the structure. Interesting effects of mutations in *Rhodobacter capsulatus* ferrocytochrome *c*2 on the exchange rates of amide protons at sites distant from the mutations have been reported by Gooley *et al.* (1991, 1992). Thus a Y75F mutation increases by factors of 3–10 the exchange rates of residues 45, 46, 57, 58, 60, 61, 82–87, 98 and 100, distributed between turns and two helices while a P35A mutation increases by factors of 5–50 the rates of residues 16–18, 20, 34, 37, 43, 45, 46 and 58, all but the last located in turn and loop regions. Changes similar to those for the Y75F mutant are seen upon change of the oxidation state in the native protein. In the three-dimensional structure the two clusters of residues lie relatively near each other, but not near enough to assume stabilization by direct contact. If the increased exchange rates are taken to signify a destabilization of the local structure the nature of this destabilization remains to be explained. A 400-fold decrease in the amide exchange rates in the B-chain helix upon binding of the allosteric ligand phenol has been found to accompany the TR transition in insulin hexamers (Hardaway *et al.*, 1994). No distinction could be drawn however between stabilization of the helix and decrease in its solvent accessibility.

An additional example of the use of hydrogen exchange to study dynamics in the small allosteric system of the *trp*-repressor from *E. coli* is discussed in the next section. In this case it has been possible to establish a clear-cut correlation between the change of backbone dynamics by mutation and a change in the functional properties of the protein.

5.5. *Observables and the Underlying Motions*

Relating the motions that are possible within a protein structure to the observables by which they can be monitored is a non-trivial problem, especially for motions of the backbone. Somewhat more straightforward is the analysis of side chain motions as long as the backbone behaves, to at least the first approximation, as a rigid unit, as exemplified in the cited study by Ernst and Ernst (1994). Even for the fast thermal motions of individual atoms in proteins that can be detected by crystallographic B-factors and NMR relaxation measurements and simulated in an MD trajectory up to the time-scale of a few nanoseconds there are discrepancies in the messages we are given by different methods. A frequent impression created by molecular dynamics simulations is that very fast motions of relatively high amplitude can occur throughout the structure, although this impression has lessened through the years as damping by water became a more prominent part of the simulations. ^{15}N and ^{13}C NMR relaxation measurements of motions along the backbone tell us unequivocally that very fast, very large-amplitude motions cannot be common, or they would be detectable as very small order parameters in the simplest analysis. The agreement between order parameters calculated from MD simulations and those measured by NMR is generally not very good (Smith *et al.*, 1995). In defining our physical picture of the events we need to give precedence to experiment.

If in the analysis of ^{15}N relaxation data we describe the internal motion of the NH vector as a "wobble-on-a-cone," we should remember that this net motion reflects multiple rotations or oscillations around bonds connecting the peptide backbone. It is possible to show by modeling experiments that multiple backbone rotations will in fact result in a motion of an affected NH vector that is well described by a wobble-on-a-cone, but it is not possible to resolve the resultant motion into its individual components (Zheng *et al.*, 1995a,b). The only firm relationship is that the time-scale of the "wobble" and that of the backbone rotations causing it are identical. The same limitation applies to the interpretation of hydrogen exchange. Only the time-scale of the event is defined, but neither the nature nor the amplitude of the motions responsible for it. In the case of either measurement it can plausibly be argued that if amide n reflects a motion, but $n-1$ and $n+1$ do not, the motion must be in the immediately adjacent bonds. Yet this is not necessarily so; to cite but one complication: the magnitude of the "exchange term" associated with slow motions invoked to explain T_2 data depends on the magnitude of the chemical shift difference between the exchanging species. This could be large for a distant proton and small for all the intervening protons equally affected by the same distant motion. A still worse problem affects the analysis of B-factors. The possible net excursion is given by the size of the B-factor and it is possible to show using models that an anisotropic or even isotropic distribution of excursions will result from multiple angular rotations of the backbone. It is not possible in crystallographic measurements to decompose the observable into its components nor, in contrast to NMR relaxation and other spectroscopic measurements, to define the time-scale on which the motions occur. Definitions of the nature of responsible motions from measurements of changes in visible, UV or fluorescence spectra is similarly impossible. Only the time-scale is rigorously defined. An understanding of the slow (micro–millisecond) and functionally important motions in proteins thus critically depends on the development of a better theoretical framework and of methods for a realistic simulation of slow motions, which is not possible with the existing tools of molecular dynamics.

VI. THE LESSONS OF THE *TRP*-REPRESSOR

The most complete picture of the relationship between the dynamics of a protein structure and protein function has been obtained on the small primitive allosteric system, the *trp*-repressor from *E. coli.* The repressor deserves to be considered a primitive allosteric device, because it manifests the essential feature of all allosteric mechanisms: a structural change upon the binding of a controlling ligand, even though it may not be allosteric in the classical sense of the term, as correctly pointed out by Perutz (1989), because the two binding sites are too near each other to separate direct and indirect (allosteric) effects. A summary of the findings on this system defines the scope and limits of what can currently be learned about dynamics and function in a single system. It also provides the framework for distinguishing gaps in our understanding of a particular system from the gaps in our understanding of the basic principles of protein dynamics and of the relationship between protein structure and function.

The *trp*-repressor is a 25 kD DNA-binding protein which regulates the transcription of at least five operons: *trpEDCBA, trpR, aroH, mtr* and *aroL* (Zubay *et al.*, 1972; Rose *et al.*, 1973; Zurawski *et al.*, 1981; Heatwole and Somerville, 1991; Heatwole and Somerville, 1992). The number may be as high as 10, as additional DNA sequences with high binding constants to the repressor have been identified (B. Hurlburt, 1996; personal communication). In all cases the amino acid tryptophan is required as a co-repressor. Only weak DNA binding is observed with the ligand free aporepressor (K_D 10^6–10^7), while the ligand-containing holorepressor manifests binding constants of the order of K_D 10^{10}–10^{11} (Carey, 1988; Chou *et al.*, 1989; Hurlburt and Yanofsky, 1990).

The structures of both the apo- and the holorepressor have been determined both by X-ray diffraction (Schevitz *et al.*, 1985; Zhang *et al.*, 1987; Lawson *et al.*, 1988) and by NMR (Arrowsmith *et al.*, 1991; Zhao *et al.*, 1993), and so have three structures of its DNA complexes (Otwinowski *et al.*, 1988; Lawson and Carey, 1993; H. Zhang *et al.*, 1994). The structure of the 1:1 DNA complex is shown diagrammatically in Fig. 9. The repressor is an intertwined dimer of two identical polypeptide chains, each containing 108 residues and six α-helices (A, B, C, D, E, F). Helices D and E form a helix-turn-helix "DNA reading head" that binds to the major groove of the DNA in the liganded holorepressor, but not in the aporepressor form. The position of the DE helix-turn-helix is shifted with respect to the ABCF core by the binding of the corepressor in both the crystal and solution structures, as shown in Fig. 2. This shift is accomplished by changing only two angles in the hinge regions on both sides of the DE helix-turn-helix, i.e. φ of T65 changes by 17° and that of K91 by 14° (Zhang *et al.*, 1987). Of great interest for an understanding of the dynamics of this molecule is the finding that changes in the relative position of the reading heads of a similar magnitude can be produced by changing the crystal form of the holorepressor (Lawson *et al.*, 1988). The original proposal (Zhang *et al.*, 1987) was that the *en-bloc* shift of the DE reading heads was the essence of the allosteric mechanism, allowing a better fit of helix E into the major groove of the DNA. The actual situation is more complicated. In addition to a net *en-bloc* shift there is a structural change within the DE helices themselves and further structural changes upon complex formation with DNA. This was most clearly seen in the NMR structures (Zhao *et al.*, 1993; H. Zhang *et al.*, 1994), but comparisons of the relevant crystal structures (Otwinowski *et al.*, 1988; Lawson and Carey, 1993) indicate similar changes.

Both the observation that the position of the DE domain can be changed not only by ligand binding, but also by crystal packing and the observation of fine structural changes within the domain itself upon binding of both the corepressor tryptophan and the DNA have given rise to the notion that this domain is unusually "flexible." This impression was further strengthened by the initial calculations of the NMR structures, when it was found that the NMR spectrum of this domain contained fewer NOEs per residue than the core of the structure, so that the domain appeared disordered in the calculated structure. The r.m.s. deviations of the solution structures from each other and from the crystal structures, (Zhao *et al.*, 1993), the crystallographic B-factors (Lawson *et al.*, 1988) and of the excursion amplitudes in an MD simulation (Howard and Kollman, 1992) all show the same profile and suggest unusual flexibility or disorder in the DNA-binding domain.

These findings and the finding that the change in the heat capacity on ligand binding is larger than that expected from the enthalpy of interaction (Spolar and Record, 1994) has led to the proposal that folding of the helix-turn-helix and DNA binding are coupled in this system (Reedstrom and Royer, 1995). Taken literally, this concept is clearly inconsistent with the structural finding that the DNA-binding domain remains helical at least 90% of the time (Zheng *et al.*, 1995b; Gryk and Jardetzky, 1996a). Nevertheless it contains a kernel of truth. Detailed studies of the dynamics of this region (Gryk *et al.*, 1995; Zheng *et al.*, 1995a,b) show that, even though the helical configuration is preserved, the structure fluctuates, so that all individual hydrogen bonds in it are accessible to exchange approximately 50% of the time. Helical and hydrogen-bonded are not synonymous. Measurements which do not have an atomic resolution, as does NMR, such as heat capacity, CD and fluorescent probe binding, represent averages over both the entire molecule and time (grand averages, in the sense of statistical mechanics). They cannot distinguish between the possibilities that (a) 50% of the structure is unfolded all of the time and (b) all of the structure is "unfolded" (not hydrogen-bonded) 50% of the time. They also cannot distinguish between the possibilities that in the second case the "unfolded" residues (i) form a contiguous segment or (ii) are interspersed throughout the helix. These distinctions can be made by studying the "protection factors" in proton exchange for individual residues in the sequence (Gryk *et al.*, 1995). If 50% of the helix-turn-helix were unfolded all of the time (and folding were coupled to ligand binding) the protection factors for the 50% of the residues in the open segment would be very small and identical for all residues, and in the remaining 50% normally large and also nearly identical for all residues. If however, the geometry of the entire structure remained helical on the average, but individual hydrogen bonds were open 50% of the time, the protection factors would have values intermediate between those of a helix and a coil throughout the structure and would vary from residue to residue. This in fact is the observation that has been made on the *trp*-repressor DNA-binding domain. The domain is clearly fragile on a millisecond time-scale, and flexible in this sense. It is geometrically helical and rigid on a nanosecond time-scale. The findings of Spolar and Record and Reedstrom and Royer are readily interpreted as reflecting an overall stabilization of the structure, rather than a complete folding of one of its parts.

On a nanosecond time-scale the entire repressor molecule behaves as a rigid anisotropic rotor with one degree of internal motional freedom for the polypeptide backbone (Zheng *et al.*, 1995a). The only exceptions to this are the N-terminal and C-terminal residues, which show much more extensive local motions. If one uses the free wobble-on-a-cone model of Howarth to describe the internal motion, the average semiangle of the cone for the single motion is found to be $20.9 \pm 5.7°$. This corresponds to an order parameter of 0.82, both values indicating that the helical core of the protein does not undergo significant fluctuations on the nanosecond time-scale. This type of experiment allows us to make a distinction between "true flexibility," implying a high degree of motional freedom, as reflected in the NMR relaxation parameters, and "apparent flexibility" of structures intact by relaxation criteria, but fragile on much longer (millisecond and second) time-scales, as reflected, for example in the rates of backbone proton exchange. With true flexibility, as in the case of TMV folding of a protein segment is indeed coupled to the formation of a complex. In the case of apparent flexibility, exemplified by the *trp*-repressor and the c-myb protein (Zheng *et al.*, 1996), relatively minor adjustments suffice to produce the required induced fit. It is not yet clear whether this distinction is reflected in functional differences, or is simply a by-product of protein architecture. Either type of flexibility allows the molecule to make fine adjustments, but it is worth noting that folding a completely disordered, highly mobile segment is energetically more expensive than making the adjustments from an almost intact, but fragile state. Several indirect lines of evidence suggest that some degree of flexibility is required for the function of the entire helix-turn-helix class of DNA-binding proteins (Pabo and Sauer, 1992; Chalepakis *et al.*, 1994; Petersen *et al.*, 1995). A comparison of the dynamics data of the *trp*-repressor obtained by different methods is shown in Fig. 10 and a flexibility map based on hydrogen exchange data (Finucane and Jardetzky, 1996) in Fig. 11.

Ligand binding stabilizes not merely the local structure, but the protein structure in its entirety (Czaplicki *et al.*, 1991; Zhao *et al.*, 1993; H. Zhang *et al.*, 1994; Zheng *et al.*, 1995a). Thus, the exchange rate of backbone protons is decreased in the core helices of the repressor, as well as in the DNA-binding domain upon binding of both the corepressor and the operator DNA. This phenomenon has been known for some time (see discussion

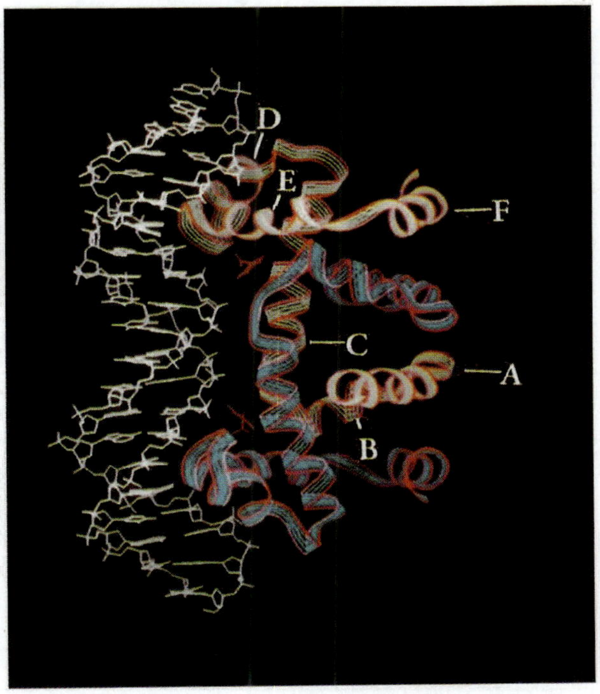

Fig. 9. Solution structure of the *trp*-repressor–consensus operator complex, *trp*-repressor at backbone representation and DNA at full atomic representation. Cofactor L-tryptophan molecules are marked in red. (Fig. 7b from H. Zhang *et al.* © 1994, reprinted by permission of the publisher, Academic Press Limited, London.)

A

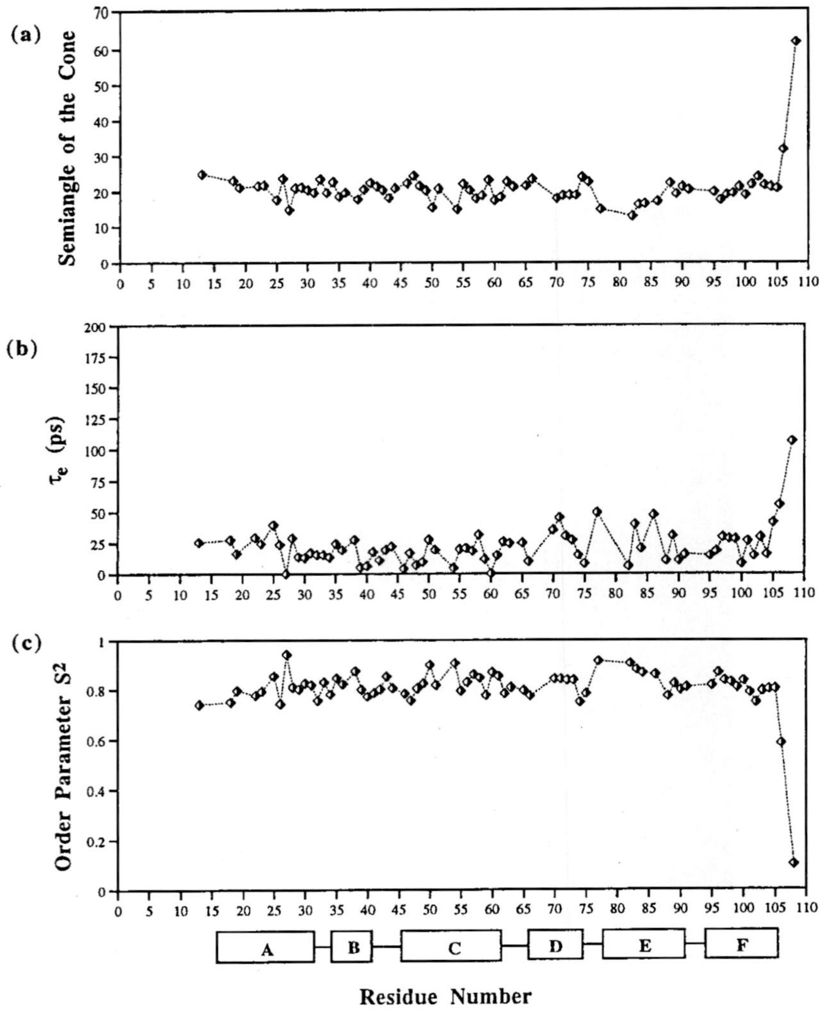

Fig. 10(A).

Fig. 10. Nanosecond and millisecond dynamics of *trp*-repressor. (A) Plots of the (a) semiangle of the cone, (b) internal correlations time and (c) order parameter as a function of residue number. The secondary structure determined in solution for *trp* repressor is indicated with boxes. (Fig. 7 from Zheng *et al.* © 1995 (b), American Chemical Society, reprinted with permission.) (B) Lifetimes ($\tau_{ex} = 1/k_{ex}$) of amide protons versus residue number of (a) *trp* aporepressor and (b) holorepressor at 35°C. (Fig. 1 from Arrowsmith *et al.* © 1991, American Chemical Society, reprinted with permission.)

of staphylococcal nuclease above), and there are many examples of stabilization against denaturation on ligand binding. Yet it is rarely pointed out that this stabilization implies an action at a distance; the transmission of a signal to distant parts of the protein chain. The mechanism by which this is accomplished remains obscure (Fig. 12).

Of particular interest in the *trp*-repressor case is the stabilization of the DNA-binding domain by point mutations in the turn of the helix-turn-helix. A series of mutants in the 77 position (AV77, AC77, etc.) also exhibit enhanced repressor activity at low concentrations of tryptophan and have been dubbed "superrepressors." The X-ray and NMR structures of the AV77 mutant are identical to those of the wild type within the

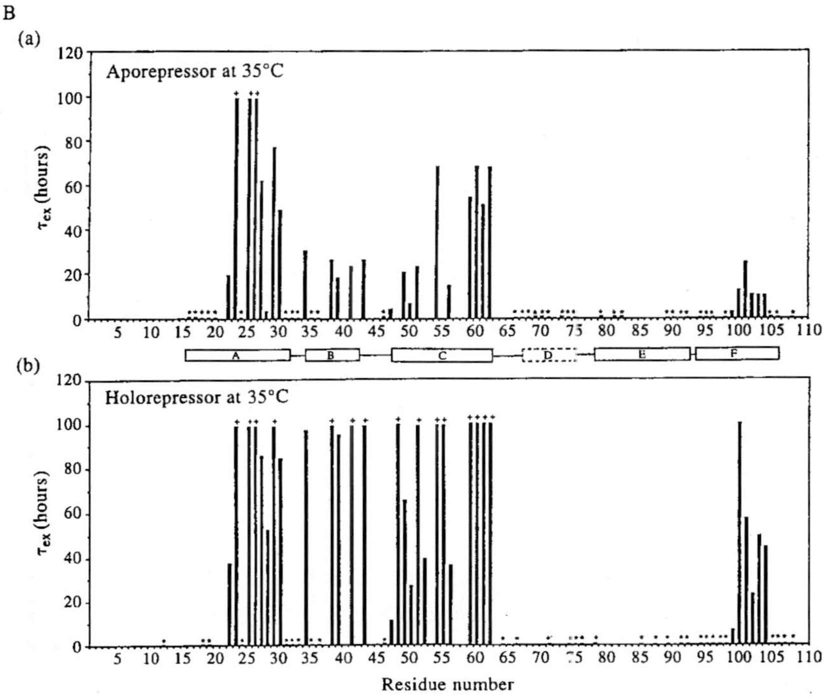

Fig. 10(B).

current limits of resolution (Zheng *et al.*, 1995b; Gryk *et al.*, 1995). The relationship between the stability and function of this mutant has been extensively investigated (Gryk and Jardetzky, 1996a,b; Gryk *et al.*, 1995, 1996; Reedstrom and Royer, 1995). The conclusion is that the increased stability of the AV77 structure leads to a decrease in the versatility of the mutant, as compared to the wild type. Whereas the wild type is uniformly effective in blocking the expression of five known operons, the AV 77 mutant shows enhanced activity in blocking the trpR operon, but decreased activity in blocking the trpL and aroH operons (Gryk *et al.*, 1996). This loss of versatility actually accounts for the superrepressor activity. Since the binding constants of the wild type and AV77 for the trp R operator DNA are identical (Hurlburt and Yanofsky, 1992), weaker binding to other operators increases its availability to the trpR operator (Arvidson *et al.*, 1994). The *trp*-repressor findings can be contrasted to the destabilization of a structure by mutation, noted above for T4 lysozyme, ferrocytochrome c and the c-myb DNA-binding domain (Ogata *et al.*, 1995). Thus, mutations can have either a stabilizing or destabilizing effect on protein dynamics. The relationship between flexibility and function is clear in the *trp*-repressor example. Decrease in flexibility by mutation makes it more difficult for the repressor to adjust to different DNA sequences by induced fit and thus limits the choice between different operator sequences.

The flexibility of the peptide backbone in the DNA-binding domain of the *trp*-repressor observed by amide hydrogen exchange can thus be related to two functions of the simple allosteric system: (1) it facilitates the allosteric transition the *en-bloc* movement of the reading head on tryptophan binding, and (2) it facilitates induced fit to each of the five different operator DNA sequences for the five operons controlled by the repressor. Changes in the flexibility of the repressor change the ability of the molecule to discriminate between different DNA sequences.

The example of the *trp*-repressor suggests that quite generally an optimum of flexibility may be essential for function. Protein function depends on the successive and reversible formation and breakage of specific atomic contacts. Rigidity is required to make good contacts, but flexibility is required to break them. Flexibility is also required to correct imperfections of geometric fit and to allow versatility for making contacts with more than one partner, for example specifically binding to more than one DNA sequence. A compromise between the conflicting requirements of (1) strong, but reversible binding and (2) high specificity, but also some versatility is essential in the design of any control mechanism. The relatively rigid bacterial and snake venom toxins make excellent irreversible inhibitors, but bind too tightly to be a useful part of a control system which must function reversibly. At the other extreme, the energetic costs of bringing a highly flexible random coil into a unique configuration suitable for specific contacts are too high for efficient function. We can expect to find a large number of different compromises in different protein systems. The compromise reflected in the *trp*-repressor system—a molecule too flexible to maintain specific contacts by itself—but stabilized sufficiently by a ligand to function effectively in the required control range, while still retaining enough flexibility to bind to more than one operator is perhaps one of the easiest to achieve. There are undoubtedly others.

VII. FLEXIBILITY AND STRUCTURE DETERMINATION

Our task of understanding the relationship between structure and function would be simpler if we could take any structure as given and examine its dynamics to establish the relation. This is, however, not always safe. The existence of flexibility in proteins has several important consequences for structure determination by either X-ray or NMR. These have to always be borne in mind and any given structure may need to be re-examined in the light of information about its dynamics.

Well known is that mobile protein segments are ill defined in any structure determination. Not as widely appreciated is the possibility that flexible segments can be ordered by crystal forces, giving rise to a crystal structure that does not accurately reflect the natural (the solution) structure of the molecule or molecular complex. Several examples of differences between the crystal structure and a solution structure that point toward this possibility are now known: calmodulin (Barbato *et al.*, 1992; Finn *et al.*, 1995; Zhang *et al.*, 1995), troponin C (Kleerekoper *et al.*, 1995), and calbindin (Kördel *et al.*, 1992). The danger that crystal forces will distort the structure of a non-covalent complex, such as a protein–DNA complex, which is held together by relatively weak forces, is particularly great (Pabo and Sauer, 1992). It may well be that such a distortion of the complex in the crystal accounts for the unusual finding in the *trp*-repressor–operator crystal structure, where no specific DNA-protein contacts, but instead several water-mediated contacts, were observed (Otwinowski *et al.*, 1988; Lawson and Carey, 1993). Distortion by crystal forces seems a much more likely explanation of the lack of specificity of the complex than the "indirect readout" postulated by the authors, especially in view of the fact that the solution structure of the same complex is compatible with direct DNA-protein contacts (H. Zhang *et al.*, 1994).

In the case of NMR structures, neglect of dynamics, especially of conformational equilibria involving whole segments or domains can easily result in the calculation of a wrong structure. The conventional calculation of NMR structures from internuclear distances reflected in the NOEs is based on the assumption of a single rigid structure. Several attempts have been made to introduce corrections taking internal dynamics into account. Some improvement of the calculated structures has been noted (Torda and Norton, 1989; Torda *et al.*, 1990), although the danger of introducing artefacts by such calculations has also been pointed out (Billeter and Wüthrich, 1991). However, these efforts have been centered on correcting for fast thermal motions, and this is not the major problem. The major problem arises if there exists a conformational equilibrium between structures that correspond to two or more energy minima. In this case a single structure calculated on the assumption that the data represent a single structure has no physical meaning. It does not even represent the (linear) physical average between the different structures, because the NMR parameters reflected in the NMR data are non-linear averages (Jardetzky, 1980; Jardetzky and Roberts, 1981). In some cases where the nature of the conformational change is known, as in tyrosine rotation, substantial improvement in accuracy can be achieved by introducing corrections for the motion (Fejzo *et al.*, 1991).

In general, the most serious problem is that the NOE data used in structure calculations contain few clues as to whether they represent a single species or an average. Additional information, some obtainable from NMR, is necessary to decide on the possible nature and number of interconverting conformers. The serious drawback is that the structures of the individual conformers cannot be independently verified, unless the exchange can be slowed down by lowering the temperature or changing the solvent to resolve their individual spectra. This has been accomplished in the study of tyrosine ring rotation, but not for a protein as a whole (Campbell and Dobson, 1975). Short of this, the "reasonableness" of the individual conformer structures has to be judged from the average spectra they yield. Even if correct, this is not as definitive as a simple straightforward structure calculation is usually claimed to be. There have been correspondingly few attempts to deal with this problem.

A prototype study was carried out by Kim and Prestegard (1989) who have developed a dynamic model for the NMR structure of acyl carrier protein in solution. Assuming that the protein existed in two different conformations they obtained a significant improvement in the fit of the calculated average distances to the cross-relaxation parameters in which the distances are reflected. Needless to say the average distances calculated from the dynamic model were not identical to the "average" distance that could be calculated from the relaxation parameters on the incorrect assumption of a single structure. The energy of the two conformers was also found to be lower than the energy of the incorrect "average" structure. The postulated two-state model involved a partial disruption of the second of the three helices and substantial motion of the connecting loops. The flexibility or fragility of this helix was more recently confirmed by amide exchange studies (Andrec *et al.*, 1995). A similar equilibrium between two conformations has been invoked to account for NMR findings in ADR1a, a zinc-finger peptide (Xu *et al.*, 1991).

It remains to be seen how many of the published structures, both X-ray and NMR, will have to be revised in the light of the growing understanding of dynamics, of the possibility that flexible structures can be distorted into artefacts by crystallization and that an NMR structure calculated on the conventional assumption that there exists a unique rigid structure will be wrong, if the structure is in fact flexible. It is already clear however that the greatest confidence can be placed in a given structure when the dynamic behavior of the molecule is examined at the same time, as has been done in the case of the several interleukins, the calcium-binding proteins calmodulin and calbindin and the *trp*-repressor. Thus understanding the dynamics of a protein is important not only for understanding its function, but also for assessing the validity of its structure (Jardetzky, 1984; Zhao and Jardetzky, 1994a; Hoogstraten *et al.*, 1995; Hoogstraten and Markley, 1996).

VIII. RELATIONSHIPS BETWEEN MOBILITY, FLEXIBILITY AND FUNCTIONALLY IMPORTANT MOTIONS

In the introduction we have noted that structural studies have defined for us what we must know to understand allosteric transitions and functionally important motions in proteins generally. However, within this framework the present state of our knowledge does not appear as a complete elegant edifice, but rather as a patchwork of disjointed observations that do not easily relate to each other or to the central question of describing functionally significant motions. The spatial and time resolution of the available information on protein dynamics is diagrammatically shown in Fig. 13.

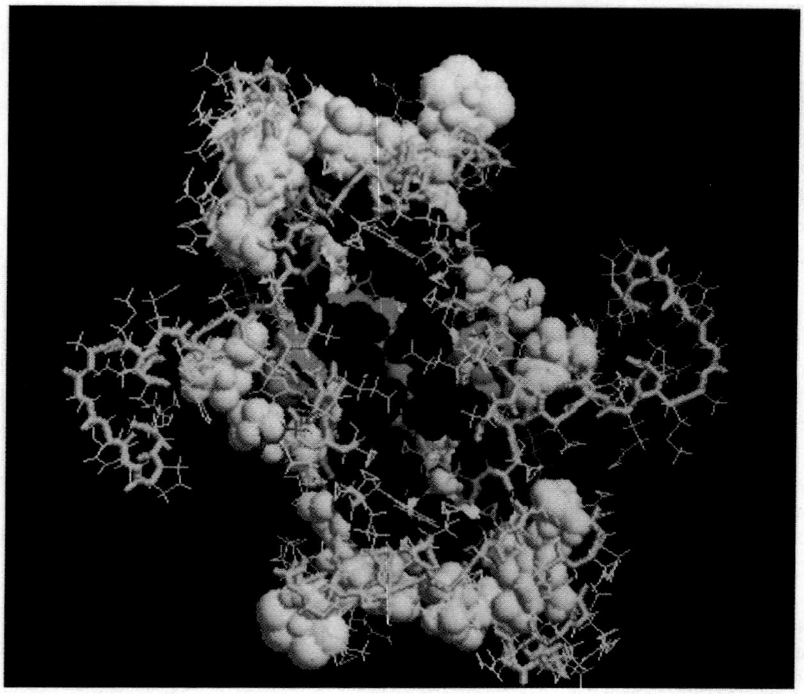

Fig. 11. Flexibility map of the *trp* repressor as it would be seen from the DNA, based on backbone proton exchange data. Amide protons at pH 7.2 that exchange between 2.5 and 23 h after dissolution of the fully protonated repressor in $^{2}H_{2}O$ are shown in cyan and those still present 23 h after dissolution are shown in dark blue. (Fig. 8c from Finucane and Jardetzky, copyright © 1996 The Protein Society, reprinted with the permission of Cambridge University Press.)

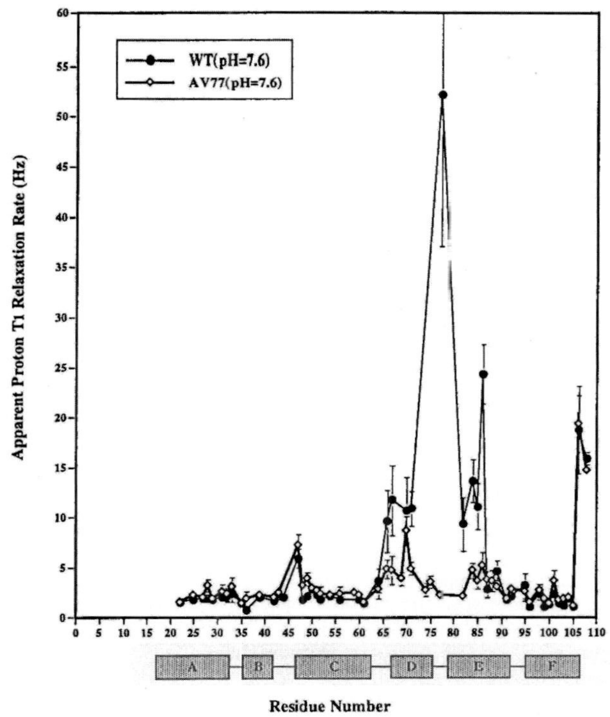

Fig. 12. Stabilization of the *trp*-repressor by the AV77 mutation: Apparent proton T_1 relaxation rates (proportional to the proton exchange rates) for the backbone amides of wild-type *trp* repressor versus amino acid sequence for the wt (fast rates in the DE helix-turn-helix region), and the mutant AV77 (slow rates) (pH = 7 6).

The overwhelming majority of data we have on protein dynamics reflect more or less random *thermal* motions which are ubiquitous in all molecular systems. Pico–nanosecond thermal motions can be detected by NMR relaxation measurements, Resonance Raman spectroscopy or the temperature dependence of Debye–Waller factors and can be modeled by molecular dynamics. Thermal motion in proteins is reasonably well described and reasonably well understood. Yet thermal motions rarely have any functional significance, though their existence does. Flexibility facilitates the in-and-out diffusion of substrates and products in enzymes, or ligands generally in proteins, the making and breaking of specific bonds, and the mutual adjustment of protein and ligand side chains—an induced fit at the side chain level.

The functional role of segmental flexibility can be similarly understood in the same simple terms. No specific sequence or pattern of motions is needed to perform a function, the mere fact that a segment is mobile or flexible suffices. In these terms we can understand the role of disordered segments that fold or adjust to form a specific set of contacts with a ligand, such as DNA, RNA or another protein, as in the case of TMV or *trp*-repressor. In the same terms we can understand the function of flexible linkers, as we find in antibodies, between kringle domains or the domains of dumb-bell-shaped calcium-binding proteins, all of which allow variation in the distance between multiple ligands. It is noteworthy that the required flexibility can manifest itself either already on the time-scale of fast thermal motions (pico–nanoseconds, "true flexibility"), as in the calcium-binding proteins and TMV, or on the much slower time-scale of msec or sec, rather as a fragility of otherwise ordered structures ("apparent flexibility"), as in the *trp*-repressor. The function of allowing induced fit, either in the coarse sense of varying the distance between two binding sites, or in a fine tuning sense of allowing adjustment of atomic positions between neighboring atoms, can be performed in either case. More or less

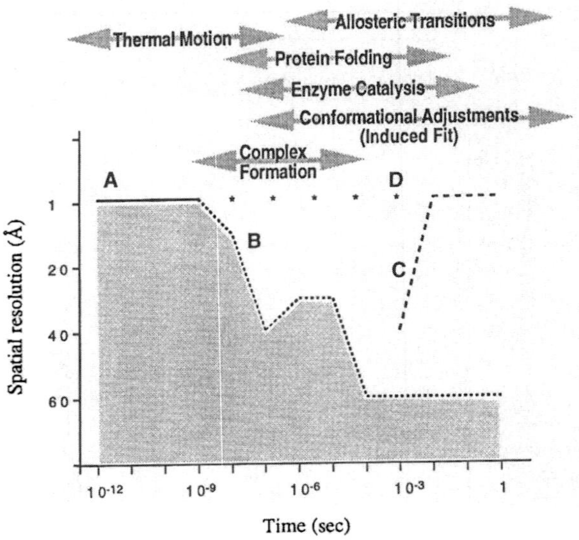

Fig. 13. Space–time resolution of information on internal motions in proteins. A: all atoms from molecular dynamics simulations, NMR relaxation and B-factor analysis, B: Motions detectable by NMR relaxation and other spectroscopic techniques, C: Backbone motions from H-exchange and D: Information potentially obtainable from NMR relaxation and chemical shift averaging experiments.

random thermal motion is in this case adequate to produce the desired result and the driving force for changing from a random to a specific fixed configuration is the formation of contacts with a specific ligand. Primitive allosteric mechanisms, such as the *trp*-repressor may require little more than the stabilization of a specific configuration from among many—rapidly or slowly—interconverting local conformations. It is conceivable, though not yet unequivocally demonstrated that allosteric control by phosphorylation similarly involves the ordering of a previously disordered segment as the key initial step (Johnson, 1994).

In contrast, hardly understood at all are the slow, collective, probably concerted motions, which can occur in highly organized systems. Vibrations of a suspension bridge, which at resonance can lead to its downfall are among the best known examples of motions engineered into an organized structure, dependent on its design and distinct from the normal random motions of the material. It is possible, even probable (though unproven), that these motions underlie the "true" allosteric transitions between two or at most a few distinct stable structures. Here we are not dealing with a fragile or disordered segment of a stable structure, which becomes ordered by thermal motion in response to a new set of possible interactions. We see a change of the structure from one well defined form to another, which is in most cases described as a change in the equilibrium distribution of the two or more interconverting, distinctly organized forms. Such transitions require no more than net movements at a few "soft spots" in a rigid body. Yet here we most often do not even know the basic facts. What are the differences of dihedral angles in the backbone that distinguish the two or more forms? How large is the number of angles changed? Are the dynamics of these angles different from those of other residues in the structure? Can these dynamics account for a perpetual interconversion that could lead to a conformational equilibrium? Can the "soft spots" be identified by their dynamic behavior?

Slow motions that can be associated with conformational transitions and protein function can easily and reliably be detected at a few individual reporter groups within a protein structure, but not yet described in atomic detail. They are therefore usually, and appropriately, described in rather vague terms; fluctuations, breathing, conformational averaging, etc. A recent partial exception to this are the "conformational exchange terms" in NMR relaxation measurements and the physical exchange of protons from the polypeptide backbone, both of which can be localized to individual amide groups. However, even here the relationship between the exchange rate and the motions responsible for it remains ill-defined. Similarly ill-defined is the terminology used for the description of motions in proteins.

To give a precise and systematic description of internal motions in proteins it is necessary to begin with the elementary observation that all motions effective in producing structural changes in proteins are rotations or oscillations about chemical bonds. These naturally fall into two classes: rotations about bonds in the polypeptide backbone, i.e. changes in the Ramachandran angles φ and ψ (with proline isomerization as a special case); and rotations about bonds in the side-chains, i.e. changes in the angles χ_1, χ_2, etc. Ramachandran angle changes are necessarily involved in all changes of secondary and tertiary structure, while changes of side-chain angles, by themselves, only affect the local conformation. Translations in a covalently linked chain are necessarily secondary to rotations about connecting bonds. Vibrations, an integral component of thermal motion and of all MD simulations are localized and will, by themselves, not result in a structural change. They can in principle, especially in organized secondary structures, serve as elements for the transmission of information. Suggestions that vibrational solitons may be carriers of information in protein structures are well known (Davydov, 1979), but conclusive experimental evidence for such effects is lacking.

A general classification of protein motions might become possible once the mechanisms are understood, if the motions leading to conformational transitions are found to fall into distinct classes, clearly distinguishable by the pattern of backbone and side-chain movements. For the present, very few distinct patterns of movements have been documented. Translocation of helices by a change of one or a few dihedral angles in the loop connecting them has been observed experimentally and it may be useful to reserve the use of the term "hinge" motion to this type of motion. Another type of motion often encountered in NMR experiments is the rapid random change of backbone angles in a longer contiguous segment of the backbone, which may be appropriately called "segmental" motion. The distinction between hinge and segmental motions lies not so much in the number of backbone angles affected (although one should beware of long hinges), as in the requirement that a hinge be bracketed by two rigid structures. Correlated two-hinge movements have been seen, e.g., in the "twisting" of β-sheets, three-hinge or multiple hinge movements in helix bundles, creation of a hinge, as in the bending of a helix and the unwinding of a helix can also be precisely described. Beyond that clear lines defining distinct categories would be difficult to draw with the experimental material at hand.

Information transfer between two points in a protein structure requires motion, but can in principle occur either with or without a change in the intervening structure. A path for the transmission of conformational information, accompanied by a change in the intervening structure, can involve either backbone rotations only or side-chain rotations only, or a combination of both. Such a path need not involve a contiguous set of angles, since the change of only one backbone angle may change the orientation of an entire rigid unit of secondary structure, such as an α helix, with the path of individual angle rotations continuing at the other end. Transmission of information can be either a cascade of events that can be resolved in time, as would be expected in a weakly coupled system, or it could be a concerted event, with all steps occurring simultaneously, as would be expected for a strongly coupled system. Alternatively, however, information could be transferred within protein structures by a modulation of their dynamics, without changes in the average structure.

The importance of the "slow" motions for information transfer in proteins is underscored by the cited repeated findings that ligand binding tends to produce at most a localized damping of the pico- and nanosecond motions, while it globally affects the slow processes, such as amide proton exchange, throughout the entire structure. Something other than an easily detectable structural change or the modulation of the very fast motions monitored by NMR relaxation and MD simulations is clearly responsible for the transmission of this information. Where no structural change is observed at good resolution, the stabilizing action at a distance must involve a coherent modulation of backbone dynamics (Finucane and Jardetzky, 1995).

A number of questions need to be asked about these "slow" motions. First—are they really slow motions or, as some molecular dynamicists prefer to think of them, rare events? The difference is not purely semantic. If one believes that the binary potentials and linear equations used in MD calculations (neglecting strong coupling that is possible in many-body interactions) are an adequate description of reality, one expects the actual motion to be fast, observable on a longer time-scale only by its rarity. If on the other hand one views the protein as a lattice in which the lattice points can be strongly coupled, the possibility of complex collective motions in which the actual motion is slow becomes real. We have as yet little to go on to make a choice between the two conceptions. The additional problem is that the mathematical treatment of the second option—of extensive coupling in an organized system as irregular and complex as a protein—is all but intractable. We thus see, however, that the lack of understanding of the slow motions is not simply the result of an insufficient number of observations, it reflects as much the lack of a clear and rigorous conceptual framework for describing collective motions in organized systems that are as highly irregular in their architecture as proteins are. An informed choice of the conceptual framework for describing the "slow" motions in proteins is one of the most urgent items on our agenda.

To achieve an understanding of information transfer and allosteric transitions in proteins we need, to begin with, to make the classical allosteric proteins accessible to detailed experimental studies of their dynamics, so that the information contained in NMR relaxation parameters and proton exchange measurements can be brought to bear on the problem. The experimental difficulties to be overcome are greater than those encountered in the prototype study of the primitive "allosteric" system—the *trp*-repressor. Most allosteric proteins have higher molecular weights, which makes the analysis of their NMR spectra more difficult because of more extensive overlap of the more numerous and broader spectral lines. Yet, with increasing use of recombinant technology and selective isotopic labeling such studies on hemoglobin and other allosteric studies of comparable size should become feasible in the near future.

Second, we need to carefully examine the theoretical basis of our description of protein motions. We need to test whether the binary potentials used in conventional MD simulations are adequate to deal with the slow and probably concerted motions and search for new ways of describing the physics of the conformational space in which these motions occur. There is reason to expect that the recently developed Monte Carlo Simulated Annealing procedure (MCDIS-MD; Monte Carlo in Dihedral Space with Molecular Dynamics) which in essence extends the time-scale of MD simulations to longer time-scales, will shed some light on this point (Zhao and Jardetzky, 1994b; 1995). In pilot experiments on crambin it has been possible to show that with the more complete exploration of conformational space afforded by MCDIS-MD, motion at some dihedral angles contributes more to the internal mobility of the peptide chain than that at others. This result is evident from the larger RMSD values for some angles subjected to MCDIS-MD, as opposed to others, as shown in Fig. 14. Also evident in the figure is a strong correlation of motions of adjacent φ and ψ angles, which has been noted before (Chandrasekhar *et al.*, 1992; Garcia, 1992), but remains largely neglected in the discussion of MD simulations and the analysis of NMR relaxation data. It is well known that even sizeable concerted changes in adjacent φ and ψ angles along a unit of secondary structure result in only a small change in the overall geometry of the structure (Ribeiro *et al.*, 1983).

Once the geometric nature of the functionally significant motions is captured by this procedure or by one of its successors, we will need to examine more seriously the possibility that they will have to be described by non-linear equations. These can more accurately represent traveling and periodic rotational motions of the backbone in

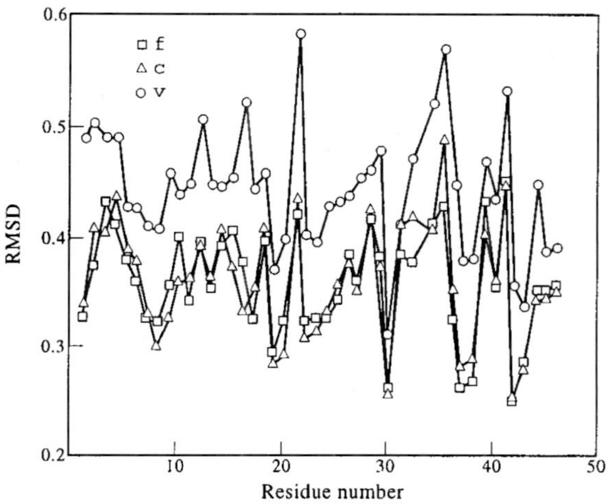

Fig. 14. RMSD of crambin structures for which one of the three dihedral angles was changed by −90° followed by energy minimization and short molecular dynamics.

organized secondary structures (Jardetzky and King, 1983). If we want to understand information transfer within proteins and allostery, we need to always bear in mind Hans Frauenfelder's dictum (Frauenfelder, 1983, p. 338)—"nature is non-linear."

ACKNOWLEDGEMENTS

The author thanks C. Brockel, W. N. Lipscomb, J. L. Markley, G. A. Petsko and R. Rigler for reading the manuscript and making valuable comments.

REFERENCES

Abraham, D. J., Safo, M. K., Boyiri, T., Danso-Danquah, R. E., Kister, J. and Poyart, C. (1995) How allosteric effectors can bind to the same protein residue and produce opposite shifts in the allosteric equilibrium. *Biochemistry* **34,** 15,006–15,020.

Ackers, G. K., Doyle, M. L., Myers, D. and Daugherty, M. A. (1992) Molecular code for cooperativity in haemoglobin. *Science* **255,** 54–63.

Akke, M., Skelton, N. J., Kördel, H., Palmer, A. G. III and Chazin, W. J. (1993) Effects of ion binding on the backbone dynamics of calbindin D_{9k} determined by ^{15}N NMR relaxation. *Biochemistry* **32,** 9832–9844.

Alber, T., Gilbert, W. A., Ponzi, D. R. and Petsko, G. A. (1983) The role of mobility in the substrate binding and catalytic machinery of enzymes. In *Mobility and Function in Proteins and Nucleic Acids*, pp. 4–24 (Ciba Foundation Symposium **93**). Pitman, London.

Allerhand, A., Doddrell, D., Glushko, V., Cochran, D. W., Wenkert, E., Lawson, P. J. and Gurd, F. R. (1971) Conformation and segmental motion of native and denatured ribonuclease A in solution. Application of natural-abundance carbon-13 partially relaxed Fourier transform nuclear magnetic resonance. *J. Am. Chem. Soc.* **93,** 544–546.

Altman, R. B. and Jardetzky, O. (1989) The heuristic refinement method for the determination of the solution structure of proteins from NMR data. *Meth. Enzymol.* **177,** 218–246.

Andrec, M., Hill, R. B. and Prestegard, J. H. (1995) Amide exchange rates in *Escherichia coli* acyl carrier protein: Correlation with protein structure and dynamics. *Prot. Sci.* **4,** 983–993.

Arata, Y. (1996) NMR of larger proteins: An approach to the structural analyses of antibody. In *Biological NMR Spectroscopy*, pp. 209–232 (eds. J. L. Markley and S. J. Opella). Oxford University Press, Oxford (*in press*).

Arnone, A. C., Bier, J. F., Cotton, F. A., Hazen, E. E. Jr., Richardson, D. C. and Richardson, J. S. (1969) The extracellular nuclease of *Staphylococcus aureus*: Structures of the native enzyme and an enzyme-inhibitor complex at 4 Å resolution. *Proc. natn. Acad. Sci. U.S.A.* **64,** 420–427.

Arrowsmith, C. H., Czaplicki, J., Iyer, S. B. and Jardetzky, O. (1991) Unusual dynamic features of the *trp*-repressor from *E. coli. J. Am. Chem. Soc.* **113,** 4020–4022.

Arvidson, D. N., Arvidson, C. G., Lawson, C. L., Miner, J., Adams, C. and Youderian, P. (1994) The tryptophan repressor sequence is highly conserved among the *Enterobacteriaceae*. *Nucleic Acids Res.* **22**, 1821–1829.

Auzat, I., Byrnes, W. M., Garel, J.-R. and Chang, S. H. (1995a) Role of residue 161 in the allosteric transitions of two bacterial phosphofructokinases. *Biochemistry* **34,** 7062–7068.

Auzat, I., Le Bras, G. and Garel, J.-R. (1995b) Hypercooperativity induced by interface mutations in the phosphofructokinase from *Escherichai coli*. *J. molec. Biol.* **246,** 248–253.

Auzat, I., Gawlita, E. and Garel, J.-R. (1995c) Slow ligand-induced transitions in the allosteric phosphofructokinase from *Escherichia coli*. *J. molec. Biol.* **249,** 478–492.

Babu, Y., Bugg, C. E. and Cook, W. J. (1988) Three-dimensional structure of calmodulin refined at 2.2 Å resolution. *J. molec. Biol.* **204,** 191–204.

Bai, Y., Milne, J. S., Mayne, L., Englander, S. W. (1993) Primary structure effects on peptide group hydrogen exchange. *Proteins: Struct. Func. Genet.* **17,** 75–86.

Barbato, G., Ikura, M., Kay, L.E., Pastor, R.W., Bax, A. (1992) Backbone dynamics of calmodulin studied by ^{15}N relaxation using inverse detected two-dimensional NMR spectroscopy: The central helix is flexible. *Biochemistry* **31,** 5269–5278.

Berglund, H., Kovács, H., Dahlman-Wright, K., Gustafsson, J.-Å. and Härd, T. (1992) Backbone dynamics of the glucocorticoid receptor DNA-binding domain. *Biochemistry* **31,** 12,001–12,011.

Bialek, W. and Onuchic, J. N. (1988). Protein dynamics and reaction rates: mode-specific chemistry in large molecules?. *Proc. natn. Acad. Sci. U.S.A.* **85,** 5908–5912.

Billeter, M. and Wüthrich, K. (1991) Time-averaged nuclear Overhauser effect distance restraints applied to tendamistat. *Chemtracts—Biochem. molec. Biol.* **2,** 32–34.

Birdsall, B., Bevan, A. W., Pascual, C., Roberts, G. C. K., Feeney, J., Bronenborn, A. and Clore, G. M. (1984) Multinuclear NMR characterization of two coexisting conformational states of the *Lactobacillus casei* dihydrofolate reductase-trimethoprim-NADP+ complex. *Biochemistry* **23,** 4733–4742.

Bourne, Y., Arva, A. S., Bernstein, S. L., Watson, M. H., Reed, S. I., Endicott, J. E., Noble, M. E., Johnson, L. N. and Tainer, J. A. (1995) Crystal structure of the cell cycle-regulatory protein sucl reveals a β-hinge conformational switch. *Proc. natn. Acad. Sci. U.S.A.* **92,** 10,232–10,236.

Browner, M. F. and Hackos, D., Fletterick, R. (1994) Identification of the molecular trigger for allosteric activation in glycogen phosphorylase. *Struct. Biol.* **1**(5), 327–333.

Brünger, A. T. (1992) *X-PLOR (Ver. 2.1) Manual.* Yale University, New Haven, CT.

Brüschweiler, R. (1992) Normal modes and NMR order parameters in proteins. *J. Am. Chem. Soc.* **114,** 5341–5344.

Brüschweiler, R. and Case, D. A. (1994) Collective NMR relaxation model applied to protein dynamics. *Phys. Rev. Lett.* **72,** 940–943.

Brüschweiler, R., Griesinger, C. and Ernst, R. R. (1989) Correlated motion monitored by NMR relaxation in the rotating frame. A source of structural and dynamic information on macromolecules. *J. Am. Chem. Soc.* **111,** 8034–8035.

Brüschweiler, R., Liao, X. and Wright, P. E. (1995) Long-range motional restrictions in a multidomain zinc-finger protein from anisotropic tumbling. *Science* **268,** 886–889.

Brzovic, P. S., Choi, W. E., Borchardt, D., Kaarsholm, N. E. and Dunn, M. F. (1994) Structural asymmetry and half-site reactivity in the T to R allosteric transition of the insulin hexamer. *Biochemistry* **33,** 13057–13069.

Buck, M., Boyd, J., Redfield, C., MacKenzie, D. A., Jeenes, D. J., Archer, D. B. and Dobson, C. M. (1995) Structural determinants of protein dynamics: Analysis of ^{15}N NMR relaxation measurements for main-chain and side-chain nuclei of hen egg white lysozyme. *Biochemistry* **34,** 4041–4055.

Campbell, I. D. and Dobson, C. M. (1975) Spin echo double resonance: A novel method for detecting decoupling in Fourier transform nuclear magnetic resonance, *J. Chem. Soc., Chem. Commun.*, 750–751.

Campbell, I. D., Dobson, C. M., Moore, G. R., Perkins, S. J. and Williams, R. J. P. (1976) Temperature dependent molecular motion of a tyrosine residue of ferrocytochrome C. *Fedn. Eur. biochem. Socs Lett.* **70,** 96–101.

Campbell, I. D., Dobson, C. M. and Williams, R. J. P. (1985) The study of conformational states of proteins by nuclear magnetic resonance. *Biochem. J.* **231,** 1–10.

Carey, J. (1988) Gel retardation at low pH resolves *trp* repressor-DNA complexes for quantitative study. *Proc. natn. Acad. Sci. U.S.A.* **85,** 975–979.

Chalepakis, G., Wijnholds, J. and Gruss, P. (1994) Pax-3–DNA interaction: Flexibility in the DNA binding and induction of DNA conformational changes by paired domains. *Nucleic Acids Res.* **22,** 3131–3137.

Chandrasekhar, I., Clare, G. M., Szabo, A., Gronenborn, A. M. and Brooks, B. R. (1992) A 500 ps molecular dynamics simulation study of Interleukin-1β in water. *J. molec. Biol.* **226,** 239–250.

Cheng, J.-W., Lepre, C. A., Chambers, S. P., Fulghum, J. R., Thompson, J. A. and Moore, J. M. (1993) ^{15}N NMR relaxation studies of the FK506 binding protein: Backbone dynamics of the uncomplexed receptor. *Biochemistry* **32,** 9000–9010.

Cho, N., Song, S. and Asher, S. A. (1994) UV resonance raman and excited-state relaxation rate studies of hemoglobin. *Biochemistry* **33,** 5932–5941.

Chothia, C. and Lesk, A. M. (1987) Canonical structures for the hypervariable regions of immunoglobulins. *J. molec. Biol.* **196,** 901–917.

Chothia, C., Lesk, A. M., Tramontano, A., Levitt, M., Smith-Gill, S. J., Air, G., Sheriff, S., Padlan, E. A., Davies, D., Tulip, W. R., Colman, P. M., Spinelli, S., Alzari, P. M. and Poljak, R. J. (1989) Conformations of immunoglobin hypervariable regions. *Nature* **342,** 877–883.

Chou, W. Y., Bieber, C. and Matthews, K. S. (1989) Tryptophan and 8-anilino-1-naphthalenesulfonate compete for binding to *trp* repressor. *J. biol. Chem.* **264,** 8309–8313.

Clore, G. M., Driscoll, P. C., Wingfield, P. T. and Gronenborn, A. M. (1990) Analysis of the backbone dynamics of interleukin-1β using two-dimensional inverse detected heteronuclear ^{15}N-^{1}H NMR spectroscopy. *Biochemistry* **29,** 7387–7401.

Cohen, G. B. and Ren, R., Baltimore, D. (1995) Modular binding domains in signal transduction proteins. *Cell* **80,** 237–248.

Connolly, M. L. (1985) Computation of molecular volume. *J. Am. Chem. Soc.* **107,** 1118–1124.

Constantine, K. L., Friedrichs, M. S., Bell, A. J., Lavoie, T. B., Mueller, L. and Metzler, W. J. (1993a) Relaxation study of the backbone dynamics of human profilin by two-dimensional ^{1}H-^{15}N NMR. *Fedn Eur. biochem. Socs 13446* **336**(3), 457–461.

Constantine, K. L., Friedrichs, M. S., Goldfarb, V., Jeffrey, P. D., Sheriff, S. and Mueller, L. (1993b) Characterization of the backbone dynamics of an anti-digoxin V_L domain by inverse detected ^1H-^{15}N NMR: Comparison with X-ray data for the Fab. *Proteins: Struct. Func. Genet.* **15**, 290–311.

Czaplicki, J., Arrowsmith, C. and Jardetzky, O. (1991) Segmental differences in the stability of the *trp*-repressor peptide backbone. *J. Biomol. NMR* **1**, 349–361.

Davydov, A. S. (1979) Solitons in molecular systems. *Phys. Scr.* **20**, 387–394.

De Staercke, C., Van Vliet, F., Xi, X.-G., Rani, C. W., Ladjimi, M., Jacobs, A., Triniolles, F., Hervé, G. and Cunin, R. (1995) Intramolecular transmission of the ATP regulatory signal in *Escherichia coli* aspartate transcarbamylase: Specific involvement of a clustered set of amino acid interactions at an interface between regulatory and catalytic subunits. *J. molec. Biol.* **246**, 132–143.

Ding, X., Rasmussen, B. F., Petsko, G. A. and Ringe, D. (1994) Direct structural observation of an acyl-enzyme intermediate in the hydrolysis of an ester substrate by elastase. *Biochemistry* **33**, 9285–9293.

Dobson, C. M. and Karplus, M. (1986) Internal motion of proteins: Nuclear magnetic resonance measurements and dynamic simulations. *Meth. Enzymol.* **131**, 362–389.

Doster, W., Cusack, S. and Petry, W. (1989) Dynamical transition of myoglobin revealed by inelastic neutron scattering. *Nature* **337**, 754–756.

Dzakula, Z., Westler, W. M. and Edison, A. S. (1992a) The "CUPID" method for calculating the continuous probability distribution of rotamers from NMR data. *J. Am. Chem. Soc.* **114**, 6195–6199.

Dzakula, Z., Edison, A. S. and Westler, W. M. (1992b) Analysis of c1 rotamer populations from NMR data by the CUPID method. *J. Am. Chem. Soc* **114**, 6200–6207.

Dzakula, Z., Westler, W. M. and Markley, J. L. (1996) Continuous probability distribution (CUPID) analysis of potentials for internal rotations. *J. magn. Reson. Ser. B* **111**, 109–126.

Elöve, G., Chaffotte, A. F., Roder, H. and Goldberg, M. E. (1992) Early steps in cytochrome c folding probed by time-resolved circular dichroism and fluorescence spectroscopy. *Biochemistry* **31**, 6876–6883.

Endicott, J. A., Noble, M. E., Garman, E. F., Brown, N., Rasmussen, B., Nurse, P. and Johnson, L. N. (1995) The crystal structure of p13-suc1, a p34-cdc2-interacting cell cycle control protein. *Eur. molec. Biol. Org. J.* **14**, 1004–1014.

Englander, S. W. and Kellenbach, N. R. (1984) Hydrogen exchange and structural dynamics of proteins and nucleic acids. *Q. Rev. Biophys.* **16**, 521–655.

Englander, S. W., Calhoun, D. B., Englander, J. J., Kallenbach, R. K., Liem, H., Malin, E. L., Mandal, C. and Rogero, J. R. (1980) Individual breathing reactions measured in hemoglobin by hydrogen exchange methods. *Biophys. J.* **32**, 577–589.

Ernst, M. and Ernst, R. R. (1994) Heteronuclear dipolar cross-correlated cross relaxation for the investigation of side-chain motions. *J. magn. Reson. Ser. A* **110**, 202–213.

Ernst, R. R. (1992) Nuclear magnetic resonance Fourier transform spectroscopy (Nobel lecture). *Angew. Chem. Int. Ed. Engl.* **31**, 805–823.

Evans, P. A., Dobson, C. M., Kautz, R. A., Hatfull, G. and Fox, R. O. (1987) Proline isomerism in staphylococcal nuclease characterized by NMR and site-directed mutagenesis. *Nature* **329**, 266–268.

Faber, H. R. and Matthews, B. W. (1990) A mutant T4 lysozyme displays five different crystal conformations. *Nature* **348**, 263–266.

Farrow, A. N., Muhandiram, R., Singer, A. U., Pascal, S. M., Kay, C. M., Gish, G., Shoelson, S. E., Pawson, T., Forman-Kay, J. D. and Kay, L. E. (1994) Backbone dynamics of a free and a phosphopeptide-complexed Src homology 2 domain studied by ^{15}N NMR relaxation. *Biochemistry* **33**, 5984–6003.

Fejzo, J., Krezel, A. M., Westler, W. M., Macura, S. and Markley, J. L. (1991) Refinement of the NMR solution structure of a protein to remove distortions arising from neglect of internal motion. *Biochemistry* **30**, 3807–3811.

Fetler, L., Tuac, P., Hervé, G., Moody, M. F. and Vachette, P. (1995a) X-ray scattering titration of the quaternary structure transition of aspartate trans arbamylase with a bisubstrate analogue: Influence of nucleotide effectors. *J. molec. Biol.* **251**, 243–255.

Fetler, L., Vachette, P., Hervé, G. and Ladjimi, M. M. (1995b) Unlike the quaternary structure transition, the tertiary structure change of the 240s loop in allosteric aspartate transcarbamylase requires active site saturation by substrate for completion. *Biochemistry* **34**, 15654–15660.

Finn, B. E., Evenäs, J., Drakenberg, T., Waltho, J. P., Thulin, E. and Forsén, S. (1995) Calcium-induced structural changes and domain autonomy in calmodulin. *Nature struct. Biol.* **2**(9), 777–783.

Finucane, M. D. and Jardetzky, O. (1995) Mechanism of hydrogen–deuterium exchange in *trp* repressor studied by ^1H-^{15}N NMR. *J. molec. Biol.* **253**, 576–589.

Finucane, M. D. and Jardetzky, O. (1996) The pH-dependence of hydrogen–deuterium exchange in *trp* repressor: The exchange rate of amide protons in proteins reflects tertiary interactions, not only secondary structure, *Prot. Sci.* **5**, 653–662.

Fox, R. O., Evans, P. A. and Dobson, C. M. (1986) Multiple conformations of a protein demonstrated by magnetization transfer NMR spectroscopy. *Nature* **320**, 192–194.

Franzen, S., Lambry, J. C., Bohn, B., Poyart, C. and Martin, J. L. (1994) Direct evidence for the role of haem doming as the primary event in the cooperative transition of haemoglobin. *Struct. Biol.* **1**(4), 230–233.

Frauenfelder, H. (1983) Summary and outlook. In *Mobility and function in proteins and nucleic acids*, pp. 329–339 (Ciba Foundation Symposium **93**). Pitman, London.

Frauenfelder, H., Petsko, G. A. and Tsernoglou, D. (1979) Temperature-dependent X-ray diffraction as a probe of protein structural dynamics. *Nature* **280**, 558–563.

Frauenfelder, H., Hartmann, H., Karplus, M., Kuntz, I. D. Jr., Kuriyan, J., Parak, F., Petsko, G. A., Ringe, D., Tilton, R. F. Jr., Connolly, M. L. and Max, N. (1987) Thermal expansion of a protein. *Biochemistry* **26**, 254–261.

García, A. E. (1992) Large-amplitude nonlinear motions in proteins. *Phys. Rev. Lett.* **68**, 2696–2699.

Gerstein, M., Lesk, A. M. and Chothia, C. (1994) Structural mechanisms for domain movements in proteins. *Biochemistry* **33**(22), 6739–6749.

Goody, R. S. (1996) A two-way structure. *Nature* **380,** 483–484.

Gooley, P. R., Caffrey, M. S., Cusanovich, M. A. and MacKenzie, N. E. (1991) A spectroscopic analysis of the Pro35→Ala mutant of *Rhodobacter capsulatus* cytochrome c_2: The strictly conserved Pro35 is not structurally essential. *Eur. J. Biochem.* **196,** 653–661.

Gooley, P. R., Caffrey, M. S., Cusanovich, M. A. and MacKenzie, N. E. (1992) Mutations Pro → Ala-35 and Tyr→Phe-75 of *Rhodobacter capsulatus* Ferrocytochrome c_2 affect protein backbone dynamics: Measurements of individual amide proton exchange rate constants by ^{1}H-^{15}N HMQC spectroscopy. *Biochemistry* **31,** 443–450.

Grasberger, B. L., Gronenborn, A. M. and Clore, G. M. (1993) Analysis of the backbone dynamics of interleukin-8 by ^{15}N relaxation measurements. *J. molec. Biol.* **230,** 364–372.

Gronenborn, A., Birdsall, B., Hyde, E. I., Roberts, G. C. K., Feeney, J. and Burgen, A. S. V. (1981) Direct observation by NMR of two coexisting conformations of an enzyme–ligand complex in solution. *Nature* **290,** 273–274.

Gryk, M. R. and Jardetzky, O. (1996a) AV77 hinge mutation stabilized the helix-turn-helix domain of *trp* repressor. *J. molec. Biol.* **255,** 204–214.

Gryk, M. R. and Jardetzky, O. (1996b) Flexibility and function of the *Escherichia coli trp* repressor. In *Biological NMR Spectroscopy*, pp. 35–57 (eds. J. L. Markley and S. J. Opella). Oxford University Press, Oxford.

Gryk, M. R., Finucane, M. D., Zheng, Z. and Jardetzky, O. (1995) Solution dynamics of the *trp* repressor. A study of amide proton exchange by T_1 relaxation. *J. molec. Biol.* **246,** 618–627.

Gryk, M. R., Jardetzky, O., Klig, L. S. and Yanofsky, C. (1996) Flexibility of DNA binding domain of *trp* repressor required for recognition of different operator sequences. *Prot. Sci.* **5,** 1195–1197.

Hansen, A. P., Petros, A. M., Meadows, R. P. and Fesik, S. W. (1994) Backbone dynamics of a two-domain protein: ^{15}N relaxation studies of the amino-terminal fragment of urokinase-type plasminogen activator. *Biochemistry* **33,** 15,418–15,424.

Hardaway, L. A., Brems, D. N., Beals, J. M. and MacKenzie, N. E. (1994) Amide hydrogen exchange of the central β-chain helix within the T- and R-states of insulin hexamers. *Biochim. biophys. Acta* **1208,** 101–103.

Haurowitz, F. (1938) Das gleichgewicht swischen hämoglobin und sauerstoff. *Hoppe-Seyler's Z. Physiol. Chem.* **254,** 266–274.

Heatwole, V. M. and Somerville, R. L. (1991) The tryptophan-specific permease gene, *mtr*, is differentially regulated by the tryptophan and tyrosine repressors in *Escherichia coli* K-12. *J. Bacteriol.* **173,** 3601–3604.

Heatwole, V. M. and Somerville, R. L. (1992) Synergism between the trp repressor and tyr repressor in repression of the *aroL* promoter of *Escherichia coli* K-12. *J. Bacteriol.* **174,** 331–335.

Henry, E. R., Levitt, M. and Eaton, W. A. (1985) Molecular dynamics simulation of photodissociation of carbon monoxide from hemoglobin. *Proc. natn. Acad. Sci. U.S.A.* **82,** 2034–2038.

Holt, J. M. and Ackers, G. K. (1995) The pathway of allosteric control as revealed by hemoglobin intermediate states. *Fedn Am. Socs exp. Biol. J.* **9,** 210–218.

Hoogstraten, C. G. and Markley, J. L. (1996) Effects of experimentally achievable improvements in the quality of NMR distance constraints on the accuracy of calculated protein structures. *J. molec. Biol.* **258,** 334–348.

Hoogstraten, C. G., Choe, S., Westler, W. M. and Markley, J. L. (1995) Comparison of the accuracy of protein solution structures derived from conventional and network-edited NOESY data. *Prot. Sci.* **4,** 2289–2299.

Howard, A. E. and Kollman, P. A. (1992) Molecular dynamics studies of a DNA binding protein. *Prot. Sci.* **1,** 1173–1184.

Howarth, O. W. (1979) Effect of internal librational motions on the ^{13}C nuclear magnetic resonance relaxation times of polymers and peptides. *J. Chem. Soc. Faraday Trans. II* **75,** 863–873.

Huang, Y. and Ackers, G. K. (1995) Enthalpic and entropic components of cooperativity for the partially ligated intermediates of hemoglobin support a "symmetry rule" mechanism. *Biochemistry* **34,** 6316–6327.

Hurlburt, B. and Yanofsky, C. (1992) The NH_2-terminal arms of *trp* repressor participate in repressor/operator association. *Nucleic Acids Res.* **20,** 337–341.

Hurlburt, B. K. and Yanofsky, C. (1990) Enhanced operator binding by *trp* superrepressors of *Escherichia coli. J. biol. Chem.* **265,** 7853–7858.

Hvidt, A. and Nielsen, S. O. (1966) Hydrogen exchange in proteins. *Adv. Prot. Chem.* **21,** 287–386.

Hynes, T. R. and Fox, R. O. (1991) The crystal structure of *Staphylococcal nuclease* refined at 1.7 Å resolution. *Proteins: Struct. Func. Genet.* **10,** 92–105.

Ishima, R. and Nagayama, K. (1995a) Protein backbone dynamics revealed by quasi spectral density function analysis of amide ^{15}N nuclei. *Biochemistry* **34,** 3162–3171.

Ishima, R. and Nagayama, K. (1995b) Quasi-spectral-density function analysis for nitrogen-15 nuclei in proteins. *J. magn. Reson. B* **108,** 73–76.

Ishima, R., Yamasaki, K. and Nagayama, K. (1995) Application of the quasi-spectral density function of ^{15}N nuclei to the selection of a motional model for model-free analysis. *J. biomol. NMR* **6,** 423–426.

Iwata, S., Kamata, K., Yoshida, S., Minowa, T. and Ohta, T. (1994) T and R states in the crystals of bacterial L-lactate dehydrogenase reveal the mechanism for allosteric control. *Struct. Biol.* **1**(3), 176–185.

Jardetzky, O. (1964) Magnetic resonance in biological systems. *Science* **146,** 552–553.

Jardetzky, O. (1970) Nuclear magnetic resonance studies of protein binding sites: The complex of Staphylococcal Nuclease with 3′,5′-thymidine diphosphate. In *Molecular properties of drug receptors*, Ciba Found. Symp., pp. 113–132 (eds. R. Porter and M. O'Connor). Churchill, London.

Jardetzky, O. (1980) On the nature of molecular conformations inferred from high-resolution NMR. *Biochim. biophys. Acta* **621,** 227–232.

Jardetzky, O. (1981) NMR studies of macromolecular dynamics. *Accs. Chem. Res.* **14,** 291–298.

Jardetzky, O. (1984) A method for the definition of the solution structure of proteins from NMR and other physical measurements: The *lac*-repressor headpiece. In *Progress in bioorganic chemistry and molecular biology*, pp. 55–63 (Proc. of the Internatl. Conference on the Frontiers of Bio-organic Chemistry and Molecular Biology, Moscow-Alma Ata, June 17–24, 1984; ed. Y. A. Ovchinnikov). Elsevier, Amsterdam.

Jardetzky, O. (1992) Condensation and Commentary on Powers *et al.*, Biochemistry 31, 9150 (1992). *Chemtracts—Biochem. molec. Biol.* **3**, 301–305.

Jardetzky, O. and Jardetzky, C. D. (1962) Introduction to magnetic resonance spectroscopy. Methods and biochemical applications. In *Methods of biochemical analysis, IX*, pp. 235–410 (ed. D. Glick). Interscience, NY.

Jardetzky, O. and King, R. (1983) Soliton theory of protein dynamics. In *Mobility and function in proteins and nucleic acids*, pp. 291–309 (Ciba Found. Symp. **93**). Pitman, London.

Jardetzky, O. and Roberts, G. C. K. (1981) *NMR in molecular biology*. Academic, New York, NY.

Jardetzky, O. and Wade-Jardetzky, N. G. (1965) On the mechanism of the binding of sulfonamides to bovine serum albumin. *Molec. Pharmacol.* **1**, 214–230.

Jardetzky, O., Akasaka, K., Vogel, D., Morris, S. and Holmes, K. C. (1978) Unusual segmental flexibility in a region of tobacco mosaic virus coat protein. *Nature* **273**, 564–566.

Jayaraman, V., Rodgers, R. T., Mukerji, I. and Spiro, T. G. (1995) Hemoglobin allostery: Resonance raman spectroscopy of kinetic intermediates. *Science* **269**, 1843–1848.

Johnson, L. N. (1994) Control by protein phosphorylation. *Struct. Biol.* 1(10), 657–659.

Joseph, D., Petsko, G. A. and Karplus, M. (1990) Anatomy of a conformation change: Hinged "lid" motion of the triosephosphate isomerase loop. *Science* **249**, 1425–1428.

Kalk, A. and Berendsen, H. J. C. (1976) Proton magnetic relaxation and spin diffusion in proteins. *J. magn. Reson.* **24**, 343–366.

Karplus, M. and Petsko, G. A. (1990) Molecular dynamics simulations in biology. *Nature* **347**, 631–639.

Kato, K., Matsunaga, C., Igarashi, T., Kim, H., Odaka, A., Shimada, I. and Arata, Y. (1991) Complete assignment of the methionyl carbonyl carbon resonances in switch variant anti-dansyl antibodies labeled with [$^{1-13}$C] methionine. *Biochemistry* **30**, 270–278.

Kay, L. E. (1993) Pulsed-field gradient-enhanced three-dimensional nmr experiment for correlating ^{13}C α-β-^{13}C' and ^{1}H-α chemical shifts in uniformly carbon-13-labeled proteins dissolved in water. *J. Am. Chem. Soc.* **115**, 2055–2057.

Kay, L. E., Muhandiram, D. R., Farrow, N. A., Aubin, Y. and Forman-Kay, J. D. (1996) Correlation between dynamics and high affinity binding in an SH2 domain interaction. *Biochemistry* **35**, 361–368.

Kay, L. W., Torchia, D. A. and Bax, A. (1989) Backbone dynamics of proteins as studied by ^{15}N inverse detected heteronuclear NMR spectroscopy: Application to Staphylococcal nuclease. *Biochemistry* **28**, 8972–8979.

Kim, H.-W., Shen, T.-J., Sun, D. P., Ho, N. T., Madrid, M. and Chien, H. (1995) A novel low oxygen affinity recombinant hemoglobin (α96Val→Trp): Switching quaternary structure without changing the ligation state. *J. molec. Biol.* **248**, 867–882.

Kim, H.-W., Shen, T.-J., Sun, D. P., Ho, N. T., Madrid, M., Tam, M. F., Zou, M., Cottam, P. F. and Ho, C. (1994) Restoring allosterism with compensatory mutations in hemoglobin. *Proc. natn. Acad. Sci. U.S.A.* **91**, 11,547–11,551.

Kim, Y. and Prestegard, J. H. (1989) A dynamic model for the structure of acyl carrier protein in solution. *Biochemistry* **28**, 8792–8797.

King, R. and Jardetzky, O. (1978) A general formalism for the analysis of NMR relaxation measurement on systems with multiple degrees of freedom. *Chem. Phys. Lett.* **55**, 15–18.

Kleerekoper, Q., Howarth, J. W., Guo, X., Solaro, R. J. and Rosevear, P. R. (1995) Cardiac troponin I induced conformational changes in cardiac troponin C as monitored by NMR using site-directed pain and isotope labeling. *Biochemistry* **34**, 13,343–13,352.

Klotz, I. M. and Hunston, D. L. (1975) Protein interactions with small molecules. Relationships between stoichiometric binding constants, site binding constants, and empirical binding parameters. *J. biol. Chem.* **250**, 3001–3009.

Kördel, J., Skelton, N. J., Akke, M., Palmer, A. G. III, Chazin, W. J. (1992) Backbone dynamics of calcium-loaded calbindin D_{9k} studied by two-dimensional proton-detected ^{15}N NMR spectroscopy. *Biochemistry* **31**, 4856–4866.

Komoroski, R. A. and Allerhand, A. (1972) Natural-abundance carbon-13 Fourier-transform nuclear magnetic resonance spectra and spin lattice relaxation times of unfractionated yeast transfer-FNA. *Proc. natn. Acad. Sci. U.S.A.* **69**, 1804–1808.

Koshland, D. E., Nemethy, G. and Filmer, D. (1966) Comparison of experimental binding data and theoretical models in proteins containing subunits. *Biochemistry* **5**, 365–385.

Kovalenko, O., Yifrach, O. and Horovitz, Q. (1994) Residue lysine-34 in groES modulates allosteric transitions in groEL. *Biochemistry* **33**, 14,974–14,978.

Krishna, N. R., Huang, D. H., Glickson, J. D., Rowan III, R. and Walter, R. (1979) Amide hydrogen exchange rates of peptides in H_2O solution by ^{1}H nuclear magnetic resonance transfer of solvent saturation method. *Biophys. J.* **26**, 345–366.

Kuboniwa, H., Tjandra, N., Grzesiek, S., Ren, H., Klee, C. B. and Bax, A. (1995) Solution structure of calcium-free calmodulin. *Nature struct. Biol.* **2**(9), 768–776.

Kull, F. J., Sablin, E. P., Lau, R., Fletterick, R. J. and Vale, R. D. (1996) Crystal structure of the kinesin motor domain reveals a structural similarity to myosin. *Nature* **380**, 550–555.

Lawson, C. L. and Carey, J. (1993) Tandem binding in crystals of a *trp* repressor/operator half-site complex. *Nature* **366**, 178–182.

Lawson, C. L., Zhang, R.-G., Schevitz, R. W., Otwinowski, Z., Joachimiak, A. and Sigler, P. B. (1988) Flexibility of the DNA-binding domains of *trp* repressor. *Proteins: Struct. Func. Genet.* **3**, 18–31.

Lefèvre, J.-F., Dayie, K. T., Peng, J. W. and Wagner, G. (1996) Internal mobility in the partially folded DNA binding and dimerization domains of GAL4: NMR analysis of the N-H spectral density functions. *Biochemistry* **35**, 2674–2686.

Linderstrøm-Lang, K. (1955) Deuterium exchange between peptides and water. *Chem. Soc. Spec. Publ.* **2**, 1–20.

Lipari, G. and Szabo, A. (1982) Model-free approach to the interpretation of nuclear magnetic resonance relaxation in macromolecules. 1. Theory and range of validity. *J. Am. Chem. Soc.* **104**, 4546–4559.

Lipscomb, W. N. (1994) Aspartate transcarbamylase from *Escherichia coli* activity and regulation. *Adv. Enzymol.* **68,** 67–152.

Lipscomb, W. N. (1991) Structure and function of allosteric enzymes. *Chemtracts-Biochem. molec. Biol.* **2,** 1–15.

Liu, Y., Zhao, D., Altman, R. and Jardetzky, O. (1992) A systematic comparison of three structure determination methods from NMR data: Dependence upon quality and quantity of data. *J. biomol. NMR* **2,** 373–388.

Loh, S. N., Prehoda, K. E., Wang, J. and Markley, J. L. (1993) Hydrogen exchange in unligated and ligated staphylococcal nuclease. *Biochemistry* **32,** 11,022–11,028.

Loll, P. J. and Lattman, E. E. (1989) The crystal structure of the ternary complex of staphylococcal nuclease, calcium, and the inhibitor deoxythymidine 3′,5′-bisphosphate, refined at 1.65 Å. *Proteins: Struct. Func. Genet.* **5,** 183–201.

Loncharich, R .J. and Brooks, B. R. (1990) Temperature dependence of dynamics of hydrated myoglobin: Comparison of force field calculations with neutron scattering data. *J. molec. Biol.* **215,** 439–455.

Mackay, J. P., Shaw, G. L. and King, G. F. (1996) Backbone dynamics of the c-Jun leucine zipper: ^{15}N NMR relaxation studies. *Biochemistry* **35,** 4867–4877.

Mandel, A. M., Akke, M. and Palmer, A. G. III (1995) Backbone dynamics of *Escherichia coli* ribonuclease HI: Correlations with structure and function in an active enzyme. *J. molec. Biol.* **246,** 144–163.

Markley, J. L. and Jardetzky, O. (1970) Nuclear magnetic resonance studies of the structure and binding sites of enzymes. XIV. Inhibitor binding to staphylococcal nuclease. *J. molec. Biol.* **50,** 223–233.

Mattevi, A., Valentini, G., Rizzi, M., Speranza, M. L., Bolognesi, M. and Coda, A. (1995) Crystal structure of *Escherichia coli* pyruvate kinate type I: Molecular basis of the allosteric transition. *Structure* **3,** 729–741.

Matthews, S. J., Jandu, S. K. and Leatherbarrow, R. J. (1993) ^{13}C NMR study of the effects of mutation on the tryptophan dynamics in chymotrypsin inhibitor 2: Correlations with structure and stability. *Biochemistry* **32,** 657–662.

Mau, T., Baleja, J. D. and Wagner, G. (1992) Effects of DNA binding and metal substitution on the dynamics of the GAL4 DNA-binding domain as studied by amide proton exchange. *Prot. Sci.* **1,** 1403–1412.

Miller, D. W. and Dill, K. A. (1995) A statistical mechanical model for hydrogen exchange in globular proteins. *Prot. Sci.* **4,** 1860–1873.

Molday, R. S., Englander, S. W. and Kallen, R. G. (1972) Primary structure effects on peptide group hydrogen exchange. *Biochemistry* **11,** 150–158.

Monod, J., Wyman, J. and Changeux, J. P. (1965) On the nature of the allosteric transitions: A plausible model. *J. molec. Biol.* **12,** 88–118.

Moy, F. E., Lowry, D. F., Matsumura, P., Dahlquist, F. W., Krywko, J. E. and Domaille, P. J. (1994) Assignments, secondary structure, global fold and dynamics of chemotaxis Y protein using three- and four-dimensional heteronuclear (^{13}C, ^{15}N) NMR spectroscopy. *Biochemistry* **33,** 10,731–10,742.

Neet, K. E. (1995) Cooperativity in enzyme function: Equilibrium and kinetic aspects. *Meth. Enzymol.* **249,** 519–567.

Nelson, D. J., Cozzone, P. J. and Jardetzky, O. (1974) ^{31}P nuclear magnetic resonance study of ATP and ATP-G-Actin: Direct evidence for P-β and P-γ involvement in nucleotide binding. In *Molecular and Quantum Pharmacology* pp. 501–513 (eds. E. Bergmann and B. Pullman). D. Reidel Publ. Col., Dordrecht-Holland.

Nelson, D. J., Opella, S. J. and Jardetzky, O. (1976) ^{13}C nuclear magnetic resonance study of molecular motions and conformational transitions in muscle calcium binding parvalbumins. *Biochemistry* **15,** 5552–5560.

Nicholson, L. K., Kay, L. E., Baldisseri, D. M., Arango, J., Young, P. E., Bax, A. and Torchia, D. A. (1992) Dynamics of methyl groups in proteins as studied by proton-detected ^{13}C NMR spectroscopy. Application to the leucine residues of staphylococcal nuclease. *Biochemistry* **31,** 5253–5263.

Ogata, K., Morikawa, S., Nakamura, H., Hojo, H., Yoshimura, S., Zhang, R., Aimoto, S., Ametani, Y., Hirata, Z., Srai, A., Ishii, S. and Nishimura, Y. (1995) Comparison of the free and DNA-complexed forms of the DNA-binding domain from c-Myb. *Nature struct. Biol.* **2,** 309–320.

Opella, S. J., Nelson, D. J. and Jardetzky, O. (1974) Carbon magnetic resonance study of the conformational changes in carp muscle calcium binding parvalbumin. *J. Am. Chem. Soc.* **96,** 7157–7159.

Orekhov, V. Y., Pervushin, K. V. and Arseniev, A. S. (1994) Backbone dynamics of (1-71)bacterioopsin studied by two-dimensional ^{1}H-^{15}N NMR spectroscopy. *Eur. J. Biochem.* **219,** 887–896.

Otwinowski, Z., Schevitz, R. W., Zhang, R.-G., Lawson, C. L., Joachimiak, A., Marmorstein, R. Q., Luisi, B. F. and Sigler, P. B. (1988) Crystal structure of *trp* repressor/operator complex at atomic resolution. *Nature* **335,** 321–329.

Pabo, C. O. and Sauer, R. T. (1992) Transcription factors structural families and principles of DNA recognition. In *Annual Review of Biochemistry*, Vol. 61, pp. 1053–1095 (ed. C. C. Richardson). Ann. Revs. Inc., Palo Alto.

Pedersen, H., Sogaard-Andersen, L., Holst, B. and Valentin-Hansen, P. (1991) Heterologous cooperativity in *Escherichia coli*: The CytR repressor both contacts DNA and the cyclic AMP receptor protein when binding to the deoP2 promoter. *J. biol. Chem.* **266,** 17804–17808.

Peng, J. W. and Wagner, G. (1992) Mapping of spectral density functions using heteronuclear NMR relaxation measurements. *J. magn. Reson.* **98,** 308–332.

Peng, J. W. and Wagner, G. (1992) Mapping of the spectral densities of N–H bond motions in eglin c using heteronuclear relaxation experiments. *Biochemistry* **31,** 8571–8586.

Perham, R., Duckworth, H. and Roberts, G. C. K. (1981) Mobility of polypeptide chain in the pyruvate dehydrogenase complex revealed by proton NMR. *Nature* **292,** 474–477.

Perutz, M. F., Shih, D.T.-B. and Williamson, D. (1994) The chloride effect in human haemoglobin. A new kind of allosteric mechanism. *J. molec. Biol.* **239,** 555–560.

Perutz, M. F. (1962) *Proteins and Nucleic Acids*. Elsevier, Amsterdam.

Perutz, M. F. (1979) Regulation of oxygen affinity in hemoglobin: influence of structure of the globin on the heme iron. *A. Rev. Biochem.* **48,** 327–386.

Perutz, M. F. (1989) Mechanisms of co-operativity and allosteric regulation in proteins. *Q. Rev. Biophys.* **22,** 139–236.

Perutz, M. F. (1990) *Mechanisms of cooperativity and allosteric regulation in proteins.* Cambridge University Press, Cambridge.

Petersen, J. M., Skalicky, J. J., Donaldson, L. W., McIntosh, L. P., Alber, T. and Graves, B. J. (1995) Modulation of transcription factor ets-1 DNA binding: DNA-induced unfolding of an α-helix. *Science* **269,** 1866–1869.

Petsko, G. A. (1996) Not just your average structures. *Nature struct. Biol.* **3,** 565–566.

Philo, J. S., Dreyer, U. and Lary, J. W. (1996) Quaternary structure dynamics and carbon monoxide binding kinetics of hemoglobin valency hybrids. *Biophys. J.* **70,** 1949–1965.

Powers, R., Clore, G. M., Stahl, S. H. J., Wingfield, P. T. and Gronenborn, A. (1992) Analysis of the backbone dynamics of the ribonuclease H domain of the human immunodeficiency virus reverse transcriptase using ^{15}N relaxation measurements. *Biochemistry* **31,** 9150–9157.

Qiu, X., Verlinde, C. L. M. J., Zhang, S., Schmitt, M. P., Holmes, R. K. and Hol, G. W. J. (1995) Three-dimensional structure of the diphtheria toxin repressor in complex with divalent cation co-repressors. *Structure* **3,** 87–100.

Radford, S. E., Buck, M., Topping, K. D., Dobson, C .M. and Evans, P. A. (1992) Hydrogen exchange in native and denatured states of hen egg-white lysozyme. *Proteins: Struct. Func. Genet.* **14,** 237–248.

Rashin, A. A. (1987) Correlation between calculated local stability and hydrogen exchange rates in proteins. *J. molec. Biol.* **198,** 339–349.

Rasmussen, B. F., Stock, A. M., Ringe, D. and Petsko, G. A. (1992) Crystalline ribonuclease A loses function below the dynamical transition at 220°K. *Nature* **357,** 423–424.

Rath, V. L. and Fletterick, R. J. (1994) Parallel evolution in two homologues of phosphorylase. *Nature struct. Biol.* **1,** 681–690.

Rayment, I., Rypniewski, W. R., Schmidt-Baese, K., Smith, R., Tomchick, D. R., Benning, M. M. and Winkelmann, D. A. (1993) Three-dimensional structure of myosin subfragment-1: A molecular motor. *Science* **261,** 50–58.

Redfield, C., Boyd, J., Smith, L. J., Smith, R. A. G. and Dobson, C. M. (1992) Loop mobility in a four-helix-bundle protein: ^{15}N NMR relaxation measurements on human interleukin-4. *Biochemistry* **31,** 10,431–10,437.

Reedstrom, R. J. and Royer, C. A. (1995) Evidence for coupling of folding and function in *trp* repressor: Physical characterization of the superrepressor mutant AV77. *J. molec. Biol.* **253,** 266–276.

Remington, S., Wiegand, G. and Huber, R. (1982) Crystallographic refinement and atomic models of two different forms of citrate synthase at 2.7 and 1.7 Å resolution. *J. molec. Biol.* **158,** 111–152.

Ribeiro, A. A., King, R. and Jardetzky, O. (1983) Rotational solitons in a folded polypeptide chain. In *Conformation in Biology*, pp. 39–47 (eds. R. Srinivasan and R. H. Sarma). Adenine, New York.

Ribeiro, A. A., King, R., Restivo, C. and Jardetzky, O. (1980) An approach to the mapping of internal motions in proteins: Analysis of ^{13}C NMR relaxation in the bovine pancreatic trypsin inhibitor. *J. Am. Chem. Soc.* **102,** 4040–4051.

Richarz, R., Nagayama, K. and Wüthrich, K. (1980) Carbon-13 nuclear magnetic resonance relaxation studies of internal mobility of the polypeptide chain in basic pancreatic trypsin inhibitor and a selectively reduced analogue. *Biochemistry* **19,** 5189–5196.

Ringe, D., Almo, S. C., Stoddard, B., Howell, P. L., Farber, G. K. and Petsko, G. A. (1990) Dynamic processes in proteins by x-ray diffraction. In *Structure and Methods. Vol. 2: DNA Protein Complexes and Proteins*, pp. 103–114 (eds. R. H. Sarma and M. H. Sarma). Adenine, New York.

Rischel, C., Madsen, J. C., Andersen, K. V. and Pulsen, F. M. (1994) Comparison of the backbone dynamics of apo- and holo-acyl-coenzyme A binding protein using ^{15}N relaxation measurements. *Biochemistry* **33,** 13,997–14,002.

Robertson, A. D. and Baldwin, R. L. (1991) Hydrogen exchange in thermally denatured ribonuclease A. *Biochemistry* **30,** 9907–9914.

Roder, H., Wagner, G. and Wüthrich, K. (1985) Amide proton exchange in proteins by EX1 kinetics: Studies on the basic pancreatic trypsin inhibitor at variable pH and temperature. *Biochemistry* **24,** 7396–7407.

Rose, J. K., Squires, C. L., Yanofsky, C., Yang, H.-Y. and Zubay, G. (1973) Regulation of *in vitro* transcription of the tryptophan operon by purified RNA polymerase in the presence of partially purified repressor and tryptophan. *Nature New Biology* **245,** 133–137.

Sablin, E. P., Kull, F. J., Cooke, R., Vale, R. D. and Fletterick, R. J. (1996) Crystal structure of the motor domain of the kinesin-related motor ncd. *Nature* **380,** 555–559.

Schevitz, R. W., Otwinowski, Z., Joachimiak, A., Lawson, C. L. and Sigler, P. B. (1985) The three-dimensional structure of *trp* repressor. *Nature* **317,** 782–786.

Schiering, N., Tao, X., Zeng, H., Murphy, J. R., Petski, G. A. and Ringe, D. (1995) Structures of the apo- and the metal ion-activated forms of the diphtheria *tox* repressor from *Corynebacterium diphtheriae. Proc. natn. Acad. Sci. U.S.A.* **92,** 9843–9850.

Schirmer, T. and Evans, P. R. (1990) Structural basis of the allosteric behaviour of phosphofructokinase. *Nature* **343,** 140–145.

Schneider, D. M., Dellwo, M. J. and Wand, A. J. (1992) Fast internal main-chain dynamics of human ubiquitin. *Biochemistry* **31,** 3645–3652.

Schumacher, M. A., Dixon, M. M., Kluger, R., Jones, R. T. and Brennan, R. G. (1995) Allosteric transition intermediates modelled by crosslinked haemoglobins. *Nature* **375,** 84–87.

Smith, F. R and , Simmons, K. C. (1994) Cyanomet human hemoglobin crystallized under physiological conditions exhibits the Y quaternary structure. *Proteins: Struct. Func. Genet.* **18,** 295–300.

Smith, P. E., van Schaik, R. C., Szyperski, T., Wüthrich, K. and van Gunsteren, W. F. (1995) Internal mobility of the basic pancreatic trypsin inhibitor in solution: A comparison of NMR spin relaxation measurements and molecular dynamics simulations. *J. molec. Biol.* **246,** 356–365.

Spera, S., Ikura, M. and Bax, A. (1991) Measurement of the exchange rates of rapidly exchanging amide protons: Application to the study of calmodulin and its complex with a myosin light chain kinase fragment. *J. biomol. NMR* **1,** 155–165.

Spolar, R. S. and Record, M. T. Jr. (1994) Coupling of local folding to site-specific binding of proteins to DNA. *Science* **263,** 777–784.

Srinivasan, R. and Rose, G. D. (1994) The T-to-R transformation in hemoglobin: A reevaluation. *Proc. natn. Acad. Sci. U.S.A.* **91,** 11,113–11,117.

Stone, M. J., Chandrasekhar, K., Holmgren, A., Wright, P. E. and Dyson, H. J. (1993) Comparison of backbone and tryptophan side-chain dynamics of reduced and oxidized *Escherichia coli* tjopredpxom isomg ^{15}N NMR relaxation measurements. *Biochemistry* **32,** 426–435.

Stone, M. J., Fairbrother, W. J., Palmer, A. G. III, Reizer, H., Saier, M. H. Jr. and Wright, P. E. (1992) Backbone dynamics of the *Bacillus subtilis* glucose permease IIA domain determined from ^{15}N NMR relaxation measurements. *Biochemistry* **31,** 4394–4406.

Stryer, L. (1975) *Biochemistry*. Freeman and Co., San Francisco.

Takahashi, H., Suzuki, E., Shimada, I. and Arata, Y. (1992) Dynamical structure of the antibody combining site as studied by ^{1}H-^{15}N sht correlation NMR spectroscopy. *Biochemistry* **31,** 2464–2468.

Tian, W. D., Sage, T., Champion, P. A., Chien, E. and Sligar, S. G. (1996) Probing heme protein conformational equilibration rates with kinetic selection. *Biochemistry* **35,** 3487–3502.

Tilton, R. F. Jr., Dewan, J. W. and Petsko, G. A. (1992) Effects of temperature on protein structure and dynamics: X-ray crystallographic studies of the protein ribonuclease-A at nine different temperatures from 98 to 320 K. *Biochemistry* **31,** 2469–2481.

Torda, A. E. and Norton, R. S. (1989) Proton NMR relaxation study of the dynamics of Anthopleurin-A in solution. *Biopolymers* **28,** 703–716.

Torda, A. E., Scheek, R. M. and Van Gunsteren, W. F. (1990) Time-averaged nuclear Overhauser effect distance restraints applied to tendamistat. *J. molec. Biol.* **214,** 223–236.

Truckses, D. M., Somoza, R. R., Prehoda, K. E. and Markley, J. L. (1996) Coupling between cis/trans proline isomerization and protein stability in staphylococcal nuclease. *Prot. Sci.,* **5,** 1906–1916.

Tuac, P., Keiser, R. T., Kantrowitz, E. R. and Vachette, P. (1994) Glu-50 in the catalytic chain of *Escherichia coli* aspartate transcarbamoylase plays a crucial role in the stability of the R quaternary structure. *Prot. Sci.* **3,** 1998–2004.

Vidugiris, G. J. A., Truckses, D. M., Markley, J. L. and Royer, C. A. (1996) High-pressure denaturation of staphylococcal nuclease proline-to-glycine substitution mutants. *Biochemistry* **35,** 3857–3864.

Wade-Jardetzky, N., Bray, R. P., Conover, W. W., Jardetzky, O., Geisler, N. and Weber, K. (1979) Differential mobility of the N-terminal headpiece in the *lac*-repressor protein. *J. molec. Biol.* **128,** 259–264.

Wagner, G. and Wüthrich, K. (1982) Amide proton exchange and surface conformation of the basic pancreatic trypsin inhibitor (BPTI) in solution: Studies with two-dimensional nuclear magnetic resonance. *J. molec. Biol.* **160,** 343–361.

Wemmer, D., Shvo, H., Ribeiro, A. A., Bray, R. P. and Jardetzky, O. (1981) High resolution ^{1}H NMR studies of the exchangeable NMR resonances of the *lac* repressor headpiece. *Biochemistry* **20,** 3351–3354.

Woessner, D. E. (1962a) Nuclear spin relaxation in ellipsoids undergoing rotational Brownian motion. *J. Chem. Phys.* **37,** 647–654.

Woessner, D. E. (1962b) Spin relaxation processes in a two-proton system undergoing anisotropic reorientation. *J. Chem. Phys.* **36,** 1–4.

Xu, R. X., Horvath, S. J. and Klevit, R. E. (1991) ADR1a, a zinc finger peptide, exists in two folded conformations. *Biochemistry* **30,** 3365–3371.

Xue, Y., Huang, S., Liang, J.-Y., Zhang, Y. and Lipscomb, W. N. (1994) Crystal structure of fructose-1,6-bisphosphatase complexed with fructose 2,6-bisphosphate, AMP and Zn^{2+} at 2.0-Å resolution: Aspects of synergism between inhibitors. *Proc. natn. Acad. Sci. U.S.A.* **91,** 12,482–12,486.

Yamazaki, T., Muhandiram, R. and Kay, L. E. (1994) NMR experiments for the measurement of carbon relaxation properties in highly enriched, uniformly ^{13}C, ^{15}N-labeled proteins: Application to ^{13}C-7α carbons. *J. Am. Chem. Soc.* **116,** 8266–8278.

Yguerabide, J., Epstein, H. F. and Stryer, L. (1970) Segmental flexibility in an antibody molecule. *J. molec. Biol.* **51,** 573–590.

Zhang, H., Zhao, D., Revington, M., Lee, W., Jia, X., Arrowsmith, C. and Jardetzky, O. (1994) The solution structures of *trp* repressor-operator DNA complex. *J. molec. Biol.* **238,** 592–614.

Zhang, M., Tanaka, T. and Ikura, M. (1995) Calcium-induced conformational transition revealed by the solution structure of apo calmodulin. *Nature struct. Biol.* **2**(9), 758–767.

Zhang, R.-G., Joachimiak, A., Lawson, C. L., Schevitz, R. W., Otwinowski, Z. and Sigler, P. B. (1987) The crystal structure of *trp* aporepressor at 1.8 Å shows how binding tryptophan enhances DNA affinity. *Nature* **327,** 591–597.

Zhang, Y., Liang, J.-Y., Huang, S. and Lipscomb, W. M. (1994) Toward a mechanism for the allosteric transition of pig kidney fructose-1,6-bisphosphatase. *J. molec. Biol.* **244**(5), 609–624.

Zhao, D. and Jardetzky, O. (1994a) An assessment of the precision and accuracy of protein structures determined by NMR—dependence on distance errors. *J. molec. Biol.* **239,** 601–607.

Zhao, D. and Jardetzky, O. (1994b) A new procedure for Monte Carlo simulated annealing on proteins. In *Abstracts of Papers (208th Nat'l Meeting of the Am. Che. Soc.).* American Chemical Society, Washington, DC, PHYS 237.

Zhao, D. and Jardetzky, O. (1995) Simulated annealing in the calculation of NMR structures. In *Adaptation of Simulated Annealing to Chemical Optimization Problems*, pp. 303–328 (ed. J. H. Kalivas). Elsevier Science, Amsterdam.

Zhao, D., Arrowsmith, C. H., Jia, X. and Jardetzky, O. (1993) Refined solution structures of the *E. coli trp* hHolo- and aporepressor. *J. molec. Biol.* **229,** 735–746.

Zheng, Z., Brockel, C., Ganter, B., Lipsick, J., Sasaki, M., Ogata, K., Nishimura, Y. and Jardetzky, O. (1996) Amide proton exchange rates in c-Myb DNA-binding domain. *Q. J. magn. Reson. Biol. Med.*, **3**, 33–45.

Zheng, Z., Czaplicki, J. and Jardetzky, O. (1995a) Backbone dynamics of *trp* repressor studied by ^{15}N NMR relaxation. *Biochemistry* **34**, 5121–5223.

Zheng, Z., Gryk, M. R., Finucane, M. D. and Jardetzky, O. (1995b) Investigation of protein amide proton exchange by ^{1}H longitudinal spin relaxation. *J. magn. Reson. B* **108**, 220–234.

Zink, T., Ross, A., Lüers, K., Cieslar, C., Rudolph, R. and Holak, T. A. (1994) Structure and dynamics of the human granulocyte colony-stimulating factor determined by NMR spectroscopy. Loop mobility in a four-helix-bundle protein. *Biochemistry* **33**, 8453–8463.

Zubay, G., Morse, D. E., Schrenk, W. J. and Miller, J. H. M. (1972) Detection and isolation of the repressor protein for the tryptophan operon of *Escherichia coli*. *Proc. natn. Acad. Sci. U.S.A.* **69**, 1100–1103.

Zurawski, G., Gunsulas, R. P., Brown, K. D. and Yanofsky, C. (1981) Structure and regulation of *aroH*, the structural gene for the tryptophan-repressible 3-deoxy-D-arabino-heptulosonic acid-7-phosphate synthetase of *Escherichia coli*. *J. molec. Biol.* **145**, 47–73.

Chapter 3

Michael G. Rossmann

Biography

I was born in Frankfurt (M), Germany on July 30, 1930. My family immigrated to England in July 1939. At the Regent Street Polytechnic I studied physics and mathematics and stayed on to obtain an M.S. degree in 1953. I then obtained a job as a lecturer in "natural philosophy" (physics) at the Royal Technical College in Glasgow, Scotland (1952–1956). At the same time I studied for a Ph.D. under J. Monteath Robertson at the University of Glasgow, graduating in 1956. I worked on the crystal structure of three aromatic hydrocarbons in Glasgow, doing all calculations by hand. Subsequently I spent two years as a postdoc and Fulbright Scholar in Bill Lipscomb's laboratory in Minneapolis, Minnesota working on the structures of some plant products using, for the first time, an electronic computer. I wrote some of the first crystallographic computer programs for a UNIVAC 1103 machine while in Minnesota.

In 1958 my wife (Audrey) and our two children returned to England to work in Max Perutz's Medical Research Council Laboratory in Cambridge (later the Laboratory for Molecular Biology). Max had started to collect three-dimensional data on horse oxyhemoglobin and the new EDSAC 2 computer had just started to become functional. After writing a three-dimensional Fourier program I had an easy and powerful solution for finding the relative y coordinates of the heavy atoms in the C2 space group of the hemoglobin crystals. We were able to solve the 5.5 resolution hemoglobin structure in the summer of 1959 and recognized the similarity to Kendrew's myoglobin structure determined a year or so earlier. These were the first protein structures to be solved. The evolutionary relationship between these structures has been a major guide for my research direction ever since. This work on the hemoglobin structure determination was also a stimulus for the development of a series of crystallographic techniques that, in collaboration with David Blow, laid the foundations of modern structural biology.

In 1964 my family and I moved to Lafayette, Indiana, where I established my own laboratory at Purdue University. I worked on lactate dehydrogenase (LDH), the structure of which was the first structure of an enzyme with a small metabolite as a substrate, and the largest structure solved to date. Three years later we determined the structure of glyceraldehyde 3-phosphate dehydrogenase (GAPDH), through which it became apparent that the building blocks of proteins were structural domains, each having had an independent evolutionary history.

In 1970 David Haas and I demonstrated the use of frozen crystals to minimize radiation damage, a technique that is universal today. A few years later I developed data processing procedures for oscillation photography as an essential component of our work on virus structure determination. Many of these procedures are used almost universally today.

After working with Bror Strandberg in Sweden on the satellite tobacco necrosis virus, I turned to some small RNA plant viruses that could be readily propagated. In 1980, this led to the structure of southern bean mosaic virus. To everyone's surprise its structure was very similar to that of tomato bushy stunt virus, which had been determined by Steve Harrison a year or so earlier, demonstrating a common origin.

I then turned my attention to animal viruses in collaboration with Roland Rueckert. This led to the structure of human rhinovirus (HRV) serotype 14 in 1985. The "canyon hypothesis" on the site of receptor binding was confirmed in 1993 for HRV14.

The accumulation of crystallographic technology opened the door for the determination of many other icosahedral viruses. Among the virus structures we published were mengovirus, canine parvovirus, bacteriophage phiX174, coxsackie virus B3, and human parvovirus B19.

In 1991 we determined the crystal structure of the nucleocapsid protein of Sindbis virus, and in 1995 we were able to publish the structure of an alphavirus using the Sindbis nucleocapsid structure to help interpret the three-dimensional cryo-electron microscopy density. The alphavirus investigation led to the development of the hybrid technology of combining crystallography with electron microscopy. In this way we determined a pseudo-atomic structure of Sindbis virus and of flaviviruses such as dengue and West Nile viruses. We were also able to determine the structure of immature dengue and yellow fever viruses, establishing the assembly process (in collaboration with Jue Chen) leading to infectious virus.

We are currently studying the structures of Mimivirus — whose genome is more like that of a bacteria, thus blurring the border between the living and the dead — and other large dsDNA icosahedral viruses, part of a group of very large DNA viruses that have evolved from the same common capsid proteins as the small icosahedral viruses mentioned above.

Personal Remembrances of the Colonel

The most important thing I learned when I was in the Colonel's lab in Minneapolis was how to program an electronic computer. The University of Minnesota had made arrangements with Univac in St. Paul to use a few hours of computing time per week. Crystallography was an obvious candidate for such privileges. We used to take university cars and drive over to the computer (this was also an opportunity for me to learn how to drive a standard gear-shift car, tutored by Bob Jacobson and Scott Mathews). The knowledge I gained in the use of computers has been an enormously powerful tool and benefit for most of the science I have done since. That is the basis of my selection of the two papers given below. The first paper, coauthored with the Colonel, was my first three-dimensional structure determination (as opposed to two-dimensional structures projected down short crystal axes). The Univac 1003 computer made this possible. I use this structure determination in my class every year to explain the way the heavy atom method works for phase determination. The second structure was the first major product of my time in Cambridge. Although the Colonel is not a coauthor of the paper about this, my time in Minnesota was essential to make this work possible. In Cambridge I spent much of my time writing programs and developing methods not only for the determination of hemoglobin but partially also for Kendrew's work on myoglobin. The results on hemoglobin have been a guide to me ever since, not only for the technical aspects but also for their biological significance. I would not have been able to do this work without the preceding two years in Minnesota.

Coming to Minneapolis was also a tremendous cultural experience. The difference between Britain and North America was far greater in those days than it is now. I had never before seen a supermarket nor driven a car. But not only was the material side of life so different, the way people behaved was also new to my wife and me. For instance, in Scotland, where I had been a graduate student, my "professor," J. Monteath Robertson, addressed my wife as "Mrs. Rossmann" and me as "Rossmann." To the Colonel we were just Audrey and Michael. I liked that! In Minnesota the car salesmen (or almost anybody) would address me with my first name, but in Britain the salespeople would invariably be super polite and always use titles to establish a good relationship.

The Colonel had been on sabbatical leave shortly before we arrived in Minneapolis. Indeed, I got to know him when he came to give a seminar in Glasgow. It was then that I decided it would be fun to work in his lab, a decision that worked out extremely well. The

Colonel must have enjoyed his time in Oxford as much as I was enjoying my time in Minneapolis. He apparently started drinking tea at 4 p.m. while in England and carried that English tradition back home. He would come down to our room in the basement of the chemistry building each afternoon and make himself a cup of tea on the Bunsen burner. At the same time he would discuss our projects with us which was usually an excellent learning experience.

I am not a musical person and, therefore, I cannot make any comments on the Colonel's enjoyment of his obviously outstanding ability with a clarinet. However, his passion for music plus his passion for science characterized him as a person who made the most out of every moment of his life. I visited him recently at Harvard and he, Jean, and I had dinner together. He had become a little frail, but the Kentucky Colonel with his string tie had not changed one little bit.

Tetrahedron, 1958, Vol. 4, pp. 275–293. Pergamon Press Ltd., London

THE SESQUITERPENOID IRESIN

M. G. ROSSMANN and W. N. LIPSCOMB
School of Chemistry, University of Minnesota, Minneapolis 14, Minn.

(*Received* 19 *May* 1958)

Abstract—The three-dimensional structure of the sesquiterpenoid iresin has been obtained from an X-ray diffraction study of the *p*-dibromobenzoate diester. The position of the angular methyl group is in agreement with the bi*cyclo*farnesol skeleton, previously found only in higher terpenes.

DJERASSI *et al.*[1] isolated a new tricyclic sesquiterpenoid $C_{15}H_{22}O_4$ from the Mexican plant "Iresine celosioides". This compound, named iresin, was shown to be tricyclic,[2,3] with the unique feature of possibly possessing the bi*cyclo*farnesol(I), or a closely related, skeleton. If so, iresin would be the "long-sought-after missing link"[2] between

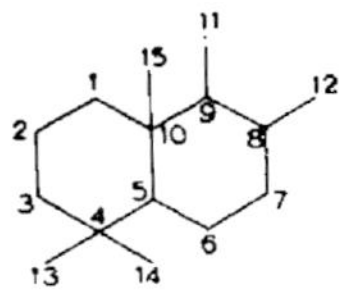

I

the higher and lower terpenes: this skeleton is present in all diterpenes, triterpenes and steroids, and its absence in the lower terpenes has been taken to indicate a fundamental difference between the biogenesis of the lower terpenes and the higher terpenes. We shall show that iresin is in fact based on the bi*cyclo*farnesol skeleton.

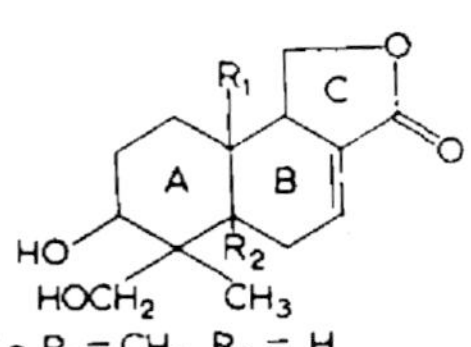

IIa $R_1 = CH_3$ $R_2 = H$
IIb $R_1 = H$ $R_2 = CH_3$

At the time we started, the detailed stereochemistry and position of the angular methyl group were unknown. Classical organic procedures had not decided between Structures (IIa) and (IIb), but the OH group on C-3, the caibon atom bearing the primary hydroxyl function, and the 5-6 bond were thought to be all *cis*. The present X-ray diffraction study has confirmed these aspects of stereochemistry, established the position of the methyl group, and established all three-dimensional aspects of

[1] C. Djerassi, P. Sengupta, J. Herran and F. Walls, *J. Amer. Chem. Soc.* **76**, 2966 (1954).
[2] C. Djerassi, W. Rittel, A. L. Nussbaum, F. W. Donovan and J. Herran, *J. Amer. Chem. Soc.* **76**, 6410 (1954).
[3] C. Djerassi and W. Rittel, *J. Amer. Chem. Soc.* **79**, 3528 (1957).

the molecular geometry except for the absolute configuration. Preliminary communications of the X-ray evidence[4] for the stereochemistry and presently available chemical evidence[5] for the stereochemistry and absolute configuration are in complete agreement.

EXPERIMENTAL DATA

A heavy atom derivative was prepared here by Dr. Paul Tomboulion who esterified the two OH groups by treating iresin with *p*-bromobenzoyl chloride in pyridine.

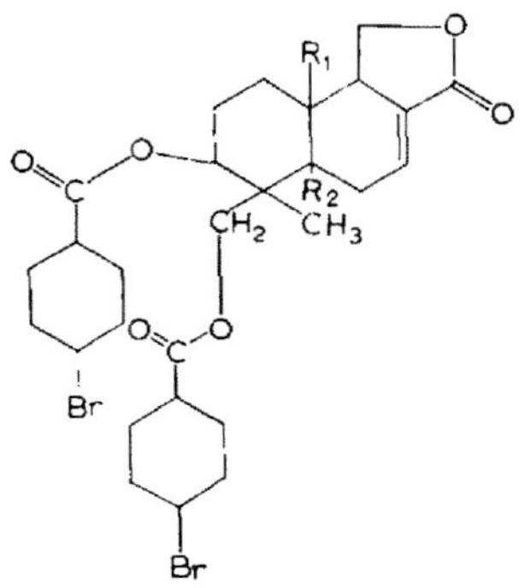

III

The product (III), recrystallized from a benzene-*cyclo*hexane mixture, melted at 211·5–212° (corrected). Infrared spectra of iresin and this derivative indicated no changes in the iresin residue other than esterification.

Very thin crystals about 0·5 × 0·5 mm² were grown from benzene solution over a period of a month. Equi-inclination Weissenberg photographs were taken of levels 0, 1, 2, 3 and 4 about the *b* and *c* axes with Cu$K\alpha$ radiation. The unit cell is monoclinic, with parameters $a = 28{\cdot}75$, $b = 7{\cdot}41$, $c = 6{\cdot}40$ Å. and $\beta = 92°$, where the *a* axis is perpendicular to the plate. Our measured density of 1·526 g cm^{-3} gives a molecular weight of 627 if two molecules are in the unit cell, in good agreement with the expected value of 632 for (III). The systematic absence only of 0*k*0 when *k* is odd indicates either $P2_1$ or $P2_1/m$, but the molecule cannot be expected to have either a mirror plane or a centre of symmetry. Hence the space group $P2_1$, later shown to be correct, was assumed.

Visual estimates of 1432 independent diffraction maxima were made with the multiple film technique in which account was taken of the variation of film factor with angle of incidence.[6] The usual Lorentz-polarization corrections were made, and corrections for absorption were estimated only on those very few reflections for which the incident or diffracted ray passed in the plane of the plate-like crystal. The very thin background band was used to indicate the magnitude of the correction for these few reflections. The use of reflections common to the several films allowed correlation to a single scale, later made absolute by comparison with trial structures.

STRUCTURE DETERMINATION

The Patterson projection along *c* yielded the Br positions which were so close to $P2_1/m$ symmetry that the remainder of the structure was obscured by a pseudo-mirror

[4] M. G. Rossmann and W. N. Lipscomb, *J. Amer. Chem. Soc.* 80, 2592 (1958).
[5] C. Djerassi, F. W. Donovan, S. Burstein and R. Mauli, *J. Amer. Chem. Soc.* 80, 2593 (1958).
[6] M. G. Rossmann, *Acta Cryst.* 9, 819 (1956).

plane. Hence, the three-dimensional Patterson function was computed with sharpened coefficients $[F_0^2/(\hat{f})^2]$ exp $[2(B - B^1)(\sin^2\theta/\lambda^2)]$, where $B - B^1$ was chosen as 1·0 Å^2 and $\hat{f} = \Sigma f_j/\Sigma Z_j$ in the usual notation. The only unusual features of the obvious Br···Br peaks were the elongation of the non-Harker peaks (Fig. 1) along y, interpreted as displacement of 0·14 A from the pseudo-mirror plane, and the elongation of the Harker peaks which later was found to be due to pronounced anisotropy of the thermal motion of the Br atoms.

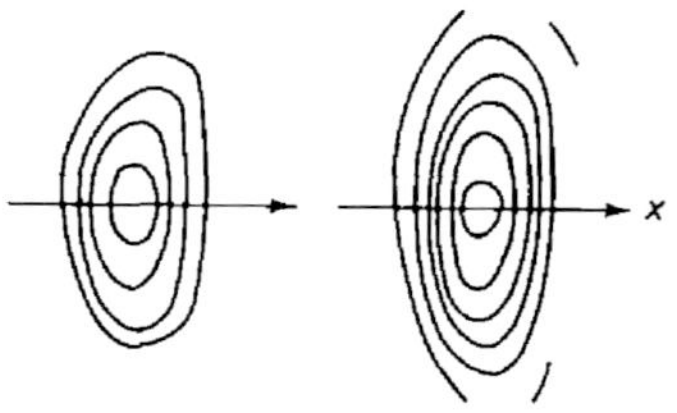

FIG. 1. (a) Single Br··· Br Harker peak. (b) General double Br··· Br peak. Contours are at 400 e^2/A^3.

The benzene rings were not obvious in the Patterson function, and hence only the positions of the Br atoms were used to compute phases for the first three-dimensional electron density map (F1). Had the Br atoms been placed exactly on the pseudo-mirror plane, the structure and its mirror image would have been superimposed, but the slight displacements of Br described above allowed the identification of the 7 C atoms (16–22) and O-1 and O-2 of one of the benzoate groups, as well as C-23, all of heights between 1 and 3 eA^{-3}. The effect of the pseudo-mirror plane was, however, still very evident. Four cycles of least-squares refinement of these atoms along with the Br atoms reduced $R = \Sigma|(|F_0| - |F_c|)|/\Sigma|F_0|$ from 0·62 to 0·41, an improvement due mostly to small changes in the Br positions and scale factor. During the last cycle of this refinement, the new phases were inserted into the computation of a new electron density map (F2).

Nearly all of the light atoms were found in F2. Those found above in F1, and therefore included in F2, had heights between 5 and 7 eA^{-3}, and those not yet introduced had heights between 1 and 3 eÅ^{-3}. A search was made for peaks at a distance of 1·5 Å, and at suitable bond angles from both the oxygen atoms that had so far been found. These peaks were marked and a similar search was then made of peaks 1·5 Å away from these, and so forth. At the same time the chemical reasonableness of the emerging structure was taken into account. In this way the molecule could be fitted to peaks in the Fourier without undue imagination except that atom C-9 did not correspond to any real peak and the bonds in the five-membered ring required a good deal of "stretching" (see Fig. 2). All other peaks in the map could be accounted for by the pseudo-mirror plane. These false peaks were generally slightly smaller in height, larger in area and of more irregular outline than the real peaks. Some atoms had no "mirror atom". There was general confusion where the second benzene ring might have been expected to turn up. Thus in the next trial structure the whole of the first benzoate group, as well as the iresin molecule itself and atoms O-4 and C-29, were included. Atom C-23 was omitted at this stage. A revision of the scattering curve for Br had been made. Unfortunately this contained an error. It effectively reduced the light atom contributions to high-order planes by a factor of three but left the low-order planes unchanged in structure factors used for F2.

Two cycles of least-squares refinement, in which R fell from 0·38 to only 0·37, were carried out, and then the third electron density map, F3, was computed. Examination of F3 indicated that atom C-23 had been incorrectly placed in F1, and atoms O-4 and C-29 had been incorrectly placed in F2. Moreover, the second benzoate

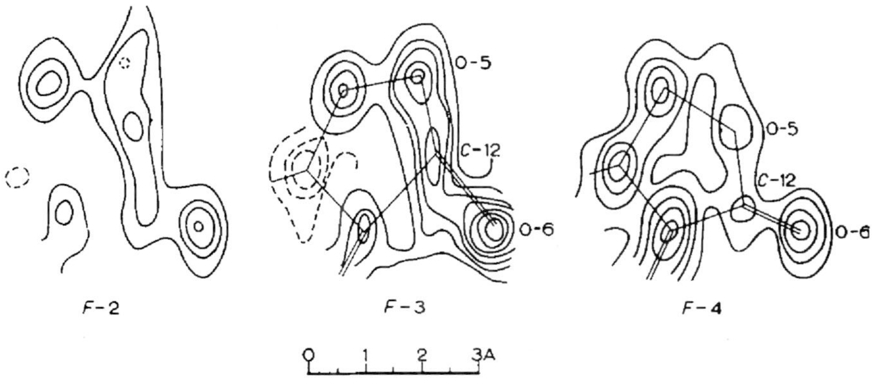

FIG. 2. Successive stages in determining the shape of the five-membered ring shown on section 14/20 c. Dotted lines show section 13/20 c and dashed line section 15/20 c. Contours are at intervals of 1 e/A^3 upwards.

group was now clearly visible, but something was wrong with the five-membered ring. Atoms C-12 and O-5 (Fig. 2) did not give the expected well-resolved spherical peak, and the apparent temperature factor shifts indicated by least squares were unusually large. Therefore, only these two atoms were omitted from the next refinement. R was found to be 0·305. Computation of F4 only in the vicinity of the five-membered ring, with phases from all atoms except C-12 and O-5, showed clearly where these two atoms really were (Fig. 2), thus completing the structure. The difficulty had been caused by a "mirror peak" in F2 which was actually higher than the true peak.

REFINEMENT OF THE STRUCTURE

All atoms but hydrogen had been located, and hence three-dimensional least-squares refinement was tried but R fell only from 0·280 to 0·277 in three cycles. In an attempt to interpret this result, a three-dimensional difference electron density map D1 was computed. This map showed no variations higher than $\pm 2{\cdot}0$ eÅ^{-3}, all of which could be interpreted as shifts of light atoms, except for peaks up to 3·5 eÅ^{-3} around both bromines, indicating strong anisotropic thermal motion (Fig. 3). Six further cycles of least-squares refinement in which six thermal parameters were introduced for each of the two bromines, produced values of $R = 0{\cdot}182$ and $r = \Sigma\omega(|F_0|^2 - |F_c|^2)^2/\Sigma\omega F_0^4 = 0{\cdot}183$. The final electron density map F5 was now computed, composite sections of which are shown in Fig. 4. Except for a very few small peaks as high as 1·5 eÅ^{-3}, the background was below 1 eÅ^{-3}. Another three-dimensional difference electron density map D2 was also computed, which indicated that the apparent thermal motions and termination of series effects around Br had been well accounted for (Fig. 3). This difference map showed no peaks higher than 1·6 eÅ^{-3}, but many peaks occurred at a height of about 1 eÅ^{-3}. When the expected parameters for hydrogen atom positions were evaluated, about three-fourths of these atoms fell on peaks greater than 1 eÅ^{-3}.

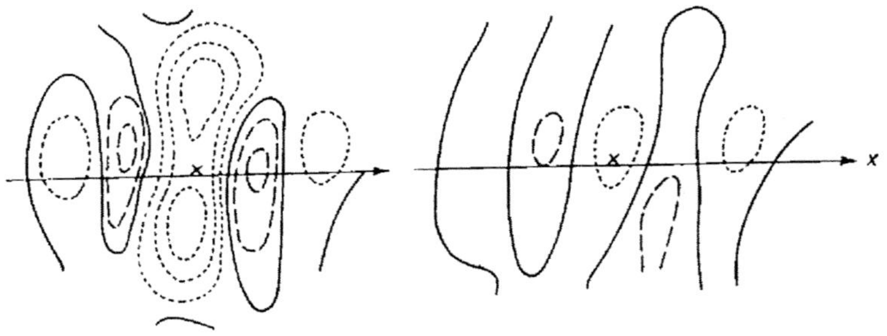

FIG. 3. Difference maps cut at 9/20 c. showing anisotropic motion of Br-2. Contours intervals of 1 e/A^3. Dashed lines are negative and dotted lines are positive contours.

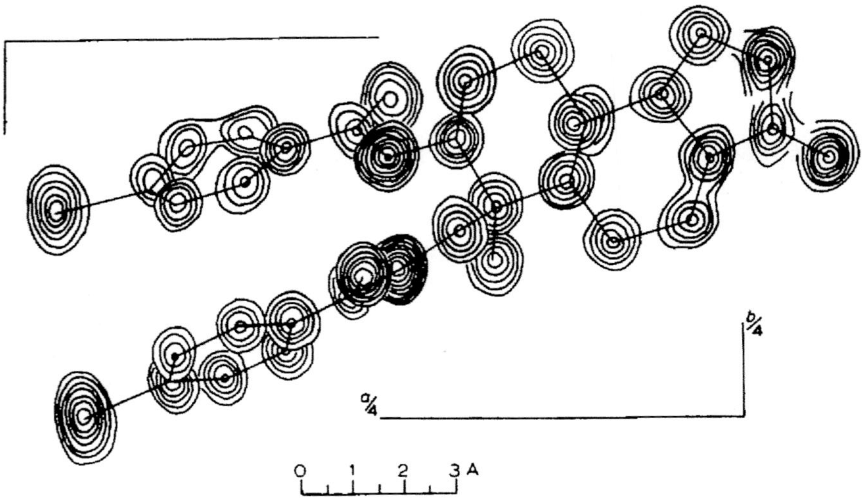

FIG. 4. Final electron density map. Contours are at intervals of 1 e/A^3 (except near bromines). The lowest contour is at a height of 2 e/A^3. Contours around bromines are at intervals of 5 e/A^3 upwards.

Further refinement of the structure is thus surely possible. The introduction of hydrogen atoms and possibly of anisotropic thermal parameters for other atoms than Br, especially atom O-3, would no doubt improve the already satisfactory agreement. In lieu of a complete table of the 1430 reflections, the agreement for different classes[7] of reflections is shown in Table 1. As a further check of the structure, a detailed examination was made of the 113 planes in Columns *a* and *b* of Table 1. Due to an error, we discovered that five of these planes, all having large indices, had

[7] R. E. Dickerson, P. J. Wheatley, P. A. Howell and W. N. Lipscomb, *J. Chem. Phys.* 27, 200 (1957).

been entered with either a wrong power of 10 or of 2 into the observed structure factor list used for refinements. However, the weighting factors for these five planes were small in the least-squares treatment, and we have shown that the maximum

TABLE 1. AGREEMENT FACTOR R FOR ALL CLASSES OF REFLECTIONS*

Class	R	n	a	b
All	0·182	1430	94	19
$0kl$	0·143	29	2	0
$h0l$	0·207	268	31	2
$hk0$	0·175	186	9	6
h odd	0·178	719	42	10
k odd	0·167	649	28	12
l odd	0·171	680	46	7
$h + k$ odd	0·180	710	42	10
$k + l$ odd	0·192	731	46	11
$h + l$ odd	0·180	721	54	13
$h + k + l$ odd	0·172	694	42	9

* Two reflections were outside the range of sin θ for which computations were made. The number of reflections in each class is n. The number of planes showing $|F_0| > 2|F_c|$ is in Column a, and the number showing $|F_0| < (\frac{1}{2})|F_c|$ is in column b.

effect on the electron density maps is less than 0·2 eÅ^{-3}, even in the unlikely event that the errors are all at their maximum at a given point. Hence the effect of these errors is negligible, and their correction reduces R to 0·176. As an additional check on the over-all agreement, we have listed R as the scattering angle increases (Table 2).

TABLE 2. VARIATION OF R WITH sin θ

Range of sin θ	R	n	$\Sigma \lvert F_0 \rvert$
0·00–0·15	0·249	14	606
0·15–0·30	0·136	91	4741
0·30–0·45	0·132	208	7283
0·45–0·60	0·140	361	7635
0·60–0·75	0·242	409	6574
0·75–0·90	0·244	242	2923
0·90–1·00	0·278	100	903

The five incorrectly listed planes were omitted. A final refinement in which all hydrogen atoms were included except those of the two methyl groups led to values of $R = 0·163$ and $r = 0·157$ and decreased the numbers in columns a and b of Table 1 by about a factor of two.

Isotropic temperature factor constants B in the formula $\exp(-B \sin^2 \theta/\lambda^2)$ are shown in Fig. 7 for each atom. We have previously observed large least-square increases of B for atoms incorrectly placed in the structure[8] at early stages of the

[8] F. L. Hirshfeld, K. Eriks, R. E. Dickerson, E. L. Lippert, Jr. and W. N. Lipscomb, *J. Chem. Phys.* **28**, 56 (1958).

TABLE 3. CO-ORDINATES OF ATOMS IN 1 MOLECULE

Atom	x/a	y/b	z/c
Br-1	0·0318	0·0445	0·0274
Br-2	0·0507	0·4981	0·4427
C-1	0·3621	0·4561	0·5750
C-2	0·3096	0·3831	0·5504
C-3	0·3047	0·2312	0·4213
C-4	0·3301	0·0463	0·5190
C-5	0·3802	0·1158	0·5619
C-6	0·4105	0·9669	0·6603
C-7	0·4616	0·0127	0·7006
C-8	0·4738	0·1783	0·7002
C-9	0·4420	0·3465	0·6312
C-10	0·3915	0·2704	0·6824
C-11	0·4711	0·5038	0·7610
C-12	0·5210	0·2607	0·7655
C-13	0·3039	−0·0163	0·7209
C-14	0·3288	0·8996	0·3514
C-15	0·3857	0·2563	0·9257
C-16	0·0961	0·1007	0·0716
C-17	0·1140	0·0627	0·2924
C-18	0·1600	0·1192	0·3204
C-19	0·1873	0·2145	0·1579
C-20	0·1670	0·2416	−0·0253
C-21	0·1213	0·2226	−0·0471
C-22	0·2361	0·2631	0·1832
C-23	0·1101	0·6021	0·5485
C-24	0·1458	0·5978	0·4290
C-25	0·1884	0·6902	0·4962
C-26	0·1917	0·7543	0·6983
C-27	0·1556	0·7445	0·8273
C-28	0·1131	0·6716	0·7611
C-29	0·2342	0·8291	0·7784
O-1	0·2560	0·1850	0·3759
O-2	0·2592	0·3427	0·0611
O-3	0·2409	0·8611	0·9749
O-4	0·2630	0·8842	0·6400
O-5	0·5188	0·4308	0·7802
O-6	0·5586	0·1735	0·7956

refinement. In the present study we found that for atoms placed between 0·15 and 0·35 Å from their correct positions, the square of the full shift in temperature factor is roughly proportional to the full shift in position. Thus our earliei experience, that the behavior of tempeiature factors is a method of picking out incorrectly placed atoms, has been confirmed.

A list of co-ordinates for atoms in one molecule is given in Table 3. The observed amplitudes of scattering are listed in Table 4.

Table 4. Observed structure factors on absolute scale

(*h*0*l*)

	$l=\bar{5}$	$l=\bar{4}$	$l=\bar{3}$	$l=\bar{2}$	$l=\bar{1}$	$l=0$	$l=1$	$l=2$	$l=3$	$l=4$	$l=5$
$h=0$	21·7	24·3	38·2	51·6	68·1	(640·0)	68·1	51·6	38·2	24·3	21·7
1	—	—	—	144·7	34·4	(cut off)	78·4	146·1	21·7	21·9	20·1
2	23·0	23·3	38·1	63·6	86·8	32·4	39·2	105·6	44·3	20·4	20·1
3	23·0	23·3	16·5	129·3	45·4	14·8	26·5	122·4	13·6	5·6	—
4	29·3	28·2	55·8	47·6	77·7	45·5	11·5	42·5	27·0	6·3	20·3
5	31·7	28·3	50·2	72·2	67·9	21·6	25·1	13·8	34·0	27·2	20·3
6	28·5	19·0	8·0	20·0	17·9	4·0	23·2	7·3	35·3	23·5	20·6
7	23·3	26·7	56·7	21·6	56·6	4·5	13·9	—	9·2	8·0	—
8	23·6	—	42·3	39·8	71·0	16·5	68·8	—	5·6	8·6	20·9
9	—	—	—	52·0	37·8	68·8	15·3	52·6	15·4	—	—
10	—	23·0	—	50·9	23·9	84·3	14·1	41·6	16·1	18·2	—
11	—	17·6	7·1	7·0	18·2	54·3	60·7	101·7	10·3	18·8	—
12	—	6·8	—	28·2	17·0	63·4	7·6	90·7	12·6	46·8	—
13	—	10·0	10·8	21·6	7·6	44·4	17·0	94·9	6·2	21·7	—
14	—	4·9	37·6	23·0	29·9	140·0	12·4	55·2	9·2	48·9	—
15	—	5·2	8·9	19·4	38·7	47·8	58·8	51·9	6·9	36·4	29·3
16	—	5·4	20·4	18·8	52·3	17·7	43·5	24·2	20·6	29·6	20·3
17	—	11·7	25·6	9·5	27·5	10·9	47·3	53·9	7·3	14·7	28·2
18	—	8·5	21·7	—	50·4	11·4	25·6	—	16·5	—	27·6
19	—	—	27·8	8·7	44·9	26·1	77·2	—	22·2	10·9	26·8
20	—	—	—	—	54·5	23·2	28·9	21·7	15·8	17·2	—
21	—	6·1	29·2	15·5	16·7	23·9	43·5	18·7	25·0	25·8	—
22	—	12·7	—	—	28·2	8·0	22·0	48·5	25·1	—	—
23	—	15·2	10·1	—	—	13·2	20·8	8·3	—	17·2	—
24	—	15·1	18·9	—	—	12·5	—	35·8	7·8	16·6	—
25	—	5·4	—	29·7	29·9	7·7	—	19·1	—	22·7	—
26	—	8·6	—	14·1	16·7	7·7	—	22·2	7·5	21·4	—
27	—	7·1	—	16·1	—	15·7	19·1	10·1	7·2	14·0	—
28	—	—	9·7	—	24·1	11·0	21·3	16·7	8·7	12·7	—
29	—	—	—	8·3	23·4	—	27·7	14·7	11·4	7·5	—
30	—	—	—	—	20·1	—	32·0	—	21·5	8·9	—
31	—	—	11·5	—	12·4	—	29·6	—	12·3	—	—
32	—	—	9·4	8·1	12·7	—	28·7	—	16·2	—	—
33	—	—	—	—	11·5	—	21·7	—	3·4	—	—
34	—	—	—	7·0	9·1	9·8	20·3	4·1	—	—	—
35	—	—	—	—	—	—	12·7	—	—	—	—
36	—	—	—	—	—	14·5	6·9	—	—	—	—

(*h*1*l*)

	$l=\bar{5}$	$l=\bar{4}$	$l=\bar{3}$	$l=\bar{2}$	$l=\bar{1}$	$l=0$	$l=1$	$l=2$	$l=3$	$l=4$	$l=5$
$h=0$	16·6	15·2	17·5	44·5	20·6	—	20·6	44·5	17·5	15·2	16·6
1	—	30·0	48·4	69·0	60·0	58·0	90·3	37·5	20·9	33·3	—
2	—	17·5	30·3	70·1	150·6	38·1	94·9	56·0	18·4	19·0	16·8
3	16·6	9·0	29·6	29·7	114·0	33·8	73·9	37·8	47·4	29·5	—
4	14·0	14·8	54·9	41·9	102·3	34·6	64·3	27·4	30·7	30·7	17·0
5	—	18·9	35·6	9·9	106·4	1·0	69·1	28·8	54·6	21·9	14·2
6	16·8	8·3	3·9	42·2	67·2	22·3	115·2	17·1	36·8	7·9	14·4
7	—	19·3	6·9	37·6	85·1	10·1	70·7	20·2	45·7	12·3	—
8	—	18·3	30·4	34·7	24·1	37·4	88·4	13·0	30·7	19·7	—
9	—	19·0	22·5	67·5	25·6	28·8	74·8	30·0	52·1	—	—
10	—	23·8	5·8	50·2	16·7	41·5	67·6	22·2	38·5	11·1	—
11	—	20·3	13·9	42·1	18·2	45·5	45·6	12·4	26·4	13·4	—
12	—	9·9	11·5	57·2	7·4	30·0	54·5	20·3	17·6	16·9	—
13	—	10·3	8·7	47·0	31·5	87·0	36·1	26·0	13·4	17·2	—
14	—	14·4	12·5	49·0	9·6	37·0	39·0	28·8	8·7	7·2	—
15	—	12·4	9·5	11·0	10·5	46·1	39·7	25·9	32·1	12·7	—
16	—	—	6·6	33·0	36·3	22·8	65·7	24·0	26·8	8·3	—
17	—	13·4	10·3	—	19·1	15·3	33·9	—	28·1	26·2	—
18	—	16·4	14·5	18·8	16·5	—	33·0	22·8	36·5	23·1	—

Table 4 (Cont'd.).

(h1l) contd.

	$l=\bar{5}$	$l=\bar{4}$	$l=\bar{3}$	$l=\bar{2}$	$l=\bar{1}$	$l=0$	$l=1$	$l=2$	$l=3$	$l=4$	$l=5$
$h=19$	—	8·3	7·3	21·0	19·8	—	40·9	12·2	39·3	8·6	—
20	—	12·0	15·4	29·2	15·5	25·9	42·1	24·0	10·3	8·6	—
21	—	13·3	15·8	37·3	—	36·0	27·0	24·3	24·6	12·5	—
22	—	—	11·7	27·5	16·0	39·9	22·5	17·7	23·9	—	—
23	—	13·1	16·2	35·8	18·4	49·0	26·8	34·4	13·6	8·2	—
24	—	—	8·0	26·2	—	40·3	11·7	29·5	13·4	11·0	—
25	—	—	—	13·6	19·8	20·6	15·8	33·5	9·8	5·8	—
26	—	—	12·3	16·2	28·0	28·8	10·5	24·3	9·5	9·2	—
27	—	—	10·0	22·3	12·9	14·5	—	30·4	7·0	9·2	—
28	—	—	9·7	—	17·9	16·6	—	23·0	8·6	8·3	—
29	—	—	12·8	8·3	9·6	12·7	—	11·6	10·8	7·2	—
30	—	—	12·2	—	9·8	10·5	—	8·3	13·4	—	—
31	—	—	10·0	—	11·5	6·3	—	—	8·1	—	—
32	—	—	9·2	—	9·8	8·5	—	9·2	9·8	—	—
33	—	—	—	7·2	7·6	5·6	—	8·1	4·2	—	—
34	—	—	—	—	—	14·5	—	19·3	—	—	—
35	—	—	—	—	—	14·5	6·7	—	—	—	—
36	—	—	—	—	—	10·9	—	—	—	—	—

(h2l)

	$l=\bar{4}$	$l=\bar{3}$	$l=\bar{2}$	$l=\bar{1}$	$l=0$	$l=1$	$l=2$	$l=3$	$l=4$	$l=5$
$h=0$	17·1	11·7	54·2	50·9	44·7	50·9	54·2	11·7	17·1	—
1	17·3	27·5	77·4	49·0	87·0	26·5	61·8	16·1	11·8	—
2	—	33·9	34·1	77·7	104·8	60·7	67·6	31·8	17·9	—
3	21·3	36·7	42·8	47·8	92·9	56·2	56·9	38·2	16·9	21·3
4	8·2	26·4	28·3	56·9	45·2	38·0	25·9	10·9	24·0	—
5	21·9	41·5	—	16·7	22·7	48·0	70·8	25·0	13·1	—
6	13·1	26·1	56·8	57·6	29·6	19·8	17·6	23·2	—	—
7	—	43·5	45·9	21·5	30·6	35·9	24·0	27·0	25·5	21·7
8	6·9	14·0	35·8	33·5	38·5	64·3	46·2	22·2	14·5	21·7
9	9·6	22·6	23·4	30·8	26·5	26·1	40·5	22·8	6·5	21·7
10	4·7	8·4	40·9	37·8	52·1	32·7	48·5	—	16·8	21·7
11	15·5	6·7	21·9	31·3	61·4	21·3	60·6	17·9	21·6	—
12	—	20·3	19·0	6·9	63·1	28·4	45·7	15·8	24·4	—
13	10·4	15·3	11·5	25·3	41·5	13·6	61·8	12·2	24·8	—
14	—	—	—	26·1	24·7	15·8	41·3	—	22·7	—
15	—	22·2	—	25·8	25·8	18·2	33·5	21·4	—	—
16	—	22·8	21·9	27·0	32·8	31·1	27·2	13·6	13·1	—
17	12·3	19·3	22·5	45·4	15·1	26·3	32·1	13·9	25·9	—
18	5·2	22·9	—	47·3	—	39·0	24·9	24·0	9·4	—
19	8·0	13·6	13·6	28·7	18·1	59·5	27·7	27·8	12·7	—
20	10·2	11·5	31·1	37·0	—	19·1	—	—	13·1	—
21	—	10·1	14·2	19·6	—	40·2	38·9	8·0	13·0	—
22	8·2	10·3	28·3	12·7	19·0	27·0	33·0	8·0	8·2	—
23	8·9	14·2	20·2	12·9	—	10·0	24·8	14·0	15·7	—
24	—	—	16·2	15·1	16·1	—	9·9	9·4	21·3	—
25	12·5	—	—	—	—	—	25·4	—	14·2	—
26	—	13·7	17·9	15·1	12·0	10·5	12·7	—	9·3	—
27	8·2	9·0	—	18·2	11·9	14·1	24·6	7·3	19·0	—
28	—	—	7·7	17·7	11·7	15·1	23·6	8·1	16·8	—
29	—	6·9	—	24·1	—	15·1	8·4	15·0	—	—
30	—	7·3	—	13·4	—	25·6	6·4	9·5	—	—
31	—	5·9	—	12·7	—	22·5	9·5	10·6	—	—
32	—	—	—	18·6	7·0	20·6	—	11·9	—	—
33	—	—	6·4	—	—	17·7	—	—	—	—
34	—	—	—	—	—	6·0	—	—	—	—
35	—	—	—	—	7·8	—	—	—	—	—

Table 4 (Cont'd.).

(*h*3*l*)

	$l=\bar{5}$	$l=\bar{4}$	$l=\bar{3}$	$l=\bar{2}$	$l=\bar{1}$	$l=0$	$l=1$	$l=2$	$l=3$	$l=4$	$l=5$
$h=0$	17·0	33·7	13·7	58·9	15·3	—	15·3	58·9	13·7	33·7	17·0
1	17·0	20·3	23·9	15·0	26·5	16·5	9·6	6·6	—	33·8	18·6
2	17·0	4·2	20·7	26·2	—	28·7	46·4	33·1	32·6	9·3	—
3	17·0	13·0	23·9	50·3	35·9	25·2	19·4	44·4	25·7	10·4	—
4	—	15·1	21·7	53·4	48·3	26·7	77·9	23·1	—	24·7	—
5	—	6·8	12·3	23·9	62·4	14·8	62·4	23·6	26·5	20·4	—
6	—	—	24·0	33·2	44·7	33·5	65·0	47·6	27·8	9·9	—
7	—	10·6	12·6	23·9	30·1	40·0	39·2	24·9	31·8	20·4	—
8	—	8·0	9·0	19·4	23·9	14·1	99·7	19·4	36·5	23·5	—
9	—	13·1	9·4	15·1	32·7	21·3	71·2	18·5	11·9	17·3	—
10	—	8·5	—	27·4	17·2	34·3	44·9	32·9	24·8	12·0	—
11	—	13·7	13·4	34·0	17·0	35·4	22·9	15·0	36·8	9·0	—
12	—	14·0	12·9	22·0	16·3	45·8	19·8	19·1	26·2	—	—
13	—	18·6	10·3	29·5	19·4	16·8	20·1	27·2	17·0	23·5	—
14	—	—	10·6	14·5	23·7	35·6	16·5	14·4	—	23·8	—
15	—	—	7·3	21·6	12·9	15·8	22·2	15·9	—	17·2	—
16	—	—	—	—	9·1	25·1	21·7	25·9	15·3	—	—
17	—	—	—	10·7	—	—	22·5	15·9	24·0	9·4	—
18	—	—	10·0	—	25·3	—	30·6	14·4	17·3	10·0	—
19	—	7·6	—	—	21·0	—	28·0	—	24·3	9·9	—
20	—	—	—	18·1	20·3	14·7	22·2	16·5	31·5	—	—
21	—	—	10·3	13·6	—	13·7	27·5	28·8	27·0	—	—
22	—	—	—	—	—	26·1	25·6	9·3	18·9	—	—
23	—	5·2	9·7	13·2	10·5	16·6	—	9·3	13·7	—	—
24	—	5·2	9·5	13·0	19·8	26·5	10·0	—	13·3	—	—
25	—	—	—	15·9	—	24·8	18·9	19·9	14·5	—	—
26	—	9·9	9·0	7·8	—	25·1	14·3	19·3	12·0	8·2	—
27	—	—	—	7·7	—	15·1	9·8	18·4	11·2	—	—
28	—	6·8	10·1	7·3	—	6·6	—	8·7	7·3	—	—
29	—	4·5	—	—	—	6·3	—	9·5	6·6	—	—
30	—	—	—	—	9·8	7·3	—	7·3	7·8	—	—
31	—	—	—	—	7·6	6·9	—	—	6·9	—	—
32	—	—	—	—	—	—	—	—	—	—	—
33	—	—	—	—	—	8·9	—	4·6	—	—	—
34	—	—	—	—	—	7·8	—	—	—	—	—
35	—	—	—	—	9·8	—	—	—	—	—	—

(*h*4*l*)

	$l=\bar{4}$	$l=\bar{3}$	$l=\bar{2}$	$l=\bar{1}$	$l=0$	$l=1$	$l=2$	$l=3$	$l=4$
$h=0$	11·7	14·2	28·2	31·1	33·4	31·1	28·2	14·2	11·7
1	24·8	12·6	18·5	20·3	37·6	28·0	36·1	21·4	23·7
2	8·6	15·8	28·0	20·6	37·8	28·7	27·8	21·5	23·8
3	11·7	8·7	7·3	38·0	30·0	21·0	33·0	7·8	11·8
4	—	6·7	33·7	43·0	12·8	44·5	21·4	30·3	16·9
5	15·8	33·7	7·5	30·8	22·8	30·4	20·8	15·4	12·1
6	7·2	11·1	6·6	37·3	17·0	7·9	21·3	—	17·2
7	7·2	24·0	6·1	17·9	21·5	37·0	14·2	19·8	—
8	7·3	—	18·5	21·0	37·3	22·5	—	20·3	12·4
9	—	—	15·5	18·2	28·9	23·2	15·0	22·6	17·8
10	—	14·8	—	17·4	20·0	17·2	36·3	20·9	14·8
11	—	8·4	10·1	12·7	27·3	20·3	15·8	—	14·8
12	10·0	11·4	8·4	—	35·8	—	12·9	21·7	14·9
13	—	8·7	8·6	—	—	15·1	35·0	15·9	9·4
14	10·1	—	10·7	18·6	16·2	—	17·6	—	—
15	14·9	11·9	8·9	12·0	16·9	15·8	19·3	22·8	—
16	—	—	7·8	19·4	17·4	—	28·0	—	11·3

Table 4 (Cont'd.).

(*h*4*l*) *contd.*

	$l=\bar{4}$	$l=\bar{3}$	$l=\bar{2}$	$l=\bar{1}$	$l=0$	$l=1$	$l=2$	$l=3$	$l=4$
$h=17$	—	—	14·5	12·4	23·0	19·4	8·7	—	11·1
18	—	—	17·9	—	—	15·8	—	—	8·9
19	—	12·2	—	15·8	13·6	15·8	8·1	—	7·1
20	—	—	8·1	15·8	9·9	19·8	18·1	12·5	6·8
21	—	—	10·4	11·7	7·0	12·9	8·0	9·7	7·9
22	—	7·6	10·4	—	—	11·7	—	—	6·2
23	—	7·5	—	—	—	11·7	—	—	11·1
24	6·1	11·1	7·8	—	9·8	—	17·7	10·1	14·4
25	7·2	7·0	—	—	—	—	26·0	—	—
26	—	8·6	7·3	—	9·3	—	6·7	—	7·3
27	—	6·2	6·9	—	9·0	—	—	—	—
28	5·4	—	—	—	10·1	9·6	20·0	—	—
29	4·4	—	—	8·4	11·5	8·8	9·6	—	—
30	—	—	—	8·6	6·6	9·3	10·3	—	—
31	—	—	—	7·6	—	—	7·8	—	—
32	—	—	—	—	—	9·3	—	—	—

(*h*5*l*)

	$l=\bar{4}$	$l=\bar{3}$	$l=\bar{2}$	$l=\bar{1}$	$l=0$	$l=1$	$l=2$	$l=3$	$l=4$
$h=0$	7·6	—	—	10·3	—	10·3	—	—	7·6
1	12·7	9·8	8·4	—	12·6	—	25·1	16·8	7·6
2	9·3	9·8	13·3	17·2	5·0	—	14·7	16·8	12·1
3	8·5	9·8	—	12·2	20·6	—	18·2	—	10·2
4	—	—	—	8·8	17·0	15·1	12·5	10·8	10·2
5	8·6	—	11·3	8·8	13·6	15·3	15·0	15·8	10·2
6	8·6	—	29·8	29·9	24·2	35·6	18·5	—	14·9
7	8·6	—	27·5	10·8	—	25·6	26·2	15·9	7·8
8	—	11·9	15·3	25·6	7·7	25·8	34·6	9·0	—
9	—	11·9	7·7	9·8	10·0	26·1	—	14·4	9·4
10	—	9·0	—	10·0	22·0	26·5	18·1	8·0	14·8
11	—	12·0	13·2	10·0	6·6	16·0	—	8·0	9·3
12	—	8·0	8·6	—	6·7	—	—	8·0	—
13	—	11·2	13·5	11·5	6·7	—	—	—	—
14	—	8·0	8·7	11·7	13·6	12·9	—	—	—
15	—	—	8·7	11·7	11·9	24·4	—	—	—
16	—	—	—	—	—	10·0	—	—	—
17	—	—	8·1	—	14·0	—	—	—	—
18	—	—	—	—	8·6	14·1	—	9·5	—
19	—	—	—	10·0	7·0	—	9·0	11·9	8·3
20	—	—	—	11·7	7·0	—	—	9·0	—
21	—	—	—	—	6·9	—	8·6	—	6·5
22	—	—	7·7	—	—	—	—	11·5	4·8
23	—	—	—	—	8·2	—	8·0	7·6	4·2
24	—	—	7·0	—	9·7	17·2	—	7·0	—
25	—	—	—	—	13·0	—	—	7·5	—
26	—	—	—	—	6·0	—	—	5·5	—
27	—	—	—	—	5·7	6·9	—	3·6	—
28	—	—	—	—	4·5	6·0	—	—	—
29	—	—	—	—	—	6·9	6·0	—	—

Table 4 (Cont'd.).

(*h*6*l*)

	$l=\bar{4}$	$l=\bar{3}$	$l=\bar{2}$	$l=\bar{1}$	$l=0$	$l=1$	$l=2$	$l=3$	$l=4$
$h=0$	—	8·0	—	14·1	—	14·1	—	8·0	—
1	—	—	8·0	—	13·2	10·3	—	—	—
2	—	—	8·0	9·8	13·2	10·3	13·2	16·2	—
3	—	—	—	9·8	—	28·9	—	11·2	7·5
4	8·5	11·9	—	16·5	23·0	10·3	16·2	8·0	9·9
5	7·5	—	—	16·5	—	11·7	18·1	11·2	8·3
6	7·5	—	10·4	—	18·7	11·7	13·2	16·2	—
7	—	—	10·4	19·6	6·8	16·5	16·2	11·1	—
8	—	—	—	—	13·8	10·5	18·1	—	—
9	—	12·9	10·6	—	13·8	10·5	—	7·8	—
10	—	11·1	—	—	14·0	16·7	—	7·8	—
11	—	9·0	8·1	12·0	—	12·0	—	7·6	—
12	—	8·9	—	—	7·0	16·0	—	—	—
13	—	10·9	9·3	—	9·5	10·5	—	8·6	—
14	—	—	8·0	10·5	8·6	—	—	—	—
15	—	—	7·2	—	—	—	—	—	—
24	—	—	—	—	—	—	—	3·4	—

(*h*7*l*)

	$l=\bar{3}$	$l=\bar{2}$	$l=\bar{1}$	$l=0$	$l=1$	$l=2$
$h=0$	—	—	—	—	—	—
1	—	—	—	—	—	—
2	—	—	10·5	—	12·9	—
3	—	—	—	—	—	—
4	—	—	—	—	—	—
5	—	—	—	—	—	9·0
6	—	—	—	7·0	—	11·0
7	—	—	—	—	—	9·0
8	—	—	—	—	—	—
9	—	—	9·8	8·6	—	—
10	8·1	—	—	7·0	—	—
11	—	8·7	—	8·5	—	—
12	—	10·7	—	—	—	—
13	—	8·4	—	—	—	—
14	—	—	—	9·2	—	—

DISCUSSION

The iresin molecule (IV) was found to have the angular methyl attached to C-10. Rings A and B are fused *trans*. The fact that the OH group on C-3, the methylene group on C-4 and the 5–6 bond are all *cis* has been confirmed. Ring A is in chair form. Rings B and C are very roughly in the same plane, except for atom C-5. However, the H on C-9 and the methyl C-15 are *trans* (Fig. 5), in agreement with the chemical results of Djerassi *et al.*[5] The shape of the molecule and its packing in the crystal can be seen in Figs. 5 and 6.

The standard deviations of the atomic position parameters, calculated from the residuals in the least-squares refinement, were essentially isotropic. Their values (Fig. 7) led to estimated standard deviations of 0·035, 0·050, 0·042 Å in the Br–C, C–C and C–O bond lengths, respectively. As an independent estimate, the mean

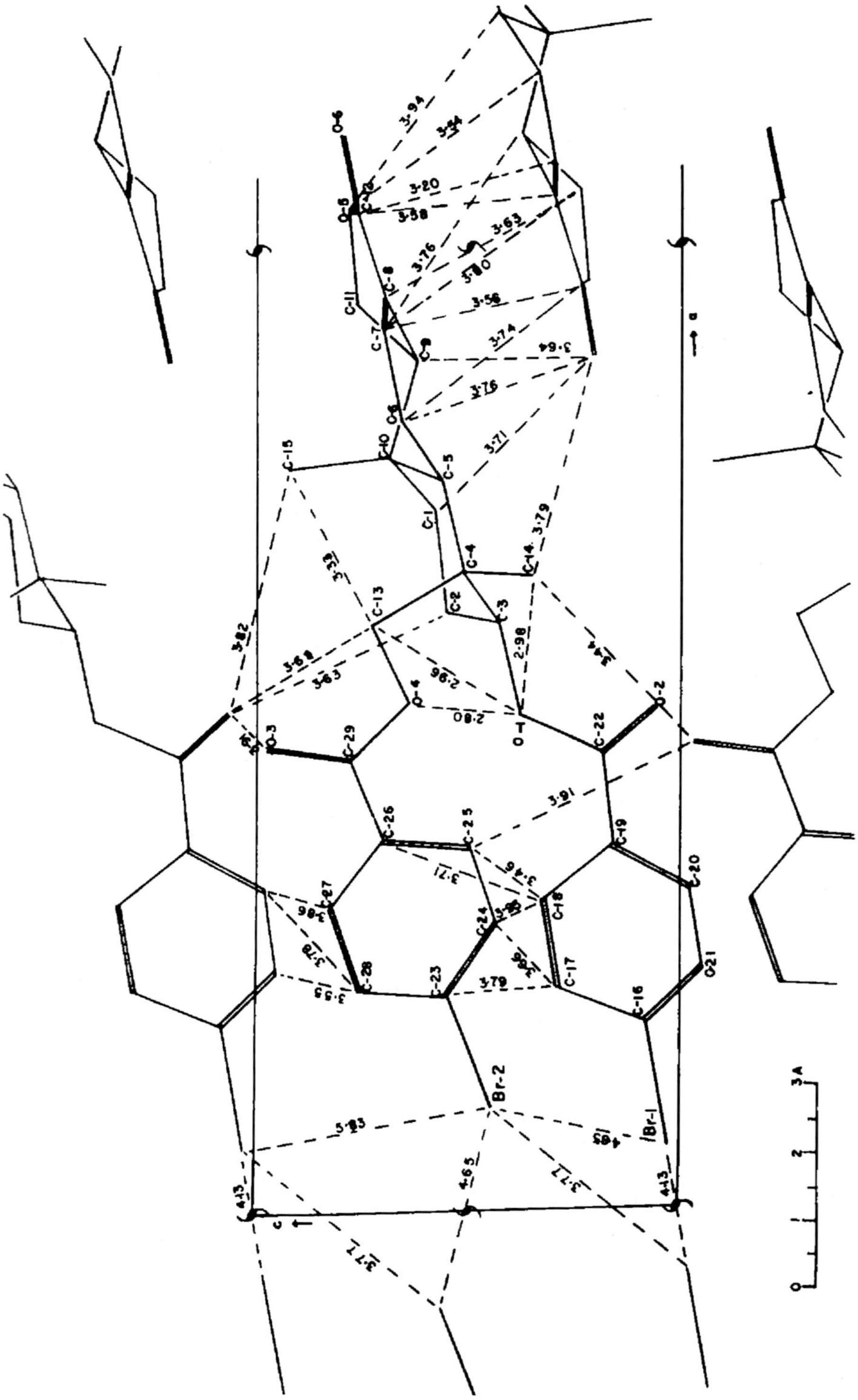

FIG. 5. Diagram showing projection of the molecule along the *b* axis, its packing and intermolecular distances.

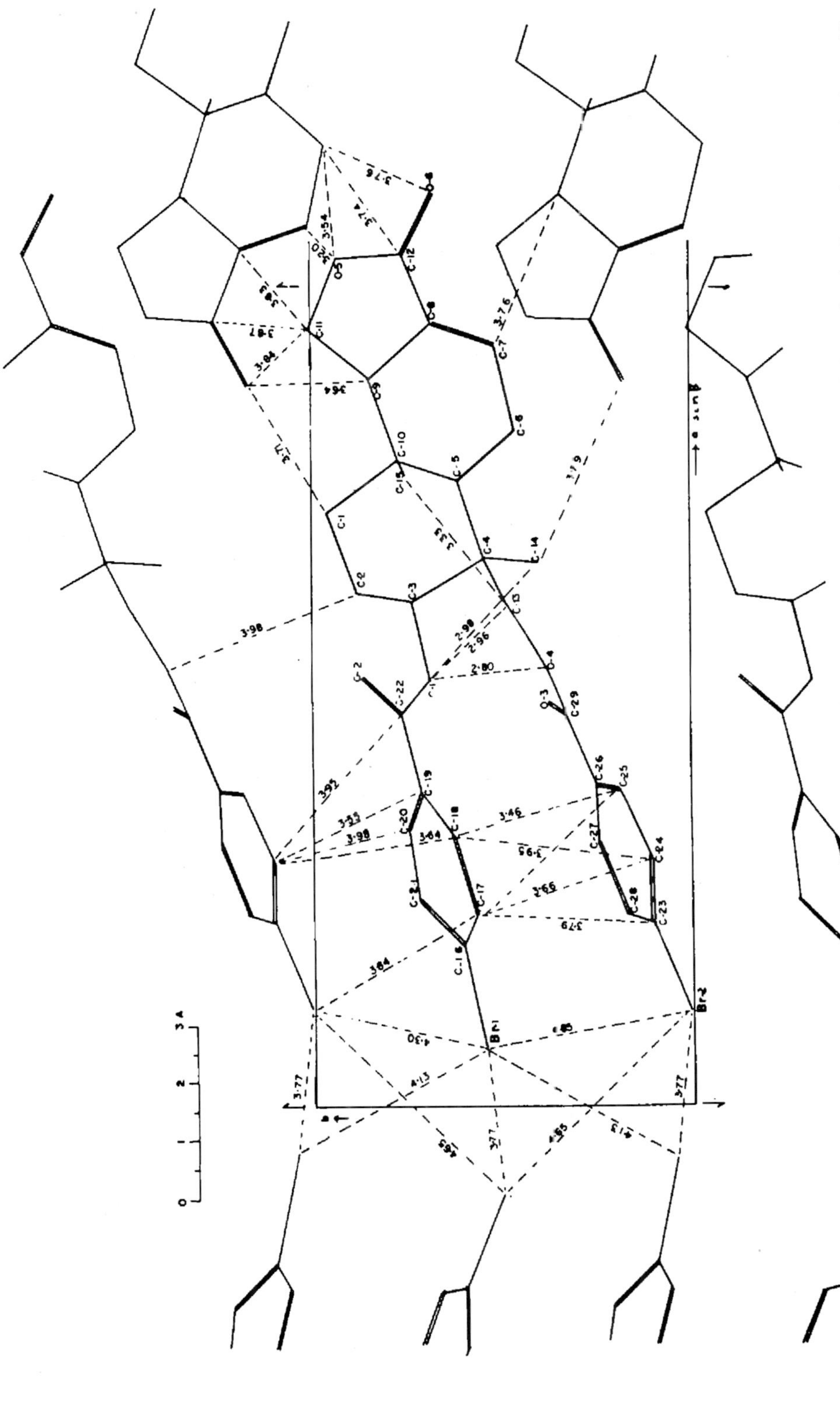

FIG. 6. Diagram showing projection of the molecule along the *c* axis, its packing and intermolecular distances.

distance of 1·395 Å for the twelve C–C bonds in the two benzene rings gave a standard deviation of 0·068 Å, quite reasonably a bit higher than the least-squares value of 0·050 Å obtained by neglect of off-diagonal terms in solving the normal equations.

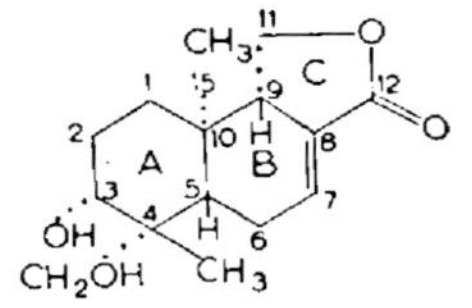

IV

Standard deviations in bond angles (Fig. 7) were about 4°. The fact that bonds C-22 · · · O-2, C-29 · · · O-3, C-29 · · · O-4, C-12 · · · O-6 and C-12 · · · O-5 are significantly shorter than a single C · · · C bond of 1·54 Å, establishes that O-2, O-3, O-4, O-5 and O-6 are oxygen atoms. Also, their higher electron density (Fig. 4) and yet sensible temperature factors (Fig. 7) further confirms this point. The shortness of the bond C-8 · · · C-7, as well as the general planarity of ring B establishes the position of the double bond in ring B.

The benzene ring attached to Br-1 and the corresponding carboxyl group and atom C-3 all lie in the same plane. The maximum distance of any one atom from this plane is 0·2 Å. In contrast, the benzene ring attached to Br-2 does not lie in the same plane as its corresponding carboxyl group. The maximum distance of any one atom from the best plane representing the benzene ring was 0·10 Å. The carboxyl group, methylene group and atom C-4 are roughly co-planar to within 0·5 Å. The angle between the normal to the benzene ring attached to Br-2 and the plane through the corresponding carboxyl group is 59°. The five-membered ring and atom O-6 was planar to within 0·09 Å, except for atom C-9, which is 0·44 Å from this mean plane. Exactly the same degree of non-planarity is found in similar five-membered rings.[9] As a general rule, one of the two carbon atoms furthest away from the oxygen is displaced by about 0·4 Å from the plane of the other four atoms. Atoms C-6, C-7 and C-10 lie on the same side of and between 0·25 to 0·35 Å away from the five membered ring plane.

Inter-molecular distances less than 4·0 Å, shown in Figs. 5 and 6, indicate no contacts significantly less than Van der Waals distances. All close C · · · C contacts are just greater than 3·4 Å, the recognized Van der Waals C · · · C distance. The close C · · · O contact of 3·20 Å between molecules related by a twofold screw is also just greater than the C · · · O Van der Waals contact distance of 3·1 Å. Within the molecule the methylene group C-13 and the methyl C-15 approach to 3·33 Å, and the oxygen O-1 approaches the methylene C-13 and methyl C-14 to 2·96 and 2·98 Å, respectively. Similarly, the two ether oxygens O-1 and O-4 approach to exactly the limit set by the sum of their radii, presumably because they are unhindered by hydrogen atoms. The angle between the normals of the two benzene rings is 50½°, which compares very favourably with about 52° in naphthalene and anthracene. Had it not been so difficult to obtain good single crystals, we would marvel at how well Nature is able to pack such an awkwardly shaped molecule.

[9] F. Jellinek, *Acta Cryst.* **10**, 277 (1957); S. Furberg, *Ibid.* 3, 325 (1950); and L. A. Beevers and W. Cochran, *Proc. Roy. Soc.* A **190**, 257 (1947).

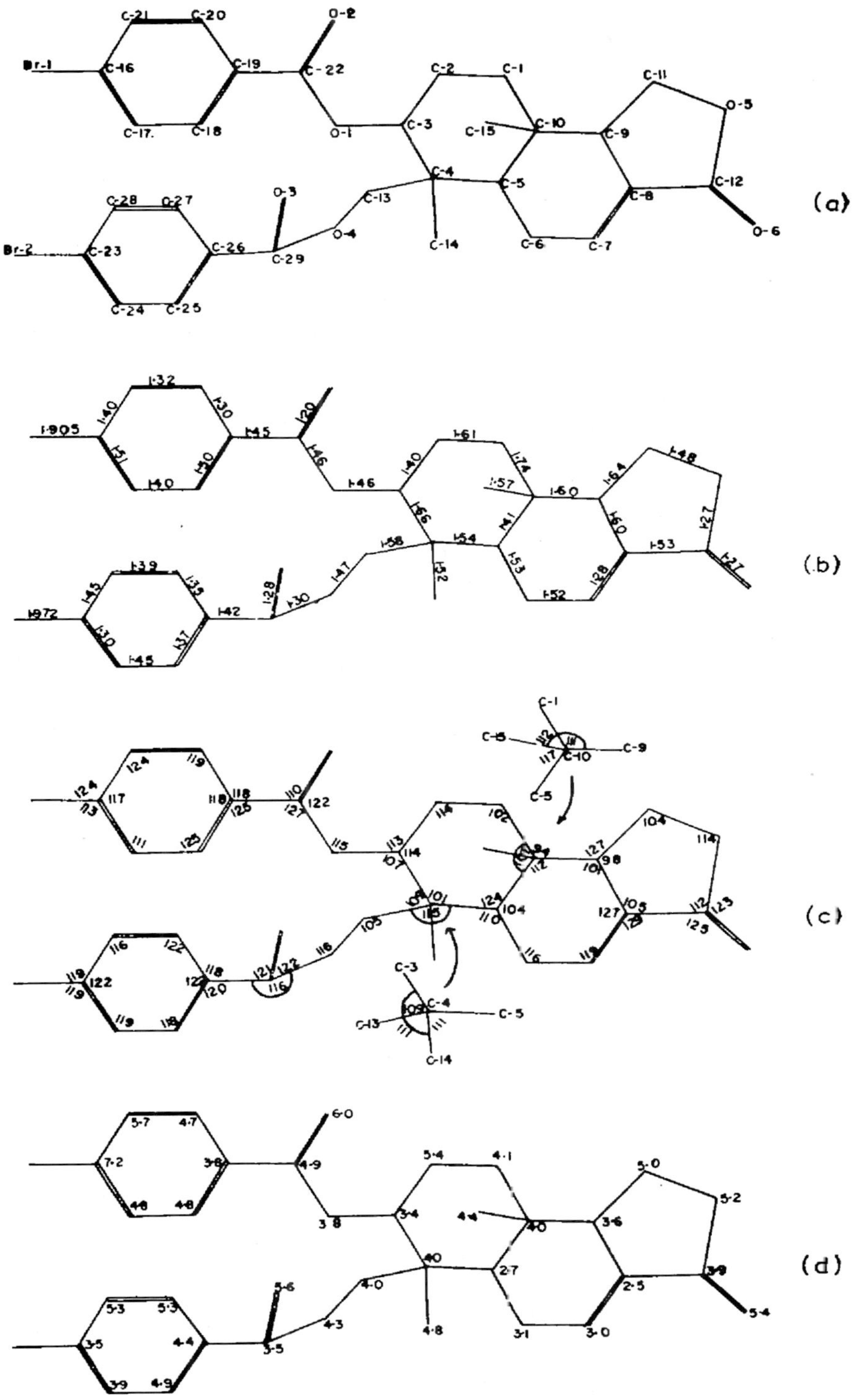

FIG. 7. Diagram (a) gives the atomic names. Diagram (b) shows bond length. Diagram (c) shows bond angles. Diagram (d) shows final isotropic temperature factors.

ANISOTROPIC THERMAL MOTION OF THE Br ATOMS

The determination of amplitudes of vibration by least-squares methods for individual atoms[10,11] is adopted here as being more suitable than the analysis of the motion of a rigid molecule[12] and to the analysis of electron density gradients in difference projections[13] or three-dimensional difference maps.[14] For an isotropic atom the atomic amplitude of scattering is reduced by $\exp(-\mathbf{B}\sin^2\theta/\lambda^2)$ where $\mathbf{B}$ is related to the mean square amplitude of vibration by $\mathbf{B} = 8\pi^2\mu^2$, and is usually expressed in units of Å². When the motion is anisotropic, the usual form is $\exp[-(\beta_1 h^2 + \beta_2 k^2 + \beta_3 l^2 + \beta_4 hk + \beta_5 kl + \beta_6 hl]$ where the β's are determined from the experimental data by the least-squares treatment. If we substitute $A = 4\beta_1/a^{*2}$, $B = 4\beta_2/b^{*2}$, $C = 4\beta_3/c^{*2}$, $D = 4\beta_4 a^*b^*$, $E = 4\beta_5/b^*c^*$ and $F = 4\beta_6/c^*a^*$, then in an orthogonal system of axes, where 2ϕ, 2ψ, 2ω is a point in reciprocal space and $\phi^2 + \psi^2 + \omega^2 = 1$, we can express $\mathbf{B}$ as

$$\mathbf{B} = A\phi^2 + B\psi^2 + C\omega^2 + D\phi\psi + E\psi\omega + F\omega\phi \qquad (1)$$

The values of ϕ, ψ and ω which define the directions of the principal axes of the ellipsoid of vibration are given by the simultaneous solutions of

$$\left.\begin{aligned} (2A-\lambda)\phi + D\psi + F\omega &= 0 \\ D\phi + (2B-\lambda)\psi + E\omega &= 0 \\ F\phi + E\psi + (2C-\lambda)\omega &= 0 \end{aligned}\right\} \qquad (2)$$

when λ is a root of

$$\lambda^3 + p\lambda^2 + q\lambda + r = 0$$

where

$$\begin{aligned} p &= -2(A+B+C) \\ q &= 4(AB+BC+CA) - (D^2+E^2+F^2) \\ r &= -2[4ABC - (AE^2+BF^2+CD^2) + DEF] \end{aligned} \qquad (3)$$

Then the three solutions of the cubic equation correspond to the three principal directions of the ellipsoid of vibration.*

The experimental data are the β's listed in Table 5 corresponding to the bromine

TABLE 5. ANISOTROPIC THERMAL PARAMETERS FOR BROMINE ATOMS

	Br-1	Br-2		Br-1	Br-2
β_1	0·0012	0·0012	A	3·9	3·9
β_2	0·0522	0·0467	B	11·5	10·3
β_3	0·0636	0·0553	C	10·4	9·0
β_4	−0·0002	−0·0036	D	−0·2	−3·0
β_5	−0·0108	−0·0193	E	−2·0	−3·6
β_6	−0·0055	−0·0050	F	−4·0	−3·6

* In the monoclinic system (b unique), A, B and D remain as defined above, but

$$C = (\csc\beta^*)[4\beta_3/c^{*2} - 4\beta_6(\cos\beta^*)/c^*a^* + 4\beta_1(\cos^2\beta^*)/a^{*2}],$$
$$E = (\csc\beta^*)[4\beta_5/b^*c^* - 4\beta_4(\cos\beta^*)/a^*b^*], \text{ and}$$
$$F = (\csc\beta^*)[4\beta_6/c^*a^* - 8\beta_1(\cos\beta^*)/a^{*2}].$$

when a, b and an axis perpendicular to a and b are chosen as orthogonal axes.

[10] J. S. Rollett and D. R. Davies, *Acta Cryst.* **8**, 125 (1955).
[11] J. Waser, *Acta Cryst.* **8**, 731 (1955).
[12] D. W. J. Cruickshank, *Acta Cryst.* **9**, 754 (1956).
[13] W. Cochran, *Acta Cryst.* **4**, 81 (1951); *Ibid.* **4**, 408 (1951).
[14] Y. C. Leung, R. E. Marsh and V. Schomaker, *Acta Cryst.* **10**, 650 (1957).

co-ordinates given in Table 3. These values were converted to $A \cdots F$ with the use of the unit cell parameters. The procedure, probably to be found useful in other similar cases, is as follows. For a given Br, substitute $A \cdots F$ into (3), solve for the roots λ_1, λ_2 and λ_3 by convenient graphical procedures, then substitute each root into (2) in order to find the proportions $\phi : \psi : \omega$, and then divide these final results by $\sqrt{\phi^2 + \psi^2 + \omega^2}$ to obtain the actual direction cosines (Table 6) which can be inserted into equation (1) to find the magnitude of the temperature factor in the direction of each principal axis (Table 6).

Table 6. Direction and magnitude of principal axes of ellipsoid of vibration

	λ	ϕ	ψ	ω	B	
Br-1	6·7	0·963	0·042	0·265	3·3	*L*
	20·7	0·230	−0·597	−0·769	10·3	*M*
	24·4	0·128	0·814	−0·571	12·1	*N*
Br-2	5·6	0·910	0·263	0·328	2·8	*L*
	17·7	0·426	−0·498	0·758	8·9	*N*
	23·1	0·034	−0·839	0·547	11·6	*M*

If we label these principal axes (Table 6) by *L*, *M*, *N* and if we name the direction of the Br ··· C bond as *P*, the normal to the benzene ring as *Q*, and an axis in the plane of each benzene ring and perpendicular to *P* and *Q* as *R*, then the angles which *L*, *M* and *N* make with *P*, *Q* and *R* are shown in Table 7.

Table 7. The angles between the principal axes of vibration with given directions in the corresponding benzene ring

Angle	Br_1	Br_2
$\angle L, P$	14°	8°
$\angle M, Q$	27°	20°
$\angle N, R$	30°	20°

Thus we see that a fairly "normal" amplitude of vibration exists in a direction parallel to the Br ··· C bonds, while the motion perpendicular to this bond is extraordinarily large. There is apparently no correlation between the plane of the benzene rings and the direction of the major axes. It does seem surprising that there is no conclusive evidence of anisotropic thermal motion in the atoms of the benzene ring in D2, as one would expect from the apparent anisotropic motion of the Br atoms, or from possible randomness in the position of the benzoate groups by about 0·35 Å perpendicular to the Br ··· C bond. From this analysis it would seem that the short Br ··· Br contact of 3·77 Å, just over the Van der Waals contact distance, is a very important feature of the molecular packing.

The standard deviation of **B** in any given direction is 0·4 Å^2 for the bromine atoms and 0·6 Å^2 for the isotropic atoms.

CALCULATIONS

All calculations were carried out on the Remington Rand UNIVAC 1103 high-speed digital computer. Phases used to compute Fourier summations were written onto magnetic tape during a least-squares calculation. These were then read back into the computer and sorted in such a way as to be suitable for our Fourier program. The *a*, *b* and *c* axes were divided into 160, 40 and 20 parts so that the intervals in these directions measured 0·18, 0·19 and 0·32 Å, respectively. The time required to punch the electron density at all the 64,000 points in the asymmetric unit was 2½ hr. One least-squares refinement cycle, when all atoms were included, and the bromine atoms were given six thermal parameters each, took 64 min. 1430 reflections were included in all Fourier summations and least-squares calculations. The Thomas–Fermi[15] scattering curve was used for bromine but McWeeny[16] curves were used for oxygen and carbon. A total of 39 hr of computing was used in this study.

Acknowledgements—We wish to express our thanks to Professor C. Djerassi who suggested this problem to us and who also supplied the sample of iresin. We are also indebted to Dr. P. Tomboulion who prepared the heavy atom derivative, and to the University of Minnesota for contributing the computing time. This research was also aided by a grant from the Upjohn Company.

[15] *Internationale Tabellen* p. 572. Zweiter Band, Berlin (1935).
[16] F. McWeeny, *Acta Cryst.* **4**, 513 (1951).

(Reprinted from Nature, Vol. 185, No. 4711, pp. 416–422, February 13, 1960)

STRUCTURE OF HÆMOGLOBIN

A THREE-DIMENSIONAL FOURIER SYNTHESIS AT 5·5-Å. RESOLUTION, OBTAINED BY X-RAY ANALYSIS

By Dr. M. F. PERUTZ, F.R.S., Dr. M. G. ROSSMANN, ANN F. CULLIS, HILARY MUIRHEAD and Dr. GEORG WILL

Medical Research Council Unit for Molecular Biology, Cavendish Laboratory, University of Cambridge

AND

Dr. A. C. T. NORTH

Medical Research Council External Staff, Davy Faraday Research Laboratory, Royal Institution, London, W.1

VERTEBRATE hæmoglobin is a protein of molecular weight 67,000. Four of its 10,000 atoms are iron atoms which are combined with protoporphyrin to form four hæm groups. The remaining atoms are in four polypeptide chains of roughly equal size, which are identical in pairs[1–3]. Their amino-acid sequence is still largely unknown.

We have used horse oxy- or met-hæmoglobin because it crystallizes in a form especially suited for X-ray analysis, and employed the method of isomorphous replacement with heavy atoms to determine the phase angles of the diffracted rays[4–7]. The Fourier synthesis which we have calculated shows that hæmoglobin consists of four sub-units in a tetrahedral array and that each sub-unit closely resembles Kendrew's model of sperm whale myoglobin[8]. The four hæm groups lie in separate pockets on the surface of the molecule.

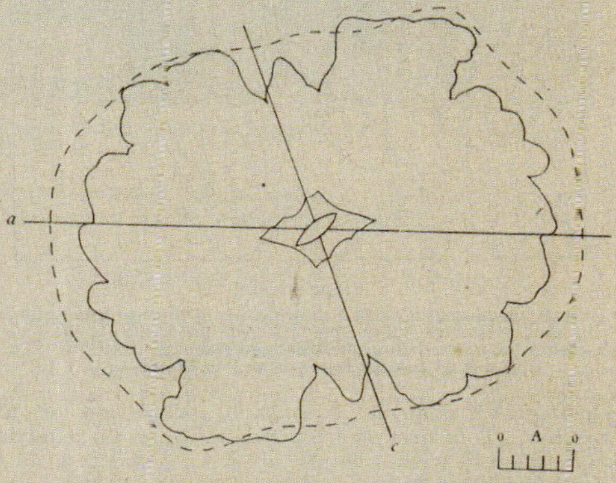

Fig. 2. External shape of the molecule. Full lines indicate the boundary derived from the contour at 0·54 electron/Å.³. The broken line shows the boundary derived by Bragg and Perutz from two-dimensional data (ref. 13). Note the hole in the middle

Method of Analysis

Horse oxyhæmoglobin, crystallized from 1·9 *M* ammonium sulphate solution at *p*H 7, has the space group *C*2 with two molecules in the unit cell which lie on dyad axes[8]. In order to determine the phase angles of the 1,200 reflexions contained in the limiting sphere of 5·5 Å.$^{-1}$, six different isomorphous heavy-atom compounds were used (ref. 9 and unpublished work). Intensities were measured photographically and by counter spectrometer (Arndt, U. W., and Phillips, D. C., unpublished work). The relative positions and shapes of the heavy-atom replacement groups were found by correlation functions based on Patterson methods[10] and refined by least squares procedures. For each reflexion the structure amplitudes of all seven compounds were combined in an Argand diagram[11], and the probability of the phase angle having a value α was calculated for $\alpha = 0, 5, 10, \ldots 355°$. The centroid of the probability distribution, plotted around a circle, was then chosen as the best vector *F* in the Fourier synthesis[12]. The results were finally plotted on 32 contour maps showing the distribution of electron density in sections spaced 2 Å. apart normal to *b* (Fig. 1). The absolute configuration of the molecule was determined from anomalous dispersion[4].

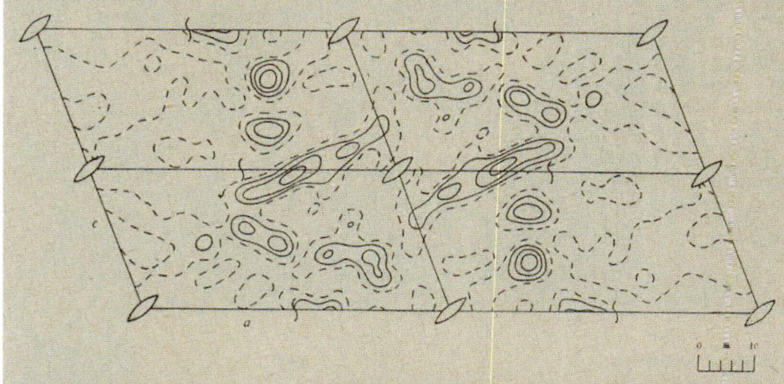

Fig. 1. Section at $y = 1/32b$. This cuts through the middle of the molecule on which the diagram is centred. 'Flat' areas indicating liquid appear on the left and right. Contours are drawn at intervals of 0·14 electron/Å.³. The broken line marks 0·4 electron/Å.³. Contours at lower levels are omitted

External Shape of the Molecule

More than half the volume of the crystals is taken up by liquid of crystallization, which mainly fills the spaces between the molecules and shows up on the contour maps in the form of flat, featureless regions (Fig. 1). The outlines of the two molecules in the unit cell can be traced by following the boundaries between these regions and the continuous electron-dense regions described below. In Fig. 2 the outline of one molecule, traced from the periphery of the 0·54

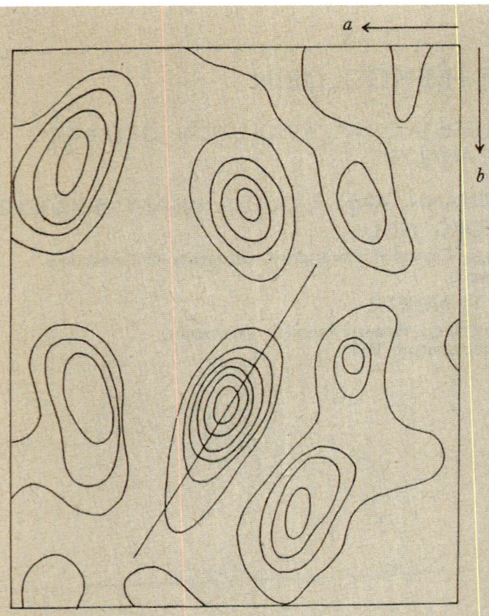

Hæm 1 and 2

Fig. 3. Section at $z = 1/4\ c$ showing one of the two hæm groups in the asymmetric unit. The straight line indicates the hæm orientation derived from electron spin resonance (ref. 14). The lowest contour shown is at 0·4 electron/Å.^{3}

electron/Å.^{3} contour, is seen in projection on the b-plane. To a first approximation it can be regarded as a spheroid with a length of 64 Å., a width of 55 Å. and a height of 50 Å. normal to the plane of the paper. Except for a slight shortening along a, this shape agrees with the earlier picture obtained from two-dimensional data, even including the dimple in the centre of the molecule (see Fig. 2 of ref. 12).

Positions of the Hæm and Sulphydryl Groups

Four peaks stand out from the rest, clearly representing the iron atoms with their surrounding porphyrin rings. A flattening of the peaks indicates the approximate orientation of the rings (Fig. 3). Fortunately, their exact orientation was already known from electron spin resonance[14], and it only remained to assign the correct one of the four alternative orientations of the hæms to each of the electron density peaks. The results are shown in Fig. 4. The iron atoms lie at the corners of an irregular tetrahedron with distances of 33·4 and 36·0 Å. between symmetrically related pairs. The closest approach between symmetrically unrelated iron atoms is 25·2 Å. (Table 1).

Table 1

	Fe_1	Fe_2	S_1
x	−6·6 Å.	12·3 Å.	5 Å.
y	7·3 Å.	−10·7 Å.	10 Å.
z	13·1 Å.	18·2 Å.	16 Å.
$Fe_1 - Fe_2$ = 33·4 Å.			
$Fe_2 - Fe_4$ = 36·0 Å.			
$Fe_1 - Fe_3$ = 25·2 Å.			
$Fe_1 - S_1$ 13 Å.			
$Fe_2 - S_1$ 21 Å.			

Horse hæmoglobin contains four cysteine residues, but only two sulphydryls combine with mercury in the native protein[5]. From the positions of the mercury atoms we inferred that each of the two sulphydryl groups is about 13 Å. away from one iron atom and 21 Å. away from another. (Fig. 4 and Table 1). The significance of this situation is discussed below.

Configuration of the Polypeptide Chains

The most prominent feature of the Fourier synthesis consists of more or less cylindrical clouds of high density, like the vapour trails of an aeroplane; they are curved to form intricate three-dimensional figures. Sections through various parts of these appear in Fig. 1. To build a model of the figures, we rolled out sheets of a thermo-setting plastic to the thickness of our sections on a scale of 2 Å. = 1 cm., and cut out the shape of each region on the contour maps where the density exceeds 0·54 electron/Å.^{3} (this corresponds to the first full contour line in Fig. 1. From this section, for example, 14 shapes would be cut). The shapes were then assembled in accordance with their positions and heights in the different sections, and the hæm groups were attached in the appropriate orientation. The model was then baked to set it permanently. For comparison, a Fourier synthesis of sperm whale myoglobin was calculated at a resolution of 5·5 Å., using the new X-ray data of Kendrew *et al.*[15], and a model of the electron density distribution was constructed by the method just described.

From the hæmoglobin Fourier synthesis there emerged four separate units which are identical in pairs. Fig. 5 shows one member of each pair together with the model of myoglobin on the left. In each unit the cloud of high density describes a complicated figure which, in the white unit (middle), can be traced from end to end by following the superimposed line. Except for two small gaps where the density sinks slightly below 0·54 electron/Å.^{3}, the black unit (right) closely resembles the white one. There are several gaps interrupting the myoglobin model where, probably due to increased thermal motion, the electron density falls below the contour-level chosen for cutting the model. However, we know from the recent work of Kendrew and his collaborators[15] that

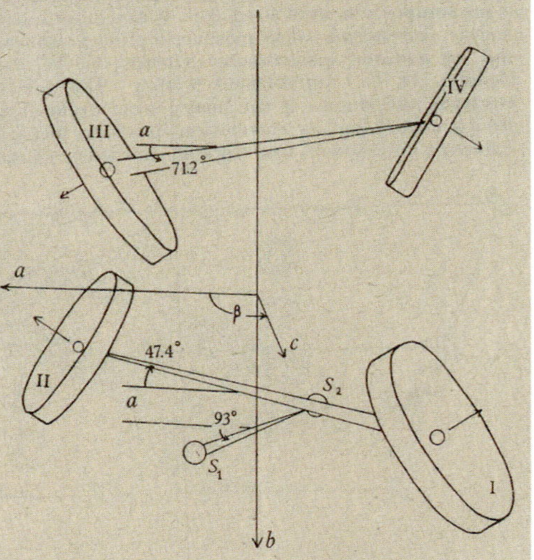

Fig. 4. Arrangement of hæm groups in hæmoglobin. Arrows indicate the reactive side of each group. The c-axis comes out of the paper towards the observer. This picture and all subsequent ones show the absolute configuration

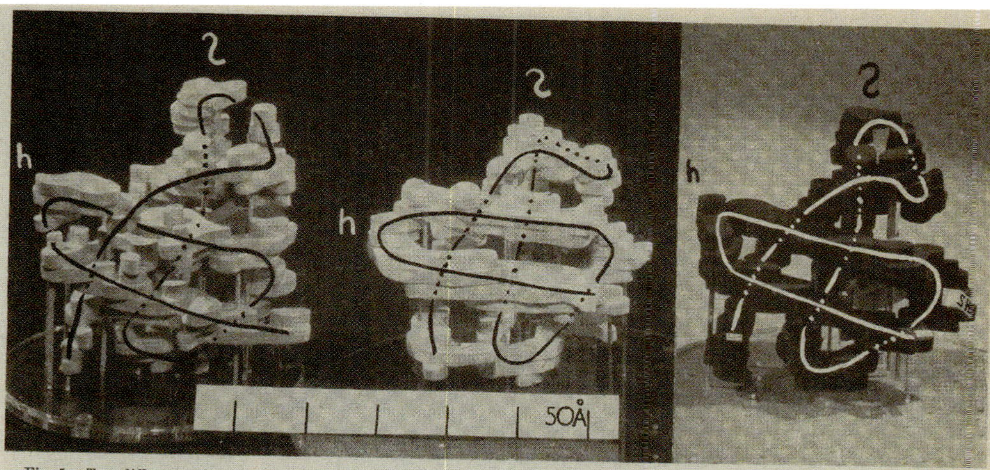

Fig. 5. Two different polypeptide chains in the asymmetric unit of hæmoglobin compared with myoglobin (left). The hæm groups are at the back of the chains

the gaps are bridged by a continuous polypeptide chain, and it is evident from Fig. 5 that apart from the gaps, the model has a configuration closely similar to the hæmoglobin units.

Clearly, the four tortuous clouds of high electron density in hæmoglobin represent the four polypeptide chains. The black and the white chains have similar, but not identical configurations. In the black chain the *S*-shaped bend at the top is more pronounced, the hæm group is lower, and the bend *h* sharper than in the white chain. These, however, are details. The most important result is their resemblance to each other and to sperm whale myoglobin.

Arrangement of the Four Sub-units

The first step in the assembly of the molecule is the matching of each chain by its symmetrically related partner (Fig. 6). It will be noted that there is comparatively little contact between the members of each pair, suggesting rather tenuous linkages. In the next step the white pair is inverted and

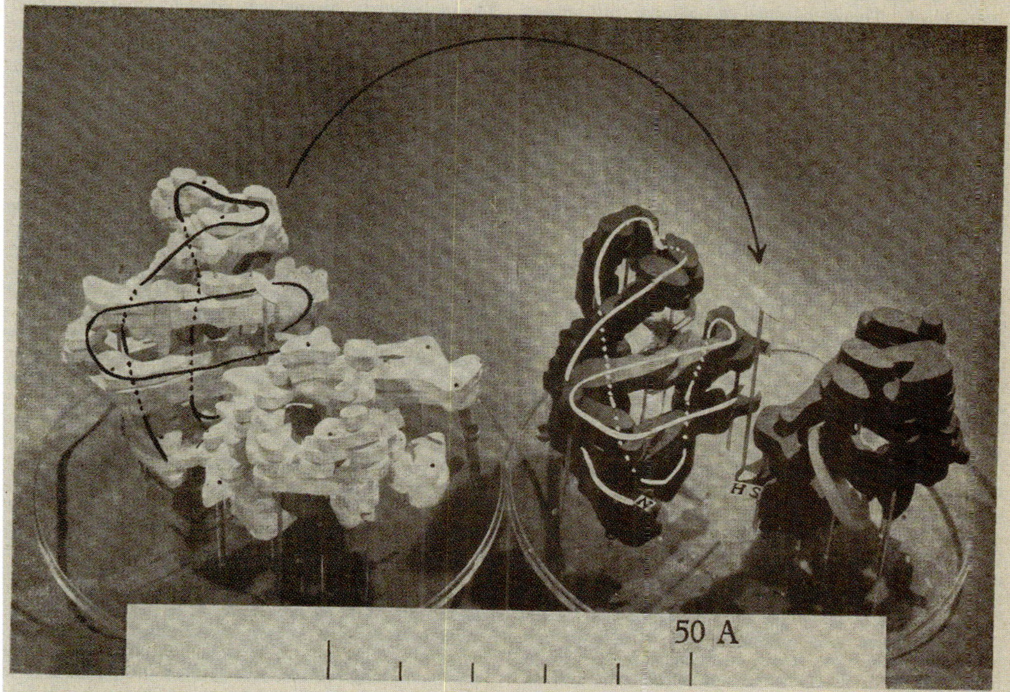

Fig. 6. Two pairs of chains symmetrically related by the dyad axis. The arrow shows how one pair is placed over the other to assemble the complete molecule

3

placed over the black pair as indicated by the arrow. Fig. 7 shows one white chain placed over the pair of black ones and Fig. 8 shows the molecule completely assembled. The resulting arrangement is tetrahedral and has almost, but not quite, the orthorhombic point group symmetry 222. It contains two 'pseudo dyads' which lie approximately at right angles to each other and to the true dyad, one emerging from the centre of Fig. 8 and the other from the centre of Fig. 9. This means that, to a first approximation, each sub-unit can be generated by a rotation of 180° from any of its neighbours. Figs. 7, 8 and 9 also show that the surface contours of the white chains exactly fit those of the black, so that there is a large area of contact between them. This structural complementarity is one of the most striking features of the molecule. Fig. 10 is a view down the true dyad axis and reveals a hole going right through the centre of the molecule, as was to be expected from the Fourier projection[13]. However, the van der Waals radii of the chains are much bigger than appears in the model, and little room may, in fact, be left for water or electrolytes to pass through. Fig. 10 also reveals a dimple at the top where the white chains meet. This is matched by a similar, but larger, hollow at the bottom where the black chains meet.

Fig. 7. Partially assembled molecule showing two black chains and one white

The hæm groups are seen to lie in separate pockets on the surface of the molecule (Fig. 8). Each pocket is formed by the folds in one of the polypeptide chains, which appears to make contact with the hæm group at four different points at least. The iron atoms in the neighbouring pockets formed by the black and the white chains are 25 Å. apart.

Information from the Fourier Synthesis of Myoglobin at 2-Å. Resolution (ref. 15)

Thanks to the similarity with myoglobin, the interpretation can be carried further than would have been possible on the basis of our results alone. Kendrew *et al.* have found that the straight stretches of rod indicated in Fig. 5 are α-helices and that the N-terminal end of the chain is at the bottom left. The ends of the two hæmoglobin chains have been labelled *N* and *C* accordingly, as seen in Figs. 7 and 10. The myoglobin Fourier synthesis also reveals the hæm-linked amino-acid side-chain, probably histidine, on one side of the (ferric) iron atom and a small peak, probably representing a water molecule, on the other side. If this information is transferred to hæmoglobin, the reactive side of the hæm group is as indicated by the arrows in Fig. 4 and by the labels O_2 in Figs. 7 and 8*a*.

Figs. 7 and 8*b* show the reactive sulphydryl group (which is absent in myoglobin) to be attached to the portion of the black chain carrying the hæm-linked histidine. It is seen that the histidine and cysteine side-chains point in roughly opposite directions, one towards and the other away from the hæm group. The sulphydryl group may possibly be in contact with the loop of the white chain which lies below the hæm group on the left. This is the situation of one pair of sulphydryl groups in the molecule. The second pair is probably attached to the white chains, but it is unreactive in the native protein and its position is still unknown.

Reliability of Results

The method of isomorphous replacement makes no assumptions about the structure of the protein, and its results suffer from none of the ambiguities which bedevil the interpretation of vector maps. Ideally, the parent protein, in combination with two different isomorphous heavy-atom compounds, should give accurate phase angles and electron density maps which are free from all except series termination errors. In practice, errors and uncertainties arise from several sources and must be minimized by using more than two heavy-atom compounds. In our case it was possible to estimate the accuracy of the phase angle for each reflexion from the measure of agreement between the angles indicated by the six different compounds. From the standard error in the vector *F*, averaged over all reflexions, the standard error in the final electron density was calculated as 0·12 electron/Å.^{3}, which amounts to 0·85 of the interval between successive contours in Figs. 1 and 3 (0·14 electron/Å.^{3}), or 0·15 of the difference in density between 'peaks' and 'valleys' (0·7 electron/Å.^{3}) (ref. 12). This calculated error is borne out by the observed fluctuations in the liquid regions between the molecules and by the difference in height between the two iron peaks (0·13 electron/Å.^{3}).

An error of at least five times this magnitude would be needed to turn a peak into a valley and thus to simulate the backbone of a polypeptide chain in a region actually occupied by side-chains.

Normally, the accuracy of a structure is checked by comparison of calculated and observed intensities. In hæmoglobin this is impossible so long as the light atoms are not resolved, but the positions of the iron atoms at least can be checked by their anomalous dispersion effect. Depending on the phase relationship between the structure factor of the four iron atoms by themselves and that of the entire molecule, the total intensity of any given reflexion $I\ (hkl)$ may be larger or smaller than $I\ (\bar{h}\bar{k}\bar{l})$. We selected 50 reflexions for which the effect of anomalous dispersion was expected to be largest and measured the intensity of each reflexion in the four symmetrically related quadrants on the counter spectrometer. Statistically significant differences between $I(hkl)+I(\bar{h}k\bar{l})$ and $I(\bar{h}\bar{k}\bar{l})+I(h\bar{k}l)$ were found in 36 reflexions. In 34 of them the signs of the differences agreed with prediction, indicating that the phase angles of the reflexions as well as the positions of the iron atoms are correct.

As a further check on the positions of the iron atoms, an attempt was made to label them with heavy atoms. This was done by allowing crystals of methæmoglobin, which are isomorphous with oxyhæmoglobin, to react with *p*-iodophenyl-hydroxylamine, which attaches itself to the iron atoms. A difference Fourier projection on the *b*-plane showed four prominent peaks which lie well within the calculated distances between the iodine and the iron atoms.

Finally, to avoid any possible bias in the interpretation of the Fourier synthesis, we have tried to construct objective models containing only those features actually found in the electron density maps. Due to series termination errors and inaccuracies in many of the phase angles, these maps must certainly contain errors of detail. Large errors, on the other hand, are unlikely because, quite apart from the checks just described, the results really prove themselves. No combination of errors could have led to the appearance of four distinct chains of roughly equal length, in agreement with chemical evidence, to the similarity between the two pairs of chains which are not related by crystal symmetry, and to the resemblance between hæmoglobin and myoglobin.

(*a*)

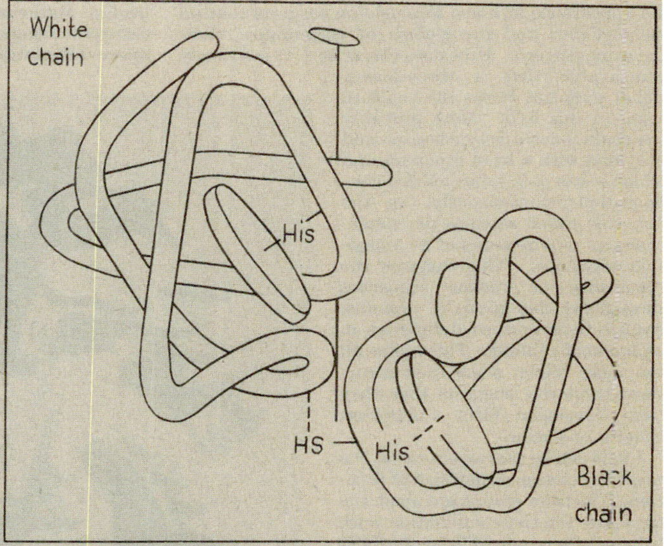

(*b*)

Fig. 8. (*a*) Hæmoglobin model viewed normal to *a*. The hæm groups are indicated by grey disks. (*b*) Chain configuration in the two sub-units facing the observer. The other two chains are produced by the operation of the dyad axis

Discussion

The polypeptide chain-fold which Kendrew and his collaborators first discovered in sperm whale myoglobin has since been found also in seal myoglobin[16].

Fig. 9. Hæmoglobin model viewed normal to *c*

structural changes detectable by X-ray analysis[17], both forms being isomorphous with the metmyoglobin normally studied. In the hæmoglobin of horse and of man, on the other hand, the oxygenated and reduced forms are crystallographically different[18–20]. The structure of reduced hæmoglobin is still unknown, but it would not be surprising if loss of oxygen caused the four sub-units to rearrange themselves relative to each other, rather than to change their individual structure to a marked degree.

Riggs has shown that blocking the sulphydryl groups reduces hæm–hæm interaction[21]. As Fig. 8 shows, these groups occupy key positions close to the hæm-linked histidines and to points of contact between two different sub-units. They may well play an important part in the transition between the oxygenated and reduced forms. Incidentally, the cysteine residue should provide a convenient marker for the peptide containing the hæm-linked histidine, and so help to determine the sequence in this important part of the chain.

A full account of this work will be published elsewhere.

We thank the Director and staff of the University of Cambridge Mathematics Laboratory for making their electronic computor *Edsac* II available to us for the many calculations involved in this work. We are

Its appearance in horse hæmoglobin suggests that all hæmoglobins and myoglobins of vertebrates follow the same pattern. How does this arise ? It is scarcely conceivable that a three-dimensional template forces the chain to take up this fold. More probably the chain, once it is synthesized and provided with a hæm group around which it can coil, takes up this configuration spontaneously, as the only one which satisfies the stereochemical requirements of its amino-acid sequence. This suggests the occurrence of similar sequences throughout this group of proteins, despite their marked differences in amino-acid content. This seems all the more likely, since their structural similarity suggests that they have developed from a common genetic precursor.

Little can be said as yet about the relation between structure and function. The hæm groups are much too far apart for the combination with oxygen of any one of them to affect the oxygen affinity of its neighbours directly. Whatever interaction between the hæm groups exists must be of a subtle and indirect kind that we cannot yet guess. A few observations of possible significance might be mentioned. Kendrew found that the combination of reduced myoglobin with oxygen involves no

Fig. 10. A view down the *b*-axis. Note the proximity of the *C* and *N* terminal ends which could serve to form links between the two white chains

also deeply grateful to the Rockefeller Foundation for its long-continued financial support and to Sir Lawrence Bragg for his unfailing enthusiasm and encouragement. Finally, we wish to thank Dr. B. R. Baker, of the Stanford Research Institute, for a gift of organic mercurials, Dr. J. Chatt of Imperial Chemical Industries, Ltd., and Mr. A. R. Powell, of Johnson Matthey, Ltd., for gifts of heavy-atom compounds, and Mrs. Margaret Allen, Miss Ann Jury and Miss Brenda Davies for assistance.

[1] Rhinesmith, S. H., Schroeder, W. A., and Pauling, L., *J. Amer. Chem. Soc.*, **79**, 4682 (1957).

[2] Braunitzer, G., *Hoppe-Seyl. Z.*, **312**, 72 (1958).

[3] Wilson, S., and Smith, D. B., *Can. J. Biochem. Physiol.*, **37**, 405 (1959).

[4] Bokhoven, C., Schoone, J. C., and Bijvoet, J. B., *Acta Cryst.*, **4**, 275 (1951).

[5] Green, D. W., Ingram, V. M., and Perutz, M. F., *Proc. Roy. Soc.*, A, **225**, 287 (1954).

[6] Kendrew, J. C., Bodo, G., Dintzis, H. M., Parrish, R. G., Wyckoff, H. W., and Phillips, D. C., *Nature*, **181**, 662 (1958).

[7] Blow, D. M., *Proc. Roy. Soc.*, A, **247**, 302 (1958).

[8] Perutz, M. F., *Proc. Roy. Soc.*, A, **225**, 264 (1954).

[9] Cullis, A. F., Dintzis, H. M., and Perutz, M. F., Conference on Hæmoglobin, National Academy of Sciences, NAS-NRC Publication 557, p. 50 (Washington, 1958).

[10] Rossmann, M. G., *Acta Cryst.* (in the press).

[11] Bodo, G., Dintzis, H. M., Kendrew, J. C., and Wyckoff, H. W., *Proc. Roy. Soc.*, A, **253**, 70 (1959).

[12] Blow, D. M., and Crick, F. H. C., *Acta Cryst.*, **12**, 794 (1959).

[13] Bragg, W. L., and Perutz, M. F., *Proc. Roy. Soc.*, A, **225**, 315 (1954).

[14] Ingram, D. J. E., Gibson, J. F., and Perutz, M. F., *Nature*, **178**, 905 (1956).

[15] Kendrew, J. C., Dickerson, R. E., Strandberg, B., Hart, R. G., Davies, D. R., Phillips, D. C., and Shore, V. C. (see following article).

[16] Scouloudi, H., *Nature*, **183**, 374 (1959).

[17] Kendrew, J. C. (private communication).

[18] Haurowitz, F., *Hoppe-Seyl. Z.*, **254**, 266 (1938).

[19] Jope, H. M., and O'Brien, J. R. P., "Hæmoglobin", 269 (Butterworths, London, 1949).

[20] Perutz, M. F., Trotter, I. F., Howells, E. R., and Green, D. W., *Acta Cryst.*, **8**, 241 (1955).

[21] Riggs, A. F., *J. Gen. Physiol.*, **36**, 1 (1952).

Printed in Great Britain by Fisher, Knight & Co., Ltd., St. Albans.

Chapter 4

Roald Hoffmann

Biography

Roald Hoffmann was born in 1937 in Złoczów, Poland. Having survived the war, he came to the US in 1949, and studied chemistry at Columbia and Harvard Universities (Ph.D. 1962). Since 1965 he has been at Cornell University, now as the Frank H. T. Rhodes Professor of Humane Letters, Emeritus. He has received many of the honors of his profession, including the 1981 Nobel Prize in Chemistry (shared with Kenichi Fukui).

"Applied theoretical chemistry" is the way Roald Hoffmann likes to characterize the particular blend of computations stimulated by experiment and the construction of generalized models, of frameworks for understanding, that is his contribution to chemistry. The pedagogical perspective is very strong in his work.

Notable at the same time is his reaching out to the general public; he participated, for example, in the production of a television course in introductory chemistry titled "The World of Chemistry," shown widely since 1990. As a writer, Hoffmann has carved out a land between science, poetry, and philosophy, through many essays and three books, *Chemistry Imagined* with artist Vivian Torrence, *The Same and Not the Same* (translated into six languages), and *Old Wine, New Flasks: Reflections on Science and Jewish Tradition*, with Shira Leibowitz Schmidt.

Hoffmann is also an accomplished poet and playwright. Four collections of his poetry have been published, as well a book of translations of his poems into Spanish. He has also co-written a play with fellow chemist Carl Djerassi, entitled *Oxygen*, and one by himself, *Should've*. Both plays have been widely produced.

Unadvertised, a monthly cabaret Roald runs at the Cornelia Street Café in Greenwich Village, "Entertaining Science," has become the hot cheap ticket in NYC.

How a Simple Molecular Orbital Method Took Shape in the Lipscomb Groupw

The extended Hückel method was the first molecular orbital method that included all valence electrons, and so could be applied to any organic or inorganic molecule. It was developed in the Lipscomb research group at Harvard, with essential contributions by L. L. Lohr, Jr., R. Stevens, W. N. Lipscomb, and myself.

In the spring of 1960, in my second year in graduate school, I was working with Martin Gouterman, trying to do a molecular orbital calculation on C_8H_8, cubane. Since people were attempting to make the molecule, I thought it would be nice to know its electronic structure. The paradigmatic model of the time, the Hückel method, wasn't directly applicable because cubane obviously didn't have a σ–π separation. Undaunted, I set up a 40×40 matrix for all the valence electrons in cubane, and using my newly-learned group theory, I reduced the matrix as much as possible. I was still left with many distinct interactions, which I labeled with a mass of Greek letters — β, γ, δ, ε, etc.

There matters lay while I went off for a year to the Soviet Union. I think I was still deciding whether I really wanted to go on in chemistry. When I came back I was committed to this wonderful molecular science and switched to working with the Colonel, who had just arrived at Harvard, on bonding in boranes. These marvelous molecules are non-planar, so the simple Hückel model wouldn't work. Still within the framework of a one-electron theory, Lipscomb advised setting the off-diagonal matrix elements of the Hamiltonian proportional to the overlap. The idea goes back to the Mulliken group at Chicago, and was used in borane calculations by Longuet-Higgins and Roberts. Earlier work by Eberhardt, Crawford, and Lipscomb on the boron hydrides was an important guide to us.

At just that time, the Lipscomb research group, including Bill Kern, Dick Stevens, Larry Lohr, Russ Pitzer, and myself, was learning computing on the IBM 604. So I programmed the vector decompositions of the overlap integrals, and adapted from others in the group the actual integral evaluation and matrix diagonalization routines. A paper by Moore, Lohr, and Lipscomb shows that pieces of the procedure were in place in 1961. Larry Lohr suggested the use of valence state ionization potentials and the Wolfsberg–Helmholtz formula, and there was the first extended Hückel program. Except we didn't call it that, yet. Lipscomb and I used the method with some success for a theoretical analysis of boron hydrides and carboranes; what you see here is the first paper in the series. Larry Lohr applied the method to XeF_4 and inorganics.

The Colonel's first graduate student at Harvard was off! From this wonderful man I learned to appreciate complexity (that's the boron hydrides), and the joys of interacting with experimentalists. I still remember the day Fred Hawthorne came by and told us of icosahedral $B_{12}H_{12}^{2-}$. It was an unforgettable, formative time at those closely-packed desks on the top floor of Gibbs.

THE JOURNAL OF CHEMICAL PHYSICS VOLUME 36, NUMBER 8 APRIL 15, 1962

Theory of Polyhedral Molecules. I. Physical Factorizations of the Secular Equation

ROALD HOFFMANN AND WILLIAM N. LIPSCOMB
Department of Chemistry, Harvard University, Cambridge 38, Massachusetts

(Received November 21, 1961)

An LCAO–MO systematization of polyhedral molecules such as B_NH_N is undertaken. Peculiarities of polyhedral systems, such as inapplicability of nearest-neighbor assumption and increased number of parameters are discussed within the framework of a Hückel type of theory. It is found that inclusion of hydrogen atoms does not affect predictions of closed shells, but is important in determining electronic transitions. Various physical factorizations of the secular equations, such as the in-surface, apex-equatorial, and ring-polar separations are critically examined. A computer program for calculations on molecules of up to 15 atoms is described and used to obtain the energy levels of a variety of polyhedral molecules.

INTRODUCTION

POLYHEDRAL molecules and ions represent a class of molecular and ionic species for which arguments based upon symmetry considerations may be expected to yield interesting results. Degeneracies in the one-electron approximation to molecular orbital descriptions often give rise to relatively large gaps in the energy level scheme, and hence relatively complex molecules may be described with some greater degree of success than is possible in molecules of lower symmetry. Such polyhedral species are prominent in boron chemistry[1–3] (B_4Cl_4, B_8Cl_8, $B_{10}H_{10}^{-2}$, $B_{12}H_{12}^{-2}$) and in intermetallic complexes and crystals. Saturated analogs are expected in carbon chemistry (tetrahedrane C_4H_4, cubane C_8H_8, dodecahedrane $C_{20}H_{20}$, etc.) and some arise in carbon-boron chemistry (the carboranes). Inorganic molecules may also form compact polyhedral structures (P_4, As_4, etc.). Some of the polyhedra discussed in this paper are shown in Figs. 1–3.

Molecular orbital descriptions have been given to the known B_4Cl_4, B_8Cl_8, $B_{10}H_{10}^{-2}$, $B_{12}H_{12}^{-2}$ species,[2,4–6] and to[7,8] $B_5H_5^{-2}$, $B_6H_6^{-2}$ and other species, in varying degrees of approximation. Our purpose here is to explore the nature of some of these approximate theories. In the first paper we investigate approximations, made for convenience in handling the secular determinants arising from linear combinations of molecular orbitals (LCAO). In the second paper, a crystal-field model is compared with the LCAO calculations, with, surprisingly, a resultant improvement in the assumptions of both types of orbital descriptions. In the third paper we give atomic charges, bond orders, and reactivity parameters for some carboranes. In the fourth contribution we present the results of a systematic survey of energy levels for a much larger variety of polyhedral species than has heretofore been investigated, having been encouraged by recent experimental developments along these lines in boron and carbon chemistry.

LCAO-MO EQUATIONS

If we consider a molecular orbital ψ built up as a linear combination of atomic orbitals ϕ,

$$\psi=\sum_i c_i\phi_i, \tag{1}$$

we obtain the following set of equations for the expan-

[1] W. N. Lipscomb, "Recent studies of the boron hydrides," in *Advances in Inorganic Chemistry and Radiochemistry* (Academic Press Inc., New York, 1959), Vol. I, p. 117.
[2] W. N. Lipscomb, A. R. Pitochelli, and M. F. Hawthorne, J. Am. Chem. Soc. **81**, 5833 (1959).
[3] A. R. Pitochelli and M. F. Hawthorne, J. Am. Chem. Soc. **82**, 3288 (1960); J. A. Wunderlich and W. N. Lipscomb, *ibid.* **82**, 4427 (1960).
[4] W. H. Eberhardt, B. C. Crawford, and W. N. Lipscomb, J. Chem. Phys. **22**, 989 (1954).
[5] R. A. Jacobson and W. N. Lipscomb (unpublished).
[6] H. C. Longuet-Higgins and M. de V. Roberts, Proc. Roy. Soc. (London) **A230**, 110 (1955).
[7] E. B. Moore, Jr., L. L. Lohr, Jr., and W. N. Lipscomb, J. Chem. Phys. **35**, 1329 (1961).
[8] H. C. Longuet-Higgins and M. de V. Roberts, Proc. Roy. Soc. (London) **A224**, 336 (1954).

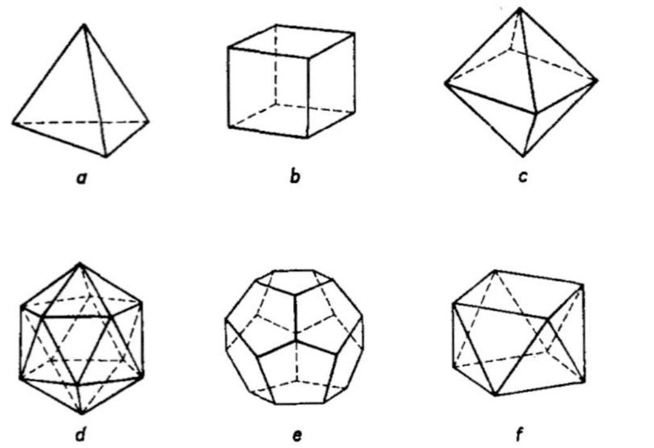

FIG. 1. a. tetrahedron, b. cube, c. octahedron, d. icosahedron, e. dodecahedron, f. archimedean antiprism. *a*, *b*, and *e* are C_NH_N possibilities.

sion coefficients:

$$(\alpha_r - ES_{rr})c_r + \sum_s{}' (\beta_{rs} - ES_{rs})c_s = 0$$

$$r = 1, 2, 3, \cdots, M, \quad (2)$$

where M = number of orbitals considered

E = energy

$S_{rs} = \int \phi_r^* \phi_s d\tau$ = overlap

$\alpha_r = \int \phi_r^* H \phi_r d\tau$ = "Coulomb integral"

$\beta_{rs} = \int \phi_r^* H \phi_s d\tau$ = "resonance integral."

H is the ubiquitous Hückel Hamiltonian, whose explicit form is hardly ever specified in the semiempirical theory. In the following, we will concern ourselves with polyhedral molecules of the formula B_NX_N, where B may be boron (or carbon) and X may be hydrogen or chlorine.

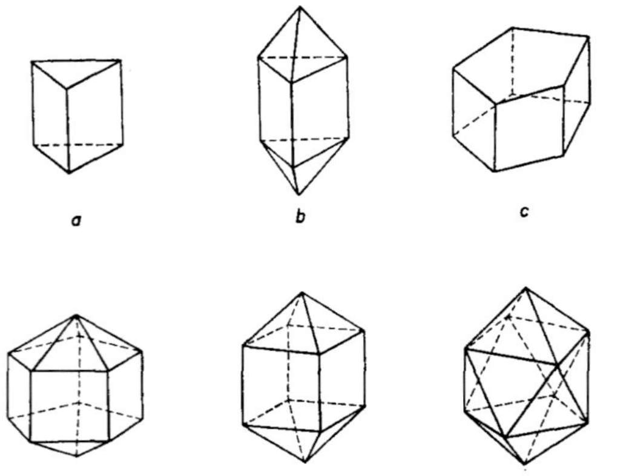

FIG. 2. a. trigonal prism, b. bicapped trigonal prism, c. pentagonal prism, d. bicapped pentagonal prism, e. bicapped cube, f. bicapped archimedean antiprism. *a*, *c* are C_NH_N possibilities.

The simplest Hückel type of treatment of these molecules differs considerably from the corresponding calculations on aromatic systems. The polyhedral frameworks are inherently three dimensional. No symmetry factorization of the σ–π type is possible, though various other physical factorizations will be discussed below. Thus the number of orbitals considered is greater for polyhedral molecules. In general, we are forced to examine the interaction of 4 orbitals on each boron (one $2s$ and three $2p$) with one orbital on each hydrogen. The secular equation is thus of degree $5N$. If we factor out the B—H bonds, we are still left with a diagonalization of a $3N$ by $3N$ matrix.

Moreover, as we will show in the next section, the equally ubiquitous tight binding assumption of aromatic theory, i.e., the neglect of all but nearest-neighbor interactions in the Hamiltonian matrix, is inapplicable to boron polyhedra. All interactions must be evaluated.

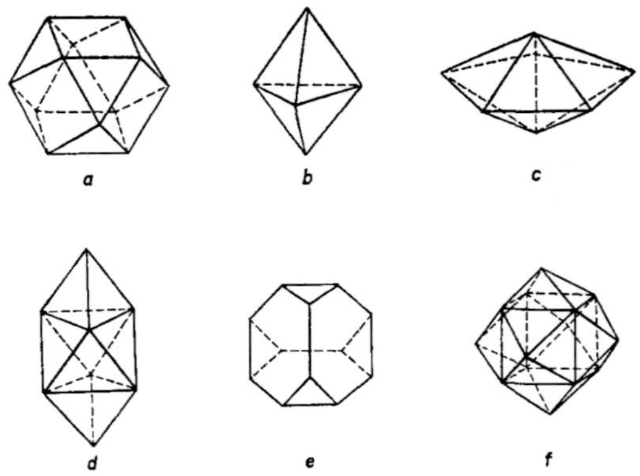

FIG. 3. a. cube-octahedron, b. trigonal bipyramid, c. pentagonal bipyramid, d. rhombic hexahedron = bicapped trigonal antiprism, e. truncated tetrahedron, f. rhombic dodecahedron = omnicapped cube. *e* is a C_NH_N possibility.

Still another difference is that, whereas there is only one Coulomb integral in the π-electron theory of homonuclear aromatics, for polyhedral molecules we are obliged to consider in the homonuclear case a $2s$ and a $2p$ Coulomb integral. In most previous treatments these have been set equal, an assumption which is not justifiable, but which reduces the number of parameters and therefore is habit forming. Arguments for setting $\alpha(2s)$ and $\alpha(2p)$ unequal will be presented in subsequent communications.[9] Since our main purpose in this paper is to examine critically various physical factorizations of the secular equation, we shall retain here the assumption of equal Coulomb integrals, in order to facilitate intercomparison of our results with those already in the literature. We shall also use a

[9] R. Hoffmann and M. P. Gouterman **36**, 2189 (1962) following paper; R. Hoffmann and W. N. Lipscomb (to be published).

Slater orbital basis (exponent 1.30 for B), though in a subsequent section we examine the consequences of using a different orbital basis. If the resonance integral is taken to be proportional to the overlap (perhaps a better assumption would be to take a different proportionality constant for each overlap type)

$$\beta_{rs}=KS_{rs} \tag{3}$$

the set of equations reduces to

$$-xc_r+\sum_s{}' S_{rs}c_s=0 \tag{4}$$

$$-x=(\alpha-E)/(K-E), \tag{5}$$

where we have chosen $S_{rr}=1$, $\alpha(2s)=\alpha(2p)$. The energy parameters are the eigenvalues of the off-diagonal overlap matrix $\mathbf{S}-\mathbf{1}$, and must all be greater than -1. From (5) we obtain

$$E_i=(\alpha+Kx_i)/(1+x_i), \tag{6}$$

and

$$E_i-E_j=(x_i-x_j)(K-\alpha)/(1+x_i)(1+x_j). \tag{7}$$

Thus the order of energy levels is preserved for all K and α (K must be taken greater than α) and we may work directly with the energy parameters x_i thus postponing a choice of K and α. The method outlined above was first introduced for boron polyhedra by Longuet-Higgins and Roberts.[6,8]

NEAREST-NEIGHBOR ASSUMPTION FOR BORON POLYHEDRA

Neglect of all but nearest-neighbor interactions in the Hamiltonian matrix is an *idée fixe* of the LCAO–MO Hückel method for aromatics. It is perhaps not widely recognized that its feasibility is merely due to the relatively small magnitude of the $p-p$ π overlap for two carbons a reasonable distance apart. As has been pointed out by Ruedenberg,[10] the tight binding approximation would fail, if this overlap, which we will call S, would exceed $\frac{1}{3}$. Consider a basis set of normalized atomic orbitals

$$\boldsymbol{\phi}=\{\phi_1,\phi_2,\cdots,\phi_N\}. \tag{8}$$

The overlap matrix $\mathbf{S}$ may be expressed as the Hilbert space inner product $\boldsymbol{\phi}^+\boldsymbol{\phi}$. From this it follows that S is positive definite and therefore its eigenvalues are all greater than zero. Now we inquire about the eigenvalues of $\mathbf{S}'$, a modification of $\mathbf{S}$ in which some of the off-diagonal elements of S have been set equal to zero. Using Hadamard's theorem,[11] we find that the eigenvalues of $\mathbf{S}'$ are bounded as follows:

$$-(M-1)\leq\lambda_i\leq M+1, \tag{9}$$

where

$$M=\max\{\sum_l{}' \mid Sk_l{}' \mid\}. \tag{10}$$

[10] K. Ruedenberg, J. Chem. Phys. **34**, 1884 (1961).

[11] See M. Parodi, *La Localisation des Valeurs Characteristiques des Matrices* (Gauthier-Villars, Paris, 1959), Chap. 2.

Thus if $M\geq1$ some of the eigenvalues of S' *may* be negative or zero; this in turn would imply linear dependence of the basis set and would lead to an incorrect energy level scheme. To ensure consistency in the theory we must have $M<1$. The maximum number of nearest neighbors in aromatics is 3, which leads to Ruedenberg's criterion that S must be less than $\frac{1}{3}$. Since $S\simeq0.25$, this is satisfied. When we turn to boron polyhedra, the maximum number of nearest neighbors is 4 or 5, moreover there is more than one orbital on each center. Even if we were to consider only one nonvanishing interaction with each center, we would still have to require that S be $<\frac{1}{4}$, $\frac{1}{5}$, respectively, whereas S actually is over 0.30. Indeed, calculations with a nearest neighbor assumption for the octahedron and cube lead to negative eigenvalues of the overlap matrix and confirm the theoretical conclusion that *for B polyhedra the assumption of only nearest-neighbor interactions leads to inconsistencies: all interactions in the polyhedron must be accounted for.*

COMPUTATION

A program for LCAO–MO calculations on polyhedral molecules has been written for the IBM 7090 computer. The overlap matrix for a single s and three p orbitals on each of up to 15 arbitrarily situated atoms is set up. Since the p orbitals transform as vectors, the coefficients (each to be multiplied by the proper overlap) of the various interactions between orbitals on the atoms i and j may be written down as:

$$C(s_i,s_j)=1,$$

$$C(s_i,p_j;\sigma)=-\mathbf{R}_{ij}\cdot\mathbf{P}_j,$$

$$C(p_i,p_j;\sigma)=-(\mathbf{R}_{ij}\cdot\mathbf{P}_i)(\mathbf{R}_{ij}\cdot\mathbf{P}_j),$$

$$C(p_i,p_j;\pi)=[\mathbf{P}_i-(\mathbf{R}_{ij}\cdot\mathbf{P}_i)\mathbf{R}_{ij}]\cdot[\mathbf{P}_j-(\mathbf{R}_{ij}\cdot\mathbf{P}_j)\mathbf{R}_{ij}]$$

$$=\mathbf{P}_i\cdot\mathbf{P}_j-(\mathbf{R}_{ij}\cdot\mathbf{P}_i)(\mathbf{R}_{ij}\cdot\mathbf{P}_j),$$

$$\mathbf{P}_i=\hat{\mathbf{i}},\hat{\mathbf{j}},\hat{\mathbf{k}} \text{ Cartesian unit vectors,} \tag{11}$$

where $\mathbf{R}_{ij}$ is the unit vector along the interatomic distance. All interactions are considered, and the full eigenvalue problem $(\mathbf{H}-E\mathbf{S})\mathbf{C}=0$ is solved, subject to the assumption that $\beta_{rs}=KS_{rs}$ for atomic orbitals. The input parameters are:

(1) Coordinates of the atoms.

(2) S and p Coulomb integrals. Provision is made for putting a different s and p Coulomb integral at each center.

(3) The parameter K.

(4) An overlap scheme. This may be a subroutine which uses Slater $2s$ and $2p$ overlaps, as was done throughout the calculations in this paper; or SCF overlaps may be read in.

The eigenvalue problem is set up in an AO basis and the energies and molecular orbitals obtained by a

TABLE I. Representations of molecular orbitals for some polyhedra.

N	Symmetry	Geometry		Representations
4	T_d	tetrahedron	$4N$	$2a_1+e+3f_2+f_1$
			out	a_1+f_2
6	O_h	octahedron	$4N$	$2a_{1g}+2e_g+3f_{1u}+f_{1g}+f_{2g}+f_{2u}$
			out	$a_{1g}+e_g+f_{1u}$
8	O_h	cube	$4N$	$2a_{1g}+2a_{2u}+e_u+e_g+f_{1g}+3f_{2g}+f_{2u}+3f_{1u}$
			out	$a_{1g}+f_{1u}+f_{2g}+a_{2u}$
12	O_h	cube-octahedron	$4N$	$2a_{1g}+a_{2g}+a_{2u}+3e_g+e_u+2f_{1g}+3f_{2g}+3f_{2u}+4f_{1u}$
			out	$a_{1g}+f_{1u}+e_g+f_{2g}+f_{2u}$
12	I_h	icosahedron	$4N$	$2a_g+3f_{1u}+f_{1g}+2f_{2u}+g_g+g_u+3h_g+h_u$
			out	$a_g+f_{1u}+h_g+f_{2u}$
10	D_{4d}	bicapped archimedean antiprism	$4N$	$5a_1+a_2+b_1+5b_2+5e_1+4e_2+5e_3$
			out	$2a_1+2b_2+e_1+e_2+e_3$

method suggested by del Re,[12] which requires only a single matrix diagonalization. When $\beta_{rs}\neq KS_{rs}$, the problem is solved by a modification of Löwdin's method,[13] requiring two matrix diagonalizations. The main time-consuming step in the calculation is the matrix diagonalization, which is carried out by Jacobi's method of two-dimensional rotations. There is provision in the program for a $g-u$ factorization when a center of symmetry is present, but otherwise no simplifications associated with molecular symmetry are introduced. It was found to be more time consuming for a computer to learn group theory than do the brute force matrix diagonalization. Approximate computation times for the solution of the $4N$ problem without a center of symmetry are 10 sec for 4 atoms, 50 sec for 8 atoms, 150 sec for 12 atoms.

To study the various physical factorizations, several subroutines are used. First sp^3 hybrids are formed at each center, with one directed axially *out* of the polyhedron center. To examine the *in-surface* separation, the $3N$ remaining orbitals are rehybridized to give an axial set of N, primarily s, and pointing *in*, and a set of $2N$ p orbitals perpendicular to the axial direction, i.e., a *surface* set. The final matrix has the form shown below.

out	I	
.	*in*	II
I	II	*surface*

To test the *out-in-surface* separation we first solve the $4N$ problem, then neglect the matrix elements in region I and solve the $3N$ secular equation. Finally we drop the matrix elements in region II and solve the separate *in* (N) and *surface* ($2N$) problems.

To examine the *apex-equatorial* factorization, we rotate the $3(N-2)$ orbitals of the equatorial atom set so that one orbital at each equatorial center is directed toward its apex atom. The secular equations for the equatorial set of $2(N-2)$ orbitals and for the single apical set of $\frac{1}{2}(N-2)+3$ orbitals are solved separately.

To examine the *polar-ring* factorization, sp^2 hybrids are formed for the ring atoms, with one pointing out. For the pyramidal molecules these outpointing orbitals are chosen arbitrarily in the plane of the polyhedron base (whereas otherwise they were chosen pointing out from the polyhedron center). The secular equation for the polar set of p_z ring, and apex orbitals is solved separately from that for the ring set.

Finally, provision is also made for doing the $5N$ problem, i.e., including a set of N hydrogen atoms.

B—H BONDS AND OUT-POINTING ORBITALS

The most complete version of the Hückel-type LCAO–MO calculation of the energy levels of a molecule B_NH_N yields an eigenvalue problem of degree $5N$. There are three Coulomb integrals: $\alpha(2s\mathrm{B})$, $\alpha(2p\mathrm{B})$, $\alpha(1s\mathrm{H})$, and 6 types of resonance integrals: $\beta(1s\mathrm{H},2sB)$, $\beta(1s\mathrm{H}, 2p\mathrm{B}, \sigma)$, $\beta(2s\mathrm{B}, 2s\mathrm{B})$, $\beta(2s\mathrm{B}, 2p\mathrm{B}, \sigma)$, $\beta(2p\mathrm{B}, 2p\mathrm{B}, \sigma)\beta(2p\mathrm{B}, 2p\mathrm{B}, \pi)$ to consider. In the polyhedra of higher symmetry considerable simplification of the eigenvalue problem may be obtained by group-theoretical methods. Thus for octahedral B_6H_6, the MO's subduce the following representations[14] of the molecu-

[12] G. del Re, Quantum Chemistry Group, Uppsala, Tech. Note, October 20, 1958.

[13] P-O. Löwdin, J. Chem. Phys. **18**, 365 (1950).

[14] The group-theoretical notation used is that of E. B. Wilson, Jr., J. C. Decius, and P. C. Cross, *Molecular Vibrations* (McGraw-Hill Book Company, Inc., New York, 1955).

Table II. Energy parameters.

5N		4N		3N	
(a) tetrahedron (T_d), B—B 1.70 A, B—H 1.21 A					
f_2	−0.854	f_2	−0.853	f_2	−0.849
f_2	−0.792	f_1	−0.556	f_1	−0.556
a_1	−0.768	f_2	−0.093	e	0.033
f_1	−0.556	e	0.033	f_2	0.685
e	0.033	a_1	0.038	a_1	2.094
f_2	0.402	f_2	0.748		
a_1	0.753	a_1	2.159		
f_2	1.022				
a_1	2.282				
(b) Octahedron (O_h), B—B 1.73 A, B—H 1.20 A					
e_g	−0.888	e_g	−0.887[a]	e_g	−0.884
f_{1u}	−0.844	f_{1u}	−0.831	f_{1u}	−0.829
e_g	−0.803	f_{1g}	−0.671	f_{1g}	−0.671
f_{1u}	−0.776	f_{2u}	−0.416	f_{2u}	−0.416
a_{1g}	−0.753	f_{1u}	−0.144	f_{2g}	0.493
f_{1g}	−0.671	e_g	−0.089	f_{1u}	1.023
f_{2u}	−0.416	a_1	0.204	a_{1g}	2.969
f_{1u}	0.337	f_{2g}	0.493		
$f_{\cdot g}$	0.493	f_{1u}	1.130		
e_g	0.652	a_{1g}	3.066		
a_{1g}	0.951				
f_{1u}	1.433				
a_{1g}	3.210				
(c) Cube (O_h), B—B 1.70 A, B—H 1.20 A					
a_{2u}	−0.938	a_{2u}	−0.938	a_{2u}	−0.937
f_{2g}	−0.883	f_{2g}	−0.879	f_{2g}	−0.876
a_{2u}	−0.816	f_{1u}	−0.729	e_u	−0.729
f_{1u}	−0.806	e_u	−0.729	f_{1u}	−0.719
f_{2g}	−0.797	f_{1g}	−0.637	f_{1g}	−0.637
a_{1g}	−0.741	f_{2g}	−0.331	f_{2g}	0.064
e_u	−0.729	a_{2u}	−0.243	f_{2u}	0.096
f_{1u}	−0.684	f_{1u}	−0.028	e_g	0.893
f_{1g}	−0.637	f_{2u}	0.096	f_{1u}	1.233
f_{2g}	−0.128	f_{2g}	0.288	a_{1g}	3.127
f_{2u}	0.096	a_{1g}	0.341		
a_{2u}	0.437	e_g	0.893		
f_{1u}	0.448	f_{1u}	1.309		
f_{2g}	0.827	a_{1g}	3.243		
e_g	0.893				
a_{1g}	1.135				
f_{1u}	1.637				
a_{1g}	3.376				
(d) Cube-octahedron (O_h), B—B 1.75 A, B—H 1.20 A					
e_g	−0.899	e_g	−0.892	f_{2u}	−0.885
f_{2u}	−0.891	f_{2u}	−0.887	e_g	−0.883
f_{1u}	−0.835	f_{1u}	−0.788	f_{2g}	−0.769
f_{2g}	−0.822	f_{2g}	−0.774	f_{1g}	−0.756
f_{2u}	−0.807	f_{1g}	−0.756	a_{2g}	−0.739
e_g	−0.794	a_{2g}	−0.739	f_{1u}	−0.739
f_{1g}	−0.756	e_u	−0.669	e_u	−0.669
a_{2g}	−0.739	f_{2u}	−0.540	f_{2u}	−0.483
f_{2g}	−0.733	e_g	−0.269	f_{1g}	−0.232
a_{1g}	−0.724	f_{1g}	−0.232	f_{1u}	0.361
f_{1u}	−0.677	f_{2g}	−0.177	f_{2g}	0.872
e_u	−0.669	f_{2u}	−0.159	a_{2u}	0.881
f_{2u}	−0.513	f_{1u}	0.154	e_g	0.991
f_{1g}	−0.232	f_{1u}	0.385	f_{1u}	1.754
e_g	0.249	a_{1g}	0.549	a_{1g}	3.616
f_{2g}	0.260	a_{2u}	0.811		
f_{1u}	0.331	f_{2g}	1.003		
f_{2u}	0.501	e_g	1.066		
f_{1u}	0.704	f_{1u}	1.805		
a_{2u}	0.881	a_{1g}	3.732		
e_g	1.240				
f_{2g}	1.326				
a_{1g}	1.404				
f_{1u}	2.156				
a_{1g}	3.889				
(e) Icosahedron, I_h, B—B 1.73 A, B—H 1.20 A					
f_{2u}	−0.887	f_{2u}	−0.886[b]	f_{2u}	−0.886
h_g	−0.873	h_g	−0.862	h_g	−0.856
f_{1u}	−0.860	f_{1u}	−0.827	f_{1g}	−0.782
f_{2u}	−0.812	f_{1g}	−0.782	f_{1u}	−0.773
h_g	−0.783	h_u	−0.678	h_u	−0.678
f_{1g}	−0.782	g_g	−0.471	g_g	−0.471
a_g	−0.718	h_g	−0.239	g_u	0.518
f_{1u}	−0.690	f_{2u}	−0.225	h_g	0.984
h_u	−0.678	f_{1u}	0.150	f_{1u}	1.907
g_g	−0.471	g_u	0.518	a_g	4.163
h_g	0.240	a_g	0.619		
f_{2u}	0.460	h_g	1.107		
g_u	0.518	f_{1u}	1.990		
f_{1u}	0.619	a_g	4.293		
h_g	1.358				
a_g	1.485				
f_{1u}	2.370				
a_g	4.469				
(f) D_{4d} bicapped antiprism, B—B 1.74 A, B—H 1.20 A					
e_2	−0.895	e_2	−0.894	e_2	−0.884
a_1	−0.875	e_3	−0.862	e_3	−0.862
e_3	−0.865	a_1	−0.859	b_2	−0.850
b_2	−0.862	b_2	−0.853	a_1	−0.812
b_2	−0.852	b_2	−0.834	b_2	−0.812
e_1	−0.851	e_1	−0.812	e_1	−0.776
e_3	−0.807	e_3	−0.754	e_3	−0.747
e_2	−0.804	a_2	−0.739	a_2	−0.739
b_2	−0.795	e_1	−0.637	e_1	−0.615
a_1	−0.786	e_2	−0.601	e_2	−0.598
e_3	−0.747	b_1	−0.590	b_1	−0.590
a_2	−0.739	e_3	−0.430	e_3	−0.327
b_2	−0.727	b_2	−0.221	e_1	0.328
e_1	−0.711	e_2	−0.198	e_2	0.756
a_1	−0.699	e_3	−0.148	e_3	0.760
e_1	−0.628	a_1	−0.120	a_1	1.004
e_2	−0.600	e_1	−0.082	e_1	1.447
b_1	−0.590	b_2	0.120	b_2	1.978
e_3	−0.363	e_1	0.379	a_1	3.877
e_1	0.197	a_1	0.489		
e_3	0.243	e_2	0.807		
e_2	0.364	e_3	0.901		
b_2	0.399	a_1	1.046		
a_1	0.446	e_1	1.584		
e_1	0.509	b_2	2.059		
b_2	0.734	a_1	3.997		
e_2	0.965				
a_1	1.095				
e_3	1.180				
a_1	1.393				
e_1	1.968				
b_2	2.375				
a_1	4.172				

[a] The energy parameters given here differ slightly from those of Longuet-Higgins and Roberts (reference 8 in text) who performed the calculation for B-B 1.72 A, Slater exponent 1.242. The latter calculations were checked with our program and an error was found in the $t_{1u}(f_{1u})$ levels, of which the authors are aware. The correct f_{1u} energy parameters are: −0.864, −0.175, 1.218.

[b] The energy parameters given here differ from those of Longuet-Higgins and Roberts (reference 6 in text) who performed the calculations for B-B 1.77 A, Slater exponent 1.271. The latter calculations were checked and found in rough agreement with the results of our program. The source of the small discrepancies here is not known.

TABLE III. Separated *in* and *surface* orbitals for some polyhedra.

		tetrahedron	
surface		in	
f_1	−0.556	f_2	−0.698
e	0.033	a_1	2.094
f_2	0.534		
		octahedron	
surface		in	
f_{1g}	−0.671	e_g	−0.884
f_{2u}	−0.416	f_{1u}	−0.400
f_{2g}	0.493	a_{1g}	2.969
f_{1u}	0.594		
		cube	
surface		in	
e_u	−0.729	a_{2u}	−0.937
f_{1g}	−0.637	f_{2g}	−0.749
f_{2g}	−0.063	f_{1u}	0.019
f_{2u}	0.096	a_{1g}	3.127
f_{1u}	0.494		
e_g	0.893		
		cube-octahedron	
surface		in	
f_{1g}	−0.756	f_{2u}	−0.848
a_{2g}	−0.739	e_g	−0.730
e_u	−0.669	f_{2g}	−0.514
f_{2u}	−0.520	f_{1u}	0.644
f_{1g}	−0.232	a_{1g}	3.616
f_{1u}	0.179		
f_{1u}	0.554		
f_{2g}	0.617		
e_g	0.838		
a_{2u}	0.881		
		icosahedron	
surface		in	
f_{1g}	−0.782	f_{2u}	−0.886
h_u	−0.678	h_g	−0.666
g_g	−0.471	f_{1u}	0.609
g_u	0.518	a_g	4.163
f_{1u}	0.526		
h_g	0.794		

lar point group O_h:

$$3A_{1g}+3E_g+4F_{1u}+F_{1g}+F_{2g}+F_{2u}.$$

After group-theoretical reduction, we are left with no worse than a quartic equation to solve. Still the number of parameters is too great, and we shall seek a simpler approach.

The B—H bonds in boron hydrides and hydride ions have the characteristics of normal covalent single bonds. The bond distance is fairly constant and does not seem to depend on the bonding of the B involved, nor on the degree of electron deficiency in the molecule. We may consider the B—H bonds formed by an sp^3 orbital from the B and the hydrogen orbital consuming an electron pair, one of which is donated by the B and the other by the hydrogen atom. Once we make this step, two approaches are possible.

In the first method,[6,8] which we will call "4*N*," the H orbitals are dropped, and the eigenvalue problem of degree 4*N*, involving only B orbitals, is solved. Upon examination of the resulting energy levels, one finds a set of orbitals near the nonbonding level which (1) have the correct symmetry for an outpointing set, and (2) concentrate electronic charge in the direction radially out from the polyhedron center at each atom. These "out" orbitals are then identified as those which form the B—H bonds. They interact with the set of H orbitals of the same symmetry to produce a bonding and an antibonding set. In filling the available electron slots, each boron and hydrogen is asked to contribute one electron to the bonding set, which is thus filled by 2*N* electrons. The remaining framework levels are populated in the usual fashion.

In the second method,[4] which we will call "3*N*," one orbital at each boron is prepared for bonding, i.e., we form an sp^3 hybrid directed out. We then assume that the interactions of these outpointing orbitals with the other framework orbitals are negligible, and then find by use of the methods given above that the secular equation becomes one of degree 3*N*.

Now once we have removed the H orbitals, we have no *a priori* rationale for giving preference to either the 3*N* or the 4*N* scheme. The 4*N* approach has the advantage that one can work with an atomic orbital basis, whereas for 3*N* we must treat hybrids. On the other hand, the size of the matrix to be diagonalized is smaller for 3*N*.

If the factorizations we have described are to have any significance, we must have a reasonable degree of coincidence between the various calculated energy level schemes. To examine this point, we have carried out the 5*N*, 4*N*, and 3*N* calculations for six molecules, with $\alpha(2sB)=\alpha(2pB)=\alpha(1sH)$. Calculations for unequal alphas give different energies, but exhibit the same kind of behavior regarding the factorizations. The molecules are the tetrahedron (T_d), octahedron (O_h), cube (O_h), bicapped archimedean antiprism (D_{4d}), cube-octahedron (O_h), icosahedron (I_h). The symmetries of the 4*N* MO's and the outpointing set are given in Table I. In Tables II(a)–II(f) we give the calculated 5*N*, 4*N*, and 3*N* energy levels.

It is seen that the 4*N* and 3*N* calculations are quite good approximations. In all methods, filled electronic shells are predicted for tetrahedral B_4H_4,[4] C_4H_4,[15] P_4, octahedral $B_6H_6^{-2}$,[8] cubical $B_8H_8^{-2}$, C_8H_8, D_{4d} $B_{10}H_{10}^{-2}$,[2] cubeoctahedral $B_{12}H_{12}^{-2}$,[16] icosahedral $B_{12}H_{12}^{-2}$.[6] Moreover, in each approximation the order of the significant bonding levels is preserved, while predicted electronic excitations vary. Indeed, the situation with respect to electronic transitions is analogous to that in certain heteronuclear aromatics, where there may be $n\rightarrow\pi^*$ and $\pi\rightarrow\pi^*$ transitions. The B—H bond orbitals (whose energy is clearly defined only by the 5*N* calculation) are the analogs of the nonbonded orbitals, while the delocalized framework MO's are analogous to the

[15] W. N. Lipscomb, Tetrahedron Letters **18**, 20 (1959).

[16] W. N. Lipscomb and D. Britton, J. Chem. Phys. **33**, 275 (1960).

TABLE IV. Energy parameters.

	4N		3N		reconstructed 3N[a]		equatorial
(a) D_{3d} rhombic hexahedron, $N=8$, B—B 1.74 A							
e_g	−0.887	e_g	−0.887	a_{2u}	−0.801	a_{2u}	−0.801
a_{1g}	−0.852	a_{2u}	−0.831	e_g	−0.795	e_g	−0.795
a_{2u}	−0.837	e_u	−0.829	a_{1g}	−0.748	a_{2g}	−0.664
e_u	−0.836	a_{2u}	−0.762	a_{2u}	−0.748	e_u	−0.618
a_{2u}	−0.819	a_{1g}	−0.736	e_g	−0.679	a_{1u}	−0.409
e_g	−0.708	e_g	−0.692	e_u	−0.679	e_g	0.287
a_{2g}	−0.664	a_{2g}	−0.664	a_{2g}	−0.664	e_u	1.004
e_u	−0.636	e_u	−0.590	e_u	−0.618	a_{1g}	2.118
e_g	−0.443	a_{1u}	−0.409	a_{1u}	−0.409		
a_{1u}	−0.409	e_g	−0.173	e_g	0.105	apex (C_{3v})	
e_u	−0.237	e_u	0.114	e_u	0.105	a_1	−0.748
a_{2u}	−0.200	e_g	0.698	e_g	0.287	e	−0.679
a_{1g}	−0.063	a_{1g}	0.830	e_u	1.004	e	0.105
a_{2u}	0.068	e_u	1.090	a_{1g}	1.896	a_1	1.896
e_g	0.153	a_{2u}	1.752	a_{2u}	1.896		
e_u	0.217	a_{1g}	3.362	a_{1g}	2.118		
a_{1g}	0.296						
e_g	0.740						
a_{1g}	0.928						
e_u	1.238						
a_{2u}	1.901						
a_{1g}	3.451						
(b) D_{4d} bicapped antiprism B—B 1.74 A							
e_2	−0.894	e_2	−0.894	a_1	−0.829	e_3	−0.828
e_3	−0.862	e_3	−0.862	b_2	−0.829	b_2	−0.812
a_1	−0.859	b_2	−0.850	e_3	−0.828	e_2	−0.740
b_2	−0.853	a_1	−0.812	b_2	−0.812	a_2	−0.739
b_2	−0.834	b_2	−0.812	e_2	−0.740	e_1	−0.608
e_1	−0.812	e_1	−0.776	a_2	−0.739	b_1	−0.590
e_3	−0.754	e_3	−0.747	e_2	−0.676	e_3	−0.034
a_2	−0.739	a_2	−0.739	e_1	−0.666	e_2	0.680
e_1	−0.637	e_1	−0.615	e_3	−0.666	e_1	1.337
e_2	−0.601	e_2	−0.598	e_1	−0.608	a_1	2.528
b_1	−0.590	b_1	−0.590	b_1	−0.590		
e_3	−0.430	e_3	−0.327	e_3	−0.034		
b_2	−0.221	e_1	0.328	e_3	0.337		
e_2	−0.198	e_2	0.756	e_1	0.337	apex (C_{4v})	
e_3	−0.148	e_3	0.760	e_2	0.680	a_1	−0.829
a_1	−0.120	a_1	1.004	e_1	1.337	b_2	−0.676
e_1	−0.082	e_1	1.447	a_1	2.163	e	−0.666
b_2	0.120	b_2	1.978	b_2	2.163	e	0.337
e_1	0.379	a_1	3.877	a_1	2.528	a_1	2.163
a_1	0.489						
e_2	0.807						
e_3	0.901						
a_1	1.046						
e_1	1.584						
b_2	2.059						
a_1	3.997						
(c) D_{5d} bicapped pentagonal antiprism (= I_h icosahedron), B—B 1.73 A.							
f_{2u}	−0.886	f_{2u}	−0.886[b]	e_{1g}	−0.835	e_{1g}	−0.835
h_g	−0.862	h_g	−0.856	a_{1g}	−0.827	a_{1u}	−0.810
f_{1u}	−0.827	f_{1g}	−0.782	a_{2u}	−0.827	e_{2u}	−0.809
f_{1g}	−0.782	f_{1u}	−0.773	a_{1u}	−0.810	a_{2g}	−0.783
h_u	−0.678	h_u	−0.678	e_{2u}	−0.809	e_{2g}	−0.689
g_g	−0.471	g_g	−0.471	a_{2g}	−0.783	a_{2u}	−0.678
h_g	−0.239	g_u	0.518	e_{1g}	−0.718	e_{1u}	−0.636
f_{2u}	−0.225	h_g	0.984	e_{1u}	−0.718	e_{1g}	−0.243
f_{1u}	0.150	f_{1u}	1.907	e_{2g}	−0.689	e_{2u}	0.377
g_u	0.518	a_g	4.163	a_{2u}	−0.678	e_{2g}	0.957
a_g	0.619			e_{1u}	−0.636	e_{1u}	1.653
h_g	1.107			e_{2g}	−0.613	a_{1g}	2.721
f_{1u}	1.990			e_{2u}	−0.613		
a_g	4.293			e_{1g}	−0.243	apex (C_{5v})	
				e_{2u}	0.377	a_1	−0.827
				e_{1g}	0.673	e_1	−0.718
				e_{1u}	0.673	e_2	−0.613
				e_{2g}	0.957	e_1	0.673
				e_{1u}	1.653	a_1	2.143
				a_{1g}	2.143		
				a_{2u}	2.143		
				a_{1g}	2.721		

[a] *Apex* and *equatorial* levels arranged in order.
[b] The correlations from I to D_5 are: $f \to a_2+e_2$, $h \to e_1+e_2+a_1$, $f_1 \to a_2+e_1$, $g \to e_1+e_2$, $a \to a_1$.

aromatic π orbitals. To preserve the analogy, we denote framework orbitals by f, f^* and the B—H bond orbitals by b, b^*. Thus the first excitation in B_4H_4, $B_6H_6^{-2}$, C_8H_8, O_h, and I_h $B_{12}H_{12}^{-2}$ is predicted as $b \to f^*$, in $B_{10}H_{10}^{-2}$ $b \to b^*$. If the $b \to f^*$ transition is symmetry allowed, it is nevertheless expected to be of low intensity since the spatial overlap of the b and f^* wavefunctions is small. Thus the spectra of these polyhedral molecules should be characterized by several weak absorptions in the near uv($b \to f^*$) followed by a strong absorption ($f \to f^*$) at higher energies. Dr. A. Kaczmarczyk of this laboratory has kindly communicated to us[17] preliminary, low-resolution uv spectral data on $B_{12}H_{12}^{-2}(I_h)$, $B_{10}H_{10}^{-2}$. Both compounds are characterized by an intense transition in the region 2000–2100 A, with no noticeable absorption at higher wavelengths. Since these spectra were taken at low concentration, further experimental study is desirable to establish if the low-intensity $b \to f^*$ transitions are indeed present.

[17] A. Kaczmarczyk (private communication).

IN-SURFACE FACTORIZATION

A further physical factorization of the $3N$ problem has been proposed.[16] One may rehybridize the 3 sp^3 hybrids at each boron so as to give an orbital (mostly $2s$) pointing "*in*" toward the polyhedron center, and two orbitals, pure $2p$, directed arbitrarily perpendicular to the axial direction, i.e., tangential to the polyhedron circumsphere. The latter set of $2N$ we call "*surface*" orbitals. One then makes the physical factorization by neglecting *in-surface* interactions, and thus factoring the $3N$ problem into one of degree N and one of degree $2N$. The levels thus obtained for 5 of the polyhedra discussed in the preceding section are given in Table III.

For the tetrahedron and octahedron there is no significant change. But for the cube, icosahedron, and cube-octahedron, the factorization has in each case produced a superfluous bonding or nonbonding f_{1u} orbital. When the interaction between *in* and *surface* orbitals is accounted for, this level becomes quite antibonding. But if we were to make predictions *entirely* on the basis of the solutions of the physically factorized

TABLE V. Energy parameters.

4N	3N	reconstructed 3N[a]	equatorial
(a) D_{3h} bicapped trigonal prism ($N=8$), B—B 1.74A			
e'' −0.895	e'' −0.895	e'' −0.821	e'' −0.821
a_2'' −0.860	a_2'' −0.817	a_2'' −0.721	a_2'' −0.717
a_1' −0.839	e' −0.804	a_1' −0.721	a_1'' −0.615
e' −0.819	e'' −0.722	a_2'' −0.717	a_2' −0.459
e'' −0.742	a_1' −0.692	e'' −0.679	e' −0.070
a_2'' −0.705	a_2'' −0.659	e' −0.679	e'' −0.060
a_1'' −0.615	a_1'' −0.615	a_1'' −0.615	e' 0.900
e' −0.484	a_2' −0.459	a_2' −0.459	a_1' 1.894
a_2' −0.459	e'' −0.298	e' −0.070	
e'' −0.420	e' −0.211	e'' −0.060	apex (C_{3v})
e'' −0.197	e' 0.338	e'' 0.084	a_1 −0.721
a_2'' −0.186	e'' 0.359	e' 0.084	e −0.679
a_1' −0.059	e' 0.991	e' 0.900	e 0.084
a_2'' 0.106	a_1' 1.025	a_1' 1.894	a_1 1.910
e' 0.153	a_2'' 1.706	a_2'' 1.910	
a_1' 0.309	a_1' 2.993	a_1' 1.910	
e' 0.338			
e'' 0.576			
e' 1.090			
a_1' 1.164			
a_2'' 1.877			
a_1' 3.065			
(b) D_{4h} bicapped cube, ($N=10$), B—B 1.74 A			
b_{1u} −0.928	b_{1u} −0.927	b_{1u} −0.855	b_{1u} −0.855
a_{2u} −0.869	e_g −0.862	a_{2u} −0.819	e_g −0.813
e_g −0.865	b_{2g} −0.862	a_{1g} −0.819	a_{1u} −0.711
b_{2g} −0.863	a_{2u} −0.860	e_g −0.813	a_{2u} −0.693
a_{1g} −0.851	a_{1g} −0.792	a_{1u} −0.711	a_{2g} −0.618
e_u −0.788	e_u −0.745	a_{2u} −0.693	e_u −0.295
e_g −0.750	e_g −0.735	b_{1u} −0.687	b_{2g} −0.192
a_{2u} −0.714	a_{1u} −0.711	b_{2g} −0.687	e_g −0.182
a_{1u} −0.711	b_{1u} −0.711	e_g −0.660	b_{2u} 0.113
b_{1u} −0.711	a_{2u} −0.685	e_u −0.660	b_{1g} 0.871
a_{2g} −0.618	a_{2g} −0.618	a_{2g} −0.618	e_u 1.193
e_u −0.474	e_u −0.407	e_u −0.295	a_{1g} 2.280
e_g −0.404	e_g −0.394	b_{2g} −0.192	
b_{2g} −0.309	b_{2g} 0.079	e_g −0.182	apex (C_{4v})
e_g −0.309	b_{2u} 0.113	b_{2u} 0.113	a_1 −0.819
b_{1u} −0.230	e_u 0.463	e_g 0.312	b_2 −0.687
a_{2u} −0.195	e_g 0.586	e_u 0.312	e −0.660
a_{1g} −0.110	b_{1g} 0.871	b_{1g} 0.871	e 0.312
e_u 0.012	a_{1g} 1.165	e_u 1.193	a_1 2.201
b_{2u} 0.113	e_u 1.301	a_{2u} 2.201	
a_{2u} 0.183	a_{2u} 2.005	a_{1g} 2.201	
b_{2g} 0.285	a_{1g} 3.519	a_{1g} 2.280	
a_{1g} 0.440			
e_u 0,492			
e_g 0.819			
b_{1g} 0.871			
a_{1g} 1.236			
a_{2u} 2.084			
a_{1g} 3.620			
(c) D_{5h} bicapped pentagonal prism, ($N=12$), B—B 1.74 A			
e_2'' −0.913	e_2'' −0.911	e_2'' −0.856	e_2'' −0.856
a_2'' −0.872	a_2'' −0.867	a_2'' −0.833	e_1'' −0.799
e_1'' −0.855	a_1' −0.847	a_1' −0.833	a_1'' −0.758
a_1' −0.853	e_1'' −0.840	e_1'' −0.799	a_2' −0.694
e_2' −0.823	e_2' −0.820	a_1'' −0.758	a_2'' −0.675
e_1' −0.800	e_1'' −0.764	e_1'' −0.713	e_1' −0.432
e_1'' −0.764	a_1'' −0.758	e_1' −0.713	e_1'' −0.298
a_1'' −0.758	e_1' −0.752	a_2' −0.694	e_2' −0.294
a_2'' −0.734	a_2' −0.694	a_2'' −0.675	e_2'' 0.088
a_2' −0.694	a_2'' −0.686	e_2'' −0.620	e_2' 0.948
e_2'' −0.684	e_2'' −0.682	e_2' −0.620	e_1' 1.480
e_1' −0.536	e_1' −0.535	e_1' −0.432	a_1' 2.452
e_1'' −0.506	e_1'' −0.499	e_1'' −0.298	
e_2'' −0.311	e_2' −0.011	e_2' −0.294	apex (C_{5v})
e_2' −0.269	e_2'' 0.122	e_2'' 0.088	a_1 −0.833
a_1' −0.217	e_1' 0.633	e_1'' 0.660	e_1 −0.713
e_1'' −0.211	e_1'' 0.894	e_1' 0.660	e_2 −0.620
a_2'' −0.201	e_2' 0.948	e_2' 0.948	e_1 0.660
e_2' 0.101	a_1' 1.136	e_1' 1.480	a_1 2.178
e_1' 0.126	e_1' 1.708	a_1' 2.178	
e_2'' 0.178	a_2'' 1.977	a_2'' 2.178	
a_2'' 0.245	a_1' 3.754	a_1' 2.452	
a_1' 0.553			
e_1' 0.637			
e_2' 1.006			
e_1'' 1.050			
a_1' 1.237			
e_1' 1.769			
a_2'' 2.031			
a_1' 3.870			

[a] *Apex* and *equatorial* levels arranged in order.

secular equation, we would conclude that cube-octahedral $B_{12}H_{12}^{-2}$ and cubical C_8H_8 still give closed shells, but with very small energy gaps separating highest filled and lowest unfilled levels. In the case of the icosahedron, we would be led to the erroneous prediction of the closed shell character of $B_{12}H_{12}$ with a minute gap, and the prediction of instability for the doubly negative anion.

The failure of the *in-surface* factorization does not imply that this simplification may not be used, with certain precautions, to obtain an approximate picture of the level ordering. The precaution to be observed is the following. If two levels of the same symmetry and near in energy are produced, one arising from the *in* set, the other from the *surface* set, then they will in general interact in such a way as to make one of them quite antibonding. Similar precautions must be kept in mind in judging the results of other physical factorizations considered below.

EQUATORIAL-APEX FACTORIZATION

This factorization has been used in treating the proposed $B_{10}H_{10}^{-2}$ structure[2] (D_{4d} bicapped antiprism) and in discussing icosahedral bonding.[7] The procedure applies to bicapped prisms and antiprisms where we distinguish two apex atoms from $N-2$ "*equatorial*" atoms. The following physical factorization of the $3N$ problem has been proposed by Lipscomb: Orient the three hybrids on each equatorial boron in such a manner that one orbital is directed toward the adjacent apex. Such an orbital set is illustrated in reference 7. Now separate the orbitals in three noninteracting groups: the $2(N-2)$ equatorial orbitals and two apical groups each composed of 3 apex orbitals and $\frac{1}{2}(N-2)$

orbitals pointing toward the apex. In Tables IV(a)–IV(c) and V(a)–V(c) we give the $4N$, $3N$, *apex*, *equatorial*, and "reconstructed" $3N$ (superimposed *apex* and *equatorial* energies, to facilitate comparison with $3N$) energies for the antiprism (D_{3d} bicapped trigonal antiprism, D_{4d} bicapped archimedean antiprism, D_{5d} bicapped pentagonal antiprism $= I_h$ icosahedron) and the prism (D_{3h} bicapped trigonal prism, D_{4h} bicapped cube, D_{5h} bicapped pentagonal prism) series.

TABLE VI. Energy parameters.

	$4N$		$3N$		reconstructed $3N$[a]		ring
(a) D_{3h} trigonal bipyramid ($N=5$), B—B 1.70 A							
e'	−0.865	e'	−0.863	a_2''	−0.858	e'	−0.834
a_2''	−0.858	a_2''	−0.858	e'	−0.834	a_2'	−0.556
a_1'	−0.853	a_1'	−0.823	e''	−0.649	e'	0.376
e''	−0.649	e''	−0.649	a_2'	−0.556	a_1'	1.473
a_2'	−0.556	a_2'	−0.556	e'	0.052		
e'	−0.388	e'	−0.301	a_1'	0.272		polar
a_1'	−0.104	e''	0.335	e''	0.335	a_2''	−0.858
a_2''	−0.063	e'	0.758	e'	0.376	e''	−0.649
e'	−0.052	a_2''	1.108	a_2''	1.108	e'	0.052
a_1'	0.124	a_1'	2.567	a_1'	1.473	a_1'	0.272
e''	0.335					e''	0.335
e'	0.898					a_2''	1.108
a_2''	1.136						
a_1'	2.615						
(b) D_{4h} tetragonal bipyramid (O_h octahedron, $N=6$), B—B 1.73 A							
e_g	−0.887	b_{1g}	−0.886[b]	b_{1g}	−0.886	b_{1g}	−0.886
f_{1u}	−0.831	a_{1g}	−0.884	a_{2u}	−0.829	e_u	−0.755
f_{1g}	−0.671	a_{2u}	−0.829	e_u	−0.755	a_{2g}	−0.671
f_{2u}	−0.416	e_u	−0.819	a_{2g}	−0.671	e_u	0.421
f_{1u}	−0.144	a_{2g}	−0.671	e_g	−0.671	b_{2g}	0.493
e_g	−0.089	e_g	−0.671	b_{2u}	−0.416	a_{1g}	1.732
a_{1g}	0.204	b_{2u}	−0.416	e_u	0.089		
f_{2g}	0.493	e_u	−0.416	a_{1g}	0.400		polar
f_{1u}	1.130	b_{2g}	0.493	e_u	0.421	a_{2u}	−0.829
a_{1g}	3.066	e_g	0.493	b_{2g}	0.493	e_g	−0.671
		e_u	0.990	e_g	0.493	b_{2u}	−0.416
		f_{1u}	1.023	a_{2u}	1.023	e_u	0.089
		a_{1g}	3.017	a_{1g}	1.732	a_{1g}	0.400
						e_g	0.493
						a_{2u}	1.023
(c) D_{5h} pentagonal bipyramid ($N=7$) B—B 1.74 A							
a_1'	−0.870	a_2''	−0.851	a_2''	−0.851	e_2'	−0.845
a_2''	−0.851	a_1'	−0.847	e_2'	−0.845	a_2'	−0.726
e_1'	−0.849	e_2'	−0.845	a_2'	−0.726	e_1'	−0.709
e_2'	−0.849	e_1'	−0.840	e_1'	−0.709	e_2'	0.494
a_2'	−0.726	a_2'	−0.726	e_1''	−0.669	e_1'	0.562
e_1''	−0.669	e_1''	−0.669	e_2''	−0.371	a_1'	1.722
e_1'	−0.553	e_1'	−0.508	e_1'	0.219		
e_2''	−0.371	e_2''	−0.371	e_2'	0.494		polar
a_2''	−0.349	e_2'	0.494	e_1''	0.524	a_2''	−0.851
e_2'	−0.187	e_1''	0.524	e_1'	0.562	e_1''	−0.669
a_1'	−0.071	a_2''	0.740	a_1'	0.704	e_2''	−0.371
e_1'	0.020	e_1'	1.420	a_2''	0.734	e_1'	0.219
a_1'	0.275	a_1'	3.273	a_1'	1.722	e_1''	0.524
e_1''	0.524					a_1'	0.704
e_2'	0.549					a_2''	0.734
a_2''	1.084						
e_1'	1.483						
a_1'	3.311						

[a] *Ring* and *polar* orbitals arranged in order.

[b] Here orbitals lose O_h symmetry when out orbitals are removed, since the latter, in this factorization, were not chosen all equivalent. The correlations $O \rightarrow D_4$ here are $e \rightarrow b_1 + a_1$, $f_1 \rightarrow a_2 + e$, $f_2 \rightarrow b_2 + e$, $a_1 \rightarrow a_1$.

TABLE VII. Energy parameters.

	$4N$		$3N$		reconstructed $3N$[b]		ring
(a) C_{3v} trigonal pyramid (T_d tetrahedron, $N=4$), B—B 1.70 A							
f_2	−0.853	e	−0.853[a]	e	−0.834	e	−0.834
f_1	−0.556	a_1	−0.840	a_2	−0.556	a_2	−0.556
f_2	−0.093	a_2	−0.556	e	−0.495	e	0.376
e	0.033	e	−0.534	a	−0.426	a_1	1.473
a_1	0.038	e	0.029	e	0.234		
f_2	0.747	e	0.639	e	0.376		polar
a_1	2.158	a_1	0.697	a_1	0.949	e	−0.495
		a_1	2.138	a_1	1.473	a_1	−0.426
						e	0.234
						a_1	0.949
(b) C_{4v} tetragonal pyramid ($N=5$), B—B 1.74 A							
b_1	−0.882	b_1	−0.881	b_1	−0.881	b_1	−0.881
a_1	−0.854	a_1	−0.853	e	−0.748	e	−0.748
e	−0.803	e	−0.794	b_2	−0.664	b_2	−0.664
b_2	−0.664	b_2	−0.664	e	−0.454	e	0.421
e	−0.542	e	−0.542	b_2	−0.409	a_2	0.492
b_1	−0.409	b_1	−0.409	a_1	−0.332	a_1	1.709
e	−0.152	e	0.106	e	0.368		
a_1	−0.126	a_2	0.492	e	0.421		polar
b_1	−0.087	a_1	0.711	a_2	0.492	e	−0.454
a_1	0.130	e	0.816	a_1	0.914	b_2	−0.409
e	0.135	a_1	2.433	a_1	1.709	a_1	−0.332
a_2	0.492					e	0.368
a_1	0.803					a_1	0.914
e	0.931						
a_1	2.458						
(c) C_{5v} pentagonal pyramid ($N=6$), B—B 1.74 A							
a_1	−0.860	a_1	−0.848	e_2	−0.845	e_2	−0.845
e_2	−0.849	e_2	−0.845	a_2	−0.726	a_2	−0.726
e_1	−0.813	e_1	−0.801	e_1	−0.709	e_1	−0.709
a_2	−0.726	a_2	−0.726	e_1	−0.373	e_2	0.494
e_1	−0.592	e_1	−0.563	e_2	−0.372	e_1	0.562
e_2	−0.371	e_2	−0.371	a_1	−0.144	a_1	1.722
a_1	−0.220	e_1	0.185	e_1	0.448		
e_2	−0.187	e_2	0.494	e_2	0.494		polar
e_1	−0.042	a_1	0.636	e_1	0.562	e_1	−0.373
a_1	0.214	e_1	1.106	a_1	0.737	e_2	−0.372
e_1	0.231	a_1	2.527	a_1	1.722	a_1	−0.144
e_2	0.549					e_1	0.448
a_1	0.842					a_1	0.737
e_1	1.173						
a_1	2.554						

[a] Here T_d symmetry is lost when out orbitals are removed. The correlations $T_d \rightarrow C_{3v}$ are $f_2 \rightarrow e + a_1$, $f_1 \rightarrow e + a_2$, $e \rightarrow e$, $a_1 \rightarrow a_1$.

[b] *Ring* and *polar* orbitals arranged in order.

On examining the closed shell possibilities from the $3N$ scheme we find $B_8H_8^{-2}$, $B_{10}H_{10}^{-2}$, $B_{12}H_{12}^{-2}$ for the antiprism series, with $B_8H_8^{+2}$ also a distinct possibility. For the prism series we have $B_8H_8^{-2}$, $B_{10}H_{10}$, $B_{12}H_{12}^{-2}$ though these would not be configurations boron would favor.

To make use of the factorization, it has been proposed that for $N=8$, 10, 12 respectively 3, 5, 7 electron pairs be assigned to the *equatorial* orbitals and 3 pairs to each *apical* set. Now the *equatorial*/*apical* orbital ratios were 12/12, 16/14, 20/16, while this electron assignment would give electron ratios of 6/12, 10/12, 14/12, respectively. We would expect the factorization to break down when these ratios differ considerably. Indeed it may be seen from Table IV that the fac-

TABLE VIII. Energy levels of the rhombic dodecahedron ($N=14$) for B—B=1.70 A, $\alpha(2s)=-15.36$ ev, $\alpha(2p)=-8.63$ ev, $K=-21.0$ ev.

Level	E(ev) Slater	E(ev) SCF-Slater
a_{1g}	62.79	48.52
f_{2g}	50.20	31.00
a_{2u}	31.34	35.29
e_g	30.88	24.86
f_{1g}	30.63	38.61
f_{1u}	26.07	42.10
f_{2g}	25.89	15.10
f_{2u}	21.88	30.10
f_{1u}	15.03	25.53
e_u	10.37	18.33
f_{1u}	−0.92	4.18
a_{1g}	−5.86	−4.08
f_{2g}	−6.56	−3.06
e_g	−8.35	−5.74
f_{1g}	−8.65	−5.46
a_{2u}	−10.15	−9.89
f_{1u}	−11.72	−10.98
f_{1u}	−12.52	−12.12
f_{2u} •	−14.08	−13.05
a_{1g}	−15.13	−14.65
f_{2g}	−16.55	−16.84
e_g	−16.92	−17.10
f_{1u}	−18.61	−18.86
a_{1g}	−19.66	−19.78

torization fails for the $N=8$ antiprism where another *equatorial* level has come below the degenerate *apical* level. It should be noted that the factorization is somewhat better for the prism series, since the prismatic arrangement makes interactions between *equatorial* orbitals smaller with respect to the more compact antiprismatic configurations. The gaps between filled and unfilled orbitals in the prism series are correspondingly smaller.

RING-POLAR SEPARATION

This physical factorization has been devised for $N=5, 6, 7$ bipyramids and $N=4, 5, 6$ pyramids.[18] The "ring" atoms are those at the base of the pyramid or in the equatorial section of the bipyramid. Let us hybridize the orbitals of the ring so that we have 3 sp^2 hybrids at each atom in the plane of the ring, and a p_z perpendicular to that plane. The hybrids are so oriented that one of them is directed radially out from the center of the ring plane. At the apices the usual set of sp^3 hybrids is formed. We reduce the problem to degree $3N$ by removing outpointing orbitals, and then attempt the physical factorization of separating for the bipyramids the $2(N-2)$ sp^2 hybrids in the plane of the ring from the $(N-2)p_z$ orbitals combined with six apex hybrids. The latter form what we call the "*polar*" set. The planar *ring* orbitals may be considered as giving rise to $N-2$ classical covalent bonds, consuming $2(N-2)$ electrons. Indeed we may note here the empirical rule, to which we have found no exception in our calculations, that whenever a molecule (such as C_4H_4, P_4, C_8H_8) or a molecular fragment (such as the B_4, B_5, B_6 rings) can be assigned a classical non-electron-deficient single-bonded structure, the LCAO–MO calculation gives a closed shell structure with a large energy gap. The LCAO's may be transformed by a unitary transformation into a set of equivalent orbitals which closely resemble the localized saturated bonds.

[18] W. N. Lipscomb, Proc. Natl. Acad. Sci. U. S. **47**, 1791 (1961).

The $N-2$ *polar* p_z orbitals yield the familiar aromatic ring arrangement. In a naive approach the interaction of this aromatic ring with the six apex orbitals is such that the lowest orbitals of the resultant *polar* set are an *a* and a doubly degenerate *e*. This may be seen in the last column of Tables VI(a)–VI(b) for the B_5 and B_6 bipyramids. Filling these orbitals and the above mentioned $N-2$ *ring* orbitals requires then $N+1$ electron pairs, or a $B_NH_N^{-2}$ ion is predicted. Unfortunately, the electron/orbital ratios in such a separation differ greatly, being 6/6, 8/8, 10/10 in the ring, 6/9, 6/10, 6/11 in the polar set for $N=5, 6, 7$, respectively. In the actual calculation, a totally symmetric orbital (a_1', a_{1g}) comes much too low in the *polar* set for each case studied, indeed for $N=7$ it goes below the *ring e* orbitals, leading to a prediction of a closed shell for B_7H_7, in disagreement with the clearcut prediction of $B_7H_7^{-2}$ by the $3N$ calculation.

For the B_N pyramid, there are $2(N-1)$ orbitals in the *ring* set, and $N+2$ *polar* orbitals. The results of the LCAO–MO calculations are given in Tables VII(a)–VII(c). Considerations similar to the above lead us to prediction of closed shells for B_NH_N or $B_NH_N^{-4}$. The electron/orbital ratios are 6/6, 8/8, 10/10 in the *ring*, 2/6, 2/7, 2/8 in the *polar* set, for B_NH_N, $N=4, 5, 6$, and as expected the factorization is bad, a *polar e* level coming close to the least bonding *ring e* level. The electron/orbital ratios for $B_NH_N^{-4}$ are 6/6, 6/7, 6/8 in the *polar* set, and the separation correspondingly good. However, such highly charged species are not expected to be stable. Their carborane analogs may well exist, pyramidal $C_4B_2H_6$, C_4BH_5, C_4H_4 (tetrahedrane).

It should be pointed out that for the *ring-polar* separation as for other physical factorizations discussed in this paper, judicious juggling of parameters whose relative, but not absolute, magnitudes are assumed, may lead, when coupled with correspondingly judicious caution, to worthwhile level schemes. The actual failure or success of an approximation is revealed only when numbers are substituted for the parameter symbols.

OVERLAP INTEGRALS

We now ask, "What is the best basis set to use in Eq. (1)? Shall we use atomic self-consistent field (SCF) functions, Morse-Young-Haurwitz orbitals, or Slater orbitals?" To examine this question, two calculations of the $4N$ type were carried out for the rhombic dodecahedron, one using Slater overlaps, the other using a mixed orbital set consisting of $2s$ Slater functions and $2p$ SCF functions for carbon. The $N=14$ polyhedron was chosen because of its symmetry (O_h)

and the large number of levels available for comparison. Overlaps for the mixed Slater-SCF sets are given by Mulliken[19]. The SCF 2*p* function has much greater spatial extent, and thus the overlaps are quite different, particularly at greater distances.

The energy levels of the polyhedron are given in Table VIII. The SCF calculation was checked against a previous calculation for this molecule by L. L. Lohr, who has suggested the use of these functions.[7,20] A $B_{14}H_{14}^{-2}$ species will yield a closed electronic shell. The levels of the SCF calculation agree quite well with the Slater levels, particularly for the bonding and nonbonding states, which, after all, are the only ones that matter. There is also the general trend that the SCF levels are more bonding or antibonding than their Slater counterparts, this being explained by the greater SCF overlap.

[19] R. S. Mulliken, J. Chem. Phys. **19**, 900 (1951).
[20] L. L. Lohr (private communication).

The conclusion we draw is that general predictions are not overly affected by the use of different overlap schemes. There is indeed no *a priori* reason why atomic SCF functions should be better basis orbitals for a molecular calculation. Slater orbitals are easy to handle, and will be used throughout our calculations.

ACKNOWLEDGMENTS

We would like to thank L. L. Lohr for valuable discussions and the U.S. Army Research Office (Durham), the Office of Naval Research, and the National Institutes of Health for financial support.

Chapter 5

Thomas A. Steitz

Biography

After completing my undergraduate studies in chemistry at Lawrence College, WI, I undertook my Ph.D. in biochemistry and molecular biology at Harvard, graduating in 1966. I then worked as a postdoctoral fellow at the MRC Laboratory of Molecular Biology in Cambridge, UK. In 1976–1977 I was a Macy Fellow at the Max-Planck Institute for Biophysical Chemistry at the University of Göttingen, Germany, and in 1984–1985 I was a Fairchild Scholar at Caltech. I have held positions at Yale University since I was first appointed as an assistant professor of molecular biophysics and biochemistry in 1970. Since 1986 I have been an investigator at the Howard Hughes Medical Institute and Sterling Professor of Molecular Biophysics and Biochemistry since 2001. From 1994 until 2001 I was Eugene Higgins Professor of Molecular Biophysics and Biochemistry.

I have over 250 publications to my name, 51 of which appeared in *Nature*, *Science*, and *Cell*. I have served as an advisor and board member for a number of research bodies including the Jane Coffin Childs Memorial Fund for Medical Research (1998–2006) and the Walter and Eliza Hall Institute of Medical Research in Australia (2001–present). In 1990 I was made a member of the National Academy of Sciences and the American Academy of Arts and Sciences. I have also been the recipient of a number of prizes and awards, including the Keio Medical Science Prize in 2006, the Gairdner International Award in 2007, and the Nobel Prize for Chemistry in 2009.

After a decade of successfully studying the structural basis of the glucose induced fit in hexokinase, the research in my laboratory has focused on obtaining structural insights into the molecular mechanisms by which the proteins and nucleic acids involved in the central dogma of molecular biology carry out gene expression from the replication and recombination of the DNA genome to its transcription into mRNA followed by the various component steps associated with the translation of mRNA into protein.

In the 1980s we established the first structure of a transcription factor, the catabolite gene activator protein and later its DNA complex, the structure of the first DNA polymerase and the first structure of an aminoacyl-tRNA synthetase bound to tRNA. Our studies of T7 RNA polymerase captured in many of its functionally important states — initiation, intermediate, elongation — as well as stages of nucleotide incorporation provide the most complete mechanistic picture of the structural basis of RNA transcription by an RNA polymerase.

Our recent structure of a replicative DNA polymerase from eubacteria showed it to be evolutionarily unrelated to the replicative DNA polymerase from eukaryotes, and thus is a potential antibiotic target and suggests that the last common ancestor of eukaryotes and eubacteria may have had a replicating DNA polymerase made of RNA. Our structure of the enzyme that copies the HIV viral RNA, HIV reverse transcriptase, into DNA was established complexed with nevirapine, which is a non-nucleotide inhibitor of HIV RT that is used to treat AIDS, and has facilitated the development of new non-nucleotide inhibitors that may also have clinical impact.

Perhaps the most significant insights have been derived from the atomic structure of the large ribosomal subunit. This structure proved that the ribosomal RNA is entirely responsible for catalyzing peptide bond formation and thus, the ribosome is a ribozyme (an enzyme made of RNA) and provided insights into how this mammoth RNA assembly is folded and functions as an enzyme. The large ribosomal subunit is the major target of antibiotics that are effective pharmaceuticals against bacterial infections. The many structures of the large subunit complexed with various different antibiotics determined at Yale have identified numerous different antibiotic binding sites near the site of protein synthesis.

This information has been enormously facilitating in the development of new antibiotics that will be effective against the rapidly increasing number of antibiotic resistant bacteria, such as MRSA. Rib-X Pharmaceuticals, Inc. in New Haven, CT (where I am cofounder and chair of the scientific advisory board) has used this structural information to develop one compound that has successfully completed phase II clinical trials and a second that should enter clinical trials shortly.

Joining the Colonel's Army: Personal Anecdotes

I came to Harvard as a graduate student to work on biophysical studies of nucleic acids, but fortunately, I went a different way and joined the Colonel's army. In the spring of 1963, I attended three Dunham lectures given by Max Perutz in which he presented the first atomic resolution protein crystal structure, that of myoglobin. He showed stereo slides, and I was stunned to see the atomic structure of myoglobin pop out in three dimensions over Max's head; this was clearly the way to understand how macromolecules carry out their biological functions. Shortly thereafter, while playing tennis with a Lipscomb graduate student (Peter Boer), I mentioned how unfortunate it was that no one was doing protein crystallography at Harvard. He said on the contrary, the Colonel, as Bill Lipscomb was referred to by his students and postdocs, had a group who was working on the crystal structure of bovine carboxypeptidase A (CPA). Shortly thereafter I made my way to the Colonel's office, a little nervous because I had already been turned down (fortunately) by another faculty member, in order to make an appointment to see him. He was standing in the office of his secretary, who was not there, and when I asked if I could make an appointment to see him, he invited me right in. After he described the CPA project, I excitedly asked if I could join, and he said yes. So, I had just received the wonderful opportunity to join the Colonel's army, and the rest is history.

The CPA team consisted of five postdocs at that time, of whom Martha Ludwig was the most important for my training. I worked with her on structural studies of substrate and inhibitor complexes of CPA as well as with the whole group on the determination of the crystal structure of the apo protein. I remember being excited at one time when we succeeded in collecting 5,000 reflections in one week using the Hilger–Watts diffractometer. Now, of course, we can collect 5,000,000 reflections from crystals of the ribosome in an hour (about a factor of 10^5 faster for an assembly that is about 80 times larger). All computer programs were written in Fortran for an IBM 7094 computer that had 32 K of memory, had no discs and used computer cards. The advantage of the latter was that we could save the used cards and sell them as used paper in order to support lab parties every few months. (I always thought we should have written a program called GENCARD to increase the party frequency.)

In retrospect, I know that it was due in to the superb team that the Colonel assembled and his encouraging management style that the project moved on well. In 1966 we published a 6 Å resolution map, including a model of the polypeptide backbone that I had over-optimistically built, but showing many of its structural features. Martha and I, with Flo Quiocho, published a low resolution map of an inhibitor complex that showed the first example of a substrate induced conformational change. In 1967 we obtained what I now realize was a superb 2.0 Å resolution electron density map of the apo-CPA that allowed us to correctly position every residue, even in the absence of an amino acid sequence. CPA was tied in 1967 with three other proteins, RNase A, RNase S, and chymotrypsin, for being the third high resolution protein structure determined.

The Colonel was an inspiring research advisor and leader who maintained constant contact with individual lab members. He would wander through the labs and offices a couple of times a day and was always there on nights and weekends, which of course inspired us to do the same. Since he engaged in a longer talk if one had a result, I tried to have something new and interesting to tell him as often as possible. He encouraged us to learn to solve our problems ourselves rather than simply telling us the answer himself. After showing him a few diffraction pictures from a crystal of methyl ethylene phosphate, he said that he knew what the space group was, and when I figured it out, I should come to his office and tell him, which I did.

Unlike PIs today, the Colonel was almost never absent from the lab to attend meetings and present seminars. He did, however, take a sabbatical in England, and while he was gone, I came up with the idea of using direct methods of phasing to phase the $F_{derivative} - F_{native}$ difference coefficients in order to calculate a projection difference Fourier map, which clearly showed the heavy atom positions. When I showed him the paper I had written on the work upon his return, he allowed me to publish the paper in *Acta Crystallographica* without his being a coauthor, also not a common practice among PIs today.

The Colonel also had an important influence on my personal life. One day while I was at the bench weighing out some platinum tetrachloride, the Colonel came up behind me and said, "I hope you won't find this improper." He then asked if I would invite Joan to come with me to a gathering of friends at his house who were to be entertained by some woodwind chamber music performed by the Colonel, a visiting theoretical chemist Longuet-Higgins, and friends of Lipscomb's from the Boston Symphony. Joan had played in a chamber group with the Colonel when he was at the University of Minnesota, so she agreed to go — to hear the Colonel. It was sort of our first "date." As a wonderful "encore," the Colonel and two of his friends from the Minneapolis Symphony played the Hayden London Trios at our wedding in Minneapolis.

The Colonel provided me with what turned out to be a great opportunity that had an important impact on my future in the summer of 1966 when he arranged for me to give a talk at the Protein Gordon Conference, which was chaired by Fred Richards. I talked about the CPA structure and the conformational change produced by substrate or inhibitor binding — induced fit. Dan Koshland was a participant and greatly appreciated this new experimental evidence for his hypothesis of substrate induced fit conformational changes. I assume that my talk and the opportunity to meet him afterward partly motivated his advocating that the Berkeley Biochemistry Department interview me and offer me a faculty position, which they did, and I accepted. Due to the reluctance of the department to consider hiring a woman (Joan) for a faculty position in 1970, I resigned my Berkeley position after two months and accepted a position at Yale offered by Fred Richards.

After Harvard and before going to Berkeley I spent three years at the MRC Laboratory of Molecular Biology in Cambridge in the group of David Blow, who was recommended to me by Hilary Muirhead, who was a postdoc with the Colonel and a former student with Max Perutz. In David's lab I worked on chymotrypsin substrate complexes with Richard Henderson. During my last year in Cambridge I started work on hexokinase, which was to occupy me and my coworkers through most of the 1970s. The question to be answered again involved induced fit: why does hexokinase not hydrolize ATP in the absence of glucose, as Koshland had asked? Subsequently, our structures of hexokinase with and without glucose determined at Yale showed there is indeed a large glucose induced conformational change. So my initial observation of substrate induced fit by CPA that I made in the Colonel's lab led to my pictures of glucose induced fit in hexokinase that are now published in dozens of textbooks, and to our more recent observation of substrate induced fit in the ribosome (Schmeing *et al.*, 2005, see below).

RESEARCH ARTICLES

The Complete Atomic Structure of the Large Ribosomal Subunit at 2.4 Å Resolution

Nenad Ban,[1]* Poul Nissen,[1]* Jeffrey Hansen,[1] Peter B. Moore,[1,2] Thomas A. Steitz[1,2,3]†

The large ribosomal subunit catalyzes peptide bond formation and binds initiation, termination, and elongation factors. We have determined the crystal structure of the large ribosomal subunit from *Haloarcula marismortui* at 2.4 angstrom resolution, and it includes 2833 of the subunit's 3045 nucleotides and 27 of its 31 proteins. The domains of its RNAs all have irregular shapes and fit together in the ribosome like the pieces of a three-dimensional jigsaw puzzle to form a large, monolithic structure. Proteins are abundant everywhere on its surface except in the active site where peptide bond formation occurs and where it contacts the small subunit. Most of the proteins stabilize the structure by interacting with several RNA domains, often using idiosyncratically folded extensions that reach into the subunit's interior.

In the last step of the gene expression pathway, genomic information encoded in messenger RNAs is translated into protein by a ribonucleoprotein called the ribosome (*1*). As in most other organisms, the prokaryotic ribosome (MW $\approx 2.6 \times 10^6$) is about two-thirds RNA and one-third protein and consists of two subunits, the larger of which is approximately twice the molecular weight of the smaller (*2*). The small subunit, which sediments at 30*S* in prokaryotes, mediates the interaction between mRNA codons and tRNA anticodons on which the fidelity of translation depends. The large subunit, which sediments at 50*S* in prokaryotes, includes the activity that catalyzes peptide bond formation—peptidyl transferase—and the binding site for the G-protein (GTP–binding protein) factors that assist in the initiation, elongation, and termination phases of protein synthesis.

Because the structures of several DNA and RNA polymerases have been determined at atomic resolution, the mechanisms of DNA and RNA synthesis are both well understood. Determination of the structure of the ribosome, however, has proven a daunting task. It is several times larger than the largest polymerase, and 100 times larger than lysozyme, the first enzyme to be understood at atomic resolution. Until now an atomic resolution structure for the ribosome has not been available, and as a result the mechanism of protein synthesis has remained a mystery.

Electron microscopy has contributed to our understanding of ribosome structure ever since the ribosome was discovered. In the last few years, three-dimensional (3D) electron microscopic images of the ribosome have been produced at resolutions sufficiently high to visualize many of the proteins and nucleic acids that assist in protein synthesis bound to the ribosome (*3*). Earlier this year, an approximate model of the RNA structure in the large subunit was constructed to fit a 7.5 Å resolution electron microscopic map of the 50*S* subunit from *Escherichia coli* as well as biochemical data (*4*).

Crystallization studies of the ribosome begun two decades ago by Yonath and Wittmann (*5*) and by the group at Pushchino (*6*) opened the possibility of using x-ray crystallography to determine the structure of the ribosome at atomic resolution. The first electron density map of the ribosome that showed features recognizable as duplex RNA was a 9 Å resolution x-ray crystallographic map of the large subunit from *Haloarcula marismortui* published 2 years ago (*7*). A year later, extension of the phasing of that map to 5 Å resolution made it possible to locate several proteins and nucleic acid sequences, the structures of which had been determined independently (*8*). At about the same time, with the use of similar crystallographic strategies, a 7.8 Å resolution map was generated of the entire *Thermus thermophilus* ribosome, showing the positions of tRNA molecules bound to its A, P, and E sites (*9*), and a 5.5 Å resolution map of the 30*S* subunit from *T. thermophilus* was obtained, which allowed the fitting of solved protein structures and the interpretation of some of its RNA features (*10*). Subsequently, an independently determined, 4.5 Å resolution map of the *T. thermophilus* 30*S* subunit was published, which was based, in part, on phases calculated from a model corresponding to 28% of the subunit mass that had been obtained with a 6 Å resolution experimental map (*11*). The interpretation of the subunit packing in the two 30*S* structures is not the same, even though the crystals used by the two groups appear to be identical.

Using a 2.4 Å resolution, experimentally phased, electron density map, we have produced an atomic structure of the *H. marismortui* 50*S* ribosomal. The model includes 2711 of the 2923 nucleotides of 23*S* ribosomal RNA (rRNA), all 122 nucleotides of its 5*S* rRNA, and structures for the 27 proteins that are well ordered in the subunit. Here, we describe the architecture of the subunit, the structure of its RNAs, and discuss the location, structures, and functions of its proteins.

The secondary structures of both 5*S* and 23*S* rRNA are remarkably close to those deduced for them by phylogenetic comparison. The secondary structure of the 23*S* rRNA divides it into six large domains, each of which has a highly asymmetric tertiary structure. The irregularities of their shapes notwithstanding, the domains fit together in an interlocking manner to yield a compact mass of RNA that is almost isometric. The proteins are dispersed throughout the structure and mostly concentrated on its surface, but they are largely absent from the regions of the subunit that are of primary functional significance to protein syntheses: the 30*S* subunit interface and the peptidyl transferase active site. The most surprising feature of many of these proteins is the extended, irregular structure of their loops and termini, which penetrate between RNA helices. The primary role of most proteins in the subunit appears to be stabilization of the 3D structure of its rRNA.

Structure determination. Several experimental approaches were used to extend the resolution of the electron density maps of the *H. marismortui* 50*S* ribosomal subunit from 5 to 2.4 Å. A back-extraction procedure was developed for reproducibly growing crystals that are much thicker than those available earlier and that diffract to at least 2.2 Å resolution. The twinning of crystals, which obstructed progress for many years (*8*), was eliminated by adjusting crystal stabilization conditions (*12*). All of the x-ray data used for high-resolution phasing were collected at the Brookhaven National Synchrotron Light Source except for two native data sets, which were collected at the Advanced Photon Source at Argonne (*13*) (Table 1). Osmium pentamine (132 sites) and iridium hexamine (84 sites) derivatives proved to be the most effective in producing isomorphous replacement and anomalous scattering phase information to 3.2 Å resolution (*14*). Intercrystal density averaging, which had contributed significantly at lower resolution, was not helpful beyond about 5 Å resolution. Electron density maps were dramatically improved, and

[1]Department of Molecular Biophysics & Biochemistry, and [2]Department of Chemistry, Yale University, and [3]Howard Hughes Medical Institute, New Haven, CT 06520-8114, USA.

*These two authors contributed equally to this work.
†To whom correspondence should be addressed.

their resolutions were eventually extended to 2.4 Å with the solvent-flipping procedure in the CNS program (*15*, *16*).

Except for regions obscured by disorder, the experimentally phased 2.4 Å resolution electron density map was of sufficient quality that both protein and nucleic acid sequencing errors could be identified and corrected. Each nucleotide could be fitted individually, and the difference between A and G was usually clear without having to refer to the chemical sequence, as was the distinction between purines and pyrimidines (Fig. 1). Only a few of the many water molecules and metal ions evident in the electron density have been positioned so far.

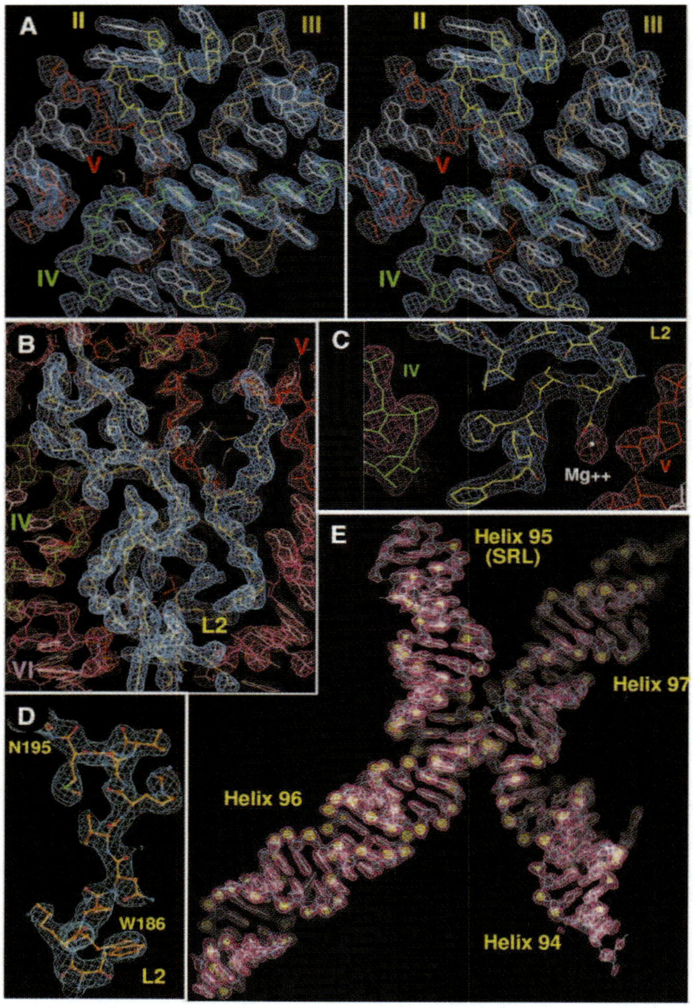

Fig. 1. Portions of the experimental 2.4 Å resolution electron density map. (**A**) A stereo view of a junction between 23*S* rRNA domains II, III, IV, and V having a complex structure that is clearly interpretable. The electron density is contoured at 2σ. The bases are white and the backbones are colored by domain as specified in Fig. 4. (**B**) The extended region of L3 interacting with its surrounding RNA, where the red RNA density is contoured at 2σ and the blue protein density is contoured at 1.5σ. (**C**) Detail in the L2 region showing a bound Mg^{2+} ion. (**D**) Detail from L2 showing amino acid side chains. (**E**) Helices 94 through 97 from domain VI. The red contour level is at 2σ, and the yellow contour at 6σ shows the positions of the higher electron density phosphate groups.

Subtraction of the atomic model from the experimental electron density map leaves no significant density except water and ions, showing that the model accounts for all the macromolecular density. Preliminary refinement of the model was achieved with experimental phase restraint in the program CNS (*16*). The model was further refined in real space against the 2.4 Å electron density map with the program TNT (*17*), which yielded a model with an *R* factor of 0.33. One additional round of mixed target refinement of both atomic positions and *B* factors with CNS led to the structure described here. The current free *R* factor is 0.26 (Table 1).

Sequence fitting and protein identification. Guided by the information available on the secondary structures of 23*S* rRNAs (*18*), the sequence of 23*S* rRNA was fit into the electron density map nucleotide by nucleotide starting from its sarcin/ricin loop sequence [A2691 to A2702 (*E. coli* numbers A2654 to A2665)] whose position had been determined at 5 Å resolution (*8*). The remaining RNA electron density neatly accommodated 5*S* rRNA. The interpretation of electron density corresponding to protein was more complicated because each protein region had to be identified chemically before the appropriate sequence could be fit into it; with the assistance of D. Klein, L. Min, S. Antolić, and M. Schmeing, ~4000 amino acid residues of 27 proteins were fit into electron density.

The *H. marismortui* 50*S* subunit appears to contain 31 proteins, and sequences for 28 of them exist in the Swiss-Prot data bank, including one called HMS6 or L7ae, which originally had been assigned to the small ribosomal subunit (*19*). The three remaining proteins were identified using the sequences of the ribosomal proteins from eukaryotes and other archaeal species as guides. No electron density was found for one of the *H. marismortui* large ribosomal subunit proteins in the sequence database, LX. Either the assignment of LX to the large subunit is in error, or LX is associated with a disordered region of the subunit. It is also possible that LX is absent from the subunits examined altogether.

The 2.4 Å resolution electron density map lacks clear electron density for proteins L1, L10, L11, and L12, the positions of which are known from earlier low-resolution x-ray and/or electron microscopic studies. These proteins are components of the two lateral protuberances of the subunit, which are both poorly ordered in these crystals. L1 is the sole protein component of one of them (*20*) and is visible in 9 Å resolution density maps of the subunit (*7*), but not at higher resolutions. L10, L11, and L12 are components of the other protuberance, which is often referred to as the L7/L12 stalk (*20*). L11 and the RNA to which it binds were located in the 5 Å resolution electron density map of the *H. marismortui* large subunit (*8*) using the independently determined crystal structures of that complex (*21*, *22*). A protein fragment (~100 residues) associated with the RNA stalk that supports the L11 complex can be seen in the 2.4 Å resolution map. On the basis of its location, the fragment must be part of L10. No electron density corresponding to L12 was seen at any resolution, but the L12 tetramer is known to be attached to the ribosome through L10, and the L10/L12 assembly is known to be flexible under some circumstances (*23*), which may explain its invisibility here.

The structures of eubacterial homologs of proteins L2, L4, L6, L14, and L22 have previously been determined in whole or in part (Table 2). L2, L6, and L14 were initially located in the 5 Å resolution map (*8*). L4 and L22 have now been identified and positioned the same way. Electron density corresponding to most of the remaining proteins was assigned by comparing chain lengths and sequence motifs deduced from the electron density map with known sequence lengths. Occasionally, these comparisons were assisted by the information available on relative protein positions (*24*) and protein interactions with 23*S* rRNA and 5*S* rRNA (*25*). Each of the protein electron density regions so identified is well accounted for by the amino acid sequence assigned to it.

The most interesting of the proteins identified by sequence similarity was L7ae, which first appeared to be L30e. The L30e identification seemed plausible because the structure of yeast L30e superimposes neatly on the electron density of L7ae, and the structure of the RNA to which L7ae binds resembles that of the mRNA element to which yeast L30e binds (*26*). Nevertheless, the sequence of HMS6, which by sequence similarity is a member of the L7ae protein family, better fits the electron density. Four of the other proteins identified by sequence similarity, L24e, L37e, L37ae, and L44e, contain zinc finger motifs. The rat homologs of L37e and L37ae were predicted to be zinc finger proteins on the basis of their sequences (*27*), and this prediction helped identify their homologs in *H. marismortui*. Even though no *H. marismortui* sequences were available for the proteins L10e, L15e, and L37ae, they could be identified using the alignments of other available archaeal sequences.

General appearance of the subunit. In its rotated crown view (Fig. 2), the large ribosomal subunit, which is about 250 Å across, presents its surface that interacts with the small subunit to the viewer with the three projections that radiate from that surface pointed up. Although the protuberance that includes L1 is not visible in the 2.4 Å resolution electron density map, the structure of L1, which has been determined independently (*28*), has been positioned approximately in lower resolution maps (*7*) and is included here to orient the reader. It is evident that, except for its two lateral protuberances, the large ribosomal subunit is monolithic. There is no hint of a division of its structure into topologically separate domains. In addition, partly because it lacks obvious domain substructure but also because it is so large, it is impossible to comprehend looking at it as a whole. To convey a sense of how it is put together, the subunit must be dissected into its chemical components.

RNA secondary structure. All the base pairs in *H. marismortui* 23*S* rRNA stabilized by at least two hydrogen bonds were identified with a computer program that searched the structure for hydrogen bond donors and acceptors separated by less than 3.2 Å. Bases linked by at least two such bonds were considered paired if the angle between their normals was less than 45° and if the angle between bonds and base normals was also less than 45°. On the basis of the results of this analysis, R. Gutell and colleagues prepared a secondary structure diagram (Fig. 3) in the format standard for 23*S*/28*S* rRNAs. The secondary structure predicted for this molecule by phylogenetic comparison was remarkably accurate, but it did not find all of the tertiary pairings and failed to identify interactions involving conserved bases. In addition to base pairs of nearly every type, the RNA contains numerous examples of well-known secondary structure motifs such as base triples, tetraloops, and cross-strand purine stacks, but no dramatically new secondary structure motifs have been identified so far.

The secondary structure of this 23*S* rRNA consists of a central loop that is closed by a terminal stem, from which 11 more or less complicated stem-loops radiate. It is customary to describe the molecule as consisting of six domains and to number its helical stems sequentially starting from the 5′ end (Fig. 4) (*29*). The division of the molecule into domains as shown in Fig. 4 deviates from standard practice with respect to helix 25, which is usually considered part of domain I. Here, it is placed in domain II because it interacts more strongly with domain II than the other elements of domain I.

There are five sequences longer than 10 nucleotides in 23*S* rRNA whose structures cannot be determined from the 2.4 Å resolution map because of disorder. Together they account for 207 out of the 232 nucleotides missing from the final model. The disordered regions are: all of helix 1, the distal end of helix 38, helix 43/44 to which ribosomal protein L11 binds, the loop end of stem-loop 69, and helix 76/77/78, which is the RNA structure to which L1 binds. For completeness, these regions are included in Fig. 3 (in gray) with the secondary structures determined for them phylogenetically.

Overall architecture of rRNA. The six domains of 23*S* rRNA and 5*S* rRNA all have complicated, convoluted shapes that fit together to produce a compact, monolithic RNA mass (Fig. 4, A and B). Thus, despite the organization of its RNAs at the secondary structure level, in three dimensions the large subunit is a single, gigantic domain. In this respect, it is quite different from the small subunit. Even in low-resolution electron micrographs the small subunit consists of three structural domains, each of which contains

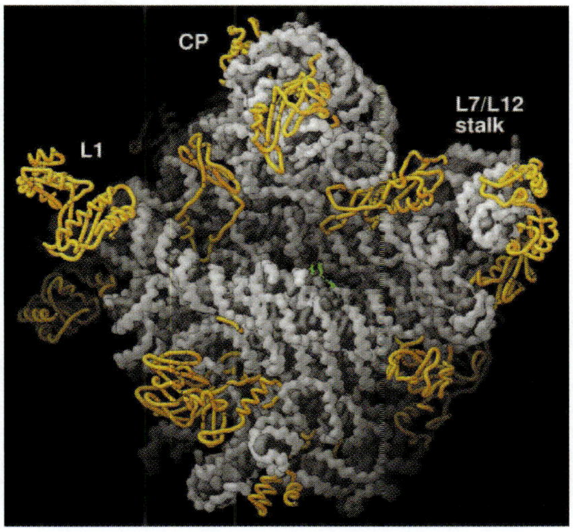

Fig. 2. The *H. marismortui* large ribosomal subunit in the rotated crown view. The L7/L12 stalk is to the right, the L1 stalk is to the left, and the central protuberance (CP) is at the top. In this view, the surface of the subunit that interacts with the small subunit faces the reader. RNA is shown in gray in a pseudo–space-filling rendering. The backbones of the proteins visible are rendered in gold. The Yarus inhibitor bound to the peptidyl transferase site of the subunit is indicated in green (*64*). The particle is approximately 250 Å across.

Secondary Structure: large subunit ribosomal RNA - 5′ half

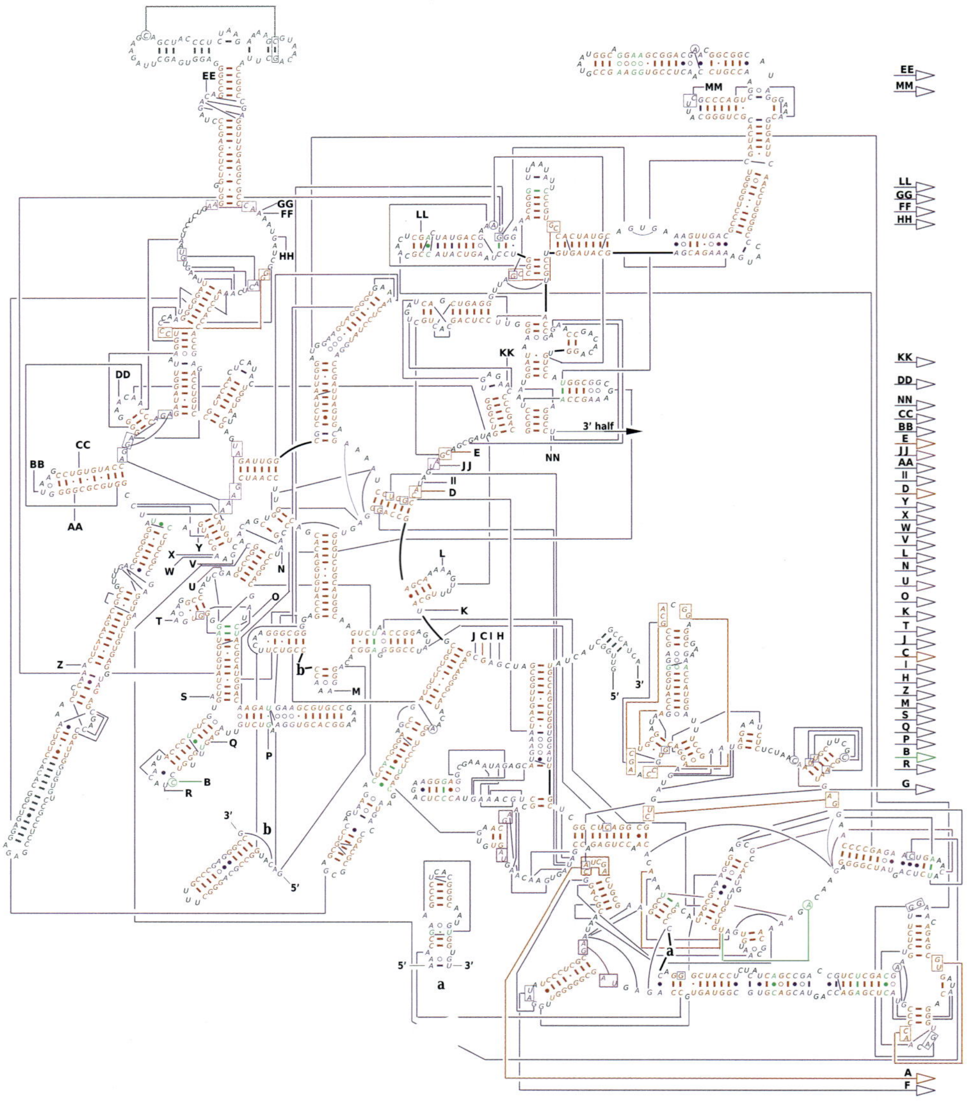

Fig. 3.

one of the three secondary structure domains of its RNA (*30*). This qualitative difference between the two subunits may reflect a requirement for conformational flexibility that is greater for the small subunit.

Domain I, which looks like a mushroom (Fig. 4E), lies in the back of the particle, behind and below the L1 region. The thin part of the domain starts in the vicinity of domain VI, which is the location of its first and last residues. Helices 1 and 25 span the particle in the back and then the domain expands into a larger, more globular structure below and behind the L1 region.

Domain II is the largest of the six 23*S* rRNA domains, accounting for most of the back of the particle. It has three protrusions that reach toward the subunit interface side of the particle (Fig. 4F). One of them (helix 42 to 44) is the RNA portion of the L7/L12 stalk, which is known to interact with elongation factors, is not well ordered in these crystals. The second domain II protrusion is helix 38, which is the longest, unbranched stem in the particle. It starts in the back of the particle, bends by about 90° and pro-

Secondary Structure: large subunit ribosomal RNA - 3′ half

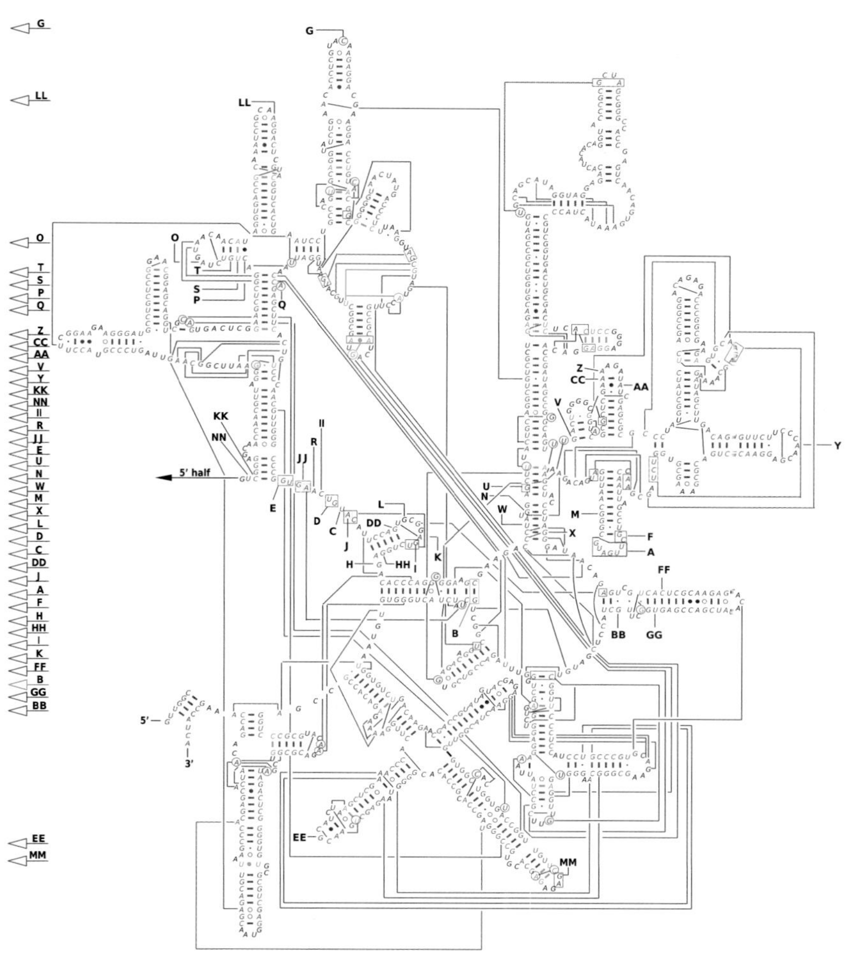

Fig. 3. The secondary structure of the 23*S* rRNA from *H. marismortui* is shown in a format made standard by R. Gutell and colleagues (*65*). It was prepared by Dr. Gutell to show all the base pairings seen in the crystal structure of the large subunit that are stabilized by at least two hydrogen bonds. Pairings shown in red were predicted and were observed. Those shown in green were predicted, but were not observed. Interactions shown in blue were observed, but were not predicted. Bases shown in black were not involved in pairing interactions. Sequences that cannot be visualized in the 2.4 Å resolution electron density map are depicted in gray with the secondary structures predicted for them.

trudes toward the small subunit between domains V and 5*S* rRNA. The third region (helix 32 to 35.1) points directly toward the small subunit and its terminus, the loop of stem-loop 34, interacts directly with the small ribosomal subunit (*31*). This loop emerges at the subunit interface between domains III and IV.

Domain III is a compact globular domain that occupies the bottom left region of the subunit in the crown view (Fig. 4G). It looks like a four-pointed star with the origin of the domain (stem-loop 48) and stem-loops 52, 57, and 58 forming the points. The most extensive contacts of domain III are with domain II, but it also interacts with domains I, IV, and VI. Unlike all the other domains, domain III hardly interacts with domain V at all; the sole contact is a van der Waals interaction involving a single base from each domain.

Domain IV accounts for most of the interface surface of the 50*S* subunit that contacts the 30*S* subunit (Fig. 4H). It forms a large diagonal patch of flat surface on that side of the subunit and connects to domains III and V in the back of the particle. Helices 67 through 71 constitute the most prominent feature of domain IV and form the front rim of the active site cleft, which is clearly visible at low resolution (Fig. 2). This is one of the few regions of the 23*S* rRNA that is not extensively stabilized by ribosomal proteins. Helix 69 in the middle of this ridge interacts with the long penultimate stem of 16*S* rRNA in the small ribosomal subunit (*9*).

Domain V, which is sandwiched between domains IV and II in the middle of the subunit, is known to be intimately involved in the peptidyl transferase activity of the ribosome (*32*). Structurally, this domain can be divided into three regions (Fig. 4, I and J). The first starts with helix 75 and ultimately forms the binding site for protein L1. The second, which consists of helices 80 through 88, forms the bulk of the central protuberance region and is supported in the back by the 5*S* rRNA and domain II. The third region, which includes helices 89 through 93, extends toward domain VI and helps stabilize the elongation factor-binding region of the ribosome.

The smallest domain in 23*S* rRNA, domain VI, which forms a large part of the surface of the subunit immediately below the L7/L12 stalk, resembles a letter X with a horizontal bar at the bottom (Fig. 4K). The most interesting region of this domain is the sarcin-ricin loop (SRL) (stem-loop 95), the structure of which has been extensively studied in isolation (*33*, *34*). The SRL is essential for factor binding, and ribosomes can be inactivated by the cleavage of single covalent bonds in this loop (*35*). As suggested by nucleotide protection data, the major groove of this loop is exposed to solvent (*36*), and its conformation is stabilized by proteins and through interaction with domain V.

Table 1. Statistics for data collection, phase determination, and model construction. HA, heavy-atom concentration; ST, soaking time; Res, resolution; λ, wavelength; Obs, observations; Redun, redundancy; Compl, completeness; (*) last-resolution shell. R_{iso}: $\Sigma|F_{PH} - F_P|/F_{PH}$, where F_{PH} and F_P are the derivative and the native structure factor amplitudes, respectively. R_{sym}: $\Sigma\Sigma_i|I_{(h)i} - I_{(h)}|/\Sigma\Sigma_i I_{(h)i}$, where $I_{(h)}$ is the mean intensity after reflections. Phasing power: rms isomorphous difference divided by the rms residual lack of closure. R_{cullis}: $\Sigma(||F_{PH} - F_P| - |F_{H(calc)}||)/\Sigma|F_{PH} - F_P|$, where F_{PH} is the structure factor of the derivative and F_P is that of the native data. The summation is valid only for centric reflection. FOM (figure of merit): mean value of the cosine of the error in phase angles. Abbreviations: MIRAS, multiple isomorphous replacement, anomalous scattering; SAD, single wavelength anomalous diffraction.

Data statistics

	MIRAS1			MIRAS2			
	Native1	$Os(NH_3)_5^{2+}$	$UO_2F_5^{3-}$	Native2	$Ir(NH_3)_6^{3+}$	$Os(NH_3)_6^{3+}$	$Ta_6Br_{12}^{2+}$
HA (mM)	–	30.0	0.5	–	20.0	4.5	3.0
ST (hours)	–	1.5	4	–	24 hours	24 hours	24 hours
Sites no.	–	132	20	–	84	38	9
Res (Å)	90–2.4	40–3.5	40–3.8	30–2.9	30–3.2	30–3.5	30–3.8
(*)	(2.5–2.4)	(3.6–3.5)	(3.9–3.8)	(3.32–3.22)	(3.27–3.20)	(3.6–3.5)	(3.97–3.80)
λ (Å)	1.00	1.14	1.30	1.00	1.075	1.14	1.255
Obs	6,089,802	1,308,703	596,166	2,832,360	1,823,861	1,646,468	1,288,524
Unique	665,928	429,761	313,863	390,770	541,488	488,275	346,745
Redun (*)	9.1 (6.5)	3.0 (2.5)	1.9 (1.6)	7.2	3.4	4.3 (4.2)	3.7
Compl (*)	95.6 (71.0)	99.4 (96.8)	92.0 (54.1)	97.1	93.8	98.1 (99.0)	99.5
$I/\sigma I$ (*)	25.5 (1.9)	13.5 (3.3)	8.9 (1.6)	18.0 (6.4)	12.0 (2.6)	10.6 (2.7)	10.8 (3.2)
R_{merge} (*)	8.6 (69.1)	7.2 (32.0)	9.1 (37.9)	11.2 (36.9)	8.5 (29.5)	12.1 (46.0)	12.1 (40.5)
χ^2 (ano) (*)	–	2.8 (1.0)	1.5 (1.0)	–	2.63 (1.48)	1.8 (1.0)	2.42 (1.18)
R_{merge} (ano)	–	6.2	8.0	–	6.7	6.9	
R_{iso} (*)	–	14.1 (22.7)	26.4 (47.0)	–	12.9 (28.1)	19.5 (39.4)	

Phasing statistics

Resolution shells (Å): ~73,200 reflections per bin

	30.0–5.1	5.1–4.0	4.0–3.5	3.5–3.2	Total
MIRAS1 (FOM)	0.52	0.31	0.14	–	0.32
$Os(NH_3)_5^{2+}$					
Phasing power	0.87	0.72	0.66	–	0.75
Phasing power (SAD)	1.40	0.58	0.26	–	0.75
R_{cullis} (centric)	0.62	0.65	0.67	–	0.65
$UO_2F_5^{3-}$					
Phasing power	0.47	0.33	0.28	–	0.36
Phasing power (SAD)	0.46	0.25	–	–	0.36
R_{cullis} (centric)	0.72	0.77	0.75	–	0.75
MIRAS2 (FOM)	0.48	0.40	0.28	0.12	0.33
$Ir(NH_3)_6^{3+}$					
Phasing power	1.02	0.92	0.78	0.66	0.89
Phasing power (SAD)	2.02	1.60	1.22	0.83	1.47
R_{cullis} (centric)	0.58	0.63	0.70	0.74	0.63
$Os(NH_3)_6^{3+}$					
Phasing power	0.62	0.57	0.58	0.58	0.59
Phasing power (SAD)	0.47	0.39	–	–	0.42
R_{cullis} (centric)	0.78	0.78	0.78	0.76	0.78
$Ta_6Br_{12}^{2+}$ (used for SAD phasing only)					
Phasing power (SAD)	2.77	0.35	0.13	–	1.19
$FOM_{(MIRAS1+MIRAS2+SAD)}$	0.76	0.51	0.31	0.14	0.37

Model statistics

		rms deviations:		Average *B* factors (Å²)	
Resolution range (Å)	90.0–2.4	Bonds (Å)	0.0064	All atoms	37.4
Reflections	577,304	Angles (°)	1.19	23*S* rRNA	32.3
R_{cryst} (%)	25.2	Dihedrals (°)	28.8	5*S* rRNA	43.2
R_{free} (%)	26.1	Impropers (°)	1.68	Minimum/Max *B* factors (Å²)	7.0/107.9

RESEARCH ARTICLES

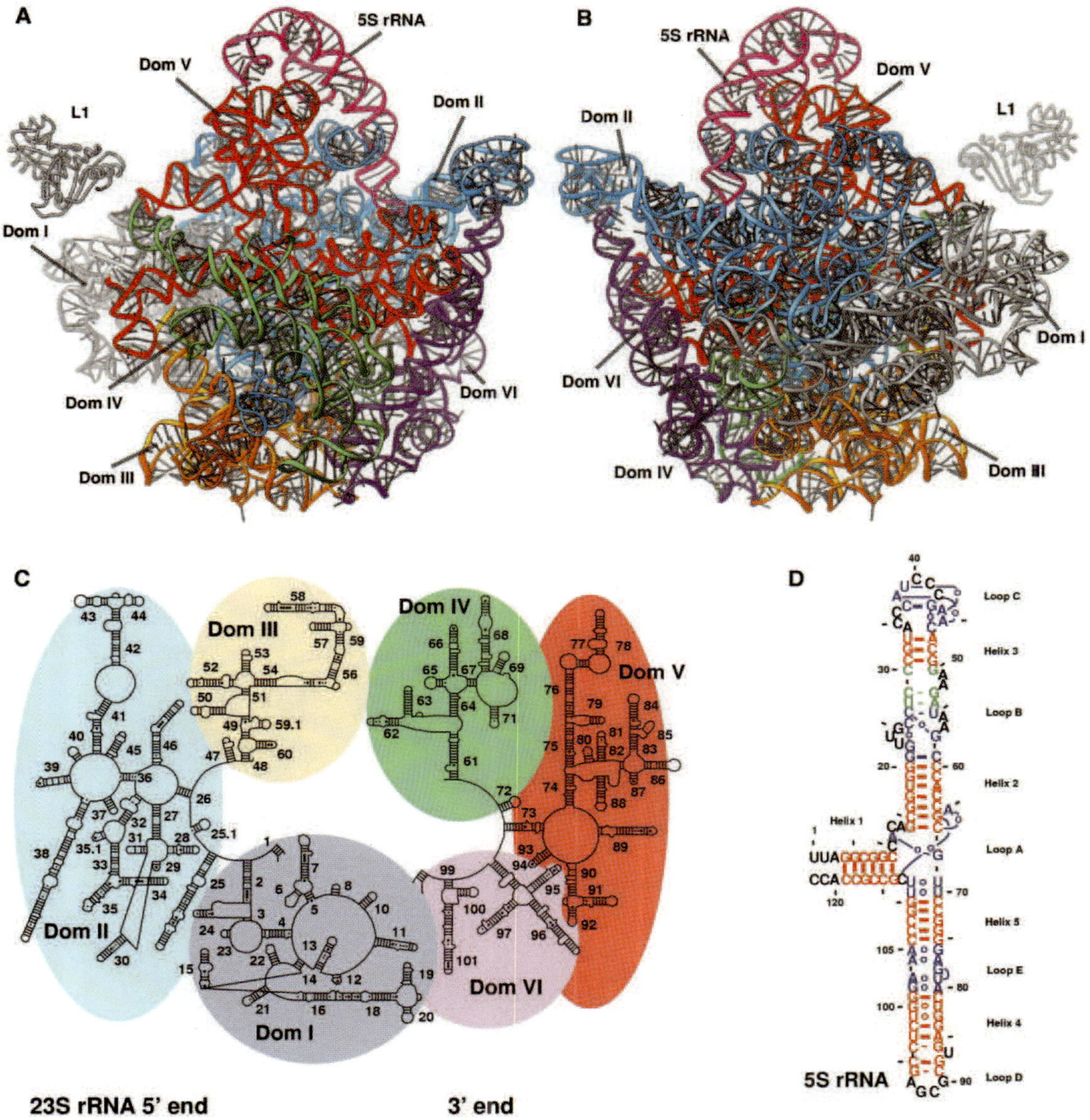

Fig. 4. The tertiary and secondary structures of the RNA in the *H. marismortui* large ribosomal subunit and its domains. (**A** and **B**) The RNA structure of the entire subunit. Domains are color-coded as shown in the schematic (**C**). (A) The subunit particle in its crown view. (B) The crown rotated by 180° about a vertical axis in the plane of the image. (C) Schematic secondary structure diagram of 23*S* rRNA with the domain coloring used throughout the figures and the helices numbered according to Leffers *et al.* (*29*). (**D**) The secondary structure of 5*S* rRNA from *H. marismortui*. Bases joined by thick lines represent Watson-Crick pairing, and those joined by a lower case "o" indicate non–Watson-Crick pairing. Bases joined by thin lines interact via a single hydrogen bond, whereas those in black are unpaired. Base pairings shown in red are phylogenetically predicted pairings that are now confirmed (*66*). Pairs shown in blue were observed but were not predicted, and pairs shown in green were predicted but were not observed. (**E** through **L**) Stereo views of the RNA domains of the 23*S* rRNA and of 5*S* rRNA. Each domain is color-coded from its 5′ end to its 3′ end to help the viewer follow its trajectory in three dimensions. The backbones are shown as ribbons and the bases as sticks. The surfaces where the most important interdomain interactions occur are shown in mono to the right. (E), Domain I; (F), domain II; (G), domain III; (H), domain IV; (I), domain V, crown view; (J), domain V, back view; (K), domain VI; and (L), 5*S* rRNA.

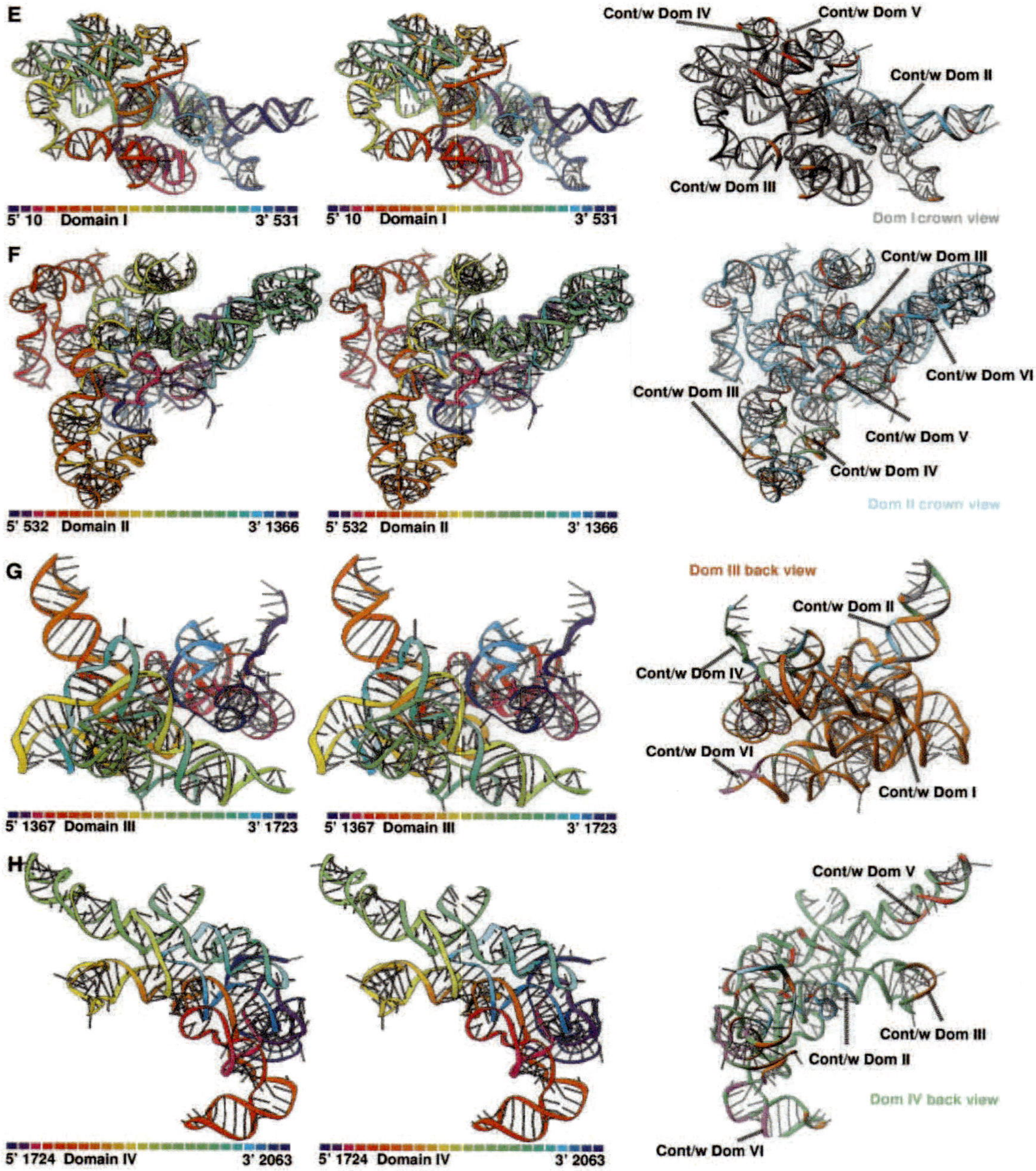

Fig. 4. (continued)

5*S* ribosomal RNA, which is effectively the seventh RNA domain in the subunit, consists of three stems radiating out from a common junction called loop A (Fig. 4D). In contrast to what is seen in the crystal structure of fragment 1 from *E. coli* 5*S* rRNA (*37*), the helix 2/3 arm of the molecule stacks on its helix 4/5 arm, not helix 1 (Fig. 4L). This arrangement results from a contorted

RESEARCH ARTICLES

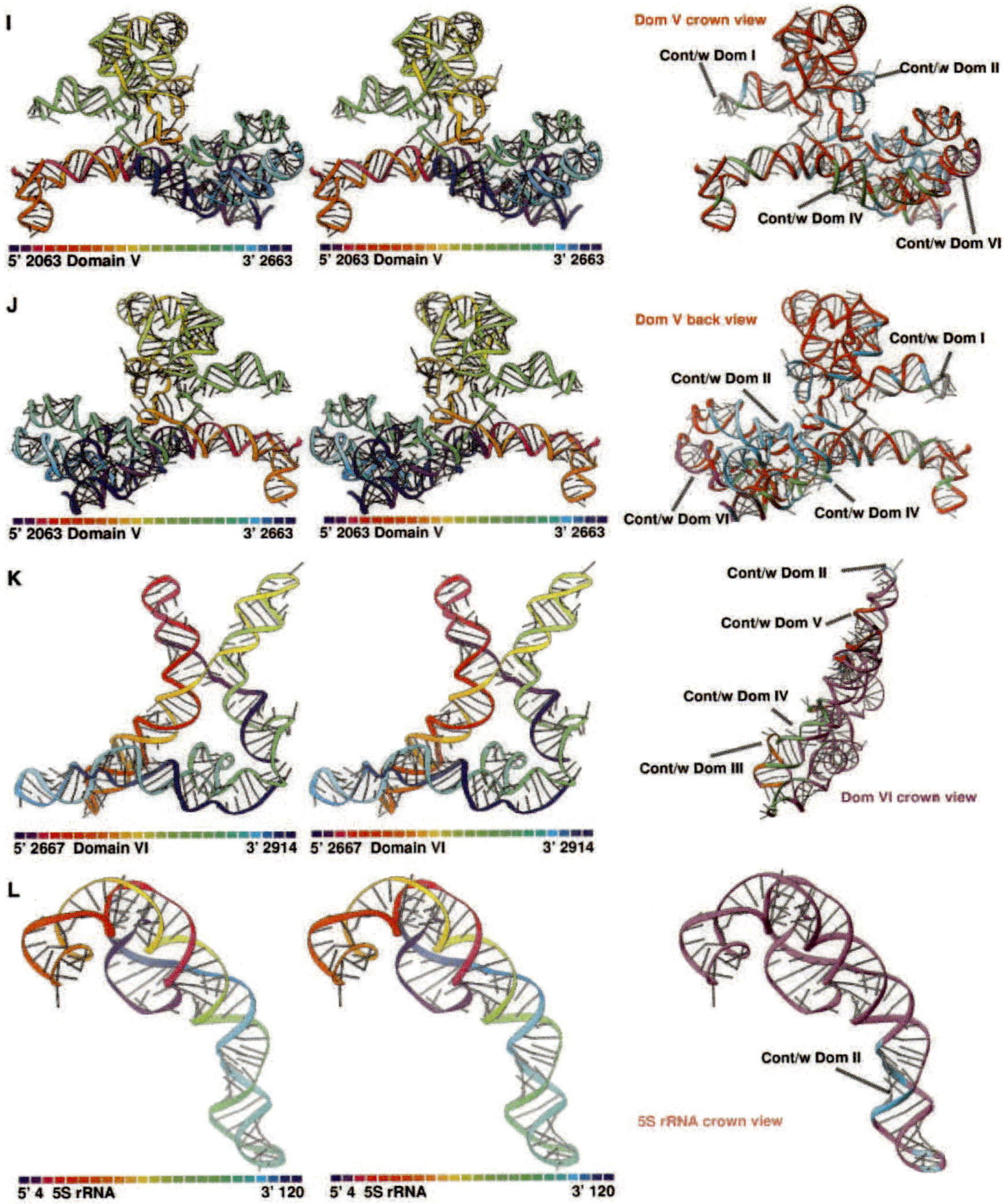

Fig. 4. (continued)

conformation of loop A residues that involves two stacked base triples. Indeed, from the secondary structure point of view, the loop A–helix 2/3 arm of 5*S* rRNA is remarkable, with a high concentration of unusual pairings leading to a convoluted RNA secondary structure.

Sequence conservation and interactions in 23*S* rRNA. Although 23*S*/28*S* rRNAs contain many conserved sequences, they also vary substantially in chain length. Shorter 23*S*/28*S* rRNAs are distinguished from their longer homologs by the truncation of, or even the elimination of, entire stem-loops, and by comparing sequences, one can identify a minimal structure that is shared by all (*38*). The expansion sequences in the 23*S* rRNA of *H. marismortui*, i.e., the sequences it contains that are larger than the minimum, are shown in Fig. 5 in green. They are largely absent from the subunit interface surface of the particle, but they are abundant on its back surface, far from its active sites. This is consistent with low-resolution electron microscopic observations, suggesting that the region of the large subunit whose structure is most conserved is the surface that interacts with the small subunit (*39*).

There are two classes of conserved sequences in 23*S* rRNA. One contains residues concentrated in the active site regions of the large subunit. The second class consists of much shorter sequences scattered throughout the particle (Fig. 5, red sequences). The SRL sequence in domain VI and the cluster of conserved residues belonging to domain V located at the bottom of the peptidyl transferase cleft are members of the first class. They are conserved because they are essential for substrate binding, factor binding, and catalytic activity. Most of the residues in the second class of conserved residues are involved in the inter- and intradomain interactions that stabilize the tertiary structure of 23*S* rRNA. Adenosines are disproportionately represented in this class. The predominance of adenosines among the conserved residues in rRNAs has been pointed out previously (*40*). Throughout the particle, adenosines are observed to participate in tertiary interactions by exploiting the smooth N1-C2-N3 face of the adenine base, which allows for very close packing and additional backbone-backbone interactions. In particular, a reoccurring pattern of two or more stacked adenosines that dock into the minor grooves of receptor helices seems to reveal a very basic principle in tertiary RNA structure formation and could be regarded as an equivalent of a hydrophobic core formation in glob-

Table 2. Large-subunit proteins from *Haloarcula marismortui*. The top block of proteins include all those known to have eubacterial homologs of the same name. The second block lists proteins found in the *H. marismortui* large ribosomal subunit that have only eukaryotic homologs (*19*). Their names are all followed by the letter "e" to distinguish them from eubacterial proteins that would otherwise have the same name. The third block are large-subunit proteins for which no *H. marismortui* sequence yet exists. They are identified by sequence homology with standard L names. [1]The structures of all or part of homologs of the following proteins were previously determined: L1 (*28*), L2 (*43*), L4 (*44*), L6 (*58*), L11 (*21, 22, 59*), L12 (*60*), L14 (*61*), L22 (*62*), and L30 (*63*). All other structures, except L10, have been newly determined in this study. [2]Rat homolog. Rat equivalents to *H. marismortui* protein are from (*26*). [3]Sequence chain length. [4]Conformation: glb, globular; ext, extension. [5]The protein interactions with the six domains of 23*S* rRNA, 5*S* rRNA, and other proteins are specified. (+) Implies that the interaction is substantial; (±) implies a weak, tangential interaction. Protein names in parentheses implies that the interactions are weak; otherwise, the interaction is substantial.

Name[1]	Hmlg[2]	Lgth[3]	Conf[4]	Interactions[5]							Proteins
				I	II	III	IV	V	VI	5*S*	
L1*	?	211	glb.					+			None
L2†	RL8	239	glb+ext		+	+	+	+			(L37ae)
L3	RL3	337	glb+ext		±		+	+	+		L14, L24e, (L13)
L4†	RL4	246	glb+ext	+	+			±			(L18e), (L24), (L37e)
L5	RL11	176	glb					+		+	L18
L6	RL9	177	glb		±			±	+		(L13)
L10*	RP0	348	glb?		+						L12
L11*	RL12	161	glb		+						None
L12*	RP1/2	115	glb								L10
L13	RL13a	145	glb		+			±	±		(L3), (L6)
L14	RL23	132	glb				+	+	+		L3, L24e
L15	RL27a	164	glb+ext	+	+			+			(L18e), (L32e)
L18	RL5	186	glb+ext		±			+		+	L5, L21e
L19	RL19	148	glb+ext		+	+	+		±		None
L22	RL17	154	glb+ext	+	±	+	+	+	+		None
L23	RL23a	84	glb	±		+					L29, (L39e)
L24	RL26	119	glb+ext	+							(L4)
L29	RL35	70	glb	+							L23
L30	RL7	154	glb		+					+	None
L18e	RL18	115	glb		+						(L4), (L15)
L21e	RL21	95	glb		+			+		±	L18
L24e	RL24	66	glb				±		+		L3, L14
L31e	RL31	91	glb			+	+		+		None
L32e	RL32	240	glb	±	+						(L15)
L37e	RL37	56	glb+ext	+	+	+	±				(L4)
L39e	RL39	49	ext	+		+					(L23)
L44e	RL36a	92	glb+ext	+			±	+			(L15e)
L7ae	RL7a	110	glb	±							L15e
L10e	RL10	163	glb		+			+		±	None
L15e	RL15	184	glb+ext	+	±	±	±	+			(L44e), L7ae
L37ae	RL37a	72	glb+ext		+	+	+				L2

*All entries so designated describe proteins that are not fully represented in the electron density maps described here. The summary information provided is derived from literature sources and is included here for completeness only. †The structure available for this protein in isolation does not include the extension(s) reported here.

From Nenad Ban, Poul Nissen, Jeffrey Hansen, Peter B. Moore, and Thomas A. Steitz. The complete atomic structure of the large ribosomal subunit at 2.4 Å resolution. *Science*, **289**, 905–920 (2000). Reprinted with permission from AAAS.

ular protein domains. Common RNA structural motifs, such as the ribose zipper and the tetraloop-tetraloop receptor interaction, depend on this principle of adenosine packing. A manuscript in preparation describes these A-dependent interactions at greater length.

In addition to its reliance on A-dependent motifs, the tertiary structure of the domains of 23*S* rRNA and their relative positions are stabilized by familiar tertiary structure elements like pseudoknots and tetraloop-tetraloop receptor motifs (*41*, *42*). Thus, in many places, base pairs and triples stabilize the interactions of sequences belonging to different components of the secondary structure of 23*S* rRNA.

5*S* rRNA and 23*S* rRNA do not interact extensively with each other. The few RNA/RNA interactions that do occur involve the backbones of the helix 4/5 arm of 5*S* rRNA and of helix 38 of 23*S* rRNA. Most of the free energy and specificity of 5*S* rRNA binding to the large ribosomal subunit appears to depend on its extensive interactions with proteins that act as modeling clay, sticking it to the rest of ribosome.

Proteins. We have determined the structures of 27 proteins found in the large ribosomal subunit of *H. marismortui* (Table 2). Twenty-one of these protein structures have not been previously established for any homologs, and the structures of the six that do have homologs of known structure have been rebuilt into the electron density map with their *H. marismortui* sequences. In addition, there are structures available for homologs of *H. marismortui* L1, L11, and L12, which cannot be visualized in the 2.4 Å resolution electron density map. Only the structure of L10 is still unknown among the 31 proteins of this subunit.

Almost all of these structures are complete. Yet, an entire domain of L5 is missing from the electron density, presumably because of disorder. Further, L32e is also noteworthy. Its NH_2-terminal 97 residues are not seen in the electron density map, and the electron density map suggests that its COOH-terminal residue may be covalently bonded to the most NH_2-terminal of its visible residues.

Of the 30 large subunit ribosomal proteins whose structures are known, 17 are globular proteins, similar in character to thousands whose structures are in the Protein Data Bank (Table 2). The remaining 13 proteins either have globular bodies with extensions protruding from them ("glb+ext") or are entirely extended ("ext"). Their extensions often lack obvious tertiary structure and in many regions are devoid of significant secondary structure as well (Fig. 6). These extensions may explain why many ribosomal proteins have resisted crystallization in isolation. The exceptions that prove the rule are L2 and L4, both of which are proteins belonging to the "glb+ext" class. Protein L2 was crystallized and its structure solved only after its extensions had been removed (*43*), and the large loop of L4 that is extended in the ribosome is disordered in the crystal structure of intact L4 (*44*).

Except for proteins L1, L7, L10, and L11, which form the tips of the two lateral protuberances, the proteins of the 50*S* subunit do not extend significantly beyond the envelope defined by the RNA (Fig. 7). Their globular domains are found largely on the particle's exterior, often nestled in the gaps and crevices formed by the folding of the RNA. Thus, unlike the proteins in spherical viruses, the proteins of the large ribosomal subunit do not form a shell around the nucleic acid with which they associate, and unlike the proteins in nucleosomes, they do not become surrounded by nucleic acid, either. Instead, the proteins act like mortar filling the gaps and cracks between "RNA bricks."

The distribution of proteins on the subunit surface is nearly uniform, except for the active site cleft and the flat surface that interacts with the 30*S* subunit. In the crown view, the proteins lie around at the periphery of the subunit (Fig. 7A), but when viewed from the side opposite the 30*S* subunit binding site (the "back side"), they appear to form an almost uniform lattice over its entire surface (Fig. 7B). Similarly, the bottom surface of the subunit, which includes the exit of polypeptide tunnel, is studded with proteins (Fig. 7C). Indeed, the six proteins that surround the tunnel exit may play a role in protein secretion because they are part of the surface that faces the membrane and the translocon when membrane and secreted proteins are being synthesized (*45*).

Although Fig. 7 shows protein chains disappearing into the ribosome interior, the degree to which proteins penetrate the body of the particle can be fully appreciated only when the RNA is stripped away. The interior of the particle is not protein-free, but it is protein-poor compared with the surface of the particle. Extended tentacles of polypeptide, many of which emanate from globular domains on the surface, penetrate into the interior, filling the gaps between neighboring elements of RNA secondary structure (Fig. 8E). The bizarre structures of these extensions are explained by their interactions with RNA. A detailed analysis of these proteins and their interactions with RNA will be presented elsewhere.

Although extended, nonglobular structures are rare in the protein database, they are not unknown. Extended protein termini often form interprotein contacts, e.g., in viral capsids, presumably adopting fixed structures only upon capsid formation (*46*). The basic

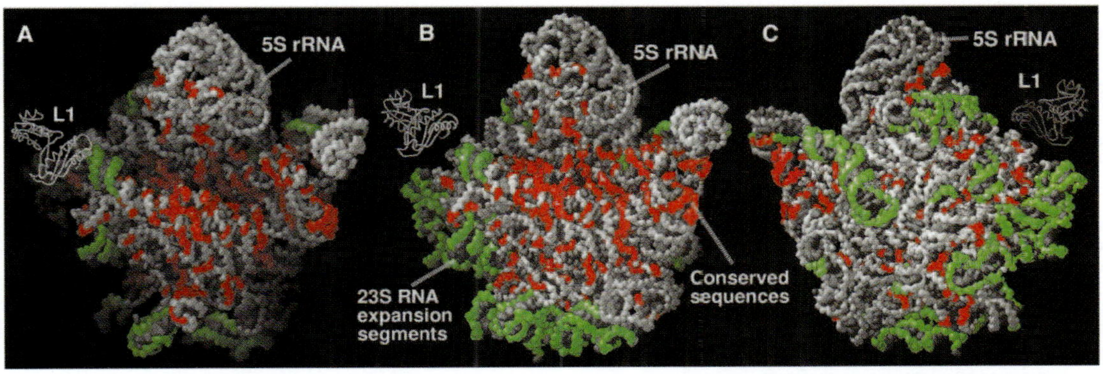

Fig. 5. Conserved residues and expansion sequences in the 23*S* rRNA of *H. marismortui*. The general, nonconserved RNA in these images is gray. Sequences that are found to be >95% conserved across the three phylogenetic kingdoms are shown in red. Sequences where expansion in the basic 23*S* structure is permitted are shown in green (*65*). (**A**) The particle rotated with respect to the crown view so that its active site cleft can be seen. (**B**) The crown view. (**C**) The back view of the particle, i.e., the crown view rotated 180° about its vertical axis.

"tails" of histones may behave the same way when nucleosomes form (*47*). The NH_2-terminal sequences of capsid proteins are often positively charged, and in virus crystal structures, the electron density for these sequences often disappears into the interior of the virus where they presumably interact with asymmetrically arranged nucleic acid. The interactions observed in the ribosome could be useful models for these viral interactions.

The interactions between extended polypeptides and RNA in the large subunit, which stabilize its massive nucleic acid structure, result in an intertwining of RNA and protein in the center of the subunit (Fig. 8, A and B). It is hard to imagine such an object assembling from its components efficiently in anything other than a highly ordered manner. Chaperones may well be required to prevent the aggregation of the extended regions of these proteins, which are likely to be disordered outside the context provided by rRNA, and to manage the folding of rRNA.

Mutations in some ribosomal proteins render bacteria resistant to certain antibiotics. One such example is a deletion of three amino acids in the β hairpin loop of protein L22 that renders bacteria resistant to erythromycin (*48*). Because this β hairpin is forming part of the surface of the tunnel wall, the mutation changes the surface properties of the polypeptide exit tunnel and may prevent the antibiotic from binding; alternatively, the mutation could be acting indirectly through RNA.

Protein and RNA interactions. Because protein permeates the large subunit extensively, there are only a few segments of the 23*S* rRNA that do not interact with protein at all. Of the 2923 nucleotides in 23*S* rRNA, 1157 make at least van der Waals contact with protein (Fig. 8D), and there are only 10 sequences longer than 20 nucleotides in which no nucleotide contacts protein. The longest such sequence contains 47 nucleotides, and is the part of domain IV that forms the ridge of the active site cleft.

The extent of the interactions between RNA and protein that occur when the large subunit assembles can be estimated quantitatively. Using the Richards algorithm (*49*) and a 1.7 Å radius probe to compute accessible surface areas, it can be shown that 180,000 $Å^2$ of surface become buried when the subunit forms from its isolated, but fully structured components. This is about half their total surface area. The average is about 6000 $Å^2$ per protein. Although this is an enormous amount compared with the surface buried when most protein oligomers form, it should be recognized that ribosome assembly must be accompanied by a large loss in conformational entropy that does not occur when most proteins oligomerize. The extended protein termini and loops of the ribosomal proteins are almost certainly flexible in isolation, and in the absence of protein, the RNA is probably quite flexible as well. Thus, the burial of a large amount of surface area may be required to provide the free energy required to immobilize the structures of these molecules.

All of the proteins in the particle except L12 interact directly with RNA, and all but 7 of the remaining 30 proteins interact with two rRNA domains or more (Table 2). The "champion" in this regard is L22, which is the only protein that interacts with RNA sequences belonging to all six domains of the 23*S* rRNA (Fig. 8C). The protein-mediated interactions between 5*S* rRNA and 23*S* rRNA are particularly extensive. Protein L18 attaches helix 1 and helix 2/3 of 5*S* rRNA to helix 87 of 23*S* rRNA. Protein L21e mediates an interaction between the same part of 5*S* rRNA and domains II and V. Protein L30 binds helix 4/5 region of 5*S* RNA to domain II. Loop C is linked to domain V by protein L5, and loop D is attached to domains II and V by protein L10e. Whatever else they may do, it is evident that an important function of these proteins is stabilization of the relative orientations of adjacent RNA domains. Several also help secure the tertiary structures of the domains with which they interact.

Because most ribosomal proteins interact with many RNA sequences and the number of proteins greatly exceeds the number of RNA domains, it can hardly come as a surprise that every rRNA domain interacts with multiple proteins (Table 2). Domain V, for example, interacts with 15 proteins, some intimately and a few in passing.

It is clear that the oligonucleotide binding experiments long relied on for information about the RNA binding properties of ribosomal proteins have underestimated their potential for interacting with RNA. The high-affinity RNA binding site identified on a protein by such an experiment may indeed be

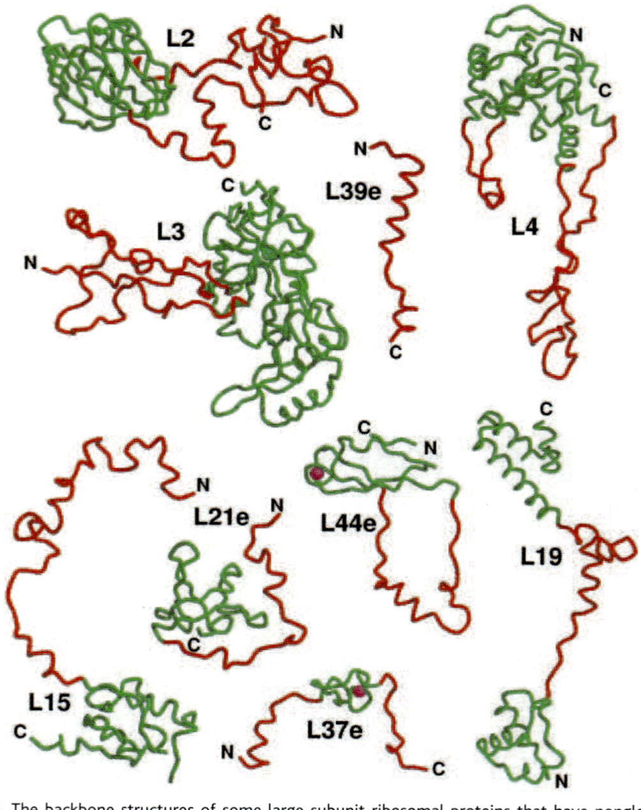

Fig. 6. The backbone structures of some large subunit ribosomal proteins that have nonglobular extensions. The globular domains of these proteins are shown in green, and their nonglobular extensions are depicted in red. The positions of the zinc ions in L44e and L37e are indicated by large dots in red.

important for ribosome assembly, but its many, weaker interactions with other sequences are likely to be missed, and they too may be vital for ribosome structure. Most ribosomal proteins crosslink RNA, and crosslinking is impossible without multiple interactions. Similar considerations may apply to proteins that are components of other ribonucleoproteins, such as the spliceosome.

Of the seven proteins that interact with only one domain, three (L1, L10, and L11) participate directly in the protein synthesis process. Rather than being included in the ribosome to ensure that the RNA adopts the proper conformation, it seems more appropriate to view the RNA as being structured to ensure the correct placement of these proteins. Another three (L24, L29, and L18e) interact with several secondary structure elements within the domains to which they bind, and presumably they function to stabilize the tertiary structures of their domains. The last of the single RNA domain proteins, L7ae, is puzzling. It cannot function as an RNA stabilizing protein because it interacts with only a single sequence in domain I, but it is far from the peptidyl transferase and factor binding sites. It is quite close to L1, however, which appears to be important for E-site function (*50*), and maybe it is involved in that activity. It could also be involved in the 70*S* assembly, because L7ae was originally assigned as a small subunit protein (HMS6).

While many ribosomal proteins interact primarily with RNA, a few interact significantly with other proteins. The most striking

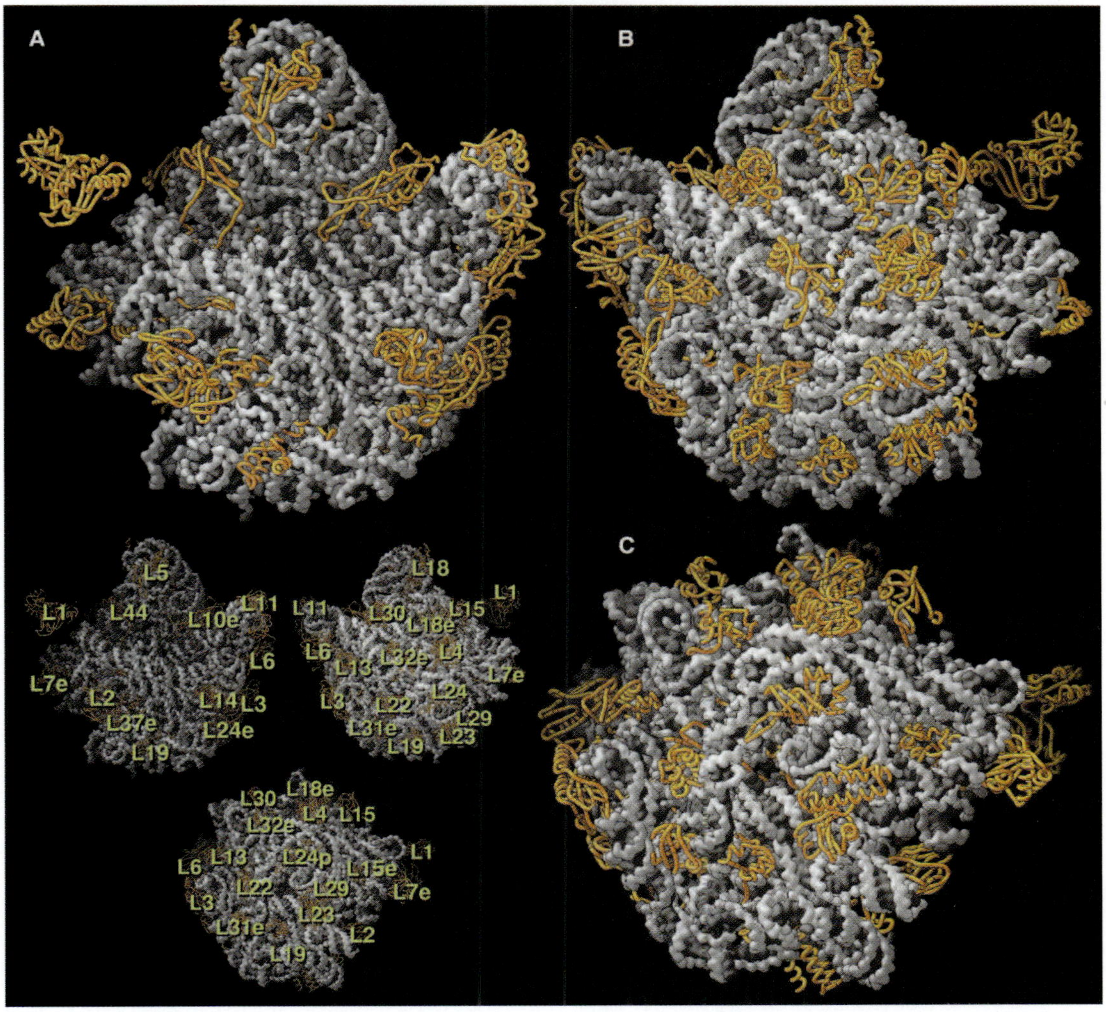

Fig. 7. Proteins that appear on the surface of the large ribosomal subunit. The RNA of the subunit is shown in gray and protein backbones are shown in gold. (**A**) The crown view of the subunit. (**B**) The back side of the subunit in the 180° rotated crown view orientation. (**C**) A view from the bottom of the subunit down the polypeptide tunnel exit which lies in the center. The proteins visible in each image are identified in the small images at the lower left of the figure. Figures were generated using RIBBONS (*67*).

RESEARCH ARTICLES

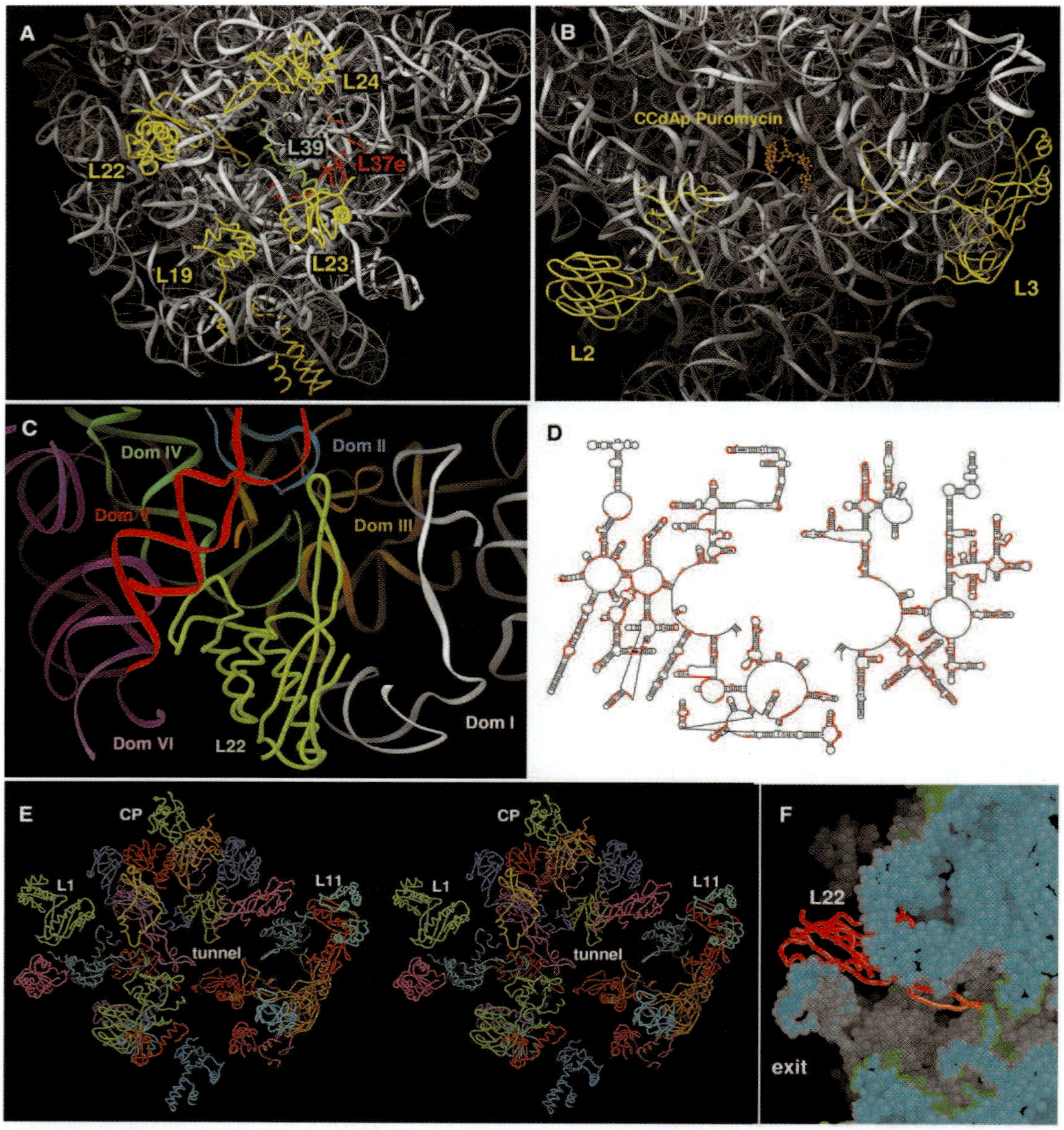

Fig. 8. The protein extensions into the RNA and multiple domain interactions. **(A)** Some of the proteins in the neighborhood of the polypeptide tunnel exit, showing the unusual extended structure of L39e (green) that enters the tunnel and L37e (red) that interpenetrates the RNA. L29, which is on top of L37e, has been removed. Protein L22 extends a long β hairpin extension inside the 23*S* rRNA. L24 has a similar extension but the entire protein is on the surface of the particle. L39 is the only protein in the subunit that lacks tertiary structure, whereas L37e has both NH_2- and COOH-terminal extensions. L19 is unique in having two globular domains on the surface of the subunit connected by an extended sequence that weaves through the RNA, shown as gray ribbons. **(B)** The nonglobular extensions of L2 and L3 reaching through the mass of 23*S* rRNA toward the peptidyl transferase site, which is marked by a CCdAp-puromycin molecule, the Yarus inhibitor (*64*). **(C)** L22 interacting with portions of all six of the domains of 23*S* rRNA. **(D)** Schematic of the 23*S* rRNA secondary structure showing the locations sequences (red) that make contact with protein. **(E)** Stereo view of the proteins of the large ribosomal subunit without the RNA. Proteins are colored as an aid to visualization only. **(F)** A cross section of the subunit in the area of the tunnel exit. Protein L22 is shown as ribbons in red, and the β hairpin loop where mutations confer erythromycin resistance is in orange. Atoms on the surface are gray, protein atoms are green, and atoms at the slice interface are blue.

structure generated by protein-protein interactions is the protein cluster composed of L3, L6, L13, L14, and L24e that is found close to the factor binding site. The surface of these proteins provides important interactions with factors. It may prove to be more generally the case that ribosomal proteins interacting primarily with RNA are principally stabilizing RNA structure, whereas some of those showing extensive protein-protein interactions may have additional binding functions.

The structure presented above illuminates both the strengths and weaknesses of approaches to complex assemblies that depend on determining the structures of components in isolation. The structures of the globular domains of homologs of the proteins in the large ribosomal subunit from *H. marismortui* are largely the same as those of the corresponding domains in the intact subunit, though adjustments in domain positions are sometimes required. Consequently, these structures were very useful for locating proteins and interpreting lower resolution electron density maps. However, for obvious reasons, the structures of the extended tails and loops of ribosomal proteins cannot be determined in the absence of the RNAs that give them structure, and the feasibility of strategies that depend on producing low–molecular weight RNA-protein complexes that have all the RNA contacts required to fix the structures of such proteins seems remote. The structures of RNA fragments also depend on their context. Whereas the sarcin/ricin loop has much the same structure in isolation (*33, 34*) as it does in the ribosome, the structure of 5*S* rRNA in isolation (*37*) differs in some respects from what is seen in the ribosome, and the structure of the isolated P loop (*51*) shows no resemblance to the structure of the P loop in the ribosome. Clearly, a "structural genomics" approach to the ribosome, which would have entailed determining the structures of all of the proteins and all possible rRNA fragments, neither would have provided the relevant structures of all of the pieces nor would it have shown their relative positions. Indeed, the structure of the large ribosomal subunit highlights the importance of structural studies of entire assemblies that show biological activity.

The analysis of the 50*S* ribosomal subunit structure presented here describes the overall architectural principles of RNA folding and its interaction with proteins, but many exciting details remain to be explored. The principles of protein-RNA interaction that should emerge from the 27 protein complexes with RNA have yet to be developed. On average, each of the 27 proteins has 3000 Å^2 of surface area in contact with RNA, which is comparable to the 2700 Å^2 of glutaminyl-tRNA synthetase that contact $tRNA^{Gln}$ (*52*), so the number of interactions between RNA and protein to be analyzed in the large subunit structure is 30 times the number in this synthetase complex. Further, because the RNA structure of the large subunit will increase the RNA structural database by a factor of 4 to 5, most of the important RNA secondary and tertiary structural motifs to be found in nature may be represented. It will be interesting to see whether a complete analysis of this RNA structural database will enable the prediction of structures for other RNA sequences. Unknown at this time is the ease with which it will be possible to model by sequence homology the 50*S* ribosomal subunit rRNA from other species and kingdoms. However, the extensive sequence conservation in the 23*S* rRNA that forms the core active site and peptide tunnel regions suggests that reasonably accurate homology modeling based on this *H. marismortui* subunit structure may be feasible. Finally, enormous numbers of monovalent and divalent metal ions as well as water molecules are visible in this map. Analysis of their interactions with RNA should elucidate their roles in the formation and stabilization of RNA structure.

References and Notes

1. R. A. Garrett *et al.*, Eds., *The Ribosome: Structure, Function, Antibiotics and Cellular Interactions* (American Society for Microbiology, Washington, DC, 2000).
2. B. Wittmann-Liebold, in *Structure, Function, and Genetics of Ribosomes*, B. Hardesty and G. Kramer, Eds. (Springer-Verlag, New York, 1986), pp. 326–361.
3. R. K. Agrawal, P. Penczek, R. A. Grassucci, J. Frank, *Proc. Natl. Acad. Sci. U.S.A.* **95**, 6134 (1998); H. Stark, M. V. Rodnina, H.-J. Wieden, M. van Hell, W. Wintemeyer, *Cell* **100**, 301 (2000).
4. F. Mueller *et al.*, *J. Mol. Biol.* **298**, 35 (2000).
5. A. Yonath *et al.*, *Biochem. Int.* **2**, 428 (1980).
6. S. D. Trakanov *et al.*, *FEBS Lett.* **220**, 319 (1987).
7. N. Ban *et al.*, *Cell* **93**, 1105 (1998).
8. N. Ban *et al.*, *Nature* **400**, 841 (1999).
9. J. H. Cate, M. M. Yusupov, G. Z. Yusupova, T. N. Earnest, H. F. Noller, *Science* **285**, 2095 (1999).
10. W. M. Clemons Jr. *et al.*, *Nature* **400**, 833 (1999).
11. A. Tocilj *et al.*, *Proc. Natl. Acad. Sci. U.S.A.* **96**, 14252 (1999).
12. *H. marismortui* (American Type Culture Collection 43049) was grown, and ribosomes were prepared from it as described previously (*7, 53*). The buffers and precipitants used for crystal growth were those described earlier (*7, 54*), but with the following modifications. A crystallization solution was obtained by back-extraction of precipitated subunit at saturation into the crystallization buffer [1.2 M KCl, 0.5 M NH_4Cl, 100 mM Kacetate, 30 mM $MgCl_2$, 7% polyethylene glycol (PEG) 6000, 15 mM tris, 15 mM MES, and 1 mM $CdCl_2$ (pH 7.1)]. The crystals that resulted had maximum dimensions of 0.5 mm by 0.5 mm by 0.2 mm and were harvested after ~2 weeks. Crystals were stabilized by gradual transfer into a solution containing 12% PEG 6000, 22% ethylene glycol, 1.7 M NaCl, 0.5 M NH_4Cl, 100 mM potassium acetate, 30 mM $MgCl_2$, and 1 mM $CdCl_2$ (pH 6.2) at 4°C. The crystals were flash-frozen in liquid propane.
13. All data, except the two native data sets, were collected at the National Synchrotron Light Source (Brookhaven National Laboratory, Upton, NY) from crystals frozen at 100 K using beamlines X12b and X25 and recorded using a 345-mm MAR imaging plate. For each heavy-atom derivative, anomalous diffraction data were collected at the wavelength corresponding to the peak anomalous scattering, except for uranium, where a low energy was used. The beam size was 100 μm by 100 μm for most data collections at beamline X25 and 200 μm by 200 μm at beamline X12b. The crystals were aligned along the long axis of the unit cell (~575 Å) so that 1.0° oscillations could be used to collect reflections out to a maximum of 2.7 Å resolution at the edge of the MAR detector. At beamline X12b, the crystal-to-detector distances were varied depending on wavelength, crystal quality, and beam divergence so that maximum resolution data could be collected while avoiding overlapping of spots. At beamline X25, the detector was positioned on a rigid platform at 480 mm, which allowed data collection to 3.2 Å for iridium and osmium derivatives with the wavelength set at the anomalous edge. Native data to 2.4 Å resolution were collected at the structural biology beamline 19ID of the Advanced Photon Source (Argonne National Laboratory, Argonne, IL) using a 3 by 3 charge-coupled device (CCD) detector, an 80 μm by 80 μm beam size, and 0.4° oscillations. Data sets were processed with DENZO and SCALEPACK (*55*).
14. Heavy-atom–based phasing was extended to 3.2 Å resolution by combining MIR (multiple isomorphous replacement) phases calculated for two different isomorphous groups of data (MIR1 and MIR2 (Table 1) with single-wavelength anomalous dispersion (SAD) phases. Weights on the individual phase sets were adjusted on the Hendrickson-Lattman coefficients to prevent phase bias due to nonisomorphism. The best two derivatives were osmium pentamine and iridium hexamine, each of which contained a large number of binding sites (Table 1). Other derivatives with smaller numbers of sites further improved map quality. All phasing was done by maximum likelihood methods implemented in CNS with the exception of the Ta_6Br_{12} derivative, which was refined in SHARP (*56*), represented as spherically averaged electron density (Table 1). Phases were improved and extended from 3.3 to 2.4 Å by solvent flipping (*15*), with a scheme that progressively refined the solvent mask from 12 to 2.6 Å detail. RNA structure was built by positioning of the individual residues, and their rotomers were picked from a library derived from high-resolution structures (M. Kjeldgaard, unpublished data). Models were built with O (*57*).
15. J. P. Abrahams and A. G. W. Leslie, *Acta Crystallogr.* **D52**, 30 (1996).
16. A. T. Brunger *et al.*, *Acta Crystallogr.* **D54**, 905 (1998).
17. D. E. Tronrud, *Methods Enzymol.* **277** 306 (1997).
18. R. R. Gutell, in *Ribosomal RNA: Structure, Evolution, Processing and Function in Protein Biosynthesis*, A. Dahlberg and R. Zimmerman, Eds. (CRC Press, Boca Raton, FL, 1996), pp. 111–128.
19. B. Wittmann-Liebold *et al.*, in *The Ribosome: Structure, Function, and Evolution*, W. E. Hill *et al.*, Eds. (American Society for Microbiology, Washington, DC, 1990), pp. 598–616.
20. M. Oakes, E. Henderson, A. Scheinman, M. Clark, J. A. Lake, in *Structure, Function and Genetics of Ribosomes*, B. Hardesty and G. Kramer, Eds (Springer-Verlag, New York, 1986), pp. 47–67; G. Stoeffler and M. Stoeffler-Meilicke, in *Structure, Function, and Genetics of Ribosomes*, B. Hardesty and G. Kramer, Eds. (Springer-Verlag, New York, 1986), pp. 23–46.
21. G. L. Conn, D. E. Draper, E. E. Lattman, A. G. Gittis, *Science* **284**, 1171 (1999).
22. B. T. Wimberly, R. Guymon, J. P. McCutcheon, S. W. White, V. R. Ramakrishnan, *Cell* **97**, 491 (1999).
23. W. Moller and J. A. Maassen, in *Structure, Function, and Genetics of Ribosomes*, B. Hardesty and G. Kramer, Eds. (Springer-Verlag, New York 1986), pp. 309–325.
24. J. Walleczek, D. Schuler, M. Stoeffler-Meilicke, R. Brimacombe, G. Stoeffler, *EMBO J.* **7**, 3571 (1988).
25. P. Ostergaard *et al.*, *J. Mol. Biol.* **284**, 227 (1998).
26. H. Mao, S. A. White, J. R. Williamson, *Nature Struct. Biol.* **6**, 1139 (1999).
27. I. G. Wool, Y.-L. Chan, A. Gluck, *Biochem. Cell Biol.* **73**, 933 (1995).
28. N. Nevskaya *et al.*, *Struct. Fold. Des.* **8**, 363 (2000).
29. H. Leffers, J. Kjems, L. Ostergaard, N. Larsen, R. A. Garrett, *J. Mol. Biol.* **195**, 43 (1987).
30. H. F. Noller *et al.*, in *The Ribosome: Structure, Function, and Evolution*, W. E. Hill *et al.*, Eds. (American Society for Microbiology, Washington, DC, 1990), pp. 73–92.

31. G. M. Culver, J. H. Cate, G. Zh. Yusupova, M. M. Yusupov, H. F. Noller, *Science* **285**, 2133 (1999).
32. R. A. Garrett and C. Rodriguez-Fonseca, in *Ribosomal RNA: Structure, Evolution, Processing and Function in Protein Biosynthesis*, R. A. Zimmerman and A. E. Dahlberg, Eds. (CRC Press, Boca Raton, FL, 1996), pp. 327–355.
33. A. A. Szewczak and P. B. Moore, *J. Mol. Biol.* **247**, 81 (1995).
34. C. C. Correll *et al.*, *Proc. Natl. Acad. Sci. U.S.A.* **95**, 13436 (1998).
35. I. G. Wool, A. Gluck, Y. Endo, *Trends Biochem. Sci.* **17**, 266 (1992).
36. D. Moazed, J. M. Roberston, H. F. Noller, *Nature* **334**, 362 (1988).
37. C. C. Correll, B. Freeborn, P. B. Moore, T. A. Steitz, *Cell* **91**, 705 (1997).
38. S. Gerbi, in *Ribosomal RNA: Structure, Evolution, Processing and Function in Protein Biosynthesis*, R. A. Zimmermann and A. E. Dahlberg, Eds. (CRC Press, Boca Raton, FL, 1996), pp. 71–88.
39. P. Dube *et al.*, *Structure* **6**, 389 (1998).
40. V. C. Ware *et al.*, *Nucleic Acids Res.* **22**, 7795 (1983).
41. P. B. Moore, *Annu. Rev. Biochem.* **68**, 287 (1999)
42. E. Westhof and V. Fritsch, *Structure* **8**, R55 (2000).
43. A. Nakagawa *et al.*, *EMBO J.* **18**, 1459 (1999).
44. M. Wahl, R. Huber, M. C. Wahl, *EMBO J.* **19**, 807 (2000).
45. R. Beckmann *et al.*, *Science* **278**, 2123 (1997).
46. M. G. Rossmann and J. E. Johnson, *Annu. Rev. Biochem.* **58**, 533 (1989); A. Liljas, *Int. Rev. Cytol.* **124**, 103 (1991).
47. K. Lugor *et al.*, *Nature* **389**, 251 (1997).
48. H. S. Chittum and W. S. Champney, *J. Bacteriol.* **176**, 6192 (1994).
49. B. Lee and F. M. Richards, *J. Mol. Biol.* **55**, 379 (1971).
50. R. K. Agrawal *et al.*, *J. Biol. Chem.* **274**, 8723 (1999).
51. E. V. Puglisi, R. Green, H. F. Noller, J. D. Puglisi, *Nature Struct. Biol.* **4**, 775 (1997).
52. M. A. Rould, J. J. Perona, D. Söll, T. A. Steitz, *Science* **246**, 1135 (1989).
53. A. Shevack, H. S. Gewitz, B. Hennemann, A. Yonath, H. G. Wittmann, *FEBS Lett.* **184**, 68 (1985).
54. K. vanBohlen *et al.*, *J. Mol. Biol.* **222**, 11 (1991).
55. Z. Otwinowski, in *Data Collection and Processing*, L. Sawyer, N. Isaacs, D. Bailey, Eds. (SERC Daresbury Laboratory, Warrington, UK, 1993), pp. 52–62.
56. E. de La Fortelle and G. Bricogne, *Methods Enzymol.* **276**, 472 (1997).
57. T. A. Jones, S. Cowan, J.-Y. Zou, M. Kjeldgaard, *Acta Crystallogr.* **A46**, 110 (1991).
58. B. L. Golden, V. Ramakrishnan, S. W. White, *EMBO J.* **12**, 4901 (1993).
59. M. A. Markus, A. P. Hinch, S. Huang, D. E. Draper, D. E. Torchia, *Nature Struct. Biol.* **4**, 70 (1997).
60. M. Leijonmarck, S. Eriksson, A. Liljas, *Nature* **286**, 824 (1980).
61. C. Davies, S. W. White, V. Ramakrishnan, *Structure* **4**, 55 (1996).
62. J. Unge *et al.*, *Structure* **6**, 1577 (1998).
63. K. S. Wilson, K. Appelt, J. Badger, I. Tanaka, S. W. White, *Proc. Natl. Acad. Sci. U.S.A.* **83**, 7251 (1986).
64. P. Nissen, J. Hansen, N. Ban, P. B. Moore, T. A. Steitz, *Science* **289**, 920 (2000).
65. R. R. Gutell *et al.*, in preparation. (Data can be found at www.rna.icmb.utexas.edu.)
66. M. Symanski, T. Specht, M. C. Barciszewska, J. Barciszewski, V. A. Erdmann, *Nucleic Acids Res.* **26**, 156 (1998).
67. M. Carson, *Methods Enzymol.* **277**, 493 (1997).

68. We thank B. Freeborn for her skilled technical assistance in preparing 50*S* ribosomal subunit material and crystals. We are indebted to D. Klein, M. Lu, S. Antoć, and M. Schmeing for their help with the fitting of protein sequences into electron density. We thank M. Kjeldgaard for providing us with a prerelease version of O adapted for RNA model building, J. Cate for contributing the iridium hexamine, R. Gutell and J. Cannone for preparing the secondary structure diagram for 23*S* rRNA, M. Wahl for sending us coordinates for L4 before their release, J. Williamson for sending us coordinates of the L30e-RNA complex before their release, R. Sweet, L. Berman, and M. Capel for their assistance with data collection at the National Synchrotron Light Source, and A. Joachimiak and the staff of 19ID at the Advanced Photon Source. Supported by grants from NIH to T.A.S. (GM22778) and P.B.M. (GM54216) and by a grant from the Agouron Institute to T.A.S. and P.B.M. N.B. is supported by a Burroughs Welcome Fund Career Award. Complete coordinates for 23*S* and 5*S* rRNAs and α carbon coordinates for the 27 proteins discussed have been deposited in the Protein Data Bank. The accession number is 1FFK for the amplitudes, experimental phases, and coordinates.

29 June 2000; accepted 24 July 2000

From Nenad Ban, Poul Nissen, Jeffrey Hansen, Peter B. Moore, and Thomas A. Steitz. The complete atomic structure of the large ribosomal subunit at 2.4 Å resolution. *Science*, **289**, 905–920 (2000). Reprinted with permission from AAAS.

nature Vol 438|24 November 2005|doi:10.1038/nature04152

LETTERS

An induced-fit mechanism to promote peptide bond formation and exclude hydrolysis of peptidyl-tRNA

T. Martin Schmeing[1], Kevin S. Huang[1], Scott A. Strobel[1,2] & Thomas A. Steitz[1,2,3]

The large ribosomal subunit catalyses the reaction between the α-amino group of the aminoacyl-tRNA bound to the A site and the ester carbon of the peptidyl-tRNA bound to the P site[1], while preventing the nucleophilic attack of water on the ester, which would lead to unprogrammed deacylation of the peptidyl-tRNA. Here we describe three new structures of the large ribosomal subunit of *Haloarcula marismortui* (*Hma*) complexed with peptidyl transferase substrate analogues that reveal an induced-fit mechanism in which substrates and active-site residues reposition to allow the peptidyl transferase reaction. Proper binding of an aminoacyl-tRNA analogue to the A site induces specific movements of 23S rRNA nucleotides 2618–2620 (*Escherichia coli* numbering 2583–2585) and 2541(2506), thereby reorienting the ester group of the peptidyl-tRNA and making it accessible for attack. In the absence of the appropriate A-site substrate, the peptidyl transferase centre positions the ester link of the peptidyl-tRNA in a conformation that precludes the catalysed nucleophilic attack by water. Protein release factors[2] may also function, in part, by inducing an active-site rearrangement similar to that produced by the A-site aminoacyl-tRNA, allowing the carbonyl group and water to be positioned for hydrolysis.

During protein synthesis, the ribosome polymerizes amino acids by repeatedly positioning the α-amino group of aminoacyl-tRNAs to react with the ester carbon of peptidyl-tRNAs[1]. The ester link of the peptidyl-tRNA must be protected from hydrolysis when the A site is unoccupied, to prevent the production of truncated protein fragments. The peptidyl transferase centre selectively activates one nucleophile, the α-amino group of the aminoacyl-tRNA, while preventing the activated attack by another, a water molecule. The questions that arise are analogous to those posed by Koshland nearly 50 years ago in explaining why hexokinase does not catalyse the hydrolysis of ATP in the absence of glucose[3]: How does the ribosome protect the peptidyl-tRNA from inopportune catalysed hydrolysis during protein synthesis when the A site is empty? Why does it not activate the nucleophilic attack of water, just as it does for the α-amino group? For hexokinase Koshland proposed[3], and Bennett and Steitz established[4], that the binding of glucose induces an enzyme conformational change that is required for the reaction, and water does not produce this change.

Early biochemical experiments found that binding of an unacylated tRNA mimic in the A site promotes the hydrolysis of a peptidyl-tRNA bound to the P site by two orders of magnitude[5]. The trinucleotide CCA, which corresponds to the acceptor end of the tRNA, promotes this hydrolysis equally well, whereas CA and A have no effect on the hydrolysis rate. These data suggest an important alteration of the active site produced by binding of tRNA to the A site, for which C74 is required. More recently, Ehrenberg and colleagues[6] have found that whereas fMet-$tRNA^{fMet}$ bound to the 70S P site is stable for nearly 6 h, the addition of unacylated $tRNA^{Phe}$ to the A site stimulates deacylation as much as 300-fold, almost to the level produced by a protein release factor. Hydrolysis at this rate could lead to one premature termination in about every five proteins produced in the cell, because the A site is mostly unoccupied during elongation as a result of rapid reaction and progression to the hybrid state[7] after rate-limiting tRNA accommodation[8].

To understand these observations, we have determined two structures of the large ribosomal subunit complexed simultaneously with both an A-site and a P-site substrate (Fig. 1) (2.58 and 2.75 Å resolution; Supplementary Table S1). One of the A-site substrates, cytidine-cytidine-hydroxypuromycin (CChPmn), includes C74; the other, ChPmn, does not. The α-amino group is replaced by a hydroxyl group, which decreases the rate of reaction 500-fold[9], allowing unreacted substrates to be revealed in electron density maps. The P-site substrate in both cases is cytidine-cytidine-adenosine-phenylalanine-caproic acid-biotin (CCApcb)[10]. We have also determined the structure (2.60 Å resolution) of the 50S subunit complexed to a novel transition-state analogue (TSA) that mimics that tetrahedral intermediate of the reaction (Fig. 1). Systematic movements of the rRNA are observed, including the rearrangement of some nucleotides that are known to be conformationally flexible[10–12].

The mechanism by which the ribosome protects the peptidyl-tRNA from hydrolysis can be understood by inspecting the conformation of the peptidyl-tRNA analogue bound to the large ribosomal subunit. The P-site ester is inaccessible to nucleophilic attack by water or the α-amino group in the structures of three complexes that include a P-site-bound peptidyl-tRNA fragment without a full A-site component (Supplementary Fig. S2). The structures of the 50S subunit complexed with CCApcb and the antibiotic sparsomycin[13] (complex 1) CCApcb bound to either the P or A sites[13] (complex 2) and CCApcb bound in the P site and ChPmn bound in the A site (complex 3) provide an excellent approximation to the structure of the ribosome bound to a P-site substrate alone. In complex 1, CCApcb is bound homogeneously to the P site. Sparsomycin sterically overlaps the A site, but it does not make the same interactions as an A-site substrate does. Complex 2 includes a significant contribution from molecules with CCApcb bound to the P site alone, but the peptidyl moiety is overlapped with density from molecules having exclusive binding of CCApcb to the A site. Simultaneous binding of the substrate to both sites is excluded by overlap between the peptidyl portions of the molecule within the exit tunnel. In complex 3, the binding of ChPmn drives CCApcb to the P site, but the lack of C74 leaves the rRNA in an uninduced state. The uninduced conformation of the rRNA in complexes 1 and 3 are fully compatible with the observed electron density for complex 2, although the mixture of A-site and P-site binding makes the electron density in this structure difficult to interpret in the absence of the other structures. Together, these structures provide clear evidence

[1]Department of Molecular Biophysics and Biochemistry, [2]Department of Chemistry, Yale University and [3]Howard Hughes Medical Institute, New Haven, Connecticut 06520, USA.

that the complex with substrate bound only to the P site is in the uninduced state.

These complexes all show that the ester group of CCApcb is both positioned and sequestered by the 23S rRNA. U2620(2585) lies on one side of the ester and A2486(2451) and C2104(2063) are positioned on the other (Fig. 2). These interactions require the carbonyl oxygen of the ester group to point towards the A site. Small-molecule studies[14] show that the optimal angle for nucleophilic attack is about 105° from the plane of the ester group. Such an attack is not possible in this ester group orientation because the required approach is blocked from both sides (Fig. 2). Furthermore, in serine proteases, a system that resembles this case more closely, the attack angle has been shown[15] to be about 90°, which would require an approach trajectory that is even more unfavourable. This conformation of rRNA around the peptidyl-tRNA could therefore protect the nascent peptide from premature hydrolysis by occluding water from appropriate attack positions and is consistent with the observation (E. Youngman and R. Green, personal communication) that binding of peptidyl-tRNA

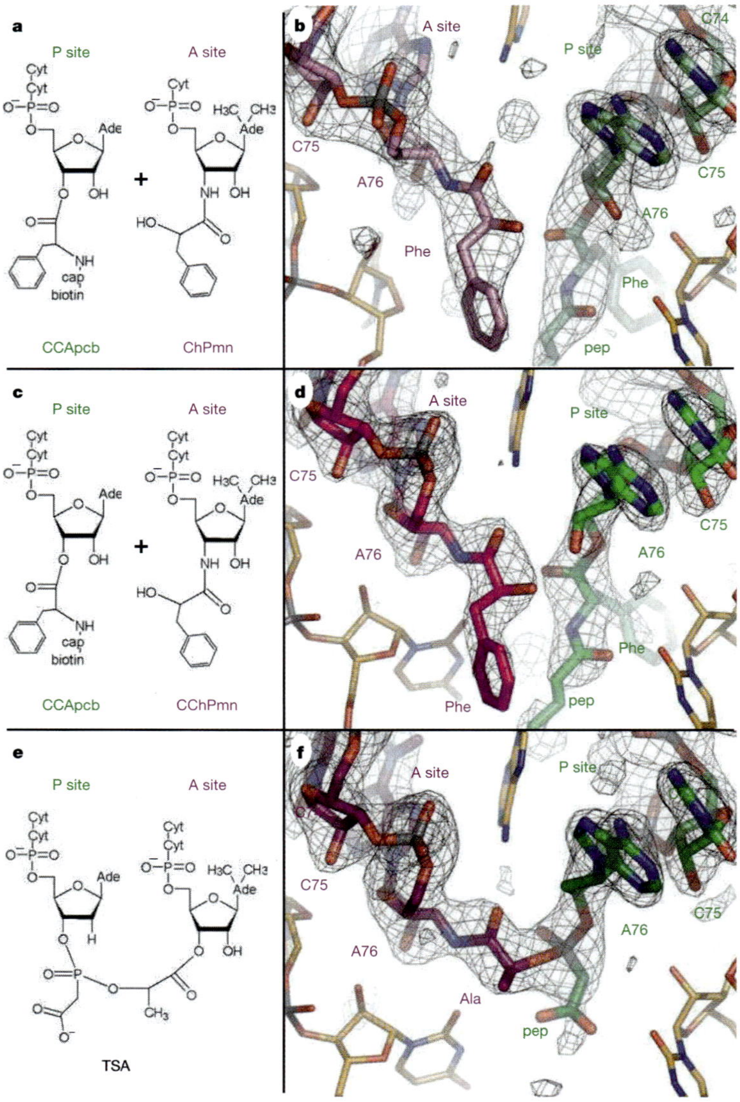

Figure 1 | Chemical structures and electron density maps. a, c, e, The chemical structures of CCApcb and ChPmn (**a**), CCApcb and CChPmn (**c**) and the TSA (**e**). **b, d, f,** Minimally biased F_o–F_c electron density maps from experiments with the respective substrates: ChPmn (pink), CCApcb (light green), and 23S rRNA (orange) displayed with a map calculated at 2.7 Å resolution, contoured at 2.5σ (**b**); CChPmn (purple), CCApcb (green), and 23S rRNA (orange) displayed with map calculated at 2.4 Å resolution, contoured at 3.0σ (**d**); the TSA (A-site portion in dark purple and the P-site portion in dark green) and 23S rRNA (orange) shown with a map calculated at 2.5 Å resolution, contoured at 3.0σ (**f**). The same view of the peptidyl transferase centre is shown in **b**, **d** and **f**.

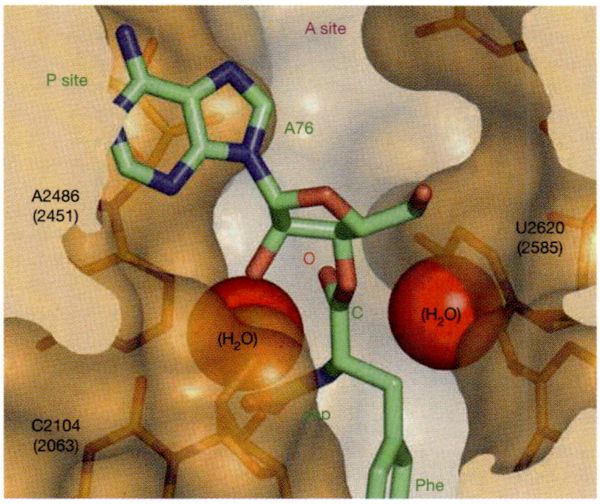

Figure 2 | Steric exclusion of water results in protection of peptidyl-tRNA from deacylation in the uninduced state. When the peptidyl transferase centre is not in the induced state, as occurs when ChPmn and CCApcb (green) are bound, the ribosome (orange surface) occludes water from positions that could attack the ester group. Theoretical water molecules (red spheres), are shown aligned for attack at 105° to the plane of the ester group, 2.8 Å away from the ester carbon. Steric clashes with A2486(2451) and C2104(2063) block the position on one side, whereas the uninduced conformation of U2620(2585) would block the other side.

to the ribosome decreases the rate of peptidyl-tRNA hydroysis by more than 20-fold.

In the complex of the 50S subunit with ChPmn and CCApcb, not only is the P-site substrate protected from attack by a water molecule, the substrates are in a conformation that is unfavourable for peptidyl transferase reaction. The carbonyl ester oxygen of CCApcb is pointing towards the α-amino group of ChPmn, and the atoms that are to react are about 4 Å apart. The conformation of the rRNA in this complex is similar to those with an empty A site[16], which include a G–U wobble pair formed by G2618(2583) and U2541(2506). Nevertheless, ChPmn makes the interactions typical of A-site binding[11], as C75 of ChPmn base pairs with G2588(2553), and A76 makes an A-minor interaction[17] with G2618(2583) (Fig. 3a).

Binding of an A-site substrate containing C74 induces a conformational change in the ribosome that exposes the ester group of the peptidyl-tRNA for attack by the aminoacyl-tRNA (Fig. 3, and Supplementary movies 1 and 2). The inclusion of C74 in CChPmn produces a continuous stacking interaction with U2590(2555) that positions the substrate closer to the centre of the active site (Fig. 3a). C75 of the A-site substrate still base pairs with G2588(2553), but G2618(2583) shifts to maintain the A-minor interaction with A76 (Fig. 3a). This shift of G2618(2583) causes its G–U wobble pair with U2541(2506) to break, and U2541(2506) swings through 90° (Fig. 3b). Furthermore, the movement of G2618(2583) is propagated along the rRNA chain, and the two adjacent bases, ^{methyl}U2619(2584) and U2620(2585), shift in the same direction. U2620(2585), which had been near the P site with its base in van der Waals contact with the ester group of the peptidyl-tRNA mimic, swings over to approach the 2′-hydroxyl of A76 in the A site and adopts a conformation previously prevented by ^{methyl}U2619(2584). This displacement of U2620(2585) creates space adjacent to the ester of the P-site substrate, allowing the peptidyl moiety and the sugar of A76 to rotate, exposing the ester to attack from the A site. The stacking of the CCA bases also lowers the α-amino group in the active site by almost 1 Å, placing it in a position that is more favourable for nucleophilic attack on the more appropriately oriented carbonyl carbon of peptidyl-CCA (Fig. 4). In this pre-reaction ground state, the hydroxyl group mimic of the α-amino forms hydrogen bonds with both N3 of A2486(2451) and the 2′-hydroxyl of A76 of the P-site substrate, a group that is critical for reaction[18] and may enhance its nucleophilicity. The conformation of the rRNA in the TSA complex is nearly identical to that of the complex with CChPmn and CCApcb (Fig. 3, and Supplementary Fig. S3), showing that the RNA rearrangements induced by the binding of CCPmn are indeed likely to be relevant to the peptidyl transferase reaction pathway.

The ribosome therefore uses a substrate-induced fit mechanism to facilitate peptide bond formation and exclude hydrolysis, similar to

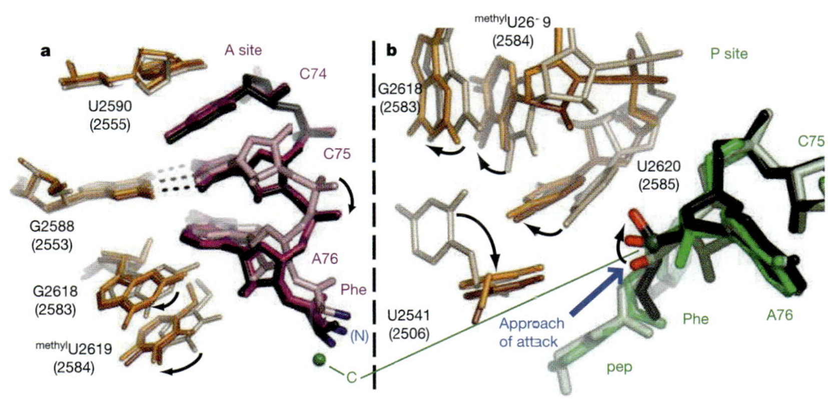

Figure 3 | Movements of rRNA and peptidyl-tRNA induced by proper binding of the A-site substrate. a, An A-site view of the three complexes. Without C74, ChPmn (pink, with rRNA in wheat colour) is positioned higher in the A site. In substrates with C74, both CChPmn (purple, with orange rRNA) and the TSA (black, with brown rRNA), C74 stacks with U2590(2555), shifting the substrates down and the α-amino group closer to the ester carbon of the P-site substrate (green). G2618(2583) shifts to maintain the A-minor interaction, causing ^{methyl}U2619(2584) to move too. **b,** P-site view of the same three complexes. The movement induced in G2618(2583) by A-site substrate binding breaks its G–U wobble pair with U2451(2506), which swings through 90°. ^{methyl}U2619(2584) and U2620(2585) shift, allowing the ester group to move from its position in the CCApcb (light green) and ChPmn complex, to that when CCApcb (medium green) is bound with CChPmn, and finally to that when it has been attacked by the A-site substrate, as shown with the TSA (black). This induced-fit mechanism is illustrated in Supplementary movies 1 and 2.

NATURE|Vol 438|24 November 2005 LETTERS

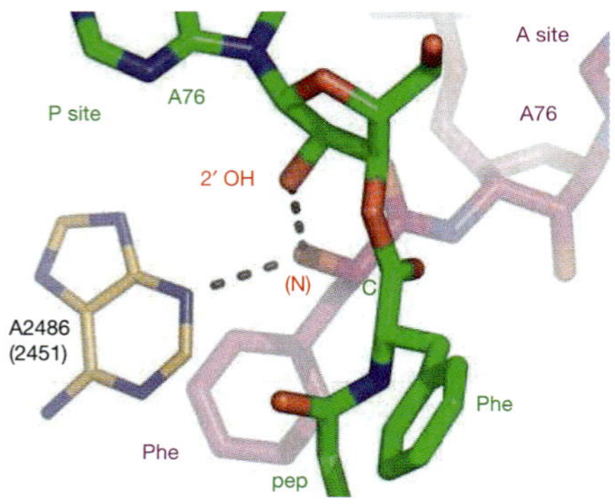

Figure 4 | Pre-attack conformation of the substrates. The hydroxyl group representing the α-amino group of the A-site substrate, CChPmn (purple) is in position to attack the ester group of the P-site substrate CCApcb (green). It is within hydrogen-bonding distance of N3 of A2486(2451) and the 2′ hydroxyl group of the P-site substrate. In this ground state, the reactive groups are 3.7 Å apart.

that proposed for several enzymes, including hexokinase[4]. The reactive faces of the P-site ester are protected from nucleophilic attack by water except when a complete CCA is bound in the A site. The ester becomes exposed and properly positioned only when aminoacyl-tRNA binds the A loop, an event that is coordinated with positioning of the α-amino group for attack. If a CCA or tRNA bound to the A site is not aminoacylated, the resulting conformational change exposes the peptidyl-tRNA to nucleophilic attack by water, which accounts for the observations that either CCA[5] or full tRNA[6] in the A site promotes the hydrolysis of peptidyl-tRNA. Indeed, the structure of the large ribosomal subunit with CCA bound simultaneously to the A, P and E sites[19] shows similar conformational changes.

Although C74 seems to be important in the proper positioning of substrate and the resultant conformational changes, the catalytic rate of peptidyl transfer is similar for cytidine-puromycin (CPmn) and CCPmn[20]. It therefore seems that the conformational change is not rate-limiting for the fragment reaction. In the electron density map for the complex of ChPmn and CCApcb (Fig. 1), the features of the ChPmn are not as clear as expected at this resolution, indicating that there might be more than one mode of binding. Presumably both the induced and uninduced conformations are present in the crystal, but the uninduced state predominates. The apparent contradiction between the ability of CPmn to react and the inability of CA to stimulate deacylation can be attributed to an important function of the interactions made by the amino acid moiety of ChPmn in shifting the conformational equilibrium towards the induced state. CPmn can access the induced conformation to react, but the deacylated CA cannot access the induced state enough to increase the rate of hydrolysis[5].

When a translating ribosome encounters a stop codon in mRNA, a protein release factor binds and promotes the deacylation of the peptidyl-tRNA[2]. A universally conserved GGQ motif in class one release factors (RFs) interacts with the peptidyl transferase centre[6,21–23]. In the context of this induced-fit model, the RF presumably promotes the conformational rearrangement of at least a subset of those rRNA nucleotides that are responsible for activation, in particular U2620(2585). Such rearrangement would allow the ester group in the P site to move out of the protected conformation and undergo attack by a water molecule. This is consistent with the result that mutation of U2620(2585) leads to a decrease in the rate of factor-catalysed peptide release[24]. The requirement for the RF to induce these conformational changes is also consistent with the similar rate of deacylation stimulated by the RF or tRNA in the presence of 20% ethanol[6]. Furthermore, because unacylated tRNA promotes hydrolysis, despite not having functional groups that could participate directly in the reaction, the primary role of the RF might be to promote the conformational change that we visualize here, and not to activate the water molecule[6].

Thus, the peptidyl transferase centre may use this induced-fit mechanism to help catalyse both essential reactions in which it participates: the peptidyl transferase reaction during elongation and the programmed deacylation of peptidyl-tRNA at termination. In addition, the structural demonstration that the peptidyl transferase centre uses an induced-fit mechanism to enhance substrate specificity shows that such mechanisms pre-date protein enzymes and were available to the ribozymes of the RNA world.

METHODS

The syntheses of ChPmn, CChPmn and the TSA are as reported elsewhere[9,25]. Crystals of the large ribosomal subunit of *H. marismortui* were grown and stabilized as described previously[26]. Crystals were soaked in stabilization solutions containing either 3 mM ChPmn and 1 mM CCApcb, 3 mM CChPmn and 1 mM CCApcb with 100 mM $SrCl_2$, or 1 mM TSA for between 1.5 and 3.5 h before being flash-frozen in liquid propane. Diffraction data sets were collected at the National Synchrotron Light Source beamline X25 and the Advanced Light Source beamline 8.2.2, using 0.4° oscillations. Data were processed with the program HKL2000 (ref. 27). Structures were determined by rigid-body refinement of the apo-subunit structure (PDB code 1S72; ref. 16) with the program CNS[28], followed by iterative rounds of gradient energy minimization, *B*-factor refinement and modelling with the program O[29] (Supplementary Table S1). The set of reflections used for the calculation of R_{free} for these structures is the same as that used for the refinement of the apo subunit[26]. The complexes were compared (Fig. 3) by the superimposition of phosphorus atoms of domain 5 of 23S rRNA, using the lsq function in the program O.

Received 13 December 2004; accepted 23 August 2005.

1. Green, R. & Lorsch, J. R. The path to perdition is paved with protons. *Cell* **110**, 665–668 (2002).
2. Kisselev, L. L. & Buckingham, R. H. Translational termination comes of age. *Trends Biochem. Sci.* **25**, 561–566 (2000).
3. Koshland, D. E. in *The Enzymes* (eds Boyer, P. D., Lardy, H. & Myrback, K.) 305–346 (Academic, New York, 1959).
4. Bennett, W. S. Jr & Steitz, T. A. Glucose-induced conformational change in yeast hexokinase. *Proc. Natl Acad. Sci. USA* **75**, 4848–4852 (1978).
5. Caskey, C. T., Beaudet, A. L., Scolnick, E. M. & Rosman, M. Hydrolysis of fMet-tRNA by peptidyl transferase. *Proc. Natl Acad. Sci. USA* **68**, 3163–3167 (1971).
6. Zavialov, A. V., Mora, L., Buckingham, R. H. & Ehrenberg, M. Release of peptide promoted by the GGQ motif of class 1 release factors regulates the GTPase activity of RF3. *Mol. Cell* **10**, 789–798 (2002).
7. Moazed, D. & Noller, H. F. Intermediate states in the movement of transfer RNA in the ribosome. *Nature* **342**, 142–148 (1989).
8. Pape, T., Wintermeyer, W. & Rodnina, M. Induced fit in initial selection and proofreading of aminoacyl-tRNA on the ribosome. *EMBO J.* **18**, 3800–3807 (1999).
9. Katunin, V. I., Muth, G. W., Strobel, S. A., Wintermeyer, W. & Rodnina, M. V. Important contribution to catalysis of peptide bond formation by a single ionizing group within the ribosome. *Mol. Cell* **10**, 339–346 (2002).
10. Schmeing, T. M. *et al.* A pre-translocational intermediate in protein synthesis observed in crystals of enzymatically active 50S subunits. *Nature Struct. Biol.* **9**, 225–230 (2002).
11. Nissen, P., Hansen, J., Ban, N., Moore, P. B. & Steitz, T. A. The structural basis of ribosome activity in peptide bond synthesis. *Science* **289**, 920–930 (2000).
12. Agmon, I. *et al.* Ribosomal crystallography: a flexible nucleotide anchoring tRNA translocation, facilitates peptide-bond formation, chirality discrimination and antibiotics synergism. *FEBS Lett.* **567**, 20–26 (2004).
13. Hansen, J. L., Schmeing, T. M., Moore, P. B. & Steitz, T. A. Structural insights into peptide bond formation. *Proc. Natl Acad. Sci. USA* **99**, 11670–11675 (2002).
14. Burgi, H. B., Dunitz, J. D. & Shefter, E. Geometrical reaction coordinates. II. Nucleophilic addition to a carbonyl group. *J. Am. Chem. Soc.* **95**, 5065–5067 (1973).
15. Radisky, E. S. & Koshland, D. E. Jr. A clogged gutter mechanism for protease inhibitors. *Proc. Natl Acad. Sci. USA* **99**, 10316–10321 (2002).
16. Klein, D. J., Moore, P. B. & Steitz, T. A. The roles of ribosomal proteins in the structure assembly, and evolution of the large ribosomal subunit. *J. Mol. Biol.* **340**, 141–177 (2004).

17. Nissen, P., Ippolito, J. A., Ban, N., Moore, P. B. & Steitz, T. A. RNA tertiary interactions in the large ribosomal subunit: the A-minor motif. *Proc. Natl Acad. Sci. USA* **98**, 4899–4903 (2001).
18. Weinger, J. S., Parnell, K. M., Dorner, S., Green, R. & Strobel, S. A. Substrate-assisted catalysis of peptide bond formation by the ribosome. *Nature Struct. Mol. Biol.* **11**, 1101–1106 (2004).
19. Schmeing, T. M., Moore, P. B. & Steitz, T. A. Structures of deacylated tRNA mimics bound to the E site of the large ribosomal subunit. *RNA* **9**, 1345–1352 (2003).
20. Seila, A. C., Okuda, K., Nunez, S., Seila, A. F. & Strobel, S. A. Kinetic isotope effect analysis of the ribosomal peptidyl transferase reaction. *Biochemistry* **44**, 4018–4027 (2005).
21. Scarlett, D. J., McCaughan, K. K., Wilson, D. N. & Tate, W. P. Mapping functionally important motifs SPF and GGQ of the decoding release factor RF2 to the *Escherichia coli* ribosome by hydroxyl radical footprinting. Implications for macromolecular mimicry and structural changes in RF2. *J. Biol. Chem.* **278**, 15095–15104 (2003).
22. Rawat, U. B. *et al.* A cryo-electron microscopic study of ribosome-bound termination factor RF2. *Nature* **421**, 87–90 (2003).
23. Frolova, L. Y. *et al.* Mutations in the highly conserved GGQ motif of class 1 polypeptide release factors abolish ability of human eRF1 to trigger peptidyl-tRNA hydrolysis. *RNA* **5**, 1014–1020 (1999).
24. Youngman, E. M., Brunelle, J. L., Kochaniak, A. B. & Green, R. The active site of the ribosome is composed of two layers of conserved nucleotides with distinct roles in peptide bond formation and peptide release. *Cell* **117**, 589–599 (2004).
25. Weinger, J. S., Kitchen, D., Scaringe, S. A., Strobel, S. A. & Muth, G. W. Solid phase synthesis and binding affinity of peptidyl transferase transition state mimics containing 2′-OH at P-site position A76. *Nucleic Acids Res.* **32**, 1502–1511 (2004).
26. Ban, N., Nissen, P., Hansen, J., Moore, P. B. & Steitz, T. A. The complete atomic structure of the large ribosomal subunit at 2.4 A resolution. *Science* **289**, 905–920 (2000).
27. Otwinowski, Z. & Minor, W. Processing of X-ray diffraction data collected in oscillation mode. *Methods Enzymol.* **276**, 307–326 (1997).
28. Brunger, A. T. *et al.* Crystallography & NMR system: A new software suite for macromolecular structure determination. *Acta Crystallogr. D Biol. Crystallogr.* **54**, 905–921 (1998).
29. Jones, T. A., Zou, J. Y., Cowan, S. W. & Kjeldgaard, M. Improved methods for building protein models in electron density maps and the location of errors in these models. *Acta Crystallogr. A* **47**, 110–119 (1991).

Supplementary Information is linked to the online version of the paper at www.nature.com/nature.

Acknowledgements We thank E. Youngman and R. Green for allowing us to use their unpublished rates of peptidyl-tRNA hydrolysis; G. Muth, D. Kitchen and S. Scaringe for assistance with synthetic chemistry; J. Kavran for collection of a diffraction data set, A. Innis for reading the text; and P. Moore for experimental advice and critical reading of the manuscript. Financial support for the National Synchrotron Light Source comes from the US DOE and the NIH. The Advanced Light Source is supported by the DOE. K.S.H.. is supported by a NIH postdoctoral fellowship. This work was supported by NIH and ACS Beginning Investigator grants to S.A.S. and an NIH grant to T.A.S.

Author Contributions T.M.S. determined and evaluated the structures of the three complexes of the 50S ribosomal subunit with substrate analogues under the supervision of T.A.S., and K.S.H. designed and synthesized the substrate analogues used under the supervision of S.A.S.

Author Information The atomic coordinates and structure factors have been deposited into the RCSB Protein Data Bank under accession codes 1VQ6, 1VQ7 and 1VQN. Reprints and permissions information is available at npg.nature.com/reprintsandpermissions. The authors declare no competing financial interests. Correspondence and requests for materials should be addressed to T.A.S. (eatherton@csb.yale.edu).

Chapter 6

Joan A. Steitz

Biography

Steitz earned her B.S. in chemistry from Antioch College in 1963. Significant findings from her work emerged as early as 1967, when her Harvard Ph.D. thesis with Jim Watson examined the test-tube assembly of a ribonucleic acid (RNA) bacteriophage (antibacterial virus) known as R17.

Steitz spent the next three years in postdoctoral studies at the Medical Research Council Laboratory of Molecular Biology in Cambridge, UK, where she used early methods for determining the biochemical sequence of RNA to study how ribosomes know where to initiate protein synthesis on bacterial mRNAs. In 1970, she was appointed assistant professor of molecular biophysics and biochemistry at Yale, becoming full professor in 1978. At Yale, she established a laboratory dedicated to the study of RNA structure and function. In 1979, Steitz and her colleagues described a group of cellular particles called small nuclear ribonucleoproteins (snRNPs), a breakthrough in understanding how RNA is spliced. Subsequently, her laboratory has defined the structures and functions of other snRNPs, such as those that guide the modification of ribosomal RNAs and several produced by transforming herpes viruses.

Steitz is an investigator of the Howard Hughes Medical Institute, and a member of the American Academy of Arts and Sciences, the American Philosophical Society, the National Academy of Sciences, and the Institute of Medicine. Her many honors include the US Steel Foundation Award in Molecular Biology (1982), the National Medal of Science (1986), the Lewis S. Rosenstiel Award (2002), the FASEB Excellence in Science Award (2003), the RNA Society Lifetime Achievement Award (2004), E. B. Wilson Medal (2005), Gairdner Foundation International Award (2006), Albany Medical Center Prize in Medicine and Biomedical Research, Connecticut Women's Hall of Fame Award, and the New York Academy of Medicine Medal for Distinguished Contributions in Biomedical Science (2008). She is the recipient of 13 honorary degrees.

Dear Bill

June 18, 2009

Dear Bill,

I have chosen the single paper that Tom and I have published together over our entire careers. The reason is obvious. Without him, my life — especially my life in science — would have been very different. And it all started with you!

I distinctly recall the night I first stopped by your lab in Gibbs as a first-year graduate student. Why I was seeking information on boron hydrides I can no longer recall. You and I looked at each other and immediately knew that we had previously interacted, in Tom Nee's chamber music group at the First Unitarian Society in Minneapolis. Shortly thereafter, I found myself invited to accompany Tom to a chamber music evening at your home, occasioned by the visit of Christopher Longuet-Higgins. It was fabulous and I was so pleased to be included! But exactly how you engineered Tom's inviting me is a story for him to tell.

Later, you came to Minneapolis and gathered together a group of musical friends to play the Hayden London Trios, and other things, at our wedding and the reception thereafter. That was so special. I do not think I can ever adequately thank you for this extraordinary gift.

But back to science. After Harvard, I did postdoctoral work at the Cambridge MRC Lab because of its "excellence in protein crystallography." Tom was fixed to work with David Blow, and Jim Watson had written to Francis Crick saying I would need a position too. When I first arrived, Francis said there was no space and maybe I should find a theoretical project to do in the library. I knew that theory was not my forte, and Mark Bretscher luckily took pity on me, offering me three feet of bench space. What emerged was our first glimpse into sequences in mRNAs that drive the initiation of protein synthesis.

The rest is history. Our later realization that bacterial ribosomes find start sites on mRNA by using RNA–RNA base pairing, in this case with the 3′-end of 16S rRNA, launched the singular focus of my entire career. That is, virtually everything my lab at Yale has done in the ensuing 30+ years has been to characterize the many ways in which bacterial and vertebrate cells use RNA–RNA base pairing to achieve specificity in the multiple reactions that underlie gene expression.

The *PNAS* paper represents a special twist on this theme. Tom's structural work on DNA polymerase and other protein-catalyzed phosphoryl transfer reactions had made it clear that a two-metal-ion active site could work for RNA catalysis as well. We ribophiles in the RNA processing field believed from the start that splicing could be RNA based, with the small nuclear RNAs (snRNAs) representing the active components of the spliceosome. Tom's and my *PNAS* paper proposed two-metal-ion models for the active sites of both the Group I self-splicing intron and the spliceosome. Later structural studies from our colleague Scott Strobel at Yale have verified this model for the Group I intron at the atomic level. Recently, there is growing evidence (from Saba Valadhkan at Case Western) that pieces of the snRNAs can indeed catalyze splicing-like reactions. Now, the huge challenge is to produce structural evidence for RNA catalysis (and presumably two-metal ions) in the spliceosome active site. I despair that this endeavor is so difficult that I may never see the answer.

The structural underpinnings with which I have always tried to view my biochemical endeavors have — I believe — allowed insights not always obvious to others. You and Tom are responsible.

Thanks so very much!

With love and congratulations on your 90th!

Joan A. Steitz

Sterling Professor of Molecular Biophysics and Biochemistry

Investigator, Howard Hughes Medical Institute

Proc. Natl. Acad. Sci. USA
Vol. 90, pp. 6498–6502, July 1993
Biochemistry

A general two-metal-ion mechanism for catalytic RNA

(phosphoryl transfer mechanism/ribozyme/group I splicing/spliceosome/group II splicing)

THOMAS A. STEITZ*† AND JOAN A. STEITZ*

Departments of *Molecular Biophysics and Biochemistry and †Chemistry and Howard Hughes Medical Institute, Yale University, New Haven, CT 06511

Contributed by Thomas A. Steitz, April 26, 1993

ABSTRACT A mechanism is proposed for the RNA-catalyzed reactions involved in RNA splicing and RNase P hydrolysis of precursor tRNA. The mechanism postulates that chemical catalysis is facilitated by two divalent metal ions 3.9 Å apart, as in phosphoryl transfer reactions catalyzed by protein enzymes, such as the 3′,5′-exonuclease of *Escherichia coli* DNA polymerase I. One metal ion activates the attacking water or sugar hydroxyl, while the other coordinates and stabilizes the oxyanion leaving group. Both ions act as Lewis acids and stabilize the expected pentacovalent transition state. The symmetry of a two-metal-ion catalytic site fits well with the known reaction pathway of group I self-splicing introns and can also be reconciled with emerging data on group II self-splicing introns, the spliceosome, and RNase P. The role of the RNA is to position the two catalytic metal ions and properly orient the substrates via three specific binding sites.

The finding that RNA can catalyze a chemical reaction (1, 2) was unexpected, in large part because RNA does not contain functional groups that have pKa values and chemical properties similar to those considered important in protein-based enzymes. How then can an RNA enzyme be an effective catalyst? Studies of protein enzymes have shown that high rates of catalytic activity result from the contributions of several factors: (*i*) the specific orientation of substrates relative to each other and to the catalytic groups of the enzyme, (*ii*) design of the active site of the enzyme to be complementary to the structure of the reactants in the transition state of the reaction, (*iii*) use of acid–base catalysis, and (*iv*) formation of a covalent enzyme–substrate intermediate (3–6). As RNA molecules show complex and specific three-dimensional structures (7–10), one can imagine how an RNA enzyme could achieve specific binding and orientation of substrates. But what chemical groups would assist in the chemistry of catalysis?

Both a general mechanism involving the expected requirement for an acid and a base (11) and specific mechanisms using one, two, or three Mg^{2+} ions have been suggested for RNA enzymes (12–14, 59). Although these proposals are consistent with available data, none has a precedent in the more extensively studied protein enzymes. Structural and biochemical studies have shown that the number of different mechanisms by which proteins achieve catalysis is much smaller than the number of enzymes catalyzing a particular category of reaction. Thus, it is reasonable to expect that one of the mechanisms of phosphoryl transfer used by protein enzymes might also be used by RNA enzymes if that mechanism does not depend on the chemical properties of protein side chains (13, 15, 16).

We now expand in detail an earlier suggestion (15, 16) regarding how a two-metal-ion mechanism might function in RNA splicing and in hydrolysis of precursor tRNA by RNase P. This mechanism is used by several protein phosphoryl-transfer enzymes and does not require that protein side chains of the enzyme participate directly in the chemistry of the reaction. Thus, the mechanism can be easily accommodated by an enzyme composed of RNA. Moreover, evidence supporting the existence of one of the two metals has been obtained for a group I intron by Cech and coworkers (17). Finally, this two-metal-ion mechanism predicts the existence of three critical substrate-binding sites on the ribozyme and is, therefore, useful in organizing the rapidly accumulating genetic and biochemical data on the several RNA splicing systems and RNase P.

The publication costs of this article were defrayed in part by page charge payment. This article must therefore be hereby marked "*advertisement*" in accordance with 18 U.S.C. §1734 solely to indicate this fact.

Mechanisms of Nucleic Acid Hydrolysis

Three general categories of mechanisms for RNA or DNA hydrolysis have emerged from structural, mutagenic, and biochemical studies of protein enzymes. RNA hydrolysis that results in a 3′-PO_4 and 5′-OH and proceeds through a 2′,3′ cyclic phosphoryl intermediate was first established for pancreatic ribonuclease A (18), which uses one lysine and two histidines to facilitate the reaction. The same 2′,3′ cyclic phosphate is formed by the "hammerhead" self-cleaving ribozyme, although how the RNA accelerates this reaction is unknown. Bovine pancreatic DNase I represents a second category of nucleases and uses a divalent cation to interact with the scissile phosphate and a histidine residue as a general base to activate an attacking water molecule (19).

A third mechanism uses two metal ions and applies to those RNA or DNA phosphoryl-transfer reactions that yield 5′-PO_4s and 3′-OHs. For example, in the 3′,5′-exonuclease of *Escherichia coli* DNA polymerase I (pol I), the hydrolytic phosphoryl-transfer reaction is facilitated by two divalent metal ions spaced 3.9 Å apart (15, 16). The same mechanism (described below) has been established for both steps of the alkaline phosphatase reaction (20), is likely for RNase H (21), and is possible for P1 nuclease (22) and phospholipase C (23). In all five enzymes, there are two divalent metal ions 3.8–4 Å apart capable of interacting with the scissile phosphate.

Two-Metal-Ion Phosphoryl-Transfer Mechanism

A two-metal-ion-catalyzed phosphoryl-transfer mechanism (Fig. 1) has been derived from the crystal structure of the DNA polymerase I 3′,5′-exonuclease domain complexed with single-stranded DNA or product (15, 16) and is consistent with mutagenic and kinetic studies (24, 25). Neither the structure nor mutagenesis implicates any protein side chain in the chemistry of catalysis. Rather, protein residues bind and correctly orient for catalysis two divalent metal ions, the 3′ terminal and penultimate residue of the DNA substrate, and the attacking water molecule. In the crystal structure the two metal ions form inner-sphere complexes with the scissile phosphate and water and, thus, facilitate formation of the attacking hydroxide ion and stabilize the transition state. Specifically, metal ion A forms a metal OH^- ion that is

Abbreviation: snRNA, small nuclear RNA.

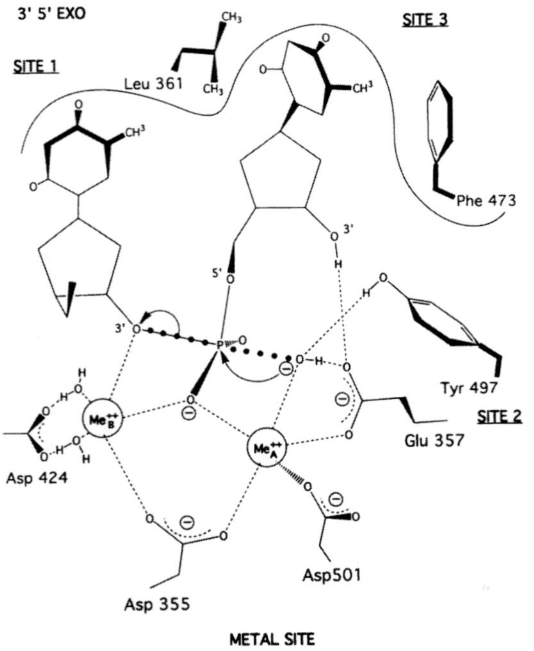

FIG. 1. The proposed transition state (or intermediate) in the mechanism of the 3′,5′-exonuclease reaction of *E. coli* DNA polymerase I (16). The 3′,5′-exonuclease activity can be supported by Mg^{2+}, Mn^{2+}, or Zn^{2+}, but the native enzyme is likely to contain Zn^{2+} in site A and Mg^{2+} in site B (15, 23). Metal ion A is proposed to facilitate formation of an attacking OH^- that has its lone pair of electrons oriented toward the phosphorus by interactions with the metal ion, Tyr-497, and Glu-357. Metal ion B is hypothesized to facilitate leaving of the 3′-OH group and to stabilize the 90° O—P—O bond angle between apical and equatorial oxygen atoms. The bases on either side of the scissile phosphate are positioned by interactions with the side chains of Phe-473 and Leu-361. Positions of all atoms in this figure were established by x-ray crystal structures, except for the positions of the phosphorous and the three equatorial oxygens, which have been moved by a few tenths of an Angstrom to construct the transition state.

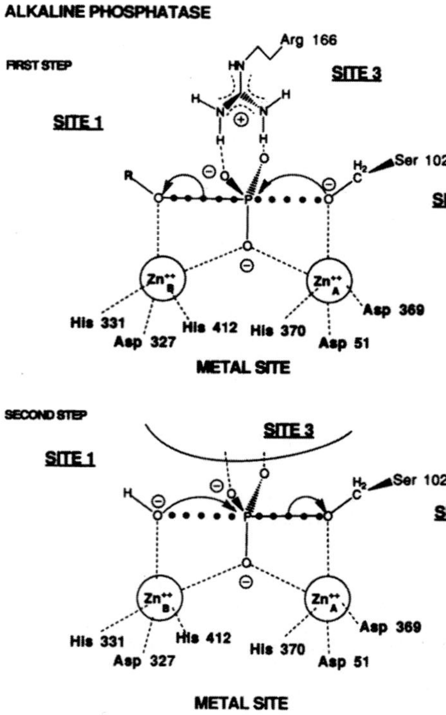

FIG. 2. The two-step mechanism of *E. coli* alkaline phosphatase. Transition states for each of the two steps are stabilized by two Zn^{2+} ions (20).

properly oriented with the help of Tyr-497 and Glu-357 for in-line attack on the phosphorus of the scissile phosphate. Metal ion B acts as a Lewis acid in facilitating the leaving of the 3′ oxyanion (26); it also serves to stabilize the pentacovalent transition state (or intermediate) by interacting with both the nonbridging and 3′ bridging oxygen atoms in a manner analogous to what has been observed with certain organophosphates, such as methylethylene phosphate (27). The bond angle between the O—P—O bridged by the ethylene group in methylethylene phosphate is distorted to 99° from the tetrahedral bond angle of 109° (28), closer to the 90° that occurs in the transition state. Westheimer (27) proposed that strain induced by the ethylene bridge accounted for the enhanced hydrolysis rate of methylethylene phosphate (some 10^6 times faster than trimethyl phosphate); a similar role is suggested for the bridging metal ions in Fig. 1.

Alkaline phosphatase provides an example of how the same active-site configuration can carry out a two-step reaction. Alkaline phosphatase contains binding sites for three closely positioned divalent metals, two of which bind phosphate as described above (20). The mechanism proposed for each stage of the reaction (Fig. 2) is comparable to that of the 3′,5′-exonuclease. In the first step, the phosphate is transferred to the OH group of Ser-102; reversal with OH^-, now occupying substrate binding site 1, achieves phosphate release. The RNase H domain of human immunodeficiency virus reverse transcriptase likewise binds two divalent metal ions 4 Å apart in the active site, implying a similar mechanism (21). The single-stranded P1 nuclease has three bound zinc ions, two of which are observed to coordinate to the 5′ phosphate of a bound nucleotide (22). Finally, phospholipase exhibits the same overall protein structure as P1 nuclease, including three Zn^{2+} atoms at the active site (23).

Although all of these five enzymes may use a two-metal-ion mechanism of phosphoryl transfer, only two (P1 nuclease and phospholipase) have obviously homologous structures. This observation implies enormous variability in the structural context capable of positioning two metal ions precisely relative to the phosphoryl group. That the intermetal ion distance is 3.9 Å in four enzymes of separate evolutionary origin implies an important chemical role for this positioning. This distance may be related to the distance (in the transition state) between the apical attacking and leaving oxygen atoms directly bonded to the metal ions. Could RNA molecules also position two Mg^{2+} ions 4 Å apart (perhaps using the phosphates of the RNA backbone rather than carboxylates) to achieve relevant interactions with a substrate phosphoryl group? Can ribozyme groups be identified that position the attacking water or sugar, the leaving ribose oxyanion, and its adjacent nucleotide?

Group I Splicing Conforms to the Two-Mg^{2+}-Ion Mechanism. The group I self-splicing intron catalyzes two phosphoryl-transfer reactions that result in intron removal and ligation of flanking exons (Fig. 3) (29). In each step of the splicing reaction, the 3′ OH of a ribose attacks the backbone phosphate of an RNA chain. The chemistry of this transesterification reaction is identical to the hydrolysis reactions

6500 Biochemistry: Steitz and Steitz Proc. Natl. Acad. Sci. USA 90 (1993)

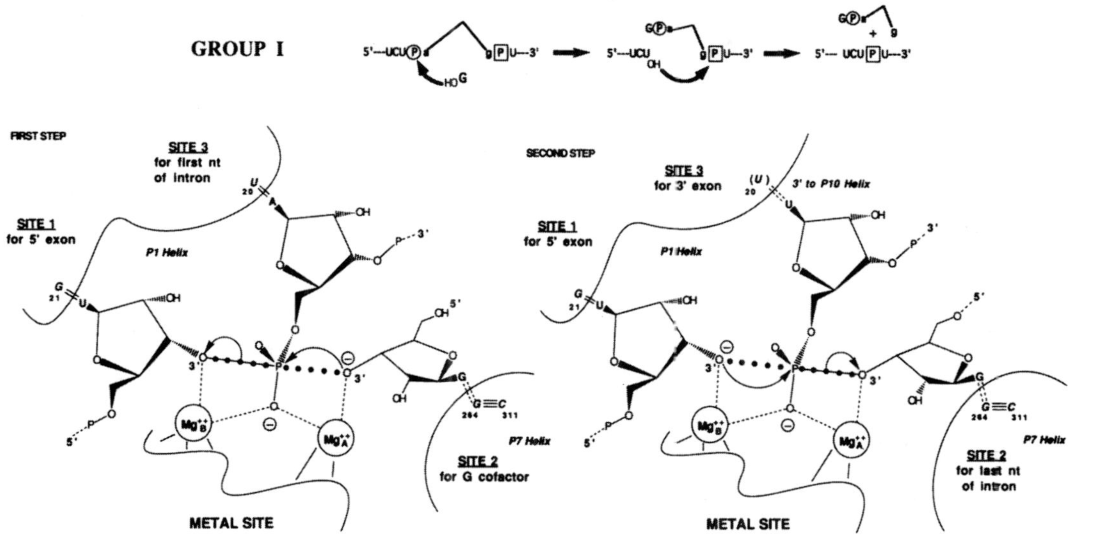

FIG. 3. Mechanism for group I self-splicing. References for assignment of *Tetrahymena thermophila* intron nucleotides and helices to substrate-binding sites 1–3 during the first and second steps of the reaction (shown above) are given in text. Intron nt 18–20 are included in helix P1 for step 1, whereas nt 13–19 pair with 3′ exon sequences to form P10 for the second step.

considered above, except that, instead of water, the 3′-OH of a nucleotide (guanosine in the first step and uridine in the second step) is activated for nucleophilic attack on the phosphoryl group. In Fig. 3, the 3′-OH of the guanosine that initiates the reaction is activated by Mg^{2+} ion A in exactly the same fashion as the attacking water in the 3′,5′-exonuclease reaction (Fig. 1). Also, as with the exonuclease, a second Mg^{2+} ion B is postulated to act as a Lewis acid to facilitate (26) leaving of the 3′-oxyanion of uridine and stabilize the pentacovalent intermediate (or transition state). Evidence for this metal ion comes from the observation that, although it is unknown whether phosphates and/or bases bind the metals, replacing Mg^{2+} by Mn^{2+} or Zn^{2+} rescues cleavage of a phosphorothioester at the 5′-splice site (17), as anticipated from the known preference of Mg^{2+} versus Mn^{2+} for coordination by oxygen versus sulfur. After step 1, the chain containing the newly added guanosine must dissociate from sites 2 and 3, allowing a new substrate, the 3′-splice site, to take its place. The newly created 3′-OH of uridine at the end of the 5′ exon remains in the same position, whereas the guanosine residue at the 3′ end of the intron occupies the same guanosine-binding site as the guanosine cofactor (site 2) and the first nucleotide of the 3′ exon now occupies site 3. Hence, in the second step, the Mg^{2+} ions reverse roles and Mg^{2+} B activates the attacking 3′-OH of uridine, whereas Mg^{2+} A facilitates the leaving oxyanion of guanosine. The mirror symmetry of the two metal ions in the catalytic center reflects the identical chemistry of the two steps of the reaction.

Not only is the chemistry of the group I intron the best understood among ribozymes, but assignment of RNA moieties that form binding sites for the substrates is most advanced. The uridine at the end of the 5′ exon (as well as several preceding residues) is base-paired to the so-called internal guide sequence and remains bound in this configuration (called helix P1) during both steps of the reaction (29) (site 1 in Fig. 3). The location of the binding site for the guanosine cofactor for the first step of the reaction (site 2) is also known (30): critical interactions are made with the G^{264}-C^{311} pair in the core of the ribozyme (helix P7). The same binding site orients the guanosine at the 3′ end of the intron for the second step of the reaction (30, 31) and also serves to bind water for ribozyme-catalyzed hydrolysis at the 5′ site (32). The third substrate-binding site, which holds the 5′ end of the intron during the first step and the 3′ exon during the second step of the reaction, has likewise been identified for the group I intron (29). A sequence adjacent to the internal guide sequence initially pairs with bases at the 5′ end of the intron (P1) but then exchanges partners to form several pairs with the 3′ exon (called P10) for the second step of the reaction. Because all three substrate-binding sites are identical for the two steps of group I splicing, inhibition of only one (the second) reaction step by an R_p phosphorothioate diasteriomer at the splice site is observed (33, 34).

Finally, critical metal-ion-binding sites have been located near position 207 (in P5) and 306 (near P7) in the catalytic core of the group I intron (35).

A Model for the Active Sites of Group II Introns and the Spliceosome. Intron removal by group II self-splicing introns (36) and by the spliceosome (37), which assembles from five small nuclear RNAs (snRNAs) (U1, U2, U4, U5, and U6) and many proteins, follows a different two-step pathway (Fig. 4, see top). In the first phosphoryl-transfer reaction, the 2′-OH group of an adenosine residue located upstream of the 3′-splice site attacks the backbone phosphate at the 5′-splice site. Magnesium ion A is proposed to activate this OH group (Fig. 4, first step) as in Figs. 1 and 3 (first step), whereas Mg^{2+} ion B again stabilizes the leaving oxyanion and pentacovalent intermediate. After step 1, the lariat structure must be replaced by the 3′-splice site in sites 2 and 3 (Fig. 4, second step). Meanwhile, the excised 5′ exon is held fixed, so that in the second step of the reaction its newly created 3′-OH can attack the phosphoryl group between the conserved residue at the 3′ end of the intron and the 3′ exon. This relationship is pictured as a reversal of the first reaction.

In contrast to group I splicing, where all three substrate-binding sites remain unchanged for the two reaction steps, our fragmentary understanding of group II introns and the spliceosome argues that two of these binding sites (2 and 3) change. Yet, in the simplest formulation of the model, site 1 (which holds the 5′ exon) and the positions of the two metal ions could remain unaltered. How can current knowledge of

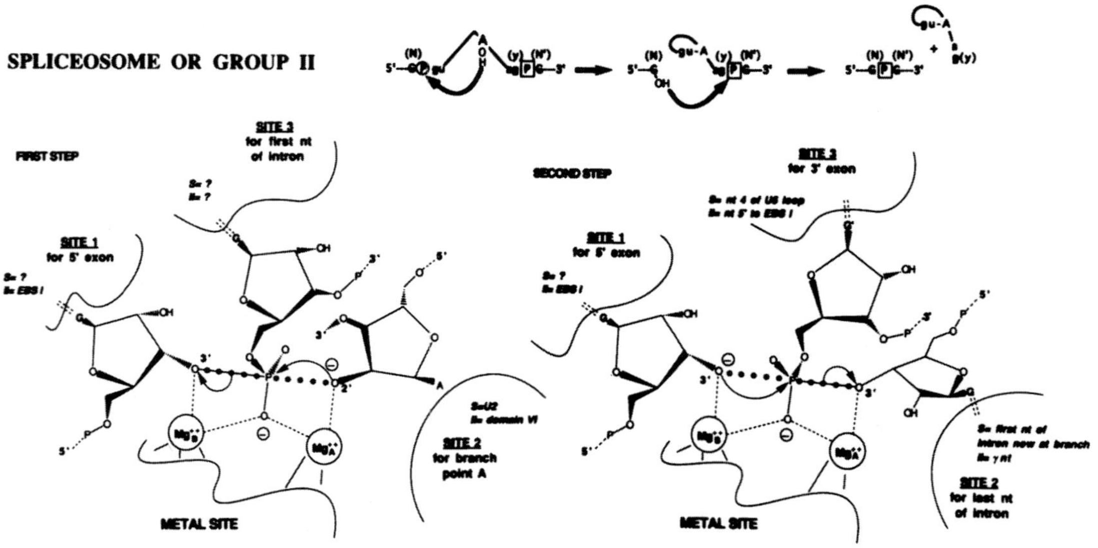

FIG. 4. Mechanism for spliceosome and group II splicing. Above is shown the two-step reaction with consensus nucleotides indicated for the spliceosome-catalyzed reaction and, in parentheses, consensus nucleotides that differ in group II self-splicing introns. In both subgroup IIa and IIb, N and the adjacent 5 nt of the 5′ exon covary with EBS1; N′ in subgroup IIa covaries with the intron nucleotide 5′ to EBS1; y indicates a pyrimidine. The two steps shown below use the spliceosome consensus sequences. Experimental evidence supporting the assignment of particular group II intron (II) or snRNA (S) nucleotides to binding sites 1–3 is cited in text. A rotation of binding site 3 by 120° away from the reader between the two reaction steps, whereas sites 1 and 2 are approximately fixed, is proposed.

functional or conserved residues in the group II intron and in the snRNAs of the spliceosome be reconciled with the model (Fig. 4)?

In group II introns, the 5′ exon is held by base-pairing to two nonadjacent sequences in domain I of the intron, termed EBS1 and EBS2 (36). The 5′-splice site is determined by its position opposite the 5′-most nucleotide in EBS1 (38, 39). Because the 5′ exon–EBS1 interaction appears to be maintained through both steps of splicing (36), it precisely fits the requirements of site 1 in Fig. 4. Domain VI of the intron (36), a double helix from which the branchpoint A residue is bulged, comprises binding-site 2 for the first step of the reaction, whereas for the second step the pairing of the last nucleotide of the intron with an intron nucleotide (located between domains II and III) termed γ is critical (36, 40). Yet, only a small displacement of domain VI between steps may be required because its structure is known to influence 3′-splice-site recognition (41). The third substrate-binding site likewise cannot be identical for the two steps of the reaction. No nucleotide has yet been identified as interacting with the first base of the intron during the first step (36, 42, 43), but intron nt 3 and 4 (and perhaps adjacent residues) base-pair with 2 nt called ϵ in intron domain I (36, 40); this pairing is maintained for the second step of the reaction (40). For step 2, introns of subgroup IIA pair the intron nucleotide 5′ to EBS1 with the first nucleotide of the 3′ exon; this interaction is important for 3′-splice-site specificity (36, 38) and, therefore, probably contributes to site 3 for the second step, but this same intron nucleotide does not appear to pair with the first base of the intron during step 1 (36). Finally, domain V is an excellent candidate for binding the two metal ions: it is the most highly conserved of the six intron domains and, when added *in trans* to domains I–III, it stimulates 5′-splice-site hydrolysis (44).

In the spliceosome, only three snRNAs (U2, U5, and U6) are believed to contribute functionally once catalysis begins (37). No RNA has yet been identified in contact with the last nucleotide of the 5′ exon during either step of the reaction, although an earlier interaction of the penultimate nucleotide of the exon with an absolutely conserved 9-nt loop sequence in U5 has been documented both genetically and biochemically (45, 46). In striking analogy to group II introns, the branch site A residue is bulged from a helical interaction with U2 snRNA, identifying binding-site 2 as containing U2 snRNA (37) for the first reaction step. Before the second step, the model (Fig. 4, second step) proposes that the invariant guanosine at the 3′ end of pre-mRNA introns replaces the branch-point A residue in site 2. A slight shift of the U2 snRNA within the spliceosome might achieve such a realignment: a non-Watson–Crick interaction between the first and last bases of pre-mRNA introns has been proposed based on genetic suppression data (47), and the first intron nucleotide is now attached to the branch-point A residue. The third substrate-binding site again would require significant remodeling between steps. The results of chemical crosslinking have suggested that intron nucleotides downstream of the 5′ splice site contact the conserved loop sequence in U5 snRNA during the first step (48), whereas genetic results argue that the first two positions of the 3′ exon interact with adjacent nucleotides in the same U5 conserved loop for the second step of the reaction (49). Thus, two different portions of the U5 conserved loop could comprise site 3 for the two reaction steps, suggesting movements that might be facilitated by known U6 contacts (48, 50) (that perhaps mimic ϵ in the group II intron) with sequences near the 5′ end of the intron. Specifically, a 120° shift in the position of site 3 (a rotation about the axial oxygens in the pentacovalent intermediate) between the two steps of pre-mRNA splicing (Fig. 4) would place the other nonbridging oxygen between the two metals. This geometry would rationalize the results of Moore and Sharp (51), who recently observed that both steps of the spliceosome reaction are inhibited by the same phosphorothioate diasteriomer, whereas neither step is affected by the opposite diasteriomer. Finally, an interaction between conserved sequences in U2 and U6 snRNAs recently identified by genetic analyses (52) in yeast exhibits uncanny structural resemblance to domain V of group II introns. This snRNA–snRNA duplex is, therefore, a most attractive candidate to provide metal-binding sites in the active spliceosome.

Ribonuclease P

Ribonuclease P catalyzes the simple hydrolysis of a specific phosphodiester bond in pre-tRNAs to create the 5′-PO_4 terminus of all mature tRNA molecules (53, 54). Although the native enzyme contains both RNA and protein, the RNA alone is active in the presence of Mg^{2+} under nonphysiological salt conditions (2). The two-metal-ion model for the RNase P active site again includes three substrate-binding sites (see Fig. 1): site 1 for the last nucleotide of the 5′ leader sequence, site 2 for the attacking OH^-, and site 3 to hold the first nucleotide of the mature tRNA. Biochemical experiments have suggested candidate residues in the *E. coli* RNase P RNA that may serve as sites 1 and 3 for substrate [C^{92} (55) and A^{246} (56)] and metal [positions 257 and 295 (57)] binding.

Perspectives

We have described how a mechanism of phosphoryl transfer documented for several protein enzymes could be used by RNA enzymes involved in splicing and RNA hydrolysis. In the protein enzymes, no side chains are directly involved in the chemistry of catalysis; rather, they serve to bind two catalytically essential metal ions and orient the substrates. These functions can equally well be assumed by RNA. The same mechanism could also be used by RNA enzymes, past or present, to catalyze other, but related, reactions, including RNA or DNA polymerization reactions (16, 58).

This paper is dedicated to the memory of Hatch Echols. We are indebted to T. Cech, C. Guthrie, D. Hershlag, M. Moore, E. Sontheimer, and O. Uhlenbeck for comments. We thank P. Eatherton for expert typing and P. Raccuia for drafting the figures.

1. Cech, T. R., Zaug, A. J. & Grabowski, P. J. (1981) *Cell* **27**, 487–496.
2. Guerrier-Takada, C. A., Gardiner, K., Marsh, T., Pace, N. & Altman, S. (1983) *Cell* **35**, 849–857.
3. Blow, D. M. & Steitz, T. A. (1970) *Annu. Rev. Biochem.* **39**, 63–100.
4. Jencks, W. P. (1975) *Advances in Enzymology and Related Areas of Molecular Biology* (Interscience, New York), Vol. 43.
5. Fersht, A. (1985) *Enzyme Structure and Mechanism* (Freeman, New York), 2nd Ed.
6. Knowles, J. R. (1980) *Annu. Rev. Biochem.* **49**, 877–913.
7. Kim, S. H., Suddath, F. L., Quigley, F. L., McPherson, A., Sussman, J. L., Wang, A. H. J., Seeman, N. C. & Rich, A. (1974) *Science* **185**, 435–440.
8. Robertus, J. D., Ladner, J. E., Finch, J. T., Rhodes, D., Brown, R. S., Clark, B. F. C. & Klug, A. (1974) *Nature (London)* **250**, 546–551.
9. Kim, S. H. & Cech, T. R. (1987) *Proc. Natl. Acad. Sci. USA* **84**, 8788–8792.
10. Michel, R. & Westhof, E. (1990) *J. Mol. Biol.* **216**, 585–610.
11. Cech, T. R. (1987) *Science* **246**, 1532–1539.
12. Guerrier-Takada, C., Heydock, K., Allen, L. & Altman, S. (1986) *Biochemistry* **25**, 1509–1515.
13. Yarus, M. (1993) *FASEB J.* **7**, 31–40.
14. Uchimaru, T., Uebayosi, M., Tanabe, K. & Taira, K. (1993) *FASEB J.* **7**, 137–139.
15. Freemont, P. S., Friedman, J. M., Beese, L. S., Sanderson, M. R. & Steitz, T. A. (1988) *Proc. Natl. Acad. Sci. USA* **85**, 8924–8928.
16. Beese, L. S. & Steitz, T. A. (1991) *EMBO J.* **10**, 25–33.
17. Piccirilli, J. A., Vyle, J. S., Caruthers, M. H. & Cech, T. R. (1993) *Nature (London)* **361**, 85–88.
18. Richards, F. M. & Wyckoff, H. W. (1971) in *The Enzymes*, ed., Boyer, P. D. (Academic, New York), Vol. 4, pp. 647–804.
19. Suck, D., Lahm, A. & Oefner, C. (1988) *Nature (London)* **321**, 620–625.
20. Kim, E. E. & Wyckoff, H. W. (1991) *J. Mol. Biol.* **218**, 449–464.
21. Davies, J. F., Hostomska, Z., Hostomsky, Z., Jordan, S. R. & Matthews, D. A. (1991) *Science* **252**, 88–94.
22. Volbeda, A., Lahm, A., Sakiyama, F. & Suck, D. (1991) *EMBO J.* **10**, 1607–1618.
23. Hough, E., Hansen, L. K., Birknes, B., Jynge, K., Hansen, S., Hordirk, A., Little, C., Dodson, E. J. & Derewenda, Z. (1989) *Nature (London)* **338**, 357–360.
24. Derbyshire, V., Freemont, P. S., Sanderson, M. R., Beese, L., Friedman, J. M., Joyce, C. M. & Steitz, T. A. (1988) *Science* **240**, 199–201.
25. Derbyshire, V., Grindley, N. D. F. & Joyce, C. M. (1991) *EMBO J.* **10**, 17–24.
26. Herschlag, D. & Jencks, W. P. (1987) *J. Am. Chem. Soc.* **109**, 4665–4674.
27. Westheimer, F. A. (1968) *An. Chem. Res.* **1**, 70–79.
28. Steitz, T. A. & Lipscomb, W. N. (1965) *J. Am. Chem. Soc.* **87**, 2488–2489.
29. Cech, T. R. (1990) *Annu. Rev. Biochem.* **59**, 543–568.
30. Michel, F., Hanna, M., Green, R., Bartel, D. P. & Szostak, J. W. (1989) *Nature (London)* **342**, 391–395.
31. Been, M. D. & Perrotta, A. T. (1991) *Science* **252**, 434–437.
32. Legault, P., Herschlag, D., Celander, D. W. & Cech, T. R. (1992) *Nucleic Acids Res.* **20**, 6613–6619.
33. McSwiggin, J. A. & Cech, T. R. (1989) *Science* **244**, 679–683.
34. Suh, E. R. & Waring, R. B. (1992) *Nucleic Acids Res.* **20**, 6303–6309.
35. Christian, E. L. & Yarus, M. (1993) *Biochemistry* **32**, 4475–4480.
36. Michel, F. K., Umesono, K. & Ozeki, H. (1989) *Gene* **82**, 5–30.
37. Guthrie, C. (1991) *Science* **253**, 157–163.
38. Jacquier, A. & Jacquesson-Breuleux, N. (1991) *J. Mol. Biol.* **219**, 415–428.
39. Müller, M. W., Schweyen, R. J. & Schmelzer, C. (1988) *Nucleic Acids Res.* **16**, 7383–7395.
40. Jacquier, A. & Michel, F. (1990) *J. Mol. Biol.* **213**, 437–447.
41. Schmelzer, C. & Müller, M. W. (1987) *Cell* **51**, 753–762.
42. Peebles, C. L., Belcher, S. M., Zhang, J., Dietrich, R. C. & Perlman, P. S. (1993) *J. Biol. Chem.*, in press.
43. Wallasch, C., Möri, M., Niemer, I. & Schmelzer, C. (1991) *Nucleic Acids Res.* **19**, 3307–3314.
44. Jarrell, K. A., Dietrich, R. C. & Perlman, P. S. (1988) *Mol. Cell. Biol.* **8**, 2361–2366.
45. Newman, A. J. & Norman, C. (1991) *Cell* **65**, 115–123.
46. Wyatt, J. R., Sontheimer, E. J. & Steitz, J. A. (1992) *Genes Dev.* **6**, 2542–2553.
47. Parker, R. & Siliciano, P. G. (1993) *Nature (London)* **361**, 660–662.
48. Wassarman, D. A. & Steitz, J. A. (1992) *Science* **257**, 1918–1925.
49. Newman, A. J. & Norman, C. (1992) *Cell* **68**, 743–754.
50. Sawa, H. & Abelson, J. (1992) *Proc. Natl. Acad. Sci. USA* **89**, 11269–11273.
51. Moore, M. J., Query, C. C. & Sharp, P. A. (1993) in *The RNA World*, eds Gesteland, R. & Atkins, J. (Cold Spring Harbor Lab. Press, Plainview, NY), in press.
52. Madhani, H. D. & Guthrie, C. (1992) *Cell* **71**, 803–817.
53. Pace, N. R. & Smith, D. (1990) *J. Biol. Chem.* **265**, 3587–3590.
54. Altman, S., Kirsebom, L. & Talbot, S. (1993) *FASEB J.* **7**, 7–15.
55. Guerrier-Takada, C., Lumelsky, N. & Altman, S. (1989) *Science* **286**, 1578–1584.
56. Burgin, A. B. & Pace, N. R. (1990) *EMBO J.* **9**, 4111–4118.
57. Kazakov, S. & Altman, S. (1991) *Proc. Natl. Acad. Sci. USA* **88**, 9193–9197.
58. Steitz, T. A. (1993) *Curr. Opin. Struct. Biol.* **3**, 31–38.
59. Smith, D. & Pace, N. R. (1993) *Biochemistry* **32**, 5273–5281.

Chapter 7

Florante A. Quiocho

Biography

Florante A. Quiocho was born in the Philippines and obtained his undergraduate degree there. He received his Ph.D. degree in biochemistry from Yale University, where he worked in the laboratory of Frederic Richards to firmly establish the catalytic activities of enzymes in the crystalline state. With the major persistent concern at the time about the relevance of protein structures to biology, the resulting work derailed the skeptics by showing catalytic activity in totally insoluble, cross-linked crystals of the enzyme carboxypeptidase A. Cross-linking of protein crystals, which was an important by-product of the work, still remains in use.

Following a postdoctoral appointment at Harvard University conducting research in protein crystallography with William Lipscomb, in 1972 Dr. Quiocho joined the faculty of the Department of Biochemistry at Rice University, rising to the rank of professor. He then moved to Baylor College of Medicine in 1986 where he is currently the Charles C. Bell Professor of Structural Biology and professor of biochemistry and molecular biology. He was also a Howard Hughes Medical Institute investigator for 18 years. Dr. Quiocho has been the recipient of fellowships from the John Simon Guggenheim Memorial Foundation and the European Molecular Biology Organization (EMBO). He has published over 200 scientific papers, including ten in the journal *Nature* (two in the News & Views section), six in *Science*, and four in *Cell*. He has served on numerous National Institutes of Health (NIH) review panels. His research interests center on the relationship of the three-dimensional structures and functions of several proteins. These studies have yielded new insights into molecular recognitions of ions, carbohydrates and capped RNA, calmodulin-regulated signal transduction processes, ABC-type active transport, and enzyme catalysis. Current studies of the Quiocho group focus on the area of membrane fusion and secretion and vesicular trafficking.

Happy 90th Birthday, Colonel!

Whatever success I may have attained in my chosen field I owe immeasurably to two illustrious and fabulous scientists and mentors: Professors Frederic M. Richards and William N. Lipscomb, Jr. (or "the Colonel"). Fred ("El Jefe" or "the Chief"), who was my doctoral research advisor at Yale University, sadly passed away in 2009. The Colonel was my postdoctoral advisor, and it is to him that I pay homage in this essay for the book dedicated to his 90th birthday.

I joined the Colonel's lab in 1966, at a time when less than a handful of protein crystal structures were being determined and long before the term "structural biology" was coined for the field. It was an exciting time; the structure determination of carboxypeptidase A, believed to be the third protein and second enzyme, was beginning to flourish in the Colonel's lab. At the same time, solving the protein X-ray structure was challenging and difficult; diffraction data were collected from a generator with a sealed tube, a diffractometer with Weisenberg geometry and a paper input and output, and computing used a very slow IBM computer (one difference Patterson projection took a couple of days, provided the IBM cards did not get chewed up by the mechanical reader). I am fortunate to be still around to see and benefit from the huge advancement in all aspects of protein crystallography.

I also worked, on the side and by myself, on concanavalin A, which attracted considerable attention then because of the protein's possible anti-proliferative feature. The concanavalin A structural analysis was beginning to gel and prove a source of satisfaction.

For reasons that will become obvious, I have chosen two papers to highlight in this book commemorating the Colonel's 90th birthday. The first is "The structure of carboxypeptidase A. V. Studies of enzyme-substrate and enzyme-inhibitor complexes at 6 Å resolution" by Thomas A. Steitz, Martha L. Ludwig, Florante A. Quiocho, and William N. Lipscomb, which was published in the *Journal of Biological Chemistry* in 1967. This first pick is particularly special and full of significance. First, it chronicled for the first time the power of X-ray structure analysis in revealing the precise location of the active site in carboxypeptidase A and the occurrence of substrate- and inhibitor-stabilized conformational change. Second, the paper signifies my association with two Nobel Laureates in Chemistry: the Colonel, in 1976, for his work on borons; and his Ph.D. student, Tom Steitz, in 2009, for his work on the ribosome. This rare privilege is a source of deep pride for me. Tom and I shared an office and lab bench. Third and finally, the paper was selected as a remembrance and tribute to the late Martha Ludwig, who more than anyone contributed to the success of the carboxypeptidase A project and story.

The second paper is "Sulphate sequestered in the sulphate-binding protein of *Salmonella typhimurium* is bound solely by hydrogen bonds" by James W. Pflugrath and Florante A. Quiocho, which was published in *Nature* in 1985. This paper, one of eight *Nature* papers from my lab, described our stunning structural and seminal discovery that the sulfate bound in the sulfate-binding protein, an initial receptor of the ABC sulfate transporter, is completely buried, dehydrated, devoid of counterions, and held in place entirely by hydrogen bonds with dipolar groups of residues with no formal charges (mostly main chain peptide and some hydroxyl side chain groups). The receptor–sulfate interactions, which we called "ion–dipole interactions," not only confer specificity but also compensate the charge of the buried isolated sulfate charges. Three prominent examples of ion–dipole interactions were later revealed in the structural analysis by the McKinnon Lab showing the binding of chloride in the chloride channel and potassium in the selectivity filter of the KcsA potassium channel, by the Brünger Lab showing the interaction of the synaptobrevin Arg56 in the SNARE complex, and by our lab demonstrating the interaction of Lys79 in the AP180, a protein involved in clathrin-mediated endocytosis.

The Colonel and El Jefe are both giants in their fields, and my stays in their laboratories were memorable, rewarding, and priceless. I am extremely fortunate and forever indebted to them.

THE JOURNAL OF BIOLOGICAL CHEMISTRY
Vol. 242, No. 20, Issue of October 25, pp. 4662–4668, 1967
Printed in U.S.A.

The Structure of Carboxypeptidase A

V. STUDIES OF ENZYME-SUBSTRATE AND ENZYME-INHIBITOR COMPLEXES AT 6 A RESOLUTION*

(Received for publication, August 2, 1967)

THOMAS A. STEITZ, MARTHA L. LUDWIG, FLORANTE A. QUIOCHO, AND WILLIAM N. LIPSCOMB
From the Department of Chemistry, Harvard University, Cambridge, Massachusetts 02138

SUMMARY

Model peptide substrates, *e.g.* glycyl-L-tyrosine, and inhibitors, *e.g.* β-(*p*-iodophenyl)propionate, are shown at 6 A resolution to bind to crystalline carboxypeptidase A_α in regions near the zinc atom. Although the question of bond formation of these added molecules to zinc must await studies at higher resolution, the present studies are not inconsistent with a direct role for the metal in enzymic reactions. Binding of glycyl-L-tyrosine to either native or acetylated crystals produces an increase in electron density in two adjacent regions (A and B) near the zinc. The main region (A) extends into the enzyme's pocket from above the zinc and is attributed to Gly–Tyr bound with its tyrosyl side chain projected into the pocket. Gly–Tyr bound to either apo- or copper-carboxypeptidase increases electron density only in Region A. Furthermore, there appears to be a structural change in the enzyme when substrates are bound to the native or modified enzyme.

As part of a continuing study of the structure of bovine pancreatic carboxypeptidase A (1–3), we are investigating the interaction of carboxypeptidase crystals with substrates and inhibitors. The structure of a glycyl-L-tyrosine complex has already been reported in a preliminary communication (2). The present paper is a full account of three-dimensional x-ray diffraction studies at 6 A resolution of a number of complexes which are formed when crystals of CPA,[1] or modified CPA, bind model peptide substrates and inhibitors. In particular, we examine the complexes of glycyl-tyrosine with native, apo-, copper, and acetylcarboxypeptidases, and of hippuryl-L-phenylalanine with acetyl-CPA. In addition, we present studies of CPA-inhibitor complexes in which the inhibitor is L-lysyl-L-tyrosineamide, β-(*p*-iodophenyl)propionate, or L-phenylalanine.

The structures of crystals containing bound inhibitors and substrates have been determined by the difference Fourier method. This procedure was first applied to the location of light atom molecules bound to proteins by Stryer, Kendrew, and Watson (4) and has subsequently been used by Johnson and Phillips (5) to study the interaction of substrate analogues with lysozyme. In difference Fourier maps the enzyme structure has been subtracted from the structure of the complex, giving an image of the added small molecule. Alterations of a small fraction of the protein structure can also be elucidated by this method (6).

In order to determine structures of enzyme-substrate complexes, a significant amount of substrate must still be bound to the molecules in the crystal at the end of the measuring time, which is approximately 2 days in our experiments. To attain a turnover time of this order, both exceedingly poor substrates and modified proteins have been used. For our purposes it is fortunate that crystallization itself decreases the enzymic activity appreciably; for example, Quiocho and Richards (7) have shown that the hydrolysis of carbobenzoxyglycyl-L-phenylalanine by crystals proceeds at $\frac{1}{300}$ the rate found in solution. As a result of the reduction in activity accompanying crystallization, the complex of Gly–Tyr with cross-linked native CPA has proven sufficiently stable for x-ray studies. Of the available modified forms of CPA (8–11) we have so far utilized the apo-, copper, and acetyl enzymes. Apo- and copper-carboxypeptidase are enzymically inactive (10, 11) but are still able to bind peptides in solution (12) and in the crystal (2). We find that acetylated CPA crystals, which have been shown to have reduced peptidase activity (13), are capable of binding peptides. The structures of such inactive substrate complexes may, of course, provide further information about the mechanism of catalysis.

Our previous studies (1) have shown that the metal atom required for enzymic activity lies in a depression in the molecular surface and is adjacent to a pocket extending into the interior. Now, we have found that both substrates and competitive inhibitors bind near the zinc and that, in one mode of binding, the pocket is occupied. The binding of substrates appears to be accompanied by conformational changes in the protein, occurring in regions adjacent to the substrate.

EXPERIMENTAL PROCEDURE

Bovine carboxypeptidase A_α crystals were prepared as described previously (1). CPA crystallizes in space group $P2_1$.

* This research was supported in part by the National Institutes of Health and by the Advanced Research Projects Agency.

[1] The abbreviations used are: CPA, carboxypeptidase A; apo-CPA, zinc-free carboxypeptidase A; IPP, β-(*p*-iodophenyl)propionate.

Unit cell parameters are $a = 51.4$, $b = 59.9$, $c = 47.2$ A, $\beta = 97°35'$. The β-(*p*-iodophenyl)propionate was prepared by iodination of β-phenylpropionate (14) and after recrystallization was found to have a melting point of 139–140°. Other reagents were obtained from the following companies: Gly–Tyr, Sigma; Lys–Tyr-amide, Cyclo Chemical Corporation; L-phenylalanine, Nutritional Biochemicals; *N*-acetylimidazole, K and K Laboratories; hippuryl-L-phenylalanine, Mann.

All experiments, including modification reactions, assays, and the collection of x-ray data, have been carried out at cold room temperatures (approximately 4°).

Crystals of apo-CPA were prepared by immersing native CPA crystals in a solution containing 5×10^{-3} M 8-hydroxyquinoline-5-sulfonate and 0.04 M tris(hydroxymethyl)aminomethane, pH 7.5, for 2 to 3 weeks (1). Crystals of copper-CPA were then obtained by placing the apo-CPA crystals in a solution of 10^{-4} M $CuCl_2$-0.02 M Tris, pH 7.5. After addition of $CuCl_2$ to the apoenzyme crystals, the intensities in the h0l zone were not visually different from those of zinc-CPA.

Assays for enzymic activity of CPA crystals were performed at 4° following either of the two methods developed in Richards' laboratory (7, 15). In the assay with a column of packed crystals (7) the substrate solution consisted of 0.02 M carbobenzoxyglycyl-L-phenylalanine, 0.2 M LiCl, and 0.02 M Tris, pH 7.5. When glycyl-L-tyrosine was used as substrate, several cross-linked CPA crystals were suspended in 0.7 ml of 0.05 M Gly–Tyr, 0.2 M LiCl, and 0.02 M Tris, pH 7.5, and rotated in a tube for 20 days, at which time hydrolysis was 10% complete. After the crystals had been allowed to settle, crystal-free substrate solution was removed, and the products, glycine and tyrosine, were analyzed with the use of an automatic amino acid analyzer. The rest of the separated substrate solution was incubated for 10 days and then analyzed as before. No increase in product concentration could be detected, indicating the absence of dissolved CPA.

Intermolecular cross-linking of CPA crystals was essential for most of our experiments, either to keep the crystals suspensions free of dissolved enzyme, which would consume substrates, or to prevent solution of crystals at moderate ionic strengths. Native crystals used in these experiments were cross-linked according to the procedure of Quiocho and Richards (16). Crystals were suspended in a solution of 0.1% glutaraldehyde, 0.1 M LiCl, and 0.02 M sodium Veronal, pH 7.5, and gently rotated. The glutaraldehyde solution was changed at least twice during a period of 6 hours. Excess aldehyde was removed after the reaction by repeated washing with 0.1 M LiCl and 0.02 M sodium Veronal, pH 7.5. Comparison of photographs of cross-linked crystals and of native CPA crystals showed only small intensity changes.

Acetylated CPA crystals were prepared by placing cross-linked native crystals into a tube containing 0.01 M *N*-acetylimidazole, 0.02 M sodium Veronal, and 0.2 M LiCl at pH 7.5, and gently rotating the tube; after several hours the solution was replaced by a freshly prepared solution of the same composition. The reaction was allowed to continue for about 12 hours, after which the crystals were washed with buffer. Acetylated crystals yielded a diffraction pattern very similar to that of native cross-linked crystals. Peptidase assays, using a column of crystals as described by Quiocho and Richards (7, 13), showed that after acetylation the crystals retained 10 to 15% of their initial activity.

All of the CPA substrate and inhibitor complexes were prepared by placing crystals into fresh substrate or inhibitor solutions for about 12 hours, after which the crystals were mounted and the data were collected. The presence of salt (0.1 to 0.2 M LiCl) is required in these experiments. When crystals are placed in buffered substrate or inhibitor solutions (ionic strength below 0.05) without added salt, disordered x-ray patterns are obtained. The preparations of the various inhibitor and substrate complexes are summarized in Table I. Substitution of mercury for zinc produces no changes in cell dimensions, and cross-linked crystals soaked in Gly-Tyr show lattice changes no greater than 0.3 to 0.4%. The maximum cell change, observed with cross-linked, acetylated crystals in the presence of hippurylphenylalanine, was 0.6% in *c* axis.

All x-ray diffraction data were collected on Supper-Pace diffractometers, with the use of copper target x-ray tubes with nickel filters. For each of the complexes reported here the 6 A three-dimensional data were obtained from a single crystal within about 2 days. Corrections for Lorentz and polarization factors and for absorption were made, after which three-dimensional difference electron density maps were computed at 6 A resolution,[2] using our best set of native protein phases, α, to compute the Fourier coefficients

$$A = m(|F_D| - |F_P|) \cos \alpha$$
$$B = m(|F_D| - |F_P|) \sin \alpha$$

where $|F_D|$ is the measured amplitude of a given reflection from a derivative crystal, $|F_P|$ is the corresponding amplitude for native CPA, and m is the figure of merit for the reflection (17). The phases from the protein structure determination are thus assumed to be valid for the amplitudes of x-ray scattering from the complex. In the absence of rigorous criteria for isomorphism, one depends ultimately on chemical sense in the interpretation, internal consistencies among related studies, smallness of cell parameter changes, and low background of the difference electron density maps in order to assess the validity of the results.

RESULTS

Relevant portions of difference maps are summarized in Figs. 1 to 4; peak positions and occupancies are given in Table I.

Glycyl-L-tyrosine on Native CPA—The binding of glycyl-L-tyrosine to crystalline native CPA yields increased electron density in two regions, A and B, near the zinc. These regions are shown as *dotted contours* on a superposition of several sections of the 6 A map (*solid contours*) in Fig. 1. The placement of Region A (*striped*) and Region B (*dotted*) in a balsa model of the 6 A contours is shown in Fig. 5, and the effect of change of concentration of Gly–Tyr from 0.05 M to 0.01 M is compared in Fig. 2 (*Parts 1* and *2*). The major region, A, extends into the pocket of the enzyme from above the zinc. Region B is approximately normal to the plane of Fig. 1, extending up toward the reader, and hence is roughly perpendicular to the long direction of Region A. Region B has a lower density than Region A (Table I). The occurrence of negative peaks near B makes the interpretation that B is a second substrate molecule somewhat uncertain. The peaks in A and B do not represent bound products of the enzymic reaction, because of difference electron density map using data taken from a cross-linked CPA crystal soaked in 0.05 M glycine

[2] X-Ray data have been deposited as Document 9642 with the American Documentation Institute, Auxiliary Publication Project, Photoduplication Service, Library of Congress, Washington, D. C. 20540. A copy may be secured by citing the document number and remitting $16.25 for photoprints or $5.00 for 35-mm microfilm. Advance payment is required. Make checks or money orders payable to: Chief, Photoduplication Service, Library of Congress.

TABLE I

Summary of difference electron density maps

Experiment	Substrate or inhibitor	Modification	Site	Coordinates of peak			Peak density[a]	Electron count[a]
				x	*y*	*z*		
							e/A^3	*electrons*
1	Gly-Tyr, 0.05 M	Cross-link	A	0.95	0.48	0.78	0.12	13
			B	0.88	0.54	0.87	0.086	5
2	Gly-Tyr, 0.01 M	Cross-link	A	0.95	0.48	0.80	0.084	5
			B	0.84	0.54	0.85	0.064	2
3	Gly-Tyr, 0.05 M	Cross-link	A	0.95	0.47	0.82	0.13	10
		Acetylate	B	0.85	0.53	0.87	0.09	4
4	Gly-Tyr, 0.01 M	Copper-CPA	A	0.96	0.47	0.86	0.12	13
5	Gly-Tyr, 0.01 M	Apo-CPA	A	0.92	0.47	0.84	0.12	11
6	Lys-Tyr-NH_2 0.15 M	Cross-link	A, B	0.083	0.52	0.13	0.17	21
7	Hippurylphenylalanine, 0.05 M	Cross-link	A	0.95	0.47	0.82	0.16	16
		Acetylate	B	0.84	0.53	0.85	0.082	8
8	IPP, 0.01 M	Cross-link	1	0.93	0.48	0.68	0.17	12
			2	0.90	0.58	0.97	0.27	15
9	IPP, 0.05 M	Cross-link	1	0.93	0.48	0.68	0.19	17
			2	0.90	0.58	0.97	0.32	28
			3	0.31	0.20	0.60	0.20	16
			4	0.97	0.53	0.97	0.13	6
10[b]	IPP, 0.05 M	Apo-CPA	1	0.90	0.48	0.75	0.10	8
			3	0.31	0.20	0.61	0.09	6
			4	0.97	0.53	0.97	0.12	10
11[b]	IPP, 0.10 M	Apo-CPA	1	0.90	0.47	0.77	0.15	12
			3	0.31	0.20	0.61	0.24	24
			4	0.97	0.53	0.97	0.21	16
12	Phenylalanine, 0.05 M	Native CPA	A	0.95	0.47	0.82	0.19	11

[a] Peak density in electrons per cubic angstrom (e/A^3) and electron counts are recorded here as twice the observed values in order to correct the scaling error introduced by use of the native phases, and have been divided by the average figure of merit of 0.89. The occupancy of the mercury which replaces the zinc was set at 50 electrons after refinement at 2.0 A resolution, in order to calibrate the absolute scale. For light atom substrates, the background density is approximately one-half of the density in Site A.

[b] No LiCl is present in Experiments 10 and 11, but 0.02 M Tris (pH 7.5) is present. In all other experiments 0.1 to 0.2 M LiCl and 0.02 M Tris (pH 7.5) were also present.

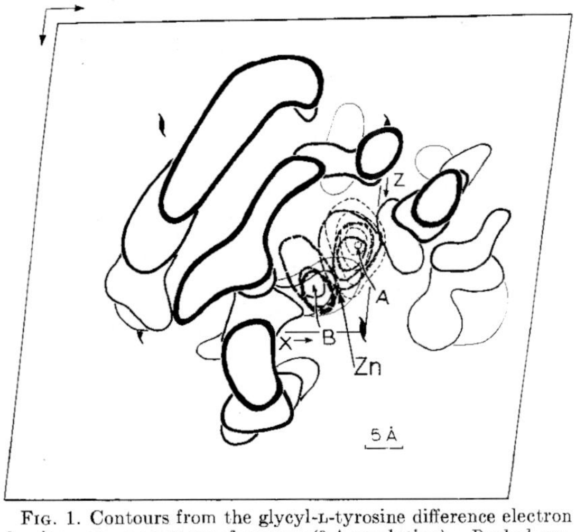

FIG. 1. Contours from the glycyl-L-tyrosine difference electron density map on contours of enzyme (6 A resolution). *Dashed contours* from sections y = 14/30 to 17/30 of the difference map calculated with data from a native crystal soaked in 0.05 M Gly–Tyr have been superimposed on the *smooth contours* of sections y = 13/30 to 17/30 of the 6 A native protein map. Neighboring protein molecules have been omitted. The two regions of added density are labeled *A* and *B*.

and saturated (0.004 M) tyrosine showed no significant peaks in either Region A or Region B. Several other peaks, with densities approximately 60% of Site A, occur in regions removed from the active site and are known to represent cross-links.

Glycyl-L-tyrosine on Acetyl-CPA—The electron density difference map (Fig. 2, *Part 3*) shows that Gly–Tyr is bound to CPA which had been acetylated with *N*-acetylimidazole. The occupancy of Gly–Tyr in this experiment is essentially the same as that found with the native enzyme (Table I), while the activity of the crystals toward carbobenzoxyglycyl-L-phenylalanine has been reduced by at least 85 to 90%. The resolution (6 A) is not sufficient to detect small shifts from the position occupied by Gly–Tyr in native CPA, nor can any new peaks attributable to acetylation be observed.

Glycyl-L-tyrosine on Copper- and Apo-CPA—The binding of Gly–Tyr to copper-CPA and apo-CPA, shown in Fig. 2 (*Parts 4* and *5*, respectively), produces increased electron density only in Region A at a concentration of 0.01 M Gly–Tyr. The peak density in Region A is shifted some 3 or 4 A in both of these examples from the position found for Gly-Tyr bound to native CPA. Region B shows no increase in electron density when 0.01 M Gly–Tyr binds to copper-CPA or apo-CPA.

L-Lysyl-L-tyrosinamide on Native CPA—An attempt was made to examine the effect of converting the charged carboxylate group to a neutral amide, since amides are known to bind to CPA (12) but are not substrates. L-Lysyl-L-tyrosinamide was chosen rather than glycyltyrosinamide because of the greater solubility conferred by the lysyl residue. The electron density difference map (Fig. 2, *Part 6*) indicates about twice as much density in Region B as in Region A, in contrast with the densities in these two regions when Gly-Tyr binds.

β-(p-Iodophenyl)propionate on CPA and apo-CPA—The competitive inhibitor β-phenylpropionate (19, 7) has been modified for x-ray study to IPP in order to locate the molecule more precisely from the heavy atom position. At 0.05 M, IPP binds to crystalline native CPA at four sites, identified in Table I by coordinates of iodine positions. Two of these sites, 1 and 2, are near Regions A and B, respectively, where Gly-Tyr binds to CPA. In apo-CPA site 2 does not bind IPP even at concentrations as high as 0.1 M IPP, while Site 1, near Region A, is shifted and reduced in height in the absence of zinc. Site 4 is reasonably near zinc (Table I), but Site 3 is far removed from the active site region.

L-Phenylalanine on Native CPA—L-Phenylalanine, a product of the hydrolysis of some CPA substrates, does bind to CPA crystals, but only in Region A (Table I, Fig. 4*a*).

Hippuryl-L-phenylalanine on Acetyl-CPA—We examined hippurylphenylalanine in an effort to study more reactive peptide substrates. X-ray photographs of the hippuryl-L-phenylalanine-

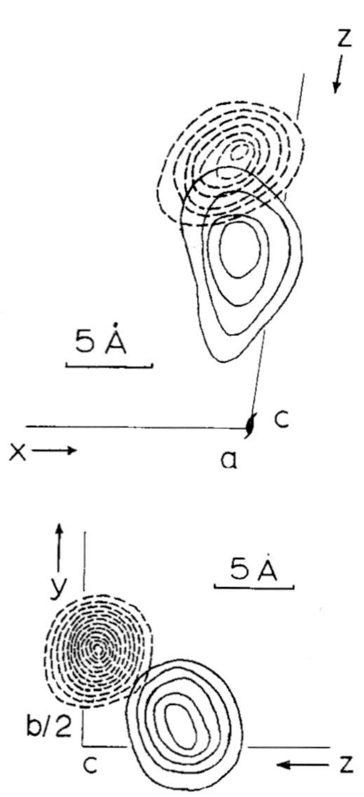

FIG. 3. *a*, a section through Gly-Tyr Region A and *p*-iodophenylpropionate Site 1 at y = 0.467 (14/30). The *smooth contours* are from the 0.05 M Gly-Tyr difference electron density map while the *dashed contours* are from the 0.05 M *p*-iodophenylpropionate on native CPA difference map. *b*, a section through Region B and Site 2 at x = 26/30. The *smooth contours* are from a 0.15 M Lys-Tyr-NH_2 difference map while the *dashed contours* are from the 0.05 M IPP difference maps.

FIG. 2. Superposition of contours from substrate difference maps. *1*, 0.05 M Gly-Tyr on native zinc-CPA; sections y = 14/30 to 17/30. *2*, 0.01 M Gly-Tyr on native zinc-CPA; sections y = 14/30 to 16/30. *3*, 0.05 M Gly-Tyr on acetylated CPA, sections y = 14/30 to 16/30. *4*, 0.01 M Gly-Tyr on copper-CPA, sections y = 13/30 to 15/30. *5*, 0.01 M Gly-Tyr on apo-CPA, sections y = 13/30 to 15/30. *6*, 0.15 M Lys-Tyr-NH_2 on zinc-CPA, sections y = 13/30 to 16/30. In each case the lowest contour is at 0.046 e/A^3 and the contour interval is 0.023 e/A^3.

CPA complex are different from those of the phenylalanine-CPA complex. In the hippuryl-L-phenylalanine difference map there is some increase in density in Region B, as well as in Region A, in contrast to the phenylalanine map (Table I, Fig. 4). However, the general similarity of the L-phenylalanine and hippuryl-L-phenylalanine difference maps does raise a question as to whether some formation and binding of product occurred during the later stages of the data collection.

Conformational Changes—In addition to Regions A and B described above we consistently observe another positive and an adjacent negative region in eight of the difference maps. In the most favorable example (Fig. 4*b*), hippuryl-L-phenylalanine is complexed with acetylated CPA; here the depth of the hole in the difference map is about 1.5 times the magnitude of the next highest unexplained feature of the map, and the centers of the peak and hole are separated by about 4 or 5 A. For comparison, sections through the L-phenylalanine difference map showing the peak and hole are presented in Fig. 4*a*. While in most other cases this peak and hole are the next largest features (after the substrate peaks themselves), they are not significantly above background. However, this peak and hole appear consistently when Region A is occupied, and the appearance of these features can be explained if a portion of the CPA molecule moves when

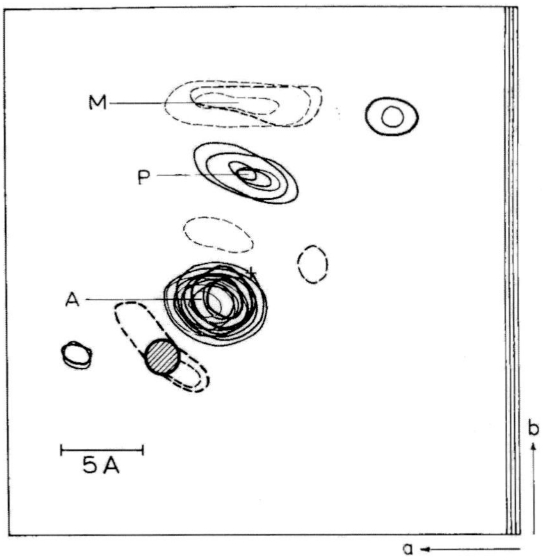

FIG. 5. Photograph of balsa model of 6 A CPA electron density map along with increase in electron density produced by 0.05 M glycyl-L-tyrosine. Region A is *striped* while Region B is *dotted*. The movement of protein which occurs when substrates are bound is indicated by the *arrow*. The *black spot* below Region A is the zinc atom. The crystallographic *b* axis is along the directions of the vertical support rods, while the *a* axis is along the front edge of the base.

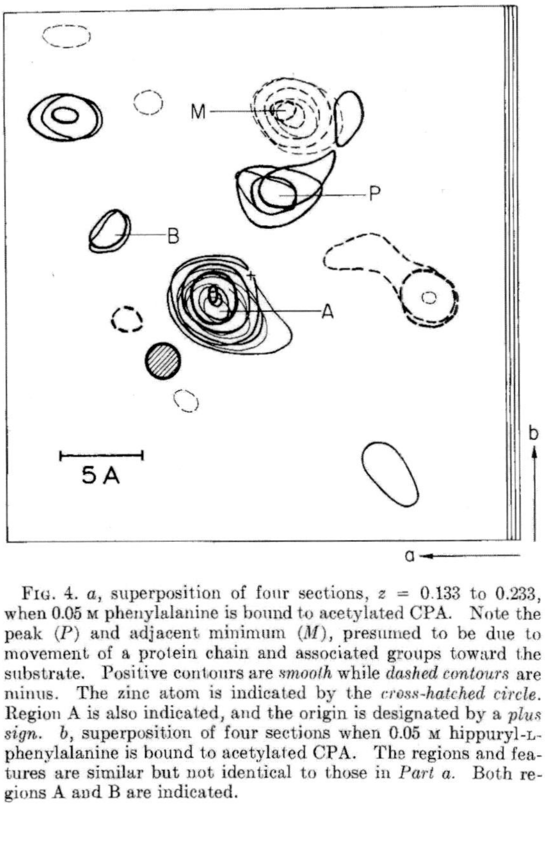

FIG. 4. *a*, superposition of four sections, z = 0.133 to 0.233, when 0.05 M phenylalanine is bound to acetylated CPA. Note the peak (*P*) and adjacent minimum (*M*), presumed to be due to movement of a protein chain and associated groups toward the substrate. Positive contours are *smooth* while *dashed contours* are minus. The zinc atom is indicated by the *cross-hatched circle*. Region A is also indicated, and the origin is designated by a *plus sign*. *b*, superposition of four sections when 0.05 M hippuryl-L-phenylalanine is bound to acetylated CPA. The regions and features are similar but not identical to those in *Part a*. Both regions A and B are indicated.

substrates are bound. This suggested movement is indicated by an *arrow* in Fig. 5.

DISCUSSION

Our present results show that model peptide substrates and competitive inhibitors are bound to crystalline CPA in regions which are sufficiently near the zinc to permit these added molecules to interact with the metal atom when it is present. We must leave a direct verification as to whether or not a bond is formed between zinc and substrate for experiments, now in progress, at higher resolution. However, the present studies provide some support for the hypothesis that the metal atom is involved at some step in the complete enzymic reaction.

Studies of the binding of β-(*p*-iodophenyl)propionate to the native enzyme have allowed us to draw some conclusions concerning the orientation of inhibitors and substrates, since the heavy atom on IPP permits us to locate one end of the bound inhibitor accurately and distinguishes the added compound from alterations in the protein structure. Of the four IPP sites, we shall discuss only Sites 1 and 2, since they are most closely related, although not identical, to Regions A and B of the Gly-Tyr maps. Assuming that the centers of the strongest peaks on the difference maps correspond to the iodine atoms, we have calculated that the iodine atoms in Sites 1 and 2 are 10.5 to 11 A from the zinc atom. This is the zinc to iodine distance expected for a fully extended IPP ion the carboxylate group of which is a zinc ligand. Our observations that the binding of IPP at Sites 1 and 2 is reduced by the removal of zinc, while very little structural change takes place in the enzyme itself when zinc is removed (1), are consistent with the proposal that the carboxylate group of the inhibitor is attached to the zinc. These results are also consistent with a nuclear magnetic resonance study (18) which indicates that a water molecule is displaced from the manganese of manganese-CPA when β-phenylpropionate is added.

Furthermore, if the carboxylate groups of IPP in both Sites 1 and 2 are zinc ligand, then binding to one site should exclude binding at the other. We have looked for evidence of competition between these sites by predicting the relative occupancies of Sites 1 and 2 at 0.05 M IPP, using K_s values derived from the occupancies observed at 0.01 M IPP. The results (Table II) tend to support competition between Sites 1 and 2, in agreement with the hypothesis based upon steric arguments.

The K_s values for Sites 1 and 2, assuming overlapping sites, are 2.2×10^{-2} M for Site 1 and 1.7×10^{-2} M for Site 2. These constants are similar to the K_I of β-phenylpropionate (10^{-2} M) in CPA crystals (7) but very different from the K_I of β-phenylpropionate (6×10^{-5} M) in solution (13). We cannot say, however, whether the K_s values of both Sites 1 and 2 change upon going from solution to crystal.

It is more difficult to give equivalent details about light atom substrates at 6 A resolution. Comparison of IPP contours in Site 1 with those of Gly–Tyr in Region A is shown in Fig. 3*a*, and of IPP contours in Site 2 with those of lysyltyrosinamide in Region B in Fig. 3*b*. Since the iodine of the IPP in Site 1 is near Region A, the phenol side chain of bound Gly–Tyr probably extends into the pocket, but a detailed model is not inconsistent with a slight displacement of the IPP phenyl group from that of Gly–Tyr. Also, the results for the IPP are not to be extrapolated to indicate that the carboxyl group of Gly–Tyr in Region A is bound to the zinc. The steric relation of iodine of IPP in Site 1 to Region A is quite reasonable. On the other hand, Region B cannot be superimposed on an IPP molecule extending from the iodine of Site 2 to the zinc, so that some doubt remains as to whether Region B is a second substrate site for dipeptides, or whether it is some evidence of conformational changes in the protein. Of course, the absence of density in Region B when Gly–Tyr interacts with apo-CPA or with copper-CPA and the observation that the largest peak in the lysyltyrosinamide difference map occurs in Region B favor the assignments of Region B as a second substrate site, perhaps slightly shifted from Site 2 of IPP. Multiple binding sites for substrates might be expected, since substrate inhibition of CPA has been demonstrated both in solution (20) and in the crystal (7). However, the kinetic data of Izumiya and Uchio (21) show no evidence for substrate inhibition by Gly–Tyr. Studies at higher resolution will be required to demonstrate conclusively whether Region B is a second substrate site.

The question of whether these dipeptide complexes are "productive" is also difficult to answer unequivocally. In lysozyme crystals, a stable complex, apparently nonproductive, is formed with tri-*N*-acetylglucosamine (6), which is actually a poor substrate in solution (21). Nevertheless, several observations suggest that the structures of the CPA complexes may be relevant to the enzyme mechanism. The apoenzyme is inactive both in solution and in the crystal (10, 13); the zinc is required for peptidase activity. If we suppose that the zinc is not involved directly with the substrate during binding or catalysis or both then it must induce activity by maintaining a tertiary protein structure which is necessary for activity. However, at least to 6 A resolution, there is no evidence of structural change when zinc is removed (1), except possibly for slight repositioning of metal ligands. Also a 2 A resolution difference map of mercury-CPA provides no indication of any protein structural changes when mercury is exchanged for zinc. Thus the "productive" binding site is probably near the zinc, which probably has a direct function in substrate binding or catalysis or both. This conclusion is also supported by the observation that the competitive inhibitor, L-phenylalanine, binds only at Site A. Higher resolution maps of both the protein and substrate complexes should help to ascertain whether the enzyme is capable of splitting a peptide bound at either A or B.

Substrate binding to acetylated CPA is discussed here on the assumption that results from experiments in solution also apply to the crystalline enzyme. It has been shown in solution that that when Gly-Phe binds to copper-CPA the rate of exchange of copper for zinc is slowed, but that after acetylation of CPA this substrate no longer affects zinc exchange; these results have been used to imply a failure of the substrate to bind to acetylated CPA (8). On the contrary, we find that Gly–Tyr binds to crystalline acetylated CPA near zinc. It therefore seems reasonable to suggest that some small structural differences do exist between CPA and acetyl-CPA complexes with Gly–Tyr, although these differences are below our limits of observation in difference electron density maps at 6 A resolution. If we are indeed looking at a productive complex between native CPA and Gly–Tyr, then the decrease in peptidase activity after acetylation must result either from changes in the binding below our limits of resolution or from specific effects on catalytic groups.

The apparent conformational change that occurs when substrate is bound in Region A is perhaps the most interesting result of our present work. It occurs to varying extents, most clearly for hippuryl-L-phenylalanine (Fig. 4*b*). The characteristic positive (*P*) and negative (*M*) peaks (Fig. 4*a*) are clearly observed when phenylalanine is bound in Region A, and in the 0.05 M iodophenylpropionate map these density changes of *P* and *M* are also present. However, no conformational changes detectable at 6 A resolution are produced by binding of lysyltyrosinamide. The separation of the positive (*P*) from the negative (*M*) region, approximately 4 to 5 A, does not necessarily mean that an entire segment of the peptide chain has moved that distance. For example, these features could result from a smaller motion of the main chain which displaces residues on opposite sides of the chain. Data on changes in ultraviolet absorption on binding of β-phenylpropionate (23) and on the absorbance of nitrotyrosyl residues after addition of inhibitor (24) may reflect the changes we have observed but cannot in themselves be uniquely interpreted as showing structural changes.

Table II

Fractional occupancy of β-(p-iodophenyl)propionate at 0.05 M

Site	Observed	Model assumed[a]	
		Competing sites	Independent sites
1	0.32	0.38	0.60
2	0.53	0.49	0.67

[a] Dissociation constants are calculated from the fractional occupancy of IPP at 0.01 M IPP, assuming competition between Sites 1 and 2 and assuming no competition. These constants are then used to predict the occupancy of IPP at 0.05 M IPP.

The details of conformational changes in proteins are now being elucidated by crystallographic studies. A large relative movement (approximately 8 A) of the β chains occurs upon deoxygenation of hemoglobin (25). Also, small structural changes (less than 1 A) in lysozyme, as a result of inhibitor binding (6), and in chymotrypsin, due to acylation of serine 195 (26), have been documented. On the other hand, the protein movement which we observe when CPA binds substrates or some inhibitors is larger (several angstroms) than that which occurs in the other single subunit enzymes which have been studied thus far. We expect that study of these and other complexes at high resolution will enable us to comment on the specific nature and function of the structural change as well as its relevance to the "induced fit" theory advanced by Koshland (27, 28).

Preliminary Results at Higher Resolution—In collaboration with Hartsuck and Reeke, we have recently computed a 2.0 Å three-dimensional electron density map of native CPA. We have also computed a 2.8 Å three-dimensional difference map between the Gly-Tyr complex with CPA and the native CPA. In the map of native CPA, we have located a region whose density is consistent with the sequence Ile-Tyr-Gln-Ala (or structurally similar residues) shown by Roholt and Pressman (29) to contain the tyrosine essential for peptidase activity. The tyrosyl side chain of this sequence is at the position of the negative region (*M* in Fig. 4) of the 6 Å difference maps attributed by us to a conformational change. Our interpretation of the 2.8 Å glycyl-tyrosine difference map places the hydroxyl oxygen of this tyrosine near the peptide bond of the substrate in Site A, a motion of the center of the tyrosyl side chain of roughly 8 Å from its position in the native enzyme. In addition to side chain motion, the peptide backbone of at least 2 residues (Tyr-Gln) moves when substrate is bound. Furthermore, a large side chain in a different sequence moves about 2 Å toward what we believe to be the carboxylate group of this bound substrate. An extension of data to 2.0 Å resolution is in progress for this complex of Gly-Tyr with CPA.

Acknowledgments—We wish to acknowledge the very considerable assistance of J. A. Hartsuck, G. N. Reeke, and H. Halsey. We also acknowledge helpful conversations with J. A. Rupley concerning the importance of salt in the preparation of crystalline substrate complexes. We are especially indebted to C. H. W. Hirs, who has provided us with pancreatic juice and procarboxypeptidase A.

REFERENCES

1. Lipscomb, W. N., Coppola, J. C., Hartsuck, J. A., Ludwig, M. L., Muirhead, H., Searl, J., and Steitz, T. A., *J. Mol. Biol.*, **19**, 423 (1966).
2. Ludwig, M. L., Hartsuck, J. A., Steitz, T. A., Muirhead, H., Coppola, J. C., Reeke, G. N., and Lipscomb, W. N., *Proc. Nat. Acad. Sci. U. S. A.*, **57**, 511 (1967).
3. Lipscomb, W. N., Ludwig, M. L., Hartsuck, J. A., Steitz, T. A., Muirhead, H., Coppola, J. C., Reeke, G. N., and Quiocho F. A., *Fed. Proc.*, **26**, 385 (1967).
4. Stryer, L., Kendrew, J. C., and Watson, H. C., *J. Mol. Biol.*, **8**, 96 (1964).
5. Johnson, L. N., and Phillips, D. C., *Nature*, **206**, 761 (1965).
6. Blake, C. C. F., Johnson, L. N., Mair, G. A., North, A. C. T., Phillips, D. C., and Sarma, V. R., *Proc. Roy. Soc. (London), Ser. B*, **167**, 378 (1967).
7. Quiocho, F. A., and Richards, F. M., *Biochemistry*, **5**, 4062 (1966).
8. Coleman, J. E., Pulido, P., and Vallee, B. L., *Biochemistry*, **5**, 2019 (1966).
9. Simpson, R. T., Riordan, J. F., and Vallee, B. L., *Biochemistry*, **2**, 616 (1963).
10. Vallee, B. L., Rupley, J. A., Coombs, T. L., and Neurath, H., *J. Biol. Chem.*, **235**, 64 (1960).
11. Coleman, J. E., and Vallee, B. L., *J. Biol. Chem.*, **236**, 2244 (1961).
12. Coleman, J. E., and Vallee, B. L., *Biochemistry*, **1**, 1083 (1962).
13. Quiocho, F. A., Bishop, W. H., and Richards, F. M., *Proc. Nat. Acad. Sci. U. S. A.*, **57**, 525 (1967).
14. Platt, J. T., Strain, W. H., and Warren, S. L., *J. Amer. Chem. Soc.*, **65**, 1273 (1943).
15. Doscher, M. S., and Richards, F. M., *J. Biol. Chem.*, **238**, 2399 (1963).
16. Quiocho, F. A., and Richards, F. M., *Proc. Nat. Acad. Sci. U. S. A.*, **52**, 833 (1964).
17. Blow, D. M., and Crick, F. H. C., *Acta Crystallogr.*, **12**, 794 (1959); Dickerson, R. E., Kendrew, J. C., and Strandberg, B. E., *Acta Crystallogr.*, **14**, 1188 (1966).
18. Shulman, R. G., Navon, G., Wyluda, B. J., Douglass, D. C., and Yamane, T., *Proc. Nat. Acad. Sci. U S. A.*, **56**, 39 (1966).
19. Elkins-Kaufman, E., and Neurath, H., *J. Biol. Chem.*, **178**, 645 (1949).
20. Lumry, R., Smith, E. L., and Glantz, R. R., *J. Amer. Chem. Soc.*, **73**, 4330 (1951).
21. Izumiya, N., and Uchio, H., *J. Biochemistry (Tokyo)*, **46**, 235 (1959).
22. Rupley, J. A., and Gates, V., *Proc. Nat. Acad. Sci. U. S. A.*, **57**, 496 (1967).
23. Fujioka, H., and Imahori, K., *J. Biol. Chem.*, **237**, 2804 (1962).
24. Riordan, J. F., Sokolovsky, M., and Vallee, B. L., *Biochemistry*, **6**, 358 (1967).
25. Muirhead H., and Perutz, M. F., *Nature*, **199**, 633 (1963).
26. Matthews, B. W., Sigler, P. B., Henderson, R., and Blow, D. M., *Nature*, **214**, 652 (1967).
27. Koshland D. E., Jr., *Cold Spring Harbor Symp. Quant. Biol.*, **28**, 473 (1963).
28. Koshland D. E., Jr., *Proc. Nat. Acad. Sci. U. S. A.*, **44**, 98 (1958).
29. Roholt, O. A., and Pressman, D., *Proc. Nat. Acad Sci. U. S. A.*, **58**, 280 (1967).

Reprinted from Nature, Vol. 314, No. 6008, pp. 257-260, 21 March 1985

Sulphate sequestered in the sulphate-binding protein of *Salmonella typhimurium* is bound solely by hydrogen bonds

J. W. Pflugrath & F. A. Quiocho

Department of Biochemistry, Rice University, Houston, Texas 77251, USA

An important question in understanding substrate binding by proteins is how charged groups are stabilized in the absence of their solvation shell. We have addressed this question here by solving the structure of the sulphate-binding protein of *Salmonella typhimurium* with bound substrate at 2.0 Å resolution. The results are remarkable in that the charged oxygen atoms of the sulphate molecule, which is buried and completely inaccessible to the solvent, are not stabilized by the formation of salt-bridges but by hydrogen bonds donated by specific residues of the protein. These hydrogen bonds are in turn coupled via peptide units to several resonating hydrogen bonding systems. These findings may be of general significance for the role of electrostatic interactions in protein structure and function.

In addition to the sulphate-binding protein, we have recently refined the molecular structures of three other binding proteins specific for L-arabinose[1], D-galactose and Leu, Ile or Val (N. K. Vyas, M. N. Vyas, J. S. Sack and F.A.Q., unpublished data; see also refs 2, 3). These periplasmic proteins, with dissociation constants of $\sim 10^{-6}$ M, belong to the family of proteins which act as initial receptors for active transport systems and chemotaxis in bacteria[4]. All four structures are ellipsoidal and composed of two similar globular domains connected by a flexible hinge.

The tertiary structure of the sulphate-binding protein at *p*H 4.3, which consists of 2,570 non-hydrogen atoms from 308 residues, 1 sulphate and 143 ordered water molecules, was refined by the method of Hendrickson and Konnert[5] to an *R*-factor of 0.14 for the 16,959 reflections from 8 to 2.0 Å. The refined structure differs from ideal bond lengths and angle distances by overall r.m.s. deviations of 0.015 Å and 0.038 Å, respectively. A more detailed description of the structure determination and protein conformation will be presented elsewhere.

Figure 1 shows the location of the binding site and Fig. 2 and Table 1 illustrate the precise details of sulphate binding. The sulphate is buried in the cleft ~7 Å below the protein surface and becomes completely inaccessible to the solvent; none of the 73-$Å^2$ accessible surface area[6] of the free sulphate is exposed. There are no positively charged residues, cations or water molecules within van der Waals' distance of the sulphate, thus precluding the formation of salt-bridges to neutralize the charged sulphate. Instead, all of the sulphate oxygens are the recipients of hydrogen bonds from seven residues distributed

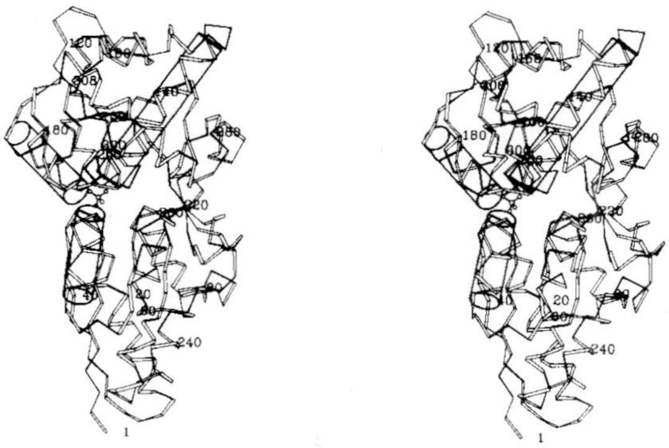

Fig. 1 Stereo view of the Cα backbone trace of the sulphate-binding protein, with the sulphate bound in the cleft between the two domains. The sulphate is shown in a ball-and-stick model and the three helices (44-54, 131-146, 172-181) close to the sulphate are further enclosed in cylinders. The residues hydrogen-bonded to the sulphate originate from both domains: the N-terminal domain supplies Asp 11 and Ser 45 and the C-terminal domain provides Ser 130-Gly 131-Gly 132-Ala 133, Ala 173 and Trp 192. Because of four likely changes to the published sequence of 310 residues[15] as a result of the present refinement, the sequence numbering scheme used in this study is based on the refined structure. However, as these changes (none in the binding site) consisted of one insertion, one deletion and two amino-acid substitutions, the total number of amino acids is unaltered. The last two residues are highly disordered. (Drawing produced by ARPLOT[16].)

Table 1 Hydrogen bonds between sulphate and binding protein

Protein atoms, donor (X)	Sulphate atoms, acceptor (Y)	Distance (Å) X···Y	Angle (°) X–H···Y
Ser 130 OG	O-1	2.76	167
Ala 173 N	O-1	2.70	151
Ser 45 N	O-2	2.84	169
Gly 131 N	O-2	2.83	170
Gly 132 N	O-3	2.77	149
Trp 192 NE1	O-3	2.82	157
Asp 11 N	O-4	2.67	152
	Overall mean:	2.76 (0.07)	159 (9)

Hydrogen bonds were determined on the basis of angle and distance criteria and hydrogen atoms were located on the basis of standard geometry (see ref. 1).

in both domains. These H-bonding residues are also buried. The sequence Ser 130-Gly 131-Gly 132-Ala 133, which forms part of a loop and the first turn of a helix, has a key role in binding, as the first three residues provide a complementary three-point docking site for O-1, O-2 and O-3 of the sulphate, respectively, and the Ala 133 peptide NH donates a hydrogen bond to the Ser 130 hydroxyl (see Fig. 2). The peptide bonds with CO groups from residues 44, 131 and 172 initiate three α-helices (Figs 1, 2*b*).

What neutralizes or stabilizes the sequestered dianion charges? Stabilization could be provided by the inducible hydrogen-bond dipoles, by the three helix macrodipoles (44-54, 131-146, 172-181) close to the sulphate (with opposite polarity) (Figs 1, 2*b*) or by a combination of the two. We propose that fixed hydrogen bond dipoles radially oriented around the sulphate, induced through polarization by the strong electric field of the dianion and further coupled (via peptide units) to several resonating or relay H-bond systems (see Fig. 2*b*), could effectively neutralize the charges. One of these systems, which is composed of two H-bonds (between O-2 of the sulphate and Gly 131 peptide NH and between Ser 130 peptide CO and Arg 134 NE), could dissipate the buried charges. Arg 134, which is 5.7 Å to the sulphate, is further involved in an inter-domain bidentate salt-link with Asp 68. Another relay system is formed by hydrogen bonding between O-4 of the sulphate and NH of Asp 11 and between CO of Tyr 10 and His 42 ND1, which is in contact with the solvent. A similar mechanism for dissipating charges could be achieved by three separate peptide bonds, each providing an NH H-bond to a sulphate oxygen and initiating via the CO group a 'zig zag' continuous sequence of alternating peptide units and H-bonds within the α-helix. The left-handed helical H-bond array has the form $O_i \cdots NH_{i+4}$, where $i = m + 3(n-1)$; m is the residue bearing the first O and $n = 1, 2, 3 \cdots$. An α-helix has three such arrays; ones of interest in SBP have m values of 44, 131 and 172.

As the Ser 130 hydroxyl is polarized by the dianion and is the recipient of a hydrogen bond from 133 NH located at the N-terminus of helix 131-146, the hydroxyl proton could be shared with the O-1 of the sulphate, yielding transient HSO_4^- species (the pK_1 and pK_2 of the sulphuric acid are ~ -3 and 1.9, respectively[7]). Proton sharing could be further facilitated by the coupling of 133 NH with another resonating H-bond array ($m = 132$).

Helix macrodipoles have been implicated in anion binding at the positive N-terminus of macrodipoles[8]. It has been calculated that binding of a single negative charge 5 Å from the helix macrodipole N-terminus involves a substantial attractive energy contribution of 12 kcal mol^{-1}, which is minimally diminished by desolvation energy[8] (this value may be overestimated significantly)[9]. The three helices in the sulphate-binding protein close to the dianion could provide charge stabilization. The effect of the macrodipoles may be diminished as the sulphate is not centred on any helix axis. These helices and several others with their N-termini converging towards the binding site cleft could also promote long-range attraction of anions preparatory to actual tight binding[8]. This would seem to be generally applicable to binding proteins except that several of the substrates are uncharged sugars and amino acid zwitterions.

The mechanism we have described above confines the function of an α-helix primarily to a specific H-bonded peptide array and takes into account resonances and polarizability of the array in dissipating buried charges. The mechanism would be more effective when the array is exposed to the solvent. This is what we observed in the amphipathic[10] helices with m values of 44 and 172; the third helix is partially buried. Further, we have made similar observations from other proteins (for example, flavodoxin) known to bind anions at N-termini of amphipathic helices (unpublished findings). In this connection, it has been shown that carbonyl oxygens within the hydrophilic side of helices are often H-bonded to solvents[11].

The proposed function of a specific H-bond helical array could, in much the same way as it might enhance the sharing of a serine hydroxyl proton with the sulphate, increase the reactivity of enzyme active-site residues or of coenzymes hydrogen-bonded to the first turn of helices.

The structure of flavodoxin resembles one of the domains of the periplasmic binding proteins (G.L.G., M.A.S. and F.A.Q., unpublished data). Although the substrate site of flavodoxin is located in a crevice at the surface[12] (coordinates obtained from Protein Data Bank[13]), we observed some similarities in the binding of the phosphate moiety of the buried flavin mononucleotide with the sulphate binding described above. (1) Of the 10 H-bond donors to the phosphate oxygens, 5 are from peptide NH groups of loops and the N-terminus of the one helix (10-26) near the phosphate dianion; (2) a Ser OH simultaneously donates a H-bond to the phosphate and accepts a H-bond from a peptide NH; (3) the one line of peptide H-bonds ($m = 10$) is exposed to the solvent. As the phosphate is not deeply buried within the flavodoxin, several relay systems as found in SBP may not be required. Nonetheless, the carbonyl oxygens of peptides bearing the NH groups H-bonded to the phosphate

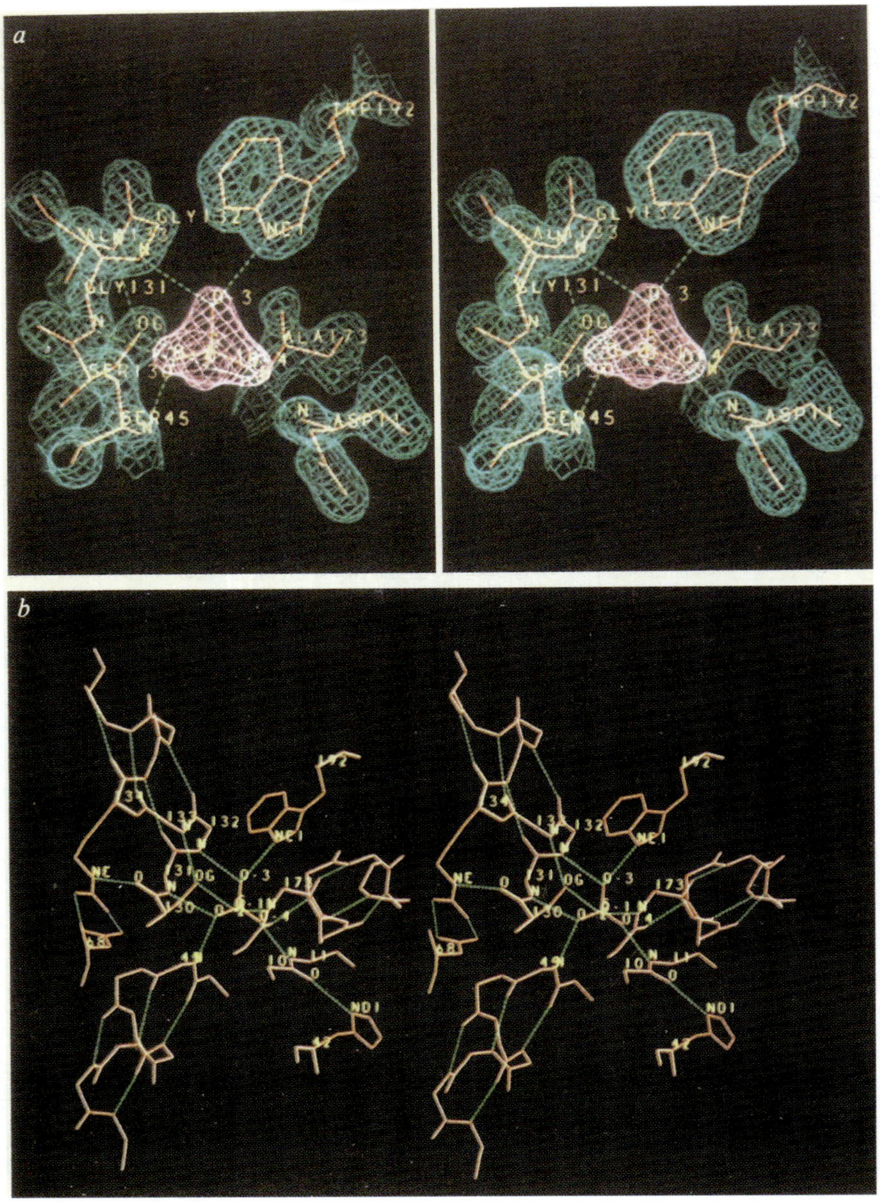

Fig. 2 *a*, Stereo view of the electron density of the tetrahedral sulphate and essential residues of the sulphate-binding protein superimposed on the refined model. The density map was calculated with coefficients $(2|F_0|-|F_c|)$ and final 2 Å refined α_c. The density surface (sulphate in purple and amino acids in blue) was contoured at 1.5 times the estimated standard deviation of the electron density and at grid-spacing of 0.4 Å. The green dotted lines represent hydrogen bonds. The H-bonds formed between the sulphate and the sulphate-binding protein are listed in Table 1. The α-carbons are identified with the amino-acid name and number. The plane formed by Ser 130 OG, Gly 131 N and Gly 132 N is parallel to and maximally superimposed on the plane of O-1, O-2 and O-3 of the sulphate. *b*, The same as *a* except that the electron density surface and residue names have been removed for clarity. Furthermore, other hydrogen-bonding systems (described in the text) and at least the first turn of three helices close to the sulphate have been added. The molecular graphics program FRODO[17] extensively modified for the E & S PS300 system[18] was used in the refinement and in generating these figures.

are in contact with the solvent.

Despite completely different substrate specificities, there are similarities in the binding sites of the L-arabinose-binding protein[1] and the sulphate-binding protein. The L-arabinose is also bound in the cleft via several H-bonds from six buried residues in both domains and is inaccessible to the bulk solvent. It is notable, especially in view of the importance of peptide bonds in anion binding, that while only side chains are involved in hydrogen-bonding the sugar in the arabinose receptor, five of the seven H-bonds to the sulphate are from the peptide backbone. (We have also observed that the charge of the solvent-inaccessible Arg 151, one of three essential charged residues in the arabinose-binding protein, is stabilized entirely by protein and sugar dipoles[1].) As H-bonds are highly directional, they confer geometrical specificity on the site and ensure correctness of fit for the substrate. It may also be important that these bonds are stable enough to provide significant binding but are of sufficiently low strength to allow rapid ligand dissociation. The estimated turnover rates of binding proteins *in vitro* are comparable to the V_{max} values for *in vivo* transport[14]. Transport could be impeded if the sulphate were bound with salt-links.

The novel finding of general significance presented here is that charges from as strong an acid as sulphate can be stabilized in an enclosed, solvent-inaccessible site within a protein by bonds other than salt-links. This particular result and other observations reported here have far-reaching importance in understanding not only other ion-binding and transport systems, but protein stability, enzyme catalysis and other biochemical processes that require charged intermediates.

J.W.P. was the recipient of an NSF predoctoral fellowship. We thank Dr N. K. Vyas for assistance. This work was supported by grants from NIH (GM21371), NSF (PCM82-14057) and the Welch Foundation (C-581).

Received 29 August; accepted 10 December 1984.

1. Quiocho, F. A. & Vyas, N. K. *Nature* **310,** 381–386 (1984).
2. Vyas, N. K., Vyas, M. N. & Quiocho, F. A. *Proc. natn. Acad. Sci. U.S.A.* **80,** 1792–1796 (1983).
3. Saper, M. A. & Quiocho, F. A. *J. biol. Chem.* **258,** 11057–11062 (1983).
4. Landick, R. & Oxender, D. L. in *Bacterial Transport* (ed. Martonosi, A. N.) 81–88 (Plenum, New York, 1982).
5. Hendrickson, W. A. & Konnert, J. H. in *Computing in Crystallography* (eds Diamond, R., Ramaseshan, S. & Venkatesan, K.) 13.01–13.23 (Indian Academy of Sciences, International Union of Crystallography, Bangalore, 1980).
6. Lee, B. K. & Richards, F. M. *J. molec. Biol.* **55,** 517–526 (1971).
7. Pauling, L. *The Nature of the Chemical Bond,* 325 (Cornell University Press, Ithaca, 1966).
8. Hol, W. G. J., van Duijnen, P. T. & Berendsen, H. J. C. *Nature* **273,** 443–446 (1978).
9. Rees, D. C. *J. molec. Biol.* **141,** 323–326 (1980).
10. Perutz, M. F., Kendrew, J. C. & Watson, H. C. *J. molec. Biol.* **13,** 669–678 (1965).
11. Blundell, T., Barlow, D., Borkakoti, N. & Thornton, J. *Nature* **306,** 281–283 (1983).
12. Smith, W. W., Burnett, R. M., Darling, G. D. & Ludwig, M. L. *J. molec. Biol.* **117,** 195–225 (1977).
13. Bernstein, F. C. *et al. J. molec. Biol.* **112,** 535–542 (1977).
14. Miller, D. M. III, Olson, J. S., Pflugrath, J. W. & Quiocho, F. A. *J. biol. Chem.* **258,** 13665–13672 (1983).
15. Isihara, H. & Hogg, R. W. *J. biol Chem.* **255,** 4614–4618 (1980).
16. Lesk, A. M. & Hardman, K. D. *Science* **216,** 539–540 (1982).
17. Jones, T. A. in *Computational Crystallography* (ed. Sayre, D.) 3303–3317 (Clarendon, Oxford, 1982).
18. Pflugrath, J. W., Saper, M. A. & Quiocho, F. A. in *Methods and Applications in Crystallographic Computing* (eds Hall, S. & Ashida, T.) 404–407 (Clarendon, Oxford, 1984).

Chapter 8

Gareth R. Eaton

Biography

Gareth R. Eaton received his A.B. at Harvard in 1962 and obtained his Ph.D. in inorganic chemistry at MIT in 1972. Since then he has been in the Department of Chemistry (now the Department of Chemistry and Biochemistry) at the University of Denver. He was promoted to professor in 1980. He served as dean of natural sciences from 1984–1988 and as vice provost for research from 1988–1989. He teaches inorganic chemistry. His research is performed jointly with Sandra S. Eaton, who received her bachelor's degree at Wellesley College in 1968 and obtained her Ph.D. in inorganic chemistry at MIT in 1972. She joined the faculty of the University of Colorado at Denver in 1973. In 1990 she moved to the Department of Chemistry and Biochemistry at the University of Denver. She teaches undergraduate and graduate classes in physical and inorganic chemistry, and is currently chair of the department.

In 1997 Gareth and Sandra Eaton jointly received the John Evans Professorship at the University of Denver. In 2001 the Electron Paramagnetic Resonance (EPR) Discussion Group of the Royal Society of Chemistry (London) awarded the Bruker Prize to them jointly. In 2008 they were honored as fellows of the International EPR/ESR Society. Their research program involves the development of instrumentation and methodology for continuous wave, rapid scan, and pulsed EPR, and applications of these methods to the study of relaxation times, spin-spin interaction, metal ions in biological systems, and EPR imaging. They jointly organized the International EPR Symposium, which is held in Colorado each summer, for 30 years. They have jointly authored more than 300 research papers, reviews, and book chapters.

Gareth Eaton's first book, *NMR Studies of Boron Hydrides and Related Compounds* with William N. Lipscomb, was originally intended as an updating of a chapter in the Colonel's *Boron Hydrides*. This was largely written the summer before Gareth Eaton entered graduate school at MIT in 1967 and was published in 1969 with help from Sandra S. Shaw, whom he married just before the book was published.

Gareth and Sandra Eaton edited the books *EPR Imaging and in Vivo EPR* jointly with K. Ohno in 1991, *Foundations of Modern EPR* jointly with Kev M. Salikhov in 1998, and three volumes in the Biological Magnetic Resonance Series: *Distance Measurements in Biological Systems by EPR in* 2000, *Biomedical EPR — Part A: Free Radicals, Metals, Medicine, and Physiology*, and *Biomedical EPR — Part B: Methodology, Instrumentation, and Dynamics* both in 2005. They edited a special issue of *Applied Magnetic Resonance* on high-field and high-frequency electron paramagnetic resonance instrumentation in 1999. With James S. Hyde, in 2006, they prepared a special thematic issue of *Concepts in Magnetic Resonance* titled "EPR at Work."

With Don C. Wiley and Oleg Jardetzky, Gareth Eaton edited *Structures and Mechanisms: From Ashes to Enzymes* in honor of the Colonel's 80th birthday. This book was published in 2002, again with substantial editing and formatting assistance by Sandra Eaton.

Another book, *Quantitative EPR*, was published in 2010.

Symmetry

The Colonel taught a graduate course in symmetry and molecular orbitals. I used the cross-registration agreement between MIT and Harvard to take this course in the spring of 1968. One day, as the Colonel was leaving the classroom, he stopped where I was sitting and said, "Aren't you glad you had Al's course first!" He was referring to the course in chemical applications of group theory that F. Albert Cotton taught at MIT. Since the Colonel covered topics many times faster than did Cotton, I was indeed glad for the background preparation.

An assignment in the course was to apply what we had learned to a topic of our choice. I proposed extending the borane treatments to the chemistry of beryllium, but the Colonel sagely cautioned me that not enough structural chemistry of Be was known to make this possible — he had tried.

My notes from the Colonel's course were my main reference book the first time I taught an analogous course at the University of Denver. The most important insight from the Colonel's course, though, was a little aside about non-rigid groups that he talked about in one of the lectures. A few years later, when Sandra Eaton was struggling to find a completeness argument about the rearrangements of trischelate complexes of transition metals, it came to me that we should look into the non-rigid groups that the Colonel had mentioned ever so briefly. There had been many previous attempts to find all possible rearrangements of trischelate complexes of transition metals. At one time, the Holm research group had a table covered with molecular models of every mechanism or intermediate anyone in the group could think of. The power of the non-rigid group approach was in the completeness argument. The result of this was our 1973 paper in the *Journal of the American Chemical Society* which is reproduced in this volume. The papers that resulted from Sandra Eaton's Ph.D. thesis research[1–4] form the basis for the modern discussion of rearrangements of four-coordinate and six-coordinate transition metal complexes, such as is presented in Chapter 14 of *Inorganic Chemistry* by Purcell and Kotz.[5] Of interesting relevance to the present context, the next section in this chapter is a discussion of the mechanism of carboxypeptidase A, based on the research of the Colonel and coworkers.

Impact of Laboratory Design on Progress

A few comments on the impact of laboratory design on progress in science may not be out of place here. During our early years in graduate school, we benefited from working in a crowded (Sandra and I shared one desk and one chair), somewhat dilapidated, former freshman chemistry lab, which was shared by about two dozen members of the Richard Holm and Alan Davidson research groups. The hood worked only if a sheet of plywood was strategically positioned over the holes in the top of the hood. Each student and postdoc had as work space two contiguous freshman lab workstations (benchtop and drawers and a sink). This was a great learning environment, because the experienced postdocs and other graduate students were within conversational range, and one could get immediate advice on almost any topic or technique. Later, MIT renovated labs in another location, and we moved into a two-person lab, just in time to work out the details of the non-rigid group approach to trischelate complexes. There, we had a quiet environment for thinking. We each started with a large sheet of paper and worked out all elements of the group independently and then checked each other's work. Different tasks require different environments.

SSE and GRE worked jointly on this project. First person wording in these comments refers to GRE.

References

1. Holm R. H. and Eaton S. S., Stereochemically non-rigid cobalt (III) complexes. *J. Am. Chem. Soc.*, **93**, 4913–4914 (1971).
2. Holm R. H. and Eaton S. S., Inversion and ligand exchange kinetics of tetrahedral bis-chelate complexes of zinc (II) and cadmium (II). *Inorg. Chem.*, **10**, 1446–1452 (1971).
3. Eaton S. S., Hutchison J. R., Holm R. H., and Muetterties E. L., Intramolecular rearrangement reactions of tris-chelate complexes. III. Analysis of the rearrangements of tris (a-isopropenyl- and a-isopropyltropolonato)-aluminum (III) and -cobalt (III). Examples of stereochemically nonrigid aluminum (III) and cobalt (III) complexes. *J. Am. Chem. Soc.*, **94**, 6411–6426 (1972).
4. Eaton S. S., Eaton G. R., Holm R. H., and Muetterties E. L., Intramolecular rearrangement reactions of trischelate complexes. IV. Further investigation of the rearrangements of tris(a-isopropenyl- and a-isopropyltropolonato)-metal (III,IV) complexes. *J. Am. Chem. Soc.*, **95**, 1116–1124 (1973).
5. Purcell K. F. and Kotz J. C., *Inorganic Chemistry*. Philadelphia: W. B. Saunders (1977).

1825

Symmetry Groups of Nonrigid Tris-Chelate Complexes

S. S. Eaton[1] **and G. R. Eaton***[1]

Contribution from the Department of Chemistry, Massachusetts Institute of Technology, Cambridge, Massachusetts 02139.
Received August 3, 1972

Abstract: Rearrangements of tris-chelate complexes are treated by the method of molecular symmetry groups for nonrigid molecules introduced by Longuet-Higgins. Full details are given for the case of three identical unsymmetrical ligands and for the case of two identical unsymmetrical ligands and a symmetrical ligand. The rearrangements derived include three not previously discussed.

Determination of the mechanisms of intramolecular rearrangement reactions of tris-chelate complexes is a long-standing problem.[2] Although a large amount of kinetic data has been obtained, only in a few recent cases has it been possible to significantly restrict the number of mechanisms[3–10] or to deduce a unique rearrangement mechanism[11] consistent with the experimental results.

The rearrangement mechanisms which have been discussed for tris-chelate complexes include[12] (a) the Bailar,[13–15] or trigonal,[5] twist about the real threefold molecular axis, (b) the Ray–Dutt,[14–16] or rhombic, twist about the imaginary threefold molecular axis, and (c) bond rupture with a five-coordinate square-pyramidal (SP) or trigonal-bipyramidal (TBP) transition state, with, in each case, dangling axial or equatorial (basal) ligands.[3] Prior analyses of rearrangements of tris chelates in terms of the foregoing mechanisms have been based largely on the apparent physical reasonableness of the postulated intermediates. Thus there is no assurance that every feasible mechanism has been considered. As the measurements become more refined[7–11] it becomes more important to ensure that no feasible rearrangement mechanism has been overlooked. In this paper we report an analysis of all of the rearrangements of tris-chelate complexes, using the molecular symmetry groups for nonrigid molecules introduced by Longuet-Higgins.[17,18]

Symmetry Groups of Nonrigid Molecules. By definition a nonrigid molecule undergoes geometrical changes within the time scale of the measurement technique considered. For example, for a tris-chelate complex the geometry might change from **1** to **2** during the mea-

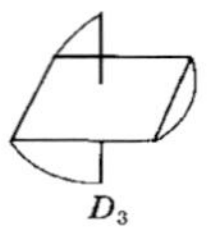

D_3
1

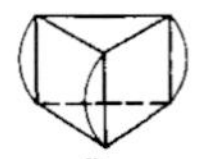

D_{3h}
2

surement time. Thus the question is raised whether the symmetry should be considered to be D_3 or D_{3h} for the purpose of analyzing the data. As another example, consider that a tris chelate containing three identical unsymmetrical ligands could exist as a mixture of the cis **3** and trans **4** isomers in rapid equilibrium. Should the

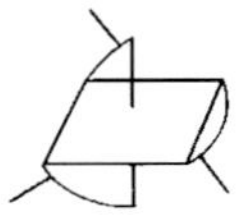

C_3
3

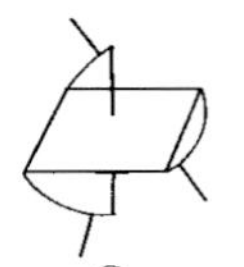

C_1
4

system be analyzed in terms of C_3 or C_1 symmetry? The answer Longuet-Higgins[17] proposed to this type of seemingly paradoxical question is that the system should not be analyzed in terms of the rigid molecule point group. Rather, nonrigid molecules should be interpreted[17] using a molecular symmetry group which is the set of (a) all feasible permutations, P, of the positions and spins of identical nuclei, including the identity E, and (b) all feasible permutation-inversions, P^*, which simultaneously permute and invert the coordinates of all particles in the center of mass. P^* is the product $PE^* = E^*P$, where E^* is the inversion of all particle positions (E^* is not always among the feasible operations.)[17]

An alternative formulation of the molecular symmetry groups of nonrigid molecules has been proposed,[19] but in its most complete form it affords results identical with those obtained by Longuet-Higgins' method.[19–22]

The method introduced by Longuet-Higgins has been applied to interpretation of vibrational and rotational aspects of the spectra of many nontransition metal

(1) Department of Chemistry, University of Denver, Denver, Colo.

(2) For recent reviews see (a) N. Serpone and D. G. Bickley, *Progr. Inorg. Chem.*, in press; (b) F. Basolo and R. G. Pearson, "Mechanisms of Inorganic Reactions," 2nd ed, Wiley, New York, N. Y., 1967, pp 300–334; (c) J. J. Fortman and R. E. Sievers, *Coord. Chem. Rev.*, **6**, 331 (1971).

(3) J. G. Gordon, II, and R. H. Holm, *J. Amer. Chem. Soc.*, **92**, 5319 (1970).

(4) J. R. Hutchison, J. G. Gordon, II, and R. H. Holm, *Inorg. Chem.*, **10**, 1004 (1971).

(5) R. C. Fay and T. S. Piper, *ibid.*, **3**, 348 (1964).

(6) A. Y. Girgis and R. C. Fay, *J. Amer. Chem. Soc.*, **92**, 7061 (1970).

(7) L. H. Pignolet, R. A. Lewis, and R. H. Holm, *ibid.*, **93**, 360 (1971).

(8) L. H. Pignolet, R. A. Lewis, and R. H. Holm, *Inorg. Chem.*, **11**, 99 (1972).

(9) S. S. Eaton and R. H. Holm, *J. Amer. Chem. Soc.*, **93**, 4913 (1971).

(10) M. C. Palazzotto and L. H. Pignolet, *J. Chem. Soc., Chem. Commun.*, 6 (1972).

(11) S. S. Eaton, J. R. Hutchison, R. H. Holm, and E. L. Muetterties, *J. Amer. Chem. Soc.*, **94**, 6411 (1972).

(12) See ref 9 for a detailed discussion.

(13) J. C. Bailar, Jr., *J. Inorg. Nucl. Chem.*, **8**, 165 (1958).

(14) C. S. Springer, Jr., and R. E. Sievers, *Inorg. Chem.*, **6**, 852 (1967).

(15) J. E. Brady, *ibid.*, **8**, 1208 (1969).

(16) P. Ray and N. K. Dutt, *J. Indian Chem. Soc.*, **20**, 81 (1943).

(17) H. C. Longuet-Higgins, *Mol. Phys.*, **6**, 445 (1963).

(18) A summary of some of the results has been reported, ref 11.

(19) S. L. Altman, *Proc. Roy. Soc., Ser. A*, **298**, 184 (1967).

(20) C. M. Woodman, *Mol. Phys.*, **19**, 753 (1970).

(21) J. K. G. Watson, *ibid.*, **21**, 577 (1971).

(22) S. L. Altman, *ibid.*, **21**, 587 (1971).

compounds including ethane,[17] H_3CBF_2,[17,23] p-fluorotoluene,[24] cyclobutane,[25] cyclopentane,[26] and methylsilylacetylene.[27] Similar applications have been reported for ferrocene[28] and octahedral metal hexaammine ions.[29] There have been no reported applications of the Longuet-Higgins method to metal chelate complexes or to the nmr spectra of metal complexes.

Tris-Chelate Complexes. Case (1). Three Identical Unsymmetrical Ligands. In the case of tris-chelate complexes containing three identical unsymmetrical ligands, $M(A-B)_3$, the set of all permutations and permutation-inversions of the ligating nuclei is a group of order 384. This result is obtained by starting with one isomer, such as **5**, and writing down all isomers which can be obtained from **5** by permuting and inverting (through the center of mass) the ligating nuclei, with the restriction that the ligands cannot span trans positions. The notation is illustrated in **5** for the cis isomer

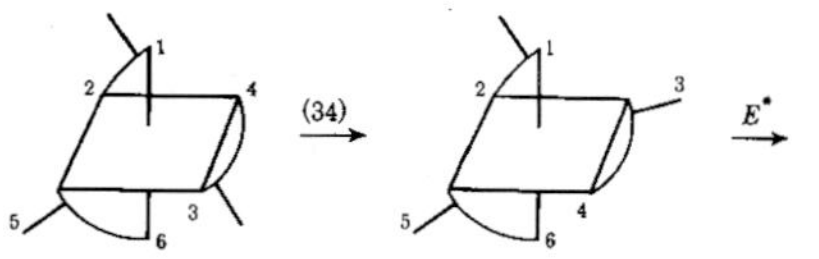

5 [135−462] **6** [145−362]

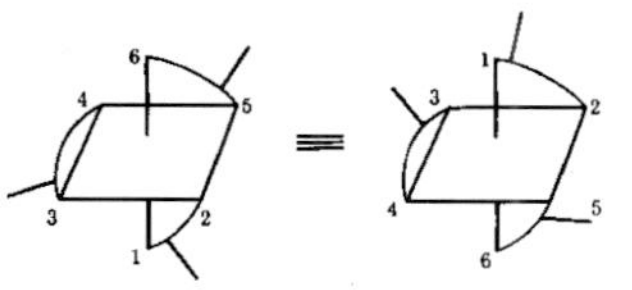

7 [154−263]

[135-462]. As the complex is viewed down the real-C_3 (or pseudo-C_3) axis with the triangular face including vertex 1 upward, the vertex labels are given clockwise for this face starting with 1, followed by the labels for the lower triangular face starting with the vertex to the right of 1. Square brackets denote isomers and parentheses denote permutations. One end of the ligand is flagged to denote its unsymmetrical nature (*e.g.*, the isopropenyl substituent in α-isopropenyltropolonate). The permutation (34) is illustrated by **5** → **6** and takes cis to trans. Isomer **6** is converted to **7** by E^*, the inversion of all ligands through the center of mass, and the operation which takes **5** to **7** is the permutation-inversion (34)*.

By inspection it is found that each isomer can be converted into 15 other isomers which cannot be superimposed on it by rigid-body rotations, but that for each of these 16 isomers there are 23 rearranged forms which can be superimposed on it by rigid body rotations. These rigid body motions are the same for each isomer and have group properties. Furthermore, there are 16 operations which interconvert the isomers within the set of 16 isomers. These 16 operations form an Abelian group. Thus the set of all permutations and permutation-inversions of the ligating nuclei is a group of order 384 which factors into a group of order 16 consisting of the rearrangements and a group of order 24 that can be regarded as consisting of the rigid-body rotations of the molecule. The 16 operations in the group of order 16, which is of primary interest here, constitute the complete group of rearrangements of stereochemically nonrigid tris-chelate complexes. It is important to note that the 16 rearrangements are expressed as permutations, *i.e.*, they merely describe the net change and not the path by which the atoms move from initial to final position.

The 16 isomers are drawn in Table I. The multiplication table for the 16 operations is given in Table II. The effect of each of the 16 operations on the isomer [135-462] is given in Table III. The notation used in these tables to indicate nonequivalent environments is illustrated in **8** for a ligand with a diastereotopic group at one end. The magnetic environments *x*, *y*, and *z* are, in principle, distinguishable for any unsymmetrical ligand whereas the environments r and s apply only to diastereotopic substituents such as the methyl resonances of α-isopropyltropolone or the benzyl CH_2 resonances of *N*-methyl-*N*-benzyldithiocarbamate.

Table I. Isomers of $M(A-B)_3$ Complexes

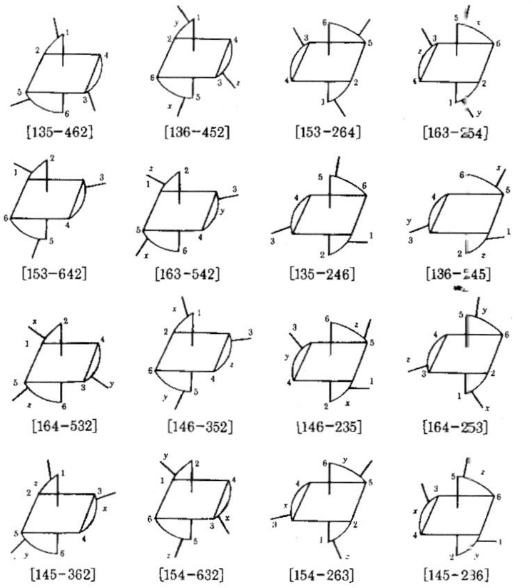

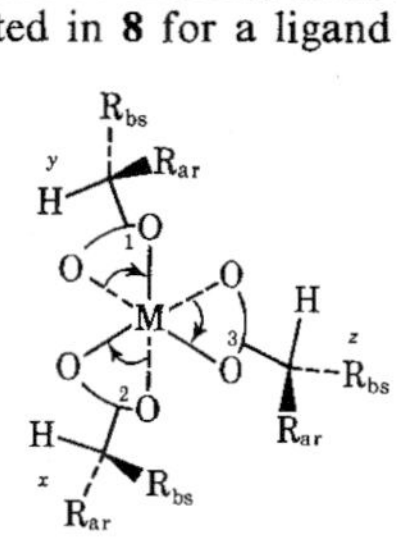

8

Since the nmr experiments detect only the net effect on all isomers, permutations which give the same net averaging pattern for the nonequivalent sites are placed

(23) P. R. Reed, Jr., and R. W. Lovejoy, *Spectrochim. Acta, Part A*, **26**, 1087 (1970).
(24) T. Cvitas and J. M. Hollas, *Mol. Phys.*, **20**, 645 (1971).
(25) J. M. R. Stone and I. M. Mills, *ibid.*, **18**, 631 (1970).
(26) I. M. Mills, *ibid.*, **20**, 127 (1971).
(27) P. R. Bunker, *ibid.*, **9**, 257 (1965).
(28) P. R. Bunker, *ibid.*, **8**, 81 (1964); *ibid.*, **9**, 247 (1965).
(29) A. J. Stone, *J. Chem. Phys.*, **41**, 1568 (1964).

together in averaging sets, A_i, although the permutations within each set A may have different effects on a particular isomer. The effect of the eight operations (averaging sets) on the pmr spectra are discussed below for ligands with nondiastereotopic and diastereotopic substituents.

Nondiastereotopic. When the substituents are not diastereotopic, the site averaging does not indicate whether a rearrangement proceeds with inversion or retention of configuration at the metal center. Thus the operations, A_i, are pairwise indistinguishable: (A_1, A_5), (A_2, A_6), (A_3, A_7), (A_4, A_8). (A_1, A_5) has no effect on the pmr spectra. (A_2, A_6) averages two trans sites (yz) reducing the signal multiplicity from four at slow exchange to three at fast exchange. Distinctions between (A_3, A_7) and (A_4, A_8) are more subtle. Both pairs effect cis–trans isomerization and thus average all four signals to one at fast exchange. The two pairs differ in the four matrix elements related to exchange between y and z out of the total of 16 matrix elements in the pmr kinetic exchange matrix.[11]

Diastereotopic. Although the substituents are diastereotopic and thus distinguish between processes which effect inversion and those which do not, the difficulties in assigning observed resonances to absolute magnetic environments precludes distinctions between some operations. A_1, the identity operation, is obviously easily distinguished. A_2 effects two pair-wise exchanges ($yr \rightleftharpoons zr$, $ys \rightleftharpoons zs$), leaving four signals unbroadened and reducing signal multiplicity from eight to six. A_3 and A_4 both interconvert cis and trans without inversion at the metal center, reducing the signal multiplicity from eight to two. However, as in the case of nondiastereotopic nuclei the exchange matrices for A_3 and A_4 are very similar, differing in eight matrix elements out of 64. A_5 and A_6 result in four pair-wise exchanges which are indistinguishable in the absence of unambiguous assignments of the magnetic environments yr, ys, zr, and zs to observed resonances (A_5 averages yr with ys and zr with zs whereas A_6 averages yr with zs and ys with zr). A_7 and A_8 average all eight resonances to a single peak but the differences again are restricted to eight matrix elements which have a relatively small effect on the overall shape of the exchange-broadened spectra.

Comparison of Results for Diastereotopic and Nondiastereotopic Substituents. If the ligand contains both diastereotopic and nondiastereotopic groups, then the combined information allows distinction between the following operations on the basis of changes in signal multiplicity alone: A_1, A_2, (A_3, A_4), A_5, A_6, (A_7, A_8). For example, by observing the nondiastereotopic CH_3 and diastereotopic CH_2 resonances in tris(*N*-methyl-*N*-benzyldithiocarbamato)iron it was concluded that A_6 is the low-temperature process.[30] In principle such a comparison could also be done for the nondiastereotopic α'-H and diastereotopic isopropyl methyls of α-isopropyltropolone. Where it is not feasible to look at diastereotopic and nondiastereotopic nuclei in the same ligand, closely related complexes can be compared, provided the assumption is made that both rearrange by the same mechanism. As is discussed in detail elsewhere,[10,31] this procedure was used to determine that A_6 is the only mechanism which fits the nmr data for the low-temperature processes of both $M(\alpha\text{-}C_3H_5T)_3$ and $M(\alpha\text{-}C_3H_7T)_3$ for M = Co, Al, and Ga.[11,31–33]

Physical Mechanisms. The following correspondences can be made between the various physical mechanisms considered previously[4,6] and the permutations derived here: p-C_3 or r-C_3 twists, A_6; i-C_3 twists, A_8; TBP-axial, A_7; TBP-equatorial, A_3; SP-axial, $A_3 + A_6 + A_8$. Stiefel has also proposed a "distorted hexagonal planar" intermediate.[34] Rotation of the ligands through a planar transition state to positions roughly perpendicular to the original ones gives as a net result A_5. If the rotation is continued through a trigonal twist motion, the net result is A_2. Various physical representations of A_i can be obtained by comparison of a starting isomer with each of the 24 rotated forms of the product isomer to suggest the ring motion necessary to effect the observed site averaging. The magnetic molecular models recently described[35] are useful for this purpose.

Tris-Chelate Complexes. Case (2). One Symmetrical and Two Identical Ligands, $M(A\text{–}B)_2(C\text{–}C)$. If the substituents on the symmetrical ligand can be studied by nmr (^{19}F, ^{1}H, . . .), the two ends may be distinguishable if they are in magnetically nonequivalent environments. Thus the permutation-inversion group is the set of permutations and permutation-inversions of six distinguishable nuclei connected in pairs of two, just as for $M(A\text{–}B)_3$. The 16 isomers for $M(A\text{–}B)_2(C\text{–}C)$ are the same as those in Table I if the ligand **5–6** is regarded as the symmetrical ligand, C–C. There are, then, three geometrical isomers and a total of four magnetic environments for a nondiastereotopic substituent on either the symmetrical ligand or unsymmetrical ligand. The notation for the isomers[7] and magnetic environments is illustrated below. Isomers ct and tc have two-

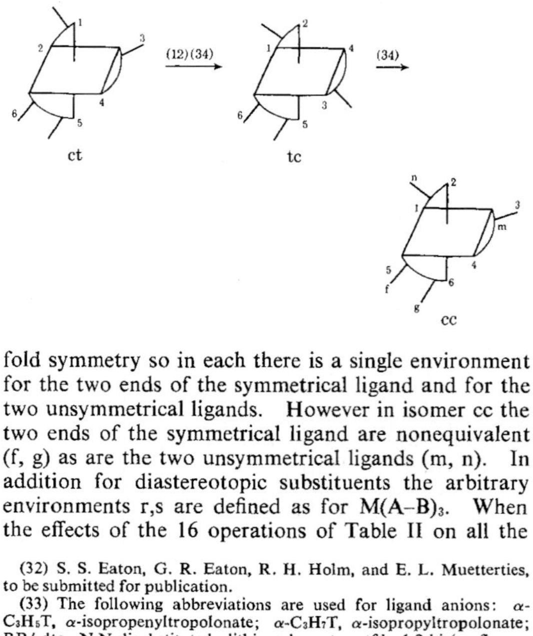

fold symmetry so in each there is a single environment for the two ends of the symmetrical ligand and for the two unsymmetrical ligands. However in isomer cc the two ends of the symmetrical ligand are nonequivalent (f, g) as are the two unsymmetrical ligands (m, n). In addition for diastereotopic substituents the arbitrary environments r,s are defined as for $M(A\text{–}B)_3$. When the effects of the 16 operations of Table II on all the

(30) D. J. Duffy and L. H. Pignolet, *Inorg. Chem.*, **11**, 2843 (1972).

(31) S. S. Eaton, Ph.D. Thesis, Massachusetts Institute of Technology, 1972.

(32) S. S. Eaton, G. R. Eaton, R. H. Holm, and E. L. Muetterties, to be submitted for publication.

(33) The following abbreviations are used for ligand anions: α-C_3H_5T, α-isopropenyltropolonate; α-C_3H_7T, α-isopropyltropolonate; RR′-dtc, N,N-disubstituted dithiocarbamate; tfd, 1,2-bis(perfluoromethyldithiolene); mnt, maleonitriledithiolene.

(34) E. I. Stiefel and G. F. Brown, *Inorg. Chem.*, **11**, 434 (1972).

(35) S. S. Eaton and G. R. Eaton, to be submitted for publication.

Table II. Multiplication Table

	E	(12)(34)(56)	(12)	(34)	(56)	(12)(34)	(34)(56)	(12)(56)
E	E	(12)(34)(56)	(12)	(34)	(56)	(12)(34)	(34)(56)	(12)(56)
(12)(34)(56)	(12)(34)(56)	E	(34)(56)	(12)(56)	(12)(34)	(56)	(12)	(34)
(12)	(12)	(34)(56)	E	(12)(34)	(12)(56)	(34)	(12)(34)(56)	(56)
(34)	(34)	(12)(56)	(12)(34)	E	(34)(56)	(12)	(56)	(12)(34)(56)
(56)	(56)	(12)(34)	(12)(56)	(34)(56)	E	(12)(34)(56)	(34)	(12)
(12)(34)	(12)(34)	(56)	(34)	(12)	(12)(34)(56)	E	(12)(56)	(34)(56)
(34)(56)	(34)(56)	(12)	(12)(34)(56)	(56)	(34)	(12)(56)	E	(12)(34)
(12)(56)	(12)(56)	(34)	(56)	(12)(34)(56)	(12)	(34)(56)	(12)(34)	E
E^*	E^*	(12)(34)(56)*	(12)*	(34)*	(56)*	(12)(34)*	(34)(56)*	(12)(56)*
(12)(34)(56)*	(12)(34)(56)*	E^*	(34)(56)*	(12)(56)*	(12)(34)*	(56)*	(12)*	(34)*
(12)*	(12)*	(34)(56)*	E^*	(12)(34)*	(12)(56)*	(34)*	(12)(34)(56)*	(56)*
(34)*	(34)*	(12)(56)*	(12)(34)*	E^*	(34)(56)*	(12)*	(56)*	(12)(34)(56)*
(56)*	(56)*	(12)(34)*	(12)(56)*	(34)(56)*	E^*	(12)(34)(56)*	(34)*	(12)*
(12)(34)*	(12)(34)*	(56)*	(34)*	(12)*	(12)(34)(56)*	E^*	(12)(56)*	(34)(56)*
(34)(56)*	(34)(56)*	(12)*	(12)(34)(56)*	(56)*	(34)*	(12)(56)*	E^*	(12)(34)*
(12)(56)*	(12)(56)*	(34)*	(56)*	(12)(34)(56)*	(12)*	(34)(56)*	(12)(34)*	E^*

Table III. Permutational Analysis of $M(A–B)_3$ Nmr Spectra

Operation	Resulting isomer[a]	Averaging set	Net configurational change	Net site interchanges (trans)
E	[135-462]	A_1	None	None
(12)(34)(56)	[153-642]	A_2	None	(yz)
(12)	[164-532]		Cis → trans	
(34)	[145-362]	A_3	Trans → $^1/_3$ cis + $^2/_3$ trans	$(xy),(xz)$
(56)	[136-452]			
(12)(34)	[163-542]		Cis → trans	
(34)(56)	[146-352]	A_4	Trans → $^1/_3$ cis + $^2/_3$ trans	$(xzy),(xyz)$
(12)(56)	[154-632]			
E^*	[153-264]	A_5	$\Delta \rightleftharpoons \Lambda$	(rs)
(12)(34)(56)*	[135-246]	A_6	$\Delta \rightleftharpoons \Lambda$	(yz),(rs)
(12)*	[146-235]		$\Delta \rightleftharpoons \Lambda$	
(34)*	[154-263]	A_7	Cis → trans	$(xy),(xz)$,(rs)
(56)*	[163-254]		Trans → $^1/_3$ cis + $^2/_3$ cis	
(12)(34)*	[136-245]		$\Delta \rightleftharpoons \Lambda$	
(34)(56)*	[164-253]	A_8	Cis → trans	$(xyz),(xzy)$, (rs)
(12)(56)*	[145-236]		Trans → $^1/_3$ cis + $^2/_3$ trans	

[a] Isomers resulting from performing the indicated operations on [135-462] are given as examples. Columns 3, 4, and 5 summarize net effects of operations on all 16 isomers.

Table IV. Permutational Analysis of $M(A–B)_2(C–C)$ Nmr Spectra

Operation	Averaging set	Net configurational change	Net site interchanges (cc isomer)	$M(A–B)_3$ analog
E	A_1'	None	None	A_1
(12)(34)(56)	A_2'	ct $\rightleftharpoons$ tc	(mn)	A_2
(56)	A_3'	None	(fg)	A_3
(12)				A_3
(34)	A_4'	tc $\rightleftharpoons$ cc $\rightleftharpoons$ ct		A_3
(34)(56)				A_4
(12)(56)				A_4
(12)(34)	A_5'	ct $\rightleftharpoons$ tc	(fg), (mn)	A_4
E^*	A_6'	$\Delta \rightleftharpoons \Lambda$	(rs)	A_5
(12)(34)(56)*	A_7'	ct $\rightleftharpoons$ tc, $\Delta \rightleftharpoons \Lambda$	(mn), (rs)	A_6
(56)*	A_8'	$\Delta \rightleftharpoons \Lambda$	(fg), (rs)	A_7
(12)*				A_7
(34)*		tc $\rightleftharpoons$ cc $\rightleftharpoons$ ct		A_7
(34)(56)*	A_9'	$\Delta \rightleftharpoons \Lambda$	(rs)	A_8
(12)(56)*				A_8
(12)(34)*	A_{10}'	ct $\rightleftharpoons$ tc, $\Delta \rightleftharpoons \Lambda$	(fg), (mn), (rs)	A_8

isomers are considered, it is found that the operations fall into ten nmr averaging sets, A_i' for $M(A–B)_2(C–C)$ (*cf.* Table IV). The changes in signal multiplicity

Table V. Changes in Signal Multiplicity for Diastereotopic and Nondiastereotopic Ligands

	A–B		C–C	
	Nondiastereotopic	Diastereotopic	Nondiastereotopic	Diastereotopic
A_1'	4 → 4	8 → 8	4 → 4	8 → 8
A_2'	4 → 2	8 → 4	4 → 3	8 → 6
A_3'	4 → 4	8 → 8	4 → 3	8 → 6
A_4'	4 → 1	8 → 2	4 → 1	8 → 2
A_5'	4 → 2	8 → 4	4 → 2	8 → 4
A_6'	4 → 4	8 → 4	4 → 4	8 → 4
A_7'	4 → 2	8 → 4	4 → 3	8 → 4
A_8'	4 → 4	8 → 4	4 → 3	8 → 4
A_9'	4 → 1	8 → 2	4 → 1	8 → 2
A_{10}'	4 → 2	8 → 4	4 → 2	8 → 4

Table VI. Distinctions Made between Averaging Sets A_i' due to Changes in Signal Multiplicity

Nondiastereotopic A–B[a] nondiastereotopic C–C	Nondiastereotopic A–B diastereotopic C–C	Diastereotopic A–B nondiastereotopic C–C	Diastereotopic A–B diastereotopic C–C
(A_1', A_6')	A_1'	A_1'	A_1'
(A_2', A_7')	A_2'[b]	(A_2', A_7', A_8')	A_2'
(A_3', A_8'	A_3'	A_3'	A_3'
(A_4', A_9')	(A_4', A_9')	(A_4', A_9')	(A_4', A_9')
(A_5', A_{10}')	(A_5', A_7', A_{10}')	(A_5', A_{10}')	(A_5', A_6', A_7', A_8', A_{10}')
	(A_6', A_8')	A_6'	

[a] The degeneracies in this case are due to the inability to distinguish processes which occur with inversion from those which occur without inversion. [b] A_2' can be distinguished from (A_6', A_8') if the signals due to C–C and A–B, respectively, can be unambiguously identified; the overall multiplicity change is 12 → 8 for all three averaging sets.

which can be deduced from Table IV for diastereotopic and nondiastereotopic ligands A–B and C–C are shown in Table V. Therefore the distinctions which can be made between averaging sets A_i' on the basis of changes in signal multiplicity for the four possible combinations of diastereotopic and nondiastereotopic ligands are shown in Table VI. Thus the maximum information is obtained by comparing the nmr averaging pattern for a complex with nondiastereotopic substituents on both A–B and C–C with a closely related complex containing diastereotopic groups on C–C. With these combined results the following averaging sets are distinguishable: A_1', A_2', A_3', (A_4', A_9'), (A_5', A_{10}'), A_6', A_7', A_8'.

The permutational analysis is simplified at the cost of

E^*	(12)(34)(56)*	(12)*	(34)*	(56)*	(12)(34)*	(34)(56)*	(12)(56)*
E^*	(12)(34)(56)*	(12)*	(34)*	(56)*	(12)(34)*	(34)(56)*	(12)(56)*
(12)(34)(56)*	E^*	(34)(56)*	(12)(56)*	(12)(34)*	(56)*	(12)*	(34)*
(12)*	(34)(56)*	E^*	(12)(34)*	(12)(56)*	(34)*	(12)(34)(56)*	(56)*
(34)*	(12)(56)*	(12)(34)*	E^*	(34)(56)*	(12)*	(56)*	(12)(34)(56)*
(56)*	(12)(34)*	(12)(56)*	(34)(56)*	E^*	(12)(34)(56)*	(34)*	(12)*
(12)(34)*	(56)*	(34)*	(12)*	(12)(34)(56)*	E^*	(12)(56)*	(34)(56)*
(34)(56)*	(12)*	(12)(34)(56)*	(56)*	(34)*	(12)(56)*	E^*	(12)(34)*
(12)(56)*	(34)*	(56)*	(12)(34)(56)*	(12)*	(34)(56)*	(12)(34)*	E^*
E	(12)(34)(56)	(12)	(34)	(56)	(12)(34)	(34)(56)	(12)(56)
(12)(34)(56)	E	(34)(56)	(12)(56)	(12)(34)	(56)	(12)	(34)
(12)	(34)(56)	E	(12)(34)	(12)(56)	(34)	(12)(34)(56)	(56)
(34)	(12)(56)	(12)(34)	E	(34)(56)	(12)	(56)	(12)(34)(56)
(56)	(12)(34)	(12)(56)	(34)(56)	E	(12)(34)(56)	(34)	(12)
(12)(34)	(56)	(34)	(12)	(12)(34)(56)	E	(12)(56)	(34)(56)
(34)(56)	(12)	(12)(34)(56)	(56)	(34)	(12)(56)	E	(12)(34)
(12)(56)	(34)	(56)	(12)(34)(56)	(12)	(34)(56)	(12)(34)	E

mechanistic information in the case where the symmetrical ligand does not function as a probe of environment in the experiments performed. When the two ends of C–C become indistinguishable (**5–6** becomes **5**–**5**), only eight distinct isomers are found and eight operations, which comprise an Abelian group, interconvert the isomers. This is a subset of the group in Table II, found by eliminating all operations which involve permutation (56). This adds the ambiguity that, for example, (12) and (12)(56) become indistinguishable. The nmr averaging sets are given in Table VII.

Table VII. Permutational Analysis of $M(A\text{–}B)_2(C\text{–}C)$ Nmr Spectra Where the Ends of C–C are Indistinguishable

Operation	Averaging set	Net configurational change	Net site interchanges (cc isomer)	$M(A\text{–}B)_3$ analog
E	A_1''	None	None	A_1
(12)(34)	A_2''	ct ⇌ tc	(mn)	A_2, A_4
(12), (34)	A_3''	ct ⇌ cc ⇌ tc		A_3, A_4
E^*	A_4''	Δ ⇌ Λ	(rs)	A_5
(12)(34)*	A_5''	ct ⇌ tc, Δ ⇌ Λ	(mn), (rs)	A_6, A_8
(12)*, (34)*	A_6''	ct ⇌ cc ⇌ tc, Δ ⇌ Λ	(rs)	A_7, A_8

For nondiastereotopic substituents on A–B, multiplicity changes distinguish (A_1'', A_4''), (A_2'', A_5''), (A_3'', A_6''). A comparison of diastereotopic and nondiastereotopic substituents could distinguish A_1'', (A_2'', A_5''), (A_3'', A_6''), A_4''. Thus the lack of information from C–C severely restricts the conclusions that can be drawn from the experiment.

Utility. Recent studies of the temperature-dependent nmr spectra of Fe(RR′-dtc)$_2$tfd[7,33] and Fe(RR′-dtc)$_2$mnt[8] illustrate the utility of the above approach. The published results[7] for the nondiastereotopic substituents are consistent with A_2' or A_7' (A_2 or A_6 in the notation for $M(A\text{–}B)_3$), and not just inversion, as was concluded[7,8] based on what was known at that time. Subsequent to ref 7 and 8 Musher has also recognized the need to consider a rearrangement of type A_2'.[36] As shown above, this ambiguity could be resolved by a study of a complex with a diastereotopic symmetrical ligand. This analysis also explicitly delineates the reasons that the study of Fe(RR′-dtc)$_2$mnt provided less information than was obtained for Fe(RR′-dtc)$_2$-tfd.[7,8]

In addition to the tropolonate complexes mentioned above, an application to $M(A\text{–}B)_3$ complexes is demonstrated by the trisdithiocarbamates, M(RR′-dtc)$_3$. Originally it was thought that the variable-temperature nmr spectra of *trans*-Fe(MePh-dtc)$_3$ supported a twist mechanism for the low-temperature rearrangement process.[10] Based on the results presented herein, it is recognized that the low-temperature process is consistent with either A_2 or A_6 and that by studying Fe(Me-PhCH$_2$-dtc)$_3$, as described above, it can be concluded that A_6 is the low-temperature process.[30]

Acknowledgments. This work was performed in the laboratory of Professor R. H. Holm, whom we thank for support under GP-18978X from the National Science Foundation.

(36) J. I. Musher, *J. Amer. Chem. Soc.*, **94**, 5662 (1972); *Inorg. Chem.*, **11**, 2355 (1972). Musher has approached the problem of a complete set of rearrangements of a tris chelate complex in an alternate fashion. However, the notation for his set of rearrangements fails to distinguish between plausible rearrangements and ones which require a ligand to span trans octahedral positions. Further, although group properties are necessary to prove completeness, his set, as written, is not a group.

Electron Spin Relaxation Times

Structures and Mechanisms

At the time of the Colonel's 80th birthday, some of us put together a book[1] that illustrated the wide range of research in which his former students and postdocs were producing cutting-edge results. The book includes an autobiographical sketch by the Colonel, and a synopsis of his science by me. The main thing the publisher did was change the location and formatting of the dedication page, and in the process changed Denver to Dover. This error was the first thing the Colonel brought up during my next visit.

From X-ray to NMR

I initially went to the Colonel's lab with the hope of doing X-ray crystallography but he redirected me to synthesis and NMR because a crystal structure was not feasible in just a year as an undergraduate. At just about that time (1961–1962), two problems confronted the Lipscomb group and I was invited to try to solve them. One was the effect of S-state transition metal ions on the ^{11}B NMR spectra of $B_{10}H_{10}{}^{2-}$ and the other was a free radical in a reaction of $B_{10}H_{10}{}^{2-}$ with $CuCl_2$.[2,3] Then, Fred Hawthorn described to the Colonel an NMR spectrum of what he thought was $B_{10}H_{10}{}^{2-}$ but the spectra were not fully congruent with the structure that the Colonel had predicted. In order to sort out the ^{11}B and ^{1}H NMR spectra, I had to learn to make my own NMR tubes and to convert the Baldeschwieler research group NMR spectrometer from ^{1}H or ^{19}F to ^{11}B each night after the graduate students had finished using the spectrometer and convert it back to the other nucleus before morning. I also ran ^{1}H spectra on one of the first Varian A60 spectrometers. Since the $B_{10}H_{10}{}^{2-}$ was soluble in water, the water peak was on top of the borane peaks in the ^{1}H spectrum. I developed a method of adding a little paramagnetic Co(II) to the solution to shift the water peaks away from the borane peaks, and thus get an interpretable ^{1}H spectrum of $B_{10}H_{10}{}^{2-}$. When the ^{11}B and ^{1}H NMR spectra were consistent with the structure that the Colonel had predicted (anyone that was in the group can recall the frequent "I predicted that!"), Roald Hoffmann became intrigued, and predicted the magnetic susceptibility. I measured the susceptibility on a Gouy balance system constructed by Richard Holm, who was then an instructor at Harvard, and with whom I would later do my Ph.D. research at MIT.

A Book Coauthored with the Colonel Because an Instrument Failed

After my five years of service in the navy, the Colonel provided me with a research position during the summer of 1967, before I entered graduate school at MIT. When the equipment that I was supposed to use for studying rearrangements of polyhedral boranes failed (back then, I did not know how to build and maintain equipment), I suggested that I revise the chapter in *Boron Hydrides* about NMR, since the Colonel was talking about revising the entire book. This effort turned into our 617-page book on NMR, *Studies of Boron Hydrides and Related Compounds*.[4] The timing of the book was lucky — we exhaustively documented the low-field CW NMR studies just as higher-field and pulsed NMR became available (spectra of $B_{10}H_{14}$ at different field strengths are shown in Fig. A.5 on p. 560).

Relaxation

One day, while struggling with some report in the boron NMR literature, I made a comment that the Colonel misinterpreted, giving me too much credit. He excitedly asked, "Have you got a new idea about relaxation? The field needs new ideas!" His comment stuck with me and remains true today. When referees sometimes say that our new ideas are crazy, or simply wrong, I occasionally think about that comment from the Colonel. During my time in the Lipscomb lab I did not work on the free radical problem beyond being introduced to EPR by Gus Maki, but since moving to Denver in 1972, most of my effort has been in EPR (about 300 papers and reviews and six books[5–10]).

Much of our current effort is to learn enough about spin relaxation so that we can apply it to the study of the structure and dynamics of biomolecules and *in vivo* imaging. Part of our recent effort has been to develop spin labels for proteins, peptides, etc. that can be used to measure distances at temperatures above that of liquid nitrogen, and as close to physiological temperature as possible. The nitroxyl radicals that are persistent enough, and chemically suitable for spin-labeling proteins, all have two pairs of gem-dimethyl groups adjacent to the N-O moiety on which the unpaired electron is substantially localized. Unfortunately, the methyl group dynamics make the electron spin phase memory time very short above about 80 K. Liquid helium has to be used to obtain the spectra that reveal the spin-spin interactions that can be interpreted in terms of distances between the spins. Another problem is that the cryoprotected aqueous solutions used for the protein solutions, similar to the solutions used for low-temperature X-ray crystallography, soften substantially below the glass transition temperatures. Consequently, we have been seeking spin labels without methyl groups, and seeking glasses that will dissolve the species we want to study but also have high glass transition temperatures. Routine empiricism would only accidentally find a good solution to the problem of finding proper conditions for these studies. As the Colonel said many times, theory can lead experiment. Maybe some of our fits to experimental data do not deserve to be called theory yet, but we are guided by our calculations. For a few cases, we have been able to extend spin label T_m to much longer than was heretofore feasible. There is promise of further development along these lines. We are working with a group that is putting proteins in high T_g glasses for the purpose of stabilizing drugs for delivery in regions of the world where refrigeration is not reliable. Maybe, working on this problem from several broad perspectives, and exploiting connections between many seemingly disconnected technologies, we will be able to do what no one so far has been able to do.

Some of our perspectives on electron spin relaxation are summarized in references 11–13. Explorations of temperature, viscosity, molecular structure, molecular size, and other variables are reported in references 14–33. Reference 33 shows the benefit of eliminating methyl groups, as mentioned above. Recalling old NMR literature about the "wiggles" that Bloembergen, Purcell, and Pound observed in 1948,[34] and the subsequent exploitation in rapid-scan correlation NMR spectroscopy in 1974,[35] we developed rapid-scan EPR as a new field of study, intermediate between CW EPR and pulsed EPR.[24] Some *in vivo* EPR imaging uses triaryl methyl radicals, so we have studied their electron spin relaxation.[19,25,26] There are extensive applications of nitroxyl radicals to biological systems. The interpretations that Robinson *et al.*[36] provided for frequency dependence of spin-lattice relaxation (T_1) of nitroxyl radicals predicted that relaxation would be too short to perform pulsed EPR imaging of nitroxyls at the low RF frequencies (e.g. 250–300 MHz) used for *in vivo* imaging. Other researchers had been unsuccessful in attempting time-domain EPR of nitroxyls at low frequency. However, our understanding of electron spin relaxation, and our development of methods such as rapid scan EPR came together to demonstrate with rapid scan EPR at 250 MHz (9 mT magnetic field) that the relaxation time of nitroxyl radicals was long enough that we then successfully performed pulsed EPR on nitroxyls at 250 MHz.[28] Once we demonstrated this, others succeeded in pulsed EPR imaging of nitroxides in mice.[37] Maybe here we could be permitted to echo the Colonel and say that we predicted this.

SSE and GRE worked jointly on this project. First person wording in these comments refers to GRE.

References

1. Eaton G. R., Wiley D. C., and Jardetzky O. (Eds), *Structures and Mechanisms: From Ashes to Enzymes.* ACS Symposium Series No. 827, Washington DC: American Chemical Society (2002).
2. Lipscomb W. N. and Kaczmarczyk A., Interactions of boranes and their ions with paramagnetic S-state ions. *P. Natl. Acad. Sci. USA*, **47**, 1796–1797 (1961).
3. Kaczmarczyk A., Dobrott R. D., and Lipscomb W. N., Reactions of $B_{10}H_{10}^{-2}$ ion, *P. Natl. Acad. Sci. USA*, **48**, 729–733 (1962).
4. Eaton G. R. and Lipscomb W. N., *N.M.R. Studies of Boron Hydrides and Related Compounds.* New York: Benjamin (1969).
5. Eaton G. R., Eaton S. S., and Ohno K. (Eds), *EPR Imaging and in Vivo EPR.* Boca Raton, FL: CRC Press (1991).
6. Berliner L. J., Eaton G. R., and Eaton S. S. (Eds), *Distance Measurements in Biological Systems by EPR.* Vol. 19 of Biological Magnetic Resonance, New York: Kluwer (2000).
7. Eaton S. S., Eaton G. R., and Berliner L. J. (Eds), *Biomedical EPR — Part A: Free Radicals, Metals, Medicine, and Physiology.* Vol. 23 of Biological Magnetic Resonance, New York: Kluwer Academic/Plenum Press (2005).

8. Eaton S. S., Eaton G. R., and Berliner L. J. (Eds), *Biomedical EPR — Part B: Methodology, Instrumentation, and Dynamics*. Vol. 24 of Biological Magnetic Resonance, New York: Kluwer Academic/Plenum Press (2005).
9. Hyde J. S., Eaton S. S., and Eaton G. R., *Concepts in Magnetic Resonance*, thematic issue, **28A**, 1–100 (2006).
10. Eaton G. R. and Eaton S. S., High-field and high-frequency EPR. *Appl. Magn. Reson.*, **16**, 161–166 (1999).
11. Eaton S. S. and Eaton G. R., Determination of distances based on T_1 and T_m in distance measurements in biological systems by EPR. *Bio. Magn. Re.*, **19**, 347–381 (2000).
12. Eaton S. S. and Eaton G. R., Relaxation times of organic radicals and transition metal ions in distance measurements in biological systems by EPR. *Bio. Magn. Re.*, **19**, 29–154 (2000).
13. Eaton S. S. and Eaton G. R., Measurement of interspin distances by EPR. *Specialist Periodical Report on Electron Spin Resonance*, **19**, 318–337 (2005).
14. Zhou Y., Mitri R., Eaton G. R., and Eaton S. S., Electron spin lattice relaxation processes for molecular S=1/2 systems in glassy matrices at temperatures between 10 and 130 K. *Current Topics in Biophysics*, **23**, 63–68 (1999).
15. Zhou Y., Bowler B. E., Eaton G. R., and Eaton S. S., Electron spin lattice relaxation rates for S=1/2 molecular species in glassy matrices or magnetically dilute solids at temperatures between 10 and 300 K. *J. Magn. Reson.*, **139**, 165–174 (1999).
16. Zhou Y., Bowler B. E., Eaton G. R., and Eaton S. S., Electron spin-lattice relaxation rates for high-spin Fe(III) complexes in glassy solvents at temperatures between 6 and 298 K. *J. Magn. Reson.*, **144**, 115–122 (2000).
17. Eaton S. S., Harbridge J., Rinard G. A., Eaton G. R., and Weber R. T., Frequency dependence of electron spin relaxation for three S=1/2 species doped into diamagnetic solid hosts. *Appl. Magn. Reson.*, **20**, 151–157 (2001).
18. Huber M., Lindgren M., Hammarstrom P., Martensson L. G., Carlsson U., Eaton G. R., and Eaton S. S., Phase memory relaxation times of spin labels in human carbonic anhydrase II: pulsed EPR to determine spin label location. *Biophys. Chem.*, **94**, 245–256 (2001).
19. Yong L., Harbridge J., Quine R. W., Rinard G. A., Eaton S. S., Eaton G. R., Mailer C., Barth E., and Halpern H. J., Electron spin relaxation of triarylmethyl radicals in fluid solution. *J. Magn. Reson.*, **152**, 156–161 (2001).
20. Harbridge J. R., Eaton S. S., and Eaton G. R., Electron spin lattice relaxation in radicals containing two methyl groups, generated by γ-irradiation of polycrystalline solids. *J. Magn. Reson.*, **159**, 195–206 (2002).
21. Harbridge J. R., Eaton S. S., and Eaton G. R., Electron spin-lattice relaxation processes of radicals in irradiated crystalline organic compounds. *J. Phys. Chem. A*, **107**, 598–610 (2003).
22. Harbridge J. R., Eaton S. S., and Eaton G. R., Electron spin relaxation of radicals in γ-irradiated malonic acid and methyl malonic acid. *Appl. Magn. Reson.*, **24**, 261–276 (2003).
23. Owenius R., Terry G. E., Williams M. J., Eaton S. S., and Eaton G. R., Frequency dependence of electron spin relaxation of nitroxyl radicals in fluid solution. *J. Phys. Chem. B*, **108**, 9475–9481 (2004).
24. Stoner J. W., Szymanski D., Eaton S. S., Quine R. W., Rinard G. A., and Eaton G. R., Direct-detected rapid-scan EPR at 250 MHz. *J. Magn. Reson.*, **170**, 127–135 (2004).
25. Owenius R., Eaton G. R., and Eaton S. S., Frequency (250 MHz to 9.2 GHz) and viscosity dependence of electron spin relaxation of triarylmethyl radicals at room temperature. *J. Magn. Reson.*, **172**, 168–175 (2005).
26. Fielding A. J., Carl P. J., Eaton G. R., and Eaton S. S., Multifrequency EPR of four triarylmethyl radicals. *Appl. Magn. Reson.*, **28**, 239–249 (2005).
27. Fielding A. J., Fox, S., Millhauser G. L., Chattopadhyay M., Kroneck P. M. H., Fritz G., Eaton G. R., and Eaton S. S., Electron spin relaxation of copper (II) complexes in glassy solution between 10 and 120 K. *J. Magn. Reson.*, **179**, 92–104 (2006).

28. Tseitlin M., Dhami A., Quine R. W., Rinard G. A., Eaton S. S., and Eaton G. R., Electron spin T_2 of a nitroxyl radical at 250 MHz measured by rapid-scan EPR. *Appl. Magn. Reson.*, **30**, 651–656 (2006).
29. Sato H., Kathirvelu V., Fielding A. J., Bottle S. E., Blinco J. P., Micallef A. S., Eaton S. S., and Eaton G. R., Impact of molecular size on electron spin relaxation rates of nitroxyl radicals in glassy solvents between 100 and 300 K. *Mol. Phys.*, **105**, 2137–2151 (2007).
30. Sato H., Bottle S. E., Blinco J. P., Micallef A. S., Eaton G. R., and Eaton S. S., Electron spin-lattice relaxation of nitroxyl radicals in temperature ranges that span glassy solutions to low-viscosity liquids. *J. Magn. Reson.*, **191**, 66–77 (2008).
31. Sato H., Kathirvelu V., Spagnol G., Rajca S., Rajca A., Eaton S. S., and Eaton G. R., Impact of electron-electron spin interaction on electron spin relaxation of nitroxide diradicals and tetraradical in glassy solvents between 10 and 300 K. *J. Phys. Chem. B.*, **112**, 2818–2828 (2008).
32. Ulyanov D., Bowler B. E., Eaton G. R., and Eaton S. S., Electron-electron distances in spin-labeled low-spin metmyoglobin variants by relaxation enhancement. *Biophys. J.*, **95**, 5306–5316 (2008).
33. Kathirvelu V., Smith C., Parks C., Mannan M. A., Miura Y., Takeshita K., Eaton S. S., and Eaton G. R., Relaxation rates for spirocyclohexyl nitroxyl radicals are suitable for interspin distance measurements at temperatures up to about 125 K. *Chem. Commun.*, 454–456 (2009).
34. Bloembergen N., Purcell E. M., and Pound R. V., Relaxation effects in nuclear magnetic resonance absorption. *Phys. Rev.*, **73**, 679–712 (1948).
35. Dadok J. and Sprecher R. F., Correlation NMR spectroscopy. *J. Magn. Reson.*, **13**, 243–248 (1974).
36. Robinson B. H., Haas D. A., and Mailer C., Molecular dynamics in liquids: spin-lattice relaxation of nitroxide spin labels. *Science*, **263**, 490–493 (1994).
37. Hyodo F., Matsumoto S., Devasahayam N., Dharmaraj C., Subramanian S., Mitchell J. B., and Krishna M. C., Pulsed EPR imaging of nitroxides in mice. *J. Magn. Reson.*, **197**, 181–185 (2009).

Appl. Magn. Reson. 30, 651–656 (2006)

Applied Magnetic Resonance

Electron Spin T_2 of a Nitroxyl Radical at 250 MHz Measured by Rapid-Scan EPR

M. Tseitlin[1], **A. Dhami**[1], **R. W. Quine**[2], **G. A. Rinard**[2], **S. S. Eaton**[1], and **G. R. Eaton**[1]

[1] Department of Chemistry and Biochemistry and [2] Department of Engineering, University of Denver, Denver, Colorado, USA

Received May 5, 2006; revised July 7, 2006

Abstract. Rapid-scan electron paramagnetic resonance (EPR) was used to measure the room temperature spin–spin relaxation time T_2 for the per-deuterated nitroxyl radical, 4-oxo-2,2,6,6-tetramethylpiperidin-1-oxyl-d_{16} in water at 250 MHz. Signals were recorded on a locally constructed system using either sinusoidal or triangular magnetic field scans. Values of T_2 were obtained by simulation of the rapid-scan response and were systematically shorter for the high-field line ($m_I = -1$) than for the $m_I = 0$ or $+1$ lines. For $m_I = +1$, the value of T_2 increased from 0.41±0.03 μs at 0.5 mM to 0.53±0.03 μs at 0.1 or 0.2 mM. For $m_I = -1$, T_2 increased from 0.35±0.03 μs at 0.5 mM to 0.42±0.03 μs at 0.1 or 0.2 mM. These values put lower bounds on the values of T_1 for this nitroxyl at 250 MHz.

1 Introduction

Rapid-scan measurements of T_2 use a method that grew out of an observation reported in the early nuclear magnetic resonance (NMR) paper by Bloembergen, Purcell and Pound [1]. The "wiggles" that they observed on the trailing edge of the signal were correctly interpreted as being due to passage through resonance at a rate that is fast relative to relaxation times. The line shapes were subsequently simulated by Jacobsohn and Wangness [2]. In 1974 Gupta et al. [3] and Dadok et al. [4] developed rapid-scan correlation NMR spectroscopy based on these principles. Similar transient responses were observed by Czoch et al. [5] in electron paramagnetic resonance (EPR). We are developing the methodology and technology of rapid-scan EPR [6–8], which has promise for in vivo application for the time regime between normal continuous-wave (CW) EPR and pulsed EPR. Nitroxyl radicals, which have been so useful for biomedical studies, are a natural target for rapid-scan EPR.

In this paper we report the first measurements of the spin–spin relaxation times T_2 of nitroxyl radicals at 250 MHz, a frequency that has been selected for in vivo studies [9–11]. Many in vivo or ex vivo low-frequency EPR studies have used triarylmethyl radicals or lithium phthalocyanine (LiPc), which have convenient relaxation properties for CW and pulsed EPR. So far, nitroxyl radicals have been studied at low frequency only by CW methods. The relaxation times reported in this paper suggest that some pulsed EPR methods could be applied to nitroxyl radicals in vivo.

The electron spin–lattice relaxation times of nitroxyl radicals are not yet fully understood, although there is a rapidly accumulating body of information on the subject [12–15]. Multifrequency EPR measurements of spin–lattice relaxation of nitroxyl radicals have been performed from 2.5 to 35 GHz [14] and interpreted with a model that includes four relaxation mechanisms [15]. The background literature reveals uncertainty about the relaxation times of nitroxyl radicals at lower resonant frequency, but there is an overall expectation that relaxation times will be shorter at lower frequency. In fluid solution, $T_1 \geq T_2$, so values of T_2 at 250 MHz put lower limits on values of T_1.

2 Methods

Rapid-scan EPR using sinusoidal [7] or triangular magnetic field scans [8] was performed as previously reported. The resonators used were analogous to those used by Joshi et al. [8], but with a shield between the resonator and the magnetic field scan coils. This type of resonator design is undergoing continuing development, and a detailed design will be reported after further improvements. As previously described, the EPR method uses direct detection of the EPR signal, without magnetic field modulation. Instead of small-amplitude 100 kHz (or other frequency) magnetic field modulation, the Zeeman field is swept rapidly relative to relaxation times over the entire spectral width. The Q of the resonator that was used for the sinusoidal scans was about 150 and the Q of the resonator that was used for the triangular scans was about 40. The resonator and scan coils are within and shielded by a large aluminum cook pot whose cover was attached with good conductivity by using copper alloy finger stock.

Aqueous solutions of 4-oxo-2,2,6,6-tetramethyl-piperidin-1-oxyl-d_{16} (CDN Isotopes, Quebec, Canada; tempone-d_{16}) were deoxygenated by bubbling with N_2 gas and flame-sealed in NMR tubes with an outer diameter of 10 mm. Solutions with concentrations of 0.5, 0.2 and 0.1 mM were prepared to check the effect of nitroxyl concentration.

3 Simulations

Simulations were performed by numerical integration of the rotating-frame representation of the Bloch equations including magnetic field scan as shown in the expressions in Eq. (1). The calculations for triangular scans are similar to those used for sinusoidal scans [7], and require only modification of the field scan function.

$$\frac{dM_u}{dt} = -\frac{M_u}{T_2} - (\Delta\omega + \Omega_m f(\omega_m t))M_v,$$

$$\frac{dM_v}{dt} = (\Delta\omega + \Omega_m f(\omega_m t))M_u - \frac{M_v}{T_2} - \gamma B_1 M_z,$$

$$\frac{dM_z}{dt} = \frac{M_0}{T_1} + \gamma B_1 M_v - \frac{M_z}{T_1}, \quad (1)$$

where γ is the electron magnetogyric ratio equal to $1.7608 \cdot 10^7$ rad/sG, $\Delta\omega = \omega - \omega_0$ is the offset of a spin packet from the center of the scan in angular frequency, Ω_m is the amplitude of the field scan in angular units and equals $0.5\gamma B_m$, where B_m is the peak-to-peak scan amplitude in gauss, $\omega_m = 2\pi\nu_m$ is the angular scan frequency, ν_m is the scan frequency in hertz, B_1 is the radio-frequency (rf) magnetic field in gauss (peak-to-peak), and $f(\omega_m t)$ is the sinusoidal or triangular scan function. The time evolution of the magnetization for a spin packet is evaluated using fourth-order Runge–Kutta numerical integration [16] performed at 40000 to 50000 points per field scan cycle. Simulations were calculated with a program written in Compaq Visual Fortran 6.5. The inhomogeneously broadened line shapes were modeled as the sum of contributions from 9 to 15 T_2-determined spin packets with differing offsets from the center of the signal. The weightings of the spin packets in the distribution were calculated using an approximation for a Voigt function [17]. The sweep frequency was measured with a Philips PM6680 frequency counter. The sweep widths were calibrated by comparing the hyperfine splitting of the nitroxyl lines in the rapid scan signals with the hyperfine splitting in a conventional CW spectrum at 250 MHz.

4 Results and Discussion

The rapid-scan signal for 0.5 mM tempone-d_{16} in water is shown in Fig. 1. The ^{13}C satellite lines were observed in the experimental signal, but were not included in the simulation. The ^{13}C hyperfine splitting of 6.0 G is in good agreement with values obtained at X-band for the same radical. The oscillations in intensity after the magnetic field has scanned through the peak of the absorption line are characteristic of the rapid-scan response. The spacing between peaks of the oscillations is determined by the rate at which the field is scanned through resonance. The scan widths that were required to simulate the signals were in good agreement with calibrated scan widths. The damping of the oscillation depends on inhomogeneous broadening and on the T_2 of the radical. The primary contribution to the inhomogeneous broadening is from unresolved deuteron couplings. The slow-scan line width, which reflects T_2 and unresolved hyperfine broadening, was measured at scan rates slow enough that there were no rapid-scan ef-

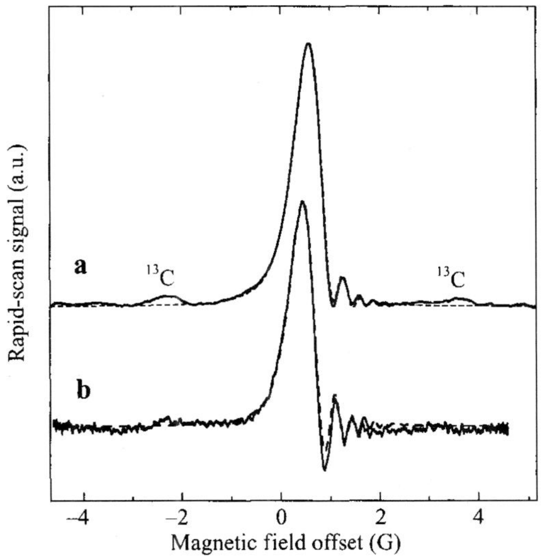

Fig. 1. Rapid-scan signals for the low-field line of tempone-d_{16} in water obtained with 40 kHz sinusoidal field sweeps and rf frequency of 245 MHz. **a** Signal for 0.5 mM solution with a sweep width of 10.0 G. The dashed line is a simulation obtained with T_2 = 0.38 μs and an inhomogeneous broadening of 15 mG. **b** Signal for 0.1 mM solution with sweep width of 9.5 G. The dashed line is a simulation obtained with T_2 = 0.56 μs and an inhomogeneous broadening of 15 mG. The ^{13}C sidebands were not included in the simulations.

fects on the line shape. The inhomogeneous broadening and value of T_2 in the simulations were adjusted until there was good agreement between simulated and experimental scans at both slow- and fast-scan rates. The estimated value of the inhomogeneous broadening was 15 to 20 mG, which is small compared with the T_2-determined contribution to the line widths. The uncertainty in the inhomogeneous broadening and noise in the experimental data result in an uncertainty of about ±0.03 μs in the values of T_2. The rapid magnetic field scans can induce eddy currents in metallic components of the resonator, which causes inhomogeneous broadening that increases with increasing scan rates [18]. The observation of consistent values of T_2 for samples examined with different resonators using either sinusoidal or triangular scans indicates that the scan-dependent contributions to inhomogeneous broadening are not causing substantial systematic errors in the T_2 measurements. For the tempone-d_{16} signals these contributions to inhomogeneous broadening are small relative to the inherent deuteron contribution.

The values of T_2 obtained by simulating the rapid-scan signals are summarized in Table 1. The T_2 values for 0.1 and 0.2 mM solutions are the same within experimental uncertainty and are significantly longer than the values for 0.5 mM solutions, which is consistent with observations in our laboratory at higher observing frequencies. The values of T_2 are longer for the center and low-field lines than for the high-field lines, which also is consistent with line width differences at 250 MHz and higher frequencies.

Table 1. T_2 values[a] (in microseconds) for tempone-d_{16} in aqueous solution at room temperature.

Concentration (mM)	$m_I = +1$	$m_I = 0$	$m_I = -1$
0.5	0.41	0.41	0.35
0.2	0.50	0.50	0.43
0.1	0.53	–	0.42

[a] Uncertainties are about ±0.03 μs.

We are not aware of prior direct measurements of nitroxyl relaxation times at 250 MHz with which to compare these results. Time-domain measurements are in process, and preliminary results are in close agreement. Analysis of the CW line shape for the nitroxyl radical 4-hydro-3-carbamoyl-2,2,5,5-tetra(perdeutero-methyl)-pyrroline-l-oxyl (CTPO-d_{12}) in low-concentration degassed water solution at 250 MHz gave $T_2 = 0.47$ μs [19], which is similar to the values obtained in this study for tempone-d_{16}. Although nitroxyl relaxation rates are strongly dependent on tumbling correlation time, tempone-d_{16} and CTPO-d_{12} have similar molar masses and presumably similar tumbling correlation times. Pulsed measurement of T_2 for the same CTPO-d_{12} sample at X-band gave 0.59 μs [19]. For Fremy's salt in degassed aqueous carbonate solution at room temperature analysis of the CW line shape gave $T_2 = 0.41$ μs [20]. This limited set of comparisons indicates weak frequency dependence of T_2 between 250 MHz and X-band.

Extrapolation from higher-frequency experiment [14] and theory [15] predicts T_1 values shorter than the T_2 values reported here, although $T_1 \geq T_2$ is expected in fluid solution. The experimental results reported here are T_2 measurements on water solutions of small radicals, whereas the most complete prior measurements at higher frequencies were T_1 measurements on larger spin-labeled molecules in lipids or small molecules in water-glycerol solutions designed to have tumbling correlation times similar to lipid samples [14, 15, 21]. Since values of T_1 become shorter as the tumbling rate increases, the discrepancy between prediction and experiment may be greater for the faster tumbling small molecules than that estimated on the basis of the models for slower tumbling radicals. More thorough analysis, and possibly additional measurements between 250 MHz and 1.5 GHz, will be needed to understand the trends in relaxation times. At 250 MHz there could be some effect of the approximations made in the models for the frequency dependence of relaxation. It is not as valid at 250 MHz to ignore hyperfine splittings relative to electron Larmor frequency as it is at 2.5 GHz and higher frequencies. Full interpretation of the T_2 values will await future studies, but the values themselves have important implications for development of new in vivo EPR methodologies.

Acknowledgments

This work was supported in part by NIH NIBIB grants EB000557 (GRE), EB002807 (GRE), and P41 EB002034 (Howard Halpern, PI), Center for EPR Imaging In Vivo Physiology. We thank Janusz Koscielniak for bringing reference 5 to our attention and thank him and Martyna Elas for translating parts of it for us.

References

1. Bloembergen N., Purcell E.M., Pound R.V.: Phys. Rev. **73**, 679–712 (1948)
2. Jacobsohn B.A., Wangness R.K.: Phys. Rev. **73**, 942 (1948)
3. Gupta R.K., Ferretti J.A., Becker E.D.: J. Magn. Reson. **13**, 275–290 (1974)
4. Dadok J., Sprecher R.F.: J. Magn. Reson. **13**, 243–248 (1974)
5. Czoch R., Duchiewicz J., Francik A., Indyka S., Koscielniak J.: Meas. Automatic Control **29**, 41–43 (1983)
6. Harbridge J.R., Rinard G.A., Quine R.W., Eaton S.S., Eaton G.R.: J. Magn. Reson. **156**, 41–51 (2002)
7. Stoner J.W., Szymanski D., Eaton S.S., Quine R.W., Rinard G.A., Eaton G.R.: J. Magn. Reson. **170**, 127–135 (2004)
8. Joshi J.P., Ballard R.L., Rinard G.A., Quine R.W., Eaton G.R.: J. Magn. Reson. **175**, 44–51 (2005)
9. Halpern H.J.: Biol. Magn. Reson. **18**, 469–482 (2003)
10. Williams B.B., Halpern H.J. in: Biological Magnetic Resonance, vol. 23, Biomedical EPR, Part A: Free Radicals, Metals, Medicine, and Physiology (Eaton S.S., Eaton G.R., Berliner L.J., eds.), pp. 283–319. New York: Kluwer Academic 2005.
11. Subramanian S., Krishna M.C.: Biol. Magn. Reson. **23**, 321–383 (2005)
12. Eaton S.S., Eaton G.R.: Biol. Magn. Reson. **19**, 29–154 (2000)
13. Owenius R., Terry G.E., Williams M.J., Eaton S.S., Eaton G.R.: J. Phys. Chem. B **108**, 9475–9481 (2004)
14. Hyde J.S., Yin J.-J., Subczynski W.K., Camenisch T.G., Ratke J.J., Froncisz W.: J. Phys. Chem. B **108**, 9524–9529 (2004)
15. Mailer C., Neilson R.D., Robinson B.H.: J. Phys. Chem. A **109**, 4049–4061 (2005)
16. Press W.H., Teukolsky S.A., Verrling W.T., Flannery B.P., in: Numerical Recipes in Fortran: The Art of Scientific Computing, pp. 704–708. Cambridge: Cambridge University Press 1992.
17. Wertheim G.K., Butler M.A., West K.W., Buchanan D.N.E.: Rev. Sci. Instrum. **45**, 1369–1371 (1974)
18. Joshi J.P., Eaton G.R., Eaton S.S.: Appl. Magn. Reson. **29**, 239–249 (2005)
19. Bowman M.K., Michalski T.J., Peric M., Halpern H.J.: Pure Appl. Chem. **62**, 271–274 (1990)
20. Kooser R.G., Volland W.V., Freed J.H.: J. Chem. Phys. **50**, 5243–5257 (1969)
21. Robinson B.H., Haas D.A., Mailer C.: Science **263**, 490–493 (1994)

Authors' address: Sandra S. Eaton, Department of Chemistry and Biochemistry, University of Denver, Denver, CO 80208 USA
E-mail: seaton@du.edu

Chapter 9

Irving R. Epstein

Biography

Irving R. Epstein is Henry F. Fischbach Professor of Chemistry, chair of the Chemistry Department, and a member of the Volen Center for Complex Systems at Brandeis University, where he has taught since 1971, served as dean of arts and sciences from 1992 to 1994 and served as provost and senior vice president for academic affairs from 1994 to 2001.

He received his A.B., M.A., and Ph.D. degrees from Harvard University in 1966, 1968, and 1971, respectively, carrying out his doctoral work in chemical physics under the direction of William N. Lipscomb. He received a diploma in advanced mathematics as a Marshall Scholar at Oxford University in 1967, working with C. A. Coulson, and was a NATO postdoctoral fellow at Cambridge University in 1971. In 1977–1978, he was a National Science Foundation professional development fellow at the Max Planck Institute for Biophysical Chemistry in Göttingen.

He has been awarded Guggenheim and Humboldt Fellowships and was a Dreyfus Foundation Teacher-Scholar. In 2006, he was named a Howard Hughes Medical Institute professor. He chaired the first Gordon Research Conference on Chemical Oscillations and Dynamic Instabilities and currently serves as associate editor of *Chaos* and as chair of the board of directors of the New England Complex Systems Institute.

His current research interests are in nonlinear chemical dynamics, particularly pattern formation, oscillating chemical reactions, chaos, and the behavior of complex networks.

He is married to Ellen Epstein, an attorney at the Massachusetts Appeals Court, and has two sons, David, a corporate attorney, and Peter, a graduate student in philosophy. His hobbies include solving puzzles of all sorts and playing tennis.

My History with the Colonel

My history with the Colonel goes back to my first day as an undergraduate in 1962. Harvard had received a grant from the National Science Foundation for an experimental curriculum in which all of undergraduate chemistry (general, organic, inorganic, and physical) would be taught to a group of 25 well-prepared students in just two years (!). The Colonel was in charge of selecting the students and teaching the first semester of the course. I met with him, and his reassuring manner somehow convinced me that I could handle such an undertaking. When I showed up for the first class, the textbook for this "freshman chemistry" course turned out to be Cotton and Wilkinson, and the first topic we covered was Hückel molecular orbital theory. The Colonel simply assumed that we'd be able to figure out anything we didn't know but needed. Remarkably, his approach worked, and the majority of us not only survived, but went on to careers in science. Forty years later, a colleague of mine at Brandeis won the Welch Award, and I was seated at the award dinner with the Colonel, who was on the selection committee. He and E. J. Corey, who was also on the committee and had taught the organic part of the course, amazed me (and shocked my wife) by recalling details from exams that I had written in the course!

Attracted by the Colonel's ability to make even the most complex concepts appear straightforward, I took his graduate group theory course as a junior and occasionally hung around his lab. When, after graduating and going off to Oxford for a year, I decided that I wanted to complete my Ph.D. in the US, returning to Harvard to work with the Colonel seemed like the natural thing to do. It was an exciting time, with the Colonel essentially running three groups — boron chemistry, structural biology, and theory, which was my area. There were perhaps 30 or 40 students and postdocs altogether, but it seemed like the Colonel was always happy to talk to me when I popped in (if he wasn't on one of his frequent lecture trips). I got extra face time with him, too, because we frequently played tennis together, often with fellow grad students Tom Koetzle and Gene Switkes.

When I asked the Colonel for suggestions about what to write my thesis on, he offered a few ideas about things that were going on in the group, but basically told me that I should work on any problem that I found interesting and not to worry too much about a unifying theme or whether it was something that I thought he was interested in (or was funded for). As it turned out, I ended up playing with a large number of only rather loosely connected problems, ranging from molecular orbital calculations of wavefunctions to the barrier to internal rotation in ethane to topological approaches to boron hydrides to Compton scattering and molecular momentum distributions. Of the 15 papers that came out of my time with the Colonel, I ended up as the sole author on six, because the Colonel felt that the ideas had come from me and he had provided only occasional advice. I don't think too many other thesis directors would have treated their Ph.D. students as magnanimously and collegially.

The Colonel's willingness to let his students move in whatever direction their intellectual interests might take them resulted in a lively atmosphere during graduate school and a remarkable diversity of career paths afterward. The theorists with whom I shared a large office in Gibbs have ended up in such fields as psychophysics, drug design, geochemistry, chemical oceanography, mushroom farming, and even crystallography. Irene Pepperberg, one of our officemates, was featured in the *New York Times* recently for her remarkable work on the intelligent behavior of which parrots are capable. I ended up studying nonlinear chemical dynamics — things like oscillating reactions, pattern formation and chaos — very far from what I did as a grad student, but it was an easy transition because what I had learned with the Colonel was not so much a particular subject as a willingness to tackle exciting problems wherever they appeared.

The period I spent in graduate school with the Colonel coincided with the height of the Vietnam War, and we were not immune to the turmoil that the political situation brought with it. Many of the graduate students signed a letter of protest when Dow Chemical, notorious for supplying napalm to the government, was scheduled to make a recruiting visit to the Chemistry Department. One member of the faculty scrutinized the letter and made noises about punishing the signers, but the Colonel made it clear that *his* students were free to express their views and that he would resist any pressures on them to "behave." As a resident tutor in one of the Harvard houses, I ran into occasional political conflicts with some of the senior faculty attached to the house. I would sometimes tell the Colonel about these incidents, and while he never said anything explicit, he made it clear that he would back me if the going got tough. When I finished my thesis work and lined up a NATO postdoc to go to England, I had not yet reached the magic age of 26, beyond which one couldn't be drafted. I was faced with the dilemma that if I turned in my thesis and started my postdoc, I was likely to end up as a soldier rather than a scientist. I talked to the Colonel about my problem, and he quickly agreed that I should put my thesis into my desk drawer and continue as a (draft-deferred) graduate student until I felt it was safe to leave. We agreed that I would learn how to do crystal structures and would attempt to solve the structure of a large borane for which I had calculated wavefunctions as part of my thesis work.

My experience in the lab convinced both of us that I was not destined to become a crystallographer or a synthetic boron chemist. One day, I walked into the lab and I noticed that Ed Wong, who had been making boranes and carboranes at a bench near mine wasn't around. Later in the day, I asked people where he was, and they replied, "Didn't you hear about the accident?" It turned out that a vacuum line had exploded the night before, sending Ed to the hospital, though fortunately his injuries were relatively minor. The fact that I had gone about my business oblivious to the broken glass and traces of Ed's blood in the area around my bench made me rethink whether I should really be doing that kind of experimental work. I've since returned to doing experiments, but of a much safer kind, and usually from a distance.

My years with the Colonel were a time of much learning and growth surrounded by outstanding people and wonderful science. He was a world-class mentor, who taught me that it's possible to do a first-rate job at many things, particularly if you're willing to work at it seven days a week.

The two chosen papers represent the breadth of problems that I worked on with the Colonel, linked by our favorite molecules, the boron hydrides. The *Journal of Chemical Physics* paper is the first in a series devoted to calculating the momentum distributions of electrons in molecules, which are experimentally accessible by measuring the line shape of Compton scattered X-rays. It's very quantitative and "physical," making use of the boron hydride molecular orbital calculations that were a focus of the theoretical group at that time. The paper from *Inorganic Chemistry* is much more qualitative and "chemical." It expands on and systematizes the three-center bond formalism that the Colonel had developed earlier in order to identify resonance structures and predict stabilities of boranes and carboranes. It uses ideas derived from much more quantitative localized molecular orbital calculations to refine the earlier qualitative topological approach.

THE JOURNAL OF CHEMICAL PHYSICS VOLUME 53, NUMBER 12 15 DECEMBER 1970

Molecular Momentum Distributions and Compton Profiles. I. General Theory and Boron Hydrides

IRVING R. EPSTEIN* AND WILLIAM N. LIPSCOMB
Department of Chemistry, Harvard University, Cambridge, Massachusetts 02138
(Received 29 July 1970)

A method is described for the calculation of momentum distributions and Compton profiles of polyatomic molecules. Distributions, profiles and momentum expectation values are calculated for the boron hydrides B_2H_6, B_4H_{10}, B_5H_9, B_5H_{11}, and B_6H_{10}. Contour maps of total and difference densities in momentum space are presented for diborane. The results are analyzed in terms of localized orbitals. The analysis shows that traditional concepts of two- and three-center electron pair bonds may find great utility in interpreting momentum space distributions.

I. INTRODUCTION

The calculation of atomic and molecular wavefunctions in momentum space has been of interest to theoretical chemists and physicists for more than 30 years. These calculations have met, however, with two major obstacles. First, the lack of good configuration space wavefunctions has hindered the application of methods which use these functions as a starting point. In addition, a scarcity of experimental results, particularly for liquids and gases, has made comparison of theory with experiment difficult, if not impossible.

Recent advances in theoretical and experimental techniques have made possible both the calculation of vastly improved molecular wavefunctions and the accurate measurement of Compton profiles of liquids and gases. In this paper we shall make use of some of these more exact configuration space wavefunctions to calculate momentum space distributions and Compton profiles for a series of boron hydrides.

The boron hydrides provide an exciting challenge to theoretical chemists because of the multicenter bonding[1,2] which accounts for the stability of these electron-deficient compounds. Recent studies in this laboratory[3–5] employing self-consistent field (SCF) and localized molecular orbital (LMO) methods have shed considerable light on the structure and bonding of the smaller boron hydrides. Although Compton scattering experiments have not yet been performed on boron hydrides, it seems likely that the first such measurements will not be long in coming. In a second paper,[6] we extend the methods developed here to a series of hydrocarbons and related molecules. In addition, we use the LMO momentum distributions introduced here to investigate the transferability of momentum space properties from small to large molecules, and to predict Compton profiles for molecules in several homologous series.

II. THEORY

A. Momentum Space Wavefunctions

There are two basic approaches to the calculation of momentum space wavefunctions. One may either solve the Schrödinger equation directly in the momentum representation or one may take a configuration space wavefunction and transform it to momentum space.

The direct approach has been applied to such simple systems as the hydrogen atom,[7] the helium atom,[8] and the hydrogen molecule ion.[9] However, solution of the momentum space Schrödinger equation for more complex atoms and molecules is a formidable task and has generally been avoided.

The less straightforward but computationally more tractable approach involves solving the Schrödinger equation to some desired degree of approximation in configuration space and then applying the Dirac transformation to the result. The transformation is given (with $\hbar=1$) by

$$\chi(\mathbf{p}_1, \mathbf{p}_2, \cdots, \mathbf{p}_n) = (2\pi)^{-3n/2} \int \prod_{i=1}^{n} \exp(-i\mathbf{p}_i\cdot\mathbf{r}_i)\psi(\mathbf{r}_1, \mathbf{r}_2, \cdots, \mathbf{r}_n) \times d\mathbf{r}_1 d\mathbf{r}_2\cdots d\mathbf{r}_n, \quad (1)$$

where there are n electrons and χ and ψ are the wavefunctions in the momentum and position representations, respectively. This transformation has the useful property that it preserves the form of the wavefunction and is independent of the spin functions. For example, if $\psi(\mathbf{r}_1, \mathbf{r}_2, \cdots, \mathbf{r}_n)$ is a single Slater determinant or a multiconfiguration sum of such terms involving orbitals $\phi_s(\mathbf{r})$ and/or geminals $\lambda_t(\mathbf{r}_1, \mathbf{r}_2)$, then $\chi(\mathbf{p}_1, \mathbf{p}_2, \cdots, \mathbf{p}_n)$ is obtained from ψ by replacing $\phi_s(\mathbf{r})$ by $\gamma_s(\mathbf{p})$ and $\lambda_t(\mathbf{r}_1, \mathbf{r}_2)$ by $\mu_t(\mathbf{p}_1, \mathbf{p}_2)$, where

$$\gamma_s(\mathbf{p}) = (2\pi)^{-3/2}\int \exp(-i\mathbf{p}\cdot\mathbf{r})\phi_s(\mathbf{r})d\mathbf{r},$$
$$\mu_t(\mathbf{p}_1, \mathbf{p}_2) = (2\pi)^{-3}\int \exp[-i(\mathbf{p}_1\cdot\mathbf{r}_1+\mathbf{p}_2\cdot\mathbf{r}_2)] \times\lambda_t(\mathbf{r}_1, \mathbf{r}_2)d\mathbf{r}_1 d\mathbf{r}_2. \quad (2)$$

This method was employed by Coulson and Duncanson[10] to study a number of molecules. These authors, while obtaining several useful results, noted that the approximations involved in their configuration space wavefunctions greatly limited the accuracy of their momentum space calculations. In the past several years Hartree–Fock wavefunctions have been used in momentum space calculations on atoms,[11] on diatomic molecules,[12] and on larger molecules representable by wavefunctions employing a single center for all the atomic basis orbitals.[13,14] We reported in a recent communication[15] that the Dirac transformation may also be conveniently applied to multicenter Hartree–Fock wavefunctions for polyatomic molecules. Here we give the details of that application and illustrate its use for molecules containing as many as sixteen atomic centers.

B. The Compton Profile

It is well known that x rays scattered by electrons at rest show an increase in wavelength, the Compton shift:

$$\lambda=(2h/mc)\sin^2(\theta/2), \quad (3)$$

where θ is the angle of scattering. Electrons in molecules are, of course, not at rest. They possess a velocity or momentum distribution determined by the momentum wavefunction $\chi(\mathbf{p})$, which may be calculated by the methods described above. The radial momentum distribution $I(p)$ is given by

$$I(p)dp=\left(\int_0^{\pi}\sin\theta_p d\theta_p\int_0^{2\pi}d\phi_p\,|\chi(\mathbf{p})|^2\right)p^2dp. \quad (4)$$

The distribution of electronic momenta gives rise to a shape or profile of the Compton line. Calculation of the Compton profile is in general a difficult undertaking, but under certain conditions the problem becomes feasible. For molecules containing only first-row atoms, the impulse approximation,[16] that the binding energy of the electrons is negligible compared with the energy imparted to them by the x ray, is a reasonable one. It can be shown[17] that in this approximation for an isotropic system the intensity J of the scattered x rays is given by

$$J(z)=\frac{1}{2}\int_{|z|}^{\infty}\frac{I(p)}{p}dp, \quad (5)$$

where $I(p)$ is given by Eq. (4). The quantity z has the dimensions of a momentum and is proportional to the wavelength displacement l from the center of the Compton line at $\lambda_0+\Delta\lambda$:

$$z=mcl/2(\lambda\lambda_0)^{1/2}\sin(\theta/2), \quad (6)$$

where λ_0 and λ are the wavelengths of the incoming and the scattered x rays, respectively. According to Eq. (5), the Compton line is symmetric about its center. Equation (6) shows that the profile at one scattering angle determines the profiles at all other angles to within a change of the wavelength scale.

III. METHOD

For the types of basis orbitals usually employed in molecular calculations, the transformation (2) to momentum space orbitals can be carried out analytically. For example, the transforms of the Slater-type orbitals (STO's)

$$\phi_{nlm\alpha}{}^S(\mathbf{r})=(2\alpha)^{n+(1/2)}[(2n)!]^{-1/2}r^{n-1}e^{-\alpha r}Y_{lm}(\theta,\phi)$$

are given by[12]

$$\gamma_{nlm\alpha}{}^S(\mathbf{p})=(n+l+1)!(-i)^l[(2n)!]^{-1/2}(2\alpha)^{n+(1/2)}p^{-1/2}$$
$$\times(\alpha^2+p^2)^{-(1/2)[n+(3/2)]}P_{n+(1/2)}{}^{-[l+(1/2)]}[\alpha/(\alpha^2-p^2)^{1/2}]$$
$$\times Y_{lm}(\theta_p,\phi_p), \quad (7)$$

where $P_{n+(1/2)}{}^{-[l+(1/2)]}[\alpha/(\alpha^2+p^2)^{1/2}]$ is a Legendre polynomial. Since our configuration space calculations employ STO's, we shall devote our attention to this case only.

Equation (7) shows that if the starting wavefunction is real, the momentum wavefunction will be complex whenever there are basis orbitals of p or f symmetry. The momentum *density*, of course, is real. Squaring (7),

TABLE I. Radial momentum distributions.

p (a.u.)	B_2H_6 $I(p)$[a]	B_4H_{10} $I(p)$[b]	B_5H_9 $I(p)$[c]	B_5H_{11} $I(p)$[d]	B_6H_{10} $I(p)$[e]
0.00	0.000	0.000	0.000	0.000	0.000
0.05	0.116	0.187	0.194	0.213	0.222
0.15	1.020	1.637	1.705	1.872	1.954
0.25	2.691	4.317	4.499	4.949	5.179
0.35	4.876	7.869	8.187	9.022	9.464
0.45	7.234	11.871	12.306	13.580	14.273
0.55	9.398	15.843	16.373	18.089	19.053
0.65	11.037	19.251	19.912	22.010	23.264
0.75	11.934	21.612	22.484	24.894	26.398
0.85	12.030	22.470	23.768	26.210	28.038
0.95	11.416	22.058	23.646	25.958	27.986
1.05	10.287	20.290	22.240	24.232	26.336
1.15	8.873	17.679	19.871	21.412	23.468
1.25	7.386	14.701	16.971	18.028	19.919
1.35	5.986	11.775	13.960	14.583	16.241
1.45	4.768	9.192	11.166	11.464	12.861
1.55	3.773	7.098	8.791	8.887	10.035
1.65	3.000	5.515	6.911	6.914	7.864
1.75	2.424	4.389	5.512	5.496	6.259
1.85	2.010	3.628	4.455	4.531	5.171
1.95	1.721	3.133	3.862	3.900	4.459
2.05	1.528	2.821	3.434	3.500	4.011
2.15	1.390	2.627	3.169	3.251	3.740
2.25	1.302	2.504	3.008	3.095	3.578
2.35	1.239	2.423	2.910	2.994	3.480
2.45	1.196	2.365	2.848	2.923	3.417
2.55	1.164	2.318	2.802	2.866	3.370
2.65	1.137	2.274	2.762	2.814	3.325
2.75	1.113	2.231	2.722	2.761	3.278
2.85	1.089	2.186	2.679	2.706	3.223
2.95	1.065	2.138	2.578	2.648	3.161
p_{max}[f] (a.u.)	0.813	0.865	0.892	0.884	0.897

[a] Normalized to 16 electrons.
[b] Normalized to 30 electrons.
[c] Normalized to 34 electrons.
[d] Normalized to 36 electrons.
[e] Normalized to 40 electrons.
[f] Momentum at which $I(p)$ reaches its maximum, calculated by quadratic interpolation.

TABLE II. Compton profiles.

z (a.u.)	B_2H_6 $J(z)$[a]	B_4H_{10} $J(z)$[b]	B_5H_9 $J(z)$[c]	B_5H_{11} $J(z)$[d]	B_6H_{10} $J(z)$[e]
0.0	8.762	15.700	17.012	18.431	19.888
0.1	8.646	15.514	16.818	18.218	19.766
0.2	8.307	14.970	16.251	17.596	19.017
0.3	7.770	14.109	15.334	16.608	17.984
0.4	7.076	12.988	14.187	15.333	16.635
0.5	6.274	11.672	12.823	13.817	15.053
0.6	5.423	10.235	11.338	12.177	13.320
0.7	4.576	8.757	9.810	10.487	11.539
0.8	3.782	7.319	8.314	8.834	9.782
0.9	3.075	5.994	6.918	7.295	8.135
1.0	2.474	4.834	5.674	5.930	6.664
1.1	1.984	3.868	4.616	4.777	5.411
1.2	1.598	3.099	3.751	3.845	4.390
1.3	1.301	2.509	3.072	3.123	3.592
1.4	1.079	2.071	2.553	2.581	2.989
1.5	0.914	1.753	2.167	2.185	2.544
1.6	0.792	1.523	1.882	1.897	2.219
1.7	0.700	1.355	1.672	1.686	1.980
1.8	0.631	1.229	1.514	1.528	1.800
1.9	0.576	1.131	1.391	1.405	1.660
2.0	0.532	1.050	1.292	1.305	1.545
2.1	0.495	0.981	1.208	1.219	1.447
2.2	0.463	0.920	1.134	1.144	1.360
2.3	0.434	0.865	1.067	1.075	1.280
2.4	0.407	0.813	1.005	1.011	1.206
2.5	0.383	0.765	0.947	0.952	1.137
2.6	0.360	0.719	0.892	0.895	1.071
2.7	0.338	0.676	0.840	0.842	1.078
2.8	0.318	0.636	0.790	0.792	0.948
2.9	0.298	0.597	0.743	0.744	0.892
3.0	0.281	0.561	0.698	0.699	0.838
half-width[f]	0.725	0.763	0.787	0.777	0.791

[a] Normalized to 16 electrons.
[b] Normalized to 30 electrons.
[c] Normalized to 34 electrons.
[d] Normalized to 36 electrons.
[e] Normalized to 40 electrons.
[f] z at which $J(z)=\frac{1}{2}J(0)$, calculated by linear interpolation.

we find that $|\chi(\mathbf{p})|^2$ is composed of terms of the form

$$\rho_{jk}f_j(p)f_k(p)Y_{l_jm_j}(\theta_p,\phi_p)Y_{l_km_k}(\theta_p,\phi_p)\exp(-i\mathbf{p}\cdot\mathbf{r}_{jk}), \tag{8}$$

where ρ is the density matrix, j and k signify atomic basis orbitals, the f's include normalizing constants and terms depending only on the radial momentum, and the interatomic vector $\mathbf{r}_{jk}$ is simply $\mathbf{r}_j-\mathbf{r}_k$.

In order to obtain the radial momentum density (4), we must calculate the integrals

$$Q_{l_jm_jl_km_k}(p,\mathbf{r}_{jk})=\int_0^{2\pi}d\phi_p\int_0^{\pi}\sin\theta_p Y_{l_jm_j}(\theta_p,\phi_p)\times Y_{l_km_k}(\theta_p,\phi_p)\exp(-i\mathbf{p}\cdot\mathbf{r}_{jk}). \tag{9}$$

We make use of this expression

$$\exp(-i\mathbf{p}\cdot\mathbf{r}_{jk})=4\pi\sum_{l,m}(-i)^l j_l(pr_{jk})Y_{lm}(\theta_p,\phi_p)Y_{lm}(\theta_{jk},\phi_{jk}), \tag{10}$$

where the j_l are spherical Bessel functions and θ_{jk} and ϕ_{jk} give the interatomic directions in configuration space. Next, we expand the product of spherical harmonics in (9) as a sum of single spherical harmonics:

$$Y_{l_jm_j}(\theta_p,\phi_p)Y_{l_km_k}(\theta_p,\phi_p)=\sum_{l'=0}^{l_j+l_k}\sum_{m'=-(|m_j|+|m_k|)}^{(|m_j|+|m_k|)}a_{l'm'}Y_{l'm'}(\theta_p,\phi_p). \tag{11}$$

For combinations of s, p, and d orbitals, we find at most seven nonzero $a_{l'm'}$.

Substituting (10) and (11) into (9), and taking account of the orthonormality of the spherical harmonics, we have

$$Q_{l_jm_jl_km_k}(p,\mathbf{r}_{jk})=4\pi\sum_{l,m}(-i)^l a_{lm}j_l(pr_{jk})Y_{lm}(\theta_{jk},\phi_{jk}), \tag{12}$$

where the a_{lm} depend upon l_j, m_j, l_k, and m_k.

When Eq. (12) is substituted into Eq. (4) we obtain an analytical expression for $I(p)$ as a finite sum of terms involving the interatomic distances and directions. The above method is applicable to any multicenter wavefunction constructed from basis orbitals which are the products of radial functions with spherical harmonics.

The problem of calculating the Compton profile has now been reduced to a semiinfinite one-dimensional integration (5) over the radial momentum p. While this integration cannot in general be done analytically, it is a straightforward matter to perform the calculation numerically, e.g., by transforming the interval $(|z|,\infty)$ to $(-1,1)$ and using Gauss–Legendre quadrature. A computer program was written for the IBM 360/65 to evaluate the radial momentum densities, to use these in calculating the expectation values of certain operators, and to perform the numerical integrations necessary to obtain the Compton profile. Sample times for the calculation of $I(p)$ and $J(z)$ at

TABLE III. Momentum expectation values[a,b] for boron hydride molecules.

	$\langle p\rangle$	$\langle p^2\rangle$	$\langle p^{-1}\rangle$	$\langle p^{-2}\rangle$
B_2H_6	1.752	6.579	1.095	2.203
B_4H_{10}	1.816	6.952	1.047	1.995
B_5H_9	1.915	7.566	1.000	1.856
B_5H_{11}	1.859	7.208	1.024	1.921
B_6H_{10}	1.933	7.698	0.994	1.831

[a] Atomic units.
[b] $\langle p^n\rangle=(\int p^n|\chi(\mathbf{p})|^2d^3\mathbf{p}/\int|\chi(\mathbf{p})|^2d^3\mathbf{p})$.

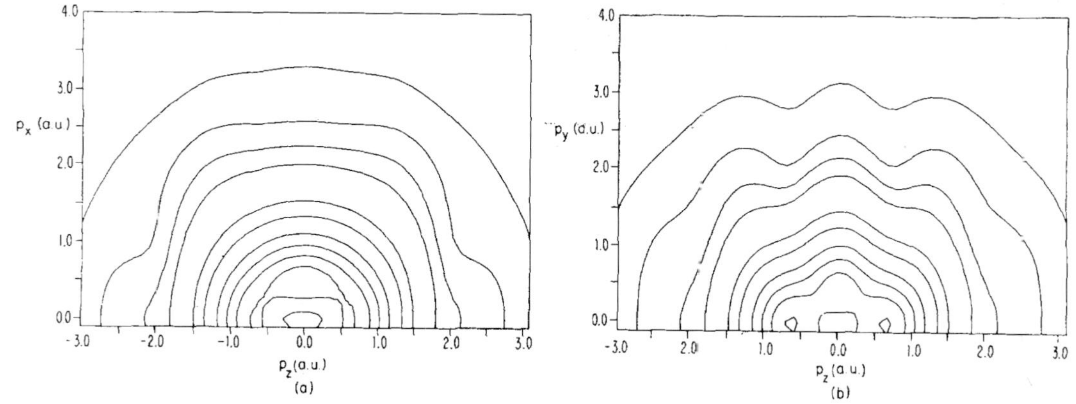

FIG. 1. Total momentum density in B_2H_6. B–B axis is along z. (a) B–H_t plane. Contours (electrons per atomic unit³) in order of intersection with the positive p_x axis are 4.00, 5.00, 5.00, 4.00, 3.00, 2.00, 1.00, 0.50, 0.10, 0.05, 0.025, 0.01. (b) B–H–B plane. Contours (electrons per atomic unit³) in order of intersection with the positive p_y axis are 4.00, 4.00, 3.00, 2.00, 1.00, 0.50, 0.10, 0.05, 0.025, 0.01. The small closed contours along the z axis represent 5.00 electrons/a.u.³

30 points are 11 sec for B_2H_6 with a basis set of 16 STO's and 97 sec for B_6H_{10} with a 40 STO basis.

IV. RESULTS

The methods described in the previous section have been applied to SCF wavefunctions for the molecules B_2H_6,[3] B_4H_{10},[4a] B_5H_9,[4a] B_5H_{11},[4a] and B_6H_{10}.[5] The wavefunctions were calculated using minimum basis sets of STO's with exponents optimized by minimizing the energy of the diborane wavefunction. Thus, our results for B_2H_6 can be expected to be somewhat more reliable than our values for the larger hydrides although, for such similar molecules, errors caused by lack of reoptimization can be expected to be small. Some of our results for B_2H_6 have been reported in an earlier communication.[15] They are reproduced here for the sake of completeness.

The radial momentum distributions, $I(p)$ [Eq. (4)], and the Compton profiles, $J(z)$ [Eq. (5)] are tabulated in Tables I and II, respectively. The values for each molecule are normalized to the total number of electrons in the molecule. While it has been common to subtract out Hartree–Fock inner shell contributions from experimental profiles,[18] this practice can be extremely misleading in comparing experimental results with calculations which employ basis sets of less than Hartree–Fock quality. The significance of the inner shell contributions to the Compton profile is discussed in detail in Paper II. We have avoided normalizing the area under the Compton profile curve since such a procedure can be justified only if the calculation is carried to high enough z so that $J(z)$ becomes negligible. In view of the great similarity among the shapes of the calculated profiles of these molecules we feel that some measure of the total intensity, for example, $J(0)$ should be one product of such a calculation. Our calculations suggest that if similar molecules, such as boron hydrides, are observed under comparable experimental conditions, experimental error may well be large enough to make them indistinguishable *on shape considerations alone.* However, intensity differences should be sufficient to distinguish even such similar species as B_5H_9 and B_5H_{11}.

Probably the most striking feature of our calculations is that the radial distributions and profiles are so similar. There is a trend in both sets of curves and in the expectation values of Table III for the larger molecules to have a greater probability density of electrons at high momenta and hence to have broader profiles, but it is a subtle trend. The fact that the profile for B_5H_9 is broader than that for B_5H_{11} shows that the broadening effect is not simply a function of the number of electrons in the molecule. One explanation for this "inverted ordering" lies in the relatively low ratio of bridge to terminal hydrogens in B_5H_{11} (3 bridge, 8 terminal in B_5H_{11}; 4 bridge, 5 terminal in B_5H_9). The localized orbital analysis presented below shows that B–H_b–B bridge LMO's contribute a somewhat broader distribution than do B–H_t terminal LMO's. The relatively narrow profile of diborane can also be accounted for to some extent by its 1:2 ratio of bridge to terminal hydrogens. However, the size effect cannot be discounted, since B_6H_{10}, which has the same number of both bridge and terminal hydrogens as B_4H_{10} has a considerably broader profile.

The expectation values in Table III give several useful pieces of information about the molecules. The kinetic energy is simply $\langle p^2 \rangle$ times the number of occupied orbitals, while $\langle p \rangle$ gives the average electronic velocity, which is related to the mean current density. The value $\langle p^{-1} \rangle$ is proportional to the maximum height of the Compton profile.

Although such quantities as the Compton profile and momentum expectation values can provide some information about the electronic structure of molecules, they give little insight into the details of chemical

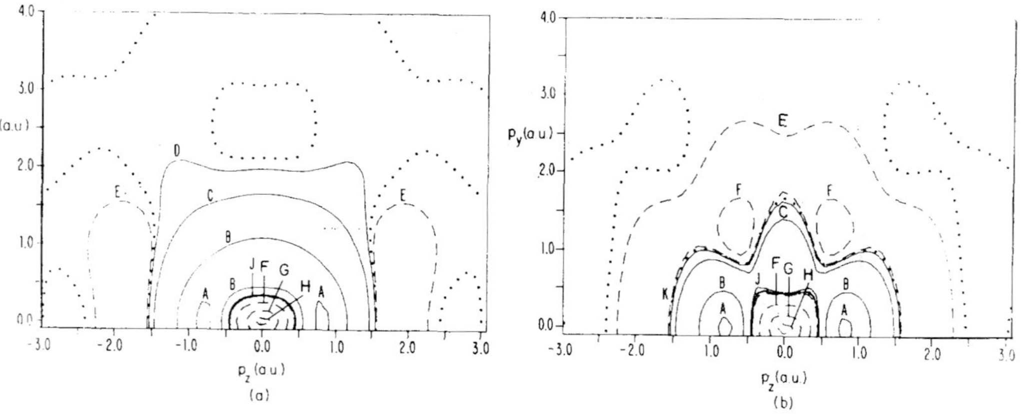

FIG. 2. Difference momentum density in B_2H_6. B–B axis is along z. The solid contours (positive difference density) are (A) 2.00, (B) 1.50, (C) 0.50, (D) 0.05. The dotted contours show zero difference density. The dashed contours (negative difference density) are (E) −0.05, (F) −0.50, (G) −1.50, (H) −2.50. The heavy lines show the near overlapping of several contours where the difference density changes rapidly. $J=C, D, E$, zero. $K=D, E$, zero. All contours in electrons per atomic unit[3]).

bonding. For this purpose it is more instructive to examine the total electron density either in position space or in momentum space before averaging over angles or performing any other integrations.

The total electron density in momentum space in the x-z (B–H terminal) and y-z (B–H bridge) planes of B_2H_6 are shown in Figs. 1(a) and (b). The corresponding difference densities, calculated by subtracting from the total density spherical atomic densities at the equilibrium positions, are shown in Figs. 2(a) and (b). The procedures involved in calculating such difference densities in position space, as well as the justification for employing spherical atoms rather than the valence state atoms used by Henneker and Cade,[12] are discussed in Ref. 4(a).

The total density maps do not show the marked tendency observed in diatomics[10,13] for momentum density to be greater at higher momenta perpendicular to the bond direction and at lower momenta along the bond direction. In polyatomic molecules such as diborane, this effect tends to be obscured by overlapping contributions from bonds oriented in different directions. The difference density maps show more clearly the effects of bond formation on the momentum distribution. We first note the regions of negative density about the origin in momentum space, which show the increase in electronic momentum which accompanies molecule formation. These areas of "momentum deficit" appear somewhat less well localized than those found in diatomics.[13] The positive contours in Fig. 2(a) which extend to regions of high p_x show more clearly than the total density maps the great increase in the component of momentum perpendicular to the bond direction. This effect is evident to a somewhat lesser extent in the B–H_b–B plane, Fig. 2(b). All contour maps show half planes since the assumption that the molecule experiences no net translation requires that the momentum density possess a center of inversion.

V. LOCALIZED ORBITALS IN MOMENTUM SPACE

Total momentum densities for the larger boron hydrides tend to be rather uninformative, owing to the interaction of large numbers of bonds, and the absence of any single plane of special value in elucidating the momentum space structure. If feasible, one would prefer to study individual bonds in isolation, and then to view the total molecular density as composed of contributions from these bonds. The localized molecular orbital formulation is ideally suited to such an approach.

The LMO's employed here have been calculated[3,4b,5] by finding the unitary transformation of the SCF orbitals which maximizes the self-repulsion energy

$$D=\sum_i \int d\tau_1 \int d\tau_2 \mid \phi_i(1) \mid^2 1/r_{12} \mid \phi_i(2) \mid^2. \quad (13)$$

These orbitals provide a conceptual link[4b,19] between SCF wavefunctions and more traditional notions of

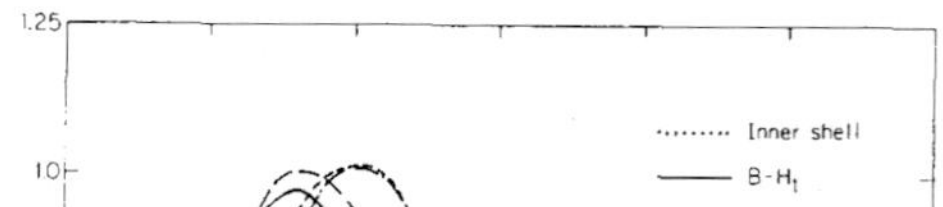

FIG. 3. Average radial momentum density as a function of momentum for boron hydride LMO's.

electron pair bonding. Thus, although determined purely from position space criteria, a LMO basis can enable us to examine momentum space distributions in terms of such intuitive concepts as inner shells and two- and three-center bonds. It is pleasing to note that radial momentum distributions and Compton profiles for a given type of localized orbital (e.g., B–H–B) are nearly independent of the molecular environment. Table IV gives momentum expectation values for all the nonequivalent LMO's in the boron hydrides studied. Figures 3 and 4 show average radial distributions and Compton profiles, respectively, for the different LMO types. The implications of the fact that LMO contributions to the Compton profile have only a weak dependence upon the molecular environment of the bond are investigated in Paper II.

As evidence that at least some of the chemical information contained in the LMO's is carried over in the transformation to momentum space, we note that the anomalous B–H bond in B_5H_{11} [4a] has an even more anomalous momentum distribution. In fact, this "B–H terminal" LMO is no longer intermediate in properties between bridge and terminal B–H bonds. In momentum space it possesses a distribution which is *even broader* than those of $B–H_b–B$ LMO's.

Such an accentuation of anomaly is not found in the momentum space analog of the partial double or short single bond in B_6H_{10}.[5] This bond, which is clearly different in position space from the normal B–B LMO's found in the same molecule and in B_4H_{10}, is quite similar to those orbitals in momentum space. This fact may suggest that the special properties of the bond are due more to the geometry of the molecule than to any great qualitative differences in electronic structure.[5]

More quantitative aspects of the momentum expectation values are just beyond present experimental accuracy, but they are worth noting because they may become observable and because they show some aspects of structural differences of conceptual importance in momentum space. Commenting only on $\langle p^{-1}\rangle$, we note that the largest structural differences among isomers will probably be bridge H atoms (1.242 to 1.29) as com-

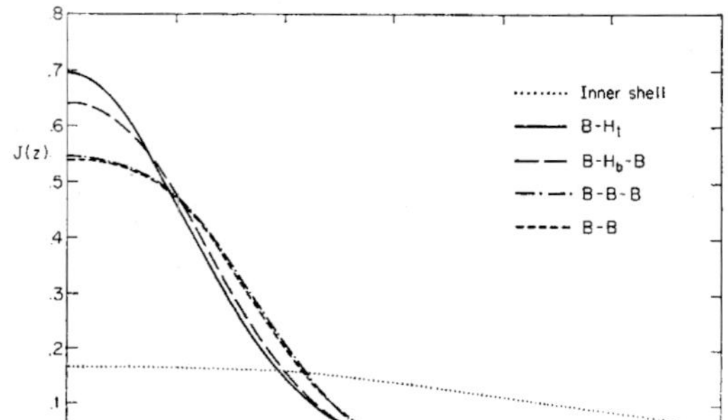

FIG. 4. Average Compton profiles of boron hydride LMO's.

TABLE IV. Momentum expectation values (atomic units) for boron hydride localized orbitals.

	$\langle p\rangle$	$\langle p^2\rangle$	$\langle p^{-1}\rangle$	$\langle p^{-2}\rangle$
B_2H_6:				
1sB	4.0457	22.3751	0.3348	0.1523
BH	0.9828	1.3178	1.3804	3.1088
BHB	0.9983	1.3063	1.2854	2.4403
B_4H_{10}[a]:				
1sB1	4.0418	22.3442	0.3356	0.1553
1sB3	4.0431	22.3570	0.3355	0.1536
B–B	1.0885	1.5546	1.1225	1.6339
B1–H1	0.9831	1.3195	1.3774	3.1094
B3–H5	0.9934	1.3380	1.3556	2.9624
BHB	1.0093	1.3305	1.2698	2.4242
B_5H_9[a]:				
1sB1	4.0411	22.3349	0.3351	0.1567
1sB2	4.0406	22.3356	0.3357	0.1551
B1, B2, B3	1.1500	1.6909	1.0652	1.5315
B1, B2, B5	1.1530	1.6949	1.0603	1.5104
B1, B3, B4	1.1531	1.6951	1.0601	1.5093
B1–H1	0.9683	1.2991	1.4148	3.3123
B2–H2	0.9801	1.3222	1.3941	3.1981
BHB	1.0017	1.3190	1.2893	2.5312
B_5H_{11}[a]:				
1sB1	4.0423	22.3437	0.3352	0.1560
1sB2	4.0425	22.3483	0.3351	0.1544
1sB4	4.0448	22.3673	0.3351	0.1532
BBB	1.1162	1.6078	1.0888	1.5616
B1–H1	0.9881	1.3285	1.3709	3.0819
B1–Hμ	1.0272	1.3640	1.2379	2.2632
B2–H3	0.9774	1.3160	1.3900	3.1480
B4–H5	0.9814	1.3164	1.3798	3.0803
B4–H7	1.0080	1.3608	1.3282	2.8500
BHB (asym)	1.0231	1.3589	1.2515	2.3530
BHB (sym)	1.0148	1.3487	1.2699	2.4731
B_6H_{10}[b]:				
1sB1	4.0414	22.3387	0.3356	0.1575
1sB2	4.0426	22.3495	0.3353	0.1543
1sB3	4.0430	22.3523	0.3352	0.1546
1sB4	4.0495	22.3995	0.3338	0.1520
B1–B2	1.1234	1.6116	1.0783	1.5190
B4–B5	1.1051	1.6587	1.1822	2.1643
BBB	1.1026	1.5849	1.1204	1.7132
B1–H1	0.9812	1.3199	1.3891	3.1860
B2–H2	0.9770	1.3152	1.3930	3.1661
B3–H3	0.9775	1.3164	1.3924	3.1615
B4–H4	0.9826	1.3246	1.3885	3.1618
B2–H23–B3	1.0111	1.3382	1.2748	2.4862
B3–H34–B4	1.0267	1.3629	1.2420	2.3028
Averages[c]:				
1sB (12)	4.043	22.354	0.335	0.155
BH[d] (14)	0.985	1.325	1.378	3.100
BHB[d] (7)	1.012	1.338	1.269	2.430
BB[e] (2)	1.106	1.584	1.100	1.577
BBB (3)	1.130	1.645	1.084	1.579

[a] See Ref. 4 (a) for numbering of atoms.
[b] See Ref. 5 for numbering of atoms.
[c] Number of nonequivalent LMO's over which average is taken are given in parentheses.
[d] Omitting anomalous B1–Hμ orbital of B_5H_{11}.
[e] Omitting anomalous B4–B5 orbital of B_6H_{10}.

pared to terminal H atoms (1.33 to 1.41). The value (1.238) for $H\mu$ in B_5H_{11} is actually outside of this range, but very similar to that (1.242) for the probably rather asymmetric[5] bridge H34 in B_6H_{10}. This type of anomaly is also shared by the endo H7 (H8) in B_5H_{11}. All of these anomalous H atoms are oriented more towards other B atoms than are H atoms showing less anomalous values of $\langle p^{-1} \rangle$. The effect of asymmetry in more nearly normal bridge H atoms is very small, 1.24–1.25 for asymmetric and 1.27–1.29 for symmetric H bridges. Also, the inner shell effect is very small; the most anomalous case is the value of 0.3338 for $B_4(B_5)$ in the very short bond of B_6H_{10}, but photoelectron spectroscopy may reveal this anomaly. The rather low values of $\langle p^{-1} \rangle = 1.060$–$1.065$ for B_5H_9 framework orbitals may contain a suggestion of delocalization in this molecule, as compared with the range 1.078–1.182 for other two-center and central three-center bonds among boron atoms in the other comparatively well localized molecules examined here.

Finally, the LMO analysis makes possible some general predictions about relative widths of Compton profiles. For example, molecules containing more bridge and fewer terminal hydrogens (B_5H_9 vs B_5H_{11}) should have broader profiles. Polyhedral ions such as $B_{10}H_{10}^{2-}$ which have a higher ratio of boron atoms to hydrogen atoms should have very broad profiles because of the greater weight given to the high momentum contributions of the inner shells. There appears to be no systematic difference between the contributions of two-center and three-center boron–boron bonds.

VI. CONCLUSIONS

One should bear in mind the limitations of the present calculations. The position space wavefunctions, while the most accurate now available for these molecules, were obtained from incompletely optimized minimum basis sets. Use of expanded basis sets including more polarizable orbitals and possibly more inner shell-type orbitals[6] might improve our results considerably. The impulse approximation and its range of applicability particularly for heavier atoms should be investigated more thoroughly. Finally, the success of our calculations rests upon the appropriateness of treating Compton scattering as a single-electron phenomenon. If it is shown that the one-electron approximation is invalid for such scattering phenomena, then our methods and others which seek to utilize Hartree–Fock wavefunctions are subject to very serious errors.

In spite of the above cautions, the prospects for obtaining reasonably accurate momentum distributions for polyatomic molecules are encouraging. Recent progress in experimental Compton scattering research lends new interest and importance to such calculations. In addition, other applications of momentum distributions to such phenomena as inelastic scattering of fast electrons[20] and positron annihilation[21] have thus far received comparatively little attention.

ACKNOWLEDGMENTS

A large number of helpful discussions with T. A. Halgren and E. Switkes are gratefully acknowledged. This research was supported in part by the U. S. Office of Naval Research.

* National Science Foundation Predoctoral Fellow.

[1] W. H. Eberhardt, B. L. Crawford, and W. N. Lipscomb, J. Chem. Phys. **22**, 989 (1954).

[2] W. N. Lipscomb, *Boron Hydrides* (Benjamin, New York, 1963).

[3] E. Switkes, R. M. Stevens, W. N. Lipscomb, and M. D. Newton, J. Chem. Phys. **51**, 2085 (1969), B_2H_6, SCF–LMO.

[4] (a) E. Switkes, I. R. Epstein, J. A. Tossell, W. N. Lipscomb, and R. M. Stevens, J. Am. Chem. Soc. **92**, 3837 (1970), B_4H_{10}, B_5H_9, B_5H_{11}, SCF; (b) E. Switkes, W. N. Lipscomb, and M. D. Newton, J. Am. Chem. Soc. **92**, 3847 (1970), B_4H_{10}, B_5H_9, B_5H_{11}, LMO.

[5] I. R. Epstein, J. A. Tossell, E. Switkes, W. N. Lipscomb, and R. M. Stevens, Inorg. Chem. (to be published). B_6H_{10}, SCF–LMO.

[6] I. R. Epstein, J. Chem. Phys. **53**, 4425 (1970) (Paper II, following paper).

[7] M. Levy, Proc. Roy. Soc. (London) **A204**, 145 (1950).

[8] R. McWeeny and C. A. Coulson, Proc. Phys. Soc. (London) **A62**, 509 (1949).

[9] R. McWeeny, Proc. Phys. Soc. (London) **A62**, 519 (1949).

[10] C. A. Coulson, Proc. Cambridge Phil. Soc. **37**, 55, 74 (1941); C. A. Coulson and W. E. Duncanson, *ibid.* **37**, 67 (1941); **38**, 100 (1941); W. E. Duncanson, *ibid.* **37**, 397 (1941); W. E. Duncanson and C. A. Coulson, *ibid.* **37**, 406 (1941).

[11] R. J. Weiss, A. Harvey, and W. C. Phillips, Phil. Mag. **17**, 241 (1968).

[12] W. H. Henneker and P. E. Cade, Chem. Phys. Letters **2**, 575 (1968).

[13] M. Roux, M. Cornille, B. Tsapline, M. Rouault, and C. Tavard, Acta Cryst. **22**, 634 (1967).

[14] M. Cornille, M. Roux, and B. Tsapline, Acta Cryst. **A26**, 105 (1970).

[15] I. R. Epstein, J. Chem. Phys. **52**, 3838 (1970).

[16] See, e.g., P. M. Platzman and N. Tzoar, Phys. Rev. **139**, A420 (1965).

[17] J. W. M. Dumond, Rev. Mod. Phys. **5**, 1 (1933).

[18] See, e.g., R. J. Weiss, J. Chem. Phys. **52**, 2237 (1970).

[19] M. D. Newton, E. Switkes, and W. N. Lipscomb, J. Chem. Phys. **53**, 2645 (1970).

[20] A. L. Hughes and M. M. Mann, Phys. Rev. **53**, 50 (1938).

[21] W. Brandt and L. Eder, Phys. Rev. **142**, 165 (1966).

Boron Hydride Valence Structures *Inorganic Chemistry, Vol. 10, No. 9, 1971* **1921**

Contribution from the Department of Chemistry,
Harvard University, Cambridge, Massachusetts 02138

Boron Hydride Valence Structures. A Topological Approach

By Irving R. Epstein and William N. Lipscomb*[1]

Received March 15, 1971

A topological approach to boron hydride three-center valence structures is presented. The method differs from earlier ones in that, on the basis of SCF-localized orbital results, open BBB bonds are excluded. Most of the boron frameworks for molecules, ions, and intermediates which are known or thought to be among the more stable species give rise to more allowed valence structures than do hypothetical topologies. A new, simpler topological computer program is described. Several charge distribution dependent weighting schemes for population analyses are considered. New structures are predicted, and use of the method in elucidating reaction pathways and more stable isomers is illustrated.

I. Introduction

The success of the three-center bond formalism[2] in describing the valence structures of the boron hydrides and of related electron-deficient molecules is now well established. The three-center, two-electron bond has played an important role in molecular orbital calculations, both semiempirical[3] and *ab initio*.[4] Its use has also served as a theoretical framework for the analysis of experimental results. Recent calculations[5] employing self-consistent field (SCF) wave functions and the localized molecular orbital (LMO) formulation[6] have put the ideas of three-center bonding on a firm quantitative footing.

One of the most interesting applications of the three-center bond formalism has been the topological or "semitopological" approach to boron hydride valence structures, developed by Dickerson and Lipscomb.[7] Capable of extension to a vast array of systems, the topological generation of boron hydride valence structures offers promise of providing a simple, systematic, pictorial description of boron hydride chemistry similar to that afforded by the dot and line drawings[8] so familiar to organic chemists. The efforts of Hoffmann and Lipscomb[3b] to program high-speed digital computers to generate boron hydride valence structures yielded a number of promising results. However, the problem is complicated, both computationally and conceptually, by the existence of two topologically distinct types of three-center boron–boron–boron bonds (see Figure 1). The SCF–LMO calculations noted, above,[5a,c,d] as well as LMO calculations[9] performed with other types of wave functions, fail to show any evidence so far for the existence of the open three-center BBB bond, Figure 1c, in the boron hydrides studied. Reformulation of the topological theory with only a single type of three-center BBB bond allows for great simplification. In this paper we describe a new computer program for the generation of all allowable boron hydride valence structures given a topological arrangement of the boron and hydrogen nuclei.

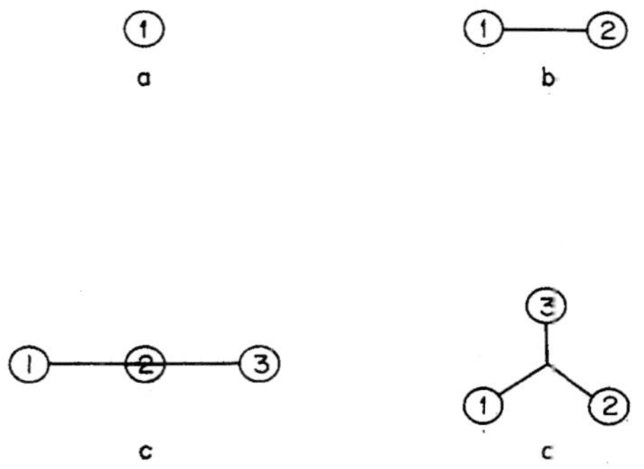

Figure 1.—Topological diagrams of orbital types found in boron hydrides.

We discuss briefly the topological approach and its application to electron-deficient molecules in general. A description of the program, with emphasis on the difference between the present version and earlier topological programs, is followed by a summary of results obtained for a number of boron hydride molecules and ions. The concluding section contains a discussion of the chemical significance and of possible extensions, modifications, and limitations of this approach.

II. The Topological Approach

If we look at molecules topologically—*i.e.*, in terms only of the connections between atoms, regardless of their identity or the distance between them (so long as we regard them as connected)—then, for our present purposes, we need to consider only the four types of orbitals shown in Figure 1. Of course, there are a few examples in which one may have to be somewhat arbitrary in deciding which atoms are connected to each other. Nevertheless, for most molecules, the complete set of connections or bonds is given by chemical or geometical intuition. If necessary one can set up quantitative criteria such as a maximum interatomic distance (dependent upon the types of atoms involved) for two atoms to be deemed "connected." The orbitals we discuss here are *idealized* charge distributions. Localized one-, two-, and three-center orbitals always

(1) National Science Foundation Predoctoral Fellow.
(2) W. H. Eberhardt, B. Crawford, and W. N. Lipscomb, *J. Chem. Phys.*, **22**, 989 (1954).
(3) (a) R. Hoffmann and W. N. Lipscomb, *ibid.*, **36**, 2179 (1962); (b) R. Hoffmann and W. N. Lipscomb, *ibid.*, **37**, 2872 (1962).
(4) F. P. Boer, M. D. Newton, and W. N. Lipscomb, *J. Amer. Chem. Soc.*, **88**, 2361 (1966).
(5) (a) E. Switkes, R. M. Stevens, W. N. Lipscomb, and M. D. Newton, *J. Chem. Phys.*, **51**, 2085 (1969); (b) E. Switkes, I. R. Epstein, J. A. Tossell, R. M. Stevens, and W. N. Lipscomb, *J. Amer. Chem. Soc.*, **92**, 3837 (1970); (c) E. Switkes, W. N. Lipscomb, and M. D. Newton, *ibid.*, **92**, 3847 (1970); (d) I. R. Epstein, J. A. Tossell, E. Switkes, R. M. Stevens, and W. N. Lipscomb, *Inorg. Chem.*, **10**, 171 (1971).
(6) C. Edmiston and K. Ruedenberg, *Rev. Mod. Phys.*, **35**, 467 (1963); *J. Chem. Phys.*, **43**, S97 (1965).
(7) R. E. Dickerson and W. N. Lipscomb, *ibid.*, **27**, 212 (1957).
(8) M. A. Couper, *C. R. Acad. Sci.*, **46**, 1157 (1858).
(9) M. D. Newton, private communication.

contain contributions from other centers which may amount to several per cent of the total electron density in the orbital.[5] A recent experimental study[10] suggests that the distinction between the two types of three-center bond (Figure 1c, d) is not nearly so clear-cut as might have been thought.

The one-center orbitals (Figure 1a) are inner shells or lone pairs. The latter make important contributions to the properties of molecules containing "electron-rich" atoms, such as nitrogen or oxygen, which have more valence electrons than available bonding orbitals.

Figure 1b shows the two-center or "normal" bond which, for many chemists, constitutes *the* chemical bond. One motivation behind our work on the boron hydrides has been the feeling that a deeper understanding of chemical bonding requires insight into the "anomalous" cases as well as the more familiar ones.

When three centers are to be linked by a single orbital, there exist two topologically distinct possibilities. We can connect them in the "open" manner of Figure 1c, in which the two end atoms are neighbors[11] of the middle one but not of each other. This is the type of bonding found in boron hydride bridge bonds (BHB), with two borons and hydrogen linked by an electron pair. Alternatively one can have a "central" three-center bond (Figure 1d), in which all three atoms are pairwise neighbors and are topologically equivalent.

The original topological formulation of boron hydride valence theory[7] allowed both types of three-center BBB bond. The starting point of our new approach is the requirement, based on our recent SCF-LMO calculations,[5] that all BBB bonds be of the central type.

Given the number of boron and hydrogen atoms in a molecule and a topological picture of how they are to be connected, the problem is to find all possible valence structures for the molecule which satisfy our topological rules. The presence of only central BBB bonds permits us to work with a somewhat simpler set of rules than that used by earlier investigators.[3b,7] We may dispense with the notions of excess connectability and connectivity,[7] and we need no longer worry about possible awkward hybridizations[3b] engendered by the crossing of two-center and open three-center bonds. We now set down our rules simply in terms of connections, orbitals, and electrons. (1) Every pair of boron atoms which are neighbors[11] must be connected by at least one B-B, BHB, or BBB bond. (2) No boron framework bond may include a pair of nonneighbor borons. (3) Each boron, since it has four valence atomic orbitals, must participate in $4 - n_i - m_i$ boron framework bonds, where n_i is the number of terminal hydrogens and m_i the number of bridge hydrogens bonded to the boron (i) in question. (4) Since each boron contributes three valence electrons and four valence orbitals, the molecule B_pH_{p+q} must have $(s - q)/2$ B-B and $p - s$ BBB bonds, where s is the total number of BHB bridges in the molecule. (5) No two borons may be bonded by both a two-center and a three-center (BBB or BHB) bond. Parenthetically, we comment that these rules for boranes are to be modified for carboranes.

The assumptions here are consistent with the "extended valence theory" of Lipscomb[12,13] in which borons may be bonded to zero, one, or two terminal hydrogens. Ionic species may be included by modifying rule 4 to allow for the appropriate number of electrons in the molecule.[14] "Heteroatoms" such as carbon or nitrogen may also be introduced by modifying rule 4. For example, the topological problem for the octahedral carborane $B_4C_2H_6$[15] is equivalent to that for the $B_6H_6^{2-}$ octahedron. Although we shall not do so here, one can also allow for multiple bonding and π donation in boron hydrides by relaxing restriction 5.[5d]

III. Computer Program

The first Hoffmann-Lipscomb topological program[3b] for computerized generation of boron hydride valence structures was modified and improved by Simpson[16] in his studies of $B_{18}H_{22}$. In its most elaborate form, Simpson's program was able to investigate many thousands of structures and to find about 425 acceptable structures per minute. Hoffmann's combinatorial subprograms were written in machine language. Simpson's later program was in FORTRAN but was longer than Hoffmann's and about 5 times the length of our present version. Simpson appears to have accepted Hoffmann and Lipscomb's hypothesis[3b] that specifying the two-center and open three-center bonds suffices to fix uniquely the remaining central three-center bonds. In Appendix I we show by (a rare) counterexample that this conjecture is false. No errors, however, appear to have resulted in earlier calculations from this assumption.

Our present program is extremely simple. It is written entirely in FORTRAN and has minimal storage requirements. Thus it can easily be used on nearly any computer. Input for a molecule consists simply of the number of borons and hydrogens, the charge, and a list of topological connections. This list is divided into B-H (B bonded to no, one, or two hydrogens and BHB groups) and B-B framework subsections.

From the input, the program calculates how many two- and three-center boron framework bonds are required for a valid structure (rule 4 above). It also constructs a canonically ordered list of allowed two- and three-center bonds. The search procedure consists in testing each possible combination of framework bonds to see if it satisfies rules 1, 3, and 5. The program operates by assembling allowable combinations of two-center bonds first and then matching them with three-center combinations. In this way, one need test only a fraction of the theoretically possible structures, since eliminating one two-center combination eliminates all possible three-center combinations with it. Our program turns out about as many acceptable structures per minute as Simpson's,[16] but it has to test far fewer.

To see the great savings brought about by the elimination of open BBB bonds, consider the decaborane molecule, $B_{10}H_{14}$. Each valence structure must have 2

(10) R. Brill, H. Dietrich, and R. Dierks, *Acta Crystallogr.*, in press.

(11) We shall refer to two atoms as neighbors if our topological framework requires them to be connected. In general the topological and geometrical notions of "neighbor" will coincide.

(12) R. Grimes, F. E. Wang, R. Lewin, and W. N. Lipscomb, *Proc. Nat. Acad. Sci. U. S.*, **47**, 969 (1961).

(13) W. N. Lipscomb, *ibid.*, **47**, 1791 (1961).

(14) One then has $y - (s/2) - (3l/2)$ two-center and $p - y + l$ three-center bonds, where l is the total charge of the ion.

(15) I. R. Epstein, T. F. Koetzle, R. M. Stevens, and W. N. Lipscomb, *J. Amer. Chem. Soc.*, **92**, 7019 (1970).

(16) (a) P. G. Simpson, Ph.D. Thesis, Harvard University, 1963; (b) P. G. Simpson and W. N. Lipscomb, *J. Chem. Phys.*, **39**, 26 (1963); (c) P. G. Simpson, K. Folting, R. D. Dobrott, and W. N. Lipscomb, *ibid.*, **39**, 2339 (1963).

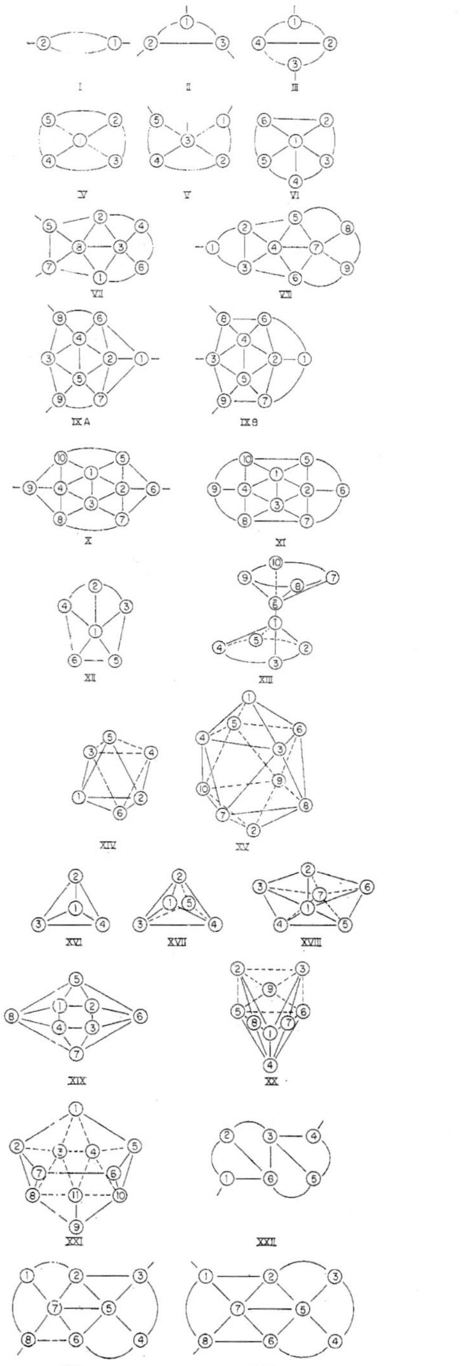

Figure 2.—Topological diagrams and numbering conventions for some known boron hydrides. ⓝ represents a BH group, ⓝ- a BH_2 group, and ⓝ a boron with no hydrogens attached. ⓝ⌒ⓜ indicates a B–H–B bridge, while ⓝ—ⓜ shows that borons *n* and *m* are neighbors in the topological sense of footnote 11.

two-center and 6 three-center boron framework bonds. There are 17 two-center, 12 central three-center, and 24 open three-center bonds topologically allowed. If we consider all possible three-center bonds, there are $[(17!)/(15!)(2!)] \times [(36!)/(30!)(6!)]$ or about 2.6×10^8 valence structures to test. Eliminating the open B–B–B bonds, as in the present program, reduces this number to $[(17!)/(15!)(2!)] \times [(12!)/(6!)(6!)]$ or about 1.3×10^5 possible structures. The savings generated are obvious. If one were willing to sacrifice the advantage of transferability among different computers, the speed could be increased even more by use of machine language combinatorial subprograms.

The final output of the program consists of a set of allowable valence structures, each one being specified by a list of its two-center and three-center bonds. The program also produces a population analysis, *i.e.*, charge and bond-order matrices obtained from the elementary distributions of Table I by averaging over

TABLE I
CONTRIBUTIONS OF BORON–HYDROGEN AND BORON–BORON ORBITALS TO POPULATION ANALYSIS

Bond type	Electrons 1	2	3	B–B overlap population 2–3	1–3	1–4
1-center	2.0	...	...	...	...	...
2-center	1.0	1.0	...	1.0	...	...
3-center open	0.5	1.0	0.5	...	0.5	...
3-center central	0.667	0.667	0.667	0.667	0.667	0.667

all the valence structures obtained. Provision can also be made for the use of various weighting factors in the population analysis, as discussed below. The program, like its predecessors,[3b,16] makes no use of molecular symmetry, since independent generation of symmetry-equivalent valence structures provides an ideal check on the reliability of the search procedure. We recognize that there are difficulties in our averaging procedures arising from nonorthogonality of these valence structures.

IV. Results

The method of analysis described above was applied to a number of boron hydride molecules and ions, both known and hypothetical. Some of the results obtained are shown in Tables II and III, with the numbering conventions used given by Figures 2 and 3.

The species in Figure 2 have all been characterized

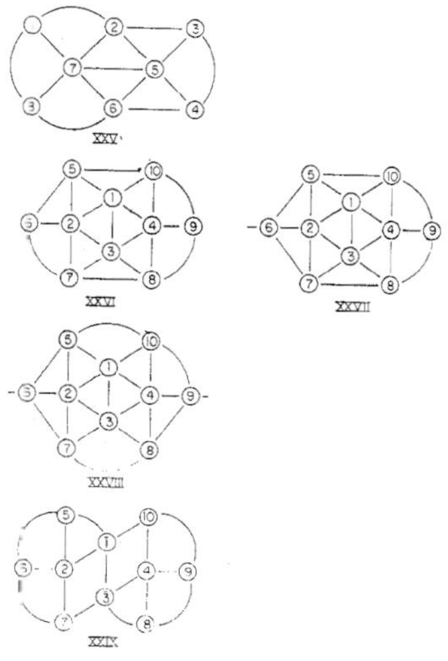

Figure 3.—Possible boron hydride topologies. Symbols are the same as those in Figure 2.

TABLE II
BORON HYDRIDE VALENCE STRUCTURES

Molecule	Topology[a]	No. of symmetry-equiv structures	B–B bonds[b]
B_2H_6	I		
1		1	...
		1	
B_3H_8	II		
1		1	2–3
		1	
B_4H_{10}	III		
1		1	2–4
		1	
B_5H_9	IV		
1		4	1–2, 2–3
		4	
B_5H_{11}	V		
1		1	...
		1	
B_6H_{10}	VI		
1		1	1–4, 2–6
2		2	1–3, 2–6
		3	
$B_8H_{13}^-$	VII		
1		2	1–7, 2–3
2		2	1–3, 5–7
3		1	3–8, 5–7
4		2	1–7, 3–4
5		2	3–4, 5–7
6		1	1–7, 2–5
		10	
B_9H_{15}	VIII		
1		1	2–3, 4–7
2		2	2–3, 5–7
3		2	3–6, 5–7
4		2	2–5, 7–8
5		1	2–5, 3–6
6		2	2–3, 7–8
		10	
$B_9H_{14}^-$	IXA		
1		2	1–6
		2	
$B_9H_{14}^-$	IXB		
1		2	2–6
2		2	6–8
		4	
$B_{10}H_{14}$	X		
1		1	...
		1	
$B_{10}H_{14}^{2-}$	X		
1		4	1–2, 3–8, 6–7
2		4	1–2, 6–7, 8–9
3		2	1–3, 5–6, 8–9
4		2	1–3, 5–6, 9–10
5		4	1–5, 3–7, 8–9
6		4	1–5, 6–7, 8–9
7		4	1–5, 6–7, 9–10
		24	
$B_{10}H_{14}$	XI		
1		2	2–5, 4–10
2		2	2–7, 4–10
3		4	4–9, 5–10
4		4	2–5, 4–9
5		4	1–2, 4–10
6		4	2–3, 4–10
7		4	1–4, 7–8
		24	
$B_{10}H_{14}^{2-}$	XI		
1		1	1–3, 2–6, 4–9, 5–10, 7–8
		1	
$B_6H_8^{2-}$ ($B_4C_2H_8$)	XII	0	...
$B_{10}H_{16}$	XIII		
1		16	1–2, 1–3, 1–6, 6–7, 6–8
		16	
$B_6H_6^{2-}$ ($B_4C_2H_6$)	XIV		
1		8	1–2, 1–3, 2–3
2		24	1–2, 1–3, 2–4
		32	
$B_{10}H_{10}^{2-}$ ($B_8C_2H_{10}$)	XV		
1		16	1–3, 1–4, 7–8
2		8	1–3, 1–4, 2–9
3		16	1–3, 1–4, 5–9
4		16	1–3, 1–4, 8–9
5		8	1–3, 2–9, 4–10
6		8	1–3, 6–9, 4–10
		72	
B_4H_4	XVI		
1		1	...
		1	
$B_5H_5^{2-}$ ($B_3C_2H_5$)	XVII		
1		2	1–2, 1–3, 1–4
		2	
$B_7H_7^{2-}$ ($B_5C_2H_7$)			
1		20	1–2, 2–3, 4–5
		20	
$B_8H_8^{2-}$ ($B_6C_2H_8$)	XIX		
1		8	1–2, 2–5, 5–8
		8	
$B_9H_9^{2-}$ ($B_7C_2H_9$)	XX		
1		12	1–7, 3–7, 2–8
2		4	1–7, 2–8, 3–9
		16	
$B_{11}H_{11}^{2-}$ ($B_9C_2H_{11}$)	XXI		
1		4	2–3, 2–8, 4–5
2		4	2–3, 4–5, 8–9
3		4	2–3, 2–8, 4–11
4		4	2–3, 4–11, 8–9
		16	
B_6H_{12}	XXII		
1		2	3–4
2		1	3–6
		3	
B_8H_{14}	XXIII		
1		2	4–5
2		1	5–7
		3	
B_8H_{14}	XXIV		
1		2	2–5
2		2	3–5
3		1	4–7
		5	
B_8H_{12}	XXV		
1		2	2–7, 4–6
2		1	2–3, 4–6
3		2	2–3, 1–7
		5	
$B_{10}H_{13}^-$	XXVI		
1		1	2–3, 4–8, 5–6
2		1	2–3, 4–10, 5–6
3		1	2–3, 5–6, 5–10
4		1	1–4, 2–5, 5–6
5		1	1–5, 3–4, 5–10
6		1	2–5, 3–4, 5–6
7		1	3–4, 5–6, 5–10
8		1	1–5, 4–9, 5–10
9		1	2–5, 4–8, 5–6
10		1	2–5, 4–9, 5–6
11		1	2–5, 4–10, 5–6
12		1	2–6, 4–8, 5–6
13		1	2–6, 5–6, 7–8
14		1	2–6, 4–10, 5–6
15		1	2–6, 5–6, 5–10
16		1	4–9, 5–6, 5–10
		16	

Boron Hydride Valence Structures *Inorganic Chemistry, Vol. 10, No. 9, 1971* 1925

Table II (*Continued*)

Molecule	Topology[a]	No. of symmetry-equiv structures	B–B bonds[b]
$B_{10}H_{13}^-$	XXVII		
1		2	4–8, 5–6
2		2	5–6, 7–8
3		2	4–10, 5–6
4		2	5–6, 5–10
		8	
$B_{10}H_{15}^-$ ($B_{10}H_{12}L_2L'$)	XXVIII		
1		1	1–3, 5–6
2		1	1–3, 6–7
3		1	1–5, 3–4
4		1	3–4, 5–6
5		1	2–3, 5–6
6		1	3–8, 6–7
7		1	3–8, 5–6
8		1	1–5, 3–7
9		1	3–7, 5–6
10		1	1–5, 6–7
		10	
$B_{10}H_{14}$	XXIX		
1		1	1–2, 2–7, 3–4, 4–10
2		2	1–2, 2–5, 3–4, 4–10
3		1	1–2, 2–5, 3–4, 4–8
4		2	2–6, 3–4, 3–7, 4–10
5		2	2–6, 2–7, 3–4, 4–10
6		2	2–6, 3–4, 3–7, 4–8
7		2	2–5, 2–6, 3–4, 4–10
8		2	2–6, 2–7, 3–4, 4–8
9		2	2–5, 2–6, 3–4, 4–8
10		2	1–10, 2–6, 2–7, 4–9
11		2	1–10, 2–5, 2–6, 4–9
12		1	2–6, 2–7, 4–9, 4–10
13		2	2–5, 2–6, 4–9, 4–10
14		1	2–5, 2–6, 4–8, 4–9
		24	

[a] See Figures 2 and 3. [b] Only the two-center bonds are given, since for the molecules listed here, the B–B–B bonds are then uniquely determined (however, see Appendix I).

experimentally and have been the subject of several theoretical studies.[5,15,17] All of them, with the exception of $B_6H_8^{2-}$, topology XII, possess one or more allowable valence structures. We have previously argued[5d] that $B_6H_8^{2-}$ (*i.e.*, $CB_5H_8^-$ or $C_2B_4H_8$) valence structures will be of the multiply bonded π-donation type. Many other topological arrangements were investigated, some of which are shown in Figure 3. In general, hypothetical topologies give rise to no allowed structures or to far fewer structures than the topologies for known hydrides with the same number of borons. Some quite plausible (but experimentally unknown) arrangements give rise to no structures at all. These include cubic B_8H_8 and $B_8H_8^{2-}$ and the icosahedral (minus one apex) $B_{11}H_{11}^{2-}$ and several other structures considered by Kettle and Tomlinson.[18]

We do not wish to suggest that the existence of many valence structures for a given topology implies the existence of a molecule with that topology or that failure to give rise to allowed structures makes a topological arrangement forbidden. Nonetheless, our three-center calculations do show a strong correlation with chemical experience and with chemical and geometrical intuition. For example, the $B_{10}H_{10}^{2-}$ ion has a strong "aromatic-like" character[17] as shown by its remarkably high diamagnetic susceptibility.[19] Since this species possesses more allowable valence structures than any other that we have investigated, we might characterize its behavior in the same resonance-structure language often applied to aromatic organic molecules. The $B_9H_{14}^-$ ion was first predicted[17] to have topology IXA, like the B_9H_{13} (CH_3CN) molecule.[20] However, a recent

(17) See W. N. Lipscomb, "Boron Hydrides," W. A. Benjamin, New York, N. Y., 1963, for a review.

(18) S. F. A. Kettle and V. Tomlinson, *J. Chem. Soc. A*, 2002 (1969).

(19) R. Hoffmann, unpublished results, 1962.

Table III

Three-Center Population Analyses[a] for Some Boron Hydrides

Molecule	Topology	Atom[b]	Charge	Bond[b]	Distance[c]	Overlap Population
$B_3H_8^-$	II	1	0.00	1–2	1.77	0.50
		2	−0.50	2–3	1.80	1.00
B_4H_{10}	III	1	0.00	1–2	1.84	0.50
		2	0.00	2–4	1.71	1.00
B_5H_9	IV	1	−0.67	1–2	1.69	0.83
		2	0.17	2–3	1.80	0.67
B_5H_{11}	V	1	−0.17	1–2	1.76	1.17
		2	0.33	1–3	1.87	0.67
		3	−0.33	2–3	1.72	0.67
				2.4	1.77	0.50
B_6H_{10}	VI	1	−0.33	1–2	1.80	0.78
		2	−0.28	1–3	1.75	0.67
		3	0.33	1–4	1.74	0.78
		4	0.22	2–3	1.74	0.94
				2–6	1.60	1.00
				3–4	1.79	0.72
$B_8H_{13}^-$ ($LB_8H_{11}NH_2$)[d]	VII	1	0.00	1–3	1.79	0.73
		3	−0.30	1–6	1.86	0.70
		4	0.27	1–7	1.92	0.77
		5	−0.60	1–8	1.74	0.80
		8	−0.03	3–4	1.82	0.73
				3–8	1.79	0.77
				4–6	2.00	0.83
				5–7	1.92	0.83
				5–8	1.73	0.80
B_9H_{15}	VIII	1	0.00	1–2	1.85	0.50
		2	−0.10	2–3	1.80	0.83
		4	−0.03	2–4	1.77	0.80
		5	0.00	2–5	1.96	0.77
		7	−0.30	4–5	1.77	0.80
		8	0.27	4–7	1.76	0.77
				5–7	1.83	0.73
				5–8	1.79	0.70
				7–8	1.75	0.73
				8–9	1.82	0.83
$B_9H_{14}^-$ ($B_9H_{13}L$)[e]	IXA	1	−0.67	1–2	1.74	0.67
		2	0.00	1–6	1.87	0.83
		3	0.00	2–4	1.76	0.67
		4	0.00	2–6	1.75	1.00
		6	0.00	3–4	1.76	1.33
		8	−0.17	3–8	1.87	0.67
				4–5	1.83	0.67
				4–6	1.73	0.67
				4–8	1.72	0.67
				6–8	1.85	0.50
$B_{10}H_{14}^{2-}$ ($B_{10}H_{12}L_2$)[f]	X	1	−0.25	1–2	1.73	0.78
		2	−0.06	1–3	1.84	0.72
		5	−0.04	1–5	1.78	0.74
		6	−0.61			
				2–5	1.75	0.81
				2–6	1.74	0.78
				5–6	1.86	0.81
				5–10	1.88	0.61
$B_{10}H_{14}$	XI	1	−0.08	1–2	1.79	0.76
		2	−0.28	1–3	1.71	0.78
		5	0.04	1–5	1.78	0.81
		6	0.28	2–5	1.78	0.74
				2–6	1.72	0.72
				5–6	1.77	0.78
				5–10	2.01	0.72
$B_{10}H_{10}^{2-}$	XV	1	−0.33	1–3	1.73	0.92
		3	−0.17	3–4	1.86	0.72
				3–8	1.81	0.74

[a] Averaged over all allowable structures with unit weights. [b] Only one of a set of symmetry-equivalent atoms or bonds is given. [c] Symmetry-average experimental distances (Å) taken from ref 11. [d] L attached at B7; NH_2 replaces bridge hydrogen H46. [e] L attached at B1. [f] L attached at B6 and B9.

(20) F. E. Wang, P. G. Simpson, and W. N. Lipscomb, *J. Chem. Phys.*, **35**, 1335 (1961), especially Figure 3.

determination[21] shows that the $B_9H_{14}^-$ structure is of topology IXB, a possibility noted by Wang, Simpson, and Lipscomb[20] and suggested by the relative numbers of allowed valence structures, 2 for IXA and 4 for IXB.

Another illustration is given by the $B_{10}H_{14}$ and $B_{10}H_{14}^{2-}$ structures (X and XI). The neutral molecule has been shown[22] to possess topology XI. A structure determination on $B_{10}H_{14}^{2-}$ is now in progress, but the best current guess of its structure is similar to that found in the $B_{10}H_{12}(NCCH_3)_2$ X-ray study,[23] *i.e.*, topology X. In both cases, Table II shows a 24:1 ratio of allowed structures in favor of the experimental topologies.

Hydrides with topologies XXII, XXIII, and XXIV were proposed by Lipscomb[24] from purely geometrical considerations. On the basis of the number of allowed valence structures, these molecules appear to be the most likely of those predicted by the geometrical theory. The eight-boron framework in topologies XXIII, XXIV, and XXV seems to offer an especially promising direction for synthetic chemists.[25]

Another example is $B_{10}H_{13}^-$, for which structures XXVI and XXVII have been assigned previously.[17] Another structure based upon this B_{10} unit has hydrogen atoms bridging the 5–6, the 7–8, and the 9–10 pairs of boron atoms. This structure has 8 valence forms, but has a more nearly uniform charge distribution (either +0.17 or −0.17 at each boron atom) than do the other structures. If the 16 resonance forms of XXVI are not to be weighted equally, the choice of the most stable structure is difficult for $B_{10}H_{13}^-$.

It would be extremely desirable if the topological theory could be used to generate, or at least to test, proposed intermediates in boron hydride reactions. Topology XXVIII is that proposed by Lipscomb[17] for the initial attack of C_2H_5OH on $B_{10}H_{12}L_2$ to begin the formation of $B(OC_2H_5)_3$ and $B_9H_{13}L$. The large number of valence structures suggests that XXVIII is indeed a plausible intermediate. However, no allowed structures result from a second step (attack on the B8–B9 bond) analogous to the first. We suggest, therefore, that intermediate XXVIII breaks down by a more complicated mechanism, possibly involving topological rearrangement of the boron framework. The transformation of $B_{10}H_{16}$ to $B_{10}H_{14}$ has been suggested[17] to occur by a least-motion path involving an intermediate with topology XXIX. While this intermediate does appear to be strongly resonance stabilized, the other proposed steps in the reaction pathway do not have allowed valence structures. Recent work by Hoffmann[26] has shown that non-least-motion paths are probably far more common than had previously been thought. We suggest that this reaction may occur by such a non-least-motion pathway, though the attractive intermediate XXIX may still play a role.

In Table III, we give population analyses for a number of experimentally characterized boron hydrides. The charge distributions agree rather well with those inferred experimentally or obtained from molecular orbital calculations. The calculated overlap populations correlate considerably less well with the observed bond distances. One source of error in these calculations is our use of equal weights for the different nonequivalent valence structures. This practice has been justified picturesquely, if overcautiously, by reference to "the symmetry of our ignorance."[3b] In fact, as recognized by Simpson,[16] our ignorance is not totally symmetric. We know at least that structures which require a large amount of charge separation will be less favorable than those in which no atom deviates greatly from neutrality.

An attempt by Simpson, *et al.*,[16c] to introduce weighting factors dependent upon total charge separation gave rise to no significant changes in the population analyses. We show here, in a preliminary study, that weighting factors can be found which will make reasonable changes in the populations. However, since there are no firm theoretical grounds for any *particular* choice of weights and since the results obtained are not, at this point, demonstrably superior to the unweighted results, we present only a brief discussion and a single example.

We investigated weighting factors of the form $v_i = \exp(-Q_i/a)$ and $w_i = (b + Q_i)^{-1}$ where $Q_i = \Sigma_{j=1}{}^{p}(q_i{}^j)^2$ and $q_i{}^j$ denotes the net charge (in fractions of an electron) on boron atom j in the ith valence structure. For example, the weighted charge, $\bar{q}^j$, is then given, using exponential weights, by $\bar{q}^j = \Sigma_i v_i q_i{}^j / \Sigma v_i$. The parameters a and b were varied to see how drastic a weighting scheme was necessary to reorder charges and overlap populations. Ultimately a and b should be expressible as functions of the number of borons and the total charges. For the exponential case, the limit $a = \infty$ gives equal weights, while a approaching zero picks out the structure(s) with least charge separation. For the reciprocal weights, large values of b tend toward equal weights, while $b = 0$ gives a Q_i-dependent weighting scheme in which structures with all atoms neutral are selected, if such exist.

We find that the order of atomic charges is almost totally insensitive to the weighting scheme or lack of one. However, the order of overlap populations can be changed significantly by weights with low values of a or b. Surprisingly, the reciprocal weights w_i appear to be somewhat more useful than the exponential weights. Table IV shows for $B_{10}H_{14}$ (XI) the charges and overlap populations obtained from equal weights, from the

TABLE IV
WEIGHTED POPULATION ANALYSIS FOR $B_{10}H_{14}$

		Weights		
		Equal	v_i ($a \to 0$)	w_i ($b = 0.01$)
Atom		Weighted charges		
1		−0.08	0.00	−0.07
2		−0.28	−0.17	−0.27
5		0.04	0.00	0.04
6		0.28	0.17	0.28
Bond	Distance,[a] Å	Weighted overlap populations		
1–2	1.78	0.764	0.833	0.768
1–3	1.77	0.778	0.833	0.768
1–5	1.76	0.806	0.833	0.814
2–5	1.79	0.736	0.667	0.729
2–6	1.72	0.722	0.833	0.740
5–6	1.78	0.778	0.667	0.760
5–10	1.97	0.722	0.833	0.734

[a] A. Tippe and W. C. Hamilton, *Inorg. Chem.*, **8**, 464 (1969).

(21) N. N. Greenwood, H. J. Gysling, and J. A. McGinnety, *Chem. Commun.*, 505 (1970).
(22) J. S. Kasper, C. M. Lucht, and D. Harker, *Acta Crystallogr.*, **3**, 436 (1950).
(23) J. Reddy and W. N. Lipscomb, *J. Chem. Phys.*, **31**, 610 (1959).
(24) W. N. Lipscomb, *Inorg. Chem.*, **3**, 1683 (1964).
(25) See, *e.g.*, J. Dobson and R. Schaeffer, *ibid.*, **7**, 402 (1968).
(26) R. Hoffmann, to be submitted for publication.

minimum Q_t structures (v_t with $a \to 0$), and from w_t weights with $b = 0.1$.

V. Discussion

The methods described here are not by any means limited to boron hydrides. The computer-assisted design of complex organic syntheses investigated by Corey and Wipke[27] draws heavily on topological techniques. When combined with geometrical considerations[24] and chemical insights, topological methods can be extremely powerful in the prediction of new molecules and of stable intermediates. Extension of the theory to the more general class of electron-deficient compounds should be straightforward. The interesting case in which more than three centers participate in a single semilocalized orbital, for example, in certain intermetallic compounds, will be more difficult to handle computationally. Nonetheless, the principles remain the same: conservation of electrons and orbitals and the requirement that all pairs of neighbors participate in at least one common bond.

We have limited our calculations here to molecules with fewer than 12 borons, but we believe that no new principles would be illustrated by the inclusion of calculations on larger structures. Calculations on larger structures are still a strain on current research budgets, even at this level of approximation. We estimate that a topological study of $B_{20}H_{16}$[28] would require about 70 min on the IBM 360/65 with the present program, though we hope to lower this considerably by further programming modifications. Much information about larger boron hydrides may be gleaned from symmetry considerations applied to combinations of smaller ones, as in the characterization[3b] of $B_{18}H_{22}$ as two symmetry-related decaborane units. It is well known[17] that most, but not all,[16b,28] known stable boron hydrides possess at least a mirror plane and often considerably higher symmetry. A more general topological approach to the study of the boron hydrides would include a systematic, preferably computerized method for generating trial topologies. This question, which becomes crucial for the study of large molecules, is under investigation. One must always bear in mind, however, that purely topological arguments which disregard steric and other important factors are likely to lead to erroneous conclusions.

Finally, we wish to take note of the peculiar cyclic path which the theoretical study of boron hydride structures has taken. The first qualitative topological calculations[2,7] inspired semiempirical quantitative studies,[3b] which in turn gave rise to more exact, nonempirical[4] and finally SCF[5] calculations. The localized orbitals from the SCF wave functions have now suggested a refinement on the topological approach. Perhaps the cycle is starting again and a new set of more accurate and more revealing studies of the boron hydrides will spring from these modest beginnings.

Acknowledgments.—We thank Dr. P. G. Simpson for providing a listing of his topological program for comparison. This work was supported in part by the Office of Naval Research.

Appendix I

Hoffmann and Lipscomb[3b] have stated that specifying the two-center bonds (and open three-center bonds if allowed) in a molecule uniquely determines the position of the central three-center bonds and hence is sufficient to specify the valence structure of the molecule. This conjecture has never, to our knowledge, been challenged. We present here a simple, though hypothetical, counterexample.

Consider the ion $B_5H_{10}^+$, derived by protonation of the apex boron of the well-known species B_5H_9. Such an ion is by no means unlikely, since calculations[5b] show the apex boron to be considerably more negative than the basal borons. The topological diagram and numbering convention are given in Figure 4a. After

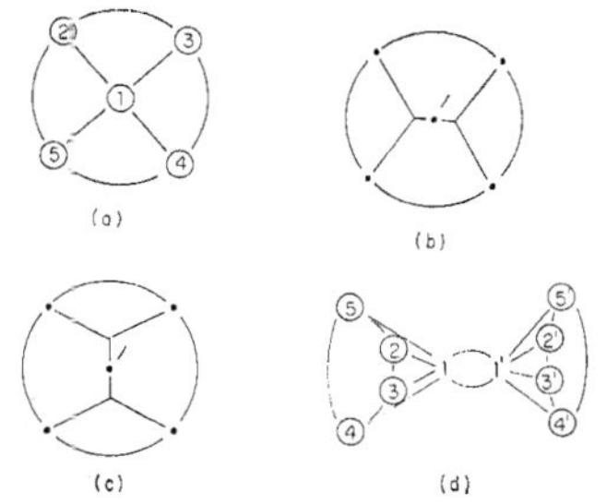

Figure 4.—$B_5H_{10}^+$ valence structures and a dimer

forming the required B–H and B–H–B bonds, we are left with four electrons and six orbitals for the boron framework. Both simple observation and our computer program show that the acceptable structures are those shown in Figure 4b and c, thus disproving the Hoffmann–Lipscomb conjecture.

We note that none of the other species studied gave such nonunique arrangements of three-center bonds. However, one can generate any number of such species by appropriately linking together $B_5H_{10}^+$ units. For example, Figure 4 shows a $B_{10}H_{20}^{2+}$ ion formed by linking two $B_5H_{16}^+$ units by a double hydrogen bridge. We note the similarity to the diborane structure and suggest the possibility of obtaining such an adduct from a B_2H_6–B_5H_9 mixture in a highly acid medium at low temperatures. An analogous B_4H_{10} type adduct may be envisioned by inserting two BH groups with a single bond between them in place of the hydrogens bridging borons 1 and 1′ in Figure 4d.

Appendix II

The closed polyhedral boranes, $B_nH_n^{l}$, have long been of great theoretical and experimental interest. In particular there has been much speculation about the source of the stability of the doubly negative ions of this series. We present in this section a simple topological argument as to the possible stable charge states for species.

The neutral molecules B_nH_n ($n > 4$) consisting of only B–H subunits have $2n$ electrons and $3n$ orbitals available for boron framework bonding. Thus their three-center valence structures must have n three-center and no two-center bonds. If we add two electrons to form $B_nH_n^{2-}$, we find that valence structures must have $n - 2$ three-center and 3 two-center bonds. If two electrons are removed to form $B_nH_n^{2+}$, we have $n - 1$ electron pairs and $3n$ orbitals. Thus, no suitable valence structures can be written for $B_nH_n^{2+}$ or for any

(27) E. J. Corey and W. T. Wipke, *Science*, **166**, 178 (1969).

(28) L. B. Friedman, R. E. Cook, and M. D. Glick, *J. Amer. Chem. Soc.*, **90**, 6862 (1968).

other B_nH_n cations employing only two- and three-center bonding. Since $3(n - 1) < 3n$, there are not enough electrons. Four- and higher-center bonds are required in order to use all available orbitals.

We may view the stability of the 2− ions as arising from a compromise between the destabilization caused by charge separation (which probably rules out 4− and 6− ions) and the added stabilization offered by the greater number of allowed valence structures. While the number of allowed valence structures mut be determined for each species by testing all possible structures for compliance with the topological rules, it seems likely that a greater number of *possible* structures will give rise to more *allowed* structures. For polyhedra in which each boron has three neighbors (*e.g.*, tetrahedron, trigonal bipyramid, octahedron) one can show that there are $3n - 6$ possible two-center and $3n - 8$ possible central three-center bonds. Thus B_nH_n will have $(3n - 8)!/[n!(2n - 8)!]$ possible structures arising from the different ways of choosing n objects from $3n - 8$. $B_nH_n^{2-}$ will have $[(3n - 8)!(3n - 6)!]/[(n - 2)!(2n - 6)!(3n - 9)!3!]$ where the additional terms arise from the possibility of choosing the two-center bonds in different ways. The ratio of the number of possible valence structures for $B_nH_n^{2-}$ to that for B_nH_n is thus $[(3n - 8)(3n - 7)(3n - 6)(n)\cdot(n - 1)]/[6(2n - 7)(2n - 6)]$, which increases from a minimum value of 120 at $n = 4$ and approaches $9n^3/8$ as n approaches infinity. Thus the 2− ions offer many more possible valence structures and, presumably, more allowable ones. Similar arguments may be given for polyhedra in which boron atoms have other numbers of nearest neighbors. Of course, the problem of symmetry degeneracy can spoil a resonance argument, and it does for tetrahedral B_4H_4, just as in organic chemistry it does for cyclobutadiene.

Chapter 10

Ada E. Yonath

Biography

Yonath, the founder of ribosomal crystallography, earned a Ph.D. degree from the Weizmann Institute of Science in 1968. She conducted her postdoctoral studies at Carnegie Mellon University and at the Massachusetts Institute of Technology (MIT). In 1970 she established the first protein crystallography laboratory in Israel, which remained the only laboratory of its kind in the country for almost a decade. Currently Yonath is a professor of structural biology at the Weizmann Institute. She is the incumbent of the Martin S. and Helen Kimmel Professorial Chair, and is the director of the Helen and Milton A. Kimmelman Center for Biomolecular Structure and Assembly. From 1998 to 2004 she was the director of the Mazer Center for Structural Biology, and from 1989 to 1994 she chaired the Departments of Structural Chemistry and Structural Biology at the Weizmann Institute. Between 1986 and 2004 she headed a Max Planck Research Unit in Hamburg, Germany, in parallel to her activities at the Weizmann Institute.

Yonath is a member of the National Academy of Sciences, the American Academy of Arts and Sciences, the Israel Academy of Sciences and Humanities, the European Academy of Sciences and Art, and the European Molecular Biology Organization. Additionally, she has honorary doctorates from Tel Aviv and Ben Gurion Universities in Israel as well as Oxford University, UK.

Her awards include the first European Crystallography Prize (2000); Certificate of Distinction from the US National Institutes of Health (2000); the Kilby International Award (2000); the most prestigious prize of the state of Israel, namely the Israel Prize in Chemistry (2002); the Harvey Prize from the Technion-Israel Institute of Technology (2002); the Cotton Medal of the US Chemical Society (2002); the Anfinsen Award of the International Protein Society (2003); the Medal of Distinction, Israeli Chemical Society (2003); Zurich University's Paul Karrer Gold Medal (2004); the University of Southern California's Massry Award and Medal for Ribosome Research (2004); the Datta Medal of the Federation of European Biochemical Societies (2005); the Fritz Lipmann Award of the German Society for Biochemistry and Molecular Biology (2005); the Louisa Gross Horwitz Prize of Columbia University in New York (2005), the Rothschild Prize for Life Sciences (2006); the

Israeli Prime Minister EMET Prize (2006); the Paul Ehrlich-Ludwig Darmstaedter Medal (2007); the Wolf Prize in Chemistry (2007); the Linus Pauling Gold Medal (2008); the UNESCO for Women in Science Prize, representing Europe (2008); the Palade Gold Medal (2008); and the "Albert Einstein" World Award of Science (2008). In 2009, Yonath shared the Nobel Prize in Chemistry with Thomas A. Steitz and Venkatraman Ramakrishnan.

In addition, she has served as a member of the Israeli National Supreme Committee for High Education (MALAG); the National Advisory Committee for Vision of Science, Israel; and the Israeli Academy Committees for Synchrotron Radiation.

William Lipscomb: My Scientific Inspiration

I was neither a student nor a postdoc of the Colonel. Yet, he inspired my research dramatically. My initial interaction with him was through attending his most illuminating course on protein structure back in the 1960s. Even then, when only a handful of structures were available, Bill delivered superb structural analysis, which was frequently "seasoned" by numerous sophisticated remarks, in-depth insights, humorous analogies, and countless philosophical comments. Furthermore, contrary to his image of being "geared up to fight," he treated the audience with unlimited patience, listened carefully to our questions, and answered our queries calmly and wisely. In the following years, when I met him at conferences, while visiting, or while serving on scientific committees, I felt extremely lucky to have the opportunity of benefiting from his guidance. In fact, many of my scientific decisions were motivated, directly or indirectly, by his opinion, experience, and advice.

I doubt that he is aware of his immense impact on my studies. Moreover, I guess that many scientists can classify themselves as Bill's spiritual children, as I do. Therefore I am using this opportunity to make sure that Bill realizes his fruitful influence on us. One paper describing research inspired by him follows.

Reports

A Tunnel in the Large Ribosomal Subunit Revealed by Three-Dimensional Image Reconstruction

A. YONATH, K. R. LEONARD, H. G. WITTMANN

A better understanding of the molecular mechanism of protein biosynthesis depends on the availability of a reliable model for the ribosome particle. The application of a diffraction technique, namely, three-dimensional image reconstruction from two-dimensional sheets of the large ribosomal subunits of *Bacillus stearothermophilus* at a resolution of 30 angstroms is described. The resulting three-dimensional model shows at least four projecting arms, arranged radially near the presumed interface with the 30*S* subunit. The projecting arms are positioned around a cleft, which turns into a tunnel with a length of 100 to 120 angstroms and a diameter of up to 25 angstroms. This tunnel spans the particle and may provide the path taken by the nascent polypeptide chain.

TO BETTER UNDERSTAND THE ROLE of ribosomes in protein biosynthesis, we have initiated structural studies of these particles by diffraction methods, and we have developed an in vitro crystallization procedure for intact ribosomal particles from bacterial sources. Three-dimensional crystals of 70*S* ribosomes from *Escherichia coli* (*1*) and of the large ribosomal subunits from *Halobacterium marismortui* (*2, 3*) and *Bacillus stearothermophilus* (*4–6*) diffracting x-rays to 6 and 13 Å, respectively, were obtained. From the latter species we were able to grow two types of well-ordered two-dimensional crystalline sheets: AL using alcohols (*7*) and ST using salt mixtures (*6–8*). These sheets are suitable for studies by image reconstruction.

Structure determination by three-dimensional image reconstruction from two-dimensional sheets is justified not only because it may yield significant results, but also because of its expected contribution to subsequent x-ray crystallographic analysis. A model obtained from image reconstruction may be used for gradual phasing of low-resolution crystallographic data. These phases, in turn, may help locate heavy atom sites for the isomorphous replacement technique.

Both AL and ST two-dimensional sheets consisted of small unit cells: $145 \pm 3 \times 311 \pm 14$ Å, $\gamma = 89° \pm 2.6°$ for the arrays grown in alcohol, and $148 \pm 10 \times 370 \pm 20$ Å, $\gamma = 109° \pm 3°$ for those grown from salts. These cell dimensions are close to those of forms 1 and 2 of our three-dimensional crystals (*6, 9*). The AL and ST sheets were well ordered, and optical diffraction patterns of electron micrographs of negatively stained specimens extended to 30 and 28 Å, respectively (Fig. 1). In both cases each unit cell contained two particles with dimensions similar to those obtained by other methods (*10*).

Because only a small fraction of the particles in the drop actually constitute the two-dimensional sheets, we could not separate them from the rest of the drop. Therefore, we tested the migration profile on sucrose gradients and the biological activity of the ribosomal particles in the crystallizing drops in the poly-U system (*11*). Particles subjected to crystallization conditions comigrated with standard particles and maintained their biological activity.

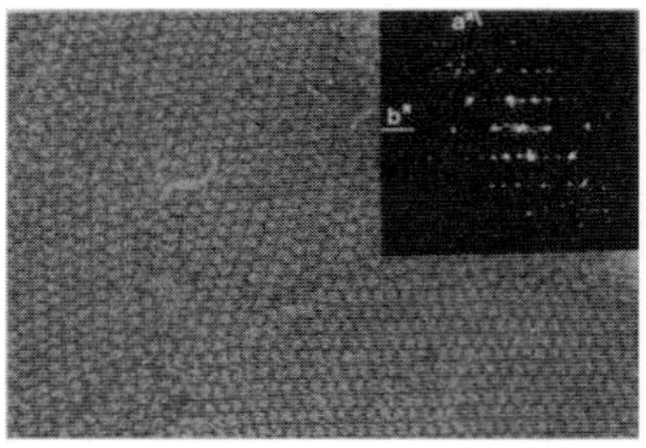

Fig. 1. A typical image of a salt-grown two-dimensional sheet stained by gold-thioglucose, and an optical diffraction pattern from a single crystalline domain containing about 20 × 10 unit cells. The a* and b* axes have been labeled. Areas for three-dimensional image reconstruction were chosen by using their optical diffractions according to the following criteria: (i) best resolution, (ii) patterns with sharpest reflections in patterns with no spurious spots, that is, including only one lattice region with the minimum mosaic spread. Only the strongest diffraction spots are shown. The others, although weak, contributed significantly to the resulting model, as shown in Fig. 5. Two-dimensional sheets were prepared as described (*7, 8*). Crystallization mixtures (7 to 10 μl) were applied to the plastic side of carbon-coated nitrocellulose grids and stained for 10 seconds with 1% (w/v) gold-thioglucose. Excess liquid was then removed by blotting. The grids were examined with a 400T Phillips electron microscope (operating at 80 kV) at an electron optical magnification of ×28,000.

We have used AL and ST sheets from three different preparations of ribosomes. We negatively stained them with either gold-thioglucose or uranyl acetate. Unlike uranyl acetate, which may interact chemically with the negatively charged components of the ribosomal particles [most likely the ribosomal RNA (rRNA)], gold-thioglucose is an inert stain that is expected to reveal the outer contour of the particles.

We have performed seven reconstructions, processed about 250 micrographs, and investigated 13 independently reconstructed particles. For each reconstruction we have used between two and four complete tilt series obtained from two-dimensional sheets. Two of these tilt series were perpendicular to each other. Between 20 and 25 crystallographically independent lattice lines, extending to a maximum resolution of 28 Å, were extracted for each reconstruction (Fig. 2).

We present here a three-dimensional model at 30 Å resolution (Fig. 3) reconstructed from three tilt series of crystalline sheets grown from salts and negatively stained with gold-thioglucose. This model is compared with those reconstructed from sheets stained with uranyl acetate.

The particles within the cells of the AL sheets were related by $p12_1$ symmetry (*7*). Although it appears that the ST sheets (Fig. 1) also contained crystallographic dimers, this was true only at lower resolution (45 Å). Three-dimensional image reconstructions were therefore carried out under the assumption of $p12_1$ symmetry for the AL crystals and $p1$ symmetry for the ST ones. In all reconstructions, the two particles in the unit cell showed the same essential features. However, in those reconstructed with $p1$ symmetry, generally one particle in the unit cell was resolved better than the other. This result may stem from preferential staining or differences in preservation.

The reconstructed particle (Fig. 3) has dimensions similar to those determined by other physical methods (*10*), and its volume (1.8×10^6 Å^3) corresponds to about 90% of the whole particle. On the basis of the known molecular weight of this particle (1.6×10^6) and of the volume obtained from the three-dimensional image reconstruction, the calculated density of a 50*S* particle is 1.3 to 1.4 g/cm^3, and the molecu-

A. Yonath, Weizmann Institute of Science, Rehovot, Israel, and Max Planck Research Unit for Structural Molecular Biology, 2000 Hamburg 52, FRG.
K. R. Leonard, European Molecular Biology Laboratory, 6900 Heidelberg, FRG.
H. G. Wittmann, Max-Planck-Institut für Molekulare Genetik, 1000 Berlin, FRG.

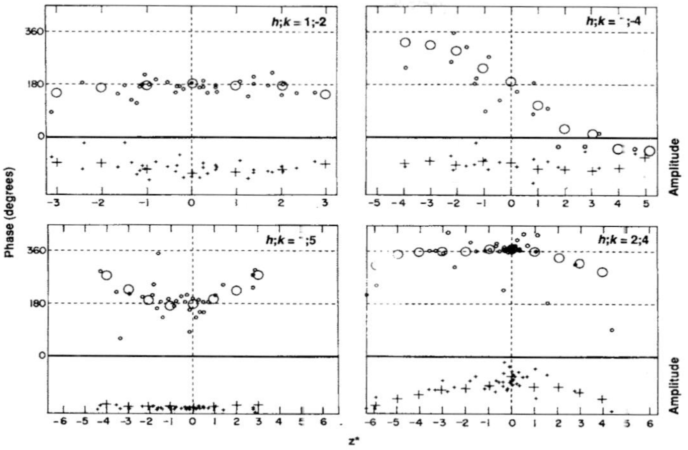

Fig. 2. Variation of phases along four typical lattice lines from which the three-dimensional model was synthesized. All amplitudes were adjusted to the same maximum value. Small circles represent the experimental values of the phases obtained by computer processing of the images, and the large ones are the interpolated values. z^* is the distance from the midpoint of the lattice line. Images were recorded at a dose of 10 electrons per square angstrom, of the specimens tilted at angles up to 55°, at intervals of 5°, with respect to the incident electron beam. A typical tilt series contained 23 images of which the first, middle, and last images were untilted. This control was essential for establishing that the specimen was not severely damaged by the electron beam during recording of the tilt series. No differences have been detected among the diffraction patterns of the untilted views taken at all stages of the experiment. The images were analyzed by standard methods (*22*) as described previously (*23*). All computational steps were carried out on a VAX 11/780. Ordered areas in the images were selected by optical diffraction and scanned with a microdensitometer (Optronics-1000) with a 25-μm raster, corresponding to 9 Å at the specimen. For Fourier transforms, optical densities of arrays (512 × 512) were used, and reciprocal lattices were fitted to the calculated diffraction patterns. Amplitudes and phases were extracted from peaks of at least 2 SD above background. Images of untilted sheets were used for the initial data in the three-dimensional analysis. The data were combined with a comparison range in z^* of 0.0016 $Å^{-1}$. The average phase error for each image based on a cumulative comparison of individual phases was 20° to 30°. Each lattice line was sampled at regular intervals of 1/300 $Å^{-1}$ along smooth curves fitted manually to the experimental values to provide terms for the Fourier synthesis. Terms along the 0,0 lattice line were not included. Fourier synthesis of 130 to 300 terms yielded the three-dimensional maps.

lar volume (*V*m) for a hypothetical crystal of a thickness of the sheets (160 to 170 Å) is 2.6 to 2.7 $Å^3$ per dalton. Both values are in good agreement with values tabulated by Matthews (*12*) and calculated for other large nucleoprotein structures (*13, 14*).

Several features were revealed by our analysis (Fig. 3). The concave surface of the particle consisted of several protrusions 25 to 30 Å in diameter. A long arm was located on the other side of the particle (bottom of the particle shown in Fig. 3A). Several projecting arms, two of which were longer than the others, were arranged radially around the edge (upper side of the particle shown in Fig. 3A). A narrow elongated cleft was formed between the projecting arms and turned into a tunnel up to 25 Å in diameter and 100 to 120 Å long (Fig. 3, B–D).

The tunnel was present in all reconstructions of the ST sheets independent of the staining material. Furthermore, indications of such a tunnel were detected in some filtered images of particular tilts (Fig. 4). In

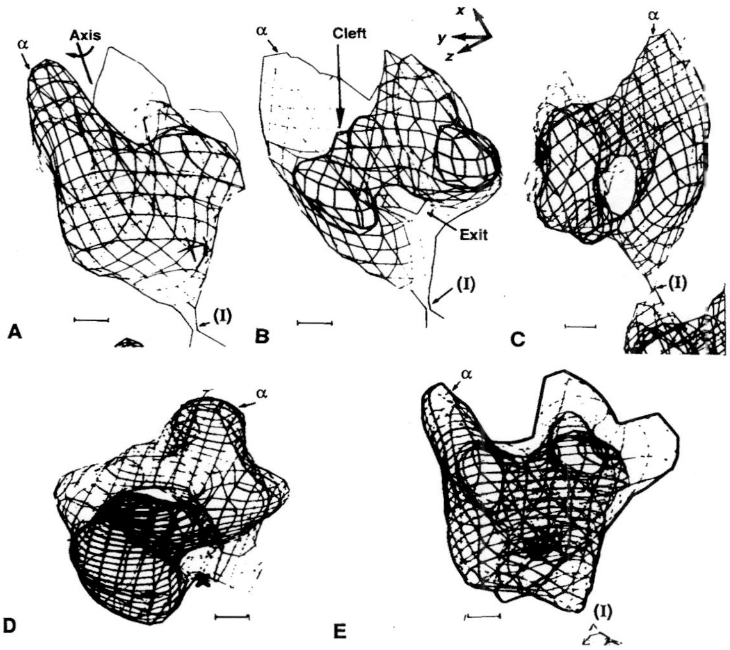

Fig. 3. Computer graphic display of the outline of the reconstructed model of the 50*S* ribosomal subunit at 30 Å resolution. The front is contoured with heavy lines, the back with lighter ones. Bar length, 20 Å. The asterisk marks the exit from the tunnel on the other side of the ribosome. Alpha (α) marks the longest projecting arm. The contour level for the particle boundary was chosen to be that at which the interparticle contacts could be resolved. For real-time display we used the computer program FRODO (*24*) combined with the GHC650 (written by G. H. Cohen, NIH) for Fourier synthesis, and FRODOMAP (written by J. L. Bhat, NIH) as well as the version, written by H. Bosshard, EMBL, Heidelberg. Views (A) to (C) were obtained from one reconstruction, and views (D) and (E) from another. (**A**) A side view of the model. The entire particle and part of a second one are shown. The arrow (I) points at the crystal contact between the two particles. (**B**) The model in (A) was rotated about the axis shown to generate the model in (B). (**C**) The model shown in (A and B) viewed into the branch of the tunnel from the exit point. (**D**) A view into the tunnel from the cleft. (**E**) The model viewed in a projection that resembles models derived from electron microscopy of single particles. The contour level is 2 to 5% higher than in previous views.

every reconstructed particle there was also a region of low density that branched off the tunnel at an angle of 60° to 70° to form a Y (or V) shape and terminated on the opposite side of the particle (Fig. 3C).

We have also reconstructed models from AL and ST sheets stained with uranyl acetate. Although the arrays used were large and well ordered, the diffraction patterns of their micrographs extended only to 32 and 35 Å, respectively. The interaction of uranyl acetate with the particles may depend on the accessibility of the stain and therefore be somewhat irregular. Consequently, the reconstructed model showed less detail. For the ST sheets, the essential features (the concave shape, the tunnel, and the projecting arms) could be resolved. In the reconstruction obtained from AL and ST crystals stained with uranyl acetate, there were two regions (on one of the long arms and on the body of the particle) where uranyl acetate, acting as a positive stain, was incorporated into the particle. In these regions the rRNA may be concentrated or easily exposed to the stain.

Several models for the 50*S* ribosomal subunits from *E. coli* have been suggested on the basis of electron microscopical visualization and averaging of single particles (*10*). Our model is more similar to those having no flat surfaces and to those in which the projecting arms are arranged radially around one side of the particle. Moreover, our model can be positioned so that its projected view resembles the usual image seen when single particles are investigated by electron microscopy (Fig. 3E). In addition, a few filtered images of two-dimensional sheets tilted by certain angles show the same shape and include the characteristic features that have been visualized by electron microscopy of single particles (Fig. 4).

However, there are also major discrepancies in the nature of the gross structural features between our model and the others, discrepancies that probably stem from the basic differences between visualization of isolated particles and crystalline sheets. There are a few preferred orientations in which the isolated particles tend to lie on grids. As a result of the contact with the flat grids, their projected views are likely to be somewhat distorted and their surfaces may appear flat rather than curved. In contrast, particles within the crystalline sheets are held together by interparticle contacts. These contacts construct a network that may stabilize the conformation of the particles and decrease, or even eliminate, the influence of the flatness of the grids. Thus, the particles within ordered arrays provide a better starting point for reconstruction studies.

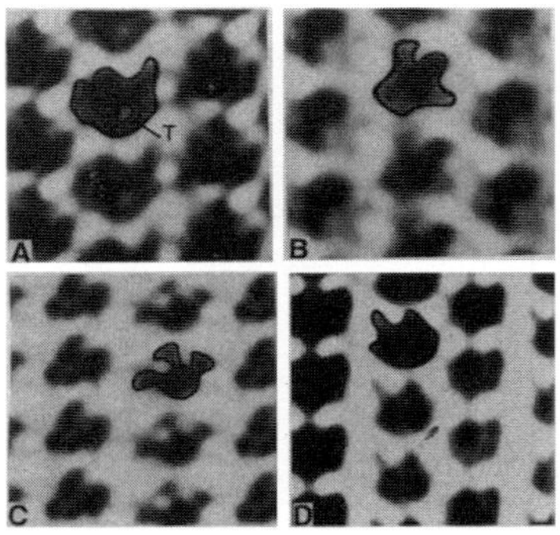

Fig. 4. Filtered images of electron micrographs in which the depicted view resembles that derived from electron microscopy of single particles. The particles are black and the background is white. (**A–C**) Filtered images of type (ST) sheets at 30°, 20°, and 35° tilt angles, respectively. The tunnel (T) can be seen. (**D**) Filtered image of type (AL) sheets at a tilt angle of 25°.

To understand why some of the features revealed in our studies have not been detected in previous investigations, we have reconstructed a model selecting from the entire set of diffraction data only the low-resolution (55 Å) terms. In the resulting model, in spite of the low resolution, the tunnel is clearly resolved. However, at this resolution the particle is almost spherical and shows only two thick and short arms instead of the elongated ones resolved in the 30 Å resolution studies (Fig. 5). Indications that at low resolution the particle may appear spherical could be detected in the low-resolution regions of several individual lattice lines (that is, $h;k = 1;-2$ or $2;-4$; Fig. 2) in which there is very little variation with tilting in amplitude or phase.

A recent model of whole ribosomes from chick embryos reconstructed at a resolution of 55 Å assigned a low-density region as a channel (*15*). However, the objects of these studies, 80*S* ribosomes, are significantly larger than ours and pack in larger unit cells (593 by 593 Å versus 152 by 350 Å). For these reasons and because of the relatively low resolution of the diffraction patterns of the two-dimensional sheets of the 80*S*, the details such as branching of the tunnel could not be revealed.

We have also reconstructed models from partial data sets (45 to 50% of the whole diffracting sphere at 30 Å) and observed that for the ST sheets these models show the main structural features: the tunnel, the concave shape, and the projecting arms. However, the partial data models appear as round objects with short arms, in contrast to the elongated oval shape of the entire particle obtained from the higher resolution set. In view of these results, it is not clear why the tunnel has not been detected in studies of negatively stained two-dimensional sheets of 50*S* subunits performed at a resolution of 40 Å (*16*).

There is a significant shortening of the long projecting arms in models reconstructed from crystals grown from alcohols. This may indicate that the arms that are positioned firmly in an environment of salts may flex when exposed to alcohols. Inherent flexibility of the long arms has also been

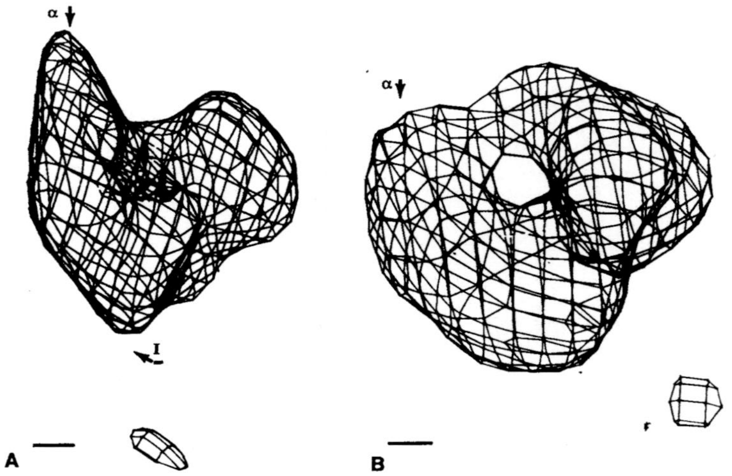

Fig. 5. The model obtained from three-dimensional image reconstruction at 55 Å resolution (part of the data used for Fig. 3A). (**A**) The view is from approximately the same direction as in Fig. 3B. Part of the second particle is shown and marked by I. (**B**) A view into the tunnel from the cleft. Bar length, 20 Å.

observed in reconstruction of rotated single particles (*17*).

The functional significance of the tunnel is still to be determined. However, it is possible that the tunnel could provide the path taken by the nascent polypeptide chain since (i) the tunnel originates at the projecting arms lying at the presumed site for protein biosynthesis (*10*) and it terminates on the other end of the particle, and (ii) it is of a diameter large enough to accommodate even the largest amino acids. Furthermore, this tunnel is long enough to accommodate and protect from proteolytic enzymes a peptide of about 40 amino acids in an extended conformation (*18–20*). Although the tunnel terminates at a location that may be compatible with that assigned by immune electron microscopy as the exit site for the growing polypeptide chain (*21*), it remains to be seen whether these two sites are identical.

Assignment of the known functional domains of the 50*S* ribosomal subunits to the various structural features of our model awaits further investigation. We hope to locate specific sites on a detailed model.

REFERENCES AND NOTES

1. H. G. Wittmann *et al.*, *FEBS Lett.* **146**, 217 (1982).
2. M. Shoham *et al.*, *ibid.* **208**, 321 (1986).
3. I. Makowski, F. Frolow, M. A. Saper, H. G. Wittmann, Y. Yonath, *J. Mol. Biol.* **193**, 819 (1987).
4. A. Yonath, H. D. Bartunik, K. S. Bartels, H. G. Wittmann, *ibid.* **177**, 201 (1984).
5. A. Yonath *et al.*, *ibid.* **187**, 633 (1986).
6. A. Yonath, M. A. Saper, H. G. Wittmann, in *Structure, Function and Genetics of Ribosomes*, B. Hardesty and G. Kramer, Eds. (Springer-Verlag, Heidelberg, 1986), pp. 112–127.
7. T. Arad, K. R. Leonard, H. G. Wittmann, A. Yonath, *EMBO J.* **3**, 127 (1984).
8. J. Piefke, T. Arad, H. S. Gewitz, A. Yonath, H. G. Wittmann, *FEBS Lett.* **209**, 104 (1986).
9. A. Yonath, *Trends Biochem. Sci.* **9**, 227 (1984).
10. H. G. Wittmann, *Annu. Rev. Biochem.* **52**, 35 (1983).
11. K. H. Nierhaus, K. Bordasch, H. E. Homann, *J. Mol. Biol.* **74**, 584 (1973).
12. B. W. Matthews, *ibid.* **33**, 491 (1968).
13. J. M. Hogle, *ibid.* **160**, 663 (1982).
14. T. Richmond, J. T. Finch, B. Rushton, D. Rhodes, A. Klug, *Nature (London)* **311**, 533 (1984).
15. R. A. Milligan and P. N. T. Unwin, *ibid.* **319**, 693 (1986).
16. M. W. Clark, K. Leonard, J. A. Lake, *Science* **216**, 999 (1982).
17. A. Verschoor, J. Frank, M. Boublik, *J. Ultrastruct. Res.* **92**, 180 (1985).
18. L. I. Malkin and A. Rich, *J. Mol. Biol.* **26**, 329 (1967).
19. G. Blobel and D. D. Sabatini, *J. Cell. Biol.* **45**, 130 (1970).
20. W. P. Smith, P. C. Tai, B. D. Davis, *Proc. Natl. Acad. Sci. U.S.A.* **75**, 5922 (1978).
21. C. Bernabeau and J. A. Lake, *ibid.* **79**, 3111 (1982).
22. L. A. Amos, R. Henderson, P. N. T. Unwin, *Progr. Biophys. Mol. Biol.* **39**, 183 (1982).
23. K. R. Leonard, P. Wingfield, T. Arad, H. Weiss, *J. Mol. Biol.* **149**, 259 (1981).
24. A. Jones, *J. Appl. Crystallogr.* **11**, 268 (1978).
25. We thank J. Piefke and T. Arad for conducting the electron microscopy experiments; J. L. Sussman for adapting the computer display programs; F. L. Hirshfeld for his interest and critical comments; B. Shaanan, H. Bosshard, M. A. Saper, and A. Levy for assistance with computer problems; and H. S. Gewitz, J. Müssig, B. Hennemann, and Y. Halfon for skillful technical assistance. Supported by NIH (GM34360), BMFT (05180 MPBO), and Minerva research grants.

12 September 1986; accepted 20 February 1987

Allelic Exclusion in Transgenic Mice That Express the Membrane Form of Immunoglobulin μ

MICHEL C. NUSSENZWEIG, ALBERT C. SHAW, ERIC SINN, DAVID B. DANNER, KEVIN L. HOLMES, HERBERT C. MORSE III, PHILIP LEDER

Antibody-producing cells display a special form of regulation whereby each cell produces immunoglobulin from only one of its two sets of antibody genes. This phenomenon, called allelic exclusion, is thought to be mediated by the product of one heavy chain allele restricting the expression of the other. Heavy chains are synthesized in two molecular forms, secreted and membrane bound. In order to determine whether it is specifically the membrane-bound form of the immunoglobulin M (IgM) heavy chain (μ) that mediates this regulation, transgenic mice were created that carry a human μ chain gene altered so that it can only direct the synthesis of the membrane-bound protein. The membrane-bound form of the human μ chain was made by most of the B cells in these animals as measured by assays of messenger RNA and surface immunoglobulins. Further, the many B cells that express the human gene do not express endogenous mouse IgM, and the few B cells that express endogenous mouse μ do not express the transgene. Thus, the membrane-bound form of the μ chain is sufficient to mediate allelic exclusion. In addition, the molecular structures recognized for this purpose are conserved between human and mouse systems.

REGULATION OF IMMUNOGLOBULIN gene expression involves genomic rearrangements at a minimum of two loci that bring regulatory and coding sequences into proximity (*1*). In addition, differential cleavage and polyadenylation of messenger RNA (mRNA) precursors generate either membrane or secreted forms of immunoglobulin heavy chains (*2–4*). Continued genomic rearrangements and splicing further generate a series of heavy chain classes (*5*). Although these changes often involve both parental sets of antibody genes, only one eventually directs the synthesis of a functional immunoglobulin in a given cell. This mechanism, allelic exclusion, is central to the process of clonal selection (*6*).

There is persuasive evidence that allelic exclusion at the light chain loci is mediated by intact immunoglobulin molecules (*7, 8*). In addition, experiments with transgenic mice suggest that expression of endogenous heavy chain alleles can be influenced by an introduced murine μ chain transgene (*9–11*). Evidence thus far indicates that the human γ-1 gene (*12*) and a version of the mouse μ gene that gives rise to a secreted form of the mouse μ chain (*11*) do not mediate this process. Since immunoglobulin M (IgM) heavy chains are produced in two forms, membrane-bound and secreted, it is reasonable to suggest that such a regulatory signal is mediated by a membrane-bound molecule (*11*) because the secreted form is sequestered in the vacuolar system. To test this theory we created transgenic mice (*13*) that carry a rearranged human μ gene altered to delete the cleavage and polyadenylation signals necessary to synthesize the secreted form of the μ heavy chain (*3*). The mutant gene (Fig. 1A) is capable of producing only the membrane-bound form of the molecule in transfected tissue culture cells.

Expression of the human transgene and the endogenous mouse μ mRNAs was assessed by ribonuclease protection experiments with probes that are species specific and can distinguish between the mRNAs encoding membrane-bound and secreted forms of the μ chain (Fig. 1). Spleen cells derived from two transgenic lines, TG SA and TG SD, were used for this analysis. The two lines expressed the membrane-bound form of the human gene but differed in the amount of expression (Fig. 1B). Densitometric analysis of data from these experiments shows that spleen cells from mice of the TG SA line produce 25 to 35

M. C. Nussenzweig, A. C. Shaw, E. Sinn, P. Leder, Department of Genetics, Harvard Medical School, and Howard Hughes Medical Institute, Boston, MA 02115.
D. B. Danner, Laboratory of Molecular Genetics, National Institute on Aging, Baltimore, MD 21224.
K. L. Holmes and H. C. Morse III, Laboratory of Immunopathology, National Institute of Allergy and Infectious Diseases, National Institutes of Health, Bethesda, MD 20892.

articles

Structural basis for the interaction of antibiotics with the peptidyl transferase centre in eubacteria

Frank Schlünzen*†, Raz Zarivach†‡, Jörg Harms*†, Anat Bashan‡, Ante Tocilj*§, Renate Albrecht§, Ada Yonath*‡ & François Franceschi§

* *Max-Planck-Research, Unit for Ribosomal Structure, Notkestrasse 85, 22603 Hamburg, Germany*
‡ *Department of Structural Biology, Weizmann Institute, 76100 Rehovot, Israel*
§ *Max-Planck-Institut für Molekulare Genetik, Ihnestrasse 73, 14195 Berlin, Germany*
† *These authors contributed equally to this work*

Ribosomes, the site of protein synthesis, are a major target for natural and synthetic antibiotics. Detailed knowledge of antibiotic binding sites is central to understanding the mechanisms of drug action. Conversely, drugs are excellent tools for studying the ribosome function. To elucidate the structural basis of ribosome–antibiotic interactions, we determined the high-resolution X-ray structures of the 50S ribosomal subunit of the eubacterium *Deinococcus radiodurans*, complexed with the clinically relevant antibiotics chloramphenicol, clindamycin and the three macrolides erythromycin, clarithromycin and roxithromycin. We found that antibiotic binding sites are composed exclusively of segments of 23S ribosomal RNA at the peptidyl transferase cavity and do not involve any interaction of the drugs with ribosomal proteins. Here we report the details fo antibiotic interactions with the components of their binding sites. Our results also show the importance of putative Mg^{+2} ions for the binding of some drugs. This structural analysis should facilitate rational drug design.

The high-resolution crystallographic structures of the ribosomal subunits revealed many of the molecular details of ribosomes, including the decoding centre at the 30S subunit and the peptidyl transferase centre at the 50S subunit[1–6]. The exact mechanism for the peptidyl transferase activity, however, is still unclear[2,7–9]. The peptidyl transferase centre is the main target site for many antibiotics and substrate analogues, such as puromycin[10].

The therapeutic use of antibiotics has been severely compromised by the emergence of drug resistance in many pathogenic bacteria. The molecular mechanisms by which bacteria become resistant usually involve drug efflux, drug inactivation, or alterations in the antibiotic target site. Although ribosomal proteins can affect the binding and action of ribosome-targeted antibiotics, the primary target of these antibiotics is rRNA[11]. Many cases of antibiotic resistance in clinical strains can be linked to alterations of specific nucleotides of the 23S rRNA within the peptidyl transferase centre[12]. However, to date the exact mode of interaction and the chemical moieties involved in binding have not been determined. The structural basis for this antibiotic action is also unknown.

Complexes of ribosomal subunits with antibiotics are essential to understand the molecular mechanisms of antibiotic binding, antibiotic resistance, and for rational drug design. Conversely, antibiotics have been extremely useful to elucidate functional aspects of protein biosynthesis and to define the structural components involved in ribosome function.

To determine ribosome–antibiotic interactions in eubacteria and to understand better the mechanism of the peptidyl transferase activity, we determined the structure of the 50S subunit of *D. radiodurans* complexed with each of the following antibiotics: chloramphenicol, clindamycin (a lincosamide), and the macrolides erythromycin, clarithromycin and roxithromycin. Analysis of the obtained X-ray structures allowed us to propose the structural principles for the action of the antibiotics examined.

Results

The 3.1 Å structure model of the 50S subunit of *D. radiodurans*[13] was used as a reference to determine the structures of the 50S–antibiotic complexes (Table 1). The electron density maps of the 50S–antibiotic complexes allowed us to unambiguously determine the binding sites of the five antibiotics. The chemical nature of the 50S–antibiotic interactions could be deduced largely from the mode of binding. On the basis of our difference electron density maps and on the available biochemical and functional data, we

Table 1 Crystallographic data

Parameter	Chloramphenicol	Clindamycin	Erythromycin*	Clarithromycin*	Roxithromycin*
Resolution (Å)	25–(3.62–3.5)	35–(3.21–3.1)	20–(3.62–3.5)	50–(3.62–3.5)	50–(3.87–3.8)
R_{sym} (%)	12.4 (57.7)	11.8 (57.0)	12.2 (45.8)	9.8 (50.6)	10.7 (24.4)
Completeness (%)	88.5 (71.0)	94.4 (82.2)	79.5 (71.4)	85.2 (78.1)	64.8 (66.8)
Redundancy	3.9	3.8	5.4	4.9	3.7
$\langle I/\sigma(I)\rangle$	9.5 (2.1)	6.5 (2.1)	7.2 (1.8)	8.1 (2.1)	6.2 (2.2)
Unit cell (Å)	171.1 × 409.3 × 696.9	170.3 × 410.1 × 697.2	169.2 × 410.0 × 695.0	169.9 × 412.7 × 697.0	170.4 × 410.7 × 694.8
R/R_{free} (%)	27.5/32.1	26.8/30.3	26.8/30.1	27.3/32.3	20.9/27.4
Bond length (Å)†	0.0038	0.0050	0.0037	0.0037	0.0035
Bond angles (°)	0.69	0.77	0.66	0.65	0.67

Values in parentheses correspond to the high-resolution bin.
* Macrolide antibiotics.
† r.m.s. deviations for bond length and bond angles. The estimated coordinate error ranges from ±0.45 Å for clindamycin to ±0.65 Å for roxithromycin.

propose the structural basis for the modes of action of the antibiotics chloramphenicol, clindamycin, erythromycin, clarithromycin and roxithromycin.

Each of the antibiotics that we analysed targets the 50S subunit only at the peptidyl transferase cavity, although the binding site from each class of antibiotic differs from each other. Each class of antibiotic interacts exclusively with specific nucleotides assigned to a multi-branched loop of domain V of the 23S rRNA in the two-dimensional structure (Figs 1a, 2a and 3a), explaining previous mutational and footprinting data[10,14,15]. None of the antibiotics show any direct interaction with ribosomal proteins. Binding of these antibiotics did not result in any significant conformational change of the peptidyl transferase cavity.

Chloramphenicol

Chloramphenicol is known to block peptidyl transferase activity by hampering the binding of transfer RNA to the A site[16,17]. Chloramphenicol has several reactive groups that can form hydrogen bonds with various nucleotides of the peptidyl transferase cavity: two oxygens of the para-nitro (p-NO_2) group, the 1OH group, the 3OH group and the 4′ carboxyl group.

One of the oxygens of the p-NO_2 group of chloramphenicol appears to form hydrogen bonds with N4 of C2431Dr (C2452Ec) (see Methods for definition), which has been shown to be involved in chloramphenicol resistance[14]. The other oxygen of the p-NO_2 group interacts with O2′ of U2483Dr (U2504Ec) (Fig. 1a–c).

The 1OH group of chloramphenicol is located at hydrogen-bonding distance (less than 4 Å) to N2 of G2044Dr (G2061Ec) of the 23S rRNA. This nucleotide has been implicated in chloramphenicol resistance in rat mitochondria[14]. A mutation of the neighbouring nucleotide, A2062Ec, confers resistance to chloramphenicol in *Halobacterium halobium*[18].

The 3OH group of chloramphenicol is fundamental for its activity, and *in vivo* modification of this group confers resistance to the drug[19,20]. The 3OH group is within hydrogen-bonding distance to 4′O of U2485Dr (U2506Ec) and to O2′ of G2484 (G2505Ec). The 3OH group of chloramphenicol is also involved in interactions coordinated through a putative Mg ion (Mg-C1, see

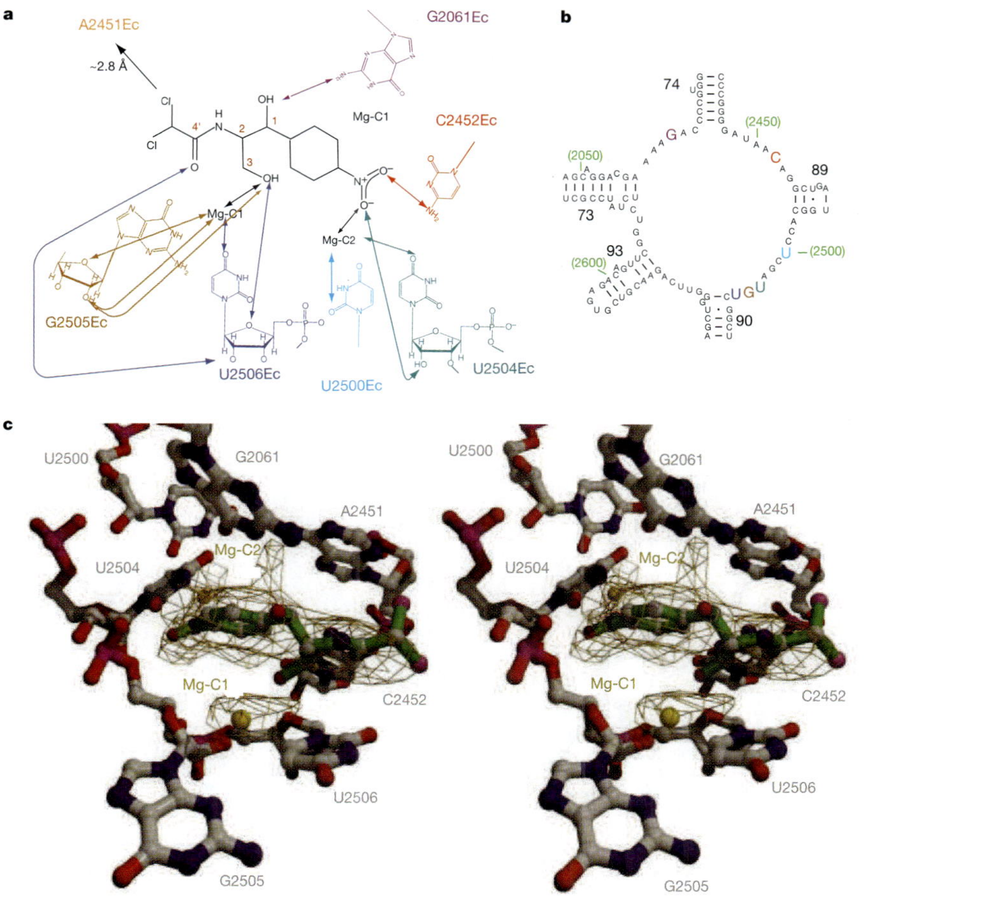

Figure 1 Interaction of chloramphenicol with the peptidyl transferase cavity. **a**, Chemical structure diagram of chloramphenicol showing the interactions (arrows) of its reactive groups with the nucleotides of the peptidyl transferase cavity. Arrows between two chemical moieties indicate that the two groups are less than 4.4 Å apart. **b**, Secondary structure of the peptidyl transferase ring of *D. radiodurans* showing the nucleotides involved in the interaction with chloramphenicol (coloured nucleotides). The colours in the secondary structure diagram match those of the chemical diagram. **c**, Stereo view showing the nucleotides interacting with chloramphenicol at the peptidyl transferase cavity of *D. radiodurans*. The difference electron density map ($2F_o - F_c$) is contoured at 1.2σ. The antibiotic is shown in green. Nucleotide numbering is according to the *E. coli* sequence. Putative Mg ions (Mg-C1, Mg-C2) are indicated.

below). The 4′ carbonyl group of chloramphenicol appears to form a hydrogen bond with the 2′OH of U2485Dr (U2506Ec) (Fig. 1a).

Mg^{2+} antibiotic interactions

In addition to the interactions between chloramphenicol and RNA nucleotides, two ions (Mg-C1 and Mg-C2) are involved in chloramphenicol binding. We propose these ions to be hydrated Mg^{2+}. However, at the present resolution we cannot exclude the possibility of the ions being either K^+ or NH_4^+. These ions are not present in the native 50S structure or in the 50S complexes with the other antibiotics that we analysed. Thus, their presence at these particular locations depends on chloramphenicol binding.

Mg-C1 mediates the interaction of the 3OH group of chloramphenicol with O4 of U2485Dr (U2506Ec) and with the 2′OH and 4′O of G2484Dr (G2505Ec). Both of these nucleotides have previously been reported as being protected by chloramphenicol[16]. Mg-C2 mediates the interaction of one of the oxygens of the p-NO_2 group with O2 of U2479Dr (U2500Ec) and O4 of U2483Dr (U2504Ec). Mg-C1- and Mg-C2-mediated interactions further stabilize the interaction of chloramphenicol with the peptidyl transferase cavity. The presence of these ions seems crucial for the interaction of chloramphenicol with the ribosome, and thus for the inhibition mechanism of this drug.

Notably, the putative Mg^{2+} (Mg101) found overlapping with the chloramphenicol location in the native structure is not observed in the chloramphenicol–50S complex, suggesting that the coordinating effect of chloramphenicol is sufficient to maintain the local structure of the 50S subunit in the absence of Mg101. The involvement of metal ions in chloramphenicol binding can partially explain why chloramphenicol, in spite of being a small molecule, has been shown by chemical footprinting to protect certain nucleotides on the peptidyl transferase ring from chemical attack, or to enhance the chemical reactivity of other nucleotides in the same region[14]. Furthermore, generation of new metal-ion-binding sites owing to antibiotic binding, as we observed for chloramphenicol, may explain the mode of action and/or binding of other drugs as well, and could be used as a tool in rational drug design.

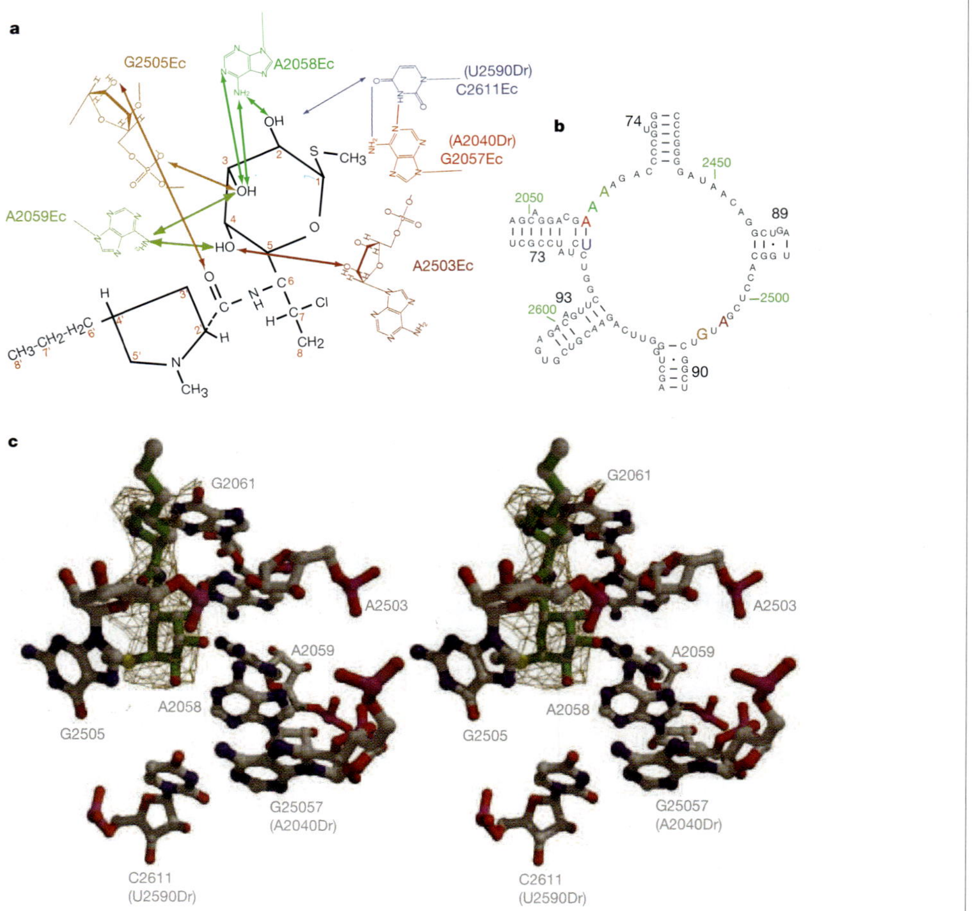

Figure 2 Interaction of clindamycin with the peptidyl transferase cavity. **a**, Chemical structure diagram of clindamycin showing the interactions (arrows) of its reactive groups with the nucleotides of the peptidyl transferase cavity. Arrows between two chemical moieties indicate that the two groups are less than 4.4 Å apart. **b**, Secondary structure of the peptidyl transferase ring of *D. radiodurans* showing the nucleotides involved in the interaction with clindamycin (coloured nucleotides). The colours in the secondary structure diagram match those of the chemical diagram. **c**, Stereo view showing the nucleotides interacting with clindamycin at the peptidyl transferase cavity of *D. radiodurans*. The difference electron density map ($2F_o - F_c$) is contoured at 1.2σ. The antibiotic is shown in green. Nucleotide numbering is according to the *E. coli* sequence.

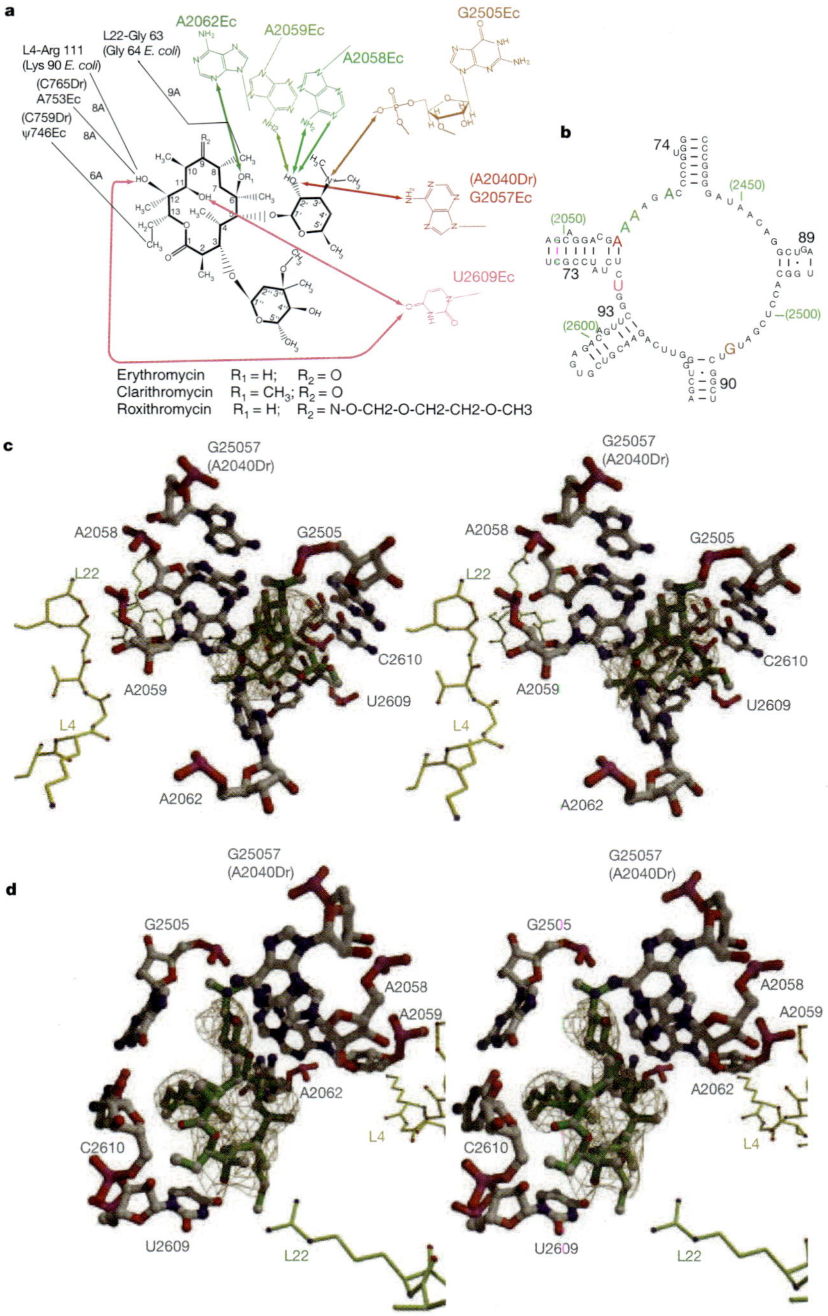

Figure 3 Interaction of macrolides with the peptidyl transferase cavity. **a**, Chemical structure diagram of the macrolides (erythromycin, clarithromycin and roxithromycin) showing the interactions (coloured arrows) of the reactive groups of the macrolides with the nucleotides of the peptidyl transferase cavity (coloured). Coloured arrows between two chemical moieties indicate that the two groups are less than 4.4 Å apart. Distance of macrolide moieties to groups implicated previously in antibiotic interaction (L4, L22 and domain II of the 23S rRNA) are shown. **b**, Secondary structure of the peptidyl transferase ring of *D. radiodurans* showing the nucleotides involved in the interaction with clindamycin (coloured nucleotides). The colours in the secondary structure diagram match those of the chemical diagram. **c**, Stereo view showing the erythromycin binding site at the peptidyl transferase cavity of *D. radiodurans*. The stereo view for clarithromycin is identical to that of erythromycin. **d**, Stereo view showing the roxithromycin-binding site at the peptidyl transferase cavity of *D. radiodurans*. For **c** and **d** the difference electron density map ($2F_o - F_c$) is contoured at 1.2σ. Green, antibiotic; yellow, part of ribosomal protein L4; light green, part of ribosomal protein L22. Nucleotides that interact with the antibiotic are shown with their chemical structure. Nucleotide numbering is according to the *E. coli* sequence.

Clindamycin

In contrast to chloramphenicol, lincosamides interact with the A and P site[21]. Although the binding site for the lincosamide clindamycin differs from that of chloramphenicol at the peptidyl transferase cavity the two sites appear to be partially overlapping (Figs 2a–c and 4). We did not localize any new metal ions involved in the binding of clindamycin; however, Mg101 was also displaced on clindamycin binding.

Clindamycin has three hydroxyl groups in its sugar moiety that can participate in hydrogen-bond formation (Fig. 2a–c). The 2OH of clindamycin appears to form a hydrogen bond with N6 of A2041Dr (A2058Ec), the pivotal nucleotide for lincosamide binding[22]. Although the 2OH group is also less than 4.0 Å away from O6 of A2590Dr (G2611Ec), additional hydrogen bonds to this nucleotide is unlikely, because mutations of U2590Dr (C2611Ec) appear to alter only the conformation of the 23S rRNA and thus affect the position of nucleotide A2041Dr (G2058Ec)[23] (Fig. 2c) (see below also).

The 3OH group can interact with N6 of nucleotide A2041Dr (A2058Ec) and with the non-bridging phosphate-oxygens of G2484Dr (G2505Ec). Thus, N6 of A2041Dr (A2058Ec) can interact with both the 2OH group and the 3OH group of clindamycin. These structural data explain in the most straightforward way why A2041Dr (A2058Ec) mutations confer resistance to clindamycin. The hydrogen bond with N6 of A2041Dr (A2058Ec) can also explain why the dimethylation of the N6 group, which disrupts the hydrogen bonds, causes resistance to lincosamides[24]. The 3OH group of clindamycin could additionally interact with N1 of A2041Dr (A2058Ec) and N6 of A2042 (A2059Ec).

The 4OH group of clindamycin could form a hydrogen bond with N6 of A2042 (A2059Ec). This interaction explains why mutations in A2042 (A2059Ec) cause clindamycin resistance in several bacterial pathogens[24]. In addition to the sugar-mediated interactions, the carbonyl group of clindamycin could form a hydrogen bond with 2′OH of G2484 (G2505Ec).

The A- and P-site character of clindamycin may be explained by the location of its proline moiety and the interaction of its 8′ carbon and its sulphur atom with the 23S rRNA. The location of the proline moiety of bound clindamycin partially overlaps that of the phenyl moiety of chloramphenicol. The proline moiety is on average 3.5 Å away from the tyrosil moiety of the puromycin-binding site described by ref. 2, thus giving an explanation to the A-site character of clindamycin. The 8′ carbon of clindamycin is located 2.5 Å away from the N3 of C2431Dr (C2452Ec), a nucleotide shown to be involved in P-site tRNA binding[25]. This location of the 8′ carbon may explain the P-site character of clindamycin. The sulphur atom of clindamycin is located around 3 Å from base G2484Dr (G2505Ec), which has been photo-crosslinked to the tRNAs bound to the A and P sites[25]. Although the two latter clindamycin groups cannot form hydrogen bonds, interactions, such as van der Waals or hydrophobic interactions, between these groups and nucleotides of the 23S rRNA may be expected.

Macrolides

In contrast to chloramphenicol and the lincosamides, macrolides of the erythromycin class do not block peptidyl transferase activity[26]. Although they bind to the peptidyl transferase ring, the erythromycin group of the macrolides, which includes clarithromycin and roxithromycin, is thought to block the tunnel that channels the nascent peptides away from the peptidyl transferase centre[2,27,28]. Our results confirm this assumption. We could unambiguously determine that the macrolides erythromycin, clarithromycin and roxithromycin all bind to the same site in the 50S subunit of *D. radiodurans*, at the entrance of the tunnel (Fig. 5). Their binding contacts clearly differ from those of chloramphenicol, but overlap those of clindamycin to a large extent (Figs 3a and 4).

Most of the 14-member-ring macrolides, which include erythromycin and its related compounds, have three structural components: the lactone ring, the desosamine sugar, and the cladinose sugar. The reactive groups of the desosamine sugar and the lactone ring mediate all the hydrogen-bond interactions of erythromycin and its second generation derivatives clarithromycin[29] and roxithromycin[30] with the peptidyl transferase cavity.

The 2′OH group of the desosamine sugar appears to form hydrogen bonds at three positions: N6 and N1 of A2041Dr (A2058Ec) and N6 of A2042Dr (A2059Ec). The hydrogen bonds between the 2′OH and N1 and N6 of A2041Dr (A2058Ec) explain why this nucleotide is essential for macrolide binding, thus shedding light on the two most common ribosomal resistance mechanisms against macrolides: the N6 dimethylation of A2041Dr (A2058Ec) by the erythromycin resistance family of methylases[31], and rRNA mutations changing nucleotide identity at this position[32]. The dimethylation of the N6 group would not only add a bulky substituent causing steric hindrance for the binding, but will prevent the formation of hydrogen bonds to the 2′OH group. The

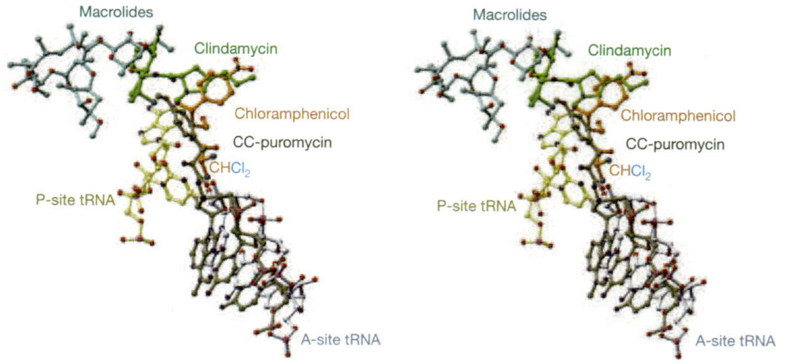

Figure 4 Relative position of chloramphenicol, clindamycin and macrolides with respect to CC-puromycin and the 3′-cytosine-adenine (CA) end of P-site and A-site tRNAs. The location of CC-puromycin was obtained by docking the position reported by ref. 2 into the peptidyl transferase centre of *D. radiodurans*. The location of the 3′-CA end of P- and A-site tRNAs was obtained by docking the position reported by ref. 42 into the peptidyl transferase centre of *D. radiodurans*. Light blue, 3′-CA end of A-site tRNA; light yellow, 3′-CA end of P-site tRNA; grey, puromycin; gold, chloramphenicol; green, clindamycin; cyan, macrolides (erythromycin). Oxygen atoms are shown in red and nitrogen atoms in dark blue. $CHCl_2$ indicates the location of the dichloromethyl moiety of chloramphenicol.

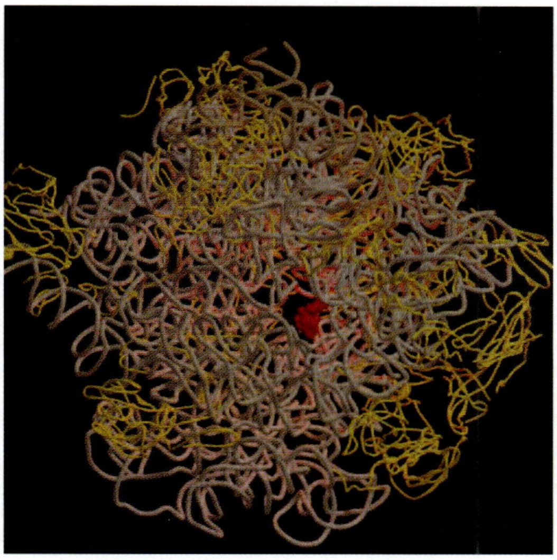

Figure 5 Top view of the *D. radiodurans* 50S subunit showing erythromycin (red) bound at the entrance of the tunnel. Yellow, ribosomal proteins; grey 23S rRNA; dark grey, 5S rRNA.

mode of interactions proposed by our structure implies that a mutation at A2041Dr (A2058Ec) will disrupt the pattern of hydrogen bonding, therefore impairing binding and rendering bacteria resistant. A2041Dr (A2058Ec) is one of the few nucleotides of the peptidyl transferase ring that are not conserved among all phylogenetic domains. Mitochondrial and cytoplasmic rRNAs of higher eukaryotes have a guanosine at position 2058Ec[33]. Therefore, the proposed mode of interaction explains the selectivity of macrolides for bacterial ribosomes.

Our results also explain the importance of the 2′OH group of the desosamine sugar for erythromycin binding known from structure activity relation (SAR) studies[34]. Changing the 2′OH group will not allow the suggested mode of interaction to take place. The 2′OH of the desosamine sugar is located 4 Å away from N6 of A2040Dr (G2057Ec), a nucleotide forming the last base pair of helix 73. Although an interaction with this group cannot be ruled out, this base pair seems to be engaged mainly in maintaining the right conformation of A2041Dr (A2058Ec)[23]. Moreover, in contrast to most eubacteria where the last base pair of helix 73 is a G·C pair, in *D. radiodurans* this base pair is an A·U pair, thus making a specific interaction of A2040Dr (G2057Ec) in the binding of macrolides unlikely.

The dimethylamino group of the desosamine sugar exists in both protonated (>96–98%) and neutral (2–4%) forms. This group, if protonated, could interact through ionic interactions in a pH-dependent manner with the backbone oxygen of G2484Dr (G2505Ec) (Fig. 3a, b). At physiological pH, both species appear to be able to bind ribosomes with the same kinetics[34]; however, the only study that revealed a strong correlation between pH and potency of inhibition of *in vitro* protein synthesis for the macrolide erythromycin concludes that the neutral macrolide form was the inhibitory species[35]. If the neutral macrolide form is in fact the inhibitory species, a hydrogen bond between the dimethylamine of the macrolides and the nucleotide G2484Dr (G2505Ec) could not be formed.

On the basis of our structural information, it will be of interest to derivatize macrolides at the dimethylamine position. These modifications could improve binding and result in the inhibition of the peptidyl transferase activity. One of the few macrolides able to inhibit peptidyl transferase is tylosin. Tylosin, which contains a mycarose moiety, has been shown to protect A2506Ec, the neighbouring nucleotide of G2484Dr (G2505Ec), from chemical modification[36]. Similarly, modification of the dimethylamine group to a reactive group that would interact with G2484Dr (G2505Ec) in a pH-independent manner may lead to more effective macrolides and structurally related compounds.

Only three of the hydroxyl moieties of the lactone ring are within hydrogen-bonding distance to the 23S rRNA: the 6-OH, the 11-OH and the 12-OH groups. The 6-OH group is within hydrogen-bonding distance to N6 of A2045Dr (A2062Ec). Although the 6-OH group has been substituted by a metoxygroup to improve acid stability in clarithromycin[29], the oxygen of the metoxy group is still at hydrogen-bonding distance to N6 of A2045Dr (A2062Ec). Our structure explains why the chemical footprinting experiments implicate this nucleotide in macrolide binding[37].

The 11-OH and 12-OH groups of the lactone ring could each form a hydrogen bond with O4 of U2588Dr (U2609Ec). Hydrogen bonds at these positions can explain substitutions of any of these two hydroxyl groups causing a moderate decrease of binding, as would be expected for groups involved in only one hydrogen-bond interaction with the same 23S rRNA nucleotide[35].

The reactive groups of the cladinose sugar are not involved in hydrogen-bond interactions with the 23S rRNA. Cladinose dispensability is confirmed by SAR studies[35] showing that the 4″-OH is dispensable for macrolide binding. A closely related group of antibiotics, the ketolides, which bind tighter to ribosomes than macrolides, do not have a cladinose sugar. Moreover, the K_d of erythromycin and the ketolide RU56006, a derivative of erythromycin that lacks the cladinose sugar, are on the same order of magnitude[37].

Although the three macrolides differ slightly in structure (Fig. 3a), we observed no differences in their binding site or mode. Our electron density map shows a small part of the etheroxime chain (R2 in Fig. 3a) of roxithromycin pointing towards the tunnel (Fig. 3d); however, this part is not involved in interactions with 23S rRNA or ribosomal proteins.

Our structure does not indicate direct interactions between erythromycin and helix 35 (H35), which has previously been implicated in the binding of erythromycin[37,38], and resistance mutations[37]. The distance between OH-11 of the lactone ring and N4 of C765Dr (A752Ec), a nucleotide of H35 proposed to interact with erythromycin[37,38], is 8.5 Å. The CH_3 group branching from position 13 of the lactone ring is located about 4.5 Å away from the O2 of C759Dr (G745Ec) of H35. Nonetheless, hydrophobic van der Waals or ion-coordinated interactions between the lactone ring and this helix loop of the 23S rRNA cannot be ruled out, suggesting that footprinting effects at A752Ec might be of an allosteric nature.

Two ribosomal proteins, L4 and L22, have been implicated in erythromycin resistance[39]. The closest distance of erythromycin (12-OH) to L4 (R111Dr/K90Ec) is 8 Å, whereas the closest distance from (8-CH_3) to L22 (G63Dr/G64Ec) is 9 Å (Fig. 3a). These distances are greater than one would expect for any meaningful chemical interaction. Therefore, the resistance of macrolides acquired by mutations in these two proteins is probably the product of an indirect effect that is produced by a perturbation of the 23S rRNA owing to the mutated proteins[40].

Overall, it is difficult to explain the high affinity of macrolides (K_d 10^{-8} M) for the ribosome solely by the proposed seven hydrogen bonds. The binding of erythromycin and its derivatives may be further stabilized by van der Waals forces, hydrophobic interactions and the geometry of the rRNA that tightly surrounds the macrolide molecules.

Overlapping binding sites

Two of the 23S rRNA nucleotides involved in the interactions with chloramphenicol are also involved in interactions with clindamycin: G2044Dr (G2061Ec) and G2484Dr (G2505Ec). Both nucleotides were previously shown by chemical footprinting to be protected by chloramphenicol or clindamycin binding[15].

The common nucleotide moieties involved in the interaction of the 23S rRNA with clindamycin and with the macrolides are N6 of A2041Dr (A2058Ec), N6 of A2042Dr (A2059Ec) and the non-bridging phosphate oxygen of U2484Dr (U2505Ec). These positions are targeted by the sugar moiety of clindamycin and the desosamine sugar of the macrolides, both of which are located in almost identical locations in the structure of the 50S–clindamycin and the 50S–macrolide complexes. The overlapping of binding sites may explain why clindamycin and macrolides bind competitively to the ribosome and why most RNA mutations conferring resistance to macrolides also confer resistance to lincosamides[15].

All examined antibiotics target nucleotide G2484Dr (G2505Ec). The importance of this nucleotide position has been established previously[41]. Although this nucleotide was shown to be protected from chemical modification on chloramphenicol, lincosamide, or macrolide binding[16,17,22], no mutation of G2484Dr (G2505Ec) conferring resistance to these antibiotics has been reported. In addition, G2484Dr (G2505Ec) is also one of the nucleotides protected on binding of peptidyl-tRNA[42]. The identity of this nucleotide is important for protein synthesis, albeit not for ribosome–antibiotic interactions where the position of its backbone oxygen or the 2′OH of the sugar appear to be the essential requirement.

Although overlapping binding sites may be coincidental, common nucleotide moieties may also point towards essential 23S rRNA nucleotides targeted by antibiotics to inhibit ribosomal function. The structural information derived from the overlapping binding sites will not only be useful to understand antibiotic action but could also have therapeutic implications as it can indicate which antibiotic combination therapies would be meaningful, and can open the door for the design of hybrid antibiotics.

Mechanism of antibiotic action

The relative binding sites for chloramphenicol, clindamycin and erythromycin at the 23S rRNA with respect to the A-site substrate analogue CC-puromycin (according to ref. 2) and the 3′ end of P- and A-site tRNAs (according to ref. 43) docked in our structure are shown in Fig. 4. From these sites it can be seen that chloramphenicol acts as a competitor of puromycin and thus as an inhibitor of the A site, consistent with previous findings[26]. In contrast to puromycin, which acts as a structural analogue of the 3′-end aminoacyl tRNA, the location of chloramphenicol suggests that this drug may act by interfering with the positioning of the aminoacyl moiety in the A site. Thus, chloramphenicol may physically prevent the formation of the transition state during peptide-bond formation. Furthermore, the carbon of the dichloromethyl moiety of chloramphenicol, a moiety shown to be important for chloramphenicol activity[44], is close to the amino acceptor group of CC-puromycin[2] (Fig. 4). The presence of such an electronegative moiety in the vicinity of the amino acceptor could also hamper or at least contribute to the inhibitory effect of chloramphenicol on peptide-bond formation.

Clindamycin clearly bridges the binding site of chloramphenicol and that of the macrolides (Fig. 4), and overlaps directly with the A and P sites. Thus, its binding position gives a structural explanation for the hybrid nature of its A and P sites. Clindamycin can interfere with the positioning of the aminoacyl group at the A site and the peptidyl group at the P site while also sterically blocking the progression of the nascent peptide towards the tunnel.

The macrolide binding site is at the entrance of the tunnel (Fig. 5). This location can allow the formation of 6–8 peptide bonds before the nascent protein chain reaches the bound macrolide. When bound, macrolides reduce the diameter of the tunnel entrance from 18–19 Å to less than 10 Å; however, as the space not occupied by erythromycin hosts a putative, hydrated Mg^{2+} ion (Mg33), the passage available for the nascent protein will be between 6 and 7 Å in diameter. Our structural results are consistent with previous biochemical findings, showing that small peptides (up to eight residues) can be produced by erythromycin-bound ribosomes[45]. It is possible that ribosomes could escape the inhibitory effects of macrolides if proteins with specific leader sequences are translated that allow the threading of the narrowed tunnel.

Overall, the binding sites of the antibiotics analysed suggest that their inhibitory action is not only determined by their interaction with specific nucleotides. These antibiotics could also inhibit peptidyl transferase by interfering with the proper positioning and movement of the tRNAs at the peptidyl transferase cavity. This sterical hindrance may be direct, as in the case of chloramphenicol, or indirect, as in the case of the three macrolides. In addition to sterical hindrance, antibiotic binding may physically link regions known to be essential for the proper positioning of the aminoacyl- and peptidyl-tRNAs, and thus prevent the conformational flexibility needed for protein biosynthesis.

The overall interaction of each analysed antibiotic together with the lack of any major conformational changes on antibiotic binding to the ribosome points towards the theory that the peptidyl transferase centre evolved as a template for the proper binding of the activated substrates, that is, the aminoacyl-tRNA and the peptidyl-tRNA[7]. However, we cannot completely exclude the presence of a catalytic nucleotide.

Discussion

Our work reveals the functional interactions involved in the binding of antibiotics to the peptidyl transferase cavity of a bacterial ribosome. All 23S rRNA nucleotides involved in interactions with these five antibiotics are part of the peptidyl transferase cavity of the 50S subunit. None of the antibiotics examined show any direct interaction with ribosomal proteins. Chloramphenicol targets mainly the A site, where it interferes directly with substrate binding. Clindamycin interferes with the A site and P site substrate binding and physically hinders the path of the growing peptide chain. Macrolides bind at the entrance to the tunnel where they sterically block the progression of the nascent peptide. Although the binding sites of the antibiotics differ from each other, they show some overlapping nucleotides.

The structural model of the peptidyl transferase centre in complex with the examined antibiotics should not only enable a rational approach for antibiotic development and therapy strategies but could be also used to identify new target sites on the eubacterial ribosome. □

Methods

Base numbering

Bases are numbered according to the *D. radiodurans* 23S rRNA sequence. A 'Dr' is added at the end of the base number for further identification. Corresponding *Escherichia coli* base numbers of the 23S rRNA sequence are given in parenthesis and have an 'Ec' added after the base number.

D. radiodurans cell growth and ribosome isolation

Cells were grown as recommended by the American Type Culture Collection (ATCC) with minor modifications to ATCC medium 679. Ribosomes were isolated as described[46].

50S crystals

Crystals of 50S belong to the space group *I*222 and contain one particle per asymmetric unit. Crystals were grown by vapour diffusion, in 10 mM $MgCl_2$, 60 mM NH_4Cl, 5 mM KCl, 10 mM HEPES, pH 7.8, using low concentrations (0.1–1%) of poly- and monovalent alcohols (dimethyl-hexandiol:ethanol) as precipitant. Co-crystallization was carried out in the presence of 0.8–3.5 mM of the antibiotics chloramphenicol, clindamycin, erythromycin and roxithromycin. Soaking of 50S crystals was performed for clarithromycin and chloramphenicol in solutions containing 0.1 mM of the respective antibiotic. Chloramphenicol data was collected from crystals grown in the presence of the antibiotic as well as from soaked crystals. Both data sets resulted in the same experimental density in the chloramphenicol difference maps and thus were merged. Difference maps for all other tested antibiotics were obtained from either co-crystallized or soaked crystals alone.

X-ray diffraction

Data were collected at 85 K from shock-frozen crystals with bright synchrotron radiation beam at ID19 at Argonne Photon Source/Argonne National Laboratory (APS/ANL), ID14/2 and 4 at European Synchrotron Radiation Facility/European Molecular Biology Laboratory (ESRF/EMBL), and at BW6 at Deutsches Elektronen-Synchrotron (DESY). Data were recorded on MAR345, Quantum 4, or APS-CCD (charge-coupled device) detectors and processed wit HKL2000 (ref. 47).

Placements and refinement

The 3.1 Å structure of the 50S subunit was refined against the structure factor amplitudes of each of the 50S–antibiotic complexes, using rigid body refinement as implemented in CNS[48]. We used SigmaA weighted difference maps for the initial manual placement of the antibiotics. Each of the 50S–antibiotic models was further refined in Refmac[49]. For the calculation of free *R*-factor (as reported in Table 1) a subset of reflections (10% of the data) was omitted from the refinement.

The experimental density allowed us the unequivocal placement of the five antibiotics that we studied. Taking into account the estimated coordinate error (Table 1) and the limits imposed by the resolution, we analysed all of the rRNA groups located less than 4.4 Å from any antibiotic moiety. Here we report all moieties from the previous group having the potential to form hydrogen bonds.

Coordinates and figures

Coordinates of chloramphenicol, clindamycin, erythromycin, roxithromycin and clarithromycin were taken from Cambridge Structural Database. Antibiotics were modelled into the difference density based on their crystal structure. Figures were produced with RIBBONS[50] or MOLSCRIPT[51] programs.

Received 3 August; accepted 27 September 2001.

1. Ban, N., Nissen, P., Hansen, J., Moore, P. B. & Steitz, T. A. The complete atomic structure of the large ribosomal subunit at 2.4 Å resolution. *Science* **289,** 905–920 (2000).
2. Nissen, P., Hansen, J., Ban, N., Moore, P. B. & Steitz, T. A. The structural basis of ribosome activity in peptide bond synthesis. *Science* **289,** 920–930 (2000).
3. Ogle, J. M. *et al.* Recognition of cognate transfer RNA by the 30S ribosomal subunit. *Science* **292,** 897–902 (2001).
4. Pioletti, M. *et al.* Crystal structures of complexes of the small ribosomal subunit with tetracycline, edeine and IF3. *EMBO J.* **20,** 1829–1839 (2001).
5. Winberly, B. T. *et al.* Structure of the 30S ribosomal subunit. *Nature* **407,** 327–339 (2000).
6. Schluenzen, F. *et al.* Structure of functionally activated small ribosomal subunit at 3.3 Å resolution. *Cell* **102,** 615–623 (2000).
7. Polacek, N., Gaynor, M., Yassin, A. & Mankin, A. S. Ribosomal peptidyl transferase can withstand mutations at the putative catalytic nucleotide. *Nature* **411,** 498–501 (2001).
8. Thompson, J. *et al.* Analysis of mutations at residues A2451 and G2447 of 23S rRNA in the peptidyltransferase active site of the 50S ribosomal subunit. *Proc. Natl Acad. Sci. USA* **98,** 9002–9007 (2001).
9. Barta, A. *et al.* Mechanism of ribosomal peptide bond formation. *Science* **291,** 203a (2001).
10. Spahn, C. M. T. & Prescott, C. D. Throwing a spanner in the works: antibiotics and the translation apparatus. *J. Mol. Med.* **74,** 423–439 (1996).
11. Cundliffe, E. in *The Ribosome: Structure, Function and Evolution* (eds Hill, W. E. *et al.*) 479–490 (ASM, Washington DC, 1990).
12. Vester, B. & Douthwaite, S. Macrolide resistance conferred by base substitutions in 23S rRNA. *Antimicrob. Agents Chemother.* **45,** 1–12 (2001).
13. Harms, J. *et al.* High resolution structure of the large ribosomal subunit from a mesophilic eubacterium. *Cell* (in the press).
14. Vester, B. & Garrett, R. A. The importance of highly conserved nucleotides in the binding region of chloramphenicol at the peptidyl transfer center of *Escherichia coli* 23S ribosomal RNA. *EMBO J.* **7,** 3577–3588 (1988).
15. Polacek, N. in *RNA-binding Antibiotics* (eds Schroeder, R. & Wallis, M. G.) 1–10 (Eurekah.Com, Georgetown, 2000).
16. Rodriguez-Fonseca, C., Amis, R. & Garrett, R. A. Fine structure of the peptidyl transferase centre on 23 S-like rRNAs deduced from chemical probing of antibiotic-ribosome complexes. *J. Mol. Biol.* **247,** 224–235 (1995).
17. Moazed, D. & Noller, H. F. Chloramphenicol, erythromycin, carbomycin and vernamycin B protect overlapping sites in the peptidyl transferase region of 23S ribosomal RNA. *Biochemie* **69,** 879–884 (1987).
18. Mankin, A. S. & Garrett, R. A. Chloramphenicol resistance mutations in the single 23S rRNA gene of the archaeon *Halobacterium halobium*. *J. Bacteriol.* **173,** 3559–3563 (1991).
19. Izard, T. & Ellis, J. The crystal structures of chloramphenicol phosphotransferase reveal a novel inactivation mechanism. *EMBO J.* **19,** 2690–2700 (2000).
20. Shaw, W. V. & Leslie, A. G. W. Chloramphenicol acetyl-transferase. *Annu. Rev. Biophys. Biophys. Chem.* **20,** 363–386 (1991).
21. Kalliaraftopoulos, S., Kalpaxis, D. L. & Coutsogeorgopoulos, C. New aspects of the kinetics of inhibition by lincomycin of peptide-bond formation. *Mol. Pharmacol.* **46,** 1009–1014 (1994).
22. Douthwaite, S. Interaction of the antibiotics clindamycin and lincomycin with *Escherichia coli* 23S ribosomal-RNA. *Nucleic Acids Res.* **20,** 4717–4720 (1992).
23. Vester, B., Hansen, L. H. & Douthwaite, S. The conformation of 23S ribosomal-RNA nucleotide A2058 determines its recognition by the Erme methyltransferase. *RNA* **1,** 501–509 (1995).
24. Ross, J. I. *et al.* Clinical resistance to erythromycin and clindamycin in cutaneous propionibacteria isolated from acne patients is associated with mutations in 23S rRNA. *Antimicrob. Agents Chemother.* **41,** 1162–1165 (1997).
25. Steiner, G., Kuechler, E. & Barta, A. Photo-affinity labeling at the peptidyl transferase center reveals 2 different positions for the A-site and P-sites in domain-V of 23S ribosomal-RNA. *EMBO J.* **7,** 3949–3955 (1988).
26. Vazquez, D. *Inhibitors of Protein Synthesis* (Springer, Berlin, 1975).
27. Yonath, A., Leonard, K. R. & Wittmann, H. G. A tunnel in the large ribosomal subunit revealed by three-dimensional image reconstruction. *Science* **236,** 813–816 (1987).
28. Milligan, R. A. & Unwin, P. N. Location of exit channel for nascent protein in 80S ribosome. *Nature* **319,** 693–695 (1986)
29. Steinmetz, W. E., Bersch, R., Towson, J. & Pesiri, D. The conformation of 6-methoxyerythromycin-A in water determined by proton NMR-spectroscopy. *J. Med. Chem.* **35,** 4842–4845 (1992).
30. Gasc, J. C., Dambrieres, S. G., Lutz, A. & Chantot, J. F. New etheroxime derivatives of erythromycin A: a structure–activity relationship study. *J. Antibiot.* **44,** 313–330 (1991).
31. Weisblum, B. Erythromycin resistance by ribosome modification. *Antimicrob. Agents Chemother.* **39,** 577–585 (1995).
32. Sigmund, C. D., Ettayebi, M. & Morgan, E. A. Antibiotic-resistance mutations in 16S and 23S ribosomal-RNA genes of *Escherichia coli*. *Nucleic Acids Res.* **12,** 4653–4663 (1984).
33. Bottger, E. C., Springer, B., Prammananan, T., Kidan, Y. & Sander, P. Structural basis for selectivity and toxicity of ribosomal antibiotics. *EMBO Rep.* **2,** 318–323 (2001).
34. Goldman, R. C., Fesik, S. W. & Doran, C. C. Role of protonated and neutral forms of macrolides in binding to ribosomes from gram-positive and gram-negative bacteria. *Antimicrob. Agents Chemother.* **34,** 426–431 (1990).
35. Mao, J. C.-H. & Puttermann, M. The intermolecular complex of erythromycin and ribosome. *J. Mol. Biol.* **44,** 347–361 (1969).
36. Poulsen, S. M., Kofoed, C. & Vester, B. Inhibition of the ribosomal peptidyl transferase reaction by the mycarose moiety of the antibiotics carbomycin, spiramycin and tylosin. *J. Mol. Biol.* **304,** 471–481 (2000).
37. Hansen, L. H., Mauvais, P. & Douthwaite, S. The macrolide-ketolide antibiotic binding site is formed by structures in domains II and V of 23S ribosomal RNA. *Mol. Microbiol.* **31,** 623–631 (1999).
38. Xiong, L. Q., Shah, S. Mauvais, P. & Mankin, A. S. A ketolide resistance mutation in domain II of 23S rRNA reveals the proximity of hairpin 35 to the peptidyl transferase centre. *Mol. Microbiol.* **31,** 633–639 (1999).
39. Wittmann, H. G. *et al.* Biochemical and genetic studies of two different types of erythromycin resistant mutants of *Escherichia coli* with altered ribosomal proteins. *Mol. Gen. Genet.* **127,** 175–189 (1973).
40. Gregory, S. T. & Dahlberg, A. E. Erythromycin resistance mutation sin ribosomal proteins L22 and L4 perturb the high order structure of 23S ribosomal RNA. *J. Mol. Biol.* **289,** 827–834 (1999).
41. Saarma, U., Spahn, C. M. T., Nierhaus, K. H. & Remme, J. Mutational analysis of the donor substrate binding site of the ribosomal peptidyltransferase center. *RNA* **4,** 189–194 (1998).
42. Moazed, D. & Noller, H. F. Sites of interaction of the CCA-end of peptidyl-transfer RNA with 23S ribosomal-RNA. *Proc. Natl Acad. Sci. USA* **88,** 3725–3728 (1991).
43. Yusupov, M. M. *et al.* Crystal structure of the ribosome at 5.5 Å resolution. *Science* **292,** 883–896 (2001).
44. Vince, R., Almquist, R. G., Ritter, C. L. & Daluge, S. Chloramphenicol binding site with analogues of chloramphenicol and puromycin. *Antimicrob. Agents Chemother.* **8,** 439–443 (1975)
45. Tenson, T., DeBlasio, A. & Mankin, A. A functional peptide encoded in the *Escherichia coli* 23S rRNA. *Proc. Natl Acad. Sci. USA* **93,** 5641–5646 (1996).
46. Noll, M., Hapke, B., Schreier, M. H. & Noll, H. Structural dynamics of bacterial ribosomes. I. Characterization of vacant couples and their relation to complexed ribosomes. *J. Mol. Biol.* **75,** 281–294 (1973).
47. Otwinowski, Z. & Minor, W. Processing of X-ray diffraction data collected in oscillation mode. *Methods Enzymol.* **276,** 307–326 (1997).
48. Brünger, A. T. *et al.* Crystallography & NMR system: a new software suite for macromolecular structure determination. *Acta Crystallogr.* D **54,** 905–921 (1998).
49. Murshudov, G. N., Lebedev, A., Vagin, A. A., Wilson, K. S. & Dodson, E. J. Efficient anisotropic refinement of macromolecular structures using FFT. *Acta Crystallogr.* D **55,** 247–255 (1999).
50. Carson, M. Ribbons. *Methods Enzymol.* **277,** 493–505 (1997).
51. Kraulis, P. J. MOLSCRIPT: a program to produce both detailed and schematic plots of protein structures. *J. Appl. Crystallogr.* **24,** 946–950 (1991).

Acknowledgements

We thank A. Hofmann for advice; A. Mankin and F. Triana for critical discussions; and T. Auerbach, H. Bartels, W. S. Bennett, H. Burmeister, C. Glotz, M. Glühmann, H. A. S. Hansen, M. Kessler, M. Laschever, S. Meier, J. Muessig, M. Peretz, M. Pioletti, B. Schmidt, A. Sitka, C. Stamer and A. Vieweger for contributing to different stages of these studies. These studies could not have been performed without the cooperation of the staff of the synchrotron radiation facilities at EMBL and Max Planck Gesellschaft at DESY, ID14/2&4 at EMBL/ESRF and ID19/APS/ANL. Support was provided by the Max-Planck Society, the US National Institute of Health, the German Ministry for Science and Technology and the Kimmelman Center for Macromolecular Assembly at the Weizmann Institute. A.Y. holds the Martin S. Kimmel Professorial Chair.

Correspondence and requests for materials should be addressed to F.F. (e-mail: Franceschi@molgen.mpg.de). Coordinates have been deposited in the Protein Data Bank under accession numbers 1JZX, 1JZY, 1JZZ, 1K00 and 1K01.

Chapter 11

Douglas C. Rees

Biography

I had the good fortune to grow up in Lexington, KY, not far from the childhood home of William N. Lipscomb. I first became aware of the Colonel (as I would later know him) when I was in high school, not through any scientific connection, but because my wife-to-be Becky took piano lessons from his sister Helen Lipscomb. My mother-in-law-to-be, who was good friends with the Colonel's mother, would mention that her friend had a son who was a famous chemist at Harvard. I had done pretty well in high school chemistry, and since I had never heard of him, with the hubris of youth I figured Lipscomb must not be too accomplished. Fortunately, I kept an open mind about this inference and have subsequently decided that there was a fundamental flaw in my logic.

The Colonel traveled from Lexington to Caltech, to Minnesota, and ultimately to Harvard in his professional career. As it happened, I made a similar journey from Lexington (with additional stops at Yale College and UCLA), but with the opposite hand. This chirality was quite fortunate, since it meant I attended graduate school at Harvard where I worked on protein crystallography with Lipscomb. Like much of my career, this event was unplanned and it was an amazing stroke of good luck to have the opportunity to study with the Colonel. (In hindsight, these developments are all the more remarkable because I went to graduate school with a negative attitude towards protein crystallography. I "saw the light," however, in a course taught by Don Wiley and Steve Harrison on protein structure and function, and this fascination was sealed with a subsequent rotation in Steve's lab. From this experience, I have become a fervent believer in the importance of rotations.)

Following postdoctoral appointments at Harvard and then at the University of Minnesota with James B. Howard (who I met at Harvard when he spent a sabbatical in the Lipscomb group), I joined the faculty of the Department of Chemistry and Biochemistry at UCLA in 1982. In 1989, I moved to the Division of Chemistry and Chemical Engineering at the California Institute of Technology, and became an investigator at the Howard Hughes Medical Institute in 1997. My research interests are in the general area of structural bioenergetics, using crystallographic and functional approaches to characterize metalloproteins and membrane proteins such as nitrogenase and ABC transporters that mediate ATP-dependent energy transduction processes.

The Metalloclusters of Nitrogenase

The enzyme nitrogenase catalyzes the conversion of atmospheric dinitrogen to the metabolically usable form of ammonia during the process of biological nitrogen fixation. One of my major research interests has been to understand how nitrogenase achieves this transformation under physiological conditions, which contrast sharply with the high temperatures and pressures of the industrial Haber–Bosch process. A central focus of our studies has been on the crystal structure determinations of the component proteins of nitrogenase (the iron (Fe-) protein and the molybdenum-iron (MoFe-) protein) and their constituent metalloclusters, particularly the iron molybdenum cofactor (FeMo-cofactor) of the MoFe-protein that provides the active site for dinitrogen reduction. The evolution in our structural understanding of the cofactor is reflected in the choice of selections for this volume — the first paper (by Kim and Rees, published in *Science* in 1992) describes the structure based on an unrefined model at 2.8 Å resolution, while the second paper (by Einsle *et al.*, published in *Science* in 2002) describes the structure at 1.16 Å resolution ("atomic resolution") that revealed the presence of a light atom ligand (likely carbon, nitrogen, or oxygen; it would be truly astonishing (and rather embarrassing) if it ultimately turned out to be boron in the center of the cofactor). The atomic identity and the mechanistic role of this interstitial ligand (if any) remain outstanding questions and I would be delighted to be able to answer these questions in a paper for a future volume honoring the Colonel.

My interest in nitrogenase grew directly from my experiences in the Colonel's group. A major influence was certainly the first hand appreciation I gained for the central role of structure in chemistry and biology. This was brilliantly exemplified, of course, by the Colonel's work on boron hydrides and enzyme structures, but it also permeated the entire atmosphere of Gibbs Laboratory through the pioneering efforts of Wiley and Harrison on the structures of cell surface proteins and viruses. In hindsight, one of the most important lessons we absorbed as graduate students in that incredibly exciting environment was the importance of identifying and solving great problems. An unexpected contributor to the intellectual excitement of that period was provided by the chemistry computer room — as the principal place for computational work, it was occupied by students and postdocs from all the structural and theoretical groups. Given the slower pace of computation at that time, there were ample opportunities while waiting for jobs to run to discuss scientific problems, personalities, and all sorts of other topics that really influenced my subsequent scientific development. (I have subsequently tried to recreate the equivalent of that computer room as a meeting room and melting pot for my group, but with the phenomenal increases in computing power and the much greater emphasis on bench work, this situation has proved to be a legacy of the past.)

As important as these general considerations were in shaping my interests in structural biology, there was also a very specific event that occurred during my time in the Colonel's group, and that was the arrival of Professor James Bryant Howard from the University of Minnesota. Jim came to spend his sabbatical working on the crystallization of the nitrogenase Fe-protein. While this did not happen, we had frequent conversations during his time in Cambridge that convinced me of three things: that Jim was an extraordinary person and scientist, that nitrogenase was an interesting system, and that I should do my postdoc with Jim to try to crystallize the nitrogenase Fe-protein. This is one of many examples of the roles of serendipity and unexpected opportunities in life, as well as illustrating the principle that these events are more likely to occur in the type of environment provided by the Colonel.

Following my time with the Colonel, I did complete my postdoc with Jim at the University of Minnesota. I was able to obtain crystals of the nitrogenase Fe-protein in 1981, and one of the first data sets was collected in the Colonel's lab at Harvard. The Fe-protein formed the core of my independent research program when I started at UCLA in 1982. Progress for the next decade was frustratingly slow; if this were to happen today, I'm positive I would have experienced what could be euphemistically termed "abortive initiation" of my professorial career. Fortunately, this did not happen, due to the support of David Eisenberg and the encouragement of Jim and the Colonel.

In 1992, we finally obtained the structure of both component proteins of nitrogenase and the associated metalloclusters. This work was the culmination of the efforts of a number of people, but particularly Millie Georgiadis, a UCLA graduate student, and Jongsun Kim, my first Caltech graduate student, who made crucial breakthroughs in producing suitable crystals and obtaining phases. One of my scientific highlights has unquestionably been establishing the unanticipated structures for the nitrogenase metalloclusters by X-ray crystallography; this has also provided me with a small appreciation for the Colonel's experiences in solving the novel structures of boron hydrides to understand their organizational principles. These examples also highlight the power of crystallography to provide these insights in a hypothesis-independent fashion, as well as the importance of high resolution diffraction data; the 1992 structural model of the FeMo-cofactor not only missed the interstitial ligand detected in the 2002 paper, but one of the central sulfides appeared to have weaker density than the other two and was initially designated "Y." In addition, the P-cluster model in the 1992 paper was based on the fusion of two Fe_4S_4 clusters with a bridging disulfide that was later shown to be the consequence of a superposition of two distinct states of the Fe_8S_7 core of the P-cluster.[1]

Of course, it should not be surprising to get to a scientific destination and find the Colonel's footprints already there. In the realm of iron-sulfur clusters, one of the first crystallographically characterized species, the black Roussinate ion ($Fe_4S_3(NO)_7^-$) was solved by the Colonel's group over 50 years ago.[2] I have a reprint of this paper inscribed by the Colonel with the comment, "An early result close to the Fe_4S_4 cluster"; it was over a decade before Fe_4S_4 clusters were actually characterized in proteins and prepared synthetically. In addition to a description of the Roussinate ion structure in this paper, there is also an analysis of the electronic structure that may be the first for an iron sulfur cluster. I believe the Colonel's interest in bonding is really at the heart of his interest in structure; I also can't escape the feeling that if I had a fraction of his insight into bonding, the shroud of mystery surrounding the mechanism of nitrogenase would be lifted.

References

1. Peters J. W., Stowell M. H. B., Soltis S. M., Finnegan M. G., Johnson M. K., and Rees D. C., Redox-dependent structural changes in the nitrogenise P-cluster, *Biochemistry*, **36**, 1181–1187 (1997).
2. Johansson G. and Lipscomb W. N., The structure of Roussin's black salt $CsFe_4S_3(NO)_7.H_2O$, *Acta Crystallographica*, **11**, 594–598 (1958).

Structural Models for the Metal Centers in the Nitrogenase Molybdenum-Iron Protein

Jongsun Kim and D. C. Rees*

Structural models for the nitrogenase FeMo-cofactor and P-clusters are proposed based on crystallographic analysis of the nitrogenase molybdenum-iron (MoFe)-protein from *Azotobacter vinelandii* at 2.7 angstrom resolution. Each center consists of two bridged clusters; the FeMo-cofactor has 4Fe:3S and 1Mo:3Fe:3S clusters bridged by three nonprotein ligands, and the P-clusters contain two 4Fe:4S clusters bridged by two cysteine thiol ligands. Six of the seven Fe sites in the FeMo-cofactor appear to have trigonal coordination geometry, including one ligand provided by a bridging group. The remaining Fe site has tetrahedral geometry and is liganded to the side chain of $Cys^{\alpha 275}$. The Mo site exhibits approximate octahedral coordination geometry and is liganded by three sulfurs in the cofactor, two oxygens from homocitrate, and the imidazole side chain of $His^{\alpha 442}$. The P-clusters are liganded by six cysteine thiol groups, two which bridge the two clusters, α88 and β95, and four which singly coordinate the remaining Fe sites, α62, α154, β70, and β153. The side chain of $Ser^{\beta 188}$ may also coordinate one iron. The polypeptide folds of the homologous α and β subunits surrounding the P-clusters are approximately related by a twofold rotation that may be utilized in the binding interactions between the MoFe-protein and the nitrogenase Fe-protein. Neither the FeMo-cofactor nor the P-clusters are exposed to the surface, suggesting that substrate entry, electron transfer, and product release must involve a carefully regulated sequence of interactions between the MoFe-protein and Fe-protein of nitrogenase.

Reduction of dinitrogen to ammonia by the nitrogenase enzyme system requires two metalloproteins, the molybdenum-iron (MoFe)–protein and the iron (Fe)–protein [reviewed in (*1*)]. The MoFe-protein is an $\alpha_2\beta_2$ tetramer with total molecular mass of ~240 kD, and the Fe-protein is a γ2 dimer. Three distinct types of redox centers are associated with these proteins: The MoFe-protein contains two types of centers, the FeMo-cofactor [reviewed in (*2*)] and the P-clusters [reviewed in (*3*)], and the Fe-protein dimer contains one 4Fe:4S cluster whose structure is described in an accompanying article (*4*). Because the active site of nitrogenase is provided by the MoFe-protein, the redox centers of this protein have attracted considerable attention. The FeMo-cofactor, first identified by Shah and Brill (*5*), most likely represents the site of substrate reduction (*6*) and has the approximate composition of one Mo atom, six to eight Fe atoms, eight to nine S atoms, and one homocitrate group (1Mo:6-8Fe:8-9S:1 homocitrate) (*2*, *7*, *8*). The P-clusters are believed to contain two 4Fe:4S clusters in close proximity (*9*–*13*), although the detailed properties of this center are distinct from better characterized proteins that con-

Division of Chemistry and Chemical Engineering 147-75CH, California Institute of Technology, Pasadena, CA 91125.

*To whom correspondence should be addressed.

tain one or more mononuclear 4Fe:4S clusters. The overall metal composition of the MoFe-protein, approximately 2Mo:30Fe:-30S, is consistent with the presence of two copies each of the FeMo-cofactor and P-clusters per MoFe-protein tetramer. Despite extensive study, the detailed structures of the FeMo-cofactor and P-clusters have not yet been established. In this report, structural models for the FeMo-cofactor and P-clusters, based on crystallographic analysis at 2.7 Å resolution of the MoFe-protein from *Azotobacter vinelandii*, are presented. Each center consists of two bridged clusters; the FeMo-cofactor has 4Fe:3S and 1Mo:3Fe:3S clusters bridged by three non-protein ligands, whereas the P-clusters contain two 4Fe:4S clusters bridged by two cysteine thiol ligands.

The MoFe-protein crystal structure was determined by the methods of multiple isomorphous replacement (MIR) and non-crystallographic symmetry (NCS) averaging within and between crystal forms. Three crystal forms of MoFe-protein, purified by protocols modified from (*14*) and (*7*) for the MoFe-proteins from *A. vinelandii* and *Clostridium pasteurianum*, respectively, were used in the structure determination. The *A. vinelandii* MoFe-protein (designated Av1) was crystallized by the microcapillary batch method (*15*) from 18% polyethylene glycol (PEG) 4000, 0.14 M NaCl, 0.10 M Na_2MoO_4, 80 mM tris-HCl, pH 8.0, and ~8 mg/ml protein. Two crystal forms of the *C. pasteurianum* MoFe-protein (designated Cp1) were obtained by crystallization from 15% PEG 4000, 0.21 M $MgCl_2$, 80 mM tris HCl, pH 8.0, and ~8 mg/ml protein (crystal form designated Mg1) and by crystallization from 18% PEG 4000, 0.3 M CsCl, and 80 mM tris HCl, pH 8.0, and ~8 mg/ml protein (designated Cs1). All three crystal forms belong to space group $P2_1$, with cell constants a = 108.4 Å, b = 130.5 Å, c = 81.5, β = 110.8°; a = 70.0 Å, b = 151.3 Å, c = 121.9 Å, β = 110.4°; and a = 87.9 Å, b = 171.4 Å, c = 73.6 Å, β = 91.5°, for the Av1, Mg1, and Cs1 forms, respectively. Each form contains one $\alpha_2\beta_2$ tetramer per asymmetric unit. The Mg1 form is similar to a crystal form previously described (*16*). Diffraction data were collected at room temperature on a Siemens multiwire area detector and processed (Table 1) with XENGEN (*17*). Heavy atom derivatives of Av1 were prepared by soaking separate crystals in solutions of 5 mM ethylmercurithiosalicylate (EMTS), 5 mM K_2PtCl_4, and 4 mM di-μ-iodobis(ethylenediamine)-di-platinum (II) nitrate (PIP). Derivatives of Mg1 and Cs1 were obtained by soaking each crystal form in 5 mM EMTS. Heavy atom positions for each form were determined from analysis of isomorphous difference Patterson maps. The required NCS relations, both within and between the different crystal forms, were determined from rotation functions (*18*) and translation functions (*19*), and were confirmed by the heavy atom locations (although EMTS binds to different sites in the Av1 and Cp1 crystals). Phases obtained from the Av1 derivative data were used to define molecular envelopes (*20*) and were then refined by iterative NCS averaging of the electron density in the different crystal forms (*21*). Although the Av1 (*22*) and Cp1 (*23*) sequences are only ~36% identical, averaging between crystal forms was invaluable in the initial stages of the analysis. Polypeptide chains were built into the averaged density with the graphics program TOM (*24*) and refined with the restrained least squares program TNT (*25*). Phases calculated from the partial models were combined with the Av1 MIR phases and further averaged. Following several iterations of averaging, building, and refinement, the current model containing ~95% of the Av1 atoms has been obtained. At present, the R factor for this Av1 model is 0.197 to 2.7 Å resolution, although the fitting of the 2026 residues in the MoFe-protein tetramer is not yet final. Although a detailed description of the complete MoFe-protein structure is forthcoming, the α and β subunits exhibit homologous polypeptide folds that consist of three domains of the α/β type. Of particular interest for the

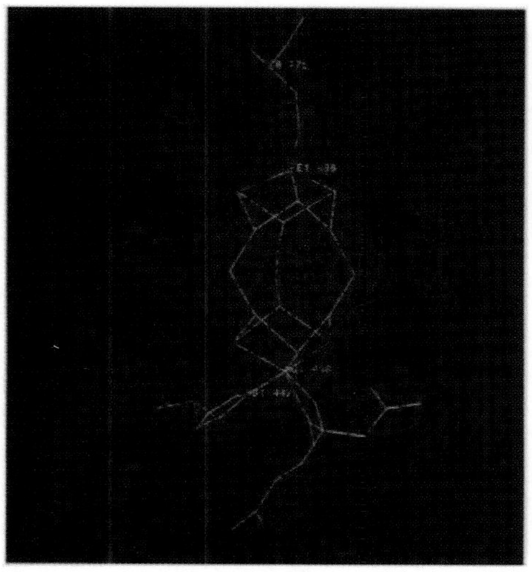

Fig. 1. Electron density map calculated at 2.8 Å resolution with superimposed FeMo-cofactor model. The map was generated by phase combination of the MIR and model phases, followed by NCS averaging in the Av1 crystal form. The blue and red densities are contoured at 2 and 4.5 times the standard deviation of the electron density map, respectively.

Table 1. Data processing and phasing statistics. Crystal form lists the type of crystal (native or derivative), with number of crystals used for data collection. Reflections with I > 0 measured to the limiting resolution were reduced to the unique data, with R_{merge} describing the agreement between symmetry-related reflections. The completeness of each data set to the limiting resolution is indicated as the percent of the theoretical number of reflections; ⟨m⟩ and pp describe the mean figure of merit and phasing power, respectively, associated with MIR phase calculations to 3.5 Å resolution.

Crystal form	Crystals (no.)	Resolution (Å)	R_{merge}	Complete reflections (%)	⟨m⟩	pp
Av1						
Native	2	2.75	0.085	94	0.62	
EMTS1	1	3.2	0.057	83		1.79
EMTS3	3	3.0	0.086	91		1.71
PTCL3	3	3.0	0.090	89		1.26
PIP3	3	3.0	0.111	92		1.17
Mg1						
Native	3	3.0	0.070	78	0.36	
EMTS	1	3.2	0.090	67		1.37
Cs1						
Native	3	3.0	0.084	74	0.35	
EMTS	1	3.2	0.087	64		1.37

present discussion, the subunits in an $\alpha\beta$ pair are approximately related by a twofold rotation, although this axis does not intersect the tetramer twofold.

The redox centers were identified as the strongest features in the electron density map and are consistent with the highest peaks in both native anomalous difference Patterson and Fourier maps. As native anomalous scattering effects were not used in the phasing of the MoFe-protein structure, the initial electron density maps are not influenced by any specific model for the centers. Models for the centers were built into the electron density maps with fragments of 4Fe:4S clusters as the basic building blocks. Metal-ligand distances were restrained to values observed in model compound structures (*26*). The FeMo-cofactor and P-cluster were identified from the positions of amino acid residues that have been deduced from mutagenesis studies to be in the cofactor environment, including residues α275 and α195 for FeMo-cofactor and residues α62, α88, α154, β70, β95, and β153 for the P-cluster (*27–31*). At 2.7 Å resolution, atoms are not resolved, so that the identities of the various sites were inferred from the available analytical and spectroscopic information. The Mo site in the FeMo-cofactor was assigned to the position in the more electron dense cluster with approximate octahedral coordination geometry, as observed in extended x-ray absorption fine structure (EXAFS) studies (*32*), although octahedral iron coordination in 4Fe:4S clusters has also been observed (*33*). It should be emphasized that at this stage the proposed cluster models are consistent with the general shape and features of the electron density map, but the unambiguous identification of individual atoms, including the possible presence of hydrides, and a more detailed description of the geometry and oxidation states would require higher resolution diffraction data of both the MoFe-protein and isolated cofactors.

The electron density map surrounding the FeMo-cofactor is illustrated in Fig. 1, with the cofactor model built into this density presented in Fig. 2. The FeMo-cofactor model contains two clusters of composition 4Fe:3S and 1Mo:3Fe:3S that are bridged by three non-protein ligands. Compounds containing 4Fe:3S cluster fragments have been described (*34*), although the terminal ligation of the Fe sites in these molecules differs significantly from that observed in the FeMo-cofactor. In the FeMo-cofactor, the Fe-Fe separation distance between bridged iron sites is about 2.7 to 2.8 Å, whereas the separation distance is ~3.8 Å between nonbridged iron sites on different cluster fragments (such as Fe2 to Fe5). Consequently, a cavity of diameter ~4 Å appears inside the cofactor between the two cluster fragments. Based on the electron density values at the bridging ligands, two are assigned as sulfur (presumably S^{2-}), while the third ligand has lower electron density and is presently modeled, perhaps somewhat fancifully, by a group containing one nitrogen. This site could be occupied by water, a dithionite derived species (*35*), a less well ordered sulfur or other possibilities, however. Homocitrate, which is an essential component of FeMo-cofactor (*8*), appears to be coordinated through hydroxyl and carboxyl oxygens to the Mo. If the threefold axes of the isolated 4Fe:3S and 1Mo:3Fe:3S clusters are superimposed, then the cluster is liganded to the protein through the metals (Fe1 and Mo) located on the threefold axis. The Fe1 and Mo sites are separated by ~7.5 Å. The protein environment around the FeMo-cofactor is primarily provided by the α subunit. $Cys^{\alpha275}$ coordinates Fe1, whereas the Mo is liganded by the side chain of $His^{\alpha442}$. Two other side chains, $His^{\alpha195}$ and $Gln^{\alpha191}$, are within 5 Å of Fe2 and Fe6, respectively, but do not appear at the present state of the crystallographic analysis to be directly liganded to the metals. The side chain of $Gln^{\alpha191}$ does interact with one of the carboxyl groups of homocitrate. $Cys^{\alpha275}$, $His^{\alpha442}$, $His^{\alpha195}$, and $Gln^{\alpha191}$ are conserved in all known MoFe-protein sequences; additionally, mutagenesis studies have implicated $Cys^{\alpha275}$ (*27, 29*) and $His^{\alpha195}$ (*28, 31*) as involved in the binding of FeMo-cofactor to MoFe-protein. Substitution of $Gln^{\alpha191}$

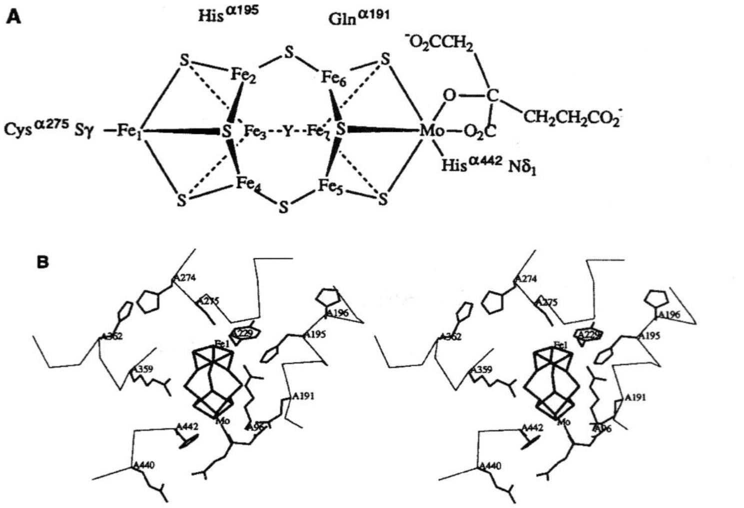

Fig. 2. (**A**) Schematic representation of the FeMo-cofactor model. Y represents the bridging ligand with relatively light electron density. (**B**) Stereoview of the FeMo-cofactor and surrounding protein environment.

Fig. 3. Electron density map calculated at 2.8 Å resolution with superimposed P-cluster model. The map was generated by phase combination of the MIR and model phases, followed by NCS averaging in the Av1 crystal form. The blue and red densities are contoured at 2 and 4.5 times the standard deviation of the electron density map, respectively.

by Lys alters the substrate reduction properties of nitrogenase (36), and the present observations support the proposal that Gln[α191] interacts with homocitrate (36).

The tetrahedral coordination geometry of Fe1 and the octahedral coordination geometry for Mo are typical of the coordination environments for these metals observed in model compounds and Fe:S proteins (3, 26). An unusual feature of the FeMo-cofactor model, however, is the approximate trigonal planar geometry of the sulfurs surrounding the remaining six Fe sites that bind the bridging ligands. Solvent molecules that might possibly serve as fourth ligands have not been observed, and hydride species could never be identified in this type of x-ray diffraction experiment. There is also no evidence in the electron density maps for a hexacoordinate sulfur atom at the center of the cluster. Trigonal coordination geometry for Fe is not unprecedented and has been described in a small molecule structure of an Fe-thiolate species (37). In this case, the coordinating thiols contained bulky substituents, and it was proposed that the low coordination number of the Fe atoms reflected the effects of steric crowding between the ligands. Although the FeMo-cofactor does not contain bulky, non-protein groups, apparently the environment provided by the MoFe-protein produces a similar effect. The sensitivity of the isolated cofactor to the solvent type used for extraction and isolation (2) may reflect the properties of this category of Fe-sites in the FeMo-cofactor.

The structural features of the FeMo-cofactor model are generally consistent with the results of analytical and spectroscopic [EXAFS, electron nuclear double resonance (ENDOR), and Mössbauer] studies of the cofactor. The composition of the non-protein part of the FeMo-cofactor model, 1Mo:7Fe:8S:1 homocitrate, is within the range of values that have been reported (2, 7, 8), although this information was used in constructing the model, particularly in the assignment of the bridging ligands. The absence of protein-bound bridging ligands between the two clusters in FeMo-cofactor is consistent with the ability to extract the intact cofactor from MoFe-protein. EXAFS studies of the Mo environment in both the MoFe-protein and isolated cofactor indicate that two to three O or N atoms and three to five S atoms are directly coordinated to Mo, with three to four Fe atoms present in the second coordination shell of Mo (32); the crystallographic model contains three O or N atoms, three S atoms, and three Fe. Studies of Fe EXAFS on the isolated cofactor (38) indicate that, on average, each Fe atom is surrounded by ~3 S and 0.8 Mo (~2.7 Å), with ~2.2 and ~1.3 Fe atoms located at ~2.7 and ~3.8 Å, respectively. The average Fe environment in the present FeMo-cofactor model bound to the MoFe-protein (including protein ligands, and assuming that ligand Y is an O- or N-containing species) contains 2.9 S, 0.4 Mo, and 0.3 N (or O atoms), with 3.4 and 1.7 Fe atoms located at ~2.7 and ~3.8 Å, respectively. ENDOR studies with ^{57}Fe (39) have indicated that five magnetically inequivalent Fe species are present; neglecting the asymmetric protein environment and the chiral homocitrate ligand, five different types of Fe sites are observed in the structural model (Fe1, Fe3, Fe7, Fe2 and Fe4, and Fe5 and Fe6). Mössbauer studies (10, 40) identified five to seven iron sites in the FeMo-cofactor that could be grouped into two sets; however, assignment of these sites to particular atoms in the FeMo-cofactor model cannot be made at present.

The electron density map surrounding a P-cluster is illustrated in Fig. 3, with the structural model built into this density presented in Fig. 4. The P-cluster model contains two 4Fe:4S clusters that are bridged by two cysteine thiol ligands (from residues α88 and β95), with the four remaining Fe liganded by singly coordinating cysteine thiols (from residues α62, α154, β70, and β153). The nonbridging cysteines coordinated to a specific 4Fe:4S cluster are from the same subunit. In addition to the cysteine ligands, Ser[β188] is close to Fe6 and may be able to coordinate this site along with Cys[β153]. The coordination environment must be distorted from ideal tetrahedral geometry for both Fe6 (due to interactions with Ser[β188]) and Fe2 (possibly due to interactions with the mainchain of Gly[α185] to obtain satisfactory fits of the 4Fe:4S clusters to the electron density. Individual replacement of any of the six cysteines by alanine eliminates diazotropic growth of the mutant strains (27, 29, 30). Interestingly, a double mutant with the two bridging cysteines, α88 and β95, both substituted by alanine exhibits activity (27), which was interpreted as indicating that these residues are located at the αβ subunit interface. Some non-alanine substitutions for Cys[α88] and Cys[β153] have been described that can still produce nitrogenase with low activity (27, 29, 30), possibly as a consequence of the substituted residue functioning as a cluster ligand, or through a structural rearrangement that allows a previously nonliganding group to coordinate the cluster.

As had been proposed from sequence comparisons and mutagenesis experiments, the P-cluster is located at the interface between the α and β subunits (27, 29). The approximate twofold symmetry of the P-cluster is reflected in the sequence similarities (41) and folding of the polypeptide chains for the α and β subunits in this region. The location of this metal center at the interface between two homologous subunits, with the coordination sphere provided by residues from both subunits, is not unique to MoFe-protein, and has been previously observed in bacterial photosynthetic reaction centers [(42), with the nonheme Fe liganded by residues from two homologous subunits] and the nitrogenase Fe-protein [(4), with the 4Fe:4S cluster liganded by residues from two identical subunits].

Mössbauer and extrusion studies indicated that the P-clusters contained 4Fe:4S clusters (9–12). The proximity of two 4Fe:4S clusters in each P-cluster was suggested from an analysis of anomalous scattering effects from Cp1 (13). The properties of the 4Fe:4S groups in the P-cluster differ in significant ways from "typical" 4Fe:4S clusters, however. The close spatial proximity of two 4Fe:4S clusters in MoFe-protein suggests that interactions between clusters might confer distinctive properties. Mössbauer studies (11) have identified three distinct classes of iron sites in P-clusters, designated Fe^{2+}, D and S, that are present in the approximate ratio of 4:10:2 in the MoFe-protein. Because the environment of each metal site is necessarily unique when coordinated to an asymmetric protein, it is not clear what effects distinguish the different Mössbauer classes, so that the assignment between P-cluster iron sites and Mössbauer classes has not yet been determined.

The protein environment in the vicinity of the FeMo-cofactor and P-clusters is illustrated in Fig. 5. As demonstrated by Bolin *et al.* (13), these centers are separated by ~19 Å, whereas the distances between centers on the other αβ subunit pair of the MoFe-protein are ~67 Å. The distance of closest approach between metal sites in FeMo-cofactor and P-clusters is ~14 Å and occurs between Fe3 of the P-cluster and the side of the FeMo-cofactor that faces the P-cluster. Interestingly, the two cysteines that can tolerate some substitutions with partial retention of activity [α88 and β153 (27, 29, 30)] are positioned on the side of the P-cluster furthest from the FeMo-cofactor. The α helix containing the P-cluster ligand Cys[α62] provides the most direct structural connection between the P-cluster and FeMo-cofactor. Both clusters are shielded from direct exposure to water by the surrounding protein.

Although a molecular-level description of dinitrogen reduction by nitrogenase cannot yet be provided, the structure and environment of the MoFe-protein metal centers suggests the following considerations are relevant to any proposed mechanism:

1) The approximate twofold symmetry

of the P-cluster and surrounding peptide could be important in binding interactions of the MoFe-protein with the symmetrical Fe-protein dimer.

2) The closest distance between the metal sites in the P-cluster and FeMo-cofactor, ~14 Å, suggests that the electron transfer rate (*43*) between these centers could be faster than the rate of nitrogenase turnover [~5 s^{-1} (*44*)].

3) The octahedral coordination of Mo in FeMo-cofactor suggests that the Mo does not directly participate in substrate binding, without the occurrence of either a change in coordination number or a change in liganding groups.

4) The importance of homocitrate to the substrate reduction mechanism (8) may arise from modulation of the redox properties of the coordinated FeMo-cofactor, by functioning in the protonation of intermediates, by participating in the electron transfer pathway between the P-cluster and FeMo-cofactor, or through a combination of all of these mechanisms.

5) The metal sites with low coordination number in FeMo-cofactor should be well suited for substrate binding and could represent sites for hydride formation (*1*). Given the rather open structure, it is possible that dinitrogen and other substrates may bind inside the FeMo-cofactor.

6) Neither the FeMo-cofactor nor the P-clusters are exposed to the surface, suggesting that substrate entry, electron transfer, and product release must involve a carefully regulated sequence of interactions between MoFe-protein, Fe-protein, and adenine nucleotides (*4*).

It is hoped that the structural models for the FeMo-cofactor and P-clusters of MoFe-protein described in this report may facilitate the development of a detailed understanding of the nitrogenase mechanism.

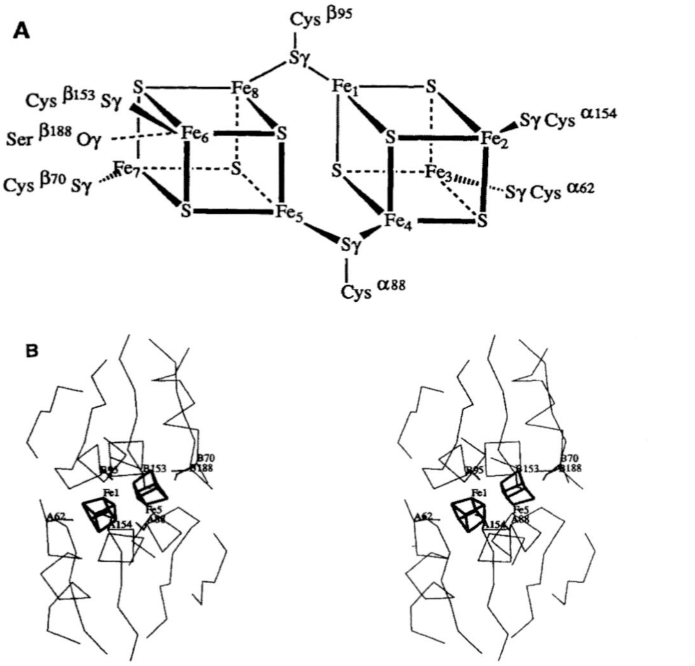

Fig. 4. (**A**) Schematic representation of the P-cluster model. (**B**) Stereoview of the P-cluster and surrounding protein model. The view is approximately along the direction of a twofold rotation axis approximately relating the α and β subunits.

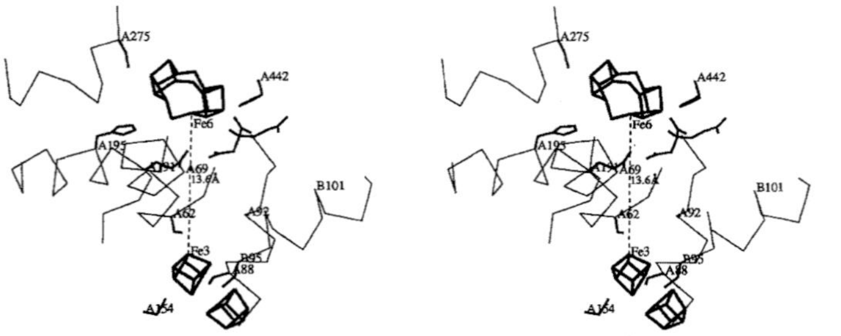

Fig. 5. Stereoview of the MoFe-protein structure in the vicinity of the FeMo-cofactor and P-cluster, including the Cα trace of the polypeptide chain and side chains of selected residues.

REFERENCES AND NOTES

1. B. K. Burgess, in *Advances in Nitrogen Fixation Research*, C. Veeger and W. E. Newton, Eds. (Nijhoff, Boston, 1984), pp. 103–114; W. H. Orme-Johnson, *Annu. Rev. Biophys. Biophys. Chem.* **14**, 419 (1985); R. H. Holm and E. D. Simhon, in *Molybdenum Enzymes*, T. G. Spiro, Ed. (Wiley-Interscience, New York, 1985), chap. 1; E. I. Stiefel *et al.*, *ACS Symp. Ser.* **372**, 372 (1988); R. H. Burris, *J. Biol. Chem.* **266**, 9339 (1991); B. E. Smith and R. R. Eady, *Eur. J. Biochem.* **205**, 1 (1992).
2. B. K. Burgess, *Chem. Rev.* **90**, 1377 (1990).
3. R. H. Holm, S. Ciurli, J. A. Weigel, H. Komiya, *Prog. Inorg. Chem.* **38**, 1 (1990).
4. M. M. Georgiadis, P. Chakrabarti, D. Woo, J. J. Kornuc, D. C. Rees, *Science* **257**, 1653 (1992).
5. V. K. Shah and W. Brill, *Proc. Natl. Acad. Sci. U.S.A.* **74**, 3249 (1977).
6. T. R. Hawkes *et al.*, *Biochem. J.* **217**, 317 (1984); J. Imperial *et al.*, *Biochemistry* **28**, 7796 (1989).
7. M. J. Nelson, M. A. Levy, W. H. Orme-Johnson, *Proc. Natl. Acad. Sci. U.S.A.* **80**, 147 (1983).
8. T. R. Hoover *et al.*, *Nature* **329**, 855 (1987); M. S. Madden, N. D. Kindon, P. W. Ludden, V. K. Shah, *Proc. Natl. Acad. Sci. U.S.A.* **87**, 6517 (1990).
9. D. M. Kurtz *et al.*, *Proc. Natl. Acad. Sci. U.S.A.* **76**, 4986 (1979).
10. R. Zimmermann *et al.*, *Biochem. Biophys. Acta.* **537**, 185 (1978); B. H. Huynh *et al.*, *ibid.* **623**, 124 (1980).
11. P. A. Lindahl *et al.*, *J. Biol. Chem.* **263**, 19412 (1988); *ibid.* **262**, 12900 (1987); K. K. Surerus *et al.*, *J. Am. Chem. Soc.*, in press.
12. W. R. Hagen, H. Wassink, R. R. Eady, B. E. Smith, H. Haaker, *Eur. J. Biochem.* **169**, 457–465 (1987).
13. J. T. Bolin *et al.*, in *Nitrogen Fixation: Achievements and Objectives*, P. M. Gresshoff, L. E. Roth, G. Stacey, W. E. Newton, Eds. (Chapman & Hall, New York, 1990), pp. 117–124.
14. B. K. Burgess, D. B. Jacobs, E. I. Stiefel, *Biochem. Biophys. Acta.* **614**, 196 (1980).
15. J. Drenth, W. G. J. Hol, R. K. Wierenga, *J. Biol. Chem.* **250**, 5268 (1975); M. M. Georgiadis, in preparation.
16. M. S. Weininger and L. E. Mortenson, *Proc. Natl. Acad. Sci. U.S.A.* **79**, 378 (1982).
17. A. J. Howard *et al.*, *J. Appl. Crystallogr.* **20**, 383 (1987).
18. M. G. Rossmann and D. M. Blow, *Acta Crystallogr.* **15**, 24 (1962); R. A. Crowther, in *The Molecular Replacement Method*, M. G. Rossmann, Ed.

(Gordon & Breach, New York, 1972), pp. 173–178.
19. R. A. Crowther and D. M. Blow, *Acta Crystallogr.* **23**, 544 (1967).
20. B. C. Wang, *Methods Enzymol.* **115**, 90 (1985).
21. M. G. Rossmann, Ed., *The Molecular Replacement Method* (Gordon & Breach, New York, 1972); G. Bricogne, *Acta Crystallogr.* **A32**, 832 (1976).
22. K. E. Brigle *et al.*, *Gene* **37**, 37 (1985).
23. S.-Z. Wang, J.-S. Chen, J. L. Johnson, *Biochemistry* **27**, 2800 (1988).
24. T. A. Jones, *Methods Enzymol.* **115**, 157 (1985).
25. D. E. Tronrud, L. F. Ten Eyck, B. W. Matthews, *Acta Crystallogr.* **A43**, 489 (1987).
26. T. Herskovitz *et al.*, *Proc. Natl. Acad. Sci. U.S.A.* **69**, 2437 (1972); D. Coucouvanis, *Acc. Chem. Res.* **24**, 1 (1991).
27. H. K. Kent *et al.*, *Biochem. J.* **264**, 257 (1989); H. M. Kent *et al.*, *Mol. Microbiol.* **4**, 1497 (1990).
28. D. J. Scott *et al.*, *Nature* **343**, 188 (1990).
29. D. R. Dean *et al.*, *Mol. Microbiol.* **4**, 1505 (1990).
30. H. D. May, D. R. Dean, W. E. Newton, *Biochem. J.* **277**, 457 (1991).
31. H. Thomann *et al.*, *J. Am. Chem. Soc.* **109**, 7913 (1987); H. Thomann, M. Bernardo, W. E. Newton, E. R. Dean, *Proc. Natl. Acad. Sci. U.S.A.* **88**, 6620 (1991).
32. M. K. Eidsness *et al.*, *J. Am. Chem. Soc.* **108**, 2746 (1986); S. D. Conradson *et al.*, *ibid.* **109**, 7507 (1987).
33. H. Lauble, M. C. Kennedy, H. Beinert, C. D. Stout, *Biochemistry* **31**, 2735 (1992).
34. C. T.-W. Chu and L. F. Dahl, *Inorg. Chem.* **16**, 3245 (1977).
35. B. Hedman *et al.*, *J. Am. Chem. Soc.* **110**, 3798 (1988).
36. D. J. Scott, D. R. Dean, W. E. Newton, in *Nitrogen Fixation: Achievements and Objectives*, P. M. Gresshoff, L. E. Roth, G. Stacey, W. E. Newton, Eds. (Chapman & Hall, New York, 1990), p. 169.
37. P. P. Power and S. C. Shoner, *Angew. Chem. Int. Ed. Engl.* **30**, 330 (1991).
38. J. M. Arber *et al.*, *Biochem. J.* **252**, 421 (1988).
39. A. E. True *et al.*, *J. Am. Chem. Soc.* **110**, 1935 (1988).
40. J. Rawlings *et al.*, *J. Biol. Chem.* **253**, 1001 (1978).
41. P. J. Lammers and R. Haselkorn, *Proc. Natl. Acad. Sci. U.S.A.* **80**, 4723 (1983).
42. J. Deisenhofer, O. Epp, K. Miki, R. Huber, H. Michel, *Nature* **318**, 618 (1985); J. P. Allen, G. Feher, T. O. Yeates, H. Komiya, D. C. Rees, *Proc. Natl. Acad. Sci. U.S.A.* **84**, 5730 (1987).
43. R. N. Marcus and N. Sutin, *Biochim. Biophys. Acta* **811**, 265 (1985); D. S. Wuttke *et al.*, *Science* **256**, 1007 (1992).
44. R. N. F. Thorneley and D. J. Lowe, *Biochem. J.* **224**, 887 (1984).
45. Discussions with J. B. Howard, J. E. Bercaw, and H. B. Gray and the assistance of M. M. Georgiadis, B. T. Hsu, D. Woo, A. J. Chirino, H. Komiya, D. Malerba, and M. K. Chan are gratefully appreciated. Supported by NSF grant DMB91-18689, with instrumentation funded in part by the Beckman Institute and the Joseph Irvine Equipment fund.

19 June 1992; accepted 18 August 1992

Nitrogenase MoFe-Protein at 1.16 Å Resolution: A Central Ligand in the FeMo-Cofactor

Oliver Einsle,[1,2] F. Akif Tezcan,[2] Susana L. A. Andrade,[1,2] Benedikt Schmid,[2] Mika Yoshida,[1,2] James B. Howard,[3] Douglas C. Rees[1,2*]

A high-resolution crystallographic analysis of the nitrogenase MoFe-protein reveals a previously unrecognized ligand coordinated to six iron atoms in the center of the catalytically essential FeMo-cofactor. The electron density for this ligand is masked in structures with resolutions lower than 1.55 angstroms, owing to Fourier series termination ripples from the surrounding iron and sulfur atoms in the cofactor. The central atom completes an approximate tetrahedral coordination for the six iron atoms, instead of the trigonal coordination proposed on the basis of lower resolution structures. The crystallographic refinement at 1.16 angstrom resolution is consistent with this newly detected component being a light element, most plausibly nitrogen. The presence of a nitrogen atom in the cofactor would have important implications for the mechanism of dinitrogen reduction by nitrogenase.

Biological nitrogen fixation provides the dominant route for the transformation of atmospheric dinitrogen into a bioavailable form, ammonia (*1–4*). This process is catalyzed by the enzyme nitrogenase, which consists of two component metalloproteins, the Fe-protein and the MoFe-protein. The homodimeric Fe-protein couples adenosine 5′-triphosphate hydrolysis to interprotein electron transfer and is the only known mechanistically competent source of electrons for the catalytically active component, the MoFe-protein. The latter is organized as an $\alpha_2\beta_2$ tetramer that contains two copies each of two unique polynuclear metal clusters designated the P-cluster and the FeMo-cofactor. Whereas the P-cluster likely participates in interprotein electron transfer, the FeMo-cofactor is the active site of substrate binding and reduction. Despite detailed structural information and a multitude of kinetic, spectroscopic, and theoretical studies, little is known about the mechanistic details of dinitrogen reduction by nitrogenase (*5*), particularly the site and mode of substrate binding.

The structures of the P-cluster and FeMo-cofactor in the MoFe-protein have been determined crystallographically at resolutions between 2.8 and 1.6 Å (*6–12*). The FeMo-cofactor, with composition [Mo:7Fe:9S]:homocitrate, is coordinated to the protein through the side chains of only two residues bound to Fe and Mo sites located at opposite ends of the cluster. Perhaps the most unusual feature of the cofactor in these structures is the trigonal prismatic arrangement of the six central iron atoms. These iron atoms lie on the surface of a sphere with a radius of 2.0 Å from the cofactor center and are each coordinated to three inorganic sulfur atoms. Furthermore, all nine sulfur atoms of the FeMo-cofactor are themselves equidistant from the center on a second sphere with a radius of 3.3 Å. The structure of the P-cluster, with composition [8Fe:7S], can be considered composed of two [4Fe:3S] subclusters that are bridged by a hexacoordinate S, with the overall assembly coordinated to the protein through six cysteine ligands.

Analysis of crystallographic structures of the MoFe-protein at resolutions up to 1.7 Å (*13*) indicated a significant (>6σ), positive $F_o - F_c$ difference density peak in the central cavity of the FeMo-cofactor. However, the corresponding $2F_o - F_c$ electron density maps did not show this feature. This contrasting behavior for the two maps suggested that the scattering properties of the whole cofactor might perturb the calculated electron density in its center through the influence of series termination effects. These are a well-known phenomenon in Fourier analyses, and in crystallography lead to resolution-dependent ripples around atomic positions (*14–16*); the effect is particularly pronounced around regions of high electron density such as metal sites. To illustrate the effect, the electron-density distribution, ρ(*r*), adjacent to an iron atom can be calculated as a function of the high-resolution limit from the expression (*16*).

$$\rho(r) = \int_0^{1/d_{max}} 4\pi s^2 f_{Fe}(s) \frac{\sin 2\pi sr}{2\pi sr} ds \qquad (1)$$

where f_{Fe} is the atomic form factor for iron; $s = 1/d$, where d is the resolution; and d_{max} is the high-resolution limit for integration. If the Fourier transform is truncated by choosing a finite integration limit instead of $1/d_{max} = \infty$, the calculated $\rho(r)$ will show resolution-dependent series termination errors (Fig. 1). At a distance of $r = 2.0$ Å from an iron atom, reminiscent of the situation in the central cavity of the FeMo-cofactor, an artificial minimum with negative electron density is created for d_{max} (resolution) between 1.6 and 2.5 Å.

To model the diffraction behavior inside the FeMo-cofactor, the influence of the entire [Mo:7Fe:9S] unit must be considered. When Eq. 1 is used to calculate the scattering contributions from the various individual components, it is apparent that the density in the central cavity is influenced mainly by the six iron atoms at 2.0 Å and the nine sulfur atoms at 3.3 Å (Fig. 2). At lower resolutions, the negative ripples surrounding these multiple iron and sulfur atoms combine to produce sufficiently negative electron density in the cofactor center to completely obscure the electron density of a light atom at this site. Consequently, this implies that the "hole" at the center of the cofactor in the $2F_o - F_c$ electron density map is an artifact, rather than the peak in the $F_o - F_c$ difference density map (*16*).

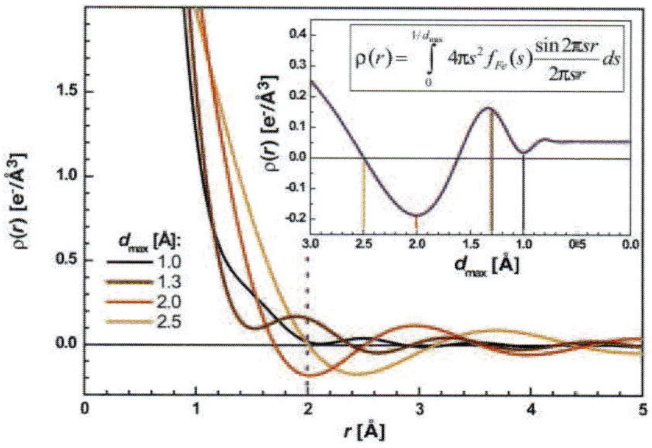

Fig. 1. The effect of series termination errors on the resolution-dependent electron density profile around an iron atom. A plot of electron density ρ(*r*) versus distance *r* (Eq. 1) shows varying effects for high-resolution limits d_{max} of 1.0 Å (black), 1.3 Å (brown), 2.0 Å (red), and 2.5 Å (orange). When ρ(*r*) is plotted versus d_{max} for the distance $r = 2.0$ Å (as found in the FeMo-cofactor), a characteristic profile is obtained with resolution-dependent maxima and minima (**inset**).

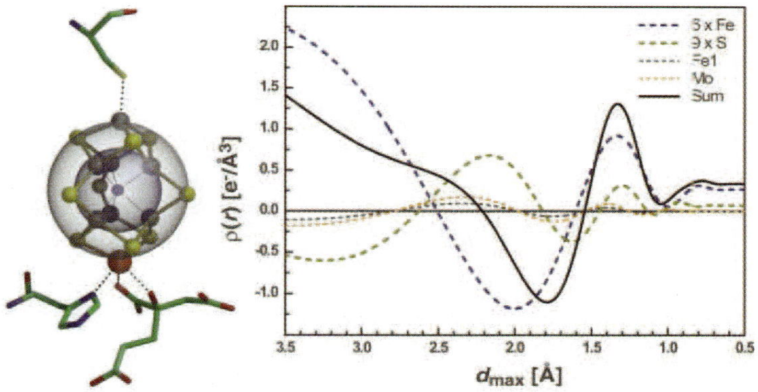

Fig. 2. Contributions of individual atom types to the resolution-dependent electron density profile in the central cavity of the FeMo-cofactor. Six iron atoms and all nine of the cluster's sulfur atoms are located on two concentric spheres. Having identical distances from the center (3.3 Å for sulfur, 2.0 Å for iron), they are the main contributors to the electron density profile there. The apical iron and molybdenum atoms exert only a minor influence. Plots of ρ(*r*) versus d_{max}, calculated analogously to Fig. 1 (inset), illustrate this effect. The curves for six iron atoms at 2.0 Å (blue), nine sulfur atoms at 3.3 Å (dark yellow), and one apical iron (Fe1, gray) and the molybdenum (orange) at 3.5 Å are shown. The sum of of all these contributions is shown in black

[1]Howard Hughes Medical Institute, [2]Division of Chemistry and Chemical Engineering, California Institute of Technology, Mail Code 147-75CH, Pasadena, CA 91125, USA. [3]Department of Biochemistry, University of Minnesota, Minneapolis, MN 55455, USA.

*To whom correspondence should be addressed. E-mail: dcrees@caltech.edu

As indicated in Fig. 2, series termination effects become less pronounced with increasing resolution. Hence we initiated the structure determination of the *Azotobacter vinelandii* MoFe-protein at a sufficiently high resolution to overcome their influence (*17*). To achieve this, we improved the quality of dithionite-reduced MoFe-protein crystals to permit data collection and refinement of the structure (Table 1) at 1.16 Å resolution to $R = 0.123$, with a diffraction component precision index [DPI (*18*)], a measure of the coordinate error, of 0.027 Å. These crystals contain two $\alpha_2\beta_2$ tetramers per asymmetric unit, and consequently four crystallographically independent copies of the FeMo-cofactor are present. Each copy clearly shows electron density at the center of the FeMo-cofactor in maps calculated with both $F_o - F_c$ and $2F_o - F_c$ Fourier coefficients (Fig. 3).

To characterize the atomic identity of the ligand at the center of the cofactor, we calculated a resolution-dependent electron density profile, analogous to that of Fig. 2, from the experimental diffraction data, which confirms the substantial impact of series termination errors (Fig. 4). The absolute values of electron density differ from those of the theoretical model (Fig. 2), owing to the absence of some low-resolution reflections, including the $F(000)$ term, in the experimental data. Consequently, the experimental electron-density minimum is even more pronounced than in the theoretical model. The electron density at the cofactor center, as determined from the observed structure factor amplitudes and phases calculated from a model lacking a central atom (black dots), is negative for all resolutions <1.55 Å. If, however, a purely theoretical curve is generated with both calculated structure factor amplitudes and phases from this same model (black line), the density at the cofactor center is substantially lower. Clearly, a central atom needs to be added to the model to reproduce the experimentally observed data. If the same model calculations are performed with an atom in the center of the cavity, the results are qualitatively similar and demonstrate that the destructive interference of the surrounding atoms is sufficient to obliterate the densities of a carbon, a nitrogen, or an oxygen atom at this position, although not that of a sulfur atom.

Unambiguous identification of an atom type solely from its electron density is problematic, even at atomic resolution. We considered carbon, nitrogen, oxygen, and sulfur as chemically plausible candidates

Table 1. Data collection and refinement statistics. For the calculation of R_{free}, 1% of the observed reflections were removed at random before refinement. a.u., asymmetric unit; ref., refinement; FOM, figure of merit; rmsd, root mean square deviation.

Resolution range	50.0–1.16 Å
Space group	$P2_1$
Unit cell	a = 108.3 Å
	b = 131.6 Å
	c = 159.2 Å
	β = 108.3°
$\alpha_2\beta_2$ per a.u.	2
Total reflections before adding partials	25,851,165
Independent reflections	1,390,520
Overall data redundancy	2.5
Completeness (outer shell)	95.6% (90.0%)
R_{merge}* (outer shell)	0.090 (0.485)
$I/\sigma(I)$ (outer shell)	10.0 (1.6)
Atoms in ref. (non-H)	37,388
Protein residues	3,708
Water molecules	5,025
Mean B value	15.6 Å^2
Overall FOM	0.923
R-factor (R_{free})†	0.123 (0.149)
rmsd bond lengths	0.020 Å
rmsd bond angles	2.251°
Estimated coordinate error (DPI)‡	0.027 Å

*$R_{merge} = \Sigma_{hkl}|\langle I\rangle - I|/\Sigma_{hkl}|I|$. †$R = \Sigma_{hkl}\|F_{obs}| - |F_{calc}\| / \Sigma_{hkl}|F_{obs}|$. ‡The diffraction component precision index (DPI) was calculated according to Cruickshank (*18*).

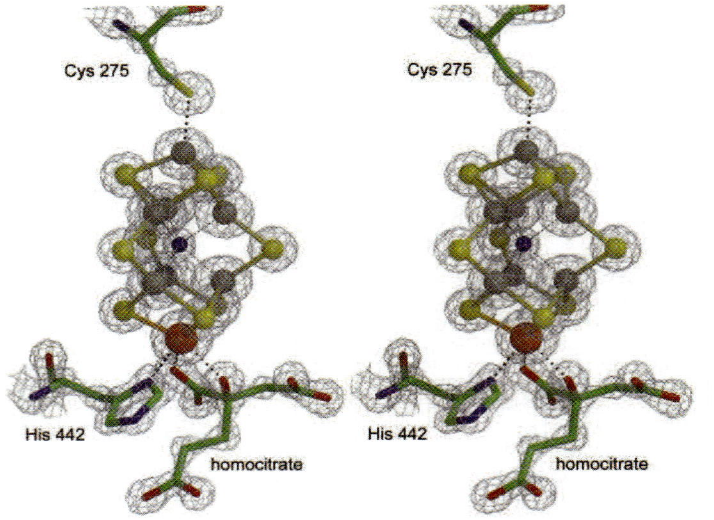

Fig. 3. Stereo representation of the FeMo-cofactor with the central ligand modeled as a nitrogen atom. The electron density map shown is a weighted $2F_o - F_c$ map of the 1.16 Å resolution structure of dithionite-reduced *A. vinelandii* MoFe-protein contoured at 3σ.

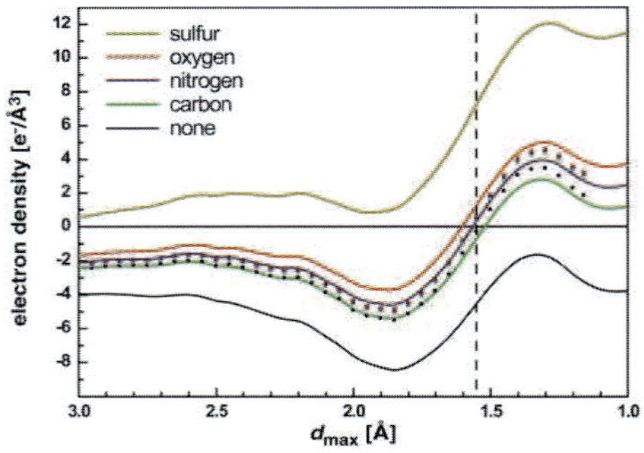

Fig. 4. Resolution-dependent electron density profiles as derived from the complete 1.16 Å resolution structure of dithionite-reduced *A. vinelandii* MoFe-protein. Solid lines are pure F_{calc} electron density values, whereas dotted curves are read from actual $F_{obs}\cdot\varphi_{calc}$ maps computed with the experimental structure factor amplitudes. The calculated density without a central ligand in the cofactor (black line) is well below the one observed in the experimental map (black dots). The latter further shows that positive density at the central position is obtained only at resolutions beyond 1.55 Å. A sulfur atom in the central cavity, however, gives a profile (dark yellow line) that at no resolution leads to negative electron density. Thus, a sulfur atom in the structure would not have been missed. If carbon (green line), nitrogen (blue line), or oxygen (red line) is placed in the center, their density still disappears, but experimental densities from structures refined with any of these atoms result in nearly identical curves (green, blue, and red dots), which agree most closely to the theoretical curve with nitrogen.

for the central atom and tested each with the available diffraction data. Because the surrounding atoms of the FeMo-cofactor are well defined and show very low temperature factor anisotropy, we assumed that the central ligand is fully occupied; partial occupancy should create at least a slight positional displacement in its environment. Although the possibility of a central sulfur atom was considered in initial models of the cofactor, of these four elements, we consider sulfur as the least likely candidate for the central atom because the observed density would allow only partial occupancy, the distances to the surrounding iron atoms are too short, and the destructive interference of the surrounding atoms is not sufficient to entirely cancel the density of the sulfur in the lower resolution structures, contrary to what is observed (Fig. 4). Refinement of a carbon, nitrogen, or oxygen atom at the central position in all cases yielded temperature factors for this site (C: 12.6 ± 2.2 Å^2; N: 14.0 ± 2.0 Å^2; O: 14.8 ± 2.5 Å^2) that correspond well to the values observed for the surrounding irons (12.2 ± 1.1 Å^2). Although the density for the ligand tends to be more anisotropic than for the inorganic components of the cofactor, it is not compatible with an ordered diatomic or larger species. Consequently, although the identity of the central ligand in the FeMo-cofactor cannot be unambiguously established from the crystallographic analysis, from the properties of the resolution-dependent electron-density profile (Fig. 4), and from the interaction of nitrogenase with dinitrogen and ammonia, we have tentatively assigned this central ligand as a fully occupied N.

Interatomic distances in the FeMo-cofactor between metals and the central nitrogen are summarized in Fig. 5. The central nitrogen ligand is hexacoordinate, with average iron-nitrogen distances of 2.00 ± 0.05 Å. This overall arrangement resembles that of a previously characterized cobalt carbonyl cluster containing interstitial nitrogen surrounded by a trigonal prismatic arrangement of metal (*19*), with an average Co-N distance of 1.94 Å. With average N–Fe–S bond angles of 102° ± 2°, the central ligand completes an approximately tetrahedral coordination environment for the six irons surrounding this group; consequently, it is no longer true that these iron sites are "three-coordinate" (*6*), at least in the dithionite-reduced form of the MoFe-protein.

The presence of a ligand in the center of the FeMo-cofactor, particularly a nitrogen, has important implications for understanding the properties of nitrogenase. One potentially relevant observation is the evidence from electron spin echo envelope modulation (ESEEM) studies for one or more nitrogen nuclei interacting with the FeMo-cofactor (*20–22*). Whereas the ESEEM signals have been assigned to nitrogen atoms of surrounding protein residues, the presence of a nitrogen atom in the cofactor suggests an alternate, nonprotein, source for at least some of the signal. Although a central nitrogen atom could be a structural component of the cofactor, it is difficult to conceive of a process whereby it is inserted without some relation to dinitrogen reduction. Indeed, a monoatomic nitrogen is consistent with the Thorneley and Lowe (*23*) kinetic model that requires the resting state of the MoFe-protein to be reduced by three electrons before dinitrogen can bind. This may reflect the need to replenish the electrons used in reducing the nitrogen to the level of nitride (N^{3-}) before it can be liberated as ammonia.

Theoretical studies of substrate binding to the FeMo-cofactor have indicated that the center of the trigonal prismatic arrangement of irons provides favorable interaction sites for dinitrogen and its reduction products (*24–26*). Furthermore, the spacing of iron atoms around this central site in the FeMo-cofactor closely parallels that of the iron surfaces used as catalysts for dinitrogen reduction in the industrial Haber-Bosch process (*27*). However, the distances between irons in the cofactor are longer (2.63 Å) than in regular metallic iron (2.47 Å), and such strained metal surfaces have been modeled to be particularly reactive as catalysts for dinitrogen dissociation (*28*). Notwithstanding the enormous disparity of reaction conditions, the parallels between the arrangement of metals in the nitrogenase FeMo-cofactor and the catalyst for the Haber-Bosch process suggest the possibility of common mechanistic elements in the reduction of dinitrogen to ammonia.

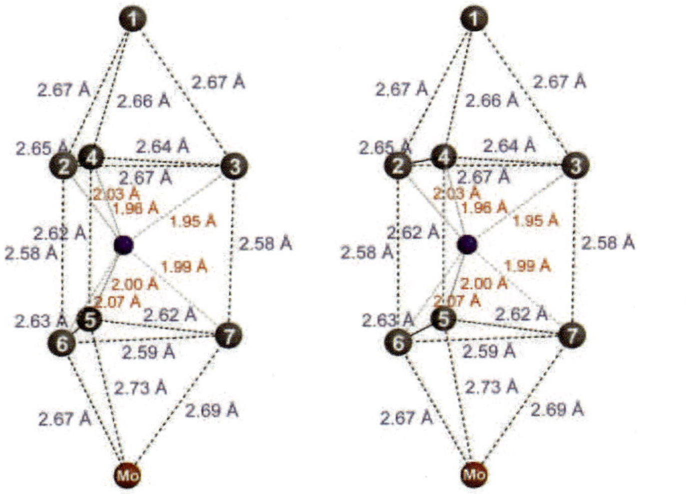

Fig. 5. Stereo representation of interatomic distances in the FeMo-cofactor. Mean values for the individual distances are the averages of the four crystallographically independent copies of the cofactor in the structure of the dithionite-reduced *A. vinelandii* MoFe-protein. Standard deviations are 0.01 Å for metal-metal and 0.03 Å for metal-nitrogen distances.

References and Notes

1. B. K. Burgess, D. J. Lowe, *Chem. Rev.* **96**, 2983 (1996).
2. J. B. Howard, D. C. Rees, *Chem. Rev.* **96**, 2965 (1996).
3. B. E. Smith, *Adv. Inorg. Chem.* **47**, 159 (1999).
4. J. Christiansen, D. R. Dean, L. C. Seefeldt, *Annu. Rev. Plant Physiol. Plant Mol. Biol.* **52**, 269 (2001).
5. D. C. Rees, J. B. Howard, *Curr. Opin. Chem. Biol.* **4**, 559 (2000).
6. J. Kim, D. C. Rees, *Science* **257**, 1677 (1992).
7. ———, *Nature* **360**, 553 (1992).
8. M. K. Chan, J. Kim, D. C. Rees, *Science* **260**, 792 (1993).
9. J. Kim, D. Woo, D. C. Rees, *Biochemistry* **32**, 7104 (1993).
10. J. T. Bolin, N. Campobasso, S. W. Muchmore, T. V. Morgan, L. E. Mortenson, in *Molybdenum Enzymes, Cofactors and Model Systems*, E. I. Stiefel, D. Coucouvanis, W. E. Newton, Eds. (ACS Symposium Series 535, American Chemical Society, Washington, DC, 1993), pp. 186–195.
11. J. W. Peters *et al.*, *Biochemistry* **36**, 1181 (1997).
12. S. M. Mayer, D. M. Lawson, C. A. Gormal, S. M. Roe, B. E. Smith, *J. Mol. Biol.* **292**, 871 (1999)
13. O. Einsle *et al.*, unpublished data.
14. W. L. Bragg, J. West, *Philos. Mag.* **10**, 823 (1930).
15. R. W. James, *Acta Crystallogr.* **1**, 132 (1948).
16. W. Cochran, H. Lipson, in *The Determination of Crystal Structures*, W. L. Bragg, Ed. (Cornell Univ. Press, Ithaca, NY, ed. 3, 1966).
17. Materials and methods are available as supporting material on *Science* Online.
18. D. W. J. Cruickshank, *Acta Crystallogr.* **D55**, 583 (1999).
19. S. Martinengo, G. Ciani, A. Sironi, B. T. Heaton, J. Mason, *J. Am. Chem. Soc.* **101**, 7095 (1979).

20. H. Thomann *et al.*, *J. Am. Chem. Soc.* **109**, 7913 (1987).
21. A. E. True, P. McLean, M. J. Nelson, W. H. Orme-Johnson, B. M. Hoffman, *J. Am. Chem. Soc.* **112**, 651 (1990).
22. H. I. Lee, K. S. Thrasher, D. R. Dean, W. E. Newton, B. M. Hoffman, *Biochemistry* **37**, 13370 (1998).
23. R. N. F. Thorneley, D. J. Lowe, in *Molybdenum Enzymes*, T. G. Spiro, Ed. (Wiley-Interscience, New York, 1985), vol. 1, pp. 221–230.
24. H. B. Deng, R. Hoffmann, *Angew. Chem. Int. Ed. Engl.* **32**, 1062 (1993).
25. K. K. Stavrev, M. C. Zerner, *Chem. Eur. J.* **2**, 83 (1996).
26. P. E. M. Siegbahn, J. Westerberg, M. Svensson, R. H. Crabtree, *J. Phys. Chem. B* **102**, 1615 (1998).
27. G. Ertl, H. Knözinger, J. Weitkamp, Eds., *Handbook of Heterogeneous Catalysis* (VCH, New York, 1997), vol. 4.
28. A. Logadottir, J. K. Nørskov, *Surf. Sci.* **489**, 135 (2001).
29. This work was supported by NIH. Stanford Synchrotron Radiation Laboratory operations are funded by the U.S. Department of Energy (DOE), Office of Basic Energy Sciences, and NIH. F.A.T. acknowledges the Burroughs Wellcome Fund and the Helen Hay Whitney Foundation for postdoctoral fellowships. We thank J. C. Peters, J. E. Bercaw, and H. B. Gray for enlightening discussions. Coordinates of the refined MoFe-protein structure have been deposited in the Protein Data Bank (accession code 1M1N).

Supporting Online Material
www.sciencemag.org/cgi/content/full/297/5587/1696/DC1
Materials and Methods
References

13 May 2002; accepted 26 June 2002

Chapter 12

Hengming Ke

Biography

I was born on August 14, 1948, in Wenling, a small town on the east coast of China. I obtained my B.S. equivalent from Zhongshan University in 1976, majoring in polymer chemistry. Until 1986 I worked as an assistant lecturer at Zhongshan University, then from 1981 to 1984 I was a visiting scholar at Harvard University. I returned to Zhongshan University as a lecturer and worked there until 1986, at which time I came back to Harvard as a research associate. I then completed my Ph.D. in biophysics at Harvard in 1989. Since 1990 I have worked at the University of North Carolina where I am a professor.

I have published some 70 articles in scientific journals as well as four book chapters. I have also given more than 60 invited lectures to academic and industrial groups as well as at international scientific conferences. I have served on the editorial board of the *Asian Journal of Biochemistry* (2006–2008) and *Current Bioactive Compounds* (2008–present), and have been the editor of the *Journal of Biological Sciences* since 2005. In addition, I have acted as an advisor to Suntory Pharmaceutical Research Laboratories, Boston, MA (2001–2003); Trimeris, Inc., in the Research Triangle Park, NC (2004–2006); and for the think tank on PDE5 inhibitory therapy for erectile dysfunction at GlaxoSmithKline and Bayer HealthCare (2002, 2003).

In our lab, our research has been focused on understanding the structure and function of medically important proteins. We have determined three-dimensional structures of ten proteins and characterized their enzymatic properties, including cyclic nucleotide phosphodiesterase (PDE), DJ-1, heat shock protein 90, DNA restriction endonuclease NaeI, ACC synthase, cyclophilin, calcineurin, fructose-1,6-bisphosphatase, chorismate mutase, and aspartate carbamoyltransferase. Recently, we have initiated the structure-based design of PDE inhibitors for the treatment of human conditions such as erectile dysfunction and Alzheimer's disease.

My Early Days Associated with the Colonel

I was the only graduate student of William N. Lipscomb (the Colonel) who came from mainland China. When China was opened to Western countries in the late 1970s, I was selected by the central government of China to study abroad. For the obvious reason, I dreamed of going to Harvard University to take advanced training. Luckily, I found Professor William Lipscomb in the Department of Chemistry at Harvard, who was happy to take me as a visiting scholar. However, after careful study, I realized that my background did not match at all with the research interests of the Colonel. I majored in polymer chemistry and was studying the crystalline morphology of polymers such as plastics by powder X-ray diffraction after I had become a junior faculty member in the Department of Chemistry at Zhongshan University, Guangzhou, China. The major research topics of the Colonel at that time were the three-dimensional structures and the enzymatic mechanisms of aspartate carbamoyltransferase and carboxypeptidase A. Obviously, my lifeboat was sailing into a surging sea of protein crystallography, a field I had no background in and had never been familiar with.

After I arrived at Harvard on June 8, 1981, I presented to the Colonel what I had done in the polymer morphology of plastics. I knew that my past work was nothing common to or overlapping with his research. However, the Colonel patiently listened and sometimes responded, "That's interesting." How wonderful I felt, a Nobel Laureate listening to my research! The Colonel's encouragement really laid down the cornerstone of my career as a protein crystallographer. It gave me great confidence that I would achieve my goal at Harvard.

With my background in general and computational chemistry, I did not have too much difficulty entering the palace of protein crystallography, but instead was faced with culture shock. As a young scholar trained in China, I was accustomed to obeying a senior, so in my first few months working with the Colonel, I felt everything was upside down. No instruction! The Colonel never told me what I should do and how it should be done, but asked me, "What's new?" and "Is there any news?" I got to learn and think for myself. When time was flying by, I realized that this is the best way the Colonel trained me. I grew quickly and have remembered forever what the Colonel did for me in the early days of my career. In fact, the two great characteristics of the Colonel, encouragement and giving freedom, have been what I have endeavored to pass on to younger generations.

Structural insight into substrate specificity of phosphodiesterase 10

Huanchen Wang*, Yudong Liu*, Jing Hou*†, Meiyan Zheng*, Howard Robinson‡, and Hengming Ke*§

*Department of Biochemistry and Biophysics and Lineberger Comprehensive Cancer Center, University of North Carolina, Chapel Hill, NC 27599-7260; and ‡Department of Biology, Brookhaven National Laboratory, Upton, NY 11973-5000

Edited by Joseph A. Beavo, University of Washington School of Medicine, Seattle, WA, and approved February 20, 2007 (received for review January 11, 2007)

Phosphodiesterases (PDEs) hydrolyze the second messengers cAMP and cGMP. It remains unknown how individual PDE families selectively recognize cAMP and cGMP. This work reports structural studies on substrate specificity. The crystal structures of the catalytic domains of the D674A and D564N mutants of PDE10A2 in complex with cAMP and cGMP reveal that two substrates bind to the active site with the same *syn* configuration but different orientations and interactions. The products AMP and GMP bind PDE10A2 with the *anti* configuration and interact with both divalent metals, in contrast to no direct contact of the substrates. The structures suggest that the *syn* configurations of cAMP and cGMP are the genuine substrates for PDE10 and the specificity is achieved through the different interactions and conformations of the substrates. The PDE10A2 structures also show that the conformation of the invariant glutamine is locked by two hydrogen bonds and is unlikely to switch for substrate recognition. Sequence alignment shows a potential pocket, in which variation of amino acids across PDE families defines the size and shape of the pocket and thus determines the substrate specificity.

crystal structure | cyclic nucleotides cAMP and cGMP

Cyclic nucleotide phosphodiesterases (PDEs) are enzymes hydrolyzing the second messengers adenosine and guanosine 3′,5′-cyclic monophosphates (cAMP and cGMP). The human genome encodes 21 PDE genes that are categorized into 11 families (1, 2). Selective inhibitors against individual PDE families have been developed as therapeutics for treatment of various human diseases (3–8). The best known examples are the PDE5 inhibitors sildenafil (Viagra), vardenafil (Levitra), and tadalafil (Cialis) that have been used for treatment of male erectile dysfunction (5). Sildenafil (Revatio) has also been approved for treatment of pulmonary hypertension (9).

PDE10 was independently identified by three groups in 1999 and shows a dual activity on hydrolysis of both cAMP and cGMP (10–12). PDE10 is highly expressed in brain striatum (13–16). Reduction of PDE10A mRNA and protein levels in striatum of transgenic mice implies a role of PDE10A in Huntington's disease (17, 18). Knockout mice experiments suggest that PDE10A is involved in regulating striatal output, possibly by reducing the sensitivity of medium spiny neurons to glutamatergic excitation (19). The PDE10 inhibitor papaverine is effective in improving executive function deficits associated with schizophrenia, and therefore inhibition of PDE10 may represent an approach to treatment of psychosis (20, 21).

PDE families contain a variable N-terminal regulatory domain and a conserved C-terminal catalytic domain. Individual PDE families show different substrate preferences. Crystal structures have been reported for the catalytic domains of seven PDE families in the unliganded form or in complex with inhibitors or products: PDE1B, PDE2A, PDE3B, PDE4B/4D, PDE5A, PDE7A, and PDE9A (22–34). However, it remains a puzzle how the conserved catalytic pocket of the PDE families selectively recognizes cAMP and cGMP. On the basis of the crystal structures of PDE4-AMP and PDE5-GMP, a "glutamine switch" mechanism was proposed for the substrate specificity (31). However, this mechanism is challenged by the mutagenesis experiments, in which the Q817A mutation in PDE5A1 did not significantly impact cAMP binding (35).

We have performed a systematic structural study on the substrate specificity of PDE10A2. Reported here are eight crystal structures of the catalytic domains of the wild-type PDE10A2 in the unliganded state or complex with products AMP and GMP, the D674A mutant and its complexes with substrates cAMP and cGMP, and the D564N mutant and its complex with cAMP. These structures reveal the conformation and interactions of substrates cAMP and cGMP in PDE10A2 and suggest a pocket that may determine the substrate specificity.

Results

Kinetic Properties of the Catalytic Domain of PDE10A2. The catalytic domain of PDE10A2 (449–789) has a K_m and k_{cat} of 56 nM and 0.33 s^{-1} for cAMP and 4.4 μM and 1.2 s^{-1} for cGMP. The k_{cat}/K_m values are 5.9 and 0.27 $s^{-1}\cdot\mu M^{-1}$, respectively, for cAMP and cGMP, and the specificity constant $(k_{cat}/K_m)^{cAMP}/(k_{cat}/K_m)^{cGMP}$ is 22. These numbers suggest a dual substrate specificity of PDE10A2, although the enzyme is more effective toward cAMP. The V_{max} values of our PDE10A2 catalytic domain are 507 nmol/min per mg for cAMP and 1,860 nmol/min per mg for cGMP and are comparable with those for the full-length PDE10A1 (10). There are two divalent metal ions at the active site of wild-type PDE10A2. The first ion coordinates with His[529], His[563], Asp[564], and Asp[674], and two waters, and has the same interactions as zinc in PDE4 (23). The second metal coordinates with Asp[564] and five water molecules.

The D674A mutant lost at least 4 orders of magnitude in the catalytic activity, as shown by an assay in which no activity was observed at 40 μg/ml D674A mutant, compared with only 5 ng/ml wild-type enzyme required for a normal assay. A consequence of the D674A mutation is loss of zinc ion, as confirmed by no electron density for the zinc site in the structures of unliganded D674A, D674A-cAMP, and D674A-cGMP. This observation implies that zinc is not critical for assembly of the PDE subdomains (23) but is essential for the catalysis. The

Author contributions: H.W. and H.K. designed research; H.W., J.H., and M.Z. performed research; Y.L. and H.R. analyzed data; and H.K. wrote the paper.

The authors declare no conflict of interest.

This article is a PNAS Direct Submission.

Abbreviations: PDE, cyclic nucleotide phosphodiesterase; S-pocket, substrate specificity pocket.

Data deposition: The atomic coordinates and structure factors have been deposited in the Protein Data Bank, www.pdb.org (PDB ID codes 2OUN, 2OUP, 2OUQ, 2OUR, 2OUS, 2OUU, 2OUV, and 2OUY).

†Present address: School of Pharmaceutical Sciences, Sun Yat-sen University, Guangzhou 510080, China.

§To whom correspondence should be addressed. E-mail: hke@med.unc.edu.

This article contains supporting information online at www.pnas.org/cgi/content/full/0700279104/DC1.

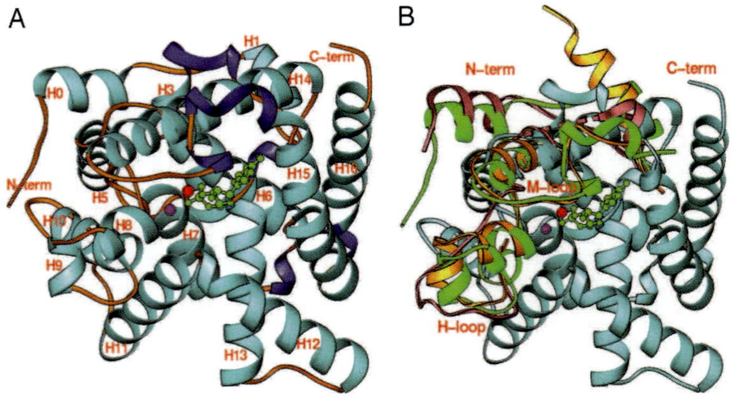

Fig. 1. Structures of the catalytic domain of PDE10A2. (*A*) Ribbon diagram. Divalent metals zinc and magnesium are shown in red and purple. Green balls–sticks represent cAMP. (*B*) Superposition of PDE10 (green ribbons) over PDE4 (gold) and PDE5 (salmon). Cyan ribbons represent comparable regions among three structures.

D564N mutant has the catalytic activity <1/1,000 of the wild-type PDE10A2, although both divalent metals are found in the D564N structures.

Architecture of PDE10 Structures. Although the PDE10A2 catalytic domain with residues 449–789 was used in the crystallization, only 449–770 can be traced unambiguously in the maps. The C-terminal residues 771–789 in all of the crystals and residues 571–586 of molecule B in the structures of PDE10A-GMP, D674A-cAMP, D674A-cGMP, and D564N-cAMP cannot be traced. Three residues from the expression vector, Ser^{446}, His^{447}, and Met^{448}, have clear electron density and are included in the structures.

The catalytic domain of PDE10A2 contains 15 α-helices and no β-strands (Fig. 1). The topological folding of PDE10A2 is the same as those of known structures of seven PDE families. The superposition of the PDE10A2 catalytic domain over the cAMP-specific PDE4D2 and the cGMP-specific PDE5A1 yielded rmsd values of 1.2 Å for the backbone atoms of 291 comparable PDE4D2 residues and 1.5 Å for 270 comparable PDE5A1 residues, indicating the overall similarity. The main difference between PDE10A2 and PDE4D2 occurs at the N terminus, in which H1 in PDE4D2 becomes a coil in PDE10A2 and H2 in PDE4D2 corresponds to a 3_{10} helix. The short helix H4 in PDE4 is missing in PDE10A2, probably resulting from the absence of 2 aa. However, PDE10A2 contains an extra helix at the N terminus, which occupies a location similar to that in PDE5A1 with a positional difference $\approx$3 Å. The H-loop of PDE10A2 (residues 571–592) contains two short α-helices and is comparable with that of PDE4D2 but not PDE5A1, in which the H-loop has variable conformations (34).

All eight crystal structures have the same space group with similar cell dimensions (Table 1) and contain two catalytic domains in their asymmetric units, which are related by a rotation axis of $\approx$100°. Because this angle differs from a twofold axis, the association of two PDE10A2 molecules appears to be crystallographic packing but is not biologically relevant. Superposition of molecules A over B yielded rmsd values of 1.4–1.6 Å for the eight structures, suggesting conformational changes of some loops. Indeed, residues 689–718 of the M-loop shift their C^{α} atoms as much as 7.5 Å, whereas other portions of the molecule remain comparable. The movements appear to be forced by the crystal lattice packing, as shown by the fact that the transposition of molecule A to B results in a clash with symmetry-related molecules.

No Significant Conformational Changes on Mutation or Substrate Binding. Structural superposition of the D564N and D674A mutants over the unliganded PDE10A2 yielded rmsd values of 0.10 and 0.12 Å for the C^{α} atoms of molecule A, 0.18 and 0.17 Å for molecule B, indicating no conformational changes on the mutations. The binding of cAMP, cGMP, or GMP caused 2.5–4.7% shrinking in the crystallographic *a* axis of D674A-cAMP, D564N-cAMP, D674A-cGMP, and PDE10A2-GMP (Table 1) and disorder of the H-loop residues 671–685 in molecule B of these structures. However, superposition of the unliganded PDE10A2 structure over PDE10A2-AMP, PDE10A2-GMP, D674A-cAMP, D564N-cAMP, and D674A-cGMP yielded rmsd values of 0.13, 0.24, 0.33, 0.28, and 0.29 for molecule A, and 0.16, 0.28, 0.72, 0.53, and 0.80 for molecule B, implying overall conformational similarity. A careful examination showed that residues Lys^{705}–Ile^{708} of the M-loop shift as much as 4 times the average rmsd in the five substrate–product complex structures. These changes appear to be the consequence of lattice contacts promoted by the hydrogen bond of cAMP/cGMP with the carbonyl oxygen of Leu^{706} of a symmetry-related molecule. In addition, the N-terminal residues 462–470 and 480–482 in molecule A of D674A-cGMP showed C^{α} positional changes 2–4 times the average difference because of the additional binding of cGMP to the N terminus. The only biologically significant change is the reorientation of the side chain of Leu^{635} of molecule A in the structures of D674A-cAMP and D564N-cAMP to avoid clashing with the ribose of cAMP.

Different Binding of Substrates cAMP and cGMP. Substrates cAMP and cGMP bind only at the active site of molecule A but not B of PDE10A2 (Fig. 2). The failure of substrate binding to B is probably because of the movement of both helix H14 and the M-loop into the active site [supporting information (SI) Fig. 5]. For example, Ile^{692} of H14 is located $\approx$4 Å to Phe^{729} in molecule B and thus blocks the interactions of the purines of cAMP and cGMP with Phe^{729} and Gln^{726}. The cAMP binding in the D674A-cAMP and D564N-cAMP structures is the same. The adenine of cAMP is sandwiched by Phe^{729} on one side and Ile^{692} and Phe^{696} on another side. N6 and N7 of adenine of cAMP form hydrogen bonds with OE1 and NE2 of Gln^{726}, respectively. The cyclic phosphate group of cAMP forms one hydrogen bond with His^{525} and three with water molecules. The ribose O5′ of cAMP hydrogen-bonds with a water molecule. O2′ forms two hydrogen bonds with a water molecule and the carbonyl oxygen of Leu^{706} from symmetry-related molecule B. In addition, cAMP interacts through van der Waals forces with residues Tyr^{524}, Leu^{635}, Leu^{675}, Val^{678}, Phe^{696}, and Met^{713} but does not directly contact the divalent metals. Substrate cAMP has a *syn* configuration and a 3′ *endo* ribose.

Three molecules of cGMP bind to the PDE10A2 catalytic

Table 1. Statistics on diffraction data and structure refinement

Data	10A native	10A-AMP	10A-GMP	D674A	D674A-cAMP	D674A-cGMP	D564N	D564N-cAMP
Space group	$P2_12_12_1$	$P2_12_12_1$	$P2_12_12_1$	$P2_12_12_1$	$P2_12_12_1$	$P2_12_12_1$	$P2_12_12_1$	$P2_12_12_1$
a, Å	51.1,	51.4	49.8	51.4	49.3	48.7	51.4	49.4
b, Å	82.0	82.0	81.9	82.0	82.3	82.0	82.2	82.3
c, Å	155.4	155.5	156.7	155.4	153.2	154.0	155.2	155.9
Resolution, Å	1.56	1.56	1.90	1.45	1.45	1.52	1.56	1.90
Reflections	85,847	87,893	48,795	110,590	101,420	95,665	91,666	47,078
Redundant	12.0	10.2	6.7	10.1	5.8	9.6	7.4	10.0
Complete, %	91.6 (58.5)*	93.3 (62.1)	94.8 (67.4)	94.4 (71.6)	90.9 (50.0)	99.9 (99.1)	97.0 (77.5)	92.3 (64.2)
Average I/σ	9.7 (2.2)	8.2 (2.0)	7.9 (2.0)	8.9 (2.1)	11.1 (2.2)	10.0 (2.5)	12.7 (3.3)	7.7 (2.1)
R_{merge}	0.067 (0.47)	0.094 (0.44)	0.071 (0.33)	0.077 (0.39)	0.073 (0.37)	0.082 (0.45)	0.064 (0.28)	0.078 (0.32)
Structure refinement								
R factor	0.197	0.200	0.213	0.209	0.218	0.204	0.209	0.217
R_{free}	0.223 (10)†	0.222 (10)	0.249 (10)	0.228 (10)	0.236 (10)	0.222 (10)	0.228 (10)	0.253 (10)
Reflections	81,787	83,987	47,053	101,832	96,206	92,068	89,422	45,261
rmsd for								
Bond, Å	0.0075	0.0064	0.0054	0.0066	0.0045	0.0063	0.0045	0.0058
Angle	1.3°	1.2°	1.2°	1.3°	1.3°	1.3°	1.1°	1.2°
Average *B* factor, Å^2								
Protein	22.9 (5,266)‡	19.5 (5,285)	27.1 (5,148)	20.6 (5,306)	19.6 (5,176)	19.7 (5,190)	19.9 (5,292)	26.9 (5,182)
Ligand		20.1 (23)	29.2 (24)		16.0 (22)	22.2 (69)		42.6 (22)
Waters	29.6 (425)‡	26.2 (399)	30.2 (322)	27.3 (425)	24.9 (437)	25.0 (366)	27.9 (458)	29.9 (320)
Zn	21.3 (2)‡	22.8 (2)	25.8 (2)				26.8 (2)	34.0 (2)
Mg	15.5 (2)‡	13.8 (2)	24.9 (2)	15.3 (2)	16.0 (2)	15.1 (2)	19.7 (2)	31.2 (2)

*The numbers in parentheses are for the highest resolution shell.
†The percentage of reflections is omitted for calculation of R_{free}.
‡The no. of atoms in the crystallographic asymmetric unit.

domain: one at the active site of molecule A and two in a pocket formed by the N-terminal residues of molecule A and symmetry-related molecules. The cGMP binding at the active site resembles that of cAMP in several aspects (Fig. 2), including the same *syn* configuration, shrinkage in *a* axis of the crystal, disorder of residues 571–586 in molecule B, and a hydrogen bond with the carbonyl oxygen of Leu^{706} from a symmetry-related molecule. However, significant differences between the binding of cAMP and cGMP are observed. The orientation of cGMP is flipped by ≈180° from that of cAMP, whereas their bases and phosphates occupy a similar location (Fig. 2). As result, the hydrogen bond of His^{525} with phosphate O2 of cAMP is swapped to O5′ of cGMP, and Gln^{726} forms only one hydrogen bond to N7 of cGMP.

An unexpected observation is that two molecules of cGMP bind to the N terminus of molecule A in D674A-cGMP. At this site, cGMP interacts with Ile^{450}–Ser^{453} of molecule A, symmetry-related residues of Leu^{547}, Leu^{654}–Leu^{656}, Gln^{743}–Pro^{747} of molecule A, and symmetry-related Leu^{466}–Lys^{470} of molecule B. The guanines of two cGMPs stack against one another and form three hydrogen bonds with Arg^{467} and Gln^{743} from symmetry-related molecules. The phosphate O2 forms two hydrogen bonds with Ser^{453} and Leu^{656}. Because this pocket does not exist in the structures in complex with cAMP, AMP, and GMP, it appears to be an artifact of the crystal packing. Nevertheless, the additional cGMP binding may lead to the positional shift of residues 462–471 and 480–483 of molecule B in D674A-cGMP, which is 2- to 4-fold the overall rmsd.

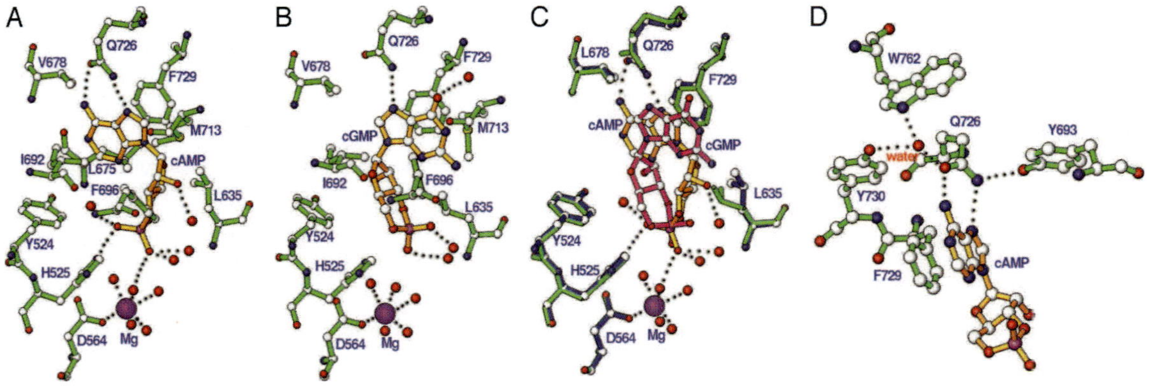

Fig. 2. Binding of cAMP and cGMP. (*A*) Interaction of cAMP (golden bonds) with PDE10A2 residues (green bonds) of the D674A mutant. The dotted lines represent hydrogen bonds. Small isolated red balls are water molecules. (*B*) Interaction of cGMP (gold) with D674A residues. (*C*) Superposition of cAMP (gold) over cGMP (pink). The residues from D674A-cAMP and D674A-cGMP are shown in blue and green, respectively. (*D*) Hydrogen bonds of Gln^{726} with Tyr^{693} and a water molecule bound to Trp^{762} and Tyr^{730}.

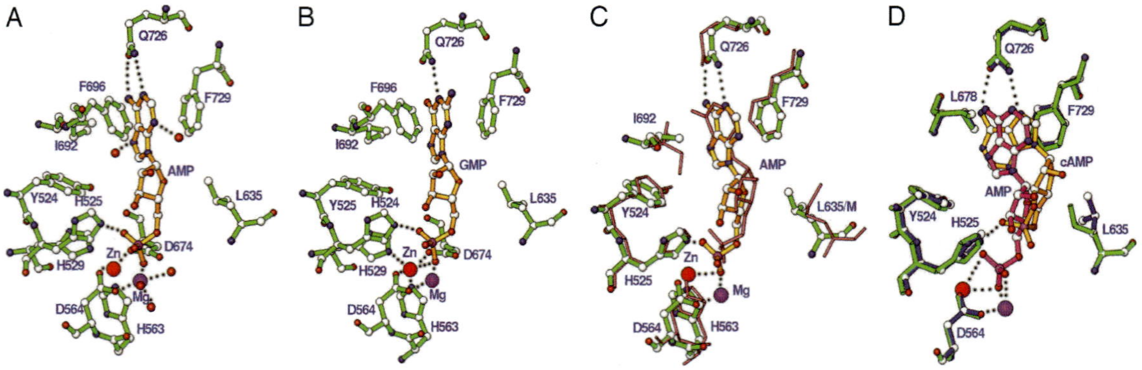

Fig. 3. Binding of products. (*A*) Interaction of AMP (gold) with PDE10A2 residues (green). (*B*) Interaction of GMP with PDE10A2 residues. (*C*) Superposition of PDE10A2-AMP over PDE4D2-AMP (salmon sticks) (27). (*D*) Superposition of AMP (pink) over cAMP (gold).

Products AMP and GMP Do Not Simulate Substrate Binding. The products AMP and GMP bind only to the active site of molecule A but not to molecule B. Both AMP and GMP have the same interactions and the same *anti* configuration and 3′ *endo* puckering (Fig. 3). The adenine of AMP and guanine of GMP are sandwiched by hydrophobic residues Phe^{729} on one side and Ile^{692} and Phe^{696} on another side. The adenine of AMP forms two hydrogen bonds with the side chain of Gln^{726} compared with only one between GMP and Gln^{726}. The interactions of AMP with PDE10A2 are the same as those in the structure of PDE4-AMP, in terms of hydrogen-bonding pattern, nucleotide conformation, and hydrophobic contact (Fig. 3*C*).

However, the contact pattern of the products AMP/GMP with the invariant glutamine is different from those of the substrates cAMP/cGMP although all of the ligands have the same stacking against Phe^{729} (Fig. 3*D*). The products have the *anti* configuration in contrast to *syn* of the substrates. Gln^{726} of PDE10A2 forms two hydrogen bonds with N1 and N6 of AMP and one with N1 of GMP. However, N6 and N7 of cAMP and N7 of cGMP are involved in hydrogen bonds with Gln^{726}. In addition, the phosphate groups of the products coordinate directly with both divalent metals and form hydrogen bonds with metal-bound residue Asp^{674} (Fig. 3). In comparison, the closest distance from the substrate atoms to the divalent metal ions is ≈4 Å, indicating no direct interaction.

Discussion

The Binding of cAMP and cGMP Simulates the Catalytic Process. Because the D674A and D564N mutants are basically inactive, it needs to be addressed whether the substrate binding in the crystal structures simulates the enzymatic process. The following observations suggest that the structural information is biologically relevant. First, the crystals of our PDE10A2 catalytic domain possess enzymatic activity, as shown by the fact that substrates cAMP and cGMP were hydrolyzed to AMP and GMP when the unmutated PDE10A2 crystals were soaked in the substrate solutions. Second, the wild-type PDE10A2 and the inactive mutants of D674A and D564N have almost identical conformations for the active-site residues, implying that substrate binding in the mutants simulates the biological process. Finally, the same conformation and interactions of cAMP in the mutant structures of D674A-cAMP and D564N-cAMP suggest that the metal ions do not directly impact the binding and conformation of the substrates. This hypothesis is supported by the fact that the closest distance of the substrate atoms to the metals is ≈4 Å in the D564N-cAMP structure. Therefore, we postulate that the substrate conformation and interaction in these crystal structures resemble those in the catalysis.

It has been reported that cAMP and cGMP exist in equilibrium between *syn* and *anti* configurations in solution, with *syn:anti* ratios of 30:70 and 95:5, respectively, for cAMP and cGMP (36). The $(F_o - F_c)$ and $(2F_o - F_c)$ maps, which were calculated from the structures before cAMP and cGMP were built in, show only the *syn* configuration of cAMP and cGMP bound to the active site of PDE10A2 (SI Fig. 6). Modeling of *anti* cAMP and cGMP into the pocket showed that the phosphate oxygen of the substrates clashes with two water molecules coordinated with zinc and magnesium, supporting that the *anti* conformers do not bind. Thus, the structures suggest that the *syn* conformer of cAMP and cGMP is the biological substrate for PDE10A2. However, it remains unknown which conformer of cAMP and cGMP is the genuine substrate for other PDE families. Early studies suggested that the preferable substrates are: *syn* cAMP and cGMP for PDE1 and PDE2, *anti* cAMP for PDE3 and PDE4, and *anti* cGMP for PDE3 and PDE5 (37, 38).

Because the substrates and products have different nucleotide configurations: *syn* for cAMP and cGMP, and *anti* for AMP and GMP, it would be interesting to know whether the enzyme takes an additional step to perform the *syn* to *anti* conversion during catalysis. To address this question, we carried out an activity assay in the presence of AMP and observed no significant inhibition by 5 mM AMP. On the basis of such poor binding of AMP, we hypothesize that the *syn* products leave the active site right after the catalysis, and the *syn* to *anti* conversion is not catalyzed by the enzyme but is an automatic process in solution. The occupancy of the *anti* AMP and GMP at the active site in the crystals may reflect the forced binding of the products at high concentration because they are predominant in solution (39). This argument is supported by the fact that only the *syn* configuration of 8-bromo-AMP was observed in the crystal when its *syn* configuration is predominant in solution (24, 39).

Implications for Substrate Specificity. The issue of the substrate specificity has not been extensively elucidated. The structures of PDE4D2-AMP, PDE4B2-AMP, and PDE5A1-GMP (24, 28, 31) show that the side chain of the invariant glutamine is fixed in opposite orientations in PDE4 and PDE5 and form two hydrogen bonds with products AMP and GMP. Because chemical formulas of AMP and GMP differ only in the phosphate portion from cAMP and cGMP, the conformation and interactions of nucleosides of the products have been assumed to simulate those of the substrates. This belief has led to a currently popular model, glutamine switch for the substrate specificity (31). In this mechanism, the glutamine is assumed to form two hydrogen bonds with cAMP but one with cGMP in cAMP-specific PDE families, and similarly cGMP-specific PDEs would have one hydrogen bond difference for the two

Table 2. Alignment of amino acids at the substrate-binding pocket

	678	685	689	692	693	713	725	726	729	730	762
PDE10A2	V	T	A	I	Y	M	G	Q	F	Y	W
PDE1B	P	H	T	L	M	L	S	Q	F	I	N
PDE2	Q	T	A	I	Y	M	L	Q	F	M	W
PDE3	P	H	T	I	V	F	L	Q	F	I	W
PDE11	V	S	A	V	T	F	L	Q	W	I	W
PDE4	P	Y	T	I	M	M	S	Q	F	I	Y
PDE7	P	S	S	V	T	L	I	Q	F	M	W
PDE8	P	C	A	I	S	V	S	Q	F	I	W
PDE5	I	Q	A	V	A	M	M	Q	F	I	W
PDE6	I	Q	A	V	A	M	L	Q	F	I	W
PDE9	E	A	V	L	L	F	A	Q	F	I	Y

substrates. For PDE families with dual substrate specificity, the side chain of the invariant glutamine would freely rotate to form two hydrogen bonds with cAMP or cGMP.

The glutamine switch mechanism is supported by the structure of dual cAMP/cGMP-specific PDE1B (31), in which the glutamine is unbonded and could be freely rotated. However, it was challenged by two lines of evidence. First, the Q817A mutation of PDE5 weakened K_m for cGMP by 60-fold but did not significantly change K_m for cAMP (35). Second, the structure of dual-specific PDE2A3 shows that the side chain of Gln^{859} is fixed by a hydrogen bond with Tyr^{827} (22). Thus, for the glutamine to switch, this preexisting hydrogen bond has to be broken, and the net gain of energy will be zero.

The glutamine switch is even less likely in view of our PDE10A2 structure, in which the side chain of Gln^{726} is locked by two hydrogen bonds with Tyr^{693} and a water bound to Tyr^{730} and Trp^{762} (Fig. 2). In addition, two hydrogen bonds between Gln^{726} and N6 and N7 of cAMP in the PDE10A2 structure are in a totally different pattern from the bidentate bonds of N1 and N6 of the model cAMP predicted by the glutamine switch (31). The single contact between Gln^{726} and N7 of cGMP (Fig. 2) differs entirely from the two predicted hydrogen bonds between the glutamine and N1 and O6 of cGMP (31). Most importantly, different conformations and interactions between the products and the substrates suggest that the products are not a reasonable model for the substrate binding, at least in PDE10, although it is unknown how similar they are in other PDE families.

The present PDE10A2 structures reveal that cAMP and cGMP bind to the same pocket but have different orientations and interactions, suggesting that the substrate specificity is determined by a different binding of the substrates in the PDE10 family. The structure-based sequence alignment of the nucleoside-binding pocket, that we tentatively term "substrate specificity pocket" or S-pocket, shows dramatic variation of amino acids across PDE families (Table 2 and Fig. 4). Apparently, the amino acid variation not only represents changes of the chemical groups in the PDE families but also defines the shape and size of the binding pocket. To match the chemical nature and shape of the pocket, substrates may bind to individual PDE families in the same or different orientations and configurations with different affinity. In some extreme cases, poor substrate binding, such as K_m = 3.9 mM for cGMP in PDE7 (33) and $K_m > 100\ \mu M$ for cAMP in PDE9 (2), would explain their preferred cAMP or cGMP specificity. The invariant glutamine in the S-pocket provides hydrogen bonds for binding of substrates but may not be a key element in distinguishing substrates. Instead, multiple elements at the S-pocket must work together to determine the substrate specificity. Each amino acid in the S-pocket may contribute to the specificity differently but should not exclusively dominate the substrate specificity.

Methods

Protein Expression and Purification of PDE10A2. The cDNA of the catalytic domain of human PDE10A2 (GenBank BAA84467) was purchased from American Type Culture Collection (Manassas, VA). The coding region for amino acids 449–789 of PDE10A2 was amplified by PCR and subcloned into the expression vector pET15b. The resultant plasmid pET-PDE10 was transferred into *Escherichia coli* strain BL21 (CodonPlus) for overexpression. The *E. coli* cell carrying pET-PDE10 was grown in 2× YT medium at 37°C to absorption A_{600} = 0.7, and then 0.1 mM isopropyl β-D-

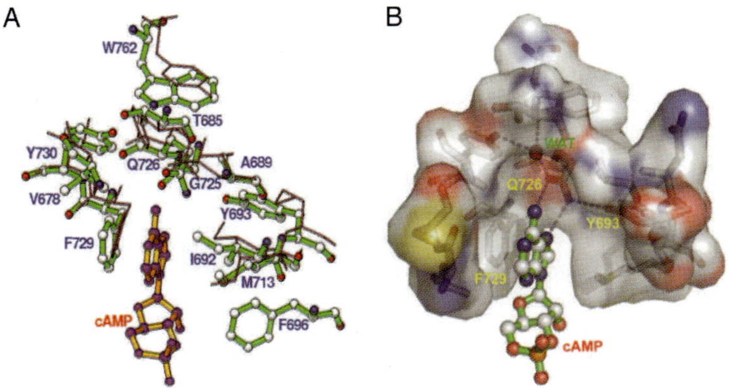

Fig. 4. A potential S-pocket. (*A*) The PDE10A2 residues (green bonds) are superimposed over the PDE4D2 residues (thinner salmon sticks). (*B*) Surface presentation of the S-pocket in PDE10.

thiogalactopyranoside was added for further growth at 20°C overnight. The recombinant PDE10 protein was purified on columns of nickel-nitrilotriacetic acid (Qiagen, Valencia, CA), Q-Sepharose, and Sephacryl S300 (Amersham Biosciences, Piscataway, NJ). A typical purification yielded ≈10 mg of PDE10A2 with a purity >95% from a 2-liter cell culture. The D674A and D564N mutants were produced by a QuikChange site-directed mutagenesis kit (Stratagene, La Jolla, CA) and verified by DNA sequencing. Overexpression and purification of the mutants used the same protocols for the wild-type protein.

Enzymatic Assay. Enzymatic activity of the catalytic domains of PDE10A2 and its mutants was assayed by using [^{3}H]cAMP or [^{3}H]cGMP (20,000 cpm per assay) as substrates in a reaction mixture of 20 mM Tris·HCl, pH 7.5/1.0 mM DTT/10 mM $MgCl_2$ at room temperature for 15 min (33). The reaction was terminated by the addition of 0.2 M $ZnSO_4/Ba(OH)_2$. Radioactivity of unreacted [^{3}H]cAMP or [^{3}H]cGMP in the supernatant was measured by a liquid scintillation counter. The turnover rate was measured at nine concentrations of substrate and at hydrolysis of 15–50% substrate. Each measurement was repeated three times. The parameters of K_m, k_{cat}, and V_{max} were calculated following steady-state kinetics.

Crystallization and Structure Determination. All crystals were grown by hanging drop and have the space group $P2_12_12_1$ with similar cell dimensions (Table 1). The unliganded PDE10A2 (449–789) was crystallized against a well buffer of 0.1 M Hepes, pH 7.5/0.2 M $MgCl_2$/50 mM 2-mercaptoethanol (2-ME)/16% PEG 3350. Crystals of the D564N and D674A mutants were grown against a well buffer of 0.1 M Hepes, pH 7.5/0.1 M $MgCl_2$/100 mM 2-ME/13% PEG 3350. The complexes of PDE10A2-AMP, PDE10A2-AMP, D674A-cAMP, D674A-cGMP, and D564N-cAMP were prepared by soaking the unliganded crystals in 20 mM cAMP or 30 mM cGMP in a buffer of 16% PEG 8000/0.1 M Hepes, pH 7.5/0.1 M $MgCl_2$/60 mM 2-ME at 4°C for 1.5–6 h. The crystallization buffer plus 20% ethylene glycol or 15% PEG 400 was used as the cryosolvent. Diffraction data were collected on beamline X29 at Brookhaven National Laboratory and processed by program HKL (Table 1) (40).

The structure of the unliganded PDE10A2 was solved by the molecular replacement program AMoRe (41), using the PDE4D2 catalytic domain as the initial model. The rotation and translation searches yielded two solutions that are obviously distinct from background and have correlation coefficients of 11.2 and 10.4 and R factors of 0.538 and 0.539 for 4543 reflections between 4 and 8 Å resolution. When the two solutions were added together, the correlation coefficient and R factor were improved to 0.21 and 0.51. The electron density map was improved by the density modification package of CCP4. The atomic model was rebuilt by program O (42) and refined by program CNS (Table 1) (43). The refined catalytic domain of the unliganded PDE10A2 was used to solve other structures.

We thank beamline X29 at National Synchrotron Light Source for collection of diffraction data, Drs. Richard Wolfenden and Charles Carter for proofreading the manuscript, and Xiuyan Qin for assistance on cell culture and protein expression. This work was supported in part by National Institutes of Health Grant GM59791 (to H.K.).

1. Bender AT, Beavo JA (2006) *Pharmacol Rev* 58:488–520.
2. Mehats C, Andersen CB, Filopanti M, Jin SL, Conti M (2002) *Trends Endocrinol Metab* 13:29–35.
3. Truss MC, Stief CG, Uckert S, Becker AJ, Wafer J, Schultheiss D, Jonas U (2001) *World J Urol* 19:344–350.
4. Liu Y, Shakur Y, Yoshitake M, Kambayashi JJ (2001) *Cardiovasc Drug Rev* 19:369–386.
5. Corbin JD, Francis SH (2002) *Int J Clin Pract* 56:453–459.
6. Lipworth BJ (2005) *Lancet* 365:167–175.
7. Castro A, Jerez MJ, Gil C, Martinez A (2005) *Med Res Rev* 25:229–244.
8. Menniti FS, Faraci WS, Schmidt CJ (2006) *Nat Rev Drug Discov* 5:660–670.
9. Galie N, Ghofrani HA, Torbicki A, Barst RJ, Rubin LJ, Badesch D, Fleming T, Parpia T, Burgess G, Branzi A, *et al.* (2005) *N Engl J Med* 353:2148–2157.
10. Soderling SH, Bayuga SJ, Beavo JA (1999) *Proc Natl Acad Sci USA* 96:7071–7076.
11. Fujishige K, Kotera J, Michibata H, Yuasa K, Takebayashi S, Okumura K, Omori K (1999) *J Biol Chem* 274:18438–18445.
12. Loughney K, Snyder PB, Uher L, Rosman GJ, Ferguson K, Florio VA (1999) *Gene* 234:109–117.
13. Seeger TF, Bartlett B, Coskran TM, Culp JS, James LC, Krull DL, Lanfear J, Ryan AM, Schmidt CJ, Strick CA, *et al.* (2003) *Brain Res* 985:113–126.
14. Kotera J, Sasaki T, Kobayashi T, Fujishige K, Yamashita Y, Omori K (2004) *J Biol Chem* 279:4366–4375.
15. Xie Z, Adamowicz WO, Eldred WD, Jakowski AB, Kleiman RJ, Morton DG, Stephenson DT, Strick CA, Williams RD, Menniti FS (2006) *Neuroscience* 139:597–607.
16. Coskran TM, Morton D, Menniti FS, Adamowicz WO, Kleiman RJ, Ryan AM, Strick CA, Schmidt CJ, Stephenson DT (2006) *J Histochem Cytochem* 54:1205–1213.
17. Hebb AL, Robertson HA, Denovan-Wright EM (2004) *Neuroscience* 123:967–981.
18. Hu H, McCaw EA, Hebb AL, Gomez GT, Denovan-Wright EM (2004) *Eur J Neurosci* 20:3351–3363.
19. Siuciak JA, McCarthy SA, Chapin DS, Fujiwara RA, James LC, Williams RD, Stock JL, McNeish JD, Strick CA, Menniti FS, Schmidt CJ (2006) *Neuropharmacology* 51:374–385.
20. Siuciak JA, Chapin DS, Harms JF, Lebel LA, McCarthy SA, Chambers L, Shrikhande A, Wong S, Menniti FS, Schmidt CJ (2006) *Neuropharmacology* 51:386–396.
21. Rodefer JS, Murphy ER, Baxter MG (2005) *Eur J Neurosci* 21:1070–1076.
22. Iffland A, Kohls D, Low S, Luan J, Zhang Y, Kothe M, Cao Q, Kamath AV, Ding YH, Ellenberger T (2005) *Biochemistry* 44:8312–8325.
23. Xu RX, Hassell AM, Vanderwall D, Lambert MH, Holmes WD, Luther MA, Rocque WJ, Milburn MV, Zhao Y, Ke H, Nolte RT (2000) *Science* 288:1822–1825.
24. Xu RX, Rocque WJ, Lambert MH, Vanderwall DE, Luther MA, Nolte RT (2004) *J Mol Biol* 337:355–365.
25. Sung BJ, Hwang KY, Jeon YH, Lee JI, Heo YS, Kim JH, Moon J, Yoon JM, Hyun YL, Kim E, *et al.* (2003) *Nature* 425:98–102.
26. Huai Q, Wang H, Sun Y, Kim HY, Liu Y, Ke H (2003) *Structure (London)* 11:865–873.
27. Huai Q, Colicelli J, Ke H (2003) *Biochemistry* 42:13220–13226.
28. Huai Q, Liu Y, Francis SH, Corbin JD, Ke H (2004) *J Biol Chem* 279:13095–13101.
29. Huai Q, Wang H, Zhang W, Colman R, Robinson H, Ke H (2004) *Proc Natl Acad Sci USA* 101:9624–9629.
30. Scapin G, Patel SB, Chung C, Varnerin JP, Edmondson SD, Mastracchio A, Parmee ER, Singh SB, Becker JW, Van der Ploeg LH, Tota MR (2004) *Biochemistry* 43:6091–6100.
31. Zhang KY, Card GL, Suzuki Y, Artis DR, Fong D, Gillette S, Hsieh D, Neiman J, West BL, Zhang C, *et al.* (2004) *Mol Cell* 15:279–286.
32. Card GL, England BP, Suzuki Y, Fong D, Powell B, Lee B, Luu C, Tabrizizad M, Gillette S, Ibrahim PN, *et al.* (2004) *Structure (London)* 12:2233–2247.
33. Wang H, Liu Y, Chen Y, Robinson H, Ke H (2005) *J Biol Chem* 280:30949–30955.
34. Wang H, Liu Y, Huai Q, Cai J, Zoraghi R, Francis SH, Corbin JD, Robinson H, Xin Z, Lin G, Ke H (2006) *J Biol Chem* 281:21469–21479.
35. Zoraghi R, Corbin JD, Francis SH (2006) *J Biol Chem* 281:5553–5558.
36. Yathindra N, Sunderalingam M (1974) *Biochem Biophys Res Commun* 56:119–126.
37. Butt E, Beltman J, Becker DE, Jensen GS, Rybalkin SD, Jastorff B, Beavo JA (1995) *Mol Pharmacol* 47:330–339.
38. Butt E, Beltman J, Becker DE, Jensen GS, Rybalkin SD, Jastorff B, Beavo JA (1995) *Mol Pharmacol* 47:340–347.
39. Lee CH, Evans FE, Sarma RH (1975) *J Biol Chem* 250:1290–1296.
40. Otwinowski Z, Minor W (1997) *Methods Enzymol* 276:307–326.
41. Navaza J, Saludjian P (1997) *Methods Enzymol* 276:581–594.
42. Jones TA, Zou J-Y, Cowan SW, Kjeldgaard M (1991) *Acta Crystallogr A* 47:110–119.
43. Brünger AT, Adams PD, Clore GM, DeLano WL, Gros P, Grosse-Kunstleve RW, Jiang JS, Kuszewski J, Nilges M, Pannu NS, *et al.* (1998) *Acta Crystallogr D* 54:905–921.

Chapter 13

Kurt Krause

Biography

Professor Kurt Krause is a structural biologist and a clinical specialist in infectious diseases with a particular interest in the pathogenicity of infectious agents such as streptococci, mycobacteria, and the HIV virus.

In 1977, Professor Krause graduated *summa cum laude* with a major in chemistry from Trinity University in San Antonio, TX. He then completed an M.D. with honors from Baylor College of Medicine and earned his M.A. and Ph.D. at Harvard University in the laboratory of Professor William N. Lipscomb, Jr. Following a short period of postdoctoral work, also at Harvard, he completed fellowships in internal medicine and infectious diseases at Baylor Affiliated Hospitals in Houston, TX. He served on the academic faculty of the University of Houston and Baylor College of Medicine and on the medical attending staff of Ben Taub General Hospital, the Methodist Hospital, and St. Luke's Hospital, before moving to New Zealand in 2005.

Professor Krause is currently the head of the Department of Biochemistry and the director of the Webster Centre for Infectious Diseases at the University of Otago in Dunedin, New Zealand. He is a member of the American Academy for the Advancement of Science, the American Chemical Society, the American Crystallography Association, and is a fellow in the Infectious Diseases Society of America.

Notes from the Colonel

From the time I first joined his research group, I loved getting notes from the Colonel. On special days at Gibbs he would leave a note for me or better yet he sometimes drew a figure for me during a meeting. He did this when he wanted to illustrate a scientific point, and since I always felt a little in awe of the Colonel during my time with him, I saved these notes. Probably I wanted to preserve a piece of my contact with him, because he represented scientific genius to me. In this tribute I want to describe four examples of these notes and their impact on my scientific life.

Smart Notes

When I first met with the Colonel to discuss possible research projects for my degree, he described several choices. This was in 1981 or 1982 and I remember that the path I followed in deciding to work with the Colonel was a bit convoluted. Originally I came to Harvard to study with Martin Karplus and Chris Dobson. I wanted to choose the popular joint NMR/theory program. But when I arrived I learned that Chris Dobson was returning to Oxford. Needing to find a mentor, I met with various faculty members in chemistry and described my interest in structural biology. No one ever mentioned the Colonel; they all mentioned Don Wiley, an ex-student of the Colonel, and another professor, Steven Harrison. It was as if the Colonel was a big secret in the hidden Gibbs Lab. Fortunately, I took a course called Enzymes from him. During his lectures he would frequently display incredible three-dimensional projections of enzymes in the classroom, and I was hooked. Seemingly without effort he would put up slides of enzymes in which the active sites would appear dangerously close to our heads. The whole thing was awesome, and I knew I wanted to work on the structural basis of enzyme action with him. Soon after this, I asked the Colonel if I could join his group. Fortunately he said yes, and he then proceeded to sketch out several projects I could pursue.

In Figure 13.1, the Colonel is illustrating a project involving the mechanism of carboxypeptidase A (CPA). Although Jean Evans was the *official* Gibbs artist, you can see perfectly the relations within the active site and how CPA was thought to interact with its substrate. Kaiser's work was mentioned along with the then recent low temperature work out of Chicago. This was enough to get me started and I plowed into a sack of potatoes to try to build on Doug Rees's potato inhibitor project.

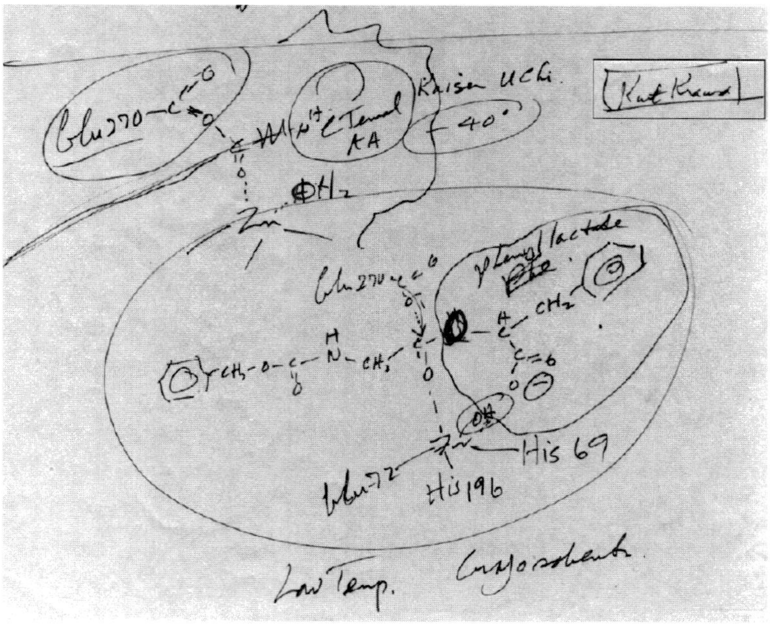

Figure 13.1 The mechanism of carboxypeptidase A (CPA).

Quick Notes

The second note from the Colonel, Figure 13.2, arose from a discussion we had about hydrogen bonds. After reviewing a treatment of hydrogen bonds in Pauling and Wilson, I was interested in the crystallographic proof of their existence. So I had one of those 9 p.m. meetings that are my most fond memory of the Colonel. Seemed to me that 9 p.m. was about the best time in the world to meet with him. He would arrive about 10 a.m. every day and immediately had meetings, classes, or phone calls all day. I never bothered him during the day, unless I had a draft of a paper to discuss. At 5 p.m. he would leave for dinner or tennis. After tennis he would return to work, and about 9 p.m. he was always available to talk.

The Colonel always seemed to know everything and could explain things in a way that a Texan like myself could understand. Probably a Southerner to Southerner thing, I bet! After a few minutes of discussing the math involved in H-bonds, he peeled off a Post-it note and drew an elegant illustration of their existence based only on the van der Waals radii of the involved atoms and the covalent radius of a nitrogen-hydrogen bond. Wolfram Sanger's book notwithstanding, this is still my most favorite figure explaining H-bonds and I use it almost every year in undergraduate lectures.

Blue Notes

Crystallography has certainly changed since I left the Gibbs Lab. I don't mean just technically, which is predictable and almost always good, but also sociologically. In the

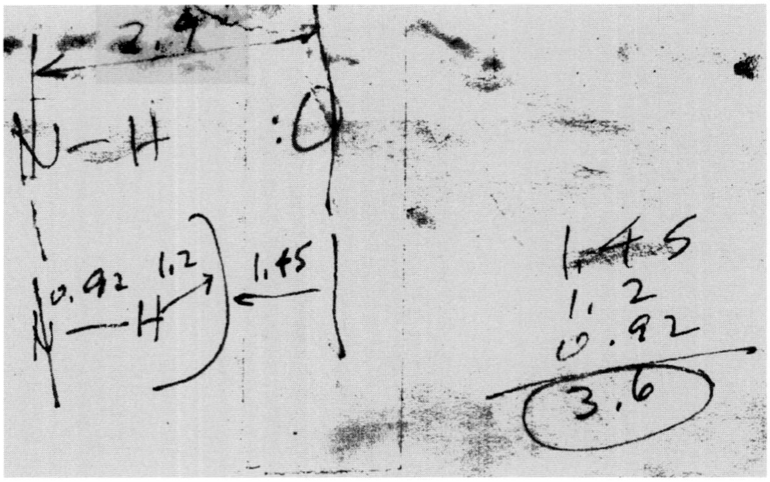

Figure 13.2 Hydrogen bonds.

1980s, it seemed like protein crystallographers comprised a small and reasonably friendly club, where everyone knew everyone. The Colonel brought visitors almost every week and often we would get to go to lunch or dinner with his guests. D. C. Phillips came by Gibbs with his camera and snapped away. Lyle Jensen and the Colonel described how far computing had come in the prior decade. They said they had to hand-wire a lot of their own computer backplanes, and apparently never had the chance to use the new "modern" Optronics units that we had at Gibbs. Another graduate student inquired on the way back to Gibbs about what "stripping beaver's lips" had to do with crystallography!

Paul Sigler and Brian Matthews were both honored guests and the latter helped set the stage for the "blue note" that can be seen in Figure 13.3. In collaboration with Woody Hastings, a bioluminescence icon at Harvard based in the biolabs, we had started a collaboration on bacterial luciferase. Before starting, the Colonel made sure I spoke to everyone known to be involved in luciferase crystallization research in order to make sure they were no longer actively pursuing that structure. That meant speaking with Paul Sigler and Brian Matthews, among others. Projects took so much effort at that time that direct competition was uncommon. Eventually we were given the go ahead on this project and by 1985, Manfred Kurfurst and I had succeeded in growing crystals that diffracted to 3Å! As we were making preparations to collect a native data set, I came to work to find a note on my desk announcing that Brian Matthews was again working on luciferase and that he had already collected native data!

We were disappointed but we stopped our work and waited for their structure, which interestingly only appeared after several more years. In structural biology today the competition is quite fierce; I sometimes feel that we have lost some of our civility.

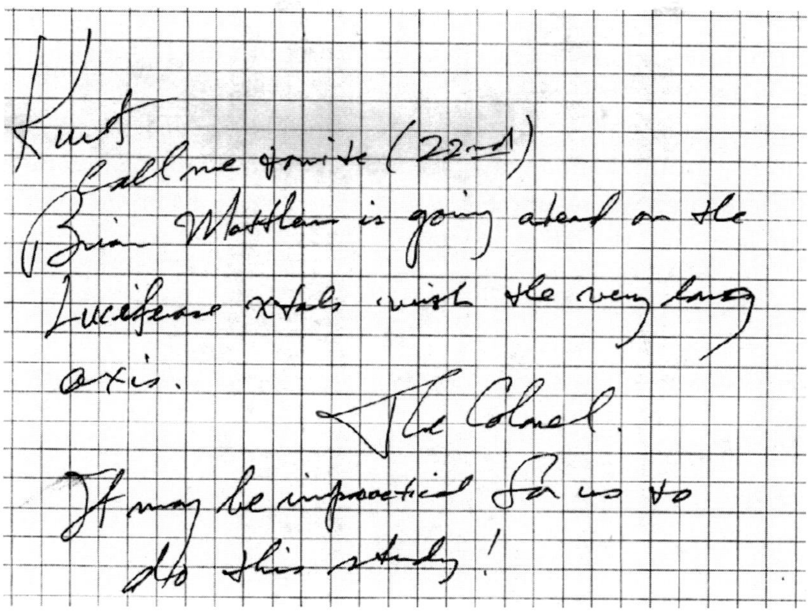
Kurt
Call me tonite (22nd)
Brian Matthews is going ahead on the
Luciferase xtals with the very long
axis.
The Colonel.
It may be impractical for us to
do this study!

Figure 13.3 Notification that Brian Matthews was again working on luciferase crystallization.

Figure 13.4 The Colonel's other kind of inspiring notes.

True Notes

No discussion of notes from the Colonel could end without mention of the truest notes he produced, those from his beloved clarinet. On many evenings after dinner and tennis, the Colonel would return to Gibbs to practice the clarinet. The notes I recall most are from Mozart's Clarinet Concerto in A Major. The Colonel was working on this piece during my tenure at Gibbs; I believe he was preparing for a concert. I suspect this piece must have been a favorite of his as well, because I remember hearing bits and pieces on and off during the evenings at Gibbs. The notes reproduced in Figure 13.4 are from a related work for clarinet and strings, also by Mozart. The Colonel played so well and with such spirit that he was inspiring. I remember that he carved all of his own reeds, and that he would head off to music camp every summer to sharpen his skills. To me this was illustrative of his nature, certainly of his virtuosity, and of his appreciation that life outside of science was important, that art was as valuable as science, or maybe that science and art were all the same thing.

As I reflect on my time at Harvard, I am very grateful to the Colonel for letting me join his group. My years at Gibbs Lab were the best years of my life. The projects were the most exciting and most important with which I have ever been involved. Whatever we try to accomplish in my group, I am always thinking about how the Colonel would approach this problem or interpret this result. He was the best mentor I could ever hope for and I think about him, Jean, and the lab every day.

J. Mol. Biol. (1987) 193, 527–553

2·5 Å Structure of Aspartate Carbamoyltransferase Complexed with the Bisubstrate Analog N-(phosphonacetyl)-L-aspartate

Kurt L. Krause, Karl W. Volz† and William N. Lipscomb

Gibbs Chemical Laboratory
Harvard University
Cambridge, MA 02138, U.S.A.

(Received 27 May 1986, and in revised form 21 August 1986)

In an X-ray diffraction study using the method of multiple isomorphous replacement, the structure of aspartate carbamoyltransferase (EC 2.1.3.2) complexed with the bisubstrate analog N-(phosphonacetyl)-L-aspartate (PALA)‡ has been solved to 2·5 Å. Ten rounds of model building and 123 cycles of restrained reciprocal space refinement have resulted in a model containing 94·4% of the theoretical atoms of the protein–inhibitor complex with an R-factor of 0·231. The fit of the model to the density is excellent, except for occasional side-chains and two sections of the regulatory chains that may be disordered.

The electron density for the PALA molecule is readily identifiable for both catalytic (c) chains of the asymmetric unit and bonding interactions with several important residues including Ser52, Arg54, Thr55, Ser80, Lys84, Arg105, His134, Arg165, Arg229 and Gln231 are apparent. The carboxylate groups of the PALA molecule are in a nearly *cis* conformation.

Gross quaternary changes between the T and R forms are noted and in agreement with earlier work from this laboratory. Namely, in the new structure the catalytic trimers move apart by 12 Å along the 3-fold axis of the enzyme and relocate by 10° relative to each other, adopting a more eclipsed position. The regulatory (r) chains in the new structure reorient about their 2-fold axis by 15°.

Large tertiary changes that include domain migration and rearrangement are also present between these two forms. In the R form both domains of the catalytic chain relocate closer to each other in order to bind to the inhibitor. The polar domain seems to bind primarily to the carbamoyl phosphate moiety of PALA, and the equatorial domain binds primarily to the L-aspartate moiety. Other changes in tertiary structure bring the 80s loop (from an adjacent catalytic chain) and the 240s loop into a position to interact with the PALA molecule.

Changes have been searched for in all interface regions of the enzyme. While the C1–C4§ and C1–R4 regions have been completely altered, most of the other interchain interfaces are similar in the T and R forms. The intrachain interfaces, between domains of the same catalytic chains, have undergone some reorganization as these domains move closer to each other when the inhibitor is bound.

This new structure allows a reinterpretation of genetic and chemical modification studies done to date. As such, the roles of Ser52, Lys84 and His134 are supported. Other residues implicated in similar studies such as Cys47, Tyr165, Lys232 and Tyr240 are too far from the inhibitor to have a direct interaction.

† Present address: Section of Biochemistry, Molecular and Cell Biology, 263 Clark Hall, Cornell University, Ithaca, NY 14833, U.S.A.

‡ Abbreviations used: PALA, N-(phosphonacetyl)-L-aspartate; c, catalytic chain; r, regulatory chain; m.i.r., multiple isomorphous replacement; 80s loop etc.: indicates a loop of residues 80 to 90 etc.

§ Capital letters, C and R, followed by a number, e.g. C1, R2, refer to a particular chain in a particular location of the dodecamer as specified in Fig. 1. Lower-case letters, c and r, in Figures and text refer to catalytic and regulatory chains as a type classification only and do not refer to a particular chain in the enzyme. Lower-case letters are frequently used in combinations with subscripts to create combinations such as: regulatory dimers, r_2: catalytic trimers, c_3, etc.

0022-2836/87/030527-27 $03.00/0

Finally, on the basis of the above results, a preliminary model for the homotropic transition is offered. In it, the change to the R-form is triggered by the binding of L-aspartate to the equatorial domain of a catalytic chain in which carbamoyl phosphate is bound to its site in the polar domain. Each C1–C4 group within the enzyme appears to be able to adopt its more active form somewhat independently, but the reorientation of the regulatory dimers that accompanies the homotropic transition serves to facilitate a T to R state transition in the other C1–C4 units. During the change from the T to R states most contact areas between r/c pairs and between r/r pairs are preserved. This is consistent with the suggestion that c-r-r-c is a functionally important unit. In addition, direct communication between C1–C4 is also consistent with the new structure.

1. Introduction

Several examples of enzymes that undergo large conformational changes induced by ligand or substrate binding have been reported recently in crystallographic studies. For example, hexokinase, liver alcohol dehydrogenase, citrate synthase, and glucose 3-phosphate dehydrogenase have all been solved in at least two forms, one with and one without substrates (Bennett & Steitz, 1978, 1980*a*,*b*; Eklund *et al.*, 1981; Remington *et al.*, 1982; Leslie & Wonacott, 1984). In many cases, the motions involved in binding can be described approximately as simple rotations of one domain of the protein relative to the other. Often large changes within the active sites occur to accommodate the presence of substrates. In each case, description of these conformational changes has helped to advance knowledge of the involved enzyme and its catalytic mechanism.

We report here the structure of another important enzyme, aspartate carbamoyltransferase, that undergoes ligand-induced conformational changes. The structure of the T form, which the enzyme adopts in the absence of substrates, has been studied since 1967 and has recently been reported in refined form (Ke *et al.*, 1984). Now the structure of the R form has been solved to 2·5 Å in the presence of the bisubstrate analog, *N*-(phosphonacetyl)-L-aspartate (PALA)†. Because of the long history of biochemical, genetic, and crystallographic investigations on this enzyme and because of its allosteric properties, aspartate carbamoyltransferase is an important subject for such a study.

Aspartate carbamoyltransferase (EC 2.1.3.2) from *Escherichia coli* is the most studied allosteric enzyme (Creighton, 1983). Its biochemical properties are well known and have been subject to review (Gerhart, 1970; Jacobson & Stark, 1973; Kantrowitz *et al.*, 1980*a*,*b*). By forming carbamoyl-L-aspartate from carbamoyl phosphate and L-aspartate, it catalyzes the first committed step in pyrimidine synthesis for many prokaryotes (Jones *et al.*, 1955; Reichard & Hanshoff, 1956). The enzyme has a molecular weight of 310,000 and is composed of two types of chains, catalytic (c) and regulatory (r), which bind substrates and nucleotide triphosphate effectors, respectively (Gerhart & Schachman, 1965). These chains combine in a c_6r_6 ± dodecameric complex that exhibits D_3 symmetry (Fig. 1; Weber, 1968; Wiley & Lipscomb, 1968).

Although the reaction mechanism of aspartate carbamoyltransferase is biochemically interesting, the enzyme has been most intensively studied because of its complex regulatory properties. The binding of its substrates, carbamoyl phosphate and L-aspartate, to the catalytic chains is regulated homotropically (Gerhart & Pardee, 1962; Bethell *et al.*, 1968). In addition, the binding of these substrates is also regulated heterotropically in the presence of nucleotide triphosphates, which bind to the regulatory chains almost 60 Å from the nearest active site. CTP, one such effector, is the final product of the pyrimidine pathway; it inhibits catalysis, presumably helping to prevent excess CTP production in the cell (Gerhart & Pardee, 1962; Bethell *et al.*, 1968). ATP, the product of a parallel pathway of purine synthesis, has a stimulatory effect on catalysis (Gerhart & Pardee, 1962; Bethell *et al.*, 1968).

Aspartate carbamoyltransferase has been shown

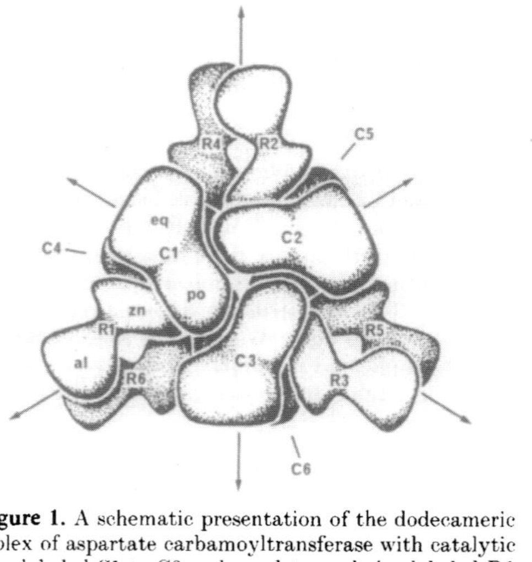

Figure 1. A schematic presentation of the dodecameric complex of aspartate carbamoyltransferase with catalytic chains labeled C1 to C6 and regulatory chains labeled R1 to R6. The domain substructures of the chains are identified as follows: polar (po), equatorial (eq), allosteric (al), and zinc (zn). The view is down the 3-fold axis with the 2-fold axes in the plane of the paper (Krause *et al.*, 1985).

† See footnote to p. 527.

to exist primarily in two major conformational states, the T (less active) form and the R (more active) form, whose properties approximate the Monod–Wyman–Changeux theory of allosterism (Gerhart & Schachman, 1968; McClintock & Marcus, 1968; Colman & Marcus, 1972; Blackburn & Schachman, 1977; Lauritzen & Lipscomb, 1980). The allosteric transition between these two forms occurs in the presence of substrates, and the nature of this transition has been studied intensively using spectroscopic (Collins & Stark, 1971; Eckfeldt *et al.*, 1970; Kempe & Stark, 1975), sedimentation (Gerhart & Schachman, 1968; Howlett & Schachman, 1977), and X-ray techniques, including low-angle scattering (Moody *et al.*, 1979) as well as single-crystal methods (Monaco *et al.*, 1978; Ladner *et al.*, 1982). These studies support the presence of large conformational changes between these two major forms of the enzyme. Other work indicates that a simple two-state model, although a good first approximation is an oversimplification of a myriad of conformational states available to the enzyme (Bethell *et al.*, 1968; Hammes & Wu, 1971; Landfear *et al.*, 1978*a*; Kerbirou & Hervé, 1972, 1973; Knier & Allewell, 1978).

Although two refined structures of the T state exist (Ke *et al.*, 1984), structural studies on the R state of aspartate carbamoyltransferase were initially hampered by substrate turnover and the instability of one of its natural substrates, carbamoyl phosphate. Most researchers wishing to probe the more active state of the enzyme have had to rely on one or more substrate analogs. The most prominent of these compounds is the bisubstrate analog PALA. It was first synthesized in 1971 (Collins & Stark, 1971) and has been used extensively in studies on the R state of the enzyme. PALA is a stable tightly bound inhibitor ($K_i = 2{\cdot}7 \times 10^{-8}$) that resembles both carbamoyl phosphate and L-aspartate (Fig. 2). It is a competitive inhibitor of carbamoyl phosphate and an uncompetitive inhibitor of L-aspartate (Collins & Stark, 1971). Upon binding PALA, aspartate carbamoyltransferase is rapidly converted to the more active R form. Most investigators have found that from four to five molecules of PALA must bind to the enzyme to complete the conformational changes associated with this transition (Howlett & Schachman, 1977; Blackburn & Schachman, 1977; Johnson & Schachman, 1980). Some recent work indicates that the transition to the R form might occur after only two molecules of PALA are bound (Foote & Schachman, 1985). On the basis of a number of experimental studies, the R state induced by this molecule is felt by most investigators to be indistinguishable from that induced by natural substrates or by carbamoyl phosphate and the L-aspartate analog succinate (Howlett & Schachman, 1977; Collins & Stark, 1971; Moody *et al.*, 1979).

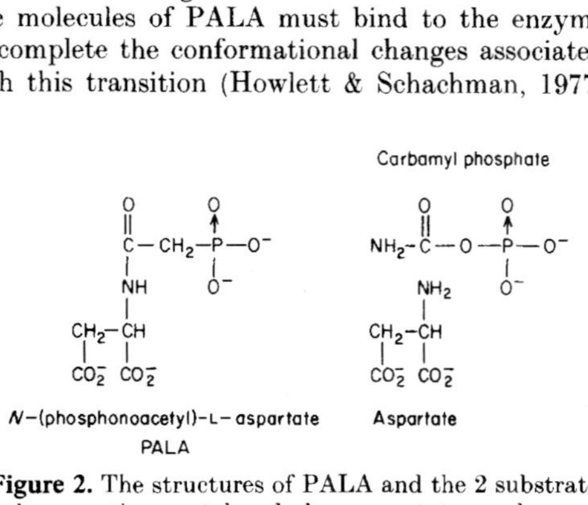

Figure 2. The structures of PALA and the 2 substrates of the reaction catalyzed by aspartate carbamoyltransferase. The compounds are arranged to emphasize their similarities (Heyde, 1976).

The crystal structure of aspartate carbamoyltransferase complexed with PALA has been pursued in order to reveal the detailed structure of the R state and to allow identification of the active site along with those residues important in substrate binding. A low resolution structure of aspartate carbamoyltransferase with PALA bound was reported by Ladner *et al.* (1982). The crystals used were of space group *P*321 and were grown in the presence of the inhibitor. The structure was solved on the basis of the unligated form of the enzyme, using a variant of the molecular replacement approach: each domain of the T state enzyme was allowed to move as a rigid unit within the unit cell of PALA-aspartate carbamoyltransferase. After each movement, the crystallographic *R*-factor and correlation coefficient between native and calculated structure factors were calculated. An attempt was made to locate the global optimum for these functions in reciprocal space as a function of domain motion. Using this method, the quaternary changes associated with PALA binding were identified. This form could not be refined adequately to higher resolution, and efforts were abandoned in favor of the multiple isomorphous replacement (m.i.r.) approach, which had been conducted in parallel.

A medium resolution structure of *P*321 PALA-aspartate carbamoyltransferase, based solely on m.i.r. methods, was described in a preliminary publication by Krause *et al.* (1985). We report here the results of further refinement of this form to 2·5 Å. A more thorough description of the quaternary changes and the tertiary changes between the T and R forms is made, as is a comparison between their active sites. The detailed binding of the PALA molecule to the R form is described. Changes in all unique interfaces within the enzyme are discussed. Genetic and chemical modification studies completed to date are reviewed with reference to the new R state structure. Finally, a tentative mechanism for the PALA-induced homotropic transition is offered.

2. Materials and Methods

(a) *Enzyme sources, crystal growth and heavy-atom preparation*

E. coli (provided by J. C. Gerhart) was grown at the New England Enzyme Center and used for the isolation and purification of aspartate carbamoyltransferase following the method of Gerhart & Holoubek (1967). After purification, the enzyme was stored at −20°C as an 18 mg/ml solution in a 1:1 (v/v) glycerol/storage buffer (40 mM-K_2PO_4, 2·0 mM-2-mercaptoethanol, 0·2 mM-EDTA (pH 7·0)) suspension. Before crystallization, 6 ml of the enzyme/glycerol suspension was dialyzed for 24 h against storage buffer alone. The solution was then diluted using storage buffer to 15 to 16 mg/ml. Subsequently, the solution was spun in an Eppendorf centrifuge for 5 min or filtered through a non-protein binding 0·22 μm filter (Millipore GV). Finally, crystallization was accomplished by dialyzing the enzyme for 1 to 2 weeks at 21°C against 50 mM-maleic acid, 1 mM-PALA, 3 mM-NaN_3, N-ethylmorpholine to pH 5·9, conductivity 3·4 mS. Temperature control was critical in obtaining a high yield of good quality crystals. The crystals grew as hexagonal bars, in space group P321 with unit cell dimensions $a = b = 122{\cdot}11$ Å, and $c = 156{\cdot}17$ Å. They diffracted to 2·3 Å resolution with essentially complete data present to 2·5 Å. The asymmetric unit contains 2 catalytic/regulatory units related by a non-crystallographic 2-fold axis of symmetry. Prior to data collection, native crystals were transferred, still within their microdialysis vials, to a 20% polyethylene glycol/crystallization buffer (w/v) solution at pH 5·8. Heavy-atom derivatives were made by transferring crystals to the polyethylene glycol/crystallization buffer solutions described above, but with either 1 mM-$K_2Pt(NO_2)_4$ for 4 to 7 days or 5·6 mM-$IrCl_3$ for 3 days. Changes in intensity on zero-level precession photographs were readily apparent for the platinum derivative, but only 1 small change for 1 reflection was noted for the iridium derivative. This difference may be a relevant point with regard to screening heavy-atom derivatives for crystals with such a high molecular weight in the asymmetric unit (a.u.) (total M_r/a.u. = 100,000).

(b) *Enzyme activity*

Aspartate carbamoyltransferase is known to be maximally active at pH 8·0 and inactive in solution at pH 5·8, the pH close to the isoelectric point at which these crystals were grown. There are, however, experimental results that suggest that, though inactive, the enzyme undergoes the same or similar conformational changes during the homotropic transition at the different pH values. These results would support the validity of using crystals grown at pH 5·8 in studies on the T and R form of the enzyme.

Lauritzen & Lipscomb (1980) found that substrate-induced difference spectra for the unligated catalytic trimer were essentially the same over a pH range of 6 to 8·5. In addition, they found a persistent differential increase in the rate of p-hydroxymercuribenzoate-induced disassociation of the T and R forms of the enzyme over this pH range, although the rate difference did increase with decreasing pH. Another investigation revealed close agreement between small-angle X-ray spectra taken at pH 8·0 on both T and R forms of the enzyme and calculated spectra derived from model co-ordinates obtained from structural studies of the T and R form at pH 5·8 (Altman et al., 1982).

Another structure of unligated aspartate carbamoyltransferase solved by X-ray diffraction methods has been reported (Young, 1984). This structure was solved from crystals grown at pH 7·0, where the enzyme retains some activity. Although this new structure is only partially refined, it shows only minor differences from the structure reported by Ke et al. (1984), incidating that no large changes have occurred in the change from pH 5·8. The initial phases, however, used in map generation for the Young structure were based on model co-ordinates from this laboratory; therefore, some bias towards finding a similar structure exists.

(c) *Data collection and analysis*

All data used for this study were collected at the Resource for Crystallography located in the laboratory of Dr Nguyen huu Xuong at the University of California, San Diego. Two multiwire proportional chambers (MARK II) (Hamlin et al., 1981) were used, together with an Elliot GX-6 rotating anode for data collection. The rotating anode was operated at 40 kV and 50 mA, and the CuKα X-ray beam was monochromatized. The rotating anode was fitted with a full-circle Nicolet goniostat, driven by a VAX 11/750 computer. Crystals of typical dimensions 0·7 mm × 0·7 mm × 1·2 mm were chosen for data collection and mounted in siliconized sealed glass capillary tubes. The protein solution from within the microdialysis wells in which the crystals were grown was used as mother liquor in mounting.

In the data collection process, the crystals were aligned on the gonionstat and the chambers were oriented at a chosen 2θ value. Crystal-to-chamber distance was chosen experimentally to maximize the simultaneous collection of diffraction data while maintaining spot separation. The average value was close to 800 mm. Helium pathways were usually employed between the chambers and crystals. Data collection proceeded by fixing values for φ and χ and scanning in 0·10° steps in ω ranging from −40° to +30°. The scanning rate was determined by the number of diffracted photons incident upon the chambers. After completion of a scan, new values were chosen for the fixed φ and χ angles and the ω scan was begun again. When sufficient replicates were collected for a given value of 2θ, the chambers were moved to a new angle and the entire process repeated until the data set was complete. By collecting high resolution data first, the same crystals could be used for low resolution data collection. An average of 2 to 3 crystals was used per data set, with total exposure times per crystal ranging up to 48 h. The intensity and σ of each observation along with all indexing and background correction were calculated simultaneously with the scanning process.

In this manner, 415,000 observations of 51,500 independent reflections were collected for native PALA-aspartate carbamoyltransferase. This data set is 97% complete to 2·5 Å and contains 4000 reflections to 2·3 Å. For the $K_2Pt(NO_2)_4$ derivative 330,000 observations of 42,600 reflections were recorded, and for the $IrCl_3$ derivative 95,000 observations of 22,000 reflections. These data sets are essentially complete to 2·7 Å and 3·4 Å, respectively. Using UCSD software (Hamlin et al., 1981), the observations were next scaled together in 5° increments of ω, averaged, and reduced by applying the usual Lorentz and polarization corrections. These scaled observations were then critiqued to uncover systematic errors in any 5° wedge of reflections. After deletions of incorrect measurements, the scaling process was repeated until convergence of the scaling R-factors occurred. Final R_{sym} measurements for the native, $K_2Pt(NO_2)_4$ and $IrCl_3$ data sets were 6·1%, 5·8% and 4·7%, respectively, based on intensities.

(d) *Heavy-atom phasing and refinement*

In preparation for the refinement of m.i.r. parameters, difference density maps produced with phases from the

Table 1
Heavy-atom parameters

Site	X	Y	Z[a]	Occupancy[b]	B (Å^2)[c]
Ir					
1	0·33333	0·66666	0·00487	0·570	64·9
2	0·39714	0·68506	0·20111	0·397	60·9
3	0·18457	0·31191	0·28629	0·655	55·6
4	0·33333	0·66666	0·48325	0·578	61·3
Pt					
1	0·79368	0·72190	0·07217	0·571	57·0
2	0·38372	0·38324	0·08497	0·497	64·9
3	0·40660	0·40472	0·10515	0·437	58·0
4	0·40899	0·52578	0·19677	0·294	120·0
5	0·38768	0·63784	0·20376	1·614	245·0
6	0·39644	0·67941	0·19878	0·415	41·9
7	0·71123	0·84937	0·28851	0·905	100·0
8	0·68919	0·87182	0·28724	0·504	50·0
9	0·81962	0·67925	0·40957	0·388	50·1
10	0·83401	0·69225	0·42771	0·517	73·5
11	0·39293	0·44436	0·43852	0·270	58·5

[a] Co-ordinates of the sites are listed as fractions of the unit cell axis along which they were measured.
[b] Fractional occupancy calculated on the basis of all reflections to 3·5 Å resolution.
[c] Isotropic temperature factor.

previously reported molecular replacement work (Ladner *et al.*, 1982) were used in determining the initial position of the heavy-atom sites. Patterson syntheses were also inspected to confirm the location of these sites. The *R*-factor, based on structure factor amplitudes, between platinum and native data sets was 11·7%, whereas the *R*-factor between the iridium and native data sets was 9·4%.

The locations, occupancies, and temperature factors of those sites located on difference density maps were next refined according to the method of Blow & Crick (1959), using either centroid or most probable phases, first to 4·0 Å and then to 3·5 Å. By using most probable phases in refinement we achieved a rapid increase in the phasing power for each derivative, an improvement in the overall figure of merit, and a large increase in the number of reflections with figure of merit greater than 0·7. Only 4 cycles of this type of refinement could be used during each set of calculations, as continued use of most probable phases allowed the iridium derivative to dominate and eventually to produce a worsening of the above statistics. Early electron density maps were generated using centroid phases based on parameters determined from either most probable refinements or centroid-weighted phase refinements. Centroid-weighted phase maps calculated from the most probable phase refinements were consistently of slightly higher quality, having a higher peak-to-noise ratio and clearer protein–solvent boundaries. As a result, all of our subsequent m.i.r. maps were calculated in this fashion.

After the initial m.i.r. refinement was completed, difference density maps revealed minor sites that were then incorporated in the refinement. Minor sites having occupancies below 0·25 were not included in the final refinement. All sites except site 4 of the platinum derivative were a part of a 2-fold related pair. The occupancies of the 2-fold related sites, however, were not usually equal.

Following the determination of the locations and occupancies for all of the sites, cautious attempts at refining each site's temperature factor, using best phases, were made. The temperature factors of all of the sites, except sites 4, 5 and 7 of the platinum derivative, refined to stable values after a few cycles. These sites eventually refined to very high, but also stable values. Final parameters for these refinements as well as final phasing statistics are shown in Tables 1 and 2.

(e) *Determination of the location of the non-crystallographic 2-fold axis*

After calculation of acceptable electron density maps based on heavy-atom phasing, determination of the location of the non-crystallographic 2-fold axis of symmetry was undertaken. An initial estimate of its location was calculated from the heavy-atom positions using a least-squares superposition of the symmetry related sites. This estimate was used as a starting point for a refinement method based on electron density (Cox, 1967). In this method, an ellipsoid of revolution was

Table 2
Phasing statistics for m.i.r. refinements

Figure of merit with resolution									
Resolution (Å)	27·3	13·84	9·27	6·99	5·59	4·66	4·00	3·50	Total
Number of observations	43	281	709	1316	2065	3111	4291	5596	17,412
Mean figure of merit	0·66	0·74	0·77	0·75	0·72	0·66	0·61	0·57	0·64
Platinum derivative statistics									
Resolution (Å)	27·3	13·84	9·27	6·99	5·59	4·66	4·00	3·50	Total
Number of reflections	40	275	701	1293	2041	3088	4259	5560	17,257
r.m.s. heavy atom *F*	597·8	441·0	357·7	243·9	183·1	144·7	114·1	88·8	164·5
r.m.s. lack of isomorphism	223·3	123·5	82·1	71·7	74·1	79·1	80·0	69·3	77·0
Phasing power	2·68	3·57	4·36	3·40	2·47	1·83	1·43	1·28	2·14
Iridium derivative statistics									
Resolution (Å)	27·3	13·84	9·27	6·99	5·59	4·66	4·00	3·50	Total
Number of reflections	37	273	702	1292	1983	2954	4066	4082	16,109
r.m.s. heavy atom *F*	301·6	282·3	262·6	228·2	192·2	157·0	121·4	92·1	155·1
r.m.s. lack of isomorphism	297·9	78·9	58·4	36·2	20·0	23·8	28·1	27·5	33·8
Phasing power	1·01	3·58	4·49	6·30	9·61	6·59	4·32	3·35	4·59

Root-mean-square (r.m.s.) heavy atom $F = (\sum f_h^2/n)^{1/2}$; r.m.s. lack of isomorphism $= [\sum (F_{ph}^{obs} - F_{ph}^{cal})^2/n]^{1/2}$, where F_{ph}^{obs} and F_{ph}^{cal} are the measured and the calculated structure factor moduli of the derivative.

$$\text{Phasing power} = \frac{\text{r.m.s. heavy atom } F}{\text{r.m.s. lack of isomorphism}}.$$

Table 3

Transformation parameters relating non-crystallographic domains of PALA-aspartate carbamoyltransferase

Parameter	Allosteric domain	Zinc domain	Polar domain	Equatorial domain
ϕ[a] (degrees)	199·9 (1·2)[b]	203·1 (2·5)	204·4 (1·2)	205·9 (1·5)
X-translation (Å)	132·4 (0·7)	134·6 (0·3)	134·8 (0·4)	134·3 (0·7)
Y-translation (Å)	22·7 (1·7)	27·4 (3·6)	29·1 (1·5)	32·0 (2·5)
Z-translation (Å)	77·2 (0·2)	79·8 (0·2)	80·1 (0·2)	80·1 (0·2)
Correlation[c] (%)	39·7	50·7	55·9	53·3

[a] The definition of Euler angles is consistent with data of Cox (1967). For the 4 parameter fit calculated here, $\theta = 180{\cdot}0°$ and $\psi = 0{\cdot}0°$.

[b] The numbers in parentheses represent standard deviations in the corresponding numerical values.

[c] Correlation $= \sum(p_1-\bar{p}_1)(p_2-\bar{p}_2)/[\sum(p_1-\bar{p}_1)^2\times(p_2-\bar{p}_2)^2]^{1/2}$, where p_1 is the density of a given domain and p_2 is the density of its symmetry-related mate.

superimposed on each of the 4 domains of a catalytic/regulatory unit: polar, equatorial, zinc, and allosteric. Twofold determinations were performed independently for each domain by minimizing the deviation between the density from an upper catalytic/regulatory domain from the density of the 2-fold related domain. Interpolation of the grid points was performed using the method of Bricogne (1974, 1976). Each new estimate for the 2-fold location was reinput into the algorithm until convergence was obtained. Table 3 reveals that the 2-fold locations for all domains except the allosteric domain show reasonable agreement within the error limitations of the measurements. As a result, the 2-fold axes determined for the other 3 domains were each weighted by their correlation coefficient and averaged to produce the final 2-fold axis used in non-crystallographic averaging. The final axis is located at (2/3, 1/3, 0·256) and is oriented 12·2° counterclockwise from the direction of the crystallographic y-axis.

(f) *Non-crystallographic averaging of m.i.r. density*

Although the refined m.i.r. maps were of good quality with readily identifiable helices and β-strands, there were areas of discontinuous density especially in the regulatory chains and in surface loops. Because of this discontinuous density it was decided, as had been done for the unligated form, to average the m.i.r. density across the 2-fold axis determined above using the methods of Bricogne (1974, 1976). In order to perform the averaging, an envelope was drawn around the asymmetric unit of the protein as displayed in the final m.i.r. electron density map. It was created using an interactive graphics program adapted for use here. This envelope was used to select portions of electron density that were to be averaged. Density outside of the envelope was set to a constant value of zero. Care was taken to draw a very generous envelope so as not to exclude any potential density either on the surface of the protein or within the central cavity.

In the averaging process, density within the upper catalytic/regulatory region was averaged with density from the lower catalytic/regulatory region. This averaged map was next back transformed to yield phases which were recombined with $2F_o-F_c$ moduli to generate an electron density map to be input for the next cycle. This iterative process was repeated until the R-factors converged (Table 4). The iterative nature of the process is necessary to correct for the loss of high resolution details that occurs due to errors in the 2-fold axis and due to local disorder. These errors are revealed in the high temperature factors that must be applied to the averaged structure factors when they are scaled with native data. The combination of these averaged structure factors with $2F_o-F_c$ moduli allows for correction of some of these deficits, as a difference map is added into the Fourier calculation. This correction is evident in the phases as the scaling temperature factor decreases with each successive averaging (Table 4).

In general, this averaging greatly improved the connectivity of the map and also lowered its noise level. Exceptions occurred in the allosteric domain, where density worsened as a result of the averaging process, and in the central cavity, where the noise level rose somewhat. It appears that the worsening of the allosteric domain could be due to the presence of a local 2-fold axis that was significantly different from that of the remaining molecule. The presence of a different 2-fold axis in this region would be in agreement with the results from the local 2-fold axis determination described in section (e), above. The selective increase in the noise level of the central cavity probably occurs because it is a large area near the 3-fold axis with little density present. It thus serves as a repository for errors accumulating from improper averaging throughout the map.

(g) *Model building and structure factor calculation*

The amino acid sequences as determined by Schachman *et al.* (1984) and by Hoover *et al.* (1983) were used primarily in fitting the density of the PALA-

Table 4

Progress in non-crystallographic symmetry averaging of the P321 *PALA-aspartate carbamoyltransferase complex*

Cycle number	R-factor[a] (%)	Correlation (%)	Mean phase change from previous cycle	Scaling[b] temperature factor
1	40·3	62·1	45·0	−25·8
2	27·5	83·8	26·8	−8·7
3	21·8	88·2	14·7	−7·1
4	19·7	90·9	9·7	−6·1

[a] $R = \sum||F_o|-|F_c||/\sum|F_o|$ here, and in Tables 5 and 6.

[b] The scaling temperature factor $= B$ in the expression $\exp[-B(\sin\theta/\lambda)^2]$. It is applied to the averaged structure factors to scale them to the native data.

aspartate carbamoyltransferase structure. Reference was also made to the protein sequences of Konigsberg & Henderson (1983) and of Weber (1968). Any discrepancies among these 4 sources of sequence information were handled as described by Ke *et al.* (1984).

The initial fitting of this sequence to the averaged density was accomplished by fitting, as rigid units, portions of each of the domains of the T form to the R form density. Next, this crude fitting was optimized on a residue by residue basis on an Evans and Sutherland PS300 system or an MPSII system using software developed by Jones (1982; FRODO) or Diamond (1982; BILDER). Modifications in this software to adapt it to our system was performed by J. Pflugrath and M. Saper in the case of FRODO and by R. Ladner in the case of BILDER. In general, side-chains with no density were left out of early models, and although PALA density was identified in the 1st difference density maps, it was not included in the phasing until several cycles of refinement and rebuilding had been completed.

Structure factors were calculated from the molecular co-ordinates with software developed by Ten Eyck (1973, 1977). In this method, an electron density map is calculated from the listing of co-ordinates, and this map undergoes Fourier transformation to yield structure factors. Temperature factor enhancement was employed only at later stages of high resolution refinement. An overall temperature factor shift of 15 was then chosen empirically, because further increases in this value did not appreciably alter the resulting phase set.

Solvent modeling was not employed. Instead, an initial R-factor determination as a function of resolution was made. Those inner shells with elevated R-factors were excluded from refinement calculations. Because PALA-aspartate carbamoyltransferase crystals were grown from solutions of low ionic strength, the low resolution data scaled well, allowing the use of data from as low as 15 Å without correction.

(h) *Restrained reciprocal space refinement*

Refinement of the model to density was performed using a modified version of the reciprocal space refinement algorithm (PROTIN and PROLSQ) developed by Hendrickson & Konnert (1981). In our system, implemented by D. Rees and M. Lewis, an electron density map was generated from the co-ordinates. This map was then back transformed to yield structure factors. Next, this information was combined with native structure factor moduli and used in making an $F_o - F_c$ difference map. An algorithm developed by Jack & Levett (1978) was then used to calculate the density gradient around each atomic position. Finally, this information along with stereochemical restraints calculated by the program PROTIN was input into a modified version of PROLSQ called GCLSQ (Rees *et al.*, 1983). GCLSQ calculated the atomic and temperature factor shifts and generated a listing of statistics on the deviation of the current model from ideality. A fraction of these shifts was applied to the atomic co-ordinates, and these new values were reinput into the next cycle. Following 8 cycles of refinement, a complete or partial refitting of the sequence to $2F_o - F_c$ maps and $F_o - F_c$ maps was performed.

In the initial fitting and refinement process, omit maps, in which phases from a portion of the model were left out of the map calculation, were used extensively. Non-crystallographic symmetry restraints were not employed in the refinement. Temperature factors were set at 20·0 until cycle 10, when they were allowed to vary under tight restraints of 1·0, 1·5, 1·5 and 2·0 for main-chain bond, main-chain angle, side-chain bond and side-chain angle, respectively. In the final refinement cycles these restraints were loosened to 1·5, 2·5, 2·5 and 4·0.

One cycle of refinement used approximately 2 h of computing time on a VAX/VMS 11/780 computer (Digital Equipment Corporation) equipped with a floating point accelerator. Temperature factor enhancement and temperature factor refinement lengthened this time appreciably. In agreement with previous reports (Remington *et al.*, 1982), we found that the largest expenditure of time was spent in the rebuilding process.

3. Results and Discussion

The structure of aspartate carbamoyltransferase complexed with the bisubstrate analog PALA represents the R form of the enzyme and is important in allowing a comparison with earlier reports of the T state structure. It also provides two additional crystallographic solutions for the catalytic and regulatory chains of aspartate carbamoyltransferase. Very significant structural differences exist between this new structure and that of the T state. It is convenient to divide these differences into quaternary changes, tertiary changes, changes at interface regions, active-site changes, and changes involved in the detailed binding of the PALA molecule itself. After a brief description of the course of refinement and the quality of the structure, each of these differences will be discussed. Following this discussion, the new structure will be used to help to reinterpret genetic and chemical modification studies done on this enzyme. Finally, a possible scenario for the mechanism of the homotropic transition will be offered.

(a) *Course of refinement and quality of structure*

To date, 123 cycles of restrained reciprocal space refinement and ten rebuilding sessions have yielded a structure containing 94·4% of the theoretical atoms of the protein–ligand complex plus 150 water molecules. The present crystallographic R-factor is 23·1% on the basis of calculations using 50,608 unique reflections; this total represents a data set 98% complete to 2·5 Å, and includes nearly 4000 reflections from 2·5 Å to 2·3 Å. The course of refinement that is outlined in Table 5 shows that we have achieved steady improvements in both R-factor and resolution. The current model has significantly better stereochemical statistics than that of the most recently reported T state model (Table 6 and Fig. 3) (Ke *et al.*, 1984).

The density is excellent throughout the catalytic chain and the zinc domain of the regulatory chain except for occasional missing or distorted side-chains. The majority of carbonyls is visible in these areas. The trace of these three domains is identical with previous reported structures; all helices and β-strands previously identified are present (Fig. 4; Honzatko *et al.*, 1982; Ke *et al.*, 1984).

Table 5
Course of refinement

Round no.	R-factor Initial	R-factor Final	Resolution (Å)	Reflections included	Atoms in model	(%) of model included
1	43·6	35·4	3·5	16,652	6435	88·5
2	35·7	31·9	3·4	18,305	6316	86·9 [a]
3	36·7	33·0	3·1	24,065	6450	90·0
4	36·7	32·7	2·9	24,460	6392	87·9
5	34·1	31·8	2·8	33,455	6548	90·1
6	33·8	31·5	2·7	37,231	6551	90·1 [b]
7	33·6	30·6	2·65	38,320	6636	91·3
8	31·3	30·0	2·55	43,917	6636	91·3 [c]
9	31·2	30·8	2·5	50,608 [d]	6636	91·3
10	32·0	30·0	2·5	50,608	6644	91·4
11	30·0	27·4	2·5	50,608	6864	94·4 [e]
12	27·4	26·5	2·5	50,608	6864	94·4
13	27·5	26·9	2·5	50,608	6864	[f]
14	27·0	27·1	2·5	50,608	6864	
15	27·1	25·4	2·5	50,608	6864	[g]
16	24·6	23·2	2·5	50,608	7012 (150) [h]	

[a] PALA phosphonate density modeled as sulfate ion.
[b] PALA molecule included in refinement.
[c] Shell scaling as a function of resolution introduced.
[d] All independent reflections included in R-factor calculation, with 98% of theoretical reflections to 2·5 Å.
[e] Refinement of isotropic temperature factors begun.
[f] Scaling and weighting changes, temperature factor enhancement.
[g] Incorporated updated versions of PROTIN and PROLSQ.
[h] Number of water molecules shown in parentheses.

The allosteric domains of the regulatory chains have, however, remained problematic throughout the refinement with the first ten to 15 residues and the 50s loop (residues 50 to 60) of each monomer lacking density. These regions were similarly problematic in the T state structure. As the refinement has progressed, difference density maps have yielded more information in these areas but the regions are still uninterpretable. At this stage, it is difficult to differentiate between true disorder in these regions and incomplete refinement. All elements of secondary structure previously reported are present (Fig. 4; Honzatko *et al.*, 1982; Ke *et al.*, 1984), but the trace is unsure from 1 to 66.

A non-crystallographic 2-fold axis is present within this unit cell and relates one upper catalytic/regulatory unit with a lower catalytic/regulatory unit within the dodecameric complex. Any two crystallographically independent catalytic/regulatory units together comprise the asymmetric unit of this crystal system. The non-crystallographic 2-fold axis used in our calculations is located at the center of the c_6r_6 complex and extends roughly in the direction of the crystallographic y-axis, except that it deviates from being parallel to this axis by 12·2°. A similar 2-fold axis for the T form deviates by 7·5° from parallelism to the crystallographic y-axis. There appear to be local variations from this 2-fold in the allosteric domain of the regulatory chains. Figure 5 illustrates on a residue-by-residue basis the deviation from an ideal 2-fold operation for both catalytic and regulatory chains.

Table 6
Initial and final refinement statistics

Geometry type [a]	Initial	Final
Distances (Å)		
Bond	0·032	0·016
Angle	0·070	0·044
Planar	0·111	0·089
Planarity (Å)	0·033	0·013
Chiral volumes (Å^3)	0·235	0·213
Non-bonded contacts (Å)		
Single contacts	0·277	0·258
Multiple contacts	0·533	0·362
H-bonding	0·528	0·302
Torsion angles (deg.)		
Planar	11·0	6·1
Staggered	24·7	25·6
Orthonormal	29·4	33·1
R-factor (%) [b]	43·6	23·2
Correlation (%) [c]	48·7	93·2

[a] As defined in Hendrickson & Konnert (1981).
[b] $R\text{-factor} = \sum ||F_o| - |F_c|| / \sum |F_o|$.
[c] $\text{Correlation} = \dfrac{\sum (F_o - \bar{F}_o)(F_c - \bar{F}_c)}{[\sum (F_o - \bar{F}_o)^2 (F_c - \bar{F}_c)^2]^{1/2}}$.

(b) *Location of heavy-atom derivatives*

With the benefit of our refined-model phases, we have investigated the locations of the iridium and platinum heavy-atom derivative sites. The iridium derivative has only four sites, two of which are located in each catalytic/regulatory unit. Two such units, related by a non-crystallographic axis of

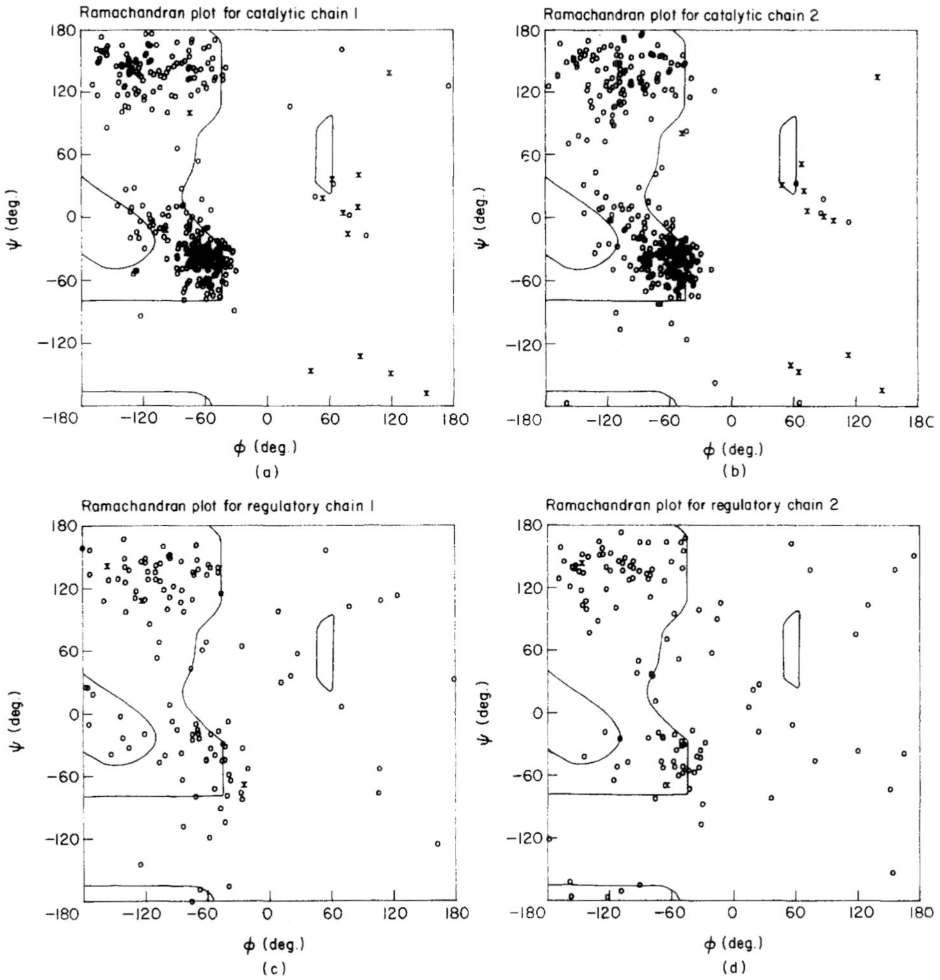

Figure 3. Ramachandran plots of the 4 chains that make up the asymmetric unit of *P*321 PALA-aspartate carbamoyltransferase crystals: (a) catalytic chain 1; (b) catalytic chain 2; (c) regulatory chain 1; and (d) regulatory chain 2. ϕ, ψ angles from unbuilt portions of the regulatory chains were excluded from the plots. Angles determined for glycine residues are indicated by the the letter "x". The wide distribution of data points in the Ramachandran plots of the regulatory chains is indicative of their greater degree of disorder relative to the catalytic chains.

symmetry, make up the asymmetric unit. Site 1 is located on the crystallographic 3-fold axis inside of the C1–C2–C3 interface of each c_3 trimer. It interacts with three His64 residues and three Ser69 residues. The density is quite clear, and there is no evidence for lack of isomorphism. This location within the c_3 trimer almost seems designed to bind to heavy-atoms. The channel is, however, too small for substrate molecules to use to reach the active site. The 2-fold related site is located in the same location of the other catalytic/regulatory unit. The next iridium site is located in the region of the 50s loop of the regulatory chain, possibly near Glu52. Its 2-fold related site is similarly situated in the other regulatory chain.

The platinum derivative contains 11 sites. All except site 4 have a 2-fold related site; each pair will be described together. Sites 1 and 11 are located near His282 in the catalytic chains. Sites 2 and 3 along with sites 9 and 10 are located at the R1–C1 interface near Asn13, Gln174, Met201 and Lys205 of the catalytic chain and Arg128, Cys137 and Glu144 of the regulatory chain. Sites 9 and 10 appear almost as three disordered sites instead of two. Site 4 has no 2-fold site. It is located in the lower catalytic/regulatory unit between helix H1 and His147 of the regulatory chain, which is between the allosteric and zinc domains. No density is present in this region in the upper catalytic/regulatory unit, although binding does not appear to be sterically hindered. Sites 5 and 6, along with sites 7 and 8, are located near the region of the 50s loops of the regulatory chains. The dyad equivalent of site 4 was inserted into the heavy-atom refinement but without success.

There are multiple heavy-atom binding sites near

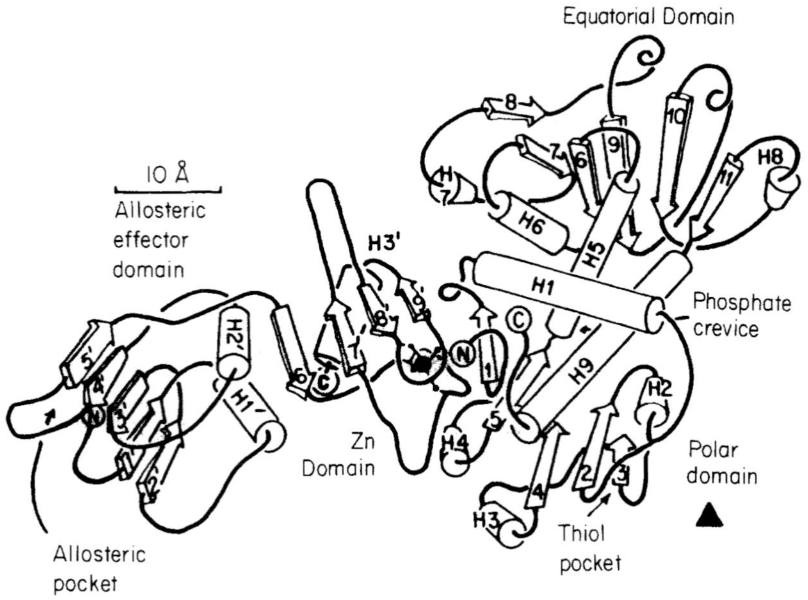

Figure 4. Schematic drawing of the trace of one catalytic/regulatory unit of aspartate carbamoyltransferase. The cylinders represent helices and the arrows designate strands of β-structure. The view is down the 3-fold axis. The PALA site is in the phosphate crevice, while the thiol pocket, containing Cys47, is not in contact with the active site.

the region of the 50s loop of the regulatory chain. Further refinement may reveal whether these heavy-atoms are responsible for the distortion evident in this region of the map.

(c) *Quaternary changes in the T to R transition*

Relative to the unligated or CTP complexed forms, the structure of aspartate carbamoyltransferase complexed with PALA has undergone extensive quaternary rearrangement (Krause *et al.*, 1985). The c_3 trimers move apart along the 3-fold axis by 12 Å and reorient counterclockwise relative to each other by 10°. This reorientation results in a more nearly eclipsed c_3–c_3 trimer interface (Fig. 6). The r_2 dimers reposition clockwise around their 2-fold axes by 15° as viewed from outside the dodecameric complex (Fig. 7). Finally, the equatorial domain as a whole relocates about 3 Å closer to the polar domain in what may be described approximately as a rigid rotation of the former toward the latter domain (Fig. 8).

With the exception of the newly identified 3 Å shift of the equatorial domain, these changes are in agreement with those described in the previously reported low-resolution structure of this system (Ladner *et al.*, 1982). They are also in accord with the theories of investigators who predicted large structural changes in the T to R transition. Gerhart (1970) was among the first to propose a rotation of each trimer relative to the other as being a part of the homotropic transition, but his suggested 60° rotation was much too large, and he did not propose a molecular expansion along the 3-fold axis. Moody *et al.* (1979) were more successful when, on the basis of small-angle X-ray scattering experiments, they correctly predicted the 12 Å elongation of the enzyme along the 3-fold axis; they also predicted an expansion along the r_2 dimers, which does not appear to take place. Monaco *et al.* (1978) were able to anticipate the expansion along the 3-fold axis in the PALA-aspartate carbamoyltransferase complex based on the corresponding expansion in the unit cell of the crystal form.

The findings of Howlett & Schachman (1977) concerning the changes in sedimentation coefficient upon PALA binding are supported by our results. They found that the enzyme sediments more slowly in the R form, which is consistent with the reported 12 Å expansion along the 3-fold axis. However, they could not distinguish between molecular expansion and other changes in shape. They also found that isolated c_3 trimers sediment more rapidly in the presence of PALA. This occurs because the catalytic chain adopts a more compact shape as the two domains relocate closer together.

The channels adjacent to each catalytic/regulatory unit are much larger in the R form. Each of these channels has more than twice the cross-sectional area compared to those in the T form. Since a hybrid form of the enzyme, c_6r_4 retains substantial homotropic and heterotropic interactions, it is unlikely that the increase in the size of these channels has catalytic relevance (Evans *et al.*, 1974).

Because of the quaternary changes described above, several of the interface regions in this

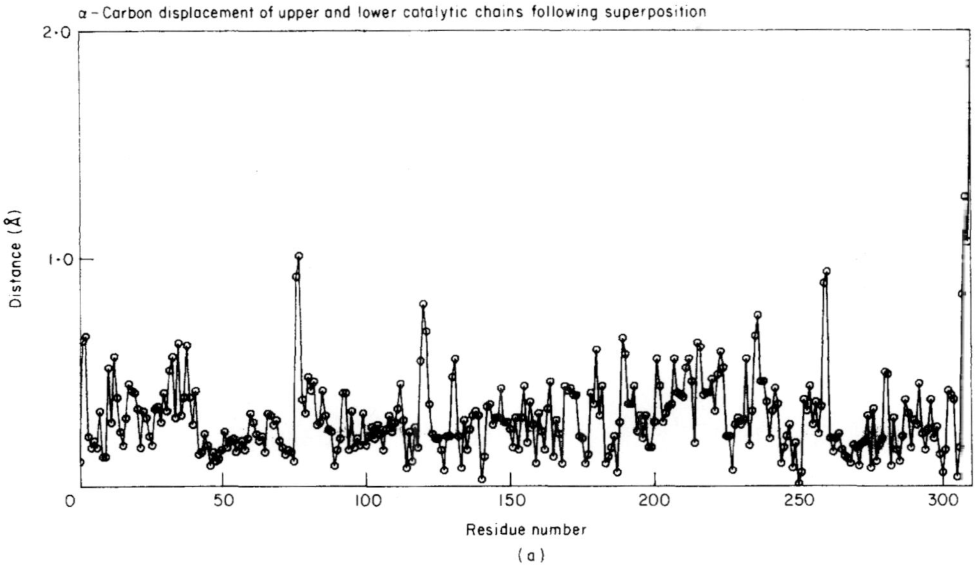

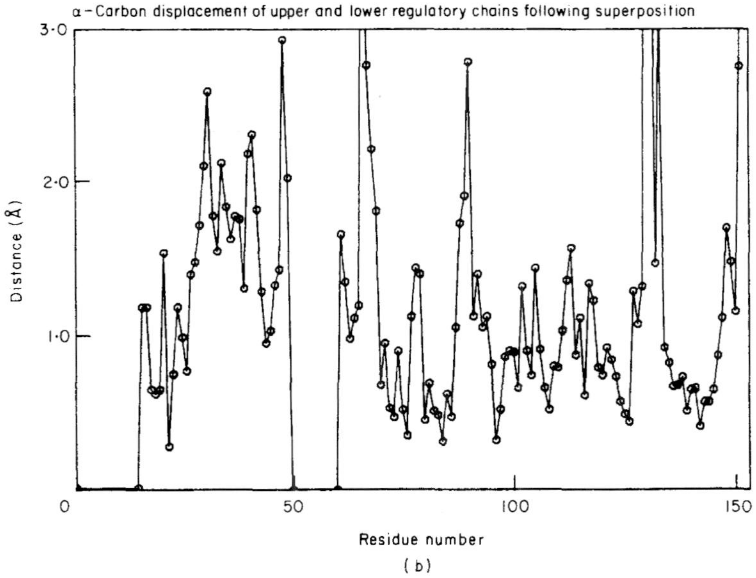

Figure 5. The α-carbon displacement between upper and lower chains of the asymmetric unit of PALA-aspartate carbamoyltransferase following superposition using the Kabsch algorithm (Kabsch, 1978). Unbuilt portions of the regulatory chains, i.e. residues 1 to 15 and 50 to 60, were left out of the superposition calculation. therefore, these distances are recorded as zero. (a) Catalytic chains; (b) regulatory chains.

structure are altered relative to the T form. These changes will be described in section (h), below.

(d) *Tertiary changes in the T and R transition*

Within the catalytic chain, the largest differences in tertiary structure between the T and R forms occur in the 80s loop and the 240s loop (Fig. 8) (Krause *et al.*, 1985). In the R form, the 80s loop is relocated about 5 Å closer to the beginning of the H2 helix of an adjacent catalytic chain bringing it close to the location of the PALA molecule (see below). The remainder of the polar domain is almost unchanged. Also in the R form, the 240s loop is repositioned much closer to the polar domain, thereby effecting a major reorganization in the C1–C4, R1–C4 interfaces. This change, described in section (h), below, results in stacking of the 240s loops on top of each other rather than next to each other as in the T form. This stacking prevents a

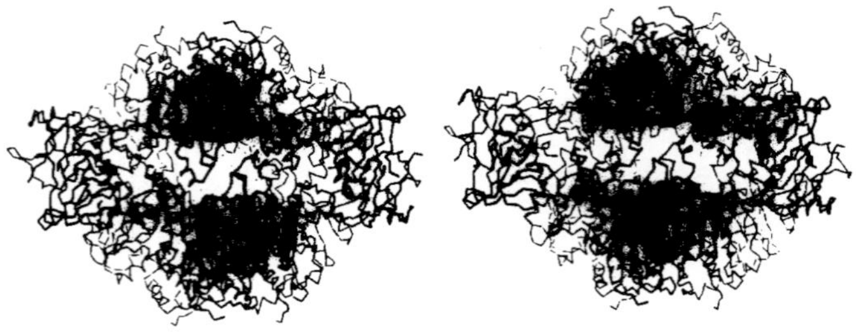

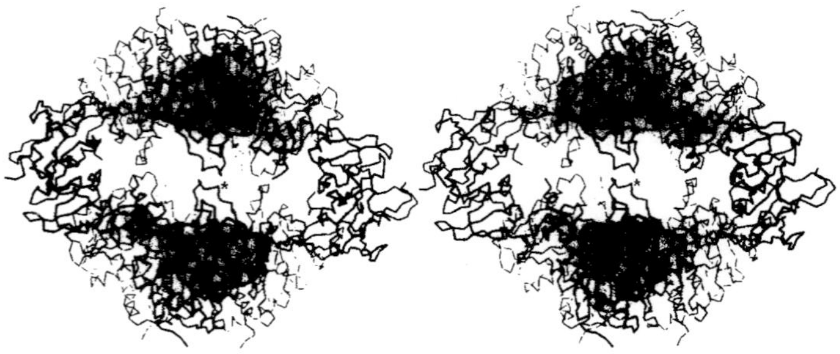

Figure 6. Stereo α-carbon tracings of the polypeptide chains of the entire dodecameric complex of aspartate carbamoyltransferase in both the T (less active) and the R (more active) forms. The view is perpendicular to the 3-fold axis, which lies vertically in the plane of the paper. One of the 3 2-fold axes projects out from the center of the enzyme directly towards the viewer. The conspicuous loop nearest the center of the diagram is the 240s loop. In the R form tracing, it is marked with an asterisk. Behind the 240s loops, in both Figures, lies the central cavity of the dodecamer.

reversion to the T form *via* a simple collapse along the 3-fold axis.

The combination of the quaternary changes and tertiary changes in the catalytic chain results in an effective domain closure on the PALA binding site of the R form. These large motions represent a new example of protein rearrangement driven by substrate binding. Several types of domain closure involving enzymes interacting with substrates or analogs have been described (Lesk & Chothia, 1984). Some, such as insulin, involve motion around hinge regions; others involve rotation of one subunit domain relative to the other. Lesk & Chothia (1984) have proposed a theory of protein domain closure that involves a shifting of helices as rigid units upon each other. The motions of the domains of aspartate carbamoyltransferase do not fit the pattern of any one of these simple cases of

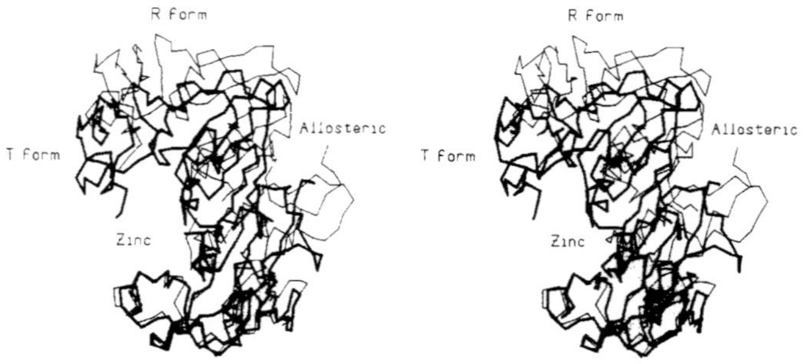

Figure 7. Stereo α-carbon tracings of the regulatory dimer as it appears in the T form (thick tracing) and R form (thin tracing). The allosteric domains are in the foreground and the zinc domains are in the background. Note the rotation of the R form dimer relative to the T form dimer. In the lower part of the Figure there is approximate superposition of the chain tracings while a large deviation is present between the 2 forms in the upper portion of the Figure. The orientation is similar to that of Fig. 6. The 2 halves of the 10 stranded β-pleated sheet are related by 1 of the molecular non-crystallographic 2-fold axes.

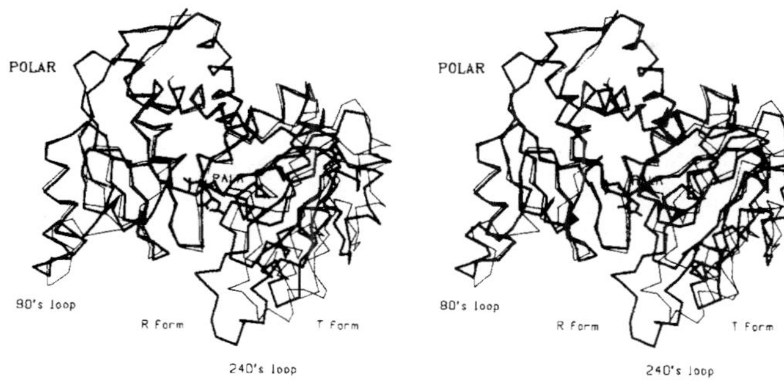

Figure 8. Stereo superposition of the catalytic chains of the T (thin line) and R (thick line) conformations. We have chosen to superimpose the polar domains of the 2 forms using the Kabsch algorithm (Kabsch, 1978). This choice reflects that, in both T and R forms, the polar domains of the dodecameric complex retain very similar contacts, suggesting that the preponderance of the motion of the homotropic transition occurs in the equatorial domains. The view is perpendicular to the 3-fold axis. The location of the PALA molecule in the R state is shown.

protein motion. Rather, a more complex rearrangement appears to take place. For example, the motion of the equatorial domain is roughly as a rigid unit around a hinge region located near the 130s region. However, the 80s loop and the 240s loop undergo extensive reorganization that is not accounted for by any simple rigid body movement.

Within the regulatory chains, the tertiary changes are more subtle. Each zinc domain and the ten strands of β-sheet found in the combined allosteric domains of a regulatory dimer have essentially the same tertiary structure in both the T and the R forms. There is, however, a slight difference in the location of the 2-fold axis for the two domains within each chain. When the zinc domains of the T and R forms of the regulatory chains are superimposed using the Kabsch algorithm (Kabsch, 1978), the structures show excellent agreement in that domain (root-mean-square deviation $=0{\cdot}45$ Å), but below residues 90 to 100, in the transition to the allosteric domain, they begin to show a small but definite deviation. Similarly, if the allosteric domains of the T and R forms of this chain are superimposed, they show good agreement that lessens as the zinc domain is approached. Taking this into consideration, the motion that takes place in the homotropic transition may be described approximately as a rigid reorientation of each regulatory dimer but with minor deviation at hinge regions between the two domains.

(e) *The PALA binding site*

The PALA molecule has been located between the polar and equatorial domains of the catalytic chain near the N terminus of helix H2†. Although this region is described as predominantly hydrophobic by Honzatko *et al.* (1982), there are numerous charged groups present. Among them are Glu50, Arg54, Arg105, Arg167, Arg229, Glu233, Arg234 and Arg296, as well as Lys84 from an adjacent catalytic chain. The relocation of the 240s loop forms a lower boundary to this area making it accessible only through the central cavity.

The phosphonate moiety of the PALA molecule is located very close the pyrophosphate and phosphate binding sites described for the T form of the enzyme (Honzatko & Lipscomb, 1982). The aspartate moiety extends toward the equatorial domain. Within the polar domain, the phosphonate moiety is bound to Ser52, Arg54, Thr55 and Arg105. Also bound to the phosphonate moiety are Ser80 and Lys84 from an adjacent catalytic chain. In addition, the main-chain nitrogens of residues 53, 54 and 55 interact with this portion of the PALA molecule.

The equatorial domain donates several residues to binding the aspartate moiety of PALA. Arg167 binds to the α-carboxylate, while the β-carboxylate is bound to Arg229 and Gln231. Arg105 also binds to the α-carboxylate group. Lys84, from an adjacent chain, binds to both carboxylate groups. His134 is within hydrogen-bonding distance of the carbonyl oxygen of PALA, as are Thr55 and Arg105. The amide NH group of PALA is linked by a hydrogen bond to the main chain carbonyl group of Leu267. The detailed binding of PALA to the enzyme is shown in Figure 9(a) to (c).

The conformation of the PALA molecule itself is somewhat different from that of the crystal structure reported by Zanotti *et al.* (1984) (Fig. 10(a)). For example, in our structure the phosphonate and carboxylate groups proceed from the *si* face of the prochiral carbonyl group, whereas, the phosphonate group and the β-carboxylate group proceed from the *re* face in their structure (Fig. 10(b)). The curl noted between these moieties is still present in our complex but in the opposite direction. In our first report, the dihedral angle between the two carboxylate groups of PALA was 110°; this value was close to the value of 132°

† This region of helix H2, with its positive dipole moment, may aid in the binding of PALA. The reference in Krause *et al.* (1985) to a binding interaction between PALA and the negative end of the dipole of helix H2 was in error.

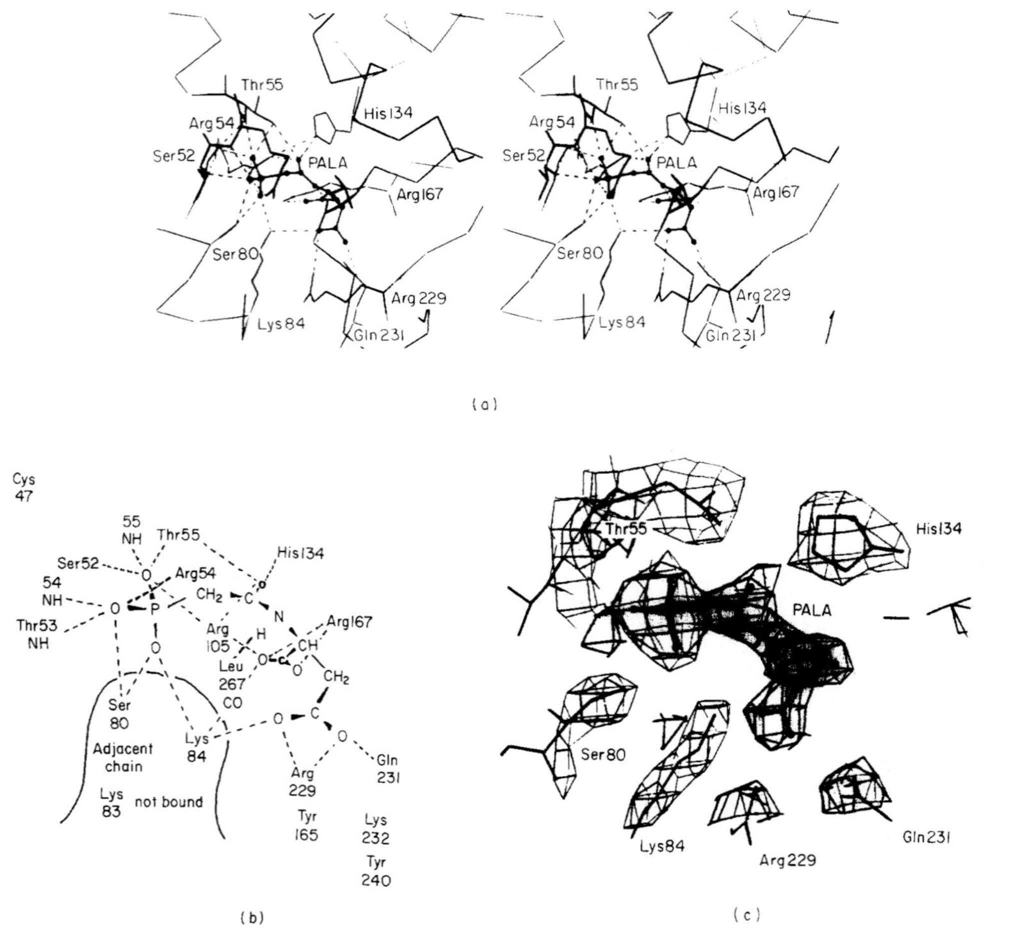

Figure 9. (a) The residues that participate in the binding of PALA are illustrated in stereo superimposed on an α-carbon tracing of the catalytic chain (see the text). The view is from the central cavity of the enzyme, roughly perpendicular to the 3-fold axis. Ser80 and Lys84 are contributed by an adjacent catalytic chain within the same catalytic trimer. Arg105 is located behind the phosphonate of PALA, and Leu267 is in front of the PALA aspartate moiety; they are unlabeled to preserve clarity. Broken lines illustrate hydrogen bonding or salt linkages. (b) Schematic illustration of the binding of PALA as shown in (a). Residues studied by other methods (Cys47, Lys83, Tyr165, Lys232 and Tyr240) are too far from PALA to have direct interactions. Hydrogen bonds, salt links, and other polar contacts less than 3·5 Å are indicated. (c) Depiction of the electron density in the active site of the R form of aspartate carbamoyltransferase as determined using $(2F_o - F_c)$ Fourier coefficients. Portions of the model are superimposed. No solvent molecules were included in the calculation.

reported by Zanotti *et al.* After further refinements, these groups appear to fit the density better in a more nearly *cis* conformation with a dihedral angle of about 60°.

The location of the PALA molecule within the catalytic chain would appear to identify unequivocally the active site of aspartate carbamoyltransferase. The direct participation of Lys84 and Ser80 from an adjacent catalytic chain lends strong support to the previously proposed active site shared between two chains (Monaco *et al.*, 1978).

The details of the binding interactions implicate several residues also cited in previous binding studies with pyrophosphate (Honzatko & Lipscomb, 1982). Among them are Ser52, Arg54, Arg105, His134 and Arg167. Many new residues have been implicated above with regard to the binding of the aspartate moiety of the bisubstrate inhibitor. Overall the binding of PALA involves several salt links that, together with bonds from polar groups and possibly from the dipoles of helices H2, H5 and H9, help to hold the molecule in place.

Some active-site residues appear to bridge both domains of the catalytic chain, thus acting to bring the substrates together and stabilize the new R state conformation. Lys84 and Arg105, for example, bind to both moieties of the PALA molecule. New salt bridges in the R form occur between Glu50 and Arg167 and between Glu50 and Arg234.

The current PALA density appears to fit a model with its carboxylates in a nearly *cis* orientation.

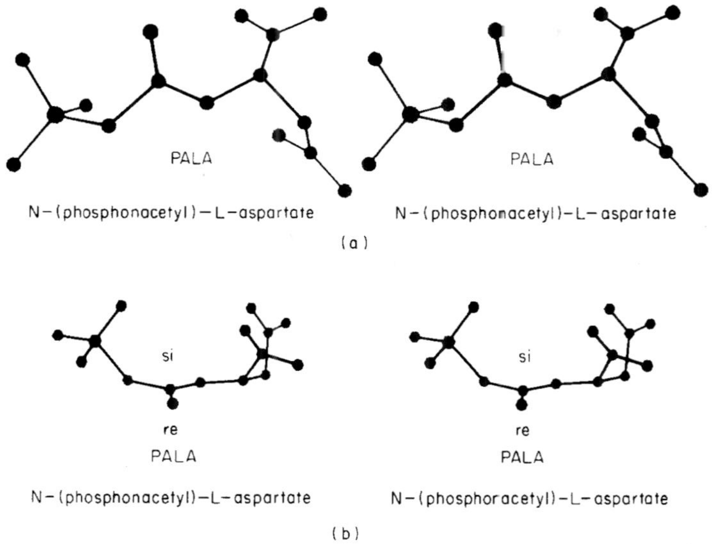

Figure 10. (a) The conformation of the PALA molecule in stereo as complexed by aspartate carbamoyltransferase. (b) The PALA molecule in stereo, as complexed by aspartate carbamoyltransferase, arranged to illustrate its stereochemistry relative to the PALA structure described by Zanotti *et al.* (1984).

This orientation is consistent with the observation of Davies *et al.* (1970) that maleate, which has its carboxylates constrained into a *cis* position, is a better inhibitor than fumarate. The basis behind this preference for substrates with *cis* carboxylates may be the orientation of residues Arg167 and Arg229. They bind to the α- and β-carboxylates of PALA, respectively, and their orientation, side by side, each projecting from the equatorial domain toward the active site, would certainly encourage the adoption of a nearly *cis* geometry in approaching dicarboxylates (Fig. 11). Although the geometry of these two residues is attractive as an explanation for this stereochemical preference, it should be remembered that an alternate, but less satisfactory, fit to the density may be made for the PALA molecule with its carboxylates in a more nearly *trans* position, as previously reported (Krause *et al.*, 1985).

Davies *et al.* (1970) also reported that stereospecific substitutions of hydroxy and certain other moieties for selected hydrogen atoms on aspartate can alter its ability to bind to the enzyme. They were able to deduce a paradigm that allowed prediction of where atomic substitutions would have the greatest effect. As yet, we can find no explanation for these findings in the current active-site geometry, although further refinements may be helpful.

Much speculation and experimentation on the mechanism of aspartate carbamoyltransferase has centered on the presence of an ordered *versus* random addition mechanism (Collins & Stark, 1971; Heyde, 1976). The structures of the T and R form do not allow a clear choice to be made between these two mechanisms; however, the following points may be made. On the basis of steric considerations only, and assuming that the con-

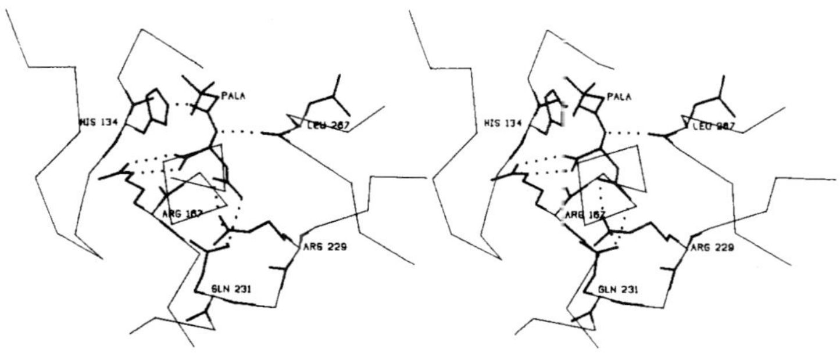

Figure 11. The interactions between the dicarboxylate moiety of PALA and selected residues of the catalytic chain are shown in stereo. The 3-fold axis is tilted away from vertical and the molecule is rotated to allow a more unobstructed view.

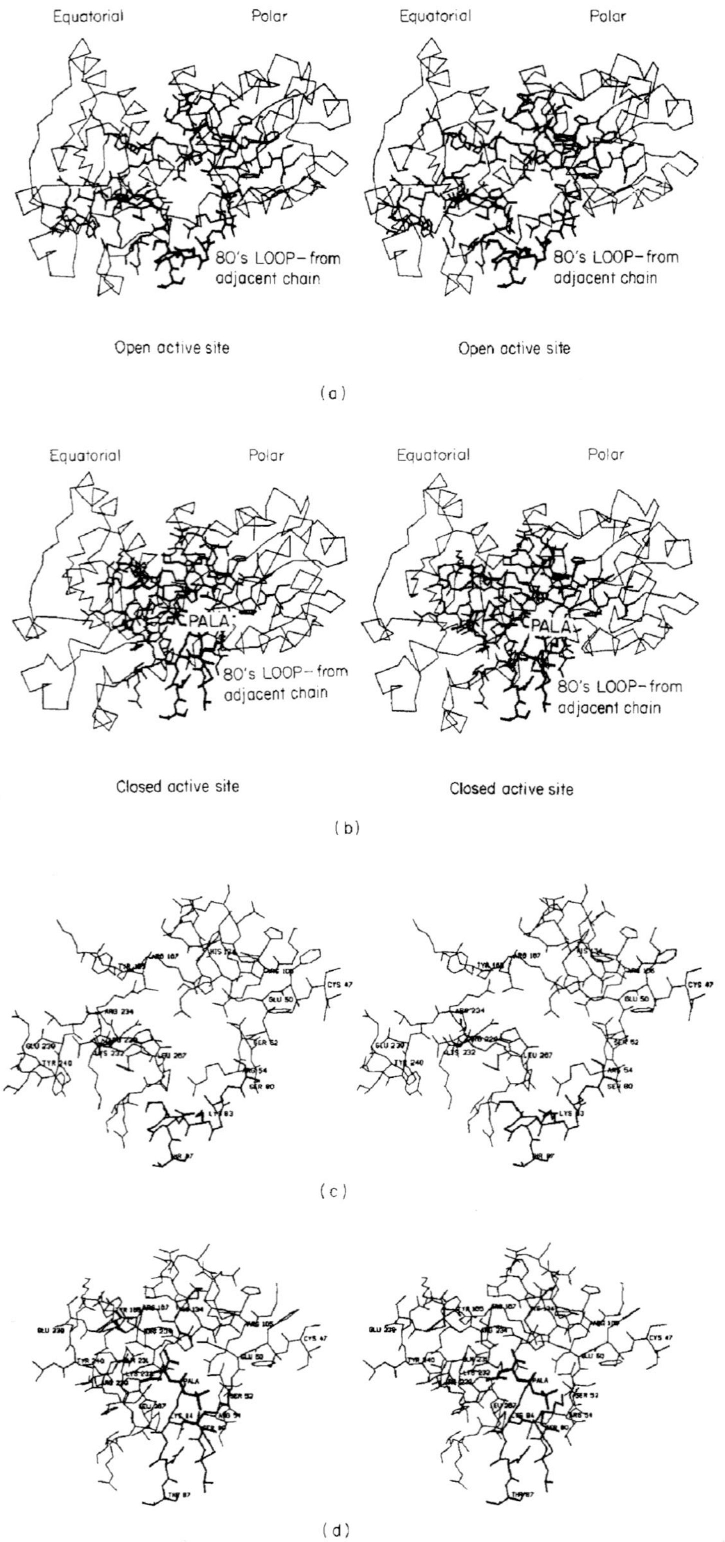
Equatorial
Polar
80's LOOP–from adjacent chain
Open active site
(a)
PALA
Closed active site
(b)
(c)
(d)

Fig. 12.

formation of PALA has relevance to the binding of natural substrates, both substrates could enter the active site of the T form and bind independently. Once bound in the R form they are trapped within the active site. Any conformational change that would free one substrate would also sterically free the other. If the equatorial domain were to move away from the phosphate binding site, it could allow the more weakly bound aspartate molecule to leave in preference to the tightly bound carbamoyl phosphate.

Another important aspect of most mechanistic theories for this enzyme is the presence of a general base within the active site to deprotonate aspartate as it prepares to attack carbamoyl phosphate (Collins & Stark, 1971). Unfortunately, there is no convincing candidate for this base as revealed by the detailed binding of PALA. Gln231 would not appear to be basic enough to serve in this capacity, nor would the carbonyl oxygen atom of Leu267, which is hydrogen bonded to the nitrogen atom of PALA. The best choice for a general base within the active site is His134. It could receive, either directly or *via* one or more water molecules, a proton from L-aspartate as it is bound. Alternatively, the phosphate could receive a proton. (J. E. Gouaux, Jr, K. L. Krause & W. N. Lipscomb, unpublished results). The water structure is still under investigation.

As noted, it appears that the carbamoyl phosphate moiety of the PALA molecule binds primarily to the polar domain while the L-aspartate moiety binds primarily to the equatorial domain. This association of a particular domain with a specific substrate is in agreement with findings on sequence homology between aspartate carbamoyltransferase and ornithine carbamoyltransferase (Houghton *et al.*, 1984). These two enzymes have one substrate in common, carbamoyl phosphate, but they each have different second substrates, aspartate and ornithine, respectively. Sequence and predicted secondary structure homology is striking for the portions of the two sequences that correspond to the polar domain of aspartate carbamoyltransferase, including the presence in both of a four-residue phosphate binding sequence . . . X-Ser-Y-(Arg, Lys) . . . present in many phosphate-binding enzymes. The sequences for the equatorial domains, which are presumably responsible for the binding of the two different amino acids, show much less homology.

(f) *Comparison of the active sites of the T and R forms*

The very large changes in quaternary and tertiary structure reported for the R state are part of, and presumably caused by, a major reorganization of the active site when PALA is bound. The differences between the structures of the active sites of the T and R forms, which occur upon binding PALA, are described below.

The active site of the T form may be characterized as being open with both polar and equatorial domains widely separated and the 80s loop of the adajcent catalytic chain on the periphery (Fig. 12(a) and (c)). Polar interactions occur between residues Glu50, Ser52 and Arg105, and between Arg105 and Arg167. There are about ten molecules of water in the active site, including a cluster of five and a cluster of three. They do not interact with any of the numerous charged residues.

The active site of the R state, on the other hand, is quite closed in appearance as both domains move in to bind to the PALA molecule (Fig. 12(b) and (d)). The 80s loop of the adjacent catalytic chain ventures deep into the active site with the C^{α} of Lys84 moving 6 Å and the side-chain moving over 10 Å in the transition from the T to the R forms.

Solvent plays a role in the active site of the R form by interacting with both the PALA molecule and with polar residues from both domains. Ten to 15 water molecules are present, depending on which part of the asymmetric unit is studied. The solvent seems to be distributed almost evenly wherever there are no steric restrictions. For convenience it may be divided into four groups. Group 1 is located beneath the dicarboxylate moiety of PALA and contains three to five water molecules with contacts to both PALA and the equatorial domain, particularly to regions near Arg167, Arg229 and Gln231. Group 2 is located near Lys84 and contains from one to three water molecules. Group 2 associations are similar to those of group 1 but shifted away from the equatorial domain. Group 3 is located above the PALA molecule and contains four to five solvent molecules near residues Arg296, Gln288 and Gln137. Finally, there is one isolated solvent molecule about 2·9 Å above His134 (group 4). It is present in both crystallographically independent active sites and may aid in proton transfers involving this residue. The electron density for several of these water molecules is somewhat weak but their temperature factors are acceptable and difference density maps are quite clean.

There are no solvent molecules closely associated with the phosphonate moiety of PALA, perhaps because the very large number of binding interactions mitigates any need for solvation.

Aside from the noted changes in water structure, the polar domain undergoes only moderate changes in the R form (Fig. 13). Certain residues, for example, His134, occupy almost identical positions in both forms, while others, such as Arg54, have 1 to 2 Å shifts. In fact the entire N-terminal region of

Figure 12. Illustrations of the active site of aspartate carbamoyltransferase in both the T and R forms in stereo; all contributing residues from adjacent chains are present. The 3-fold axis is approximately perpendicular to the plane of the Figure. (a) and (b) Overall view of the 2 forms with an α-carbon tracing of the catalytic chain included. The views are from within the central cavity. (c) and (d) Magnified view containing the active-site residues only.

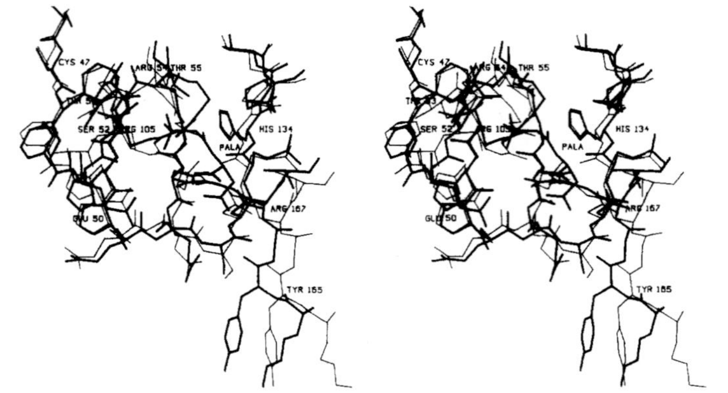

Figure 13. The polar domains of the T and R forms of aspartate carbamoyltransferase in stereo superimposed using the Kabsch algorithm (Kabsch, 1978). Residues discussed within the text as illustrative of changes between these 2 forms are identified. The light trace is of the T form, while the dark trace is of the R form. The 3-fold axis is vertical and lies in the plane of the paper.

helix H2 shifts in the R form to allow a closer association of several residues to PALA. Also in the R form, the side-chain of Ser52 orients around its C^{α}–C^{β} bond away from Arg105 to allow both of these residues to bind to PALA. Glu50 and Arg167 change from being more than 5 Å apart in the T state to 3·2 Å apart in a salt link in the R state.

The equatorial domain undergoes a more significant reorientation in the homotropic transition (Fig. 14(a) and (b)). First, as noted previously, the entire domain rotates about 3 Å closer to the polar domain in order to bind the PALA molecule. Second, the 240s loop of the same domain moves an additional 3 Å into the active site, as well as laterally 5 Å perpendicular to the active site. Arg167 binds to one carboxylate moiety of PALA, as well as to Glu50; it no longer interacts with Arg105. Arg229 and Gln231 bind to the other carboxylate moiety of PALA. To make this binding possible, Gln231 moves 3 Å toward the PALA molecule, while Arg229 actually has to move laterally away from the aspartate moiety to make room for it to bind. Arg234 moves about 4 Å toward the PALA molecule but stops short of binding to it and, instead, binds closely to Glu50. Two significant links occur between Tyr165, Lys164 and Glu239 to stabilize the new position of the 240s loop (Fig. 14(c) and (d)).

One possible scenario to explain this major reorganization would be as follows. The aspartate moiety of PALA binds to Arg167 and attracts Arg229 and Gln231. They rearrange to bind to the inhibitor and indirectly force Arg234 toward the active site pocket. Arg234 then becomes involved in a salt link with Glu50. The motion of Arg234 toward Glu50 brings the 240s loop in toward the active site until it finds a stable location when it binds to Lys164 and Tyr165. In this way, the large rearrangement is composed of several small steps, each stabilized by the formation of a polar bond or salt link.

Compression of substrates has been invoked by Collins & Stark (1971) as being part of the catalytic mechanism of aspartate carbamoyltransferase. The catalytic chain in the R form is "contracted" as compared to the T form, but there is no need to

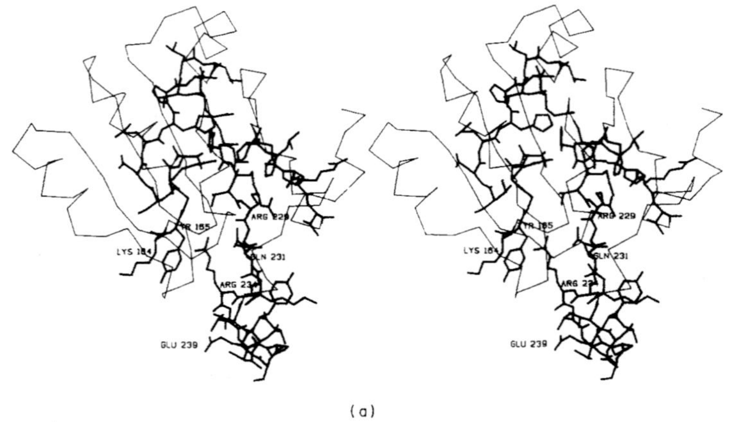

(a)

Fig. 14.

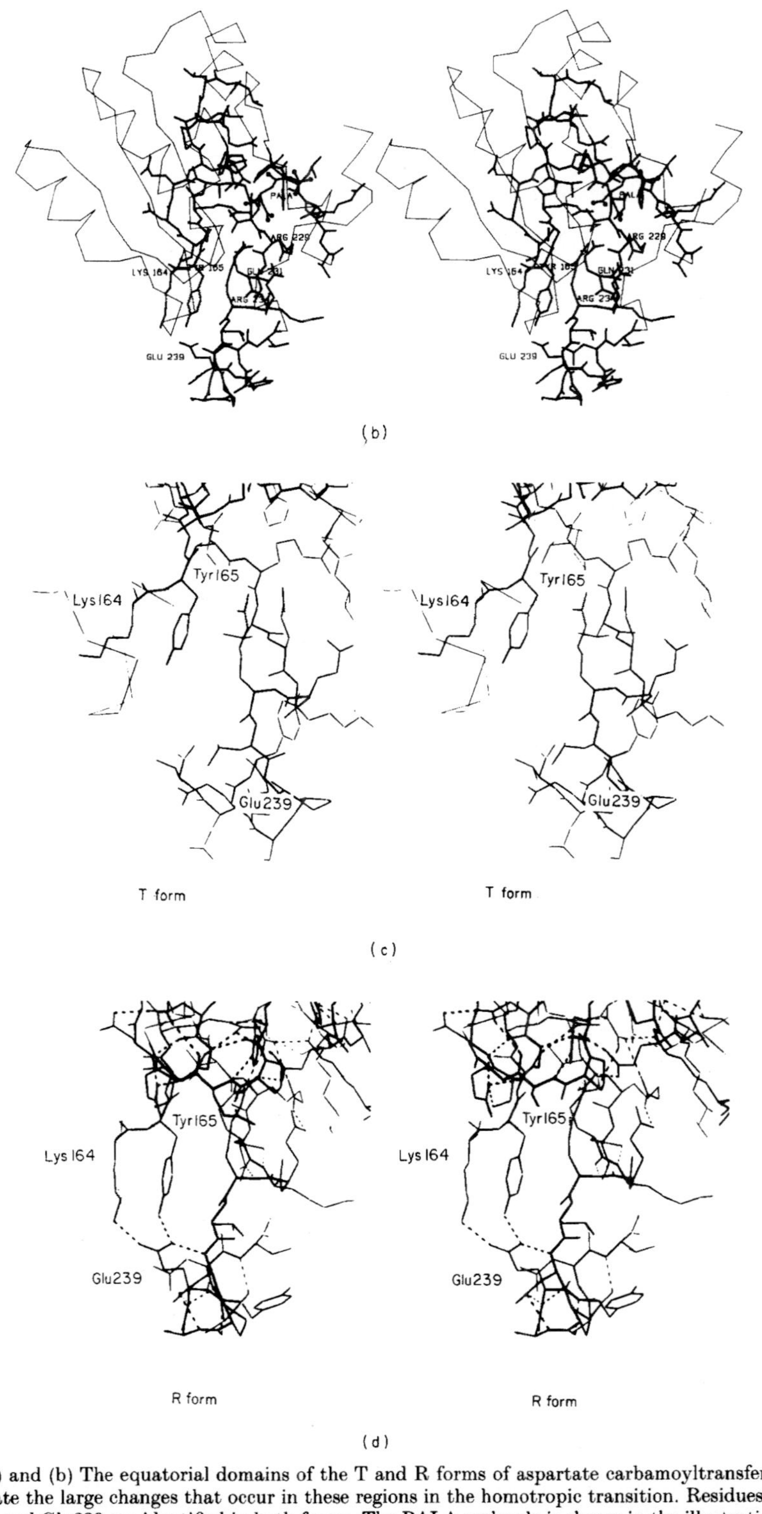

Figure 14. (a) and (b) The equatorial domains of the T and R forms of aspartate carbamoyltransferase, displayed in stereo to illustrate the large changes that occur in these regions in the homotropic transition. Residues Lys164, Tyr165, Arg229, Arg234 and Glu239 are identified in both forms. The PALA molecule is shown in the illustration of the R form. The 3-fold axis is nearly vertical, lying in the plane of the paper. (c) and (d) A detailed view in stereo of the changes involving residues Lys164, Tyr165 and Glu239 that occur in the homotropic transition. Emphasis is on the new bonding interaction between these residues in the R form. The 3-fold axis is nearly vertical, lying in the plane of the paper.

infer that this contracted chain actually "compresses" the two substrates, aspartate and carbamoyl phosphate together. There is little biochemical precedent for compression by proteins to catalyze enzymatic reactions by expending energy, as has been suggested, to force substrates into violation of their respective van der Waals' radii. There is ample biochemical precedent, however, for domain closure in proteins upon substrate binding (Lesk & Chothia, 1984). Domain closure could help to catalyze a bimolecular organic reaction by juxtaposing both substrates and increasing the probability of a productive collision. This type of closure would seem to apply in the case of aspartate carbamoyltransferase as a "contracted" catalytic chain brings aspartate and carbamoyl phosphate into position to react. Further catalytic effects could occur *via* the activation of the carbonyl of carbamoyl phosphate by a general acid and by the deprotonation of L-aspartate by a general base.

(g) *Non-covalent intersubunit interactions in the R state and the T state*

Honzatko *et al.* (1982) described a series of non-covalent interactions between side-chains of different regions of the dodecameric enzyme complex. Taken together they form the bulk of the interfaces between all the chains and domains of the enzyme. They are, therefore, presumably important in the stabilization of quaternary structure. For comparison to this early report, a table of non-covalent interactions is reported here (Table 7). Interactions present in the R but not the T form are specially identified. Criteria similar to, but slightly less stringent than, those used by Honzatko *et al.* (1982) were used in the selection of these contacts. Most reported interactions involve atoms approaching each other within 3·5 Å or less. Some interactions, in particular those involving non-polar groups, were reported for atoms approaching each other within 4·0 Å, especially if the same distance was 3·2 Å or less in the other catalytic/regulatory unit.

Complete, residue-by-residue comparison with the previously reported T form structure (Honzatko *et al.*, 1982) is difficult because the sequences used have been revised. All residue numbers in Table 7 reflect the current sequence. Final comparison with data of Honzatko *et al.* (1982) is postponed pending further refinement of both the T and R form structures.

Burley & Petsko (1985) recently reported a study of a large number of non-polar interactions in proteins involving two or more aromatic residues. They found that these residues frequently adopt a characteristic distance separation and a perpendicular geometry that helps to stabilize protein structure. Inspection of the R form of aspartate carbamoyltransferase has revealed six aromatic pairs within the asymmetric unit. Of these pairs, four meet the criterion outlined by Burley & Petsko. In the upper and lower catalytic chains, such interactions occur between Phe49 and Phe73 as well as between Tyr285 and Phe286. In the regulatory chains an interaction occurs between Phe33 and Tyr77 in the lower chain and between Phe27 and its 2-fold rleated residue in the opposite

Table 7

Non-covalent intersubunit interactions in PALA-aspartate carbamoyltransferase

Region of subunit I	Region of subunit II	Type of interaction	Comments
A. *C1–C2*			
Ala51	Asn78	Non-polar	a
Ser52	Asn78	Polar	
Thr53	Thr79	Non-polar	a
Thr53	Ser80	Polar	
Arg54	Gln86	Polar	
Arg54	Val94	Non-polar	a, b
Arg54	Tyr98	Polar	b
Arg56	Gly72	Polar	a
Leu57	Ile95	Non-polar	
His64	Val70	Polar	a
Arg65	His41	Polar	b
Arg65	Tyr98	Polar	b
Arg65	Asp100	Polar	
Asp75	Asn78	Polar	b
Lys84	Arg105	Polar	a, b
Lys84	Arg229	Mixed	a, b
Asp90	Arg269	Polar	
Val94	Phe286	Non-polar	
Tyr98	Thr61	Non-polar	a, b
Tyr98	Ile293	Non-polar	a
Arg269	Asp90	Polar	
Gly290	Thr97	Non-polar	
Ile293	Thr97	Non-polar	a, b
B. *C1–C4*			
Ser238	Ala241	Mixed	b
Tyr240	Pro237	Non-polar	b
C. *C1–R1*			
Ser11	Glu142	Polar	
Asn13	Lys137	Polar	
Thr87	Glu119	Non-polar	
Leu88	Glu119	Polar	
Ala89	Glu119	Mixed	
Pro107	Asn113	Polar	
Gln108	Asn113	Mixed	a
Glu109	Asn111	Polar	a
Glu109	Asn113	Polar	b
Glu109	Ile115	Polar	
Leu114	Glu119	Non-polar	a, b
Leu114	Val121	Non-polar	a
Glu117	Lys139	Polar	
Glu117	Lys140	Polar	
Ser131	Lys143	Polar	a
Asn132	Glu142	Polar	
D. *C1–R4*			
	None		
E. *R1–R6*			
Phe27	Phe27	Non-polar	
Leu30	Phe27	Non-polar	b
Asp39	Asn47	Mixed	
Gln40	Asn47	Polar	
Asn47	Asp39	Mixed	b
41–47	41–47	Polar	c

a, interaction found in only one catalytic/regulatory unit; b, interaction not present in the T form structure; c, antiparallel β-sheet.

chain. Neither of these latter interactions fits the geometrical criterion described above.

(h) *Interface regions of the hexamer: R state* versus *the T state*

In aspartate carbamoyltransferase, information concerning effector or substrate binding is communicated through intersubunit and intrasubunit interfaces and probably involves propagation of conformational changes. Honzatko *et al.* (1982) provided a detailed account of these interface regions for the T state of aspartate carbamoyltransferase. A newer, but as yet unpublished, listing of interchain interactions for the T form is available (Ke, Heng-Ming, personal communication). We present here a comparison between the interface regions of these two forms.

(i) *The C1–C4 and C1–R4 interface*

This interface is the most complicated of the T form and, as predicted by some investigators (Honzatko *et al.*, 1982; Moody *et al.*, 1982), it undergoes the most extensive rearrangement in the homotropic transition. Several important residues such as Tyr240 and Tyr165 are involved in these changes. Due to the rotation of the c_3 trimers, the C1–R4 contacts have completely disappeared in the PALA-bound enzyme. The C1–C4 contacts are also restructured. In the R form, the only contact between these subunits, and thus the two trimers, is through residues 238 to 240, which now stack on top of each other (Fig. 6). In the T form, these residues stack side by side (Fig. 6). This stacking prevents a reversion to the T form *via* a simple collapse along the 3-fold axis.

In the T form, the C1–C4 juxtaposition is stabilized by intersubunit hydrogen bonding between Glu239 and Asp236 of C4 with Tyr165 of C1. (The previously reported link with Glu246 is no longer present because of the insertion to correct the amino acid sequence.) In the R form, these same residues are still involved in hydrogen bonding, but now they interact with each other within the *same* subunit. Perhaps the initial and final positions of the 240s loops in the T and R forms are stabilized by these similar bonding patterns. Also the hydroxyl group of Tyr240 makes a hydrogen bond to Asp231 in the T form but this link is absent in the R form. Access to the zinc atom of both regulatory chains is increased as a result of this new interface architecture; this increased access is consistent with the observed increase in sulfydryl reactivity as the enzyme assumes the R form (Gerhart & Schachman, 1968).

Because of the nature of this interface in both forms, homotropic co-operativity could be communicated *via* $c_3{:}c_3$ interactions as well as through c-r-r-c interactions, as suggested by Chan (1975) and by Ladner *et al.* (1982). The structures do not suggest that one pathway is superior to the other. In fact, both pathways could be utilized simultaneously.

(ii) *The C1–C2–C3 interface*

This interface is very similar in the T and the R form of the enzyme because the c_3 trimers (especially the polar domains) untwist somewhat as a rigid unit. All salt links and most polar links reported for the T form, including that between Glu86 and Arg54, are still present in the R form. (N.B. All earlier references to Gln86 in the catalytic chain are probably in error. *Glu86* is correct.) Several new links, however, are present in the R form structure, for example, a link between Ser80, Lys84 and the PALA molecule in the adjacent c_3 trimer. Other new links are identified in Table 7.

(iii) *The R1–R6 interface*

This interface also is left intact in the R form because the r_2 dimers move roughly as a unit around their 2-fold axes in the homotropic transition. Thus, the antiparallel contacts between strand S2′ and helix H1′ and their respective 2-fold mates are present. New interactions do take place between Phe27 and residues 30, 31 and 36 of the opposite regulatory domain, but these need further refinement for confirmation. Further refinement is also needed to confirm an interaction between Asn47 and the backbone of residues 38 to 40 from the 2-fold related chain. Any possible interactions between the N termini cannot be detected because of poor density in this area.

(iv) *The C1–R1 interface*

The C1–R1 interface is quite complicated in the R form. It involves a very extensive array of primarily polar links that occur between the polar domain of the catalytic chains and the zinc domains of the regulatory chains. A listing of these interactions may be found in Table 7. Although suggested by the shifts in domains between the T and R forms, there appear to be no new chain links between the equatorial domain and the zinc domain in this new form. This absence of new links between these domains is in contrast to what was predicted by Krause *et al.* (1985). There are potential interactions between Arg128 and Lys129 of the regulatory chains and the equatorial domain of the catalytic unit, for example, with residues Asp200 and Glu204. At this stage of refinement, however, the distances are too long, and there is no density for the side-chains of residues 129 to 133 of the regulatory chains. This portion of the interface is dominated by an intrachain link between residues 128 and 144 both of the regulatory chain (Fig. 15).

Overall, despite minor differences, this interface appears to be nearly the same in both forms It is certainly the same within the accuracy of the model co-ordinates.

(i) *Amino acids implicated in catalysis*

Genetic and chemical modification studies along with other experimental work have identified a subset of amino acid residues implicated as being important in the catalytic mechanism of aspartate

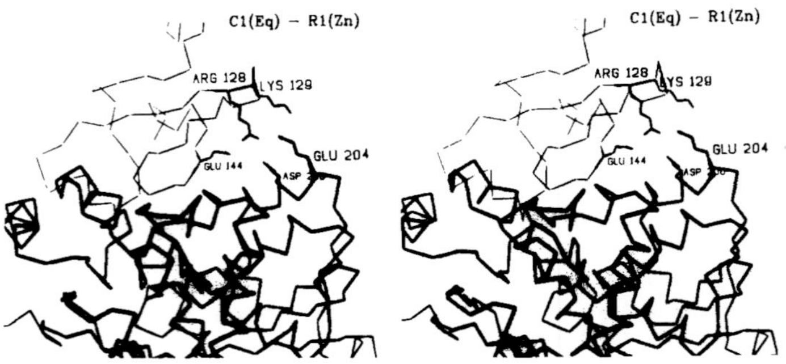

Figure 15. The α-carbon tracing of the C1–R1 interface in stereo, with emphasis on the region between the equatorial domain of the catalytic chain and the zinc domain of the regulatory chain. No contacts between these areas are present although potential interactions are described within the text. Residues Arg128, Lys129 and Glu144 of the regulatory chain (light tracing) and residues Asp200 and Glu204 of the catalytic chain (dark tracing) are identified. The 3-fold axis is perpendicular to the plane of the Figure.

carbamoyltransferase. These residues have been discussed relative to the unligated structure (Honzatko *et al.*, 1982). On the basis of the location and the binding of the PALA molecule in the R form, we can now re-examine this list.

(i) *Cys47*

Cys47, located in the polar domain of the catalytic chain, was among the first residues to be identified in aspartate carbamoyltransferase as catalytically important. This identification was made primarily through the use of 2-chloromercuri-4-nitrophenol, a compound that reacts selectively with this thiol group (Evans *et al.*, 1972). Upon reaction, the enzyme is found to lose catalytic activity in direct proportion to the number of modified cysteines. After all six residues of the dodecameric complex have reacted, no activity remains (Evans & Lipscomb, 1979). On the basis of this result, investigators concluded that this residue was near the active site. Some modifications of this residue, such as the *S*-cyano derivative, have been made that do not interfere with enzyme activity (Vanaman & Stark, 1970).

Despite these early but conflicting suggestions of catalytic importance, Cys47 would seem to be located too far from the active site in the R form to have a direct role in catalysis (Fig. 16). Indirect effects involving electrostatic or steric factors might be responsible for the observed inactivation of the enzyme.

(ii) *Ser52*

Ser52 was changed to phenylalanine in a site-specific mutagenesis experiment performed by Schachman *et al.* (1984). The mutant enzyme was found to be inactive. Because of the involvement of Ser52 and contiguous groups in the binding of the PALA phosphonate, it seems consistent that the introduction of such a large hydrophobic residue into this region would interfere with catalytic activity (Fig. 16).

(iii) *Lys83, Lys84 and Lys232*

Lysines 83, 84 and 232 are three other residues identified as having potential catalytic importance based on chemical modification studies (Fig. 16). Lauritzen & Lipscomb (1982) found that modification of any of these residues causes loss of catalytic activity. In another study, Greenwell *et al.* (1973) found that lysine residues of the catalytic subunit could be reacted with pyridoxal

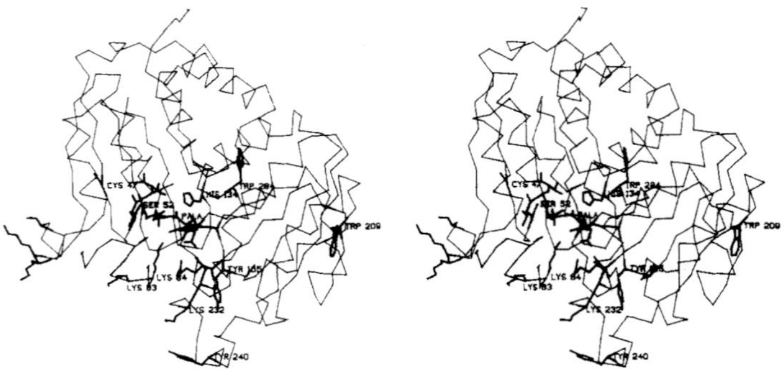

Figure 16. The α-carbon tracing of the R form of the catalytic chain in stereo, with selected residues superimposed (see the text). Lysines 83 and 84 are present on this chain, but they are also shown superimposed on a portion of an adjacent catalytic chain where along with Ser80 they contribute to the active site.

5′-phosphate to produce an inactive enzyme that could still bind PALA. Also, Lys84 and Cys47 have been modified simultaneously by the bifunctional reagent 3-acetoxymercuri-5-nitrosalicyladehyde (Landfear, 1978). This modification produces an inactive enzyme that does not bind aspartate (McMurray *et al.*, 1972).

The results of these studies seem to be consistent with the structure of the R form of aspartate carbamoyltransferase in which Lys84 is involved directly in binding both the phosphonate and aspartate moieties of the PALA molecule. It is understandable that any modification of Lys84 could, therefore, interfere with both activity and aspartate binding. It is also understandable that Lys84 may not be essential for PALA binding, for which a number of other residues share responsibility.

In the change from the T to R form, the 80s loop of the enzyme undergoes a large migration into the active site of an adjacent catalytic chain. This suggests that it possesses great flexibility. Chemical modification studies are in agreement, since the simultaneous reaction of Lys84 and Cys47 as described above would also suggest a very mobile 80s loop.

Not all results of the chemical modification studies can be rationalized in terms of the new R form structure. For example, when modification of lysine residues by bromosuccinate is carried out in the presence of carbamoyl phosphate, Lys83 is protected from reaction. This is in conflict with the role of Lys84 in binding the phosphonate of PALA. One explanation might be that Lys83 binds to carbamoyl phosphate within the central cavity and helps to convey it to the active-site pocket where it then binds to Lys84.

Chemical modification studies also suggest a binding role for Lys232 that is not supported by the structure of the R form. This residue is located near other residues that bind to the PALA molecule, such as Arg229 and Gln231. Therefore, conformational change here could interfere with aspartate binding. The reaction of Lys232 with bromosuccinate might interfere with activity by causing a link to occur between this modified lysine and Arg229. In addition, a water molecule is located in the region between this lysine and Lys84 from an adjacent chain; it is bound to both residues. Communication of conformational changes might occur *via* this pathway.

(iv) *His134*

His134 is a strong candidate to activate the carbonyl group of carbamoyl phosphate and thus fill an important role in the mechanism of aspartate carbamoyltransferase. Its participation in substrate binding has been supported by several experimental studies (Greenwell *et al.*, 1973; Roberts *et al.*, 1976; Honzatko & Lipscomb, 1982). One of these studies (Roberts *et al.*, 1976) suggests that, in solution, the PALA molecule, when bound to the enzyme, is near an aromatic residue or that its carbonyl is protonated (Roberts *et al.*, 1976). On the basis of the crystal structure, His134 is the most likely candidate to be this residue. In the active site of the R form there is a probable hydrogen bond or salt link between His134 and the carbonyl of the PALA molecule. If carbamoyl phosphate is similarly bound, the link would serve to activate its carbonyl for nucleophilic attack. A neutral His134 could also serve as a general base by receiving a proton, either directly or *via* a water molecule from L-aspartate as it binds to the active site (Volz *et al.*, 1986).

Some site-specific mutagenesis studies have also strongly implicated His134 as being catalytically important (Wall *et al.*, 1979; Wall & Schachman, 1979). Also, substitution of aspartate for Gly128 in the catalytic chain resulted in loss of activity for aspartate carbamoyltransferase. Honzatko *et al.* (1982) postulated that this mutant was inactive due to the disruption of the positively charged active site with a negative residue. Recently, a 3·0 Å partially refined structure of this mutant has become available (Young, 1984). In it, the loss of activity is linked to a conformational change that places His134 outside of the active-site cavity. It is not clear if this mutant will still bind substrates. Therefore, it is difficult to decide if its inactivity is due to the loss of an essential group from the active site, or rather to a disruption of the active site that blocks substrate binding. A crystal structure or binding study of this mutant in the presence of carbamoyl phosphate or PALA would be instructive in this regard.

(v) *Tyr165, Tyr240*

Nitration of Tyr165 or substitution of serine for tyrosine at this site drastically reduces but does not eliminate catalytic activity (Lauritzen *et al.*, 1980; Robey & Schachman, 1984). Nitration of Tyr240 preserves activity but eliminates homotropic cooperativity (Landfear *et al.*, 1978*b*). This latter effect may be rationalized in view of the critical structural role of this residue in forming the C1–C4 interface of the R form. Presumably, alteration at this site might prevent the homotropic transition from occurring. The former results are much more difficult to understand since Tyr165 is so far from the active site (Fig. 16) that it apparently does not participate in the binding of either the L-aspartate or carbamoyl phosphate moiety of PALA. A rotation of 180° about the C^{α}–C^{β} bond could bring it into the active-site region; but density for other atoms is very good, and there is no density to support such a rearrangement.

One possibility is that the role of Tyr165 lies not in direct involvement in catalysis, but in facilitating the T to R transition. Its bonding pattern, *inter*subunit bonding with residue 239 in the T form, but *intra*subunit bonding with residue 239 in the R form, may be important in stabilizing the homotropic transition. Altered enzymes may be locked into an inactive quaternary structure and effectively prevented from undergoing the T to R transition, thus explaining at once the decrease in catalytic activity and the lack of homotropic co-operativity. The decrease in rate reported for modified catalytic trimers could occur because the portion of the 240s loop involved in aspartate binding is not stabilized in these altered forms by binding with Tyr165, thereby preventing a needed conformational change. The increase in K_d for carbamoyl phosphate is not necessarily similarly explained. However, it is possible that if the enzyme is prevented from assuming its most active form, then the migration of the 80s loop into the active site may be altered. Any alteration here could have an effect on the K_d for carbamoyl phosphate.

(vi) *Trp209 and Trp284*

Foote *et al.* (1980) reported that the 7-aza substitution of Trp209 and Trp284 stabilizes a form of aspartate carbamoyltransferase with decreased affinity for aspartate. Although Honzatko *et al.* (1982) found that no definitive choice could be made, on the basis of the T structure, with regard to which of the tryptophans was important in aspartate binding, they did speculate that a polar link between residue 209 and Asp203 might be involved. Further support of this view comes from unpublished results of Smith *et al.*, which indicate that increased interaction between these two sites in the form of a salt link created by genetic means stabilizes an enzyme with prominent T state properties. They also indicate that decreased interaction between residues 209 and 203 stabilizes an enzyme with some R state properties.

There is evidence from difference spectroscopy that a conformational change in tryptophan residues occurs during the T to R transition (Collins & Stark, 1969). This evidence has led to speculation that the 203 to 209 link might be broken in the R state and that indirect propagation of these conformational changes to the active site is responsible for the reduced affinity for aspartate.

Comparison of the R state, as determined by our complex of aspartate carbamoyltransferase with PALA, with the T state structure (Ke *et al.*, 1984) has shown no discernible differences in the environments of Trp209 and Trp284. The same 203 to 209 link is prominent in the R state as in the T state (Fig. 17). Correction of the catalytic chain sequence has resulted in a repositioning of Trp284 relative to the structure reported by Honzatko *et al.* (1982). It is now located at the beginning of helix H9 and it interacts with Thr144, Lys262 and Gln287 (Fig. 18). There is no longer an interaction with Lys232.

Thus, neither form of the enzyme allows a determination of which tryptophan is involved in the effects on aspartate binding. The studies of Smith *et al.* mentioned above, however, would seem to suggest strongly that Trp209 is the responsible residue. The same indirect explanation for this effect as that offered by Honzatko *et al.* (1982) will apply until a determination of the crystal structure of the mutants can be made.

(j) *Proposed mechanism for the homotropic transition*

Given that the structures of aspartate carbamoyl-transferase, unligated, and in the presence of PALA, represent only two extreme points on the continuum between the T and R states, caution is important in any scheme to explain the homotropic transition. Some investigators even question whether the R form induced by PALA is completely equivalent to that induced by natural substrates (Heyde, 1976). As long as these caveats are kept in mind, it is tempting to speculate on a mechanism to explain the dramatic differences between the two forms. Also, it should be noted that the validity of the suggested scheme does not affect the veracity of the determination of the R structure.

(1) Carbamoyl phosphate enters the central cavity *via* the channels next to the regulatory chains and approaches a catalytic chain binding to the flexible loop containing Lys83 and Lys84. The active sites of the enzyme may only be entered through this central cavity. Van der Waals' contacts prevent any substrates from diffusing down and through the C1–C2–C3 interface. The binding of carbamoyl phosphate by lysine residues

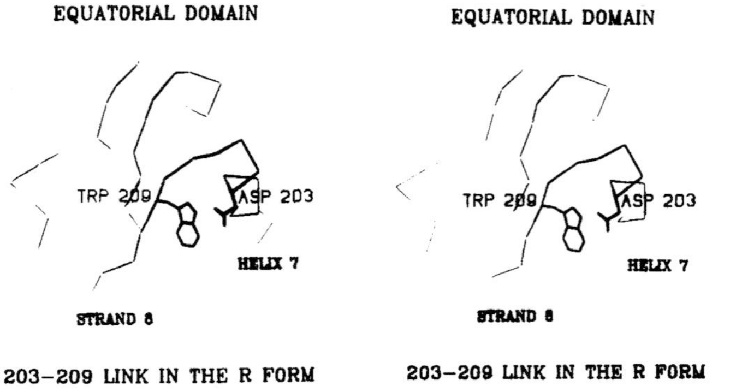

Figure 17. A portion of the equatorial domain of a catalytic chain is presented in stereo, with emphasis on the Asp203–Trp209 link.

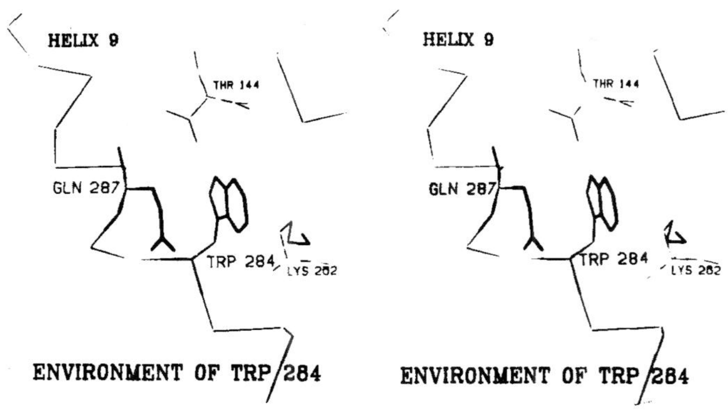

Figure 18. A portion of the equatorial domain of a catalytic chain near residue Trp284 is presented in stereo.

83 and 84 would explain the protection by carbamoyl phosphate of bromosuccination of Lys83.

(2) Lys83 and Lys84 convey carbamoyl phosphate to the active site, where it is bound by positively charged residues in the polar domain. This substrate would be drawn to the active site by the dipoles of helices H2, H5 and H9, as well as by the numerous positively charged and polar residues nearby. The needed degree of flexibility is demonstrated for the 80s loop by its large motion in the T to R transition and also by the chemical modification studies showing that it is possible to modify both Cys47 and lysines 83 and 84 with a single bifunctional reagent.

(3) Three carbamoyl phosphate molecules bind to one c_3 trimer. Experimental evidence suggests that only three sites are available for binding in the hexamer (Suter & Rosenbusch, 1977). No explanation for this effect has been found in the T or R structures to date.

(4) One aspartate molecule enters the active site of a catalytic chain that has already bound carbamoyl phosphate; it then binds to the equatorial domain. The changes occurring within the equatorial domain that allow binding to the aspartate molecule lead to the rearrangement of the bonding between the upper and lower catalytic chains. This rearrangement facilitates the movement of the equatorial domains in toward the carbamoyl phosphate binding site. Upon completion of this motion, the 240s loops stack on top of each other. Collapse to the T form along the 3-fold axis is prevented by this stacking. This motion produces a net untwisting of the c_3 trimers. It is communicated to the other catalytic subunits *via* the r_2 dimers, which move roughly as a unit with the untwisting of the trimers. This motion of the r_2 dimers helps to facilitate the T to R transition for the entire dodecameric complex. The C1–R1 and C6–R6 interfaces remain largely unchanged.

(5) L-Aspartate and carbamoyl phosphate now are bound to the equatorial and polar domains, respectively. Lys84 and Arg105 bind to both molecules and help to stabilize the R state. The salt links between Arg167 and Glu50, Arg234 and Glu50, and Glu239 and Lys164 also stabilize this new conformational state. The carbonyl group of carbamoyl phosphate is activated for nucleophilic attack by His134 and perhaps by Thr55 and Arg105. The amino group of L-aspartate is activated by proton removal possibly involving the phosphate group or active-site water molecules and His134. The energy of activation is thus lowered by a combination of general acid and base catalysis. The juxtaposition of the substrates increases the probability of properly oriented collisions, and the reaction proceeds.

(6) The formation of carbamoyl-L-aspartate occurs within the active site. As the products separate, the polar and equatorial domains move apart allowing for product release and for the binding of new substrates. Complete collapse back to the T-form does not seem to be needed here. Each of the three pairs of upper and lower catalytic chains may assume the T and R state configuration somewhat independently, although any change in their conformation is certainly communicated to the other catalytic chains *via* the r_2 dimers. The presence of some degree of independence between the active sites would explain why some studies indicate that as many as four molecules of PALA may need to bind to bring about the complete homotropic transition (Howlett & Schachman, 1977).

4. Conclusions

The structure of aspartate carbamoyltransferase complexed with PALA adds much to our understanding of this enzyme. It has allowed identification of the active site as well as those residues involved in substrate binding. In addition, it has allowed for a description of the R (more active) form of the enzyme and for a model of the homotropic allosteric transition that occurs between the T and R forms. Some work on this form remains to be done, for example, further modeling of solvent. Crystal studies involving

PALA and other substrate analogs, along with CTP and ATP, are in progress.

We thank the National Institutes of Health, grant GM06920, for support of this research. We also thank N. h. Xuong, R. Hamlin and C. Nielson for use of the data collection Resource for Crystallography at the University of California, San Diego (grant RR01644). R. Honzatko, University of Iowa, is acknowledged for his generous help with refinement software and strategy. Finally, we thank R. Hubbard of York University and D. States of the University of California, San Diego for HYDRA and PLT2, respectively. Most of the Figures in the text were generated with these programs. We also acknowledge use of the GRIP-75 molecular graphics program at the University of North Carolina, Department of Computer Science, including thanks to E. G. Britton, J. S. Lipscomb, M. E. Pique, W. Siddal, J. E. McQueen, Jr, J. Hermans, W. V. Wright and F. P. Brooks, Jr (Principal Investigator). This graphics program was supported by the National Institutes of Health, the National Science Foundation, the Atomic Energy Commission and the IBM Corporation.

References

Altman, R. B., Ladner, J. E. & Lipscomb, W. N. (1982). *Biochem. Biophys. Res. Commun.* **108**, 592–595.
Bennett, W. S., Jr & Steitz, T. A. (1978). *Proc. Nat. Acad. Sci., U.S.A.* **75**, 4848–4852.
Bennett, W. S., Jr & Steitz, T. A. (1980*a*). *J. Mol. Biol.* **140**, 183–210.
Bennett, W. S., Jr & Steitz, T. A. (1980*b*). *J. Mol. Biol.* **140**, 211–230.
Bethell, M. R., Smith, K. E., White, J. S. & Jones, M. E. (1968). *Proc. Nat. Acad. Sci., U.S.A.* **60**, 1442–1449.
Blackburn, M. N. & Schachman, H. K. (1977). *Biochemistry*, **16**, 5084–5091.
Blow, D. M. & Crick, F. H. C. (1959). *Acta Crystallogr.* **12**, 794–802.
Bricogne, G. (1974). *Acta Crystallogr. sect. A*, **30**, 395–405.
Bricogne, G. (1976). *Acta Crystallogr. sect. A*, **32**, 832–847.
Burley, S. K. & Petsko, G. A. (1985). *Science*, **229**, 23–28.
Chan, W. W.-C. (1975). *J. Biol. Chem.* **250**, 668–674.
Collins, K. D. & Stark, G. R. (1969). *J. Biol. Chem.* **244**, 1869–1877.
Collins, K. D. & Stark, G. R. (1971). *J. Biol. Chem.* **246**, 6599–6605.
Colman, P. D. & Marcus, G. (1972). *J. Biol. Chem.* **247**, 3829–3837.
Cox, J. M. (1967). *J. Mol. Biol.* **28**, 151–156.
Crawford, J. L. (1977). Ph.D. thesis, Harvard University.
Creighton, T. E. (1983). In *Proteins – Structures and Molecular Principles*, pp. 459–466, W. H. Freeman & Co., New York.
Davies, G. E., Vanaman, T. C. & Stark, G. R. (1970). *J. Biol. Chem.* **245**, 1175–1179.
Diamond, R. (1982). In *Computational Crystallography* (Sayre, D., ed.), pp. 318–325, Oxford, London.
Eckfeldt, J., Hammes, G. G., Mohr, S. C. & Wu, C.-W. (1970). *Biochemistry*, **9**, 3353–3362.
Eklund, H., Samama, J. P., Wallen, L., Branden, C. I., Akeson, A. & Jones, T. A. (1981). *J. Mol. Biol.* **146**, 561–587.
Evans, D. R. & Lipscomb, W. N. (1979). *J. Biol. Chem.* **254**, 10679–10685.
Evans, D. R., McMurray, C. H. & Lipscomb, W. N. (1972). *Proc. Nat. Acad. Sci., U.S.A.*, **69**, 3638–3642.
Evans, D. R., Pastra-Landis, E. C. & Lipscomb, W. N. (1975). *J. Biol. Chem.* **250**, 3571–3583.
Foote, J. & Schachman, H. K. (1985). *J. Mol. Biol.* **186**, 175–184.
Foote, J., Ikeda, D. M. & Kantrowitz, E. R. (1980). *J. Biol. Chem.* **255**, 5154–5158.
Gerhart, J. C. (1970). *Curr. Top. Cell. Regul.* **2**, 275–325.
Gerhart, J. C. & Holoubek, H. (1967). *J. Biol. Chem.* **242**, 2886–2892.
Gerhart, J. C. & Pardee, A. B. (1962). *J. Biol. Chem.* **237**, 891–896.
Gerhart, J. C. & Pardee, A. B. (1964). *Fed. Proc. Fed. Amer. Soc. Exp. Biol.* **23**, 727–735.
Gerhart, J. C. & Schachman, H. K. (1965). *Biochemistry*, **4**, 891–896.
Gerhart, J. C. & Schachman, H. K. (1968). *Biochemistry*, **7**, 538–552.
Greenwell, P., Jewett, S. L. & Stark, G. R. (1973). *J. Biol. Chem.* **248**, 5994–6001.
Hamlin, R., Cork, C., Howard, A., Nielsen, C., Vernon, W., Matthews, D. & Xuong, Ng. h. (1981). *J. Appl. Crystallogr.* **14**, 85–93.
Hammes, G. G. & Wu, C.-W. (1971). *Science*, **172**, 1205–1211.
Hendrickson, W. A. & Konnert, J. (1981). In *Biomolecular Structure, Function, Conformation and Evolution* (Srinivasan, R., ed.), vol. 1, pp. 43–47, Pergamon Press, London.
Heyde, E. (1976). *Biochim. Biophys. Acta*, **452**, 81–88.
Honzatko, R. B. & Lipscomb, W. N. (1982). *J. Mol. Biol.* **160**, 265–286.
Honzatko, R. B., Crawford, J. L., Monaco, H. L., Ladner, J. E., Edwards, B. F. P., Evans D. R., Warren, S. G., Wiley, D. C., Ladner, R. C. & Lipscomb, W. N. (1982). *J. Mol. Biol.* **160**, 219–263.
Hoover, T. A., Roof, W. D., Folterman, K. F., O'Donovan, G. A., Bencini, D. A. & Wild, J. R. (1983). *Proc. Nat. Acad. Sci., U.S.A.* **80**, 2462–2466.
Houghton, J. E., Bencini, D. A., O'Donovan, G. A. & Wild, J. R. (1984). *Proc. Nat. Acad. Sci., U.S.A.* **81**, 4864–4868.
Howlett, G. J. & Schachman, H. K. (1977). *Biochemistry*, **16**, 5077–5083.
Jack, A. & Levitt, M. (1978). *Acta Crystallogr. sect. A*, **34**, 931–935.
Jacobson, G. R. & Stark, G. R. (1973). In *The Enzymes* (Boyer, P. D., ed.), vol. 9, pp. 225–308, Academic Press, New York.
Johnson, R. S. & Schachman, H. K. (1980). *Proc. Nat. Acad. Sci., U.S.A.* **77**, 1995–1999.
Jones, M. E., Spector, L. & Lipmann, F. (1955). *J. Amer. Chem. Soc.* **77**, 819–820.
Jones, T. A. (1982). In *Computational Crystallography* (Sayre, D., ed.). pp. 303–317, Oxford, London.
Kabsch, W. (1978). *Acta Crystallogr. sect. A*, **34**, 827–828.
Kantrowitz, E. R., Pastra-Landis, S. C. & Lipscomb, W. N. (1980*a*). *Trends Biochem. Sci.* **5**, 124–128.
Kantrowitz, E. R., Pastra-Landis, S. C. & Lipscomb, W. N. (1980*b*). *Trends Biochem. Sci.* **5**, 150–153.
Ke, H. M., Honzatko, R. B. & Lipscomb, W. N. (1984). *Proc. Nat. Acad. Sci., U.S.A.* **81**, 4037–4040.
Kempe, T. D. & Stark, G. R. (1975). *J. Biol. Chem.* **250**, 6861–6869.
Kerbiriou, D. & Hervé, G. (1972). *J. Mol. Biol.* **64**, 379–392.
Kerbiriou, D. & Hervé, G. (1973). *J. Mol. Biol.* **78**, 687–702.
Knier, B. L. & Allewell, N. M. (1978). *Biochemistry*, **17**, 784–790.

Konigsberg, W. H. & Henderson, L. (1983). *Proc. Nat. Acad. Sci., U.S.A.* **80**, 2467–2471.

Krause, K. L., Volz, K. W. & Lipscomb, W. N. (1985). *Proc. Nat. Acad. Sci., U.S.A.* **82**, 1643–1647.

Ladner, J. F., Kitchell, J. P., Honzatko, R. B., Ke, H. M., Volz, K. W., Kalb(Gilboa), A. J., Ladner, R. C. & Lipscomb, W. N. (1982). *Proc. Nat. Acad. Sci., U.S.A.* **79**, 3125–3128.

Landfear, S. M. (1978). Ph.D. thesis, Harvard University.

Landfear, S. M., Evans, D. R. & Lipscomb, W. N. (1978*a*). *J. Biol. Chem.* **253**, 3988–3996.

Landfear, S. M., Evans, D. R. & Lipscomb, W. N. (1978*b*). *Proc. Nat. Acad. Sci., U.S.A.* **75**, 2654–2658.

Lauritzen, A. M. & Lipscomb, W. N. (1980). *Biochem. Biophys. Res. Commun.* **95**, 1425–1430.

Lauritzen, A. M. & Lipscomb, W. N. (1982). *J. Biol. Chem.* **257**, 1312–1319.

Lauritzen, A. M., Landfear, S. M. & Lipscomb, W. N. (1980). *J. Biol. Chem.* **255**, 602–607.

Lesk, A. M. & Chothia, C. (1984). *J. Mol. Biol.* **174**, 175–191.

Leslie, A. G. W. & Wonacott, A. J. (1984). *J. Mol. Biol.* **178**, 743–772.

McClintock, D. K. & Marcus, G. (1968). *J. Biol. Chem.* **243**, 2855–2862.

McMurray, C. H., Evans, D. R. & Sykes, B. D. (1972). *Biochem. Biophys. Res. Commun.* **48**, 572–577.

Monaco, H. L., Crawford, J. L. & Lipscomb, W. N. (1978). *Proc. Nat. Acad. Sci., U.S.A.* **75**, 5276–5280.

Moody, M. F., Vachette, P. & Foote, A. M. (1979). *J. Mol. Biol.* **133**, 517–532.

Rees, D. C., Lewis, M. & Lipscomb, W. N. (1983). *J. Mol. Biol.* **168**, 367–387.

Reichard, P. & Hanshoff, G. (1956). *Acta Chem. Scand.* **10**, 548–566.

Remington, S., Wiegand, G. & Huber, R. (1982). *J. Mol. Biol.* **158**, 111–152.

Roberts, M. F., Opella, S. J., Schaffer, M. H., Phillips, H. M. & Stark, G. R. (1976). *J. Biol. Chem.* **251**, 5976–5985.

Robey, E. A. & Schachman, H. K. (1984). *J. Biol. Chem.* **259**, 1180–1183.

Schachman, H. K., Pauza, C. D., Navre, M., Karels, M. J., Wu, L. & Yang, Y. R. (1984). *Proc. Nat. Acad. Sci., U.S.A.* **81**, 115–119.

Suter, P. & Rosenbusch, J. P. (1977). *J. Biol. Chem.* **252**, 8136–8141.

Ten Eyck, L. F. (1973). *Acta Crystallogr. sect. A*, **29**, 182–191.

Ten Eyck, L. F. (1977). *Acta Crystallogr. sect. A*, **33**, 486–492.

Vanaman, T. C. & Stark, G. R. (1970). *J. Biol. Chem.* **245**, 3565–3573.

Volz, K. W., Krause, K. L. & Lipscomb, W. N. (1986). *Biochem. Biophys. Res. Commun.* **136**, 822–826.

Wall, K. A. & Schachman, H. K. (1979). *J. Biol. Chem.* **254**, 11917–11926.

Wall, K. A., Flatgaard, J. E., Gibbons, I. & Schachman, H. K. (1979). *J. Biol. Chem.* **254**, 11910–11916.

Weber, K. (1968). *Nature (London)*, **218**, 1116–1119.

Wiley, D. C. & Lipscomb, W. N. (1968). *Nature (London)*, **218**, 1119–1121.

Young, T.-S. (1984). Ph.D. thesis, Duke University.

Zanotti, G., Monaco, H. L. & Foote, J. (1984). *J. Amer. Chem. Soc.* **106**, 7900–7904.

Edited by R. Huber

doi: 10.1016/S0022-2836(02)00051-7 available online at http://www.idealibrary.com on IDEAL J. Mol. Biol. (2002) 318, 503–518

JMB

The Crystal Structure of *Trichomonas vaginalis* Ferredoxin Provides Insight into Metronidazole Activation

Chetlen R. Crossnoe[1,2], Juris P. Germanas[3], Pierre LeMagueres[2], Gabriela Mustata[2] and Kurt L. Krause[1,2,4]*

[1]*Structural and Computational Biology and Molecular Biophysics Program Baylor College of Medicine Houston, TX 77030, USA*

[2]*Department of Biology and Biochemistry, University of Houston, Houston TX 77204, USA*

[3]*Department of Chemistry Southern Methodist University Dallas, TX 75235, USA*

[4]*Section of Infectious Diseases Department of Medicine Baylor College of Medicine Houston, TX 77030, USA*

**Corresponding author*

Crystallographic studies revealing the three-dimensional structure of the oxidized form of the [2Fe–2S] ferredoxin from *Trichomonas vaginalis* (TvFd) are presented. TvFd, a member of the hydrogenosomal class of ferredoxins, possesses a unique combination of redox and spectroscopic properties, and is believed to be the biological molecule that activates the drug metronidazole reductively in the treatment of trichomoniasis. It is the first hydrogenosomal ferredoxin to have its structure determined. The structure of TvFd reveals a monomeric, 93 residue protein with a fold similar to that of other known [2Fe–2S] ferredoxins. It contains nine hydrogen bonds to the sulfur atoms of the cluster, which is more than the number predicted on the basis of the spectroscopic data. The TvFd structure contains a large dipole moment like adrenodoxin, and appears to have a similar interaction domain. Our analysis demonstrates that TvFd has a unique cavity near the iron–sulfur cluster that exposes one of the inorganic sulfur atoms of the cluster to solvent. This cavity is not seen in any other [2Fe–2S] ferredoxin with known structure, and is hypothesized to be responsible for the high rate of metronidazole reduction by TvFd.

Keywords: ferredoxin structure; metronidazole; *Trichomonas vaginalis*; protein crystallography; [2Fe–2S] cluster

Introduction

TvFd is a [2Fe–2S] ferredoxin from the pathogenic protozoan *Trichomonas vaginalis*, the cause of the common sexually transmitted disease, trichomoniasis. Trichomoniasis affects millions of women worldwide each year and is a risk factor for other STDs.[1] TvFd is found in the hydrogenosome, an unusual organelle that provides trichomonads with the ability to reduce pyruvate anaerobically for the generation of ATP.[2] The biological function of TvFd is to transfer electrons from pyruvate:ferredoxin oxidoreductase (PFOR) to hydrogenase, which in turn passes these electrons to protons to form H_2.

TvFd is a [2Fe–2S] ferredoxin, an important class of iron–sulfur proteins that facilitate electron transfer in a variety of biological systems. The [2Fe–2S] ferredoxins have traditionally been divided into two subclasses; the photosynthetic (or plant-type) ferredoxins and the oxidase (or vertebrate-type) ferredoxins.[3] Several crystal and NMR structures of photosynthetic and oxidase ferredoxins exist.[4–23]

However, TvFd does not fit readily into either of these classes. Hydrogenosomal ferredoxins have distinct spectroscopic properties relative to other [2Fe–2S] ferredoxins. In the absorbance spectra of the oxidized proteins, for instance, the band at 420 nm that is prominent in other [2Fe–2S] ferredoxin spectra is not present in the hydrogenosomal ferredoxin spectra.[24,25] Circular dichroism (CD) spectra of hydrogenosomal ferredoxins are clearly different from those of oxidase ferredoxins,

Abbreviations used: CD, circular dichroism; TvFd, *Trichomonas vaginalis* ferredoxin; STD, sexually transmitted disease; PFOR, pyruvate:ferredoxin oxidoreductase; v-AnFd, *Anabaena* ferredoxin, vegetative form; PDB, Brookhaven Protein Data Bank; rmsd, root-mean-square distance; MAD, multi-wavelength anomalous dispersion; MIR, multiple isomorphous replacement.

E-mail address of the corresponding author: kkrause@uh.edu

Table 1. Comparison of TvFd sequence identity and rmsd's to other ferredoxins

[2Fe–2S] Ferredoxin	Sequence identity with TvFd	rmsd with TvFd (C^{α} atoms)
Putidaredoxin	0.25	2.33
Adrenodoxin	0.23	2.51
Haloarcula	0.18	3.47
Synechocystis	0.18	5.54
Synechococcus	0.24	3.14
Petroselinum	0.25	3.18
Equisetum	0.27	3.13
Aphanothece	0.27	3.00
Spirulina	0.26	3.24
Spinacia	0.23	3.25
Anabaena (heterocyst)	0.18	3.06
Chlorella	0.23	3.17
Anabaena (vegetative)	0.27	3.06

By comparison, putidaredoxin and adrenodoxin have a sequence identity of 0.33 and rmsd of 2.24 Å between their structures; the known photosynthetic ferredoxins have sequence identities of 0.45 to 0.75 and rmsd values of 0.70–1.64 Å. The rmsd between the three molecules of the TvFd asymmetric unit was very low (0.070 for A and B, 0.069 for A and C) due to the use of backbone NCS restraints during refinement. The rmsd values were calculated by rms superpositioning of C^{α} atoms in QUANTA (version 98, Molecular Simulations Inc. Waltham, MA).

which show little variability. For instance, two positive Cotton-effect bands at 472 and 434 nm in the oxidized *T. vaginalis* spectrum are distinguishable, while these bands are coincident in other [2Fe–2S] protein spectra.[26–28] The CD spectrum of reduced *T. vaginalis* ferredoxin has a characteristic appearance as well, with two well-separated negative extrema at 460 and 508 nm, while a single band is present in oxidase ferredoxin CD spectra. In resonance Raman spectra, the frequencies of bands of the *T. vaginalis* ferredoxin are generally downshifted relative to those for analogous bands of oxidase ferredoxins.[27,28] Hyperfine NMR spectra of hydrogenosomal ferredoxins are distinct from those of other [2Fe–2S] ferredoxins. While *T. vaginalis* ferredoxin displays six resolved sets of downfield-shifted signals at 280 K and 600 MHz, oxidase ferredoxins display four resolved clusters in the same region, and plant-type ferredoxins display a large unresolved band at 40 ppm at the same temperature and field strength.[29,30]

The oxidase ferredoxins of known structure (putidaredoxin and adrenodoxin) have a sequence identity of 0.33, and the photosynthetic ferredoxins with known structures are even more internally similar, with identities between structures ranging from 0.50 to 0.75. TvFd has a lower sequence identity with both groups, but is more similar to the oxidase class (Table 1). However, in terms of redox potential, TvFd (−350 mV) is more similar to the photosynthetic ferredoxins, which range from −325 to −440 mV, than to the oxidase ferredoxins, with redox potentials of −230 mV for putidaredoxin and −273 mV for wild-type intact adrenodoxin.[21] TvFd has, therefore, been considered by some workers as belonging to a new class of [2Fe–2S] ferredoxins, the hydrogenosomal ferredoxins,[31] others maintain that this group fits best as a subclass of the oxidase ferredoxins. Other known hydrogenosomal ferredoxins include those from *Tritrichomonas foetus*[25] and *Psalteriomonas lanterna*.[32] No structure of a hydrogenosomal ferredoxin is available.

TvFd is of pharmacological interest because of its role in the activation of metronidazole, the most important drug available for the treatment of trichomoniasis as well as many other infections caused by anaerobic bacteria and protozoa.[33] The observation that metronidazole inhibits the production of H_2 by *T. vaginalis* originally inspired the incorrect hypothesis that the drug acted by fatally disrupting hydrogenase by preferentially accepting electrons from TvFd.[34] It was later recognized that instead of blocking the H_2-generating pathway of the organism, the drug itself becomes highly reactive upon accepting an electron and proceeds to damage DNA and other cellular components.[35–38] Electron spin resonance (ESR) spectroscopy then confirmed that the reduction of metronidazole causes the formation of nitro anion radical intermediates,[39,40] which are believed to be the toxic species that damage DNA.[2]

The precise role of TvFd in the reduction of metronidazole has been the subject of debate.[41] Many researchers believe that TvFd reduces metronidazole directly in a non-enzymatic electron transfer reaction.[42–46] This was the dominant hypothesis early in the investigation of the mechanism, and is supported by evidence that a loss of TvFd gene expression is associated with metronidazole resistance,[47,48] and by the observation that addition of TvFd to hydrogenosomal extracts causes a large increase in metronidazole reduction.[25,43,49,50] Other researchers, however, believe that the drug is reduced enzymatically by a "metronidazole reductase" or "nitroreductase",[51] proposed to be the hydrogenosomal hydrogenase.[41] As this enzyme is the recipient of electrons from TvFd, this hypothesis would not necessarily conflict with the evidence demonstrating the necessity of TvFd for the reduction of metronidazole.

Table 2. Refinement statistics for TvFd structure

R-factor (%)	24.8
Free *R*-factor (%)	31.5
Resolution (Å)	100–2.2
No. reflections	15012
Completeness to 2.2 Å (%)	96.2
Completeness of outer shell (2.20–2.28 Å) (%)	81.0
Rmsd bonds (Å)	0.007
Rmsd angles (deg.)	1.136
Space group	$P2_12_12_1$
Molecules/asymm. unit	3
No. atoms/mol.	685
No. residues/mol.	93
No. water molecules	50
Average *B*-factor (Å^2)	40.0

We have addressed this problem by studying the *in vitro* reduction of metronidazole and other nitroimidazoles by TvFd using stopped-flow spectrophotometry. TvFd does reduce nitroimidazoles rapidly *in vitro*, and a complex has been demonstrated between TvFd and some nitroimidazoles, although not metronidazole. These experiments showed that TvFd transfers electrons to nitroimidazoles 100 times faster than *Anabaena* ferredoxin (vegetative form, v-AnFd). The greater rate of electron transfer by TvFd occurred despite the greater driving force from v-AnFd, which has a redox potential of −440 mV as compared to a redox potential of −350 mV for TvFd. (The redox potential of metronidazole is −345 mV.) These data will be presented in a subsequent publication.

Here, we report the 2.2 Å crystal structure of TvFd determined by multi-wavelength anomalous dispersion (MAD) phasing using the Fe atoms of the [2Fe–2S] cluster. This is the first structure of a hydrogenosomal ferredoxin to be reported. Although the fold of TvFd is similar to that of other [2Fe–2S] ferredoxins, it does contain several unique features. We will compare TvFd to other [2Fe–2S] ferredoxins, with particular emphasis on differences in their tertiary structure. Furthermore, we will describe a unique cavity near the iron–sulfur cluster that could allow direct interaction with nitroimidazole drugs and other small molecules. An increased understanding of the interaction between TvFd and metronidazole on a structural level may open the door to future drug-design efforts for pharmaceuticals with modes of action similar to metronidazole.

Results

After refinement, the model of TvFd consisted of 2055 protein atoms and 50 solvent water molecules (Table 2). The crystallographic *R*-factor was 24.8% based on 13,481 reflections collected from 100–2.2 Å and used in refinement. The final R_{test} was 31.5% for 1531 reflections used in cross-validation. Bond distance and angle rms deviations were 0.007 Å and 1.1°, respectively. Average *B*-factor values for the main-chain, side-chain, and solvent atoms were 39.17, 41.16, and 36.80 Å^2, respectively; molecules A, B, and C had average *B*-values of 34.9, 36.1, and 47.4 Å^2, respectively, for all protein atoms (Figure 1). Average real-space correlation coefficients calculated with O were 0.857, 0.851, and 0.846 for molecules A, B, and C, respectively. Ramachandran plots of each molecule of the asymmetric unit showed only one residue, the cluster-binding Cys47, that was outside the most-favored region.

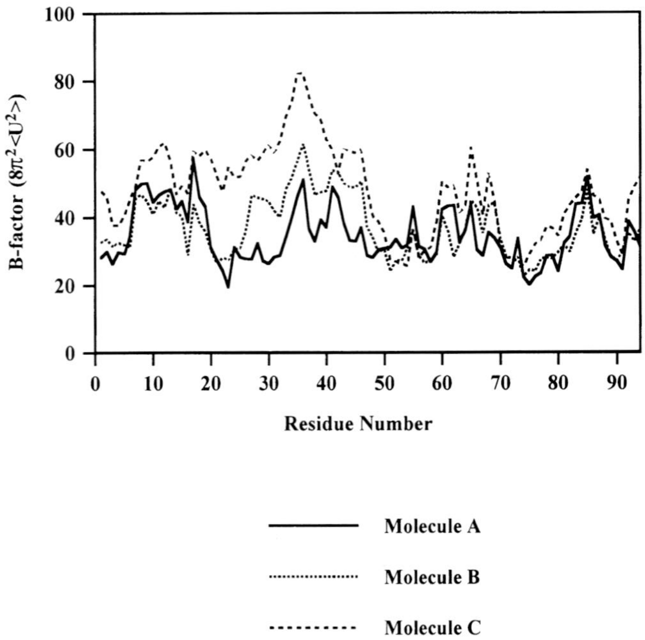

Figure 1. Temperature factor plotted as a function of residue number for the C$^\alpha$ atoms of *T. vaginalis* ferredoxin. Different line styles depict the three independent molecules A, B, and C found within the asymmetric unit.

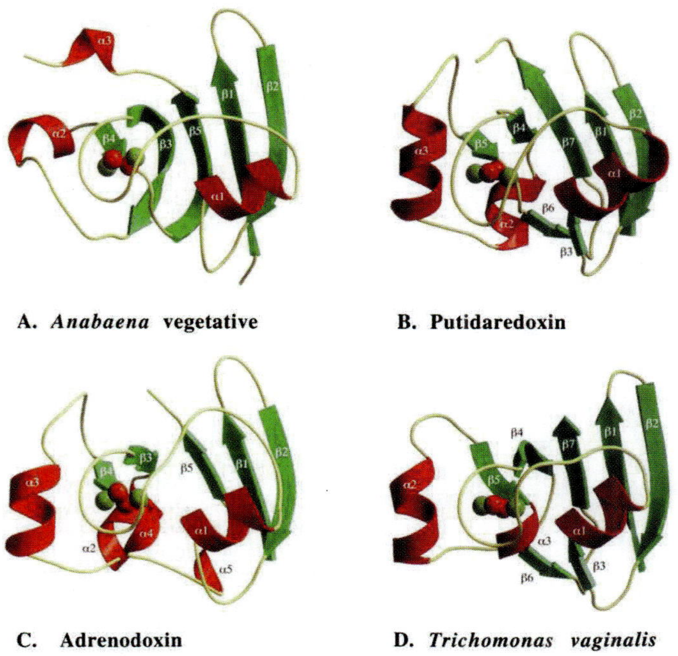

Figure 2. [2Fe–2S] ferredoxins: (a) vegetative-form *Anabaena* ferredoxin; (b) putidaredoxin; (c) adrenodoxin (residues 4–108); and (d) *T. vaginalis* ferredoxin. Iron atoms are shown in orange, inorganic sulfur atoms are green. The figure was made with MOLSCRIPT[80] and Raster3D.[81,82]

Secondary structure and other features

The most prominent features of TvFd are a five-stranded β-sheet, two α-helices, and the [2Fe–2S] cluster (Figure 2(d)). Other short segments of secondary structure are present (Table 3). The molecule begins with the first strand of the β-sheet (β_1), which extends from residues 1–7. Residues 7–10 form a type I′ reverse turn[52] leading into strand β_2 (residues 10–16). Residues 17–20 form a type II reverse turn, and residues 20–21 form the short strand β_3. Residues 22–30 form α_1, the prominent helix that is found in all known [2Fe–2S] ferredoxins. The backbone then stretches from residues 31 to 46 with no defined secondary structure except for a type II turn (34–37). Residues 33–47 form the loop covering the [2Fe–2S] cluster and contain three of the coordinating cysteine residues (Cys38, Cys44, and Cys47). Residues 47–53 then form β_4, which structurally is the fourth strand of the β-sheet; this strand is somewhat kinked and could be considered as two separate, short strands (residues 48–51 and 51–53). Residues 55–59 adopt no defined secondary structure, which is interesting because putidaredoxin and adrenodoxin, the closest homologues of TvFd, contain an α-helix in this region. Residues 60–67 form α_2, which is analogous to helix α_3 in putidaredoxin and adrenodoxin. Glu63 forms a salt-link on the interior of the protein with Arg75. Residues 67–70 and 71–74 form two more type I reverse turns, followed by a short β-strand (β_5, residues 74–76), an α-helix (α_3, residues 76–80), and another short β-strand (β_6, residues 80–82). Cys78 is the fourth coordinating cysteine residue of the cluster. Residues 83–86 and 86–89 form two more reverse turns of type III and II, respectively, and finally residues 89–93 form β_7 (Table 3).

Table 3. Secondary structure of TvFd

Residues	Structure type
1–7	β-Sheet
7–10	Type I′ reverse turn
10–16	β-Sheet
17–20	Type II reverse turn
20–21	β-Sheet
22–30	α-Helix
34–37	Type II reverse turn
38–41	Type III′ reverse turn
47–53	β-Sheet
60–67	α-Helix
67–70	Type I reverse turn
71–74	Type I reverse turn
74–76	β-Sheet
76–80	α-Helix
80–82	β-Sheet
83–86	Type III reverse turn
86–89	Type II reverse turn
89–93	β-Sheet

Comparison of independent molecules of the asymmetric unit

The *R*-factor and R_{test} of the TvFd structure are high because of disorder in molecule C of the asymmetric unit. Contoured at 1.4σ, molecule A has no main-chain break in electron density, and molecule B has only one, at Asp36. By comparison, molecule C has main-chain breaks at Gln20, Phe23,

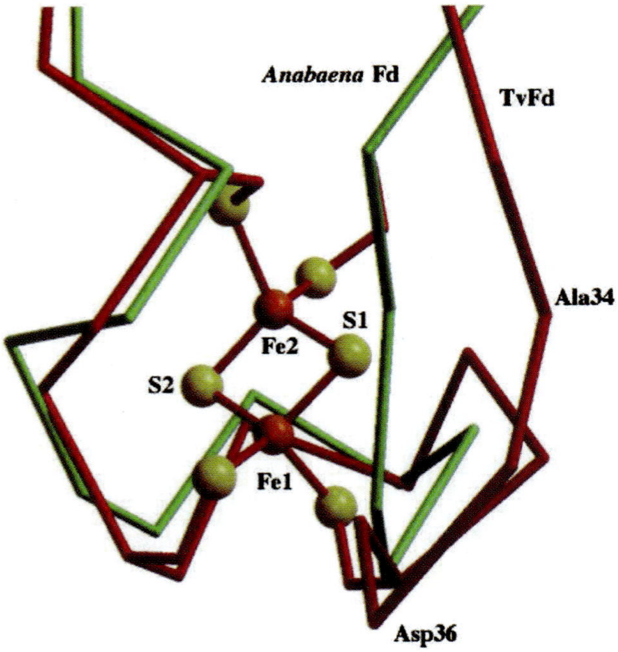

Figure 3. The backbone of TvFd (shown in red) follows a different course than other [2Fe–2S] ferredoxins at the beginning of the cluster-binding loop near Ala34, exposing the iron–sulfur cluster. The corresponding residues of other [2Fe–2S] ferredoxins (v-AnFd is shown in green) remain closer to the opposite side of the loop, covering the cluster. The figure was made with MOLSCRIPT[80] and Raster3D.[81,82]

Gly30, Leu31, Asp35, Gly40, Asn41, and Lys42. Several side-chains throughout all three molecules have poor density, but there are significantly more such residues in molecule C. Residues 30–42, comprising much of the loop covering the [2Fe–2S] cluster, are weak in molecule C. This section, however, has good density in molecules A and B, except for the side-chain of Asp36, which extends into the solvent and may not adopt a definite position in any of the three molecules. Essentially, all side-chains that are weak in one molecule are in density in at least one other molecule.

The iron–sulfur cluster region

The iron–sulfur cluster of TvFd is covered by a cluster-binding loop with a similar structure to that of all other [2Fe–2S] ferredoxins, with the exception of residues 33–36, which swing away from the body of the molecule to a much greater extent than in other ferredoxins (Figure 3). Three of the coordinating cysteine residues (Cys38, Cys44, Cys47) are contained in this loop; the fourth (Cys78) approaches the cluster from the interior of the molecule and is supplied by the α_3 helix. The binding site of the cluster is amphipathic, with hydrophobic residues (Phe23, Phe66, Leu76) forming the back and sides of the binding site (using the orientation shown in Figure 2), and hydrophilic residues (Asp36, Thr37, Gln39, Asn41, Lys42, Lys46) forming the front, top, and bottom, which are exposed to solvent. There are a total of nine hydrogen bonds to the sulfur atoms of the cluster, all from main-chain amide hydrogen atoms.

Comparison with other ferredoxin structures

The level of sequence identity of TvFd with all other [2Fe–2S] ferredoxins of known structure does not exceed 0.27 (Table 1). A detailed review of these sequences suggested more similarity to the photosynthetic ferredoxins in the amino and carboxy thirds of the molecule, but more similarity to the oxidase class in the region of the iron–sulfur cluster and the area around TvFd helix α_2. In fact, a hybrid sequence with an identity of 0.40 to TvFd can be constructed in this way from *Aphanothece sacrum* ferredoxin and putidaredoxin. Analysis of superimposed X-ray structures reveals that TvFd is more similar to the oxidase ferredoxins than to the photosynthetic ferredoxins (Table 1). This is seen most easily in the area around helix α_2. This helix is in a position analogous to that of helix α_3 of the oxidase ferredoxins, which is nearly parallel with the β-sheet, as opposed to helix α_2 of the photosynthetic ferredoxins, which is perpendicular to the β-sheet (Figure 2). In adrenodoxin, this region has been shown to be the primary interaction domain of the molecule with adrenodoxin reductase.[53] Due to the structural and electromagnetic similarities of these regions, suggesting a possible functional analogy (see below), we have preserved the term "interaction domain" when discussing the region of TvFd surrounding helix α_2. Neither TvFd nor the oxidase ferredoxins have the terminal helix typical of photosynthetic ferredoxins.

There are, also, several significant structural details of TvFd that differ from both the photosynthetic and the oxidase class of ferredoxins.[19–22] The first is in the region just prior to the

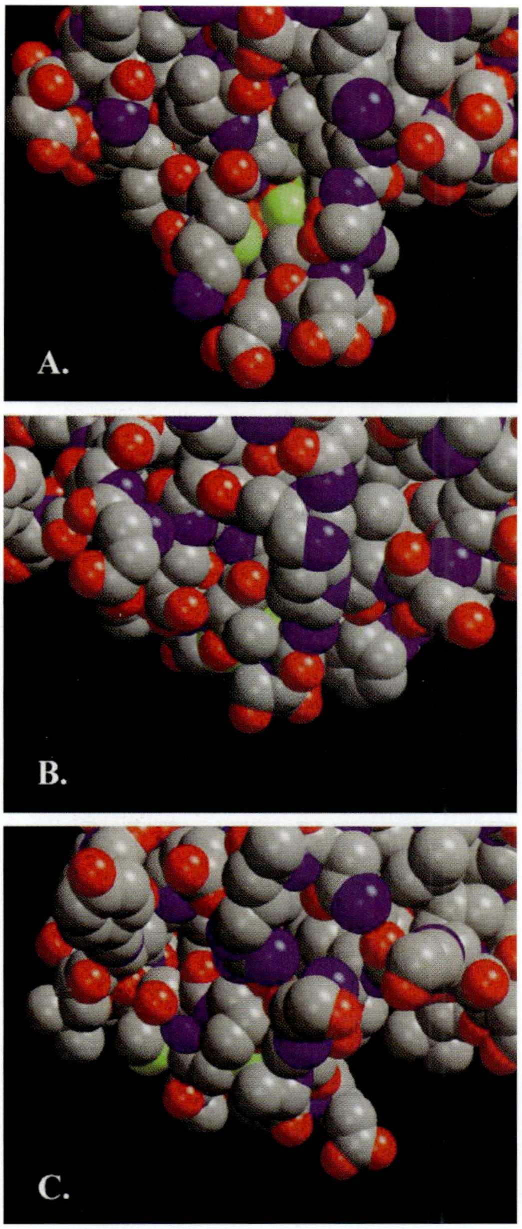

Figure 4. (a) A view of the cluster-binding loop of TvFd depicting the cavity that exposes the iron–sulfur cluster. Carbon atoms are shown in grey, oxygen in red, nitrogen in blue, sulfur in green, and iron in orange. (b) and (c) Similar views of v-AnFd and adrenodoxin, respectively, are depicted from the same orientation. These CPK figures were made with MOLSCRIPT[80] and Raster3D,[81,82] using the default atomic radii.

cluster-binding loop, and it involves the position of the backbone of TvFd around Asp35 and Asp36. In TvFd, this section of the backbone veers away from the body of the protein, forming a type II reverse turn, whereas in all other known ferredoxin structures the backbone remains closer to the body of the protein (Figure 3). This turn opens a deep crevice in the TvFd structure above the cluster, and one of the inorganic sulfur atoms of the cluster forms the floor of the crevice, something not seen in any other [2Fe–2S] ferredoxin with known structure (Figure 4). This exposure of the cluster may have implications for the reduction of metronidazole, as will be discussed. Another difference between the oxidase ferredoxins and TvFd is that the oxidase ferredoxins have a helix comprised of residues 55–61 (α_2 using the putidaredoxin numbering scheme) that is not present in TvFd (Figure 2).

The C terminus of TvFd occurs within strand β_7 of the β-sheet, rather than extending some distance past it, as it does in other [2Fe–2S] ferredoxins. The truncated adrenodoxin structure[21] terminates with Pro108, which is an essential residue for the structural integrity of adrenodoxin and has been shown to be highly conserved among vertebrate ferredoxins.[54] TvFd terminates at residue 93, which would correspond to residue 107 in adrenodoxin, suggesting an alternative method of protein stabilization in TvFd compared to adrenodoxin.

TvFd has no noticeable hydrogen-bonding network stabilizing the interaction domain, as there appears to be in the oxidase ferredoxins. The interaction domain of adrenodoxin and putidaredoxin are thought to be stabilized by hydrogen bonding from His56 (His49 in putidaredoxin), at the beginning of the interaction domain, to Tyr82 and Ser88 at the end of the interaction domain.[20,21] This linkage is proposed to transmit conformational changes upon redox activity from the cluster to the interaction domain, because one of the cluster-coordinating cysteine residues is Cys55, immediately adjacent to the histidine residue. There is no residue in TvFd analogous to His56; the interaction domain appears to be stabilized internally only by hydrophobic interactions between residues Ile48, Leu76, and Leu67. Similar hydrophobic interactions occur in both adrenodoxin and putidaredoxin.

In contrast to the differences between TvFd and other ferredoxins that are detailed above, the iron–sulfur cluster region of TvFd is in a position very similar to that of adrenodoxin. The minimum distances from the iron atoms to the surface of TvFd are 3.7 and 4.6 Å for Fe1 and Fe2, respectively, as compared to values of 3.7 and 4.4 Å for adrenodoxin.[54] Fe2 is somewhat more buried in both the putidaredoxin model and the photosynthetic ferredoxins, with minimum distances to the surface of 5.0–5.9 Å. This may be of interest because the close proximity of Fe2 to the molecular surface in adrenodoxin leads to additional possible pathways of electron transfer in that protein as compared to other ferredoxins.[54] The thinner protein shell around the adrenodoxin cluster was suggested to be responsible, in part, for the less negative redox potential of that protein relative to

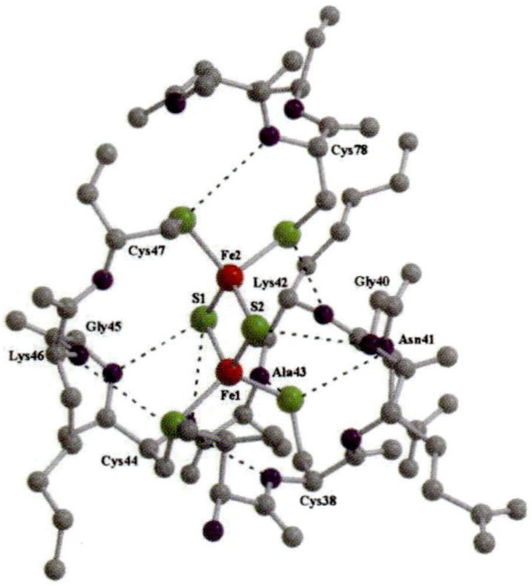

Figure 5. Stereodiagram of the iron–sulfur cluster of *T. vaginalis* ferredoxin revealing hydrogen bonds to the sulfur atoms of the cluster. Nine hydrogen bonds are identified and discussed within the text.

the photosynthetic ferredoxins. This pattern does not hold with the TvFd structure, which also has a thin protein shell but a redox potential closer to that of the photosynthetic ferredoxins.

TvFd has an overall charge due to side-chains (not including the −2 charge on the oxidized iron–sulfur cluster) of −7, which is less negative than any other [2Fe–2S] ferredoxin with known structure. The oxidase ferredoxins each have a charge of −8, the photosynthetic ferredoxins −12 to −18, and *Haloarcula marismortui* ferredoxin an impressive −28.[23] The dipole moment of TvFd, however, is large: 365 D (1 Debye = 3.335×10^{-30} Coulomb metre), close to the 385 D value for adrenodoxin,[54] which has the largest dipole moment of the [2Fe–2S] ferredoxins with known structure. The dipole of TvFd points toward the β-sheet and directly away from a cluster of negative residues on helix α_2 consisting of Glu60, Asp61, Asp62, Glu65, and Glu68. This is similar to the situation in adrenodoxin, in which the molecular dipole also points away from a cluster of negatively charged residues (Asp72, Glu73, Asp 76, and Asp79) on helix α_3. Müller *et al.*[54] suggest that for adrenodoxin and other ferredoxins the molecular dipole is important in the electrostatic steering of redox partners toward their interaction domains. Similarly, the molecular dipole of TvFd would tend to steer the electrostatically positive region of any redox partner toward this ridge along helix α_2.

In contrast to v-AnFd, TvFd does not have a water channel approaching the iron–sulfur cluster.[54,55] In TvFd, the sulfur atom of Met32 occupies the position in which the deepest of these *Anabaena* water molecules is found. The other water molecules of the channel are not seen in TvFd, although the solvent is not as well defined in this structure. However, like adrenodoxin, TvFd does have a small number of water molecules clustered on the surface depression in this area.

Discussion

Previous structural hypotheses

A number of hypotheses about the structure of TvFd have been proposed.[28,30,56] First, deuterium-dependent shifts in the resonance Raman spectrum suggested that TvFd has fewer or weaker hydrogen bonds to the sulfur atoms of the cluster than putidaredoxin.[28] The issue of the importance of the number, kind, and length of hydrogen bonds to the cluster sulfurs of [2Fe–2S] ferredoxins has been debated.[57–59] At this point, only the NMR structure of putidaredoxin is available,[19,20] and the details of the region around the cluster must be regarded as less well determined due to paramagnetic line-broadening. The interpretation of hydrogen bonds in the TvFd structure must be somewhat cautious due to the relatively low resolution of this structure. TvFd has nine hydrogen bonds from amide hydrogen atoms to the sulfur atoms of the cluster (Figure 5). By comparison, adrenodoxin has seven such hydrogen bonds, and putidaredoxin has one. This discrepancy between the structural and spectroscopic data may be related to the factors mentioned above.

Prior to our structural determination, two lines of evidence, analysis of the ionic strength dependence of the protein's redox potential, and a comparison of rates of electron transfer to cobalt coordination compounds, suggested that the cluster of TvFd was more exposed than that of

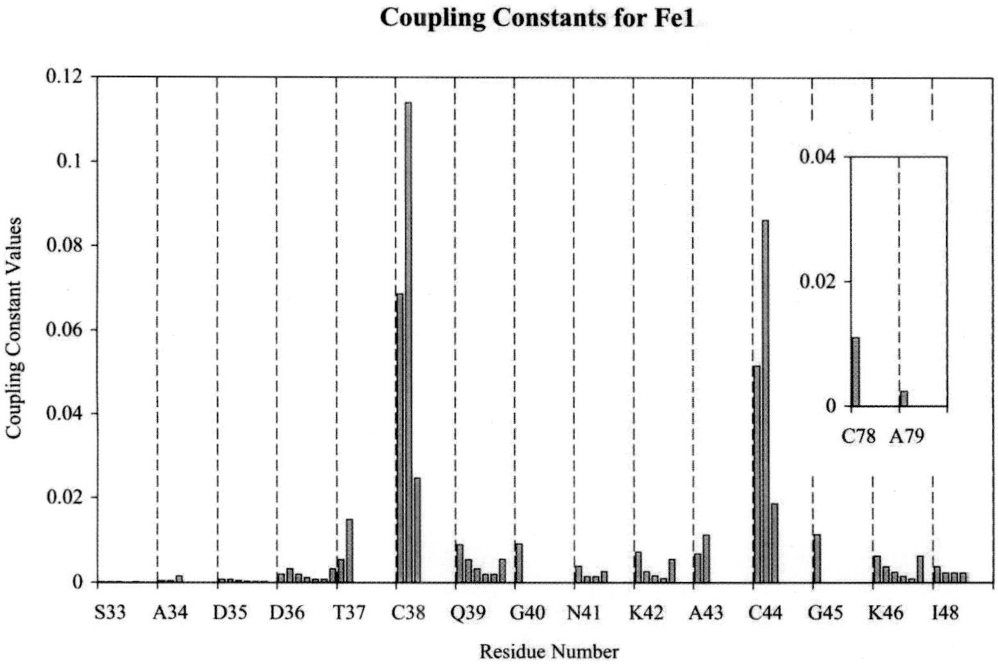

Figure 6. Coupling factors for electron transfer for the residues in the cluster-binding loop near Fe1 in *T. vaginalis* ferredoxin as calculated using PATHWAYS II[64]. Each line corresponds to an atom within the residue. The inset depicts coupling through an alternate pathway involving Cys78. To be included, atoms had to be within 15 Å of Fe1 and contain >5 Å^2 of surface area.

Anabaena ferredoxin (AnFd).[56] Viewing the molecules face-on (as shown in Figure 2) orients atom Fe1 and its coordinating cysteine residues closest to the molecular surface. From this position, most often considered the point of electron transfer, the AnFd cluster is, if anything, more exposed than the TvFd cluster. However, viewing the cluster from above (Figure 4), a cavity is present in TvFd that is not present in AnFd or any other [2Fe–2S] ferredoxin. This cavity is due to the unique backbone conformation of TvFd. The floor of the cavity is formed, in part, by the inorganic sulfur atom S2 of the cluster. The opening of the cavity is 3.3 Å wide (from the van der Waals surface of the carbonyl oxygen atom of Ser33 to the β carbon atom of Lys46) and 6.5 Å long (from the carbonyl oxygen atom of Asp 35 to the ε carbon atom of Met32). From this opening, the cavity descends 4.2 Å to the surface of sulfur atom S2 with only slight attenuation of cross-sectional area. This degree of exposure of the cluster is not seen in any other [2Fe–2S] ferredoxin, and may help explain the more rapid activation of the nitroimidazoles and other small molecules such as cobalt coordination compounds by TvFd compared to AnFd.[56]

In the comparison of TvFd to AnFd, it was hypothesized that TvFd has a less highly charged, more compact group of negative residues near the cluster.[56] It can be seen from a map of the electrostatic potential at the molecular surface that the region around the cluster is very highly negatively charged in AnFd, and substantially less so in TvFd. This is not surprising, as AnFd has a much higher molecular charge than TvFd (−17 compared to −7); what may be more surprising is the much higher molecular dipole of TvFd compared to AnFd (365 D to 177 D). This may be important in the electrostatic steering of TvFd into its association with its physiological redox partners [54]

Finally, it was hypothesized that TvFd has a somewhat different conformation in the cluster-binding loop compared to other [2Fe–2S] ferredoxins.[30] When the loops of adrenodoxin and TvFd are superimposed, they show little difference, with the exception of the backbone conformation around Asp35 and Asp36 (which has been discussed) and the peptide bond connecting Cys44 and Gly45, where the carbonyl group is pointed in opposite directions. The conformation of adrenodoxin, with the carbonyl group pointed toward the cluster, is seen in all other ferredoxin structures with the conspicuous exception of the reduced form of *Anabaena* ferredoxin.[60] Indeed, the flip of this peptide from a "CO in" position to "CO out" was the primary difference between the oxidized and reduced form of v-AnFd. Thus, the oxidized form of TvFd adopts a conformation at this position seen only in the reduced form of any other ferredoxin. This would imply that the

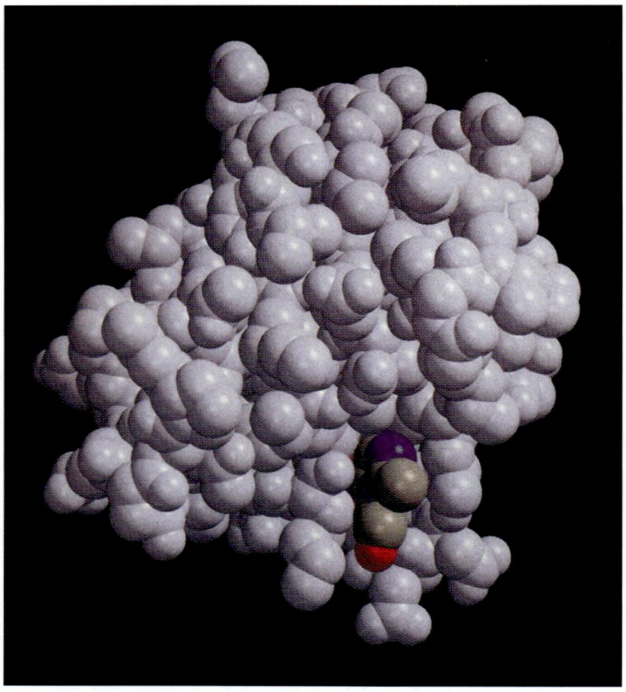

Figure 7. The structure of *T. vaginalis* ferredoxin in CPK style oriented as in Figure 4(a). Within the exposed cleft, metronidazole has been modelled as described in the text.

carbonyl flip may not be an essential part of the interconversion between the oxidized and reduced states, at least for TvFd.

Modulation of cluster redox potential

One of the most interesting questions in the study of [2Fe–2S] ferredoxins is what structural factors contribute to the redox potential and electron transfer rate. These two properties are distinct. In fact, it has been shown experimentally by the observation that variations in redox potential caused by mutations do not correspond directly to changes in electron transfer rate.[61] Marcus & Sutin demonstrated that the rate of electron transfer is dependent on the coupling between the oxidant and the reductant, the driving force behind the reaction (i.e. the redox potential), and the rearrangement energy cost of the reaction.[62]

The primary determinant of electronic coupling is the effective distance between the two species ($\kappa(R) \approx \exp(-\beta R)$).[62,63] This suggests that one of the primary structural determinants of electron transfer rate between a protein and its redox partner, be it a small molecule or protein, will be the distance between the reactants. This effective distance may be modulated by the specific components of the pathway between the species, whether they be covalent bonds, hydrogen bonds, or through space.[64] Modeling programs (such as PATHWAYS II[64]) have been designed to determine the effective distance between two reactants. The application of this program to various photosynthetic and oxidase ferredoxins demonstrated a variety of potential pathways.[54] Most significantly, adrenodoxin showed additional transfer pathways due to its thinner protein shell about Fe2 not found in the photosynthetic ferredoxins.

To investigate the electron transfer pathways for TvFd, the program PATHWAYS II[64] was used to analyze our model of TvFd. All protein atoms selected for the calculations were located within 15 Å of one or both of the iron metal ions and had at least 5 Å of solvent-accessible surface. The identification of the most favorable electron pathway from the electron donating ions to each of these electron-accepting protein atoms was calculated from a variety of possible electron transfer routes combining covalent bonds as well as hydrogen-bonded and through space (van der Waals) interactions. The highest coupling constant values were obtained with Fe1 as electron donor and are reported in Figure 6. These calculations affirm the logical conclusion that the most favored pathways are conducted through the cluster-coordinating cysteine residues and follow a path of adjacent residues to the surface of the molecule. As pointed out by Müller *et al.*,[54] as in all other ferredoxins, none of these pathways is conducted through the third coordinating cysteine residue (Cys47). Furthermore, the pathway found by that group through the fourth coordinating cysteine residue appears to exist also in TvFd (Figure 6).

The driving force behind the electron transfer reaction, the redox potential, is itself potentially modulated by a number of factors. In the past, it

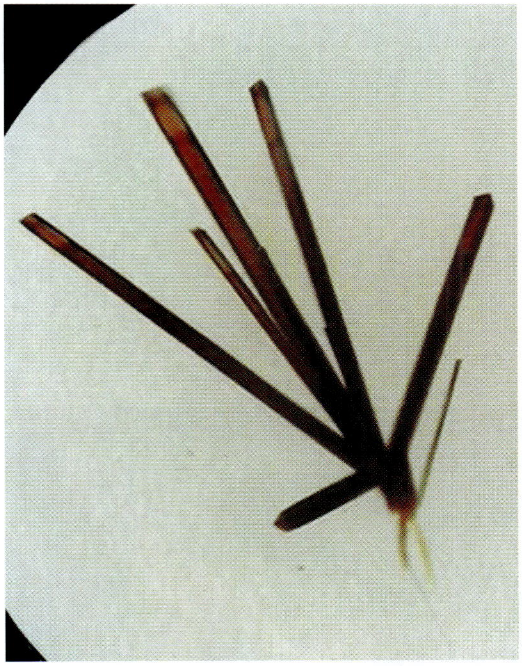

Figure 8. Crystals of *T. vaginalis* ferredoxin. The longest of the needles has dimensions of roughly 0.1 mm × 0.1 mm × 2.0 mm. The interiors of the ends of the crystals are hollow. Typical crystals used for diffraction have dimensions of 0.15 mm × 0.15 mm × 0.6 mm.

was suggested that the number and strength of hydrogen bonds to cluster sulfur atoms by backbone amide hydrogen atoms (NH···S) was a key determinant of the redox potential of the cluster.[57,58] However, further structural work has not supported this hypothesis,[59] and the determination of redox potential is believed to be more complex.[55,65,66] In any case, TvFd does have a large number (nine) of these bonds and, like all [2Fe–2S] ferredoxin structures, has a greater number to the sulfur atoms ligating Fe1 (which is believed to hold the additional charge upon reduction) than to the sulfur atoms ligating Fe2.

Reduction of metronidazole

In this study of the ferredoxin from *T. vaginalis*, an additional important goal has been to help explain the mechanism of action of a major modern pharmacological agent, metronidazole. Specifically, we have sought to understand the structural basis for the electron transfer from TvFd to metronidazole, and why this reaction appears to show a high level of specificity that is not strictly dependent on the magnitude of the redox potential difference between the two species. The explanation for this phenomenon must almost certainly be a structural one, involving either a specific binding site for nitroimidazoles, a greater accessibility to the cluster when compared with other ferredoxins, or perhaps a different pattern of rearrangement of TvFd upon oxidation.

In the previous section, it was observed that TvFd has a surface cavity that exposes one of the sulfur atoms of the cluster to a degree not found in other [2Fe–2S] ferredoxins. Hypothetical modeling shows that a metronidazole molecule can fit into this cavity if the side-chains of Thr37, Lys45, and Met32 are rotated about their $\chi 1$ bonds (Figure 7). This greater exposure of the [2Fe–2S] cluster may explain why TvFd reduces nitroimidazoles and other small molecules more rapidly than other [2Fe–2S] ferredoxins. Since *T. vaginalis* is an anaerobic protozoan, this hypothetical speed advantage in redox reactions would come with only a low risk of metabolically wasteful oxidation of the cluster by oxygen.

The common factor among all organisms treated by metronidazole is that they are anaerobic. It may be that anaerobes as a rule have more exposed electron transport machinery than aerobes. Although it is generally thought that the selectivity of metronidazole is based on the fact that oxygen inactivates metronidazole radicals,[2,46] another contributing factor might be that the electron transport machinery of aerobes must be shielded against oxygen. This shielding would block interaction with metronidazole and other nitroimidazole drugs. Further insight into the electron transfer from TvFd to metronidazole will have to await more experimentation, perhaps including co-crystallization of the protein and the drug.

Materials and Methods

The gene for TvFd, pET-3a-TvFd, was prepared as described.[50] This plasmid was digested with *Nde*I and *Hin*dIII, and inserted into a pET-26 expression vector. The resulting pET-26-TvFd was transformed into *Escherichia coli* strain BL21(DE3) containing the pLysS plasmid, and isolated by growth on LB plates containing 30 μg/ml of chloroamphenicol and 100 μg/ml of kanamycin.

TvFd was then expressed and isolated as described[28] but with the following changes: 100 μg/ml of kanamycin and 30 μg/ml of chloroamphenicol were used as selective antibiotics rather than ampicillin, and ferrous sulfate was used rather than ferrous ammonium sulfate as the iron source for ferredoxin expression. Cells were lysed using three alternating freeze/thaw cycles. This procedure typically yielded 10 mg of protein per liter of expression culture.

Crystals of oxidized TvFd (Figure 8) were grown using the sitting-drop method with a reservoir solution of 25% (w/v) polyethylene glycol (PEG) 8000, 0.2 M sodium acetate, and 0.1 M sodium cacodylate (pH 6.5) and a protein concentration of 19.0 mg/ml. Microseeding increased greatly the speed of crystallization and the number of usable crystals formed. The crystals are in space group $P2_12_12_1$, with unit cell dimensions $a = 35.59$ Å, $b = 66.33$ Å, $c = 132.5$ Å, $\alpha = \beta = \gamma = 90°$.

Table 4. Collection parameters for data sets used in the solution of the TvFd structure

Data set	Detector	R_{merge} (%)	Mosaicity (deg.)	Resolution (Å)	Complete(%)	λ (Å)	Anomalous signal (%)
MADa	Quantum 1 CCD	7.6	1.505	3.5	99	1.738	4.31
MADb		6.6		3.5	99	1.742	4.56
MADc		7.6		3.5	99	1.676	5.09
MADd		6.2		3.5	99	1.771	3.45
Native	Quantum 4 CCD	7.0	1.176	2.2	96	1.000	N/A

These crystals were unusual in that they began as 10–150 μm hollow needles, which gradually filled in from the base. Unit cell lengths *b* and *c* were found to vary by as much as 3 Å (roughly 3%) between crystals, which interfered with a structure solution using multiple isomorphous replacement (MIR). These crystals contain three molecules per asymmetric unit, giving an inverse packing density (Matthews coefficient, or V_M) of

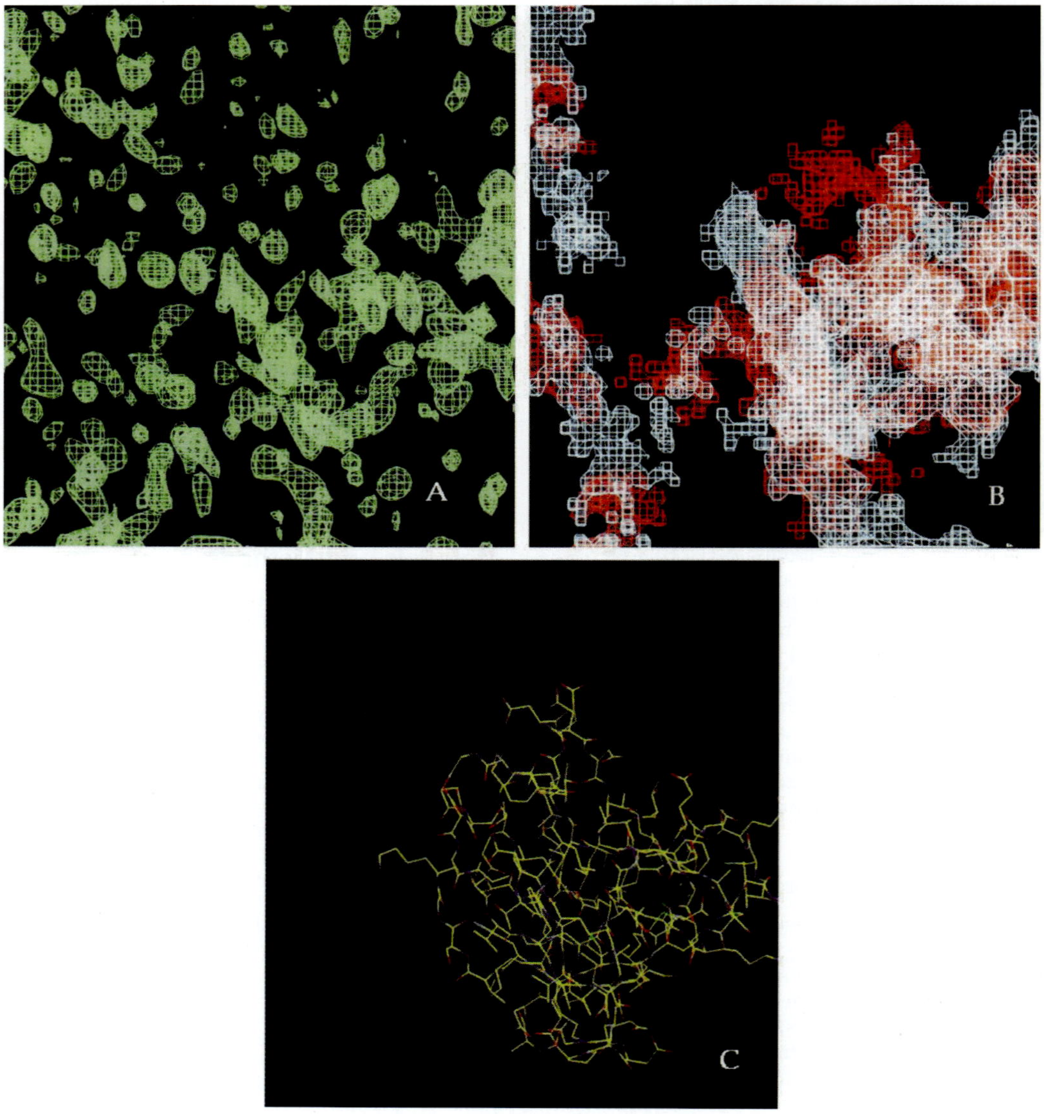

Figure 9. Maps demonstrating the use of the ESSENS programs. (a) The original MAD-phased Fourier map from SOLVE. (b) ESSENS correlation map between the MAD-phased Fourier map and a five residue helix model (red) and a five residue β-sheet model (blue). (c) The position of the final model (molecule A) in the same orientation.

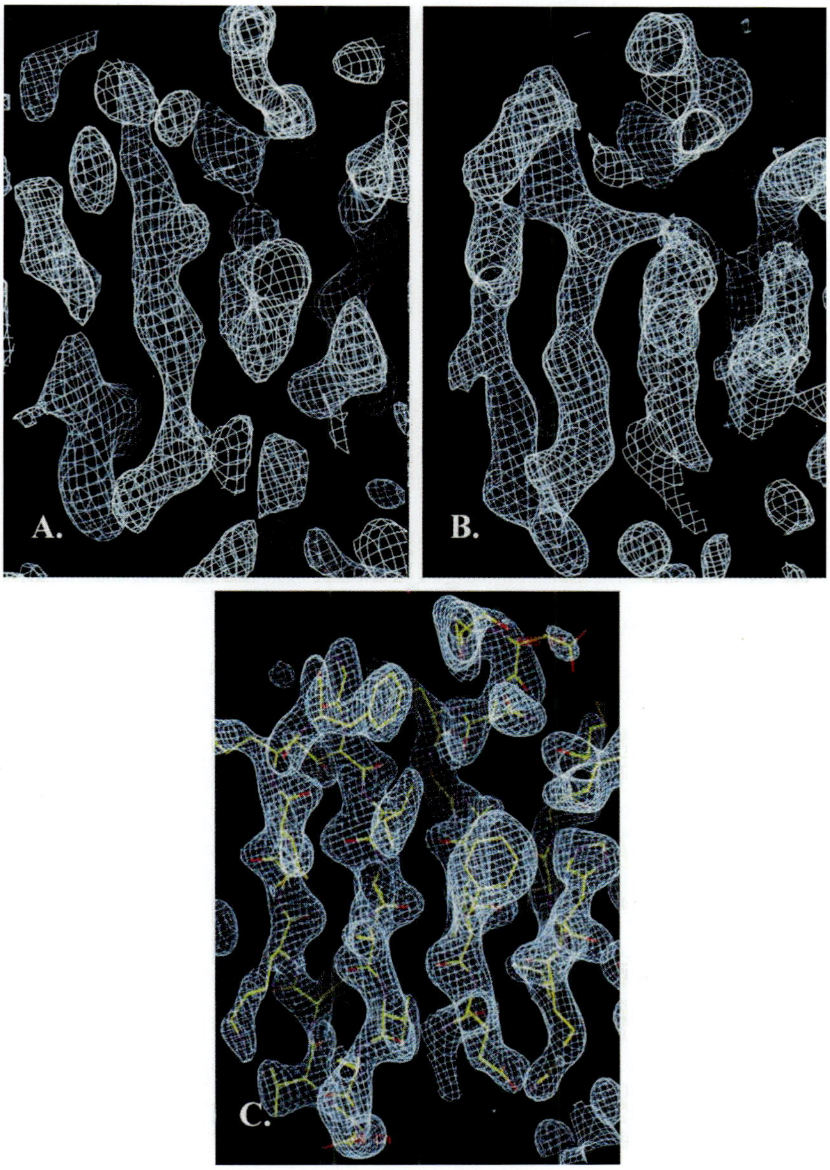

Figure 10. Electron density maps demonstrating the effectiveness of NCS averaging in the solution of the TvFd structure: (a) The MAD-phased map before NCS averaging; (b) the same map after 11 cycles of NCS averaging; and (c) the final refined $2F_o-F_c$ map.

2.56 Å^3/Da, corresponding to 52% (v/v) solvent in the crystal.[67]

A four-wavelength MAD data set was collected at 170 K on a single crystal on a Quantum-1 CCD detector at BioCARS beamline 14-BMD at the APS (Table 4). The wavelengths collected were 1.7383 Å (peak), 1.7415 Å (inflection point), 1.6755, and 1.7712 Å (remote wavelengths). At each of these wavelengths, two 110° scans were collected 180° apart to ensure collection of Friedel pairs. The data were integrated using DENZO and scaled and reduced in SCALEPACK.[68] All of the MAD data sets were better than 99% complete to 3.5 Å, with merging *R*-factors of 7.6, 6.6, 7.6, and 6.2%, respectively, and had mosaicities of approximately 1.5°.

Native data from a second crystal was collected at 170 K on a Quantum-4 CCD detector at BioCARS beamline 14-BMC at the APS (Table 4). The wavelength was set to 1.0 Å, and two 90° scans were collected 180° apart. The data were processed with DENZO and SCALEPACK, yielding a final data set that was 93.3% complete to 2.16 Å, with a merging *R*-factor of 7.0% and a mosaicity of 1.17°.

The data sets from four different wavelengths collected in the MAD experiment (X22) were phased using the program SOLVE.[69] Unfortunately, the Fourier maps after MAD phasing were not immediately interpretable (Figures 9(a) and 10(a)). Close inspection revealed unconnected electron density consistent with a small portion of the β-sheet and an α-helix (α_1) of one molecule of the asymmetric unit. The protein-solvent boundary of the remaining two molecules was only slightly discernible.

The program ESSENS[70] was used to further elucidate the approximate positions of the molecules. Two correlation maps (Figure 9(b)) were generated using five-residue helix and strand search models. Both of these correlation maps showed considerable signal in all three regions suspected of containing a protein molecule, with little noise outside these regions. Using both the Fourier map and the ESSENS correlation maps, a mask was drawn around the most distinct of the suspect regions, using putidaredoxin (1PUT) as a model. Drawing the mask was the only influence of any previous structure on the solution of TvFd. The program NCS6D[71] was then used to find the NCS operator between the first and second suspect regions, giving a correlation coefficient of 0.326 between them. The NCS operator was improved using the program IMP[68] to give a correlation coefficient of 0.381. The same procedure was used to find the NCS operator between the first and third regions, with a final correlation coefficient of 0.425.

The Fourier map was NCS-averaged using the programs AVE and RAVE,[71] for a total of 11 averaging steps. At this point the map was interpretable (Figure 10(b)). Roughly 80% of the molecule, notably excluding the iron–sulfur cluster and the loop that covers it, was built as a polyalanine model using O.[72] After positional refinement in X-PLOR[73] using strict NCS symmetry between the three molecules, this model had an *R*-factor of 44.9% and a free *R*-factor of 47.7%.

All model building was done in O, and all refinement in X-PLOR. For the first few steps of building, calculated phases were combined with MAD phases from SOLVE using the program PHASES.[74] The first several cycles of building involved manual building followed by conjugate-gradient refinement and simulated annealing.[75] Once the backbone of the model was placed with reasonable certainty, we switched from refining against the MAD data to refining against native data, gradually adding reflections to 2.2 Å. Overall *B* refinement was used until all side-chains had been placed at least tentatively, at which point we began refining individual *B*-factors. Simulated annealing omit maps were used to place problematic residues. Once all side-chains were placed as well as possible, solvent molecules and a bulk solvent mask were added. Refinement was done using NCS restraints on all backbone atoms; when these restraints were removed, the *R*-factor worsened.

During analysis of the model, the secondary structure was determined using the programs QUANTA (version 98, Molecular Simulations Inc., Waltham, MA) and PROCHECK.[76] The identities of β-turns were confirmed using manual inspection and published criteria.[77–79] Hydrogen bonding to the sulfur atoms of the cluster were counted under the criteria that the S–H distance must be less than 2.85 Å, and that the donor-H–S angle be greater than 120°. The same criteria were used to count hydrogen bonds to the clusters of the available structures of adrenodoxin and putidaredoxin.

The rmsd values between TvFd and other ferredoxins were calculated using rms superpositioning of C^{α} atoms in QUANTA using a three-step process. First, the molecules were aligned using only sequence information and superpositioned by a least-squares fit, matching only identical residues. Second, the molecules were realigned using equal weights on sequence information and C^{α}–C^{α} distance (5.0 Å cutoff), and another superposition performed using this alignment matching close residues (again with a 5.0 Å cutoff). Finally, these superimposed structures were used to determine the rmsd of the C^{α} atoms of all aligned residues.

Protein Data Bank accession codes

All coordinates will be deposited for release upon publication in the Protein Data Bank under accession code 1L5P.

Acknowledgments

Use of the Advanced Photon Source was supported by the U. S. Department of Energy, Basic Energy Sciences, Office of Science, under contract no. W-31-109-Eng-38. Use of the BioCARS Sector 14 was supported by the National Institutes of Health, National Center for Research Resources, under grant number RR07707. We acknowledge Professor George Phillips for the use of the data collection facilities at Rice University. We also thank the Institute of Molecular Design and Molecular Simulations, Inc. for the use of computational resources. This publication was supported, in part, by a training fellowship from the Keck Center for Computational Biology (National Science Foundation GRT grant BIR-92-56580) and by grants from the W. M. Keck Foundation, the Robert A. Welch Foundation and the National Institutes of Health (GM061412).

References

1. Rein, M. F. (1995). Trichomonas vaginalis. In *Principles and Practice of Infectious Diseases* (Mandell, G. L., Bennett, J. E. & Dolin, R., eds), 4th edit, pp. 2493–2497, Churchill-Livingstone, New York.
2. Müller, M. (1983). Mode of action of metronidazole on anaerobic bacteria and protozoa. *Surgery*, **93**, 165–171.
3. Matsubara, H. & Saeki, K. (1992). Structural and functional diversity of ferredoxins and related proteins. In *Iron–Sulfur Proteins* (Cammack, R., ed.), vol. 38, pp. 223–280, Academic Press, New York.
4. Sticht, H. & Rösch, P. (1998). The structure of iron–sulfur proteins. *Prog. Biophys. Mol. Biol.* **70**, 95–136.
5. Rypniewski, W. R., Breiter, D. R., Benning, M. M., Wesenberg, G., Oh, B.-H., Markley, J. L., Rayment, I. & Holden, H. M. (1991). Crystallization and structure determination to 2.5 Å resolution of the oxidized [2Fe–2S] ferredoxin isolated from *Anabaena* 7120. *Biochemistry*, **30**, 4126–4131.
6. Jacobson, B. L., Chae, Y. K., Markley, J. L., Rayment, I. & Holden, H. M. (1993). Molecular structure of the oxidized, recombinant, heterocyst [2Fe–2S] ferredoxin from *Anabaena* 7120 determined to 1.7 Å resolution. *Biochemistry*, **32**, 6788–6793.
7. Fukuyama, K., Hase, T., Matsumoto, S., Tsukihara, T., Katsube, Y., Tanaka, N., Kakudo, M., Wada, K. & Matsubara, H. (1980). Structure of *S. platensis*

[2Fe–2S] ferredoxin and evolution of chloroplast-type ferredoxins. *Nature*, **286**, 522–524.
8. Tsukihara, T., Fukuyama, K., Nakamura, M., Katsube, Y., Tanaka, N., Kakudo, M., Wada, K., Hase, T. & Matsubara, H. (1981). X-Ray analysis of a [2Fe–2S] ferredoxin from *Spirulina platensis*. Main chain fold and location of side chains at 2.5 Å resolution. *J. Biochem.* **90**, 1763–1773.
9. Fukuyama, K., Ueki, N., Nakamura, H., Tsukihara, T. & Matsubara, H. (1995). Tertiary structure of [2Fe–2S] ferredoxin from *Spirulina platensis* refined at 2.5 Å resolution: Structural comparisons of plant-type ferredoxins and an electrostatic potential analysis. *J. Biochem.* **117**, 1017–1023.
10. Tsutsui, T., Tsukihara, T., Fukuyama, K., Katsube, Y., Hase, T., Matsubara, H., Nishikawa, Y. & Tanaka, N. (1983). Main chain fold of a [2Fe–2S] ferredoxin I from *Aphanothece sacrum* at 2.5 Å resolution. *J. Biochem.* **94**, 299–302.
11. Tsukihara, T., Fukuyama, K., Mizushima, M., Harioka, T., Kusunoki, M., Katsube, Y., Hase, T. & Matsubara, H. (1990). Structure of the [2Fe–2S] ferredoxin I from the blue-green alga *Aphanothece sacrum* at 2.2 Å resolution. *J. Mol. Biol.* **216**, 399–410.
12. Ikemizu, S., Bando, M., Sato, T., Morimoto, Y., Tsukihara, T. & Fukuyama, K. (1994). Structure of [2Fe–2S] ferredoxin I from *Equisetum arvense* at 1.8 Å resolution. *Acta Crystallog. sect. D*, **50**, 167–174.
13. Binda, C., Coda, A., Aliverti, A., Zanetti, G. & Mattevi, A. (1998). Structure of the mutant E92 K of [2Fe–2S] ferredoxin I from *Spinacia oleracea* at 1.7 Å resolution. *Acta Crystallog. sect. D*, **54**, 1353–1358.
14. Bes, M. T., Parsini, E., Inda, L. A., Saraiva, L. M., Peleato, M. L. & Sheldrick, G. M. (1999). Crystal structure determination at 1.4 Å resolution of ferredoxin from the green alga *Chlorella fusca*. *Structure*, **7**, 1201–1211.
15. Lelong, C., Setif, P., Bottin, H., Andre, F. & Neumann, J.-M. (1995). ^{1}H and ^{15}N NMR sequential assignment, secondary structure, and tertiary fold of [2Fe–2S] ferredoxin from *Synechocystis* sp. PCC 6803. *Biochemistry*, **34**, 14462–14473.
16. Baumann, B., Sticht, H., Schärpf, M., Sutter, M., Haehnel, W. & Rösch, P. (1996). Structure of *Synechococcus elongatus* [Fe_2S_2] ferredoxin in solution. *Biochemistry*, **35**, 12831–12841.
17. Hatanaka, H., Tanimura, R., Katoh, S. & Inagaki, F. (1997). Solution structure of ferredoxin from the thermophilic cyanobacterium Synechococcus elongatus and its thermostability. *J. Mol. Biol.* **268**, 922–933.
18. Im, S.-C., Liu, G., Luchinat, C., Sykes, A. G. & Bertini, I. (1998). The solution structure of parsley [2Fe–2S] ferredoxin. *Eur. J. Biochem.* **258**, 465–477.
19. Pochapsky, T. C., Ye, X. M., Ratnaswamy, G. & Lyons, T. A. (1994). An NMR-derived model for the solution structure of oxidized putidaredoxin, a 2-Fe, 2-S ferredoxin from *Pseudomonas*. *Biochemistry*, **33**, 6424–6432.
20. Pochapsky, T. C., Jain, N. U., Kuti, M., Lyons, T. A. & Heymont, J. (1999). A refined model for the solution structure of oxidized putidaredoxin. *Biochemistry*, **38**, 4681–4690.
21. Müller, A., Müller, J. J., Muller, Y. A., Uhlmann, H., Bernhardt, R. & Heinemann, U. (1998). New aspects of electron transfer revealed by the crystal structure of a truncated bovine adrenodoxin, Adx (4–108). *Structure*, **6**, 269–280.
22. Pikuleva, I. A., Tesh, K., Waterman, M. R. & Kim, Y. (2000). The tertiary structure of full-length bovine adrenodoxin suggests functional dimers. *Arch. Biochem. Biophys.* **373**, 44–55.
23. Frolow, F., Harel, M., Sussman, J. L., Mevarech, M. & Shoham, M. (1996). Insights into protein adaptation to a saturated salt environment from the crystal structure of a halophilic 2Fe–2S ferredoxin. *Nature Struct. Biol.* **3**, 452–458.
24. Gorrell, T. E., Yarlett, N. & Müller, M. (1984). Isolation and characterization of *Trichomonas vaginalis* ferredoxin. *Carlsberg Res. Commun.* **49**, 259–268.
25. Marczak, R., Gorrell, T. E. & Müller, M. (1983). Hydrogenosomal ferredoxin of the anaerobic protozoon, *Tritrichomonas foetus*. *J. Biol. Chem* **258**, 12427–12433.
26. Han, S., Czernuszewicz, R. S., Kimura, T., Adams, M. W. W. & Spiro, T. G. (1989). Fe_2S_2 protein resonance Raman spectroscopy spectra revisited: Structural variations among adrenodoxin, ferredoxin, and red paramagnetic protein. *J. Am. Chem. Soc.* **111**, 3505–3511.
27. Fu, W., Drozdzewski, P. M., Davies, M. D., Sligar, S. G. & Johnson, M. K. (1992). Resonance Raman and magnetic circular dichroism studies of reduced [2Fe–2S] proteins. *J. Biol. Chem.* **267**, 15502–15510.
28. Vidakovic, M. S., Fraczkiewicz, G. & Germanas, J. P. (1996). Expression and spectroscopic characterization of the hydrogenosomal [2Fe–2S] ferredoxin from the protozoan *Trichomonas vaginalis*. *J. Biol. Chem.* **271**, 14734–14739.
29. Ratnaswamy, G. & Pochapsky, T. C. (1993). Characterization of hyperfine-shifted ^{1}H resonances in oxidized and reduced putidaredoxin, an Fe2S2 ferredoxin from *Pseudomonas putida*. *Magn. Reson. Chem.* **31**, S73–S80.
30. Liu, H. Y. & Germanas, J. P. (1998). NMR spectroscopic studies of the hydrogenosomal [2Fe–2S] ferredoxin from *Trichomonas vaginalis*: Hyperfine-shifted ^{1}H resonances. *J. Inorg. Biochem.* **72**, 127–131.
31. Johnson, P. J., d'Oliveira, C. E., Gorrell, T. E. & Müller, M. (1990). Molecular analysis of the hydrogenosomal. ferredoxin of the anaerobic protist *Trichomonas vaginalis*. *Proc. Natl Acad. Sci. USA*, **87**, 6097–6101.
32. Brul, S., Veltman, R. H., Lombardo, M. C. P. & Vogels, G. D. (1994). Molecular cloning of hydrogenosomal ferredoxin DNA from the anaerobic amoeboflagellate *Psalteriomonas lanterna*. *Biochim. Biophys. Acta*, **1183**, 544–546.
33. Finegold, S. M. & Mathisen, G. E. (1995). Metronidazole. In *Principles and Practice of Infectious Disease* (Mandell, G L., Bennett, J E. & Dolin, R., eds), pp. 329–334, Churchill Livingstone, New York.
34. Edwards, D. I. & Mathison, G. E. (1970). The mode of action of metronidazole against *Trichomonas vaginalis*. *J. Gen. Microbiol.* **63**, 297–302.
35. Ings, R. M. J., McFadzean, J. A. & Ormerod, W. E. (1974). The mode of action of metronidazole in *Trichomonas vaginalis*and other micro-organisms. *Biochem. Pharmacol.* **23**, 1421–1429.
36. Lloyd, D. & Kristensen, B. (1985). Metronidazole inhibition of hydrogen production in vivo in drug-sensitive and resistant strains of *Trichomonas vaginalis*. *J. Gen. Microbiol.* **131**, 849–853.
37. Plant, C. W. & Edwards, D. I. (1976). The effect of tinidazole, metronidazole, and nitrofurazone on nucleic acid synthesis in *Clostridium bifermentans*. *J. Antimicrob. Chemother.* **2**, 203–209.

38. Edwards, D. I. (1986). Reduction of nitroimidazoles *in vitro* and DNA damage. *Biochem. Pharmacol.* **35**, 53–58.
39. Moreno, S. N. J., Mason, R. P., Muniz, R. P. A., Cruz, F. S. & Docampo, R. (1983). Generation of free radicals from metronidazole and other nitroimidazoles by *Tritrichomonas foetus*. *J. Biol. Chem.* **258**, 4051–4054.
40. Lloyd, D. & Pederson, J. Z. (1985). Metronidazole radical anion generation in vivo in *Trichomonas vaginalis*: oxygen quenching is enhanced in a drug-resistant strain. *J. Gen. Microbiol.* **131**, 87–92.
41. Church, D. L., Rabin, H. R. & Laishley, E. J. (1988). Role of hydrogenase 1 of Clostridium pasteurianum in the reduction of metronidazole. *Biochem. Pharmacol.* **37**(8), 1525–1534.
42. Edwards, D. I., Dye, M. & Carne, H. (1973). The selective toxicity of antimicrobial nitroheterocyclic drugs. *J. Gen. Microbiol.* **76**, 135–145.
43. Lindmark, D. G. & Müller, M. (1976). Antitrichomonad action, mutagenicity, and reduction of metronidazole and other nitroimidazoles. *Antimicrob. Agents Chemother.* **10**, 476–482.
44. Müller, M. & Lindmark, D. G. (1976). Uptake of metronidazole and its effect on viability in trichomonads and *Entamoeba invadens* under anaerobic and aerobic conditions. *Antimicrob. Agents Chemother*, **9**, 696–700.
45. Müller, M. (1986). Reductive activation of nitroimidazoles in anaerobic microorganisms. *Biochem. Pharmacol.* **35**, 37–41.
46. Upcroft, P. & Upcroft, J. A. (2001). Drug targets and mechanisms of resistance in the anaerobic protozoa. *Clin. Microbiol. Rev.* **14**, 150–164.
47. Quon, D. V. K., d'Oliveira, C. E. & Johnson, P. J. (1992). Reduced transcription of the ferredoxin gene in metronidazole-resistant *Trichomonas vaginalis*. *Proc. Natl Acad. Sci.* **89**, 4402–4406.
48. Land, K. M., Clemens, D. L. & Johnson, P. J. (2001). Loss of multiple hydrogenosomal proteins associated with organelle metabolism and high-level drug resistance in trichomonads. *Exp. Parasitol.* **97**, 102–110.
49. Moreno, S. N. J., Mason, R. P. & Docampo, R. (1984). Distinct reduction of nitrofurans and metronidazole to free radical metabolites by *Tritrichomonas foetus* hydrogenosomal and cytosolic enzymes. *J. Biol. Chem.* **259**, 8252–8259.
50. Yarlett, N., Gorrell, T. E., Marczak, R. & Müller, M. (1985). Reduction of nitroimidazole derivatives by hydrogenosomal extracts of *Trichomonas vaginalis*. *Mol. Biochem. Parasitol.* **14**, 29–40.
51. Goldstein, B. P., Vidal-plana, R. R., Cavalleri, B., Zerilli, L., Carniti, G. & Silvestri, L. G. (1977). The mechanism of action of nitro-heterocyclic antimicrobial drugs. Metabolic activation by micro-organisms. *J. Gen. Microbiol.* **100**, 283–298.
52. Creighton, T. E. (1993). *Proteins: Structures and Molecular Properties*, W.H. Freeman and Company, New York.
53. Müller, J. J., Lapko, A., Bourenkov, G., Ruckpaul, K. & Heinemann, U. (2001). Adrenodoxin reductase–adrenodoxin complex structure suggests electron transfer path in steroid biosynthesis. *J. Biol Chem.* **276**, 2786–2789.
54. Müller, J. J., Müller, A., Rottman, M., Bernhardt, R. & Heinemann, U. (1999). Vertebrate-type and plant-type ferredoxins: crystal structure comparison and electron transfer pathway modelling. *J. Mol. Biol.* **294**, 501–513.
55. Holden, H. M., Jacobson, B. L., Hurley, J. K., Tollin, G., Oh, B., Skjeldal, L., Chae, Y. K., Cheng, H., Xia, B. & Markley, J. L. (1994). Structure–function studies of [2Fe–2S] ferredoxins. *J. Bioenerg. Biomembr.* **26**, 67–88.
56. Vidakovic, M. & Germanas, J. P. (1996). Electrostatic effects in electron transfer reactions of [2Fe–2S] ferredoxins with inorganic reagents. *Protein Sci.* **5**, 1793–1799.
57. Adman, E., Watenpaugh, K. D. & Jensen, L. H. (1975). NH···S hydrogen bonds in *Peptococcus aerogenes* ferredoxin, *Clostridium pasteurianum* rubredoxin, and *Chromatium* high potential iron protein. *Proc. Natl Acad. Sci.* **72**, 4854–4858.
58. Carter, C. W. (1977). New stereochemical analogies between iron–sulfur electron transport proteins. *J. Biol. Chem.* **252**, 7802–7811.
59. Backes, G., Mino, Y., Loehr, T. M., Meyer, T. E., Cusanovich, M. A., Sweeney, W. V., Adman, E. T. & Sanders-Loehr, J. (1991). The environment of Fe4S4 clusters in ferredoxins and high-potential iron proteins. New information from X-ray crystallography and resonance Raman spectroscopy. *J. Am. Chem. Soc.* **113**, 2055–2064.
60. Morales, R., Chron, M.-H., Hudry-Clergeon, G., Pétillot, Y., Norager, S., Medina, M. & Frey, M. (1999). Refined X-ray structures of the oxidized, at 1.3 Å, and reduced, at 1.17 Å, [2Fe–2S] ferredoxin from the cyanobacterium *Anabaena* PCC7119 show redox-linked conformational changes. *Biochemistry*, **38**, 15764–15773.
61. Hurley, J. K., Weber-Main, A. M., Hodges, A. E., Stankovich, M. T., Benning, M. M., Holden, H. M., Cheng, H., Xia, B., Markley, J. L., Genzor, C., Gomez-Moreno, C., Hafezi, R. & Tollin, G. (1997). Iron–sulfur cluster cysteine-to-serine mutants of *Anabaena* [2Fe–2S] ferredoxin exhibit unexpected redox properties and are competent in electron transfer to ferredoxin:NADP+ reductase. *Biochemistry*, **36**, 15109–15117.
62. Marcus, R. A. & Sutin, N. (1985). Electron transfers in chemistry and biology. *Biochim. Biophys. Acta*, **811**, 265–322.
63. Vidakovic, M.S. (1996). Structure–function studies of (2Fe–2S) ferredoxins. Doctoral dissertation, University of Houston, TX.
64. Beratan, D. N., Betts, J. N. & Onuchic, J. N. (1991). Protein electron transfer rates set by the bridging secondary and tertiary structure. *Science*, **252**, 1285–1288.
65. Beck, B. W. (1997). *Theoretical Investigations of Iron–sulfur Proteins*, Department of Biochemistry and Biophysics, Washington State University, Washington, DC.
66. Swartz, P. D., Beck, B. W. & Ichiye, T. (1996). Structural origins of redox potentials in Fe–S proteins: Electrostatic potentials of crystal structures. *Biophys. J.* **71**, 2958–2969.
67. Matthews, B. W. (1968). Solvent content of protein crystals. *J. Mol. Biol.* **33**, 491–497.
68. Otwinowski, Z. & Minor, W. (1997). Processing of X-ray diffraction data collected in oscillation mode. *Methods Enzymol.* **276**, 307–326.
69. Terwilliger, T. C. & Berendzen, J. (1999). Automated structure solution for MIR and MAD. *Acta Crystallog. sect. D*, **55**, 849–861.
70. Kleywegt, G. J. & Jones, T. A. (1997). Template convolution to enhance or detect structural features in

macromolecular electron-density maps. *Acta Crystallog. sect. D*, **53**, 179–185.
71. Kleywegt, G. J. & Read, R. J. (1997). Not your average density. *Structure*, **5**, 1557–1569.
72. Jones, T. A., Zou, J.-Y., Cowan, S. W. & Kjeldgaard, M. (1991). Improved methods for binding protein models in electron density maps and the location of errors in these models. *Acta Crystallog. sect. A*, A**47**, 110–119.
73. Brünger, A. T. (1992). *XPLOR Version 3.1–A System for X-ray Crystallography and NMR*, Yale University Press, New Haven, CT.
74. Furey, W. & Swaminathan, S. (1997). PHASES-95: a program package for processing and analyzing diffraction data from macromolecules. *Methods Enzymol.* **277**, 590–620.
75. Brünger, A. T., Krukowski, A. & Erickson, J. W. (1990). Slow-cooling protocols for crystallographic refinement by simulated annealing. *Acta Crystallog. sect. A*, **46**, 585–593.
76. Laskowski, R. A. *et al.* (1993). PROCHECK: a program to check the stereochemical quality of protein structures. *J. Appl. Crystallog.* **26**, 283–291.
77. Chou, P. Y. & Fasman, G. D. (1977). β-turns in proteins. *J. Mol. Biol.* **115**, 135–175.
78. Wilmot, C. M. & Thornton, J. M. (1988). Analysis and prediction of the different types of β-turn in proteins. *J. Mol. Biol.* **203**, 221–232.
79. Richardson, J. S. (1981). The anatomy and taxonomy of protein structure. *Advan. Prot. Chem.* **34**, 167–340.
80. Kraulis, P. J. (1991). MOLSCRIPT: a program to produce both detailed and schematic plots of protein structure. *J. Appl. Crystallog.* **24**, 946–950.
81. Bacon, D. J. & Anderson, W. F. (1988). A fast algorithm for rendering space-filling molecule pictures. *J. Mol. Graph.* **6**, 219–220.
82. Merritt, E. A. & Murphy, M. E. P. (1994). Raster3D Version 2.0—a program for photorealistic molecular graphics. *Acta Crystallog. sect. D*, **50**, 869–873.

Edited by R. Huber

(Received 18 January 2002; received in revised form 18 January 2002; accepted 22 January 2002)

Chapter 14

Rodney J. Bartlett

Biography

Rod Bartlett obtained his B.S., with a double major in chemistry and mathematics, at Millsaps College, MS in 1966, then in 1971 was awarded his Ph.D. in quantum chemistry from the University of Florida. From 1971 to 1972 he was a National Science Foundation postdoctoral fellow at Aarhus University in Denmark and from 1972 to 1974 a postdoctoral research associate at Johns Hopkins University.

After postdoctoral work, Dr. Bartlett spent seven years at Battelle Memorial Institute, first at the Department of Energy's Pacific Northwest National Laboratory and then at Battelle in Columbus, OH. In 1981 he moved to the University of Florida as professor of chemistry and physics. He was named Graduate Research Professor in 1987, a rank held by only a handful of faculty.

In 2007 Bartlett received the American Chemical Society (ACS) Award in Theoretical Chemistry and in 2008 the Schrödinger Medal of the World Association of Theoretical and Computational Chemists. In 2009 he was awarded the Boys–Rahman Prize of the Royal Society of Chemistry. He was named the Southern Chemist of the Year for 2010 by the Memphis section of the ACS. He won the Florida ACS Award in 2000.

Professor Bartlett is a fellow of the International Academy of Quantum Molecular Sciences (1991), the American Physical Society (1986), the Guggenheim Foundation (1986), and the ACS (2010). He was an E. T. S. Walton Fellow in Ireland (2003). He served as chairman of the subdivision of theoretical chemistry of the ACS (1987). He serves on the advisory boards of the New Zealand Institute of Advanced Study, the Tyndall Research Institute, Cork, Ireland, and Jackson State University, Jackson, MS.

Barlett pioneered the development of coupled-cluster theory in quantum chemistry to offer highly accurate solutions of the Schrödinger equation for molecular structure and spectra. His group is responsible for the widely used ACES program system. Other research topics include the search for metastable, high-energy density molecules like N_5^-; non-linear optics; carbon clusters; NMR coupling constants; new correlated quantum chemical methods for polymers; *ab initio* density functional theory; and the "transfer Hamiltonian" for large scale quantum mechanical simulations of materials.

He maintains a group of a dozen people. He has published over 500 papers and book chapters, presented over 200 invited lectures at major meetings, edited six books, and coauthored an advanced text on many-body methods. According to the Institute for Scientific Information, he is among the most cited chemists in the world (see ishighlycited.com). His h-index is 90.

Beryllium Borohydride: A Detective Story

The Guggenheim Foundation gave me a fellowship in 1986, and I used part of it to spend a semester at Harvard. While at Harvard, in between suffering through the inevitable Red Sox failure to become world champions (at the time they had last won when Babe Ruth was a pitcher!) and freezing through "The Game" (i.e. Harvard vs. Yale), I was involved in resolving a 50-year-old problem in molecular structure for one of the Colonel's favorite molecules, $Be(BH_4)_2$.[1] The resolution of this particular structure question, which is fundamental to understanding chemical bonds involving "electron deficient atoms," had elements of a mystery story complete with a cast of suspects, an excess of clues, "red herrings," and conflicting "expert" opinions. Consequently, William Lipscomb, John Stanton (then Lipscomb's graduate student who is now a professor at the University of Texas), and I decided to use quantum mechanics and computer programs developed in my research group at the Florida Quantum Theory Project to try to resolve this outstanding question. The ACES I program was based upon the then new methods of coupled-cluster theory and many-body perturbation theory (CC/MBPT), that my group was pursuing. This is also an example of the evolution toward "predictive" quantum chemistry, now largely represented by CC theory.

Our problem originated with the determination of molecular structure via X-ray analysis.

Professor Lipscomb obtained his degree in the area of X-ray structure determination under Linus Pauling. Pauling's Nobel Prize in Chemistry (1954) was awarded for his X-ray elucidation of the structure of proteins. Lipscomb's Nobel Prize in Chemistry (1976) dealt with another class of curious molecules, the boranes.

Boranes are composed of units involving boron and hydrogen atoms that bond together in unexpected ways. Since boron is "electron deficient," the electrons sometimes have to be shared over several atoms. Consequently, when two BH units are brought together, the molecule diborane is formed, held together by "bridged" units as shown in Figure 14.1a. These B-H-B bonds are sometimes called "banana" bonds, where two electrons are shared by three atoms. When another BH_2 unit is added we obtain B_3H_6 as shown in Figure 14.1b. As more and more units are added, we can obtain increasingly interesting and complex structures.

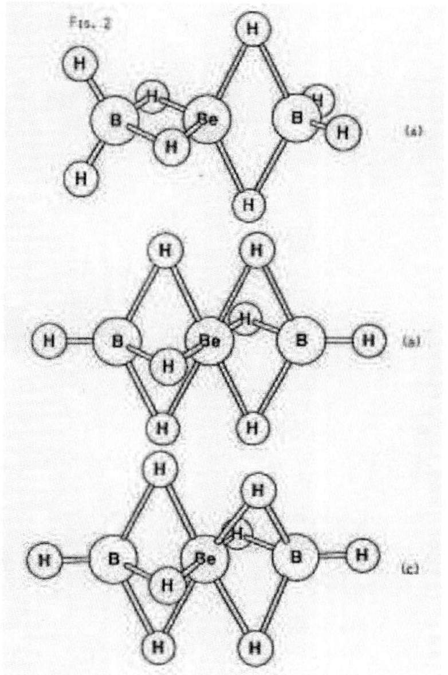

Figure 14.1 (a) Diborane, formed when two boron hydrogen units are brought together. (b) When another BH_2 unit is added, B_3H_6 is formed.

The beryllium atom, Be, is one step lighter than B, causing it to have some of the same properties, except that while B has five electrons, Be has four, making the bonding patterns potentially quite different. By combining Be with BH_4 units, we obtain the beryllium borohydride molecule, $Be(BH_4)_2$ which had been known for more than 50 years even in 1986. However, what had not been known is the mystery of how the atoms are bonded together.

Building upon an extensive knowledge of chemical bonding, different "experts" proposed more than 12 different structures for this molecule. Some of these "suspects" are shown in Figure 14.2. The aesthetically pleasing structure in Figure 14.2a contains doubly-bridged bonds between B and Be similar to those encountered in diborane. If two bridges are good, why not three? This proposed structure is shown in Figure 14.2b. Some said that the triply-bridged structure should be augmented by a different Be-B distance on one side of the molecule from the other, as in Figure 14.2c (the bond on the right is actually shorter!). Breaking symmetry in chemistry is often of fundamental importance to several phenomena. Additional proposed structures devoid of a linear arrangement of B-Be-B are shown in Figures 14.2d and 14.2e.

For almost any interesting structural chemistry problem there are feasible alternative molecular structures. However, the preferred structure of several possibilities can usually be identified by *experimental* analysis. Furthermore, experiments are usually synonymous with fact, while "theory" frequently has a dubious connotation. As Sherlock Holmes said, "It is a capital mistake to theorize before one has data."[2]

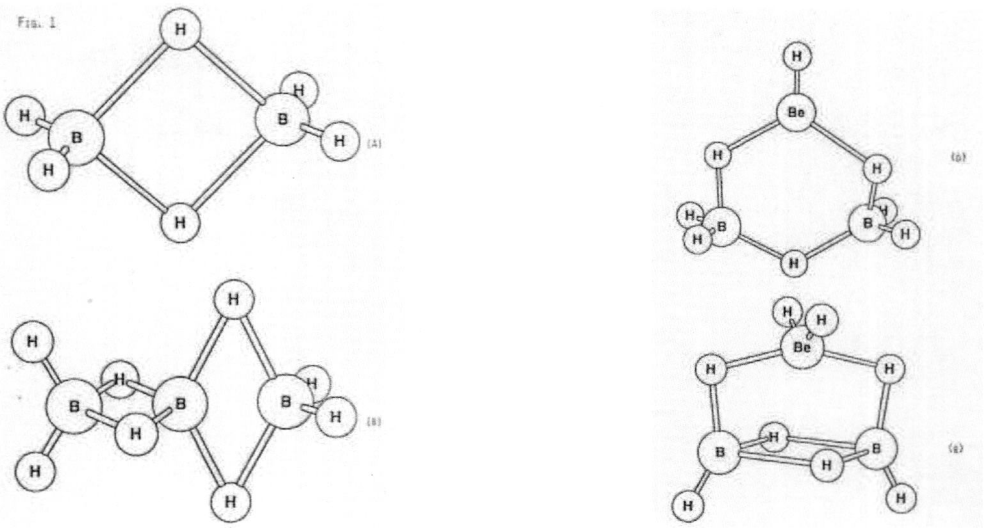

Figure 14.2 Bonds between boron and beryllium atoms that are (a) double-bridged and (b) triple-bridged. Breaking symmetry in the arrangement of the bonds in (c), (d), and (e).

In this mystery story, however, several different experimental procedures point to entirely different suspects. How can this be? X-ray analysis would be possible in the crystalline form of this molecule, but the problem here is that we want the structure of the molecular $Be(BH_4)_2$ units, and for this case the solid structure is likely to be quite different. The solid structure consists of a helix of units that have double bridges as in Figure 14.2a, however.

For gas species, if we can make the molecule we should be able to characterize it with vibrational spectroscopy. The difficulty with such an experiment occurs when we do not have a pure sample of the molecule we want to study. If the molecule is not very stable decomposition products might be present. The molecule might react with these products or itself forming new species, or impurities might cause problems, or if the molecule really exists in several forms like that suggested in Figure 14.2, more than one might exist in the sample. Any of these possibilities complicate the fingerprints that would tell us the actual form of the molecule.

Such an infra-red experiment was performed. One conclusion was that the molecule had a dipole moment, which meant it did not have a center of symmetry like the possibilities in Figures 14.2a or 14.2b. Hence, it was proposed that the suspect pictured in Figure 14.2c would offer the best explanation. Additional spectroscopic evidence indicated that there might be more than one structure present, however.

A third way to study molecular structure is electron diffraction. Two such studies proposed the structures in Figures 14.2d and 14.2e, while a third supported Figure 14.2a.

Hence, we have a real dilemma! Several different experts using several different experimental methods pointed to at least five different suspects. As Hercule Poirot might say, "Can the little gray cells do better?" In this case we mean can high-level quantum *theory* identify the right suspect, in lieu of the failures of the experimental methods? The solutions to the Schrödinger equation will tell us the molecular structure, the energy differences between the various structures, and the spectral "fingerprints." The problem is we cannot solve the Schrödinger equation exactly, so we have to obtain very good approximations to be able to select from the variety of structures of $Be(BH_4)_2$ supported by conflicting experiments.

Today, quantum chemistry can often *quantitatively predict* properties of molecules known or not; and this may be used to offer another opinion to resolve experimental discrepancies. Accurate quantum mechanical calculations, however, are extremely demanding. They frequently require powerful supercomputers, but even with supercomputers, better methods for solving the Schrödinger equation must be constantly sought.

Developing techniques for providing better solutions to the Schrödinger equation for atoms, molecules, and solids has been my primary objective. Consequently, we have developed the coupled-cluster methods that, today, are well-recognized as providing the best solutions for the largest number of problems in the field.[3] This fact was just emerging in 1986 when we applied them to the $Be(BH_4)_2$ problem.

We first calculate all the bond lengths and angles for interesting structures as in Figure 14.2 and then their energies. We find that the structure with the lowest energy is the triply-bonded form in Figure 14.2b, although the double-bridge (Figure 14.2a) is nearly as stable. The form in Figure 14.2c is always found to prefer the equal Be-B distances as in Figure 14.2b. The isomers in Figures 14.2d and 14.2e are substantially less stable, discounting their candidacy. However, only Figures 12.2c, 14.2d, and 14.2e have a dipole moment, as observed experimentally. However, when we compute their moments we get very different values, so none of these species explain that observation.

To definitively characterize our suspects we compute their fingerprints (i.e. IR spectra). By comparing with the observed spectra, we see exceptional agreement not for the suspects in Figures 14.2a or 14.2b, but for a mixture of *both* — always an effective ploy in a mystery. Furthermore, our calculations fully explain the observed fingerprints, providing (nearly) conclusive evidence that Figures 14.2a and 14.2b are indeed our culprits. Since we compute these two molecules to be equally stable, it is entirely reasonable that they would coexist in the IR experiment.

What about the dipole moments? Pondering that question in Cheers one evening, it was concluded that the dipole moment must be a "red herring," i.e. an *experimental error* caused by some impurity species. After all, no good mystery story would be without one!

References

1. Stanton J. F., Lipscomb W. N., and Bartlett R. J., Structure, energetics and vibrational spectra of beryllium borohydride isomers, *J. Chem. Phys.*, **88**, 5726–5734 (1988).
2. Doyle A. C., A scandal in Bohemia, *The Adventures of Sherlock Holmes*. New York: Oxford University Press (1993). p. 8.
3. Bartlett R. J. and Musial M., Coupled-cluster theory in quantum chemistry, *Rev. Mod. Phys.*, **79**, 291–352 (2007).

Structure, energetics, and vibrational spectra of beryllium borohydride isomers

John F. Stanton[a)] and William N. Lipscomb
Gibbs Chemical Laboratory, Harvard University, Cambridge, Massachusetts 02138
Rodney J. Bartlett[b)]
Quantum Theory Project, Departments of Physics and Chemistry, University of Florida, Gainesville, Florida 32611

(Received 22 December 1987; accepted 19 January 1988)

Several structures of beryllium borohydride (BeB_2H_8) have been studied with techniques based on many-body perturbation theory (MBPT). Energies calculated at the MBPT(4) level with a basis set of 85 contracted Gaussian functions support the two structure hypothesis of Nibler, although a qualitative MBPT(4) asymmetric stretching potential indicates that the C_{3v} double-minimum structure is unstable with respect to a more highly symmetric D_{3d} model with two sets of equivalent triple hydrogen bridges. The theoretical free energy difference between the D_{3d} structure and a diborane-like model having D_{2d} symmetry is 1.1 kcal/mol, with the latter predicted to be more stable. Uncertainties in the theoretical force field and those resulting from the finite nature of the basis set, however, may change this value by several kcal/mol. Principal evidence for the two-structure proposal comes from MBPT(2) harmonic frequencies and infrared absorption intensities, which are found to be consistent with the published vibrational spectra. These are used in a tentative assignment of most of the observed features. Unfortunately, the experimentally observed dipole moment of 2.06 D cannot be explained at present.

I. INTRODUCTION

One of the more persistent problems in structural inorganic chemistry has been the determination of the molecular structure of beryllium borohydride (BeB_2H_8).[1] First synthesized by Schlesinger nearly fifty years ago,[2] this curious substance has been the subject of a number of experimental studies.[3] An early infrared investigation was carried out by Price *et al.*,[4] who interpreted the spectrum in terms of a diborane-like structure having D_{2d} symmetry with two double hydrogen bridges (structure I in Fig. 1). A subsequent electron diffraction (ED) study by Silbiger and Bauer[5] was initially found to support a linear B–Be–B skeleton with two equivalent sets of triple hydrogen bridges having D_{3d} symmetry (structure III in Fig. 1). Later refinements, however, revealed that the D_{2d} model with an expected asymmetry in the Be–H–B bridge bonds also provided an acceptable fit to the observed diffraction patterns.[6] Consequently, the D_{2d} structure became widely accepted and was not questioned for nearly twenty years.

More recent work, however, has clouded the issue. Haaland and collaborators found that radial distribution curves calculated from their ED intensities implied a triangular arrangement of first row atoms with nearly equal Be–B and B–B distances.[7] This idea gained favor when Dyke and Nibler found that a species having a molecular weight consistent with BeB_2H_8 exhibited a dipole moment of 2.06 ± 0.12 D in an electric deflection experiment.[8]

During this period, two groups reported vibrational spectra for isotopically substituted beryllium borohydride in which the data were interpreted in very different ways. Cook and Morgan[9] assumed that certain peaks in the infrared spectrum represented partially resolved rotational contours, while Nibler[10] treated all observed features as distinct vibrational transitions. In the latter study, which included spectra of gas phase and matrix isolated BeB_2H_8 and its perdeuterated analog, a strong coincidence of infrared and Raman features formed part of a persuasive argument against the interpretation of Ref. 9. In addition, Nibler pointed out that one of the peaks crucial to Cook and Morgan's interpretation was a misassigned fundamental of diborane. Perhaps

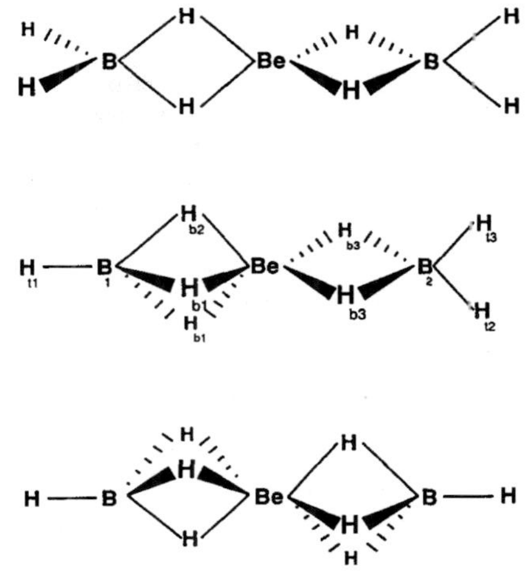

FIG. 1. Beryllium borohydride isomers I–III. Isomer IV (C_{3v} symmetry) is obtained from III via an A_{2u} distortion along the B–Be–B bond axis.

[a)] AT&T Foundation Doctoral Fellow.
[b)] Guggenheim Fellow, Harvard University, Fall 1986.

5726 J. Chem. Phys. 88 (9), 1 May 1988 0021-9606/88/095726-09$02.10

the most important result reported in this paper was that certain peaks in the vapor spectra were systematically absent or had substantially reduced intensities in the matrix. Furthermore, frequencies characteristic of both triple and double hydrogen bridges were observed in the gas phase, while only those expected from the former were prominent in the matrix spectra. On this basis, Nibler reasoned that two distinct isomers of BeB_2H_8 were present in the vapor. In the cold environment of the matrix, however, conditions favored the structure with three hydrogen bridges. Two pieces of evidence argued against the conclusion that structure III was the triply-bridged species. First, the coincidence of peaks in the gas phase (and matrix) infrared and Raman spectra implicated a structure without a center of inversion. In addition, the matrix spectra of BeB_2D_8 contained two absorptions of comparable intensity near 2200 cm^{-1}. These bands, thought to result from a BeB_2HD_7 contaminant, were assigned to stretching modes of two hydrogen atoms belonging to distinct triple bridges. To account for these observations, Nibler proposed a structure with C_{3v} symmetry (structure IV), characterized by a double-minimum asymmetric B–Be–B stretching potential, and went on to assign many of the observed features. No hypothesis was made regarding the identity of the second isomer, only the suggestion that it probably contained double hydrogen bridges.

Other electron diffraction work by Gundersen, Hedberg, and Hedberg (GHH)[11] did not clarify the issue. Diffraction patterns collected by these investigators were consistent with structures III, IV, or an unlikely Be–B–B linear arrangement with triple bridges; no acceptable fit could be found for structure I. Apparent discrepancies between their data and that of Silbiger and Bauer (which had supported structure I, as well as III), were attributed to the latter's optical scanning of the intensity profile. Using modern microphotometric methods, subtleties that were not detected visually became conspicuous. Disturbingly, the diffraction patterns of Almenningen *et al.*,[7] which favored the triangular structure, bore little resemblance to those obtained by GHH. In a largely unnoticed follow-up paper,[12] Haaland's group showed that the diffraction pattern of beryllium borohydride depended upon the condition of the sample. By varying the time between sample preparation and data collection, patterns reminiscent of their earlier data as well as those favoring the linear structures could be produced. Another important observation was that the two diffraction patterns were characterized by distinctly different background atomic scattering, indicating that the stoichiometry of the two samples was not the same.

Despite the flurry of experimental studies of BeB_2H_8 that began with the diffraction work of Almenningen *et al.*, the issue of its molecular structure has not been resolved. Strangely, no experimental work in this area has been reported since 1977, when a gas phase proton magnetic resonance study produced only one peak, indicating that the molecule was highly fluxional but with a linear B–Be–B framework.[3]

In addition to the experimental work summarized above, a few theoretical studies have been undertaken. A thorough analysis was carried out by Marynick and Lipscomb,[13] who investigated eleven different forms of BeB_2H_8 with a minimal basis of Slater orbitals (STOs) at the self-consistent field (SCF) and correlated levels of theory. Using geometrical parameters which minimized the SCF energy, CISD (configuration interaction with single and double substitutions) calculations predicted that three linear structures I, II (see Fig. 1), and III had relative electronic energies of 0.0, 6.5, and 7.3 kcal/mol, respectively. All other forms (including various triangular arrangements of the nonhydrogen atoms) were found to have significantly (>25 kcal/mol) higher energies. Attempts to optimize the structure of IV with minimal and extended[14] basis sets led directly to the D_{3d} conformer. Another investigation was reported by Ahlrichs,[15] who used a set of Gaussian lobe functions and calculated correlation energies with the independent electron pair approximation (IEPA). Using only crudely optimized geometries, structure I was found to have the lowest energy at the SCF level. Introduction of electron correlation, however, shifted the ordering of structures I and III (relative electronic energies (IEPA) in kcal/mol—I: 0.0, II: 1.1, III: -4.9). Qualitatively similar results were recently obtained by Ortiz and Lipscomb, who reported many-body perturbation theory (MBPT)[16] energy calculations for these structures.[17] Using the 6-311G* basis and geometries optimized at the SCF level with a modest split-valence basis set, they found relative energies (kcal/mol) of I: 0.0, II: 1.4, III: -1.1, using second-order MBPT [MBPT(2)]. None of the authors above, however, felt that their calculations were sufficiently accurate to discriminate between the stabilities of the three isomers.

The puzzle of the molecular structure(s) of beryllium borohydride is clearly one of the longest outstanding problems in main group inorganic chemistry. Although a number of experimental methods have been applied in an effort to solve the gas phase structure, no clear solution has emerged. Consequently, we felt it worthwhile to carry out a detailed theoretical investigation of this problem, exploiting powerful *ab initio* theoretical techniques. In this paper, we report many-body perturbation theory (MBPT) results obtained with fully polarized basis sets to study the energetics, structure, and vibrational spectra of the four isomers I–IV.

II. METHODS

Structures of beryllium borohydride isomers were fully optimized at the MBPT(2) level with a polarized contracted basis set consisting of 85 functions formed from a set of 149 primitive Gaussians. Convergence in the optimizations was defined as the point at which all atomic Cartesian forces fell to less than 0.0002 atomic units. Basis sets for boron ($3s2p1d$) and hydrogen ($2s1p$) were those of Ref. 18. For beryllium, the $3s2p$ set of Dunning and Hay[19] was used with a polarization exponent ($\zeta(d) = 0.308$) which minimizes the SDQ-MBPT(4) (fourth-order MBPT excluding the effects of triple excitations from the reference determinant) energy of beryllium hydride (BeH_2). Six Cartesian d functions were used on all nonhydrogen atoms. In the remainder of this paper, we will refer to this basis as (321/21).

Harmonic vibrational frequencies were calculated for structures I–III at the SCF level with the 3-21G[20] and (321/21) basis sets, using the respective theoretical equilbrium geometries. All second derivatives were evaluated analytically and frequencies were obtained by diagonalization of the mass-weighted Cartesian force constant matrix. For structures I and III, MBPT(2) harmonic force constants were obtained from numerical differentiation of analytic first derivatives using a step size of 0.01 bohr. Full exploitation of molecular symmetry was made in the MBPT(2) calculations; evaluation of the gradient at 14 and 11 distinct molecular geometries was sufficient to determine the complete harmonic force fields of I and III, respectively. Infrared absorption intensities were calculated within the double-harmonic approximation by the point charge scheme outlined in Ref. 21. In a recent study, we have found that intensities calculated at the MBPT(2) level with comparably sized basis sets are usually within 30% of experimental results.[22]

Dipole moments presented in Sec. III D were evaluated at the MBPT(2) level by finite difference of energies calculated in the presence of electric fields (0.005 a.u.) which were explicitly incorporated into the one-electron molecular Hamiltonian.

III. RESULTS AND DISCUSSION

A. Structures

Structures I–III were found to be stationary points on the SCF and MBPT(2) surfaces, while optimization of the C_{3v} structure (IV) led to III at all levels of theory. Internal coordinates resulting from MBPT(2) optimizations with the (321/21) basis are presented for isomers I–III in Table I, along with thermally averaged (r_g)[23] distances for III derived from the electron diffraction data of Ref. 11.

One feature common to the electron diffraction experiments which have implicated a linear arrangement of first row atoms is the presence of a prominent radial distribution peak near 1.75 Å. This observation clearly does not support the proposal that isomer I is the predominant species present under the conditions of the experiments. Although the ED studies do not rule out the possibility that trace quantities of this form may be present, the calculated MBPT(2) equilibrium distance of 1.86 Å is probably no more than 0.05 Å in error,[24] while the experimental value of r_g is significantly shorter. Furthermore, simple considerations of the effects of vibrational anharmonicity predict that equilibrium distances are shorter than the corresponding values of r_g. If diborane serves as an appropriate example, mean Be–H_b, B–H_b, and Be–B distances are likely to be 0.02–0.04 Å longer than those in the equilibrium structure, while a more modest (~0.01 Å) lengthening is expected for the B–H_t distances.[25]

It is more difficult to assess the plausibility of structure II based on the diffraction data. Superficially, one would expect the two distinct Be–B distances, which differ by 0.15 Å, to produce a doublet feature in the radial distribution function. Unfortunately, the soft vibrations characteristic of molecules containing hydrogen bridges lead to broad peaks, and closely spaced distances are often difficult to resolve. For example, features corresponding to two of the B–H distances in diborane [Δr_g(B–H) = 0.13 Å] are unresolved, and a careful analysis of the corresponding peak is required to show that its shape is different from that expected for a single internuclear distance.[26] Since GHH did not consider II in their least-squares refinement, the possibility that it is responsible for the observed diffraction pattern may not be discarded on these grounds.

Agreement between theory and experiment are more satisfactory for III, however. Application of empirical corrections of −0.01 Å for r_g(B–H_t) and −0.03 Å for the

TABLE I. Internal coordinates for the three isomers of BeB_2H_8 pictured in Fig. 1. Theoretical parameters were determined by minimization of the MBPT(2) energy with the (321/21) basis; corresponding SCF values are listed in parenthesis. Experimental results derived from the data presented in Ref. 11. All distances in Angstroms, angles in degrees.[a]

I		II[b]		III		III (expt)[c]	
r(Be–B)	1.857 (1.875)	r(Be–B_1)	1.725 (1.739)	r(Be–B)	1.734 (1.746)	r(Be–B)	1.792
r(Be–H_b)	1.478 (1.478)	r(Be–B_2)	1.875 (1.896)	r(Be–H_b)	1.607 (1.657)	r(Be–H_b)	1.676
r(B–H_b)	1.288 (1.302)	r(Be–H_{b1})	1.659 (1.685)	r(B–H_b)	1.251 (1.254)	r(B–H_b)	1.314
r(B–H_t)	1.195 (1.194)	r(Be–H_{b2})	1.565 (1.558)	r(B–H_t)	1.183 (1.181)	r(B–H_t)	1.193
r(H_t–H_t)	2.076 (2.043)	r(Be–H_{b3})	1.511 (1.512)				
		r(B_1–H_{b1})	1.247 (1.247)				
		r(B_1–H_{b2})	1.269 (1.280)				
		r(B_2–H_{b3})	1.273 (1.286)				
		r(B_1–H_{t1})	1.182 (1.180)				
		r(B_2–H_{t2})	1.197 (1.195)				
		r(B_2–H_{t3})	1.197 (1.196)				
		r(H_{b1}–H_{b1})	1.958 (1.966)				
		r(H_{b3}–H_{b3})	2.042 (2.043)				
		∠(Be–B_1–H_{t1})	175.8 (174.1)				
		∠(Be–B_2–H_{t2})	118.4 (118.9)				
		∠(Be–B_2–H_{t3})	121.3 (120.9)				
		∠(B_2–Be–B_1)	179.5 (179.0)				

[a] t and b refer to terminal and bridging hydrogen atoms, respectively.
[b] Designation for atoms is given in Fig. 1.
[c] See Ref. 23.

remaining distances brings the experimental internuclear separations into fair agreement with the theoretical equilibrium values. We are hesitant to draw any conclusions, however. Although the nonequilibrium conditions of the diffraction experiment may have preferred only one structure, the vibrational studies of Nibler[10] point to the need for additional models in the interpretation of the diffraction data. In particular, a simultaneous refinement of I and III with the molar composition as a supplementary parameter would prove informative, as would refinements including structure II.

The finding that optimization of structure IV led directly to III warrants special discussion at this point. Although the D_{3d} form is preferred at the MBPT(2) level with our basis set of 85 Gaussian functions, a larger basis set or more detailed accounting of electronic correlation could conceivably stabilize the double-minimum form. It is well documented that molecules containing beryllium and boron atoms represent formidable challenges to standard single-reference quantum chemical procedures,[27–29] due to the near degeneracy of the corresponding 2*s* and 2*p* atomic orbitals. In particular, the existence of two distinct minima in the potential curve of the diatomic Be_2 has been the subject of intense theoretical investigation.[30] At the self-consistent field level, the interaction between the two atoms appears to be repulsive at all distances, while the shallow van der Waals minimum near 4.5 Å is revealed when simple correlation corrections are applied. Characterization of the inner minimum, however, is extremely sensitive to the size of the basis and the treatment of correlation. It has been suggested[31] that inclusion of pure quadruple excitations (T_4 in the language of coupled-cluster theory) might be necessary to account quantitatively for the stability associated with this inner minimum. Clearly, such sophisticated calculations are not practical in the present case. An attempt to optimize the structure of IV at very high levels of theory, although certainly desirable, is beyond the limits of our current computational resources. As an alternative, we have carried out energy optimizations of IV at the MBPT(2) level subject to constraints on the Be–B distances. Two arrangements of the heavy atom nuclei have been chosen for study: the Be–B distances (r_a) obtained by GHH using the C_{3v} structural model in their ED refinement[11] (configuration A), and another [B; r_1 (Be–B) = 2.0 Å and r_2 (Be–B) = 1.5 Å], which represents our estimate of the limits of plausible beryllium–boron distances. At these partially optimized geometries, energies were calculated at the MBPT(4) level and are documented in Table II, along with the MBPT(2) dipole moments.

TABLE II. Electronic energies and MBPT(2) dipole moments for isomer IV. Energies are relative to structure III, and were calculated at the MBPT(4) level with the (321/21) basis. Geometries are described in Sec. III A.

Configuration	*E* (kcal/mol)	μ (D)
A	2.2	0.46
B	23.4	2.18

If the triply-bridged species observed by Nibler was the C_{3v} double-minimum structure (IV), the apparent absence of vibrational term splitting suggests an appreciable barrier to inversion $[E(D_{3d}) - E(C_{3v})]$.Clearly, our results are not consistent with this idea. Rather, they provide compelling evidence for the conclusion that III represents the lowest energy form of beryllium borohydride within the corresponding C_{3v} subspace of possible nuclear configurations.

B. Spectra

At the optimized SCF geometries, harmonic force field calculations were carried out for structures I–III using the 3-21G and (321/21) basis sets. With the larger basis, I and III represent local minima on the BeB_2H_8 hypersurface, while III was found to be unstable with respect to the E_u bridge bending mode (ν_{17}) when the 3-21G set was used. The nuclear motion associated with this mode [see Fig. 2(b)] effectively lowers the coordination number of beryllium, and suggests that polarization functions are required to provide an adequate description of the unusual bonding in this conformer. Calculations with both basis sets, however, predict that structure II represents a saddle point on the potential surface. Inspection of the transformation matrix between Cartesian and normal coordinates reveals that the corresponding transition vector is dominated by a rotation of the triply-bridged borohydride group which interconverts bridge and terminal hydrogens of I. Following these preliminary calculations, MBPT(2) harmonic force fields and infrared absorption intensities were evaluated for the two stable isomers, I and III, and are presented in Tables III and IV, respectively.

Due to the large number of observed peaks, the variability of many features, and the probable presence of overtone and combination bands in the matrix spectra, we have chosen to compare theoretical spectra of I and III with the gas phase data of Ref. 10. Infrared and Raman frequencies observed for BeB_2H_8 and BeB_2D_8 in the vapor are presented in Table V, along with a tentative assignment of many of the observed features. A cursory inspection of this assignment shows that all prominent features which were absent or substantially weaker in the matrix spectra have been attributed to structure I, while the persistent absorptions have been assigned to III.

Perhaps the most difficult part of the assignment lies in the region near 2100 cm^{-1}. Here, four absorptions are seen in both the infrared and Raman spectra. In the vapor Raman spectrum of BeB_2H_8, all four features are polarized, indicating that they correspond to totally symmetric transitions. Unfortunately, our calculations indicate that only two fundamentals belonging to totally symmetric representations are expected in this region: ν_2 of structure I and ν_2 of structure III, which are assigned to the peaks at 2075 and 2175 cm^{-1}, respectively. Also acceptable is assignment of the latter to the weaker band at 2240 cm^{-1}. The other observed features are more difficult to explain. We have assigned the band at 1995 cm^{-1} to the degenerate bridge stretch (ν_{15}) in structure I. Although no polarization is expected for the corresponding fundamental, the presence of the very weak bridge bending mode (ν_{20}) at 119 cm^{-1} indicates that a

TABLE III. Harmonic frequencies, infrared absorption intensities, and approximate mode descriptions of the normal vibrations of isomer I (D_{2d} symmetry), calculated at both SCF and correlated levels. (ip—in phase; op—out of phase).

		Harmonic frequency (cm^{-1})			Infrared intensity (km/mol)		
Symmetry	Designation	3-21G SCF	(321/21) SCF	(321/21) MBPT(2)	(321/21) SCF	(321/21) MBPT(2)	Approximate mode description
A_1							
	ν_1	2714	2666	2641	···	···	BH_2 sym str (ip)
	ν_2	2132	2122	2143	···	···	Bridge sym str (ip)
	ν_3	1709	1695	1648	···	···	Bridge sym deform (ip)
	ν_4	1253	1242	1181	···	···	BH_2 scis (ip)
	ν_5	506	511	512	···	···	B–Be–B sym str
A_2							
	ν_6	1237	1196	1151	···	···	BH_2 twist (op)
B_1							
	ν_7	1244	1204	1156	···	···	BH_2 twist (ip)
	ν_8	342	331	308	···	···	BH_4 torsion
B_2							
	ν_9	2711	2663	2639	157	109	BH_2 sym str (op)
	ν_{10}	2110	2096	2122	362	348	Bridge sym str (op)
	ν_{11}	1761	1735	1722	1485	1028	Bridge sym deform (op)
	ν_{12}	1246	1234	1177	106	47	BH_2 scis (op)
	ν_{13}	1000	995	1004	93	101	B–Be–B asym str
E							
	ν_{14}	2796	2748	2733	245	142	BH_2 asym str
	ν_{15}	1983	2005	2097	379	216	Bridge asym str
	ν_{16}	1632	1627	1603	20	9	Bridge asym deform
	ν_{17}	1158	1123	1034	6	0	BH_2 rock
	ν_{18}	919	911	890	12	4	BH_2 wag
	ν_{19}	419	406	364	16	24	B–Be–B bend
	ν_{20}	160	150	119	36	32	Bridge bend

substantial fraction of the molecules are not in the ground state of this level at room temperature. Resulting difference bands, such as $\nu_{15} + \nu_{20} - \nu_{20}$ and $\nu_{15} + 2\nu_{20} - 2\nu_{20}$, do contain a totally symmetric component, however, and might exhibit observable polarization. We are unable to assign the absorption at 2240 cm^{-1} (or 2175 cm^{-1}, if the alternative assignment is made for ν_2 of structure III). It is conceivable that this is an overtone or combination band of structure III in resonance with the nearby ν_2 transition. Possibilities include $2\nu_{12}$ and $2\nu_{15}$. The presence of this feature in the matrix argues against assignment to structure I.

Calculated MBPT(2) intensities also serve to lend plausibility to our assignment of the infrared spectrum. With the exception of the ν_{16} and ν_{18} modes of structure III, which have been assigned to weak absorptions, only those fundamentals with theoretical intensities in excess of 100 km/mol are seen in the experimental infrared spectrum, provided our assignment is correct. Due to uncertainties in both theoretical and visually estimated experimental intensity ratios, an accurate assessment of the amounts of I and III in the equilibrium vapor cannot be made, but Nibler's empirical observation that they are present in "roughly comparable amounts" appears to be correct.

Apparent relative band strengths of observed fundamentals are generally consistent with those predicted from the MBPT(2) absolute intensities. In particular, assignments of the pronounced absorptions at 1548 and 1050 cm^{-1} to ν_{11} (structure I) and ν_9 (structure III) are supported by their large calculated intensities. The only significant discrepancy occurs for the infrared band at 1000 cm^{-1}, which has been assigned to the asymmetric B–Be–B stretching mode of the D_{2d} isomer. Here, the calculated intensity of 101 km/mol is surprisingly low for this mode, and seems inconsistent with an assignment to the "very strong" absorption. While it is possible that a more extensive basis set might better describe the separation of charges brought about by this distortion and lead to a significantly higher predicted intensity, the proximity of this peak to the very intense asymmetric stretching fundamental of III in spectra of both BeB_2H_8 and BeB_2D_8 makes an accurate experimental assessment of its intensity difficult.

Although a close correspondence of infrared and Raman band positions was cited by Nibler in favor of structure IV, the weak coupling between vibrations of the respective borohydride groups in I and III nevertheless leads to nearly coincident spectral features in these molecules. Certainly, the double-minimum structure would help to explain the presence of the two peaks attributed to BeB_2HD_7 by Nibler; in view of the evidence against structure IV cited in this paper, however, we believe that alternative explanations should be sought for this observation.

By analogy with diborane, which has an extraordinarily

TABLE IV. Harmonic frequencies, infrared absorption intensities, and approximate mode descriptions of the normal vibrations of isomer III (D_{3d} symmetry), calculated at both SCF and correlated levels. (ip—in phase; op—out of phase).

		Harmonic frequency (cm^{-1})			Infrared intensity (km/mol)		
Symmetry	Designation	3-21G SCF	(321/21) SCF	(321/21) MBPT(2)	(321/21) SCF	(321/21) MBPT(2)	Approximate mode description
A_{1g}							
	ν_1	2878	2805	2782	···	···	BH str (ip)
	ν_2	2379	2358	2350	···	···	Bridge sym str (ip)
	ν_3	1366	1443	1364	···	···	Bridge sym deform (ip)
	ν_4	579	603	605	···	···	B–Be–B sym str
A_{1u}							
	ν_5	256	218	199	···	···	BH_4 torsion
A_{2u}							
	ν_6	2877	2805	2783	141	82	BH str (op)
	ν_7	2364	2339	2335	348	256	Bridge sym str (op)
	ν_8	1312	1382	1303	241	128	Bridge sym deform (op)
	ν_9	1039	1077	1118	895	786	B–Be–B asym str
E_g							
	ν_{10}	2295	2295	2333	···	···	Bridge asym str (ip)
	ν_{11}	1418	1413	1377	···	···	Bridge asym deform (ip)
	ν_{12}	1240	1206	1142	···	···	BH wag (ip)
	ν_{13}	270	390	441	···	···	Bridge sym bend
E_u							
	ν_{14}	2340	2345	2372	530	303	Bridge asym str (op)
	ν_{15}	1403	1394	1347	16	4	Bridge asym deform (op)
	ν_{16}	1235	1215	1152	54	26	BH wag (op)
	ν_{17}	$178i^a$	404	453	4	14	Bridge asym bend
	ν_{18}	331	266	287	64	50	B–Be–B bend

[a]The structure is unstable with respect to this distortion.

complicated infrared spectrum with intense overtone and combination features,[32] a complete assignment of the vibrational spectra of beryllium borohydride is likely to be difficult. By tentatively assigning most of the features in the published spectra, we hope to point future investigators in what we believe to be the correct direction.

C. Energetics

At the optimized MBPT(2) geometries, full fourth order MBPT calculations were performed for structures I–III. All 11 electron pairs were explicitly correlated into the full space of 74 virtual functions. The results, which predict that the D_{2d} structure (I) has the lowest electronic energy at both SCF and correlated levels, are given in Table VI. Consistent with our earlier studies in borane chemistry,[29,33] introducing correlation favors the conformers with the largest number of hydrogen bridge bonds. Nevertheless, at the MBPT(4) level, structure I remains the minimum on the (321/21) electronic energy surface.

Zero-point and thermal effects were calculated for the two local minimum structures, I and III, according to the standard methods of idealized gas phase statistical thermodynamics; results are presented in Table VII. As expected, inclusion of vibrational energies calculated from the MBPT(2) frequencies in Tables III and IV scaled by a factor of 0.95, which approximately corrects for differences between fundamental and harmonic frequencies as well as deficiencies in the theoretical force field, results in an additional lowering of the relative energy of III. Introduction of entropic effects, however, strongly favors I. This effect is primarily due to the substantial contribution of the weak bridge bending mode (ν_{20}) to the vibrational partition function of I, and accounts for a large fraction ($\sim 80\%$) of the total free energy difference at 298 K. The magnitude of this difference, however, is too small for us to draw any conclusions regarding the relative stabilities of I and III. While experience indicates that reaction and isomerization energies calculated for the related boron hydride systems are reasonably well converged at the MBPT(4) level[34,35] [as compared with highly correlated CCSD + T(CCSD) calculations[36]], basis set improvements could change ΔE_e by several kcal/mol.[18,37–39] Due to the greater demands placed on the basis by structure III, we believe that an augmentation would favor this conformer relative to I. This idea is supported by the MBPT(2) study of Ortiz and Lipscomb,[17] who found that expansion of the basis from 6-31G* ($3s2p1d/2s$) to 6-311G* ($4s3p1d/3s$) resulted in a lowering of III relative to I of 1.2 kcal/mol. In the present case, however, the magnitude of this effect cannot be predicted with any reliability. Furthermore, the free energy difference is sensitive to the frequencies of the weak vibrational modes of these structures. Even if our calculations were perfect (which they are not), the fundamental frequencies associated with these modes may differ substantially from those cal-

TABLE V. Tentative assignment of gas-phase infrared and Raman spectral features listed by Nibler for BeB_2H_8. The designation of the fundamentals is consistent with Tables III and IV; the isomer number is given in parentheses. Frequencies are in cm^{-1}, calculated infrared intensities in km/mol. The qualitative experimental intensities[b,c] are taken from Ref. 10. Prominent features ("medium" or higher intensity) which were much weaker or absent in the matrix spectra are marked with an asterisk.

Assignment	ν_H (obs)	ω_H (calc)[a]	ν_D (obs)	ω_D (calc)	I(obs)	I(calc)
Infrared						
ν_6 (III)	2624	2783	2000	2075	m	82
ν_{14} (I)	2550*	2733	1935	2044	m	142
ν_9 (I)	2500*	2639	1820	1916	m	109
ν_{14} (III)	2225	2372	1650	1745	m(sh)	303
ν_7 (III)	2167	2335	1580	1659	s	256
ν_{10} (I)	2071*	2122	1470	1514	s	348
ν_{15} (I)	2000*	2097	1520	1548	w(sh)	216
	1650				m(sh)	
ν_{11} (I)	1548*	1722	1166	1294	s(br)	1028
ν_8 (III)	1247	1303	1075	1144	w	128
ν_{16} (III)	1130	1152	?	861	w(sh)	26
ν_9 (III)	1050	1118	895	932	vs(br)	786
ν_{13} (I)	1000*	1004	830	847	vs(br)	101
ν_{18} (III)	287	287	275	260	w	50
Raman						
ν_1 (III)	2627	2782	?	2081	m(p)	...
	2616				w	
ν_{14} (I)	2550*	2733	?	2044	m	...
ν_1 (I)	2490*	2641	1826	1916	s(p)	...
	2352				w	
	2240				m(p)	
ν_2 (III)	2175	2350	1578	1667	s(p)	...
ν_2 (I)	2075*	2143	1471	1527	m(p)	...
ν_{15} (I)	1995	2097	1520	1548	w(p)	...
ν_4 (I)	1150	1181	?	883	vw(br)	...
	588				w[d]	
ν_4 (III)	535	605	478	537	s(p)	...
ν_{13} (III)	384	441	?	325	vw	...
	357				vw[d]	

[a] Results of MBPT(2) calculations with the (321/21) basis set.
[b] Broad peaks and shoulders designated by (br) and (sh), respectively.
[c] Intensities are those assigned to BeB_2H_8.
[d] Variable features attributed to impurity.

culated from the harmonic force field. It can only be said that our results do not contradict the possibility that I and III may coexist at equilibrium.

It is interesting to note that the predicted stability of III relative to the other structures differs substantially from the results of Ref. 13. A likely explanation for this is that the minimal STO basis set used by Marynick and Lipscomb, like the 3-21G set, is unable to satisfactorily account for the hexacoordination of the beryllium atom.

D. Dipole moment

While forbidden by symmetry for structures I and III, dipole moments for the C_s form (II) and Nibler's proposed C_{3v} structure (IV) are both expected to be significant. Al-

TABLE VI. Relative electronic energies (kcal/mol) for beryllium borohydride isomers I–III pictured in Fig. 1. Unless specified otherwise, all calculations used the (321/21) basis.

	I	II	III
SCF (3-21G)	0.0	7.4	11.2
SCF	0.0	4.7	5.6
MBPT(2)	0.0	1.5	0.2
MBPT(3)	0.0	1.9	0.9
SDQ-MBPT(4)	0.0	2.2	1.4
MBPT(4)	0.0	1.9	0.9

TABLE VII. Electronic[a] and vibrational contributions to the energy of the isomerization reaction I→III. Also included are the relative enthalpy and free energy differences, evaluated at 298 K. All values in kcal/mol.

ΔE_e	0.9
ZPC[b]	−0.6
ΔE_0	0.3
ΔH (298 K)	0.2
ΔG (298 K)	1.1

[a] Calculated from MBPT(4) energies.
[b] Calculated from the harmonic frequencies listed in Tables III and IV scaled by a factor of 0.95.

though IV does not represent a minimum on the MBPT(2) energy surface with the (321/21) basis, we nevertheless felt it instructive to calculate its dipole moment using the Be–B distances and partially optimized MBPT(2) geometries discussed in Sec. III A. The results, which are presented in Table II, suggest that the observed dipole moment could be achieved near our plausible limits of Be–B distances, but at a cost of more than 20 kcal/mol in energy. Using the least-squares distances from Ref. 11, the calculated moment is more than four times smaller than the value of 2.06 D reported in Ref. 8. Hence, if a double-minimum structure is responsible for the apparent dipole moment of beryllium borohydride, it does not constitute the majority of the substance probed in the diffraction experiments.

The MBPT(2) dipole moment for II is found to be 1.18 D, roughly one-half the experimental value. Although the present study does not rule out the possibility that II is a local minimum on the MBPT(2) potential energy surface, this result indicates that this isomer is not responsible for the experimentally observed dipole moment. This, along with the apparent success of our vibrational assignment in terms of structures I and III, provides strong evidence for the assertion that II does not exist in observable concentrations at equilibrium.

IV. CONCLUSIONS

In this paper, we have presented the results of a quantum chemical study of several possible structures of beryllium borohydride using powerful modern computational techniques. Provided both are assumed present at equilibrium, the two structures found to be local minima at the SCF level (I and III) can satisfactorily account for a large part of the experimental data. Highly correlated MBPT(4) calculations predict that the energy separation between these two isomers is indeed small, and suggest that they could coexist under ambient conditions. Principal support comes from correlated harmonic force fields and infrared absorption intensities of I and III, which are used to assign the published gas phase vibrational spectra. In complete accord with the two-structure hypothesis advanced by Nibler, features observed in the gas phase which persisted at much lower temperatures were assigned to the isomer with triple hydrogen bridges, while those only present in the vapor were attributed to I. Nevertheless, there is an important distinction between Nibler's interpretation and the results of our study: we find no evidence to support the contention that the triply-bridged species has a symmetry lower than D_{3d}. Consequently, we are unable to account for the dipole moment attributed to at least one isomer of BeB_2H_8 in the gas phase. While we believe that I and III represent the most stable forms of this molecule, the possibility remains that a transient intermediate or impurity may be responsible for this curious observation.

By no means has our study cleared up all questions concerning the structure(s) of this unusually enigmatic molecule. Of additional interest is the activation barrier for interconversion of I and III. Knowledge of the potential energy surface associated with this process would be valuable in further evaluation of the spectroscopic and electron diffraction studies reviewed in this paper. A likely channel for interconversion is provided by the respective bridge bending modes of I and III (see Fig. 2). Unfortunately, our results indicate that the frequencies of these modes, and by consequence, the potential energy surface associated with interconversion, is very sensitive to both basis set and correlation effects. Hence, location of the transition structure at the correlated level is demanded for quantitative accuracy. Such an effort would involve a considerable investment of computing resources and is left to the future.

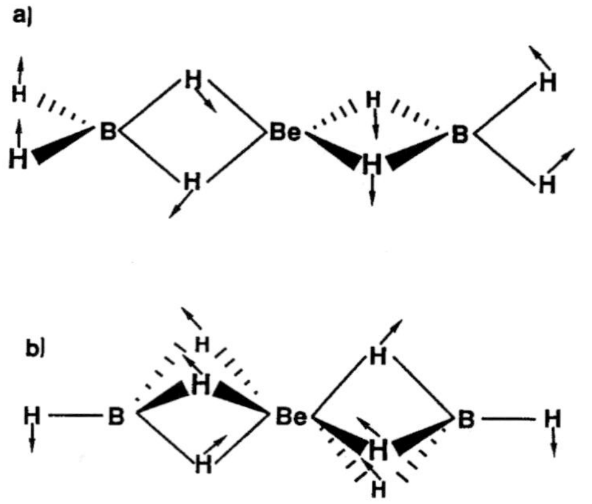

FIG. 2. Approximate form of individual components of the doubly degenerate bridge bending modes for structure I (ω_{20}) and structure III (ω_{17}).

It is our sincere hope that this report will reawaken the interest of the experimental community, which has ignored this interesting problem for a decade.

ACKNOWLEDGMENTS

Aspects of this research were supported by the National Science Foundation under Grant No. CHE85-15347 and the Office of Naval Research. We are grateful to the staff of the Pittsburgh Supercomputing Center, where many of the calculations were performed, for their courteous and prompt assistance with technical problems. We also thank J. Eric Gouaux for preparing the figures.

[1]In view of the interest in the molecule $BeBH_5$ (see Refs. 15 and 28), which could also be called "beryllium borohydride," it might alleviate confusion if a standard was developed for naming these systems. By analogy with conventions for the boron hydrides, we suggest the following names: beryllium borane(5) for $BeBH_5$ and beryllium diborane(8) for BeB_2H_8. We have chosen to adhere to tradition in this manuscript, however.

[2]A. Burg amd H. I. Schlesinger, J. Am. Chem. Soc. 62, 3425 (1940).
[3]D. F. Gaines, J. L. Walsh, and D. F. Hillenbrand, J. Chem. Soc. Chem. Commun. 1977, 224.
[4]W. C. Price, H. C. Longuet-Higgins, B. Rice, and T. F. Young, J. Chem. Phys. 17, 217 (1949).
[5]G. Silbiger and S. H. Bauer, J. Am. Chem. Soc. 68, 312 (1946).
[6]S. H. Bauer, J. Am. Chem. Soc. 72, 622 (1950).
[7]A. Almenningen, G. Gundersen, and A. Haaland, Acta Chem. Scand. 22, 859 (1968).
[8]J. W. Nibler, and T. Dyke, J. Am. Chem. Soc. 92, 2920 (1970).
[9]T. H. Cook and G. L. Morgan, J. Am. Chem. Soc. 91, 774 (1969).

[10]J. W. Nibler, J. Am. Chem. Soc. **94**, 3349 (1972).
[11]G. Gundersen, L. Hedberg, and K. Hedberg, J. Chem. Phys. **59**, 3777 (1973).
[12]K. Brendhaugen, A. Haaland, and D. P. Novak, Acta Chem. Scand. **29**, 801 (1975).
[13]D. S. Marynick and W. N. Lipscomb, J. Am. Chem. Soc. **95**, 7244 (1973).
[14]D. S. Marynick, J. Chem. Phys. **64**, 3080 (1976).
[15]R. Ahlrichs, Chem. Phys. Lett. **19**, 174 (1973).
[16]R. J. Bartlett, Ann. Rev. Phys. Chem. **32**, 359 (1981), and references therein.
[17]J. V. Ortiz and W. N. Lipscomb, in *AIP Conference Proceedings*, edited by D. Emin, T. Aselage, C. L. Beckel, I. A. Howard, and C. Wood (American Institute of Physics, New York, 1986).
[18]L. T. Redmon, G. D. Purvis, and R. J. Bartlett, J. Am. Chem. Soc. **101**, 2856 (1979).
[19]T. H. Dunning and P. J. Hay, in *Methods of Electronic Structure Theory*, edited by H. F. Schaefer (Plenum, New York, 1977), Vol. 3.
[20]J. S. Binkley, J. A. Pople, and W. J. Hehre, J. Am. Chem. Soc. **102**, 939 (1980).
[21]L. J. Schaad, C. S. Ewig, B. A. Hess, and D. Michalska, J. Chem. Phys. **83**, 5348 (1985).
[22]J. F. Stanton, D. H. Magers, and R. J. Bartlett (to be published).
[23]The published values of r_a have been corrected to correspond to mean internuclear distances (r_g).
[24]D. J. DeFrees, K. Raghavachari, H. B. Schlegel, and J. A. Pople, J. Am. Chem. Soc. **104**, 5576 (1982).
[25]J. L. Duncan, Mol. Phys. **51**, 371 (1984). Approximate differences between equilibrium and thermally averaged structures have been estimated using values of r_e from this reference and mean distances (r_g) from Ref. 26.
[26]L. S. Bartell and B. L. Carroll, J. Chem. Phys. **42**, 1135 (1965).
[27]W. N. Lipscomb, Pure and Appl. Chem. **55**, 1431 (1983).
[28]D. J. DeFrees, K. Raghavachari, H. B. Schlegel, J. A. Pople, and P. v. R. Schleyer, J. Phys. Chem. **91**, 1857 (1987).
[29]J. F. Stanton, R. J. Bartlett, and W. N. Lipscomb, in *IMEBORON VI Proceedings* (World Scientific, Singapore, 1988).
[30]R. J. Harrison and N. C. Handy, Chem. Phys. Lett. **123**, 321 (1986), and references therein.
[31]Y. S. Lee and R. J. Bartlett, J. Chem. Phys. **80**, 4371 (1984).
[32]J. L. Duncan, D. L. McKean, I. Torto, and G. D. Nivellini, J. Mol. Spectrosc. **85**, 16 (1981).
[33]M. L. McKee and W. N. Lipscomb, Inorg. Chem. **21**, 2846 (1982).
[34]In Ref. 29, the SDQ-MBPT(4) isomerization energy between D_{2d} and C_{2v} isomers of diborane(4) was mistakenly listed as the full MBPT(4) result. The true MBPT(4) value is 0.85 kcal/mol, and agrees well with the CCSD + T(CCSD) result of 1.29 kcal/mol.
[35]J. F. Stanton, R. J. Bartlett, and W. N. Lipscomb, Chem. Phys. Lett. **138**, 525 (1987).
[36]A proper assessment of the rate of convergence of the perturbation expansion requires a knowledge of the full configuration interaction (CI) result. Since these are unavailable for the borane systems we have studied, comparisons are made with CCSD + T(CCSD) [Y. S. Lee, S. A. Kucharski, and R. J. Bartlett, J. Chem. Phys. **81**, 5906 (1984); **82**, 5761E (1985)] calculations. In all cases studied to date [S. J. Cole and R. J. Bartlett, J. Chem. Phys. **86**, 873 (1987); M. Urban, J. Noga, S. J. Cole, and R. J. Bartlett, J. Chem. Phys. **83**, 4041 (1985)], basis set correlation energies computed with this model for molecules at or near their equilibrium geometries have been within 0.6% of the full CI result. Hence, as stated in Ref. 35, it is unlikely that basis set correlation errors in CCSD + T(CCSD) reaction energies for B_1 and B_2 systems exceed 1 kcal/mol. Basis set effects, however, can be significantly larger.
[37]R. J. Bartlett and G. D. Purvis, Phys. Scr. **21**, 255 (1980).
[38]G. F. Adams, G. D. Bent, G. D. Purvis, and R. J. Bartlett, Chem. Phys. Lett. **81**, 461 (1981).
[39]W. J. Hehre, *Ab Initio Molecular Orbital Theory*, edited by W. J. Hehre, J. A. Pople, R. v. R. Schleyer, and L. Radom (Wiley, New York, 1986).

The Saga of Ozone and Its Role in Developing Coupled-Cluster Theory: A Lipscomb Legacy Issue

In addition to the $Be(BH_4)_2$ story in this volume, another significant piece of work done jointly between Lipscomb's group at Harvard and mine at the University of Florida's Quantum Theory Project pertains to the ozone molecule, O_3.[1] The characteristic of most coupled-cluster methods is that they start from a single determinant reference function, Φ_0. This can be the Hartree–Fock minimum energy determinantal wavefunction, or any other single determinant like that due to Brueckner, which has maximum overlap with the exact wavefunction; or the determinant composed of the first n natural orbitals, which gives the best possible approximation to the one-particle density matrix; or the Kohn-Sham determinant that is supposed to exactly represent the density (not the density matrix) for a molecule. Once such a choice is made, the exact, correlated wavefunction written in coupled-cluster form is $\Psi_{CC} = \exp(T)\Phi_0$. The operator $T = T_1 + T_2 + T_3 \ldots$ generates all single, double, triple, ...etc. replacements of the first n-occupied orbitals, $i, j, k, \ldots$ in Φ_0 by unoccupied ones, $a, b, c, \ldots$ In the limit of all n-fold replacements the *exact* wavefunction in a finite, atomic orbital basis set is obtained. This is called the full CI.

The power of CC theory is such that one rapidly exhausts the space of determinants even with just T_1, T_2, and some limited consideration of T_3, as manifested in CCSD(T). This is primarily due to the nature of the exponential wavefunction, since even a restriction to just T_2 gives CCD and by virtue of $1/2(T_2^2)$ which provides "quadruple" replacements, and $1/3!(T_2^3)$, which puts in some "hextuple" ones, etc., *many* higher replacements are naturally introduced into the wavefunction, but with operationally no more difficulty than that for doubles in configuration interaction, CID. Once we add T_1 and T_3 it is apparent that *many* more higher replacements will be further added, providing rapid convergence to the exact results.

The primary weakness in the single-reference CC theory occurs when more than one reference function is particularly important in the wavefunction. This is often called a "multi-reference" situation. Some are well known as in bond breaking when using a restricted Hartree–Fock (RHF) reference function. For H_2, e.g. the RHF function is the determinant σ_g^2, while at complete separation, the determinant, σ_u^2 is equally important. Attempting to get the exact full CI solution based upon the RHF function means the second determinant will be gradually introduced by the T_2 operator until its coefficient has to be equal to that of the first; and this is not a small correction. CCSD can obviously overcome this problem because it is the full CI for two-electrons. But if we have more electrons, then even higher-order T_p operators might have a critical role in a multi-reference problem.

Bond breaking of this sort is a well-known weakness, but one that now has many partial solutions.[2] However, O_3 is an example that has two important determinants even in its zeroth-order description, making a single Φ_0 less appropriate. We can think of these four determinants as a^2, b^2, and ab and ba. If a and b are of different symmetries, then the last two single excitations would not contribute, but the a^2 and b^2 still would. We would like to do two-determinant CC for this, but multi-reference methods are not as easy to do in CC theory as they are in CI.[2] Such MR methods also introduce new complications for the user, since the results will strongly depend upon what choice of reference determinants are considered sufficient. Also, since CC theory is so much more powerful than CI, notably having the essential property of correct scaling with the number of electrons (size-extensivity),[2] the hope is that for even an MR problem like this, standard SR-CC can handle it. That contributes to CC's "black-box" application that is one of its very important properties. There are no choices in SR-CC except the basis set and level of correlation, and this lends itself to detailed calibration.

Again, this work was mostly carried out by John Stanton, now a professor at UT Austin, formerly a student of the Colonel's and later a postdoc of mine. We were also assisted by David Magers, a graduate student of mine and a postdoc of the Colonel's. David is now a professor at Mississippi College.

In determining the vibrational frequencies of O_3, John found something unexpected. CCSD was qualitatively fine for the structure and the frequencies, but not as good as they usually are at this level.[1] Fourth-order MBPT theory, MBPT(4), which consists of single, doubles, triples, and quadruples, was also comparatively good, attesting to the role of triple excitations neglected in CCSD.[1] However, we had previously introduced a non-iterative triples correction, CCSD+T(CCSD),[3] which means we added the non-iterative fourth-order triple correction to CCSD built upon the converged CCSD solutions for T_1 and T_2. Using this, we obtained a structure for O_3 that was not even C_{2v}! Clearly, there is a delicate balance involved. Through fourth-order SDQ-MBPT(4)~CCSD, and the triples in MBPT(4), T4~T(CCSD)=$\langle T_2^{\dagger} W T_3\rangle$, W means the two-electron interaction, and $T_3 = WT_2$. The main difference is that this reflects the infinite-order sums in CCSD in the non-iterative triples evaluation using the converged T_2 from CCSD. This approximation had to be misbalanced compared to MBPT(4).

Using this example, Raghavachari, *et al.*[4] used the slightly modified form, (T)=T(CCSD)+ $\langle T_1^{\dagger} W T_3\rangle$ instead of T(CCSD), and, through my former graduate student Trucks, our CCSD program (the only one in existence then) to do their analysis. From the CC/MBPT viewpoint this did not seem logical at the time. T(CCSD) is the pure fourth-order correction to CCSD for a Hartree–Fock reference function, with the other term arising from single excitations, T_1. This made the latter term fifth-order, and, as such, it was only one of many with comparable magnitudes. It also had only a small numerical effect compared to T(CCSD). But it worked better, and did not break the symmetry of O_3.

Later we offered a general derivation[5] that showed that $(T)=\langle T_2^{\dagger}WT_3\rangle + \langle T_1^{\dagger}WT_3\rangle + \langle T_2^{\dagger}vT_3\rangle$, where v in this term means the single-excitation part of the Hamiltonian. At this level, everything makes perfect sense, since this is the non-Hartree–Fock fourth-order correction. Measured that way, all three terms are indeed fourth-order as one would expect. Furthermore, this most general form admits any kind of reference determinant like restricted open shell Hartree–Fock, or the Brueckner, Natural, or KS determinants mentioned above. This general form also demands a semi-canonical transformation to keep CCSD(T) non-iterative. Only with this feature and the flexibility of reference function could the *many* applications be made to enable CCSD(T) to now be called the "gold-standard" in quantum chemistry!

I might add that extensive studies of ozone's antisymmetric vibrational frequencies continue to be made in the field, as the above work identified this as a test problem that continues to pay dividends to method development.

References

1. Stanton J. F., Lipscomb W. N., Magers D. H., and Bartlett R. J., Highly correlated single-reference studies of the O_3 potential surface. I. Effects of high order excitations on the equilibrium structure and harmonic force field of ozone, *J. Chem. Phys.*, **90**, 1077–1082 (1989).
2. Bartlett R. J. and Musial M., Coupled-cluster theory in quantum chemistry, *Revs. of Modern Phys.*, **79**, 291–352 (2007).
3. Urban M., Noga J., Cole S. J., and Bartlett R. J., Towards a full CCSDT model for electron correlation, *J. Chem. Phys.*, **83**, 4041–4046 (1985).
4. Raghavachari K., Trucks G. W., Pople J. A., Head-Gordon M., A fifth-order perturbation comparison of electron correlation theories, *Chem. Phys. Lett.*, **157**, 479–483 (1989).
5. Watts J. D., Gauss J., and Bartlett R. J., Coupled-cluster methods with noniterative triple excitations for restricted open-shell Hartree–Fock and other general single determinant reference functions. Energies and analytical gradients, *J. Chem. Phys.*, **98**, 8718–8733 (1993).

Highly correlated single-reference studies of the O_3 potential surface. I. Effects of high order excitations on the equilibrium structure and harmonic force field of ozone

John F. Stanton[a)] and William N. Lipscomb
Gibbs Chemical Laboratory, Harvard University, Cambridge, Massachusetts 02138

David H. Magers and Rodney J. Bartlett
Quantum Theory Project, Departments of Chemistry and Physics, University of Florida, Gainesville, Florida 32611

(Received 29 July 1988; accepted 11 October 1988)

The potential energy surface of ozone in the vicinity of the equilibrium geometry is investigated by single-reference many-body perturbation theory (MBPT) and coupled-cluster (CC) methods. As expected from the known inadequacies of the independent-particle picture of O_3, analysis of the CCSDT-1 wave function reveals considerable mixing between the [core$\cdots$]$4b_2^2 6a_1^2 1a_2^2$ and [core$\cdots$]$4b_2^2 6a_1^2 2b_1^2$ configurations. Smaller, but still significant, contributions come from other configurations involving redistribution of electrons within the out-of-plane π orbital framework. As expected, the equilibrium structure and harmonic force field computed at the SCF level of theory are in considerable error. When allowance is made for electron correlation effects, the discrepancies between theory and experiment for the equilibrium structure and totally symmetric force field are significantly reduced, and the MBPT(4), CCSD, CCSD + T(CCSD) and CCSDT-1 results are in reasonable agreement with accepted values. Asymmetric stretching frequencies, however, are found to oscillate wildly [2373 cm^{-1} at MBPT(2), 1547 cm^{-1} at MBPT(4), and 680 cm^{-1} at CCSDT-1] about the experimental value of 1089 cm^{-1}, suggesting that the C_s subspace of the potential surface poses a great challenge to theory. A possible cause of this behavior is discussed, along with implications for future studies of the global O_3 potential surface.

I. INTRODUCTION

It is widely appreciated that the traditional independent-particle model of electronic structure provides an inadequate description of the ground electronic state of ozone.[1] For example, the highly complex and controversial photoelectron spectrum of O_3 (Ref. 2) contains a number of unusually strong "shake-up" bands[2] (which are due to electron correlation effects[3]), and the first two ionization potentials predicted by Koopmans' theorem are *qualitatively* in error. This breakdown of the standard molecular orbital picture is largely due to strong configuration mixing between the lowest energy $^1A_1$ determinant and those obtained from it by redistributing electrons in the allyl-like (out-of-plane) π framework. The importance of the entire π system in the description of electron correlation in ozone has very recently been pointed out by Decleva *et al.*,[4] confirming results we have obtained in this research.

Our purpose in this paper is to explore the effects of high level correlation corrections on the predicted equilibrium structure and harmonic force field of ozone, in order to establish what level of theory is required to give a qualitatively realistic description of the ground state potential surface near the C_{2v} minimum. In the future, we will present results of an ongoing study of the dissociation and isomerization (to the equilateral triangular structure) pathways, as well as anharmonic potential constants for both isomers. We feel that the results obtained in this preliminary study are sufficiently interesting in their own right to warrant separate publication.

II. CALCULATIONS

Calculations reported in this paper were carried out with the ACES program system, developed by Bartlett and collaborators.[5] Two basis sets were used: the standard DZ Dunning–Hay basis set,[6] augmented by six *d*-type polarization functions with our usual oxygen exponent of 1.211 (Ref. 7); and a larger [5s3p2d] basis, which is described in Ref. 8. Using single determinant self-consistent-field (SCF) reference functions formed from these basis sets, correlation corrections were computed by means of many-body perturbation theory (MBPT) and various levels of the coupled-cluster (CC) approximation. Detailed discussions of both methods can be found elsewhere in the literature,[9] and will not be restated here. In correlated calculations, the three *K*-shell molecular orbitals and the three highest virtual functions ($\epsilon > 47$ a.u.) were omitted. The effect of this neglect on the parameters under study is apparently small, as observed in the MBPT(2) all-electron and frozen core results listed in Table I.

At the SCF and MBPT(2) levels of theory, minimum energy structures were determined by Newton–Raphson optimization, using analytic gradients of the energy and approximate second derivatives; at higher levels of theory, they were obtained by successive interpolation of energies calculated at various grid points. These numerical optimizations

[a)] AT&T Foundation Fellow. Current address: Department of Chemistry, University of Chicago, Chicago IL 60637.

TABLE I. Theoretical equilibrium structural parameters, harmonic force constants and harmonic vibrational frequencies calculated at various correlated levels of theory. Bond lengths and angles in angstroms and degrees, respectively; force constants (in symmetry coordinates) in atomic units, and frequencies in cm^{-1}.

	R_e	θ_e	F_{11}	F_{22}	F_{12}	F_{33}	ω_1	ω_2	ω_3
SCF	1.199	119.2	0.933	0.117	0.040	0.496	1547	859	1428
MBPT(2)[a]	1.293	116.7	0.529	0.089	0.040	1.386	1167	740	2368
MBPT(2)[b]	1.294	116.7	0.528	0.089	0.040	1.393	1167	739	2373
SDQ-MBPT(4)	1.261	117.7	0.630	0.093	0.031	0.444	1288	756	1345
MBPT(4)	1.297	117.6	0.465	0.081	0.023	0.588	1123	695	1547
CCSD	1.263	117.4	0.600	0.092	0.032	0.379	1256	748	1240
CCSD + T(CCSD)	1.293	117.0	0.443	0.079	0.024	−0.004	1097	685	128i
CCSD + T(CCSD)[c]	1.291	117.1	···	···	···	0.026	···	···	327
CCSDT-1	1.295	116.6	0.429	0.076	0.026	0.114	1076	674	680
Experiment[d]	1.272	116.8	0.498	0.083	0.036	0.293	1135	716	1089

[a] All electrons correlated.
[b] *K*-shell molecular orbitals and three highest virtual functions omitted.
[c] Calculated with [5s3p2d] basis set. Only F_{33} was evaluated.
[d] Harmonic frequencies from Ref. 14.

were carried out to high precision; errors in reported bond lengths and angles are believed to be < 0.0002 Å and 0.03°, corresponding to residual stretching and bending forces of $\approx 10^{-4}$ hartrees/bohr. Harmonic force constants in terms of the symmetry coordinates recommended by Cyvin[10] were also obtained numerically using double-sided numerical differentiation and the following relations (S_e denotes the minimum energy configuration):

$$F_{ii} \approx \frac{E(S_e + \delta_{S_i}) + E(S_e - \delta_{S_i}) - 2E(S_e)}{\delta_{S_i}^2} \quad (i = 1,2) \tag{1}$$

and

$$F_{12} \approx \frac{E(S_e + \delta_{S_1} + \delta_{S_2}) + E(S_e - \delta_{S_1} - \delta_{S_2}) - 2E(S_e) - F_{11}\delta_{S_1}^2 - F_{22}\delta_{S_2}^2}{2\delta_{S_1}\delta_{S_2}}. \tag{2}$$

Computation of the asymmetric force constant made use of the expression[11]

$$F_{33} = \frac{F_{yy}\csc^2(\theta/2)}{2}, \tag{3}$$

where F_{yy} is the second derivative of the energy with respect to displacement of the apical oxygen atom perpendicular to the rotational axis and in the molecular plane. Evaluation of F_{yy} via Eq. (1) required evaluation of the energy at only one point (aside from the equilibrium structure). Consequently, seven calculations of the energy at configurations displaced from the minimum energy structure are sufficient to determine the entire harmonic force field, and only one of these calculations need be carried out in a symmetry lower than C_{2v}. Harmonic frequencies were calculated from the force constants and equilibrium structures via the *FG* matrix formalism of Wilson, Decius, and Cross.[12] Of course, numerical determination of force constants was not necessary in the SCF and MBPT(2) calculations reported below, but these values are reported for the sake of consistency. In all cases, differences between analytic and numerical force constants at these levels of theory[13] were well below 1%, suggesting a comparable precision in the more sophisticated MBPT and CC calculations which are the focus of this paper.

The SCF configuration for ozone obtained with the DZP basis is

$$[\text{core}]3a_1^2 2b_2^2 4a_1^2 5a_1^2 3b_2^2 1b_1^2 4b_2^2 6a_1^2 1a_2^2 2b_1^0. \tag{4}$$

TABLE II. Self-consistent-field orbital energies for occupied and low-lying virtual orbitals in ozone. Eigenvalues are in atomic units, and were calculated at the optimum CCSDT-1 geometry with the DZP basis set. Orbitals belonging to the a_2 and b_1 representations constitute the out-of-plane π framework.

	Orbital	ϵ_{SCF}
Occupied		
	$1a_1$	−20.904 73
	$1b_2$	−20.709 69
	$2a_1$	−20.709 63
	$3a_1$	−1.736 44
	$2b_2$	−1.424 48
	$4a_1$	−1.101 25
	$5a_1$	−0.824 90
	$3b_2$	−0.794 62
	$1b_1$	−0.771 02
	$4b_2$	−0.567 28
	$6a_1$	−0.556 59
	$1a_2$	−0.485 62
Virtual		
	$2b_1$	−0.059 69
	$7a_1$	0.262 55
	$5b_2$	0.341 68
	$8a_1$	0.699 12
	$6b_2$	0.747 09
	$3b_1$	0.765 97
	$9a_1$	0.793 22
	$7b_2$	0.845 97
	$10a_1$	0.878 11
	$2a_2$	0.890 94

Orbital eigenvalues obtained at the CCSDT-1 minimum energy geometry are given in Table II.

III. RESULTS

A. Finite order MBPT calculations

Theoretical equilibrium structures, quadratic force constants, and harmonic vibrational frequencies calculated at various correlated levels are presented along with the corresponding experimental values[14] in Table I. In agreement with a number of earlier investigations,[1] errors in the SCF harmonic force field are large, with the two stretching force constants (F_{11} and F_{33}) overestimated by nearly a factor of 2. When electron correlation is incorporated at the level of second order MBPT, the symmetric stretch and bend (F_{22}) constants are brought into reasonable agreement with experiment, but the asymmetric stretching constant becomes unrealistically large. The resulting 117% error in the B_2 stretching frequency is undoubtedly one of the largest ever obtained in a correlated calculation. The problems associated with theoretical determination of F_{33} have previously been recognized by Lee *et al.*,[1(a)] who noted that "it would seem that the major difficulty is in the *ab initio* prediction of ω_3, and not ω_1." In their calculations, they found that configuration interaction calculations including singly and doubly substituted configurations (CISD) brought ω_1 into reasonable agreement with experiment, but continued to overestimate ω_3 by $\sim 30\%$, even when a two-reference (2R) CISD calculation was performed. The observation that their CISD and 2R-CISD frequencies were considerably better than those calculated from our MBPT(2) harmonic force field led us to speculate that singly (and, to a lesser extent, quadruply) excited determinants were required to describe the behavior of ozone as it becomes distorted from its C_{2v} equilibrium symmetry. One of the simplest levels of theory which includes these effects is the SDQ-MBPT(4) model, which treats single, double, and quadruple excitation contributions to the fourth order MBPT energy, while neglecting triply substituted determinants. From Table I, it can be seen that this improvement of the theoretical model leads to a more reasonable value of F_{33}. Coupled with the results of Lee *et al.*, this finding constitutes an apparent confirmation of the importance of single (and quadruple) excitations in this problem. Although the value of F_{11} is somewhat worse than the MBPT(2) value, the agreement of the latter is almost certainly fortuitous.

In the exact fourth order MBPT treatment, where the effects of triple excitations first enter the perturbation expansion of the correlation energy, a curious behavior is noted. The inclusion of triply excited configurations significantly reduces the totally symmetric harmonic frequencies [relative to the SDQ-MBPT(4) values], bringing them into excellent agreement with the experimental values. The mean absolute error for ω_1 and ω_2 is 2.1%, nearly a 15-fold improvement over the SCF results and representative of the accuracy obtained at this level for other systems. By contrast, the inclusion of triple excitations leads to an *increase* in the asymmetric stretching frequency. At this level of theory, the predicted ω_3 (1547 cm^{-1}) is 42% above the experimental value of 1089 cm^{-1}, and is even higher than the SCF result. This finding complements the observations of Lee *et al.*, confirming their speculation that this region of the potential represents a difficult theoretical problem. Clearly, the higher order corrections included in CC models are needed to properly describe O_3 as it is displaced along the S_3 stretching coordinate.

Predicted equilibrium geometries are also very sensitive to electron correlation effects. At the SCF level, the overestimation of the O–O bond stiffness results in a minimum energy internuclear distance which is considerably shorter than the equilibrium value inferred from experiment [14(a)] Following the pattern observed in most molecules, incorporation of electron correlation acts to make the bonds less rigid, resulting in a less compact structure.[15] In ozone, the bond lengthening is principally due to the strong mixing of the $2b_1$ virtual orbital into the molecular wave function, an effect which acts to deplete the electron density in the classical bonding regions. Although the structural parameters obtained at the SDQ-MBPT(4) and MBPT(4) levels are reasonably close to the experimental values, discrepancies are nevertheless much larger than those typically found in most molecular systems. This result, together with the poor agreement between MBPT(4) and experiment for ω_3 noted above, led us to extend our study to the level of coupled-cluster theory, where many contributions to the exact electronic wave function are treated exactly.

B. Coupled-cluster calculations

Also listed in Table I are results obtained with various coupled-cluster models. In CC theory, the correlated wave function is given by

$$|\Psi\rangle = \exp(T)|\Phi_0\rangle, \tag{5}$$

where $|\Phi_0\rangle$ is the reference function (the single SCF determinant in the present study), and T is an excitation operator usually expressed as $T = T_1 + T_2 + T_3 + \cdots$, where T_j is a j-particle excitation operator. A complete CC calculation provides a formally exact solution to the electronic Schrödinger equation within the specified basis set, and is equivalent to full configuration interaction. In practice, however, one usually works with a truncated excitation operator $T = T_1 + T_2 + \cdots + T_n$. In this paper, we consider three models: the CCSD approximation[16] (in which $T = T_1 + T_2$), CCSDT-1 [$\exp(T) \approx \exp(T_1 + T_2) + T_3$],[17] and CCSD + T(CCSD),[18] in which T(CCSD) is the first order perturbative correction to the CCSD energy. Due to the exponential nature of the excitation operator, determinants corresponding to all levels of excitation are incorporated into the correlated wave function, although the higher order terms are treated approximately. For example, CCSD treats singly and doubly excited determinants to infinite order in perturbation theory, while product terms introduce higher levels of excitation. Most important among the product terms is $T_2^2/2$, the disconnected quadruple excitation term which has an established importance in molecular systems.[19] The CC models which incorporate triple excitations [CCSD + T(CCSD), CCSDT-1, and the complete CCSDT model] provide a nearly quantitative accounting of the correlation energy; differences between results obtained with these methods and full CI calculations are typically less than 2%, even for systems exhibiting significant multireference character.[17,18,20]

The CC results in Table I beautifully illustrate the striking qualitative difference between the behavior of the totally symmetric and asymmetric regions of the O_3 potential surface in the vicinity of the minimum energy configuration. Structural parameters and totally symmetric force constants obtained with the various CC models are in good agreement with their fourth order MBPT counterparts (CCSD/SDQ-MBPT(4); CCSDT-1 [or CCSD + T(CCSD)]/MBPT(4)).[21] The asymmetric stretching force constants differ significantly, however, particularly those obtained with the models incorporating triple excitation effects. As noted in the previous section, the infinite-order effects included in CC calculations appear to play a particularly important role in the description of the asymmetric stretching coordinate. Note that while the inclusion of triple excitations at fourth order MBPT results in an increase in F_{33}, the opposite effect is observed in the CC calculations. Most bizarre is the (incorrect) implication of the DZP CCSD + T(CCSD) calculation—that the accepted C_{2v} structure of ozone is actually a transition state for interconversion of two equivalent C_s structures. Few other cases are known in which a highly sophisticated calculation has produced such a grossly incorrect answer for a ground state molecule near its equilibrium geometry.[22] Expansion of the basis from DZP to [5s3p2d] leads to the correct equilibrium point group symmetry, but still predicts an absurdly small resistance to asymmetric distortion. Some of the troubles with the CCSD + T(CCSD) method in this example can be traced to the large changes in the T_2 amplitudes which result when T_3 is included in the CC equations (see Table III). Since T(CCSD) is a perturbative measure of the first order correction to the CCSD model, it is only suitable when the changes due to the iterative inclusion of T_3 into the CC equations are not important, a condition satisfied by most molecules we have studied previously. In the present case, however, the dominant $1a_2^2 \rightarrow 2b_1^2\, T_2$ amplitude increases by nearly 50% when T_3 is included at the CCSDT-1 level, suggesting that the failure of the T(CCSD) model may be blamed, in part, on this effect. The contribution $\partial^2 T(\mathrm{CCSD})/\partial S_3^2$ to the asymmetric force constant has the proper sign, but its magnitude is much too large. Evidently, this breakdown of the T(CCSD) model is not so serious when one deals with the totally symmetric region of the potential surface.

Our most extensive calculations were those carried out with the CCSDT-1 model, which does include some effects of T_3 iteratively. As mentioned above, the symmetric stretch and bend (ω_1 and ω_2) harmonic frequencies are close to (but somewhat worse than) the finite order MBPT(4) values. Disappointingly, however, the asymmetric stretching frequency obtained at this level of theory is only slightly better than the [5s3p2d] CCSD + T(CCSD) result and is more than 400 cm^{-1} below experiment. Here, the nondynamical nature of the correlation problem in ozone is particularly conspicuous. Inspection of the T_1 and T_2 amplitudes in the CCSDT-1 wave function at asymmetrically displaced struc-

TABLE III. Dominant T_1 and T_2 amplitudes contributing to the CCSD and CCSDT-1 wave functions, and total molecular energies calculated at various correlated levels. Energies are in atomic units, and were computed at the CCSDT-1 minimum energy geometry.

T_1 amplitudes			
i	a	t_i^a(CCSD)	t_i^a(CCSDT-1)
$1b_1$	$2b_1$	0.0824	0.0881
$6a_1$	$7a_1$	0.0628	0.0663
$4b_2$	$5b_2$	0.0458	0.0488
$1a_2$	$2a_2$	0.0320	0.0320

T_2 amplitudes					
i	j	a	b	t_{ij}^{ab}(CCSD)	t_{ij}^{ab}(CCSDT-1)
$1a_2$	$1a_2$	$2b_1$	$2b_1$	0.2163	0.3206
$1b_1$	$1b_1$	$2b_1$	$2b_1$	0.0771	0.0886
$1a_2$	$6a_1$	$2b_1$	$5b_2$	0.0385	0.0504
$1a_2$	$4b_2$	$2b_1$	$7a_1$	0.0348	0.0491
$1a_2$	$1a_2$	$2b_1$	$3b_1$	0.0487	0.0463

Level of theory	Energy
SCF	− 224.295 920
MBPT(2)	− 224.931 924
MBPT(3)	− 224.888 104
D-MBPT(4)	− 224.918 096
SD-MBPT(4)	− 224.935 573
SDQ-MBPT(4)	− 224.907 514
MBPT(4)	− 224.952 783
CCSD	− 224.906 346
CCSD + T(CCSD)	− 224.944 831
CCSDT-1	− 224.945 859

tures provides some insight into the nature of the problem. Already, at $S_3 \approx 0.03$ Å, two of the four most important singly excited non-SCF contributions to $\Psi_{\text{CCSDT-1}}$ correspond to configurations which cannot interact with the ground state SCF wave function in C_{2v} symmetry ($1a_2 \rightarrow 2b_1$ and $6a_1 \rightarrow 5b_2$). The contribution of these determinants changes in first order along the B_2 symmetry coordinate, and the precipitious rise in the corresponding T_1 amplitudes (>1 Å^{-1}) suggests that these configurations become heavily mixed with Φ_0 in the C_s part of the O_3 potential surface. Consequently, for asymmetric structures of O_3, at least four configurations appear to make substantial contributions to Ψ. As a result, determinants which are excited with respect to these strongly contributing non-SCF configurations become important as well, and the high order excitation effects which are only treated approximately at this level of theory can be significant. If this analysis is true, considerably greater demand is placed on the theoretical approach than in the totally symmetric subspace, where the form of the exact molecular wave function is dominated by "only" two configurations. Evaluation of F_{33}, which requires a balanced treatment of correlation in both the C_{2v} and C_s subspaces, appears to present a formidable problem to single-reference CC theory.

The totally symmetric force constants, by contrast, depend only on the C_{2v} potential surface. Here, the CCSDT-1 model appears to be sufficiently robust to recover most of the correlation effects from the relatively poor single configuration starting point, and predicts ω_1 and ω_2 with a reasonable degree of accuracy. This result is consistent with studies of single-bond breaking,[23] a process which can be qualitatively described by two reference functions of the Heitler–London variety, in which single-reference CCSDT-1 calculations have been shown to provide a realistic description of the potential function over a wide range of internuclear distances.

Clearly, it is necessary to go beyond CCSDT-1 to obtain a reliable asymmetric stretching frequency for ozone. Related problems were observed in a study of the beryllium dimer,[24] where models which do not include T_3 in the triple excitation CC equation (such as CCSDT-1) yielded only a single minimum on the potential curve. The inner well of the double minimum potential appeared only when these effects were included.

IV. CONCLUSIONS

It appears that highly correlated models of electron correlation which incorporate triple excitation effects are capable of predicting the totally symmetric harmonic frequencies of ozone to within a few percent, even when a single determinant reference function is used. The only results in the literature that are of comparable accuracy are those obtained in a complete-active-space SCF (CASSCF) study by Adler-Golden *et al.*[25] While it could be argued that calculations based on a single determinant are not sufficiently rigorous to characterize a system known to be poorly described by the independent-particle approximation, we wish to point out that our CCSDT-1 symmetric stretch and bend harmonic frequencies and minimum energy structure are in very good agreement with the CASSCF calculations of Ref. 25 (see Table IV), and are significantly different from the double-reference CISD result of Lee *et al.*[1(a)] Both of these studies used DZP basis sets differing from ours only in the choice of polarization exponents; consequently, the comparison is appropriate. Furthermore, results obtained in the CASSCF and present studies are generally closer to experimental values than are the 2R-CISD results, which compare favorably with those obtained at the CCSD level. Hence, it appears that although the SCF and $1a_2^2 \rightarrow 2b_1^2$ configurations are the principal contributors to the molecular wave function, a CISD calculation based on these two configurations is not adequate to properly treat the correlation problem in ozone. Consequently, it would appear that a larger number of reference functions[26] need to be included in truncated CI treatments.

All of the properties mentioned above, however, are independent of the C_s region of the potential surface. As discussed in the previous section, our single-reference based CC/MBPT calculations fail to predict the asymmetric stretching frequency with any accuracy. The CASSCF value of Ref. 25 is much closer to the experimental value, but is still nearly 10% too low. The superior result obtained in the CASSCF study attests to the powerful treatment of nondynamic correlation which can be achieved in multireference (MR) calculations which use a large configuration space, and suggests that MR-CC methods might be required to accurately characterize asymmetric structures of ozone. If all of the strongly contributing configurations are included, the exponential wave operator should treat the remaining dynamic correlation effects adequately, yielding better results. It is also possible that CC methods which make fewer approximations in the treatment of triple excitations might provide a more realistic view of the C_s potential surface. For example, CCSDT-2[18] includes the effect of T_2^2 on the T_3 amplitudes, as well as that of T_1T_3 on the T_2 amplitudes. Both additional contributions are fifth order in perturbation theory, and the possible importance of the first is suggested by the very large $1a_2^2 \rightarrow 2b_1^2\, T_2$ amplitude, which can combine with many other double excitation contributions to produce unusually large values of $t_{1a_21a_2}^{2b_12b_1} t_{ij}^{ab}$. Consequently, we believe that this improvement of the theoretical model may help to overcome some of the glaring inadequacies of the CCSD + T(CCSD) and CCSDT-1 treatments. However, as suggested by the study of Be_2, it might be necessary to include the direct effect of T_3 on the T_3 amplitudes, using the full CCSDT model. Results of MR and more extensive sin-

TABLE IV. Comparison of CCSD and CCSDT-1 minimum energy structures, dipole moments, and harmonic frequencies with CASSCF and 2R-CISD calculations.[a]

Property	2R-CISD	CASSCF	CCSD	CCSDT-1	Expt.
R_e (Å)	1.271	1.296	1.263	1.295	1.272
θ_e (deg)	116.2	116.5	117.4	116.6	116.8
ω_1	1234	1098	1256	1076	1135
ω_2	745	689	748	674	716
ω_3	1352	989	1240	680	1089

[a] Values taken from Ref. 1(a).

gle-reference calculations will be reported in a subsequent paper.

Whether the remaining discrepancies in the symmetric harmonic force field and, in particular, the theoretical equilibrium structure are due to basis set deficiency or residual correlation effects is unclear. Expansion of the basis from DZP to [5s3p2d] has an insignificant effect on the CCSD + T(CCSD) "equilibrium" (C_{2v}) geometry.[27] Hence, it is possible that higher angular momentum functions need to be included. On the other hand, the inadequacy of the SCF wave function coupled with the pronounced sensitivity of the predicted molecular structure and properties to triple excitation effects suggests that higher order excitations should have great importance in ozone. In view of the great significance of ozone in atmospheric chemistry, we believe that it is important to establish what level of theory is required to describe this molecule quantitatively. In Table III, we present dominant T_1 and T_2 amplitudes, as well as MBPT and CC energies computed at the CCSDT-1 minimum energy geometry, in the hope that others will become interested in this problem. In particular, a full CI calculation of the properties discussed in this paper would be extremely useful. Although it would certainly present a formidable technical challenge, the results would serve to clarify what demands this difficult molecule places on the theoretical model.

Although the present work deals solely with O_3 in the vicinity of the equilibrium geometry, the information presented in this paper serves as a guide to more global studies of the potential surface. We have shown that MBPT and CC calculations appear to provide an adequate description of the totally symmetric subspace of the O_3 potential surface, but are not as reliable when the molecular symmetry is lowered to C_s. Consequently, it appears that the rearrangement pathway between the open isomer of O_3 and the D_{3h} ring structure present a more tractable theoretical problem than does the unimolecular dissociation to O_2 and O. In the future, we will present results of ongoing research on both reaction systems, using sophisticated theoretical methods.

ACKNOWLEDGMENTS

Aspects of this research have been supported by the National Science Foundation (Grant No. CHE85-15347) and the Office of Naval Research. We also thank the Pittsburgh Supercomputing Center for a generous allocation of time on the Cray XMP/48, on which most of our calculations were performed.

[1](a) T. J. Lee, W. D. Allen, and H. F. Schaefer, J. Chem. Phys. **87**, 7063 (1987); (b) M. V. Rama Krishna and K. D. Jordan, Chem. Phys. **115**, 423 (1987); (c) Y. Yamaguchi, M. J. Frisch, T. J. Lee, H. F. Schaefer, and J. S. Binkley, Theor. Chim. Acta **69**, 337 (1986); (d) W. D. Laidig and H. F. Schaefer, J. Chem. Phys. **74**, 3411 (1981); (e) G. Karlström, S. Engström, and B. Jönsson, Chem. Phys. Lett. **57**, 390 (1978); (f) P. J. Hay and T. H. Dunning, J. Chem. Phys. **67**, 2290 (1977); (g) P. J. Hay, T. H. Dunning, and W. A. Goddard, *ibid.* **62**, 3912 (1975); (h) W. A. Goddard, W. J. Hunt, and P. J. Hay, Acc. Chem. Res. **6**, 398 (1973); (i) R. J. Blint and M. D. Newton, J. Chem. Phys. **59**, 6220 (1973); (j) P. J. Hay, T. H. Dunning, and H. F. Schaefer, Chem. Phys. Lett. **23**, 457 (1973); (k) S. Rothenberg and H. F. Schaefer, Mol. Phys. **21**, 317 (1971); and citations included in these references.

[2](a) T. Cvitaš, L. Klasnic, and B. Kovač, Int. J. Quantum Chem. **29**, 657 (1986); (b) J. M. Dyke, L. Golob, N. Jonathan, A. Morris, and M. Okuda, J. Chem. Soc. Faraday Trans. **70**, 1828 (1974); (c) C. R. Brundle, Chem. Phys. Lett. **26**, 25 (1974); (d) D. L. Frost, S. T. Lee, and C. A. McDowell, *ibid.* 24, 149 (1974); (e) T. N. Radwan and D. W. Turner, J. Chem. Soc. A **85** (1966).

[3]For a detailed discussion of electron correlation effects in photoelectron spectra, see G. Vendlin, in *Structure and Bonding* (Springer, Berlin, 1981), Vol. 45, pp. 1–117.

[4]P. Decleva, G. DeAlti, and A. Lisini, J. Chem. Phys. **89**, 367 (1988).

[5]R. J. Bartlett, G. D. Purvis, G. B. Fitzgerald, R. J. Harrison, Y. S. Lee, W. D. Laidig, S. J. Cole, D. H. Magers, E. A. Salter, G. W. Trucks, C. Sosa, M. Rittby, and S. Pal, ACES (Advanced Concepts in Electron Structure)-An Ab Initio Program System.

[6]T. H. Dunning and P. J. Hay, in *Methods of Electron Structure Theory*, edited by H. F. Schaefer (Plenum, New York, 1977), Vol. 3.

[7]L. T. Redmon, G. D. Purvis, and R. J. Bartlett, J. Am. Chem. Soc. **101**, 2856 (1979).

[8]J. F. Stanton and D. H. Magers, J. Chem. Phys. **88**, 7650 (1988).

[9]For extensive reviews as well as a discussion of the theories, see (a) R. J. Bartlett, Annu. Rev. Phys. Chem. **32**, 359 (1981) (MBPT); (b) R. J. Bartlett, C. E. Dykstra, and J. Paldus, in *Advances Theories and Computational Approaches to the Electronic Structure of Molecules*, edited by C. E. Dykstra (Reidel, Dordrecht, 1983).

[10]S. Cyvin, *Molecular Vibrations and Mean-Square Amplitudes* (Elsevier, Amsterdam, 1968), p. 111.

[11]Equation (3), which holds rigorously at a stationary point on the potential surface, is easily derived from the transformation matrix relating internal and Cartesian coordinates. Elements of this matrix have been given explicitly in the literature (see Ref. 10). We should point out that while F_{33} is calculated in terms of rectilinear symmetry coordinates, elements of the totally symmetric harmonic force field were computed in terms of curvilinear coordinates. At a stationary point, *harmonic* force constants calculated in the two coordinate systems are identical. In the present case, residual forces are quite small, and differences between curvilinear and rectilinear quadratic constants are negligible.

[12]E. B. Wilson, J. C. Decius, and P. C. Cross, *Molecular Vibrations* (Dover, New York, 1981).

[13]Methods used for MBPT(2) second derivative evaluation in the ACES program system (see Ref. 5) are described in R. J. Harrison, G. B. Fitzgerald, W. D. Laidig, and R. J. Bartlett, Chem. Phys. Lett. **124**, 291 (1986); N. C. Handy, R. D. Amos, J. F. Gaw, J. E. Fice, E. D. Simandras, T. J. Lee, R. J. Harrison, W. D. Laidig, G. B. Fitzgerald, and R. J. Bartlett, in *Geometrical Derivatives of Energy Surfaces and Molecular Properties*, edited by P. Jørgensen and J. Simons (Reidel, Dordrecht, 1986), p. 179.

[14](a) T. Tanaka and Y. Morino, J. Mol. Spectrosc. **33**, 538 (1974) (equilibrium structure); (b) A. Barbe, C. Secroun, and P. Jouve, *ibid.* **49**, 171 (1974) (harmonic force field).

[15]For representative examples, see D. J. DeFrees, K. Raghavachari, H. B. Schlegel, and J. A. Pople, J. Am. Chem. Soc. **104**, 5576 (1980).

[16]G. D. Purvis and R. J. Bartlett, J. Chem. Phys. **76**, 1910 (1982).

[17]Y. S. Lee and R. J. Bartlett, J. Chem. Phys. **80**, 4371 (1984); Y. S. Lee, S. A. Kucharski, and R. J. Bartlett, *ibid.* **81**, 5906 (1985); **82**, 5761 (1985).

[18]M. Urban, J. Noga, S. J. Cole, and R. J. Bartlett, J. Chem. Phys. **83**, 4041 (1985).

[19]W. J. Hehre, J. A. Pople, P.v.R. Schleyer, and L. Radom, *Ab initio Molecular Orbital Theory* (Wiley, New York, 1986), and references therein.

[20]S. J. Cole and R. J. Bartlett, J. Chem. Phys. **86**, 873 (1987).

[21]SDQ-MBPT(4) and MBPT(4) energies are obtained from the low-order iterations of the CC equations [see Ref. 9(b)].

[22]For a discussion of some other pathological cases, see E. R. Davidson, Int. J. Quantum Chem. Symp. **15**, 65 (1981).

[23]W. D. Laidig, P. Saxe, and R. J. Bartlett, J. Chem. Phys. **86**, 887 (1987).

[24]C. Sosa, J. Noga, and R. J. Bartlett, J. Chem. Phys. **88**, 5974 (1988).

[25]S. M. Alder-Golden, S. R. Langhoff, C. W. Bauschlicher, and G. D. Carney, J. Chem. Phys. **83**, 255 (1985).

[26]Rama Krishna and Jordan [Ref. 1(b)] have studied the dissociation energy of ozone in a [$5s3p1d$] basis using 8R-CISD calculations. Their result for the dissociation energy (0.59 eV) is in good agreement with the CCSDT-1 value of 0.61 eV [J. F. Stanton, D. H. Magers, and R. J. Bartlett (unpublished)]. Both values, however, are far below the experimental dissociation energy of 1.13 eV.

[27]The actual CCSD + T(CCSD) equilibrium structure has C_s symmetry.

Chapter 15

Stephen K. Burley

Biography

Stephen K. Burley is an expert in structural biology, structural genomics, structure-/fragment-based drug discovery, and clinical medicine/oncology. He received his undergraduate education at the University of Western Ontario, where he studied physics and mathematics. As a Rhodes Scholar, elected from Ontario, Canada (Exeter College, 1980), Burley received a D.Phil. in molecular biophysics from Oxford University. He carried out his doctoral thesis research on molecular arrangement in viruses in the laboratory of the late Sir David C. Phillips under the direction of Dr. Andrew Miller. In 1983, Burley emigrated to the US to attend Harvard Medical School, where he received the degree of M.D. *magna cum laude* in the joint Harvard–Massachusetts Institute of Technology Health Sciences and Technology program. While a medical student, he also carried out postdoctoral research with Professor Gregory A. Petsko in the Department of Chemistry at MIT. Burley remained in Boston for internship/residency training in internal medicine at the Brigham and Women's Hospital. As a member of the "hemi-doc" program, he divided his time between clinical medicine and postdoctoral research with Nobel Laureate Professor William N. Lipscomb in the Department of Chemistry at Harvard University. With Lipscomb, he determined the three-dimensional structure of the first aminopeptidase, leucine aminopeptidase from bovine lens.

In late 1990, Burley joined the Rockefeller University as an assistant professor and an assistant investigator in the Howard Hughes Medical Institute, and was promoted to full professor and full investigator in 1994. At Rockefeller, he focused his research efforts on the structural biology of eukaryotic gene expression, determining three-dimensional structures of numerous proteins involved in transcription, mRNA splicing, and translation. In 1997, Burley was awarded a professorial chair endowed by Richard M. and Isabel P. Furlaud. He also served as chief academic officer of the university between 1999 and 2002. In 2000, Burley helped to establish the New York Structural Genomics Research Center, and led a large-scale, high-throughput structure determination facility funded by the National Institutes of Health–National Institute of General Medical Sciences (NIH–NIGMS) Protein Structure Initiative.

In early 2002, Burley left academe to become the chief scientific officer at SGX Pharmaceuticals, Inc., a San Diego structure-guided/fragment-based drug discovery company focused on oncology. The company conducted an initial public offering in early 2006, and took two "home-grown" small-molecule kinase inhibitors into the clinic in 2007 (SGX523, an exquisitely selective c-MET inhibitor for various solid tumor indications) and 2008 (SGX393, a BCR-ABL inhibitor for imatinib-resistant chronic myeloid leukemia). In August 2008, SGX was acquired by Eli Lilly and Company. Throughout his tenure at SGX, Burley continued to lead the NIH-funded Structural Genomics Research Center, which was renamed the New York SGX Research Center for Structural Genomics.

In September 2008, Burley joined Lilly Research Laboratories as a Distinguished Lilly Research Scholar. He has authored or coauthored more than 200 scholarly scientific articles and deposited more than 800 experimental protein structures into the Protein Data Bank. He is a fellow of the Royal Society of Canada and of the New York Academy of Sciences, a recipient of numerous awards and prizes, and serves on various editorial and not-for-profit scientific advisory boards.

Remembering the Colonel

My tenure in the Lipscomb laboratory (a.k.a. Gibbs), June 1987–December 1990, was one of the most rewarding periods of my scientific career. I divided my time between a postdoctoral fellowship with the "Colonel" focused on determining the structure of bovine lens leucine aminopeptidase (LAP) and internship/residency training at the Brigham and Women's Hospital in Boston. Neither was easy, but both provided unparalleled opportunities for intellectual growth. The work on LAP, a collaboration with Drs. Peter R. David and Allan Taylor, was plagued by many of the challenges of protein crystallography before the modern age. The protein had to be purified from the natural source, crystals were small, fragile, and scarce, heavy atom derivatives yet more elusive, and access to X-ray instrumentation problematic. After much struggling, a beautiful physiologic hexamer with 32 symmetry emerged from the experimental electron density map, resulting in the *Proceedings of the National Academy of Sciences* publication reprinted in this volume. It was the first aminopeptidase structure, the first de novo protein structure determination of my career, and definitely the most challenging of the many hundreds I have subsequently deposited into the Protein Data Bank.

My relationship with the Colonel was that of both postdoctoral fellow and consulting physician. An interesting combination to be sure, but my unwavering commitment to patient confidentiality stops me from going further. Management in the Lipscomb lab was also very interesting. The son of a highly successful psychiatrist, the Colonel took great pains to understand what made his charges tick. Exploiting such insights, he worked hard to motivate each lab member, exhorting them to succeed with a mix of encouragement and not always so subtle, yet always well-tailored, psychological pressure. I never knew I had it in me to work as hard as I did on LAP!

Training is all about lessons. Two important lessons from the Lipscomb laboratory have stayed with me. Above all, I solved a "hard problem." In tackling various challenges in my academic and industrial careers, I always felt that I had the advantage of knowing I could handle whatever nature was going to throw at me. Second, the Colonel explained how he made difficult decisions, saying, "I determine what is best for science, and I follow that course without deviation." This too stood me in good stead as I navigated my way through The Rockefeller University, then the biotechnology industry, and now a large pharmaceutical company.

Thank you Colonel, I would not have wanted it any other way.

Happy 90th birthday!

Proc. Natl. Acad. Sci. USA
Vol. 87, pp. 6878–6882, September 1990
Biochemistry

Molecular structure of leucine aminopeptidase at 2.7-Å resolution

(x-ray crystallography/zinc enzyme/exopeptidase/bestatin)

STEPHEN K. BURLEY*†, PETER R. DAVID*, ALLEN TAYLOR‡, AND WILLIAM N. LIPSCOMB*

*Gibbs Chemical Laboratory, Harvard University, Cambridge, MA 02138; †Department of Medicine, Brigham and Women's Hospital, Boston, MA 02115; and ‡U.S. Department of Agriculture Human Nutrition Research Center on Aging, Tufts University, Boston, MA 02111

Contributed by William N. Lipscomb, June 26, 1990

ABSTRACT **The three-dimensional structure of bovine lens leucine aminopeptidase (EC 3.4.11.1) complexed with bestatin, a slow-binding inhibitor, has been solved to 3.0-Å resolution by the multiple isomorphous replacement method with phase combination and density modification. In addition, the structure of the isomorphous native enzyme has been refined at 2.7-Å resolution, and the current crystallographic *R* factor is 0.169 for a model that includes the two zinc ions and all 487 amino acid residues comprising the asymmetric unit. The enzyme is physiologically active as a hexamer, which has 32 symmetry and is triangular in shape with a triangle edge length of 115 Å and maximal thickness of 90 Å. The monomers are crystallographically equivalent and each is folded into two unequal α/β domains connected by an α-helix to give a comma-like shape with approximate maximal dimensions of 90 × 55 × 55 Å^3. The secondary structural composition is 40% α-helix and 19% β-strand. The N-terminal domain (160 amino acids) mediates trimer–trimer interactions and does not appear to participate directly in catalysis. The C-terminal domain (327 amino acids) is responsible for catalysis and binds the two zinc ions, which are 2.88 Å apart. The pair of metal ions is located near the edge of an eight-stranded, saddle-shaped β-sheet. One zinc ion is coordinated by carboxylate oxygen atoms of Asp-255, Asp-332, and Glu-334 and the carbonyl oxygen of Asp-332. The other zinc ion is coordinated by the carboxylate oxygen atoms of Asp-255, Asp-273, and Glu-334. The active site also contains two positively charged residues, Lys-250 and Arg-336. The six active sites are themselves located in the interior of the hexamer, where they line a disk-shaped cavity of radius 15 Å and thickness 10 Å. Access to this cavity is provided by solvent channels that run along the twofold symmetry axes.**

The aminopeptidases form a group of exopeptidases that catalyze removal of amino acids from the N terminus of a protein. These enzymes are ubiquitous in nature and are of critical biological and medical importance because of their key role in protein degradation and in the metabolism of biologically active peptides. Whereas the mechanism of action and the three-dimensional structures of carboxypeptidases and endopeptidases are known in significant detail, our understanding of the aminopeptidases is much less well developed.

Leucine aminopeptidases (LAPs; EC 3.4.11.1) are widely distributed cytosolic exopeptidases that catalyze the hydrolysis of amino acids from the N terminus of polypeptide chains (1, 2). As their name implies, the LAPs cleave leucyl substrates, although substantial rates of enzymatic cleavage are seen with most amino-terminal amino acids. Bovine lens LAP is a hexameric enzyme of molecular weight 324,000, which consists of six identical subunits (3) of molecular weight 54,000 and 12 zinc ions (4). The amino acid sequence of the protein has been determined by both protein (5) and DNA (B. Wallner and A.T., unpublished data) sequencing methods. In addition, a potent, slow-binding inhibitor of LAP, bestatin or [(2*S*,3*R*)-3-amino-2-hydroxy-4-phenylbutanoyl]leucine, was isolated from culture filtrates of *Streptomyces olivoreticuli* (6) and was shown to have a $K_i = 2 \times 10^{-8}$ M for bovine lens LAP (7).

Although there is considerable interest in LAP and in other aminopeptidases, no structural data are available for any aminopeptidase at high resolution. Two electron microscopic studies of bovine lens LAP have shown that the enzyme is a hexamer with 32 symmetry (8, 9). Previous crystallographic studies have been limited to crystallization of bovine lens LAP (10), hog kidney LAP (11), and *Escherichia coli* methionine aminopeptidase (12) and a description of an electron-density map of bovine lens LAP determined by single isomorphous replacement and solvent flattening at 5-Å resolution (13).

We have solved the three-dimensional structure of the bestatin-inhibited enzyme by using the multiple isomorphous replacement (MIR) method at 3.0-Å resolution with phase combination and density modification and have refined the structure of the isomorphous native enzyme at 2.7-Å resolution. The active site has been identified and the environments of the two zinc ions have been characterized. Here, we report the general structural features of the enzyme; its secondary, tertiary, and quaternary structures; and the structure of the active site.§ A detailed analysis of the mechanism of inhibition of the enzyme by bestatin and the results of refinement of both the native and bestatin-inhibited enzymes at high resolution will be published elsewhere.

MATERIALS AND METHODS

Purification and Crystallization. LAP from bovine lens was purified according to a modification of the method of Carpenter and coworkers (14) by one of us (A.T.) and was stored in a buffer consisting of 50 mM Tris Cl (pH 7.8) and 50 μM $ZnSO_4$. The enzyme was inhibited by incubation with its slow-binding inhibitor bestatin at a concentration of 10 μM. Isomorphous crystals of the native or bestatin-inhibited enzyme were grown by one of us (A.T.) using the hanging-drop method, by vapor diffusion against the storage buffer with 1–2 M Li_2SO_4 as the precipitant. Under these conditions, crystals in the form of hexagonal bars with maximum diameter of about 0.25–0.3 mm could be obtained in 1–2 months. They show systematic absences of the hexagonal space group $P6_322$, with a = 132 Å and c = 122 Å (10). The unit cell contains two hexamers with the asymmetric unit consisting of one promoter of molecular weight 54,000, and the solvent occupies 57% of the crystal volume.

The publication costs of this article were defrayed in part by page charge payment. This article must therefore be hereby marked "*advertisement*" in accordance with 18 U.S.C. §1734 solely to indicate this fact.

Abbreviations: LAP, leucine aminopeptidase; MIR, multiple isomorphous replacement.
§The atomic coordinates and structure factors have been deposited (August 1, 1990) in the Protein Data Bank, Chemistry Department, Brookhaven National Laboratory, Upton, NY 11973 (references 1LAP, R1 LAPSF).

Table 1. Heavy-atom derivative preparation

Compound	Conc.	pH	Time
Ethylmercury phosphate (Hg1)	125 μM	7.8	5 days
Mercuric chloride (Hg2)	250 μM	7.8	5 days
Iridum hexachloride (Ir1)	50 mM	7.8	5 days
Iridum hexachloride (Ir2)	250 mM	7.8	5 days
Sodium diuranate (DU1)	½ sat.	6.4	8 days
Sodium diuranate (DU2)	½ sat.	6.8	8 days

sat., Saturation.

Structure Determination. The three-dimensional structure determination was carried out by three of us (S.K.B., P.R.D., and W.N.L.). Heavy-atom derivatives of the inhibited enzyme were prepared by soaking crystals after equilibration of the heavy-atom solutions and the crystals by vapor diffusion (see Table 1 for details of the soaking conditions; the inhibited enzyme was chosen for the MIR study in the hopes of minimizing interactions between the active-site metals and heavy-atom reagents).

Diffraction data for the native enzyme were collected at the Resource for Crystallography located in the laboratory of N. h. Xuong (University of California, San Diego), where two multiwire proportional chambers (15) were used. Diffraction data for the bestatin-inhibited enzyme and six heavy-atom derivatives were measured to 3.0-Å resolution using a Xentronics-type area detector (16) with the Harvard data-collection software (17). Small crystals and the moderately large unit cell required the use of double Franks mirrors (18, 19). Reduction to intensities was performed using the BUDDHA package (17) and reduction to structure-factor amplitudes and local scaling was performed using the algorithm of Fox and Holmes (20). Table 2 gives the statistics for data collection for the native enzyme, the enzyme–inhibitor complex, and its heavy-atom derivatives.

Isomorphous difference Patterson syntheses revealed a common, major heavy-atom binding site in the two mercurial derivatives, and the atomic position was refined by the method of Terwilliger and Eisenberg (21). Despite repeated measurement attempts, anomalous scattering from mercury and uranium heavy atoms was not detected by either of the two Xentronics-type area detectors available in our laboratory and supplied by Siemens Analytical X-ray. However, anomalous scattering from the ethylmercury phosphate derivative was successfully measured using the FAST detector in the laboratory of J. Pflugrath (Cold Spring Harbor Laboratory) (data not shown). Difference Fourier syntheses confirmed the location of the common, major mercury binding site and allowed identification of two independent minor mercury binding sites, one in each of the two mercury derivatives. In addition, these preliminary MIR phases allowed identification of heavy-atom binding sites in the diuranate- and iridium-soaked crystals. Further improvement in the accuracy of the phase data resulted from combining Wang's density modification method implemented by Wolfgang Kabsch (22, 23) with the Hendrickson–Lattman coefficients obtained from the MIR method (24). After density modification, each of the derivatives was reexamined by difference Fourier synthesis to identify yet more minor heavy-atom binding sites (see Table 2 for the results of heavy-atom refinement). The final average figure-of-merit obtained from the MIR method at 3.0-Å resolution was 0.65.

The initial electron-density maps showed some well-defined right-handed α-helices and β-strands, but the majority of the loop regions were uninterpretable. A discontinuous, unrefined polyalanine model consisting of 347 of a possible 487 amino acids was used for the first round of phase combination and density modification (25, 26). Subsequent rounds of phase combination/density modification used refined, discontinuous models obtained from manual rebuilding and chain extension using the molecular graphics package FRODO (27, 28) with an Evans & Sutherland Picture System 390, and positional refinement using XPLOR (29) with a DECstation 3100. The amino acid sequence was aligned by identifying three cysteine residues from the refined positions of the three mercury binding sites. In all, eight rounds of phase combination/model building/partial structure refinement were required to reveal the locations of all 487 amino acids and two zinc ions that comprise the asymmetric unit. This iterative process was carried out at 3.0-Å resolution to give a final *R* factor of 0.248 and root-mean-square deviations of 0.016 Å and 3.7° for bond lengths and angles, respectively. The active site showed additional electron density adjacent to the zinc ions that is consistent with the structure of the slow-binding inhibitor bestatin, but no attempt was made to establish the precise orientation of the bestatin at this resolution. Recently, x-ray diffraction data to 2.25-Å resolution from bovine lens LAP complexed with bestatin were obtained at the National Synchrotron Light Source in collaboration with R. M. Sweet. Refinement of these data is proceeding and the structure of the bestatin-inhibited enzyme will be published elsewhere.

Refinement of the isomorphous native structure at 2.7-Å resolution was carried out using the structure factors obtained with the Xuong–Hamlin area detector system. The entire model was examined using a series of unrefined delete maps that spanned the polypeptide chain, and minor corrections to side-chain and backbone positions were made. Further model building and positional and temperature-factor refinement reduced the *R* factor to 0.169, with root-mean-

Table 2. Data collection statistics

Crystal	Resolution, Å	No. unique/no. collected	R_{symm}*	ΔR†	$R_{centric}$‡	Phasing power§
Native	20–2.7	16,334/275,554	0.102			
Bestatin	20–3.0	11,388/57,147	0.057			
Hg1	20–3.0	10,345/63,489	0.076	0.12	0.58	1.35
Hg2	20–3.0	10,562/43,299	0.068	0.11	0.55	1.45
Ir1	20–3.0	10,890/41,656	0.060	0.11	0.57	1.31
Ir2	20–3.25	8,620/30,301	0.070	0.15	0.62	1.13
DU1	20–4.0	4,056/12,474	0.076	0.21	0.66	1.04
DU2	20–4.0	4,282/21,806	0.066	0.19	0.67	1.24

*$R_{symm} = \Sigma|I - \langle I \rangle|/\Sigma I$, where I is the observed intensity and $\langle I \rangle$ is the average intensity obtained from multiple observations of symmetry-related reflections.

†$\Delta R = \Sigma||F_{PH}| - |F_P||/\Sigma|F_P|$, where $|F_P|$ is the protein structure-factor amplitude and $|F_{PH}|$ is the heavy-atom-derivative structure-factor amplitude.

‡$R_{centric} = \Sigma||F_{H(obs)}| - |F_{H(calc)}||/\Sigma|F_{H(obs)}|$, where $|F_{H(obs)}|$ is the observed heavy-atom structure-factor amplitude, where $|F_{H(calc)}|$ is the calculated heavy-atom structure-factor amplitude.

§Phasing power = rms $(|F_H|/E)$, where $|F_H|$ is the heavy-atom structure-factor amplitude and E is the residual lack of closure.

square deviations of 0.015 Å and 3.19° for bond lengths and angles, respectively.

A Ramachandran plot (30) revealed only 7 nonglycine residues for which the (ϕ, ψ) values fell outside the allowed regions (data not shown). No simulated annealing was performed and no water molecules have been included in the refinement, although there are features in the ($|F_{obs}| - |F_{calc}|$) difference Fourier syntheses that are consistent with water molecules in the first shell of hydration.

The connectivity of this final model was checked by phase combination/density modification using the Hendrickson–Lattman coefficients obtained from the MIR method and a partial structure model consisting of 2217 of 3723 nonhydrogen atoms, which was prepared by deleting the two zinc ions and all the residues not found in either α-helices or β-strands. The resulting electron-density map reproduced the entire trace, with the exception of a surface loop region with poor electron density at positions 12–17, which was not well defined in any of the electron-density maps.

RESULTS AND DISCUSSION

Molecular Structure. The 487 amino acids and two zinc ions comprising the monomer are folded into two α/β-type quasispherical globular domains to give a comma-like shape, with approximate maximal dimensions of 90 × 55 × 55 $Å^3$. Fig. 1*A* shows the molecule as a ribbon drawing and Fig. 1*B* illustrates a stereodrawing of the α-carbon backbone with the two zinc ions. The secondary structure composition of the monomer is 40% α-helix and 19% β-strand.

The N-terminal 150 residues fold to give a five-stranded β-sheet sandwiched between four α-helices. The N terminus itself occurs as the middle strand of the sheet, which has four parallel strands and one antiparallel strand. There is a long loop connecting an α-helix and the fifth β-strand comprising the sheet. This loop contains the site between residues Arg-137 and Lys-138, where trypsin cleaves the hexameric enzyme but leaves the enzyme intact and biochemically active (33). There is no evidence of proteolytic cleavage of this or any other peptide bond in the protein.

An α-helix runs from residue 151 to 170 and connects the N-terminal domain with the C-terminal domain. This region of the molecule is dominated by a central, eight-membered, saddle-shaped β-sheet, which is sandwiched between groups of α-helices and constitutes the hydrophobic core of the C-terminal domain. In addition, there is a smaller three-stranded β sheet, which is located on the monomer surface and is involved in interaction with other members of the enzyme hexamer. The two zinc ions and the active site are entirely located within the C-terminal or catalytic domain. This finding is consistent with the results of photoaffinity labeling experiments, which localized the active site to the larger of the two trypsin-cleavage fragments (A.T., unpublished data).

Hexamer Structure. Fig. 2 *A* and *B* illustrate the α-carbon backbone of the LAP trimer and hexamer, respectively. The two crystallographic **32** hexamers found in the unit cell are centered at (⅔, ⅓, ¼) and (⅓, ⅔, ¾). When viewed down its threefold symmetry axis the hexamer is triangular in shape with a triangle edge length of 115 Å. The maximum extent of the hexamer along the threefold axis is 90 Å. The view of hexamer shown in Fig. 2*B* compares favorably with similar electron microscopic views (8, 9). In addition, the low-resolution models based on the results of electron microscopy (8, 9) are also in good agreement with the distribution of protein within the hexamer. The catalytic domains are clustered around the threefold axis, and the upper and lower trimers are related to one another by a twofold rotation. The N-terminal domains extend outward from the catalytic domains and are located far from the center of the hexamer. It is remarkable that our model of the LAP hexamer does not agree with the results of Vasil'ev *et al.* (13), who used single isomorphous replacement and solvent flattening to calculate an electron-density map for the hexamer at 5-Å resolution. Their model of the hexamer, derived from a "solvent flattened" electron-density map (see figure 2 in ref. 13), does

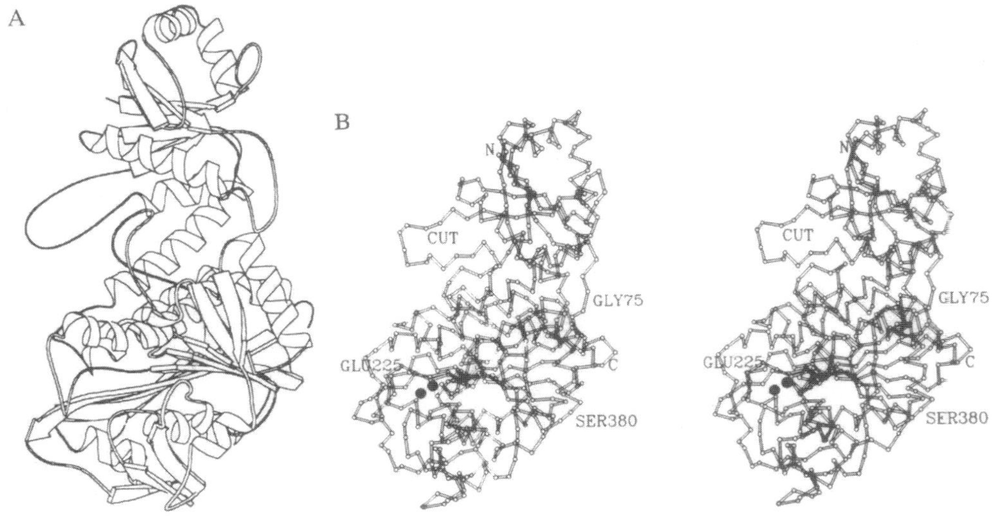

FIG. 1. (*A*) Schematic ribbon drawing of the enzyme monomer (31, 32). α-Helices are shown as ribbon helices, β-strands are shown as ribbons with arrows, and loops are drawn as single lines. This view should be compared to that shown in the stereodrawing illustrated in *B*, which is labeled with appropriate landmarks. (*B*) Stereodrawing of the α-carbon backbone of the enzyme monomer. The α-carbon atoms are denoted by small open circles and the two zinc ions located in the active site are represented by the large black circles. The N and C termini are labeled N and C, respectively. The trypsin cleavage site is indicated with the word CUT, and another four residues are labeled for the reader's convenience.

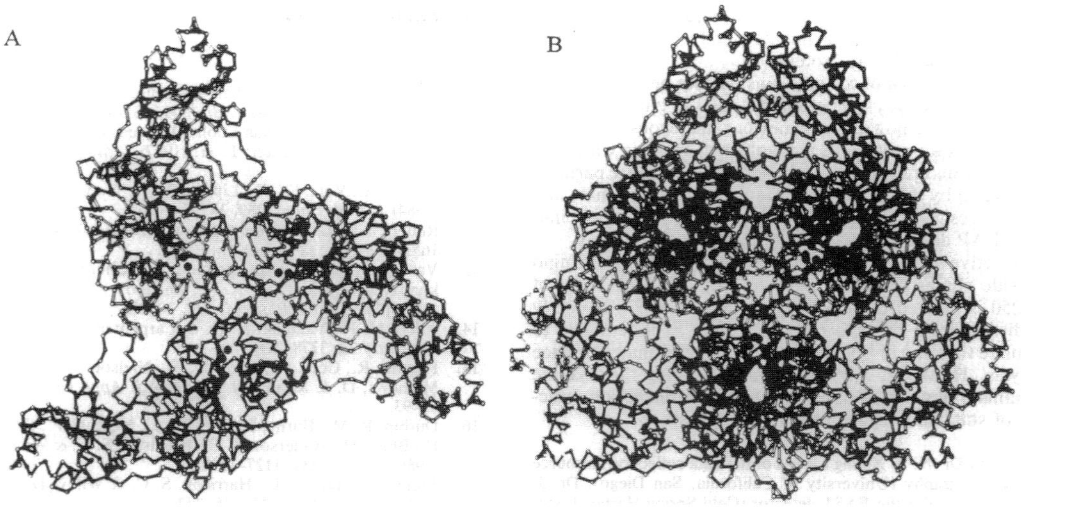

FIG. 2. (*A*) α-Carbon backbone of the enzyme trimer; view is down the threefold axis. The α-carbon atoms are shown as small open circles and the zinc ions are shown as large black circles. One monomer is drawn with shaded bonds connecting α-carbon atoms. (*B*) α-Carbon backbone of the enzyme hexamer with the same view and the same atom and bond shading as in *A*.

not show electron-density features corresponding to the N-terminal domains.

Although LAP consists of only a single polypeptide chain and has no known cooperativity, the hexameric enzyme complex illustrated in Fig. 2*B* bears a striking resemblance to the gross features of the structure of the multimeric regulatory enzyme aspartate transcarbamoylase (ATCase; aspartate carbamoyltransferase, EC 2.1.3.2), which is being actively studied in one of our laboratories (34). In LAP the catalytic domain is analogous to the catalytic or C chain of ATCase, and the LAP N-terminal domain is analogous to the regulatory or R chain of ATCase.

Active Site. The enzyme active site is illustrated in Fig. 3, which shows the electron density of the active-site residues and the two zinc ions. The active site is located near the edge of the central eight-stranded β-sheet (see Fig. 1*B*). The six active sites are themselves located in the interior of the hexamer, where they line a disk-shaped cavity of radius 15 Å and height 10 Å. Access to this interior cavity is via solvent channels that run along the crystallographic twofold axes, which are found in the planes $z = 0$ and $z = \frac{1}{2}$.

The zinc ions are 2.88 Å apart, which is slightly more than twice their covalent radii (1.25 Å) and considerably more than twice their ionic radii (0.75 Å). As suggested by earlier biochemical studies, the metal environments are not equivalent (4). One of the zinc ions appears to be more tightly bound (site 2) than the other. The metal in site 2 is coordinated by one O^{δ} atom of Asp-255, one O^{δ} atom of Asp-332, the carbonyl oxygen of Asp-332, and one O^{ϵ} atom of Glu-334, in an approximately tetrahedral arrangement. The other zinc ion appears less tightly bound to the protein (site 1) and is coordinated by one O^{δ} atom of Asp-273, one O^{δ} atom of Asp-255, and one O^{ϵ} atom of Glu-334. No fourth ligand was seen in the electron-density difference maps. At this limit of resolution and stage in the refinement, there are no density features consistent with water molecules visible in the active site. Earlier metal-substitution studies of bovine lens LAP showed that replacement of zinc in either site 1 or site 2 with

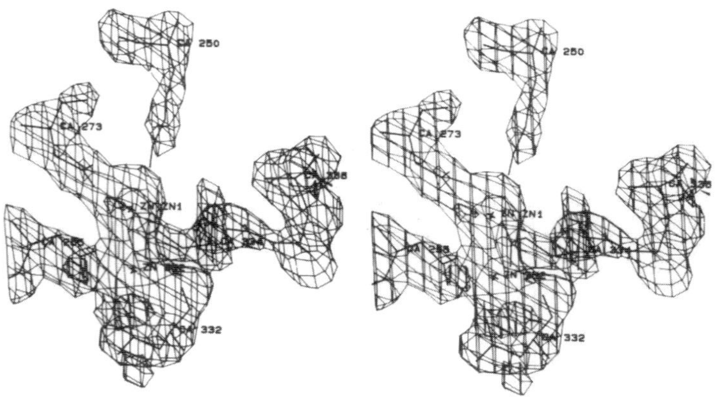

FIG. 3. Stereodrawing of the enzyme active site. This map was calculated by deleting the two zinc ions and residues Lys-250, Asp-255, Asp-273, Asp-332, Glu-334, and Arg-336 from the coordinates used to calculate the phases. The coefficients of Fourier map were $(|F_{obs}| - |F_{calc}|)$. The deleted residues are shown as stick figures and each α-carbon atom is labeled. The two zinc ions are shown as crosses and are labeled by their site number. The chicken-wire mesh represents the difference electron density contoured at 3σ.

magnesium, manganese, or cobalt ions affected both k_{cat} and K_m (14), thereby suggesting that the metal ions influence each other directly or via the protein.

This arrangement of the zinc ions has not been previously observed in x-ray crystallographic studies of zinc-containing proteases. It is, however, somewhat analogous to the arrangement of the metal ions in myohemerythrin, in which the two iron atoms are separated by about 3.23 Å and are partially coordinated by the carboxylate side chains of glutamate and aspartate residues (35). Unlike myohemerythrin, metal binding by LAP does not involve any histidine side chains.

The active site also includes two positively charged amino acid side chains (Lys-250 and Arg-336). The side chain of Lys-250 is located closest to zinc site 1. Arg-336 is nearly equidistant from the two metal-ion sites. It is tempting to speculate that one or both of these side chains may influence catalysis. However, any comments regarding the enzyme mechanism must await refinement of high-resolution structures of suitable enzyme–ligand complexes.

We thank Dr. N. h. Xuong for use of the data collection Resource for Crystallography (University of California, San Diego); Dr. J. Pflugrath for use of the FAST detector (Cold Spring Harbor Laboratory); and Dr. J. E. Gouaux for help with native data collection. We acknowledge the contributions of Drs. P. Moody and G. S. Shoham, who helped with preliminary aspects of this investigation. In addition, we thank Drs. M. Blum, H. Ke, J. Kuriyan, J. Y. Liang, G. A. Petsko, J. Pflugrath, M. Saper, and J. H. Wang for their useful discussion. S.K.B. thanks Drs. Eugene Braunwald and Marshall Wolf of the Brigham and Women's Hospital for their support and encouragement. We thank the National Institutes of Health for Grant GM06920 (to W.N.L.) and the Medical Foundation, Inc., for fellowship support of S.K.B. Purification and crystal growth by A.T. were supported by United States Department of Agriculture Contract 53-3K06-5-10 and grants from the Massachusetts Lions Eye Research Fund, Inc., the Research Corporation, and the Daniel and Florence Guggenheim Foundation.

1. Smith, E. L. & Hill, R. L. (1960) in *The Enzymes*, eds. Boyer, P. D., Lardy, H. & Myrback, K. (Academic, New York), 2nd Ed., Vol. 4, Part A, pp. 37–62.
2. Hanson, H. & Frohne, M. (1977) *Methods Enzymol.* **45**, 504–521.
3. Carpenter, F. H. & Harrington, K. T. (1972) *J. Biol. Chem.* **247**, 5580–5586.
4. Carpenter, F. H. & Vahl, J. M. (1973) *J. Biol. Chem.* **248**, 294–304.
5. Cuypers, H. T., van Loon-Klaassen, L. A. H., Vree Egberts, W. T. M., de Jong, W. W. & Bloemendal, H. (1982) *J. Biol. Chem.* **257**, 7077–7085.
6. Umezawa, H., Aoyagi, T., Suda, H., Hamada, M. & Takeuchi, T. (1976) *J. Antibiot.* **29**, 97–99.
7. Peltier, C. Z. & Taylor, A. (1986) *Fed. Proc. Fed. Am. Soc. Exp. Biol.* **45**, 1857.
8. Taylor, A., Carpenter, F. H. & Wlodawer, A. (1979) *J. Ultrastruct. Res.* **68**, 92–100.
9. Kiselev, N. A., Stelmaschuk, Tsuprun, V. L., Ludewig, M. & Hanson, H. (1977) *J. Mol. Biol.* **115**, 33–43.
10. Jurnak, F., Rich, A., van Loon-Klaassen, L., Bloemendal, H., Taylor, A. & Carpenter, F. H. (1977) *J. Mol. Biol.* **112**, 149–153.
11. Taylor, A., Volz, K. W., Lipscomb, W. N. & Takemoto, L. J. (1984) *J. Biol. Chem.* **259**, 14757–14761.
12. Roderick, S. L. & Mathews, B. W. (1988) *J. Biol. Chem.* **263**, 16531.
13. Vasil'ev, D. G., Reshetnikova, L. S., Chernaya, M. M., Nikonov, S. V., Federov, A. A., Zhdanov, A. S. & Andreeva, N. S. (1989) *Dokl. Biophys.* **304**, 20–22.
14. Allen, M. P., Yamada, A. H. & Carpenter, F. H. (1983) *Biochemistry* **22**, 3778–3783.
15. Hamlin, R., Cork, C., Howard, A., Nielson, C., Vernon, W., Mathews, D. & Xuong, N. h. (1981) *J. Appl. Crystallogr.* **14**, 85–93.
16. Durbin, R. M., Burns, R., Moulai, J., Metcalf, P., Freymann, D., Blum, M., Anderson, J. E., Harrison, S. C. & Wiley, D. C. (1986) *Science* **232**, 1127–1132.
17. Blum, M., Metcalf, P., Harrison, S. C. & Wiley, D. C. (1987) *J. Appl. Crystallogr.* **20**, 235–242.
18. Harrison, S. C., Winkler, F. K., Schutt, C. E. & Durbin, R. M. (1985) *Methods Enzymol.* **114**, 211–237.
19. Phillips, W. C. & Rayment, I. (1985) *Methods Enzymol.* **114**, 316–329.
20. Fox, G. C. & Holmes, K. C. (1966) *Acta Crystallogr.* **20**, 886–891.
21. Terwilliger, T. C. & Eisenberg, D. (1983) *Acta Crystallogr. A* **39**, 813–817.
22. Wang, B.-C. (1985) *Methods Enzymol.* **115**, 90–112.
23. Read, R. J. (1986) *Acta Crystallogr. A* **42**, 140–149.
24. Hendrickson, W. A. & Lattman, E. A. (1970) *Acta Crystallogr. B* **26**, 136–143.
25. Remington, S., Wiegand, G. & Huber, R. (1982) *J. Mol. Biol.* **158**, 111–152.
26. Robbins, A. H. & Stout, C. D. (1989) *Proteins* **5**, 289–312.
27. Jones, T. A. (1985) *Methods Enzymol.* **115**, 157–171.
28. Jones, T. A. & Thirup, S. (1986) *EMBO J.* **5**, 819–822.
29. Brunger, A. T., Kuriyan, J. & Karplus, M. (1987) *Science* **235**, 458–460.
30. Ramachandran, G. N. & Sasisekharan, V. (1968) *Adv. Protein Chem.* **23**, 283–437.
31. Richardson, J. S. (1985) *Methods Enzymol.* **115**, 359–380.
32. Priestle, J. P. (1988) *J. Appl. Crystallogr.* **21**, 572–576.
33. van Loon-Klaassen, L., Cuypers, H. Th. & Bloemendal, H. (1979) *FEBS Lett.* **107**, 366–370.
34. Kantrowitz, E. R. & Lipscomb, W. N. (1988) *Science* **241**, 669–674.
35. Sheriff, S., Hendrickson, W. A. & Smith, J. L. (1987) *J. Mol. Biol.* **197**, 273–296.

Chapter 16

Lou Massa

Biography

Born August 4, 1940 in Jamesville, NY, I was one of five brothers born to my Italian immigrant parents who were originally from Ceccano in Frosinone, Italy. I am married to Mary Gee Massa, and have two daughters, Antonia and Angelina, and a son James from a former marriage.

I gained my B.S. in chemical physics from Le Moyne College, NY, in 1961, and my M.S. from Clarkson University, NY, in 1962. In 1966 I was awarded my Ph.D. in chemical physics from Georgetown University. From 1966 to 1969 I worked as a postdoc research scientist at the Brookhaven National Laboratory. In 1969 I took a position as assistant professor of chemistry at Hunter College and the Graduate School of the City University of New York (CUNY); in 1972 I became associate professor and in 1975 full professor. I have also held visiting appointments at Harvard University, the Naval Research Lab (NRL), the University of Bordeaux, the University of London, IBM-Watson Research Lab, the University of North Carolina, the University of New Orleans, Grumman Aerospace, and the Naval Surface Warfare Center.

A member of the American Physical Society, the American Chemical Society, and the American Crystallographic Association, I have been awarded the Hunter College President's Award for Excellence in Research and the Naval Research Lab's Berman Award for Outstanding Science Paper. I have also been the American Society for Engineering Education (ASEE) Distinguished Visiting Professor at the NRL. In addition to some 120 published papers, I have given a number of lectures and more than 50 interviews on "Science and the Written Word," a CUNY-TV program.

Current research in our lab concerns quantum crystallography (QCr), which refers to the combination of structural crystallographic information with quantum-mechanical theory. The objective is to facilitate computational chemistry calculations and thereby enhance the information that may be derived from a crystallographic experiment. This concept has a long history and in recent years has been finding increased attention because of the advances in both theory and computers. Our method for obtaining quantum mechanical molecular energy involves the use of parts of a whole molecule, which in our formalism are called kernels. The individual calculations based on kernels are relatively small compared to those which would be required to treat an entire molecule all at once. Subsequently, we sum kernel contributions to obtain the energy for a whole molecule. In so doing we simplify the formidable task of obtaining a true quantum energy calculation for very large molecules. The saving of computational time is significant. The theoretical background for our approach to quantum crystallography may be found in our publications.

Origin of the Idea for the Boron Fullerenes

My greatest fun working with the Colonel had to do with the conception of the boron fullerenes. He already had the idea but he let me develop it with him. For this privilege I will always be grateful. I will come to the circumstances of our work with these molecules later, but preceding that, I do have memories of interactions with him which flood my mind.

When I was a graduate student in theoretical physics, I first heard of Professor Lipscomb from my now great friend and then mentor, Bill Clinton. That would be the professor of physics at Georgetown University and not the former president of the United States, although we all three, the two Clintons and myself, were contemporaries on the Georgetown University campus in the 1960s. I first met Professor Lipscomb at the University of New Mexico at an ACA meeting in 1972. He was already a legendary figure in the field, although this preceded his Nobel Prize by several years. I introduced myself as a student of Bill Clinton and told him of our interest in applying quantum mechanics to analysis of X-ray scattering data. He generously discussed this with me at length, standing in a public space outside the meeting halls, and he told me among other things it would be good to attach Debye–Waller factors to atomic orbitals to account for the effect of temperature upon scattering. In that same year, 1972, Clinton and I submitted a note to *Physical Review Letters* on the same subject, i.e. quantum mechanics used to describe X-ray scattering data. Over the course of a year five referees turned it down, with us arguing back after each new negative report. Finally, the journal turned it over to a super-referee that decided in our favor saying we were handling the quantum mechanics of the problem correctly. I have never discussed this with him, but I have always assumed that that referee was Professor Lipscomb. He and Jones had earlier written a paper calling attention to molecular form factors consistent with quantum mechanics. Some time later he gave a talk at Hunter College and with no prompting from me, volunteered the opinion that we were handling the quantum mechanics of X-ray scattering properly. This not only sent my stock higher in the Hunter Chemistry Department, but it confirmed for me my prejudice that he indeed had refereed our *Physical Review Letters* paper.

In 1985 I gave a talk at an IUC meeting in China on the same subject, and in the discussion period Professor Lipscomb asked me how we took care of temperature. I had the tremendous fun of being able to feign surprise and reply something along the lines of, "Why Professor Lipscomb, by attaching Debye–Waller factors to each atomic orbital, just as you recommended I should do, back in 1972 at that ACA meeting in New Mexico." He accepted my answer with very good cheer but I thought I detected some little surprise at the specificity of my reply. But then, he had made a great impression on me so I was not surprised to have remembered the when and where. And also, the advice turned out to be good. We had breakfast the next morning and reviewed progress in the application of quantum mechanics to X-ray scattering factors.

When my wife Mary Gee decided to take a Masters degree at Harvard in 1991–1992, I asked Professor Lipscomb if I could spend a sabbatical year with him, and he kindly consented. It proved to be a memorable year for me and my family. In addition to a new degree that year, we also had a new baby girl Antonia with us when we arrived in Cambridge in September 1991. We thought we would need a nanny to help with the baby while we were both at school. Mary had identical twin 17-year-old cousins that were willing to take a break between high school and college. They were inseparable so we brought them both along to Cambridge as nannies for Annie. As it turned out Bill Clinton had a sabbatical that year too, and he also joined us in Professor Lipscomb's lab. We had wonderful talks that year, especially about superconductivity. And, there were many occasions for laughter. One that stayed with me was the Clintons' first visit to our home on Fresh Pond Parkway at the end of Brattle Road. We introduced the twins to Joan Clinton who was heard to say quite simply, "Two nannies for one baby." This is considerably funnier if one knows Joan is the mother of nine children all raised, as far as I know, without benefit of even a single nanny. (Although, I was fond of babysitting for the Clintons whenever I could as a graduate student at Georgetown.)

We come now to the boron fullerenes. The origin of a research idea, as we know, happens in different ways at different times. Often a good idea is the result of a lot of hard work, upon which follows some essential insight. But the boron fullerenes came to me free of all effort, more or less as a Christmas gift. By December 1991 we had been at Harvard about four months and we were feeling quite settled. Everything was running smoothly. I had developed the habit of dropping into the Colonel's office to try out some ideas on him (I won't label them as exactly "crack pot" ideas). I recall, for example, wanting to calculate the wavefunction of an entire protein. I also toyed with the idea of making extended Hückel calculations "exact." But all that ended one morning when I arrived at my desk to find a Post-it note scribbled in pencil attached to a picture of $B_{32}H_{32}^{2-}$ from a paper the Colonel had published with some of his students,[1] years earlier than the discovery of the symmetry related compound C_{60} (Figure 16.1).

The note simply said, "Lou, see the Colonel", and it was dated December 20, 1991. It was unsigned but I assume it had been left by the Colonel's secretary, Donna. The picture of the 32 vertex structure of icosahedral symmetry indicated to me the Colonel had something very specific in mind, but I did not know what. Without pausing for a moment I wandered across the hall at Gibbs and into the Colonel's office carrying the note attached to the picture. He explained to me, forming a fullerene cage with his hands, and pointing in the air at imaginary faces, that if boron atoms were placed at the center of every C_{60} face,

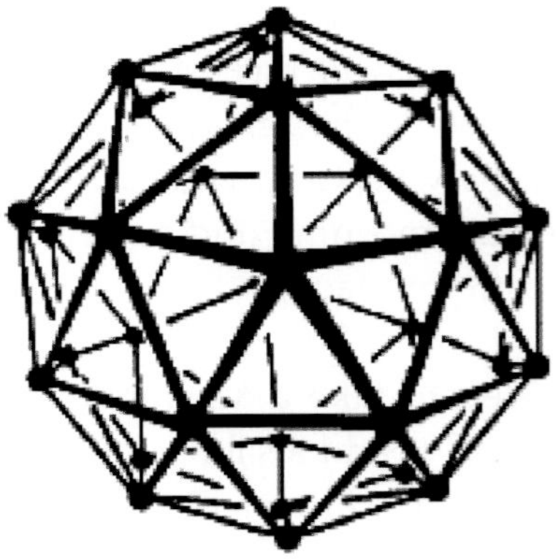

Figure 16.1 The 32-vertex structure of I_h symmetry.

and then all connected by triangular chemical bonds, one would have the boron 32 vertex icosahedral cage structure whose picture I was holding in my hand.

He introduced me to a geometrical relation that might have been discovered by the Greeks, but as it happened was first brought to light by Descartes, and later independently by Euler. It was immediately clear the Descartes–Euler relationship (see Figure 16.2) set up a one-to-one correspondence between all possible carbon fullerenes and their boron analogs (duals). We worked the correspondence out for a complete set of examples (see Table 1 in "Examples of large closo boron hydride analogs of carbon fullerenes" reprinted below).

In the process of working out the correspondence we had occasion to count the chemical connections and numbers of electrons. The arch-type buckyball, for example, has 90 connections and 240 valence electrons. The excess electrons beyond those needed for spin-pair single bonds make it a highly aromatic molecule. The Colonel, noticing at one point I was not properly accounting for the valence of carbon, asked me to remind him of the field of my Ph.D. degree. Although I suspected he knew, I politely replied, "theoretical physics." Upon this he noted wryly, "that explains something." We had a fabulous time. The science of the problem was worked out almost immediately. As a jazz musician might say, it was brief but intense.

The only difficult thing was making pictures of the molecules for publication. Mary and I tried drawing them by hand. We tried making physical models and then photographing the models. But none of it was good. Finally we went over to Cambridge Soft to get a more or less free copy of Chem3D which at the time would only run on a MAC, and thus we rented one at home for a period of a month. It took three weeks, day and night, to master the program, the machine, and the pictures, but then the paper was ready to go out to *Inorganic Chemistry*.

We did get around to sending it out sometime later, and I must say it is the only paper I've ever been associated with that was accepted the same day it was received at the journal. It's possible the Colonel's reputation in the field of boron chemistry played some role.

Abstract

Suggestion for Large Closo Boron Hydride Analogs of Carbon Fullerenes

There exists a geometrical correspondence between possible large closo boron hydrides and analog carbon fullerenes. $B_{32}H_{32}$, of icosahedral symmetry, is the analog of the Buckminster fullerene C_{60}.

C_{60}

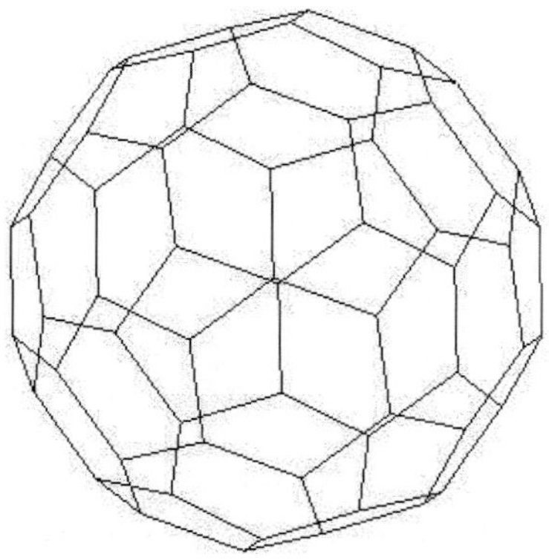

$B_{32}H_{32}$

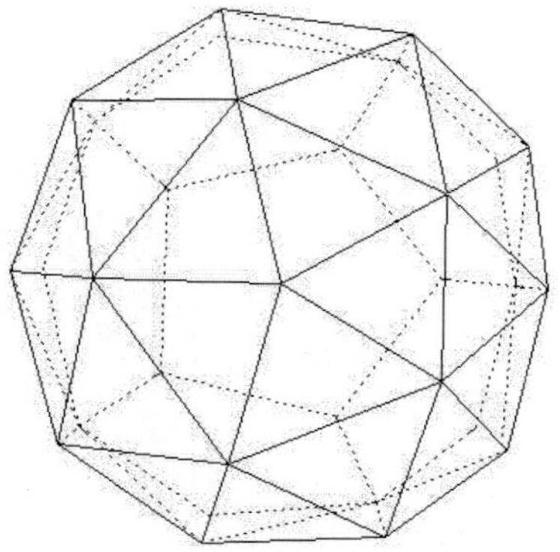

The correspondence referred to is based upon: symmetry; number equality of: (i) carbon faces and boron vertices, (ii) boron faces and carbon vertices, (iii) carbon contacts and boron contacts; and sphericity. The Descartes–Euler formula

$$P + F = C + 2 \tag{1}$$

is applied to both sets of polyhedra to relate within one molecule the number of vertices, P, the number of faces, F, and the number of contacts, C. The possibility of laser ablation of "graphitic" MB_2 (e.g. $M = Ca^{2+}$) in an atmosphere of H_2 to search for $B_{32}H_{32}$, the geometrical analog of Buckminster fullerene, is suggested.

William N. Lipscomb
Lou Massa#
Gibbs Chemical Laboratory
Harvard University
Cambridge, MA 02138

#Visiting Scholar
Hunter College and the Graduate School
City University of New York
New York, NY 10021

Figure 16.2 Abstract for "Examples of large closo boron hydride analogs of carbon fullerenes" paper.

The very day it was published I received two phone calls. One was from someone at Oak Ridge hoping to synthesize the boron fullerenes, and the other was from a reporter at *Science News* hoping to interview us about the beautifully symmetric molecules. One hope was the easier to fulfill. We still await the synthesis.

Figure 16.3 Going away presents at the end of our sabbatical.

Figure 16.4 The Lipscombs and the research group in the back yard on Garden Avenue.

The sabbatical year with the Colonel was a joyous one for me and my family. Among the times we enjoyed most were the parties that occurred intermittently at the Lipscombs' home on Garden Avenue. The families of the Lipscomb research group, including all the children, filled up the various rooms of the house. I recall many enjoyable conversations there in the big kitchen talking with the Colonel, especially about the history of various important scientific discoveries, while opening various French wines. One such party (see Figures 16.3 and 16.4) was occasioned by the end of our sabbatical in June 1992. We were sad to go. We retain our fond memories of that year with the Colonel, his family, and his research group at Harvard. Our work together on the boron fullerenes continues.[2–7]

References

1. Bicerano J., Marynick D. S., and Lipscomb W. N., Molecular orbital studies on large closo boron hydrides. *Inorg. Chem.*, **17**, 3443–3453 (1978).
2. Lipscomb W. N., Massa L., Examples of large closo boron hydride analogs of carbon fullerenes. *Inorg. Chem.*, **31**, 2297–2299 (1992).
3. Derecskei-Kovacs A., Dunlap B. I., Lipscomb W. N., Lowrey A., Marynick D. S., and Massa L., Quantum chemical studies of boron fullerene analogs. *Inorg. Chem.*, **33**, 5617–5619 (1994).
4. Lipscomb W. N. and Massa L., Closo boron hydrides and carbon fullerenes. In Hosmane N.and Holmes R. (Eds), *Phosphorous Sulfur, and Silicon, and Related Elements.* Langhorne, PA: Gordon and Reach, pp. 125–128 (1994).
5. Lipscomb W.N. and Massa L., Conjuncto closo boranes and relationships to dual structures. *Inorg. Chem.*, **33**, 5155–5156 (1994).
6. Gindulytė A., Krishnamachari N., Lipscomb W. N., and Massa, L., Quantum chemical calculations of proposed multicage boron fullerenes. *Inorg. Chem.*, **37**, 6546–6548 (1998).
7. Gindulytė A., Lipscomb W. N., and Massa L., Proposed boron nanotubes. *Inorg. Chem.*, **37**, 6544–6545 (1998).

Inorg. Chem. **1992**, *31*, 2297–2299 **2297**

Examples of Large Closo Boron Hydride Analogs of Carbon Fullerenes

There exists a geometrical correspondence between the large closo boron hydrides and their analog carbon fullerenes. Thus, the aromatic molecule $B_{32}H_{32}$, of icosahedral symmetry, the structure of which has been proposed some time ago[1] is the analog

(1) Bicerano, J.; Marynick, D. S.; Lipscomb, W. N. *Inorg. Chem.* **1978**, *17*, 3443.

0020-1669/92/1331-2297$03.00/0

2298 *Inorganic Chemistry, Vol. 31, No. 12, 1992* Communications

Table I. Carbon Compounds and Their Boron Analogs

diagram	molecule	*P*	*F*	*C* + 2	sym	*P*	*F*	molecule[a]	diagram
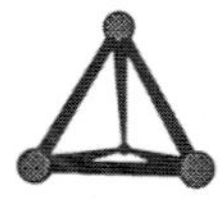	C_4	4	4	8	T_d	4	4	B_4H_4	
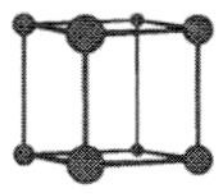	C_8	8	6	14	O_h	6	8	B_6H_6	
	C_{20}	20	12	32	I_h	12	20	$B_{12}H_{12}$	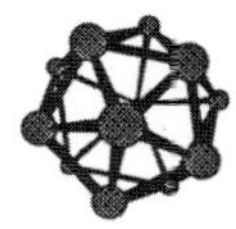
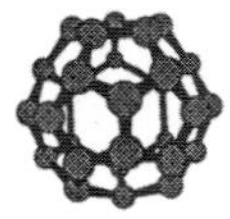	C_{28}	28	16	44	T_d	16	28	$B_{16}H_{16}$	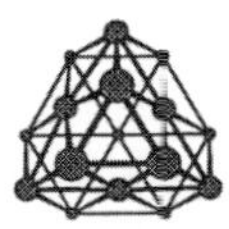
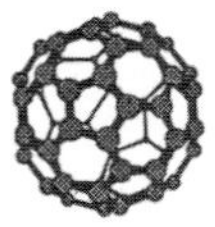	C_{60}	60	32	92	I_h	32	60	$B_{32}H_{32}$	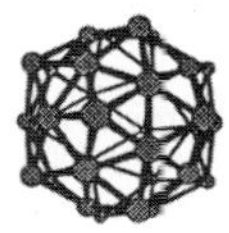
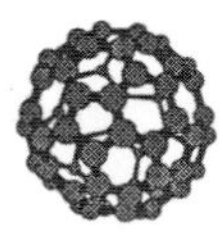	C_{70}	70	37	107	D_{5h}	37	70	$B_{37}H_{37}$	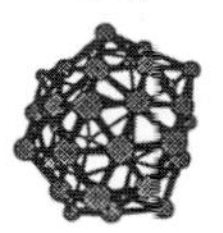
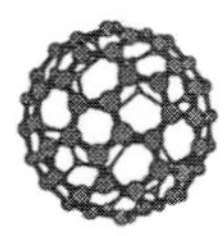	C_{80}	80	42	122	I_h	42	80	$B_{42}H_{42}$	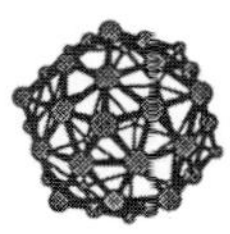

[a] B_4H_4 is predicted to be neutral, B_6H_6 and $B_{12}H_{12}$ are −2 ions, and probable charges of the larger boron species are yet to be examined using MO methods.

of the Buckminster fullerene molecule C_{60}. The correspondence referred to is based upon the following: (a) symmetry; (b) number equality of (i) carbon faces and boron vertices, (ii) boron faces and carbon vertices, and (iii) carbon contacts and boron contacts; (c) sphericity.

The correspondence is illuminated by application of the Descartes–Euler formula[2]

$$P + F = C + 2 \tag{1}$$

to both sets of polyhedra to relate within one molecule the number of vertices, *P*, the number of faces, *F*, and the number of contacts, *C*.

In Table I we list a few examples of the geometrical correspondence which prevails between closo boron hydrides and the carbon fullerenes. In particular, one may note how the geometrical structure of Buckminster fullerene maps into that of the 32 vertex closo boron hydride, $B_{32}H_{32}$. Both molecules of symmetry I_h display correspondence of the geometrical centers of the 32 carbon (polygon) faces to the 32 boron vertices, the 60 boron faces to the 60 carbon vertices, and the 90 carbon contacts to the 90 boron contacts. The sphericity[3] of each compound is maximal within its class. In accordance with eq 1, for both molecules the sum of the vertices and faces exceeds by 2 the number of contacts. A cursory review of Table I indicates that similar correspondences prevail for all the examples listed. Clearly the list of examples may be readily extended.

We mention in passing that eq 1 may be used to predict quickly which closo boron hydride is the analog of each experimentally known fullerene. Thus, for a given fullerene, if the number of carbons is multiplied by $^3/_2$ (a formal carbon contact number) to give the total number of carbon contacts, one may then calculate the number of carbon faces whose equality with the number of boron vertices yields immediately the molecular formula desired.

In view of the successes achieved in producing macroscopic quantities of the fullerenes,[4] analogous methods might be used to synthesize their closo boron hydride analogs. For this purpose we suggest the possibility of laser ablation "graphitic" MB_2 (e.g. M = Ca^{2+}) in an atmosphere of H_2 to search for $B_{32}H_{32}$, which may or may not contain an internal charge-compensating metal ion.

As has occurred with the fullerenes,[5] it is reasonable to suppose a remarkable variety of molecular properties might be associated

(2) Coxeter, H. S. M. *Introduction to Geometry*; John Wiley & Sons: New York, 1961.

(3) A numerical measure of sphericity may be taken to be $S = 1 - [\langle R^2\rangle - \langle R\rangle^2]^{1/2}/\langle R\rangle$. This yields a limiting value for a sphere of (1.0) and for the five platonic solids the values (0.966), (0.924), (0.899), (0.828), and (0.723), respectively, for a dodecahedron, icosahedron, cube, octahedron, and tetrahedron. See also related discussion in ref 1.

(4) (a) Kratschmer, K.; Lamb, L. D.; Fostiropoulos, K.; Huffman, D. R. *Nature* **1990**, *347*, 354. (b) The fullerenes were made in lesser quantity earlier: Kroto, H. W.; Heath, J. R.; O'Brien, S. C.; Curl, R. F.; Smalley, R. E. *Nature* **1985**, *318*, 162. (c) To the best knowledge of the authors, the earliest suggestion that C_{60} should be a stable, "spherical", super-conjugated molecule was made by: Eiji Osawa; Osawa, E. *Chemistry* **1970**, *25* (9), 854 (in Japanese). Yoshida, Z.; Osawa, E. *Aromaticity*; Kagakudojin: Kyoto, Japan 1971; pp 174–178 (in Japanese).

(5) Koshland, D. E., Jr. *Science* **1991**, *254*, 1705.

Inorg. Chem. 1992, 31, 2299–2300 2299

with the large closo boron hydrides, which are their geometrical duals.[6]

Acknowledgment. We thank M. Gee for model building, J. Wolff of Cambridge Scientific Computing, L. Huang and S. Chen for computer drawing, and R. C. Stevens for discussions.

(6) Coxeter, H. S. M. *Regular Polytopes*; Dover: New York, 1973. Note: Dual polyhedra are also known as reciprocals.
(7) Visiting Scholar, Hunter College and the Graduate School, City University of New York, New York, NY 10021.

Gibbs Chemical Laboratory
Harvard University
Cambridge, Massachusetts 02138

William N. Lipscomb*
Lou Massa[7]

Received March 25, 1992

Stereochemistry of Six-Coordinated Silicon Complexes. 2. Optical Resolution of $[Si(\alpha\text{-diimine})_3]^{4+}$ (α-Diimine = bpy, phen)

Six-coordinated complexes of silicon(IV) are not very familiar as compared to four-coordinated tetrahedral compounds. Though several six-coordinated Si(IV) complexes have already been synthesized, there are few studies of the isomerism.

In the present paper, we report the complete optical resolution of $[Si(phen)_3]^{4+}$. $[Si(phen)_3]^{4+}$ was synthesized from SiI_4 and 1,10-phenanthroline following the literature method[1] and chromatographically purified. The resolution was performed by a chromatographic method using an SP-Sephadex C-25 column (ϕ = 1.1 cm × 46 cm) as an adsorbent and a 0.08 M aqueous solution of sodium ((+)-tartrato)antimonate(III),[2] $Na_2[Sb_2\{(+)\text{-tart}\}_2]$ as an eluent. Figure 1 shows the elution curve. The two elution bands are completely separated, and complete resolution is effectively achieved. Each enantiomer is so stable in aqueous solution that the absorption spectrum and optical rotation were unchanged after a month at room temperature. There are some examples of enantiomeric six-coordinated silicon(IV) complexes,[3–6] but all of them are unstable in aqueous solutions. This is the first example of stable (in water) enantiomers of six-coordinated octahedral silicon(IV) complexes. The absorption and circular dichroism (CD) spectra are shown in Figure 2. According to exciton theory,[7] the absolute configuration of the faster-moving $(+)_{589}$ isomer can be assigned to Λ. Though we have already attempted the resolution of $[Si(bpy)_3]^{4+}$ by a chromatographic method, the separation of enantiomers was insufficient under the same conditions. Sodium (+)-tartrate was found to be a less effective eluting agent for both complexes.

The elution order (Λ moves faster than Δ) of the enantiomers for both the bpy and phen complexes is the same as that observed for the corresponding cobalt(III) complexes.[2]

The ^{1}H NMR spectra of $[Si(bpy)_3]^{4+}$ and $[Si(phen)_3]^{4+}$ were measured in the presence and absence of $[Sb_2\{(+)\text{-tart}\}_2]^{2-}$, in order to elucidate the mechanism of the chromatographic resolution, i.e. the stereoselective interaction between the complex ions and (tartrato)antimonate(III) ions. Figures 3 and 4 show the ^{1}H NMR spectra of an ca. 0.02 M D_2O solution of $[Si(bpy)_3]^{4+}$ and an ca. 0.004 M D_2O solution of $[Si(phen)_3]^{4+}$, respectively. The largest

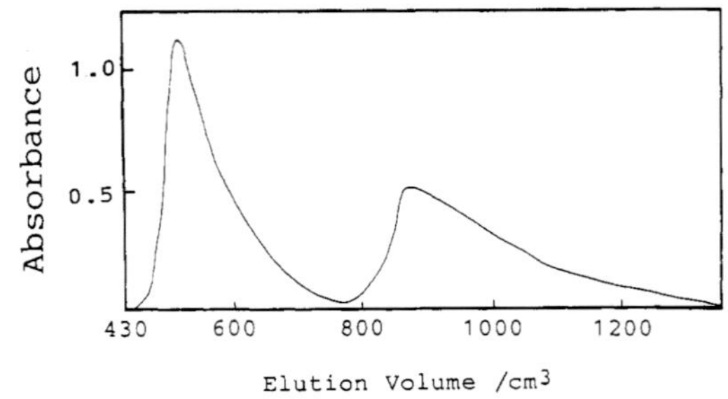

Figure 1. Elution curve of $[Si(phen)_3]^{4+}$ on an SP-Sephadex column. Chromatographic conditions: column, SP-Sephadex C-25, ϕ = 1.1 cm × 46 cm; eluent, $Na_2[Sb_2\{(+)\text{-tart}\}_2]$ (0.08 M); flow rate, 0.8 cm^3 min^{-1}; UV detection, 290 nm.

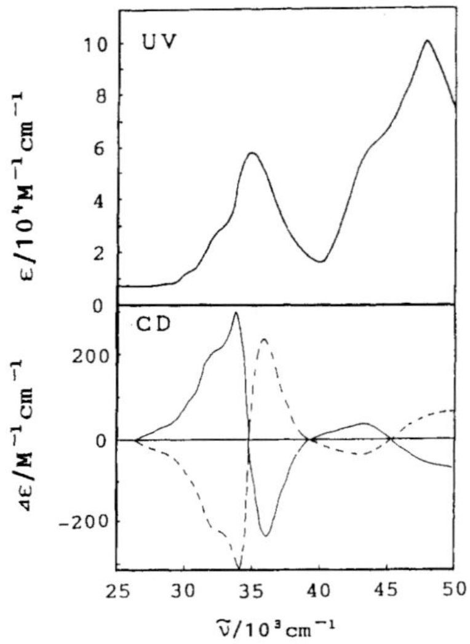

Figure 2. Absorption and CD spectra of $(+)_{589}$ (Λ)- (—) and $(-)_{589}$ (Δ)-$[Si(phen)_3]^{4+}$ (--).

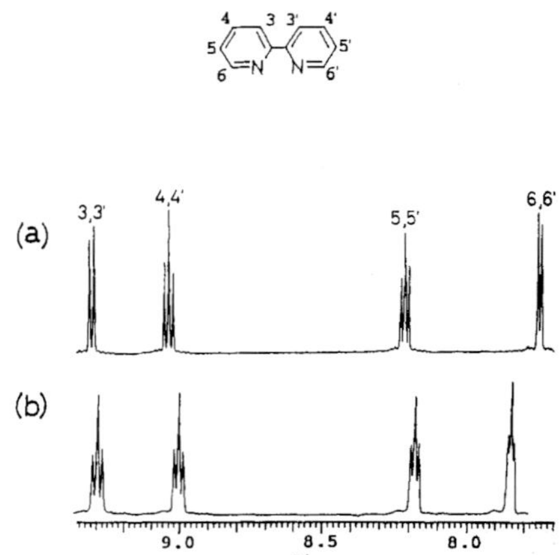

Figure 3. ^{1}H NMR spectra (500 MHz) of $[Si(bpy)_3]^{4+}$ (ca. 0.02 M) in D_2O, (a) in the absence and (b) in the presence of $Na_2[Sb_2\{(+)\text{-tart}\}_2]$ (0.4 M).

(1) Kummer, D.; Gaisser, K. E.; Seifert, J.; Wagner, R. *Z. Anorg. Allg. Chem.* **1979**, *459*, 145.
(2) Yoshikawa, Y.; Yamasaki, K. *Coord. Chem. Rev.* **1979**, *28*, 205.
(3) Dhar, S. K.; Dorn, V.; Kirschner, S. *J. Am. Chem. Soc.* **1958**, *80*, 753.
(4) Larsen, E.; Mason, S. F.; Searle, G. H. *Acta Chem. Scand.* **1966**, *20*, 191.
(5) Inoue, T.; Fujita, J.; Saito, K. *Bull. Chem. Soc. Jpn.* **1975**, *48*, 1228.
(6) Inoue, T. *Inorg. Chem.* **1983**, *22*, 2435.
(7) (a) Mason, S. F.; Norman, B. J. *Inorg. Nucl. Chem. Lett.* **1967**, *3*, 285. (b) Mason, S. F. *Fundamental Aspects and Recent Developments in Optical Rotatory Dispersion and Circular Dichroism*; Heyden & Son: Philadelphia, PA, 1973; pp 196–239.

0020-1669/92/1331-2299$03.00/0

6544 *Inorg. Chem.* **1998,** *37,* 6544–6545

Proposed Boron Nanotubes†

Asta Gindulytė,[‖,‡] **William N. Lipscomb,***[,§] **and Lou Massa**[‖,‡]

Department of Chemistry, Hunter College, 695 Park Avenue, New York, New York 10021, The Graduate School, City University of New York, 33 West 42nd Street, New York, New York 10036, and Gibbs Chemical Laboratory, Harvard University, Cambridge, Massachusetts 02138

Received May 19, 1998

Carbon nanotubes are cylindrical structures related to carbon fullerene structures. Indeed, carbon nanotube cylinders are often "capped" at their ends with "hemispherical" carbon fullerenes, illustrating the close relation of the two types of structure. Nanotube structures are of great interest because of their mechanical and one-dimensional electrical properties.[1–4]

In earlier papers[5–7] we explored the dual (reciprocal) geometries of boron and carbon fullerenes. These dual relationships were based upon the Descartes–Euler formula

$$P - C + F = 2 \qquad (1)$$

where in a polyhedral type of molecule, *P*, *C*, and *F* are counts of points (vertexes), connections (bonds), and faces, respectively. The molecules considered in refs[5–7] were based upon consideration of highly spherical compounds, either singly or compounded as "multicage" molecules. The molecules considered here, nanotubes, are based upon the geometry of a cylinder. In the search for boron nanotubes, the duals of carbon nanotubes, we employ a generalization of the above equation, viz., the Euler–Poincare formula for a cylinder,

$$P - C + F = 0 \qquad (2)$$

where *P*, *C*, and *F* retain their previous meanings.

Of course there is any number of different possible nanotubes. In this paper, for purposes of efficiency, we restrict our discussions to a few types. They are sufficiently general that the analysis shown here can be adapted afterward to any other case. In Figures 1–3 we show selected examples of carbon nanotubes and their boron duals derived from the Euler–Poincare formula, eq 2. The boron duals may be imagined as arising from a correspondence between a boron atom and the center of each of the carbon faces in the carbon nanotubes. If the formula for a given carbon nanotube is taken to be C_P, then the formula for its boron dual nanotube is B_FH_F. In the Figures, the hydrogen atoms, which extend radially outward from each boron atom, are not shown in order to simplify the drawing. Moreover, it may be possible to synthesize boron nanotubes (and boron fullerenes) as pure boron compounds, i.e., without bonds to hydrogen atoms. We would expect the basic boron geometrical structure to be the same with or without the hydrogen atoms attached. But, we recognize the electronic structure associated with the boron geometry will be greatly modified by the formation of BH bonds.

In Table 1 we display a count of *P*, *C*, and *F* for each of the molecules displayed in Figures 1–3. It may be seen how the Euler–Poincare formula is satisfied in each case. Notice in particular how the roles of *P* and *F* are interchanged in going

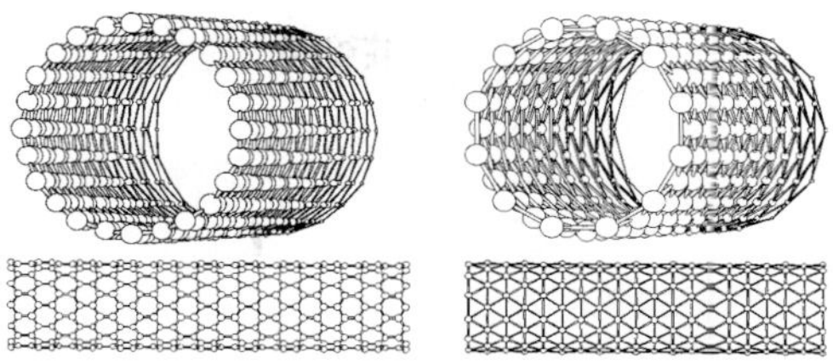

Figure 1. Carbon (left) and boron (right) nanotubes, shown as end-on (top) and side (bottom) views. The case shown is the "zigzag" geometry corresponding to (*n*, *m*) = (12,0).

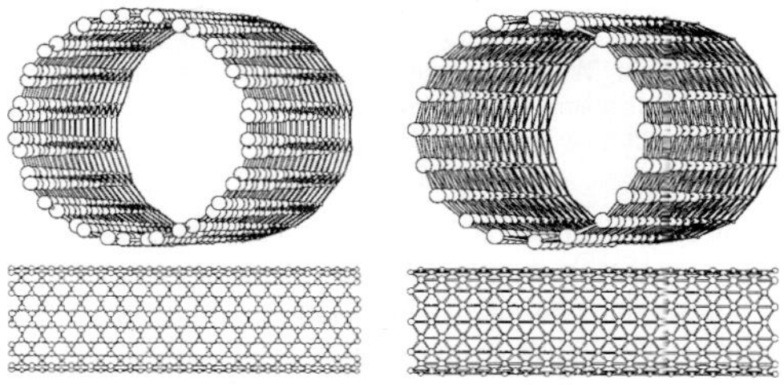

Figure 2. Carbon and boron nanotubes viewed as in Figure 1; the "armchair" geometry corresponds to (*n*, *m*) = (10,10).

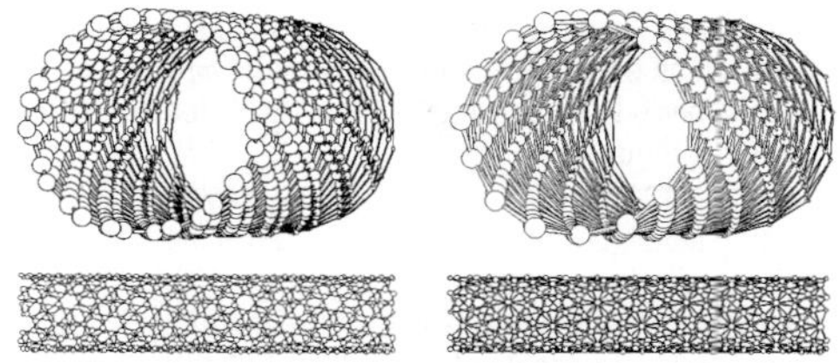

Figure 3. Carbon and boron nanotubes viewed as in Figure 1; an example chiral geometry corresponding to (*n*, *m*) = (10,5).

Table 1. Carbon and Boron Dual Nanotubes; Listed Are the Factors of the Euler–Poincare Formula for a Cylinder, viz. $P - C + F = 0$

formula	*P*	*F*	*C*	(*n*, *m*)	tube type	*P*	*F*	formula
C_{480}	480	240	720	(12, 0)	zigzag	240	480	$B_{240}H_{240}$
C_{720}	720	360	1080	(10, 10)	armchair	360	720	$B_{360}H_{360}$
C_{600}	600	300	900	(10, 5)	chiral	300	600	$B_{300}H_{300}$

† This paper is dedicated to M. Frederick Hawthorne on the occasion of his 70th birthday.
* Corresponding author.
‖ Hunter College.
‡ City University of New York.
§ Harvard University.

(1) Dresselhaus, M. S.; Dresselhaus, G.; Eklund, P. C. *Science of Fullerenes and Carbon Nanotubes*; Academic: San Diego, 1996.
(2) Mintmire, J. W.; Dunlap, B. I.; White, C. T. *Phys. Rev. Lett.* **1992**, *68*, 631.
(3) Bethune, D. S.; Kiang, C.-H.; de Vries, M. S.; Gorman, G.; Savoy, R.; Vazquez, J.; Beyers, R. *Nature* **1993**, *363*, 605.
(4) Iijima, S.; Ichihashi, T. *Nature* **1993**, *363*, 603.
(5) Lipscomb, W. N.; Massa, L. *Inorg. Chem.* **1992**, *31*, 2297–2299.
(6) Dereckei-Kovacs, A.; Dunlap, B. I.; Lipscomb, W. N.; Lowrey, A.; Marynick, D. S.; Massa, L. *Inorg. Chem.* **1994**, *33*, 5617–5619.
(7) Lipscomb, W. N.; Massa, L. *Inorg. Chem.* **1994**, *33*, 5155–5156.

10.1021/ic980559o CCC: $15.00
Published on Web 11/19/1998

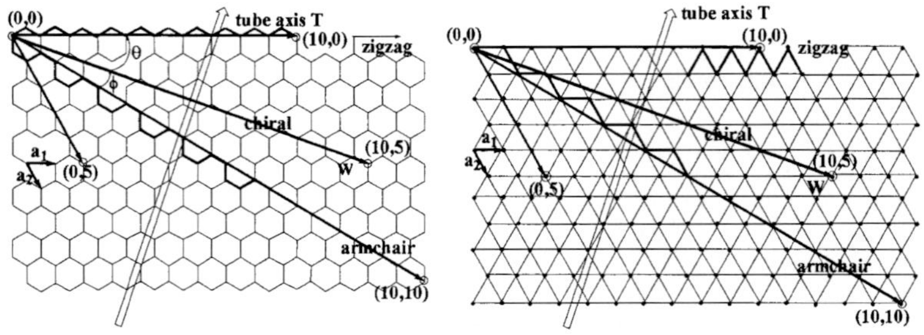

Figure 4. Relation between nanotube and lattice for carbon (left) and boron (right), illustrating lattice vectors $\mathbf{a}_1$ and $\mathbf{a}_2$, and the "wrapping" vector $\mathbf{W} = n\mathbf{a}_1 + m\mathbf{a}_2$. The extreme cases of "zigzag" (*n*, 0) and "armchair" (*n*, *n*) are indicated. The chiral angle, ϕ, is that subtended by the general wrapping vector, **W**, and the armchair vector. The sum of the angles ϕ and θ is 30°. The tube axis vector, **T**, is orthogonal to the wrapping vector **W**. The tube diameter, $d = |\mathbf{W}|/\pi$. Notice the boron lattice is mathematically obtained from the carbon lattice by the device of placing a boron point at the center of a carbon face.

from a carbon nanotube to its dual boron nanotube partner. Notice also that the number of bonds is common to the carbon and boron nanotube duals. Obtaining the count of the bonds requires consideration of all dangling bonds in both the carbon and boron cases. Dangling bonds are associated with dangling faces, and these too must be counted to obtain the values listed in the table. We mention in passing that the nanotube cylinders (including their dangling bonds) are topologically related to a torus, for which the Euler–Poincare formula, eq 2, also applies. The nanotubes also map into a flat sheet lattice structure, illustrated in Figure 4.

Following the explanation in ref 1, one may specify a single wall carbon nanotube "by bisecting a C_{60} molecule at the equator and joining the two resulting hemispheres with a cylindrical tube one monolayer thick and with the same diameter as C_{60}". Bisecting the C_{60} molecule along different directions will yield carbon nanotubes of differing symmetry. Thus, the "zigzag" and "armchair" carbon nanotubes of Figures 1 and 2 are related to bisections of C_{60} normal to a 3-fold axis and a 5-fold axis, respectively. Analogously, if "spherical" $B_{32}H_{32}$, the geometrical dual[5] of C_{60}, is bisected normal to a 5-fold and 3-fold axis one specifies thereby the dual "zigzag" and "armchair" boron nanotubes respectively, as shown in Figures 1 and 2. Additional chiral nanotubes, examples of which are illustrated in Figure 3, can be defined in relation to more general "hemispherical" caps.

Again following the explanation of ref 1, Figure 4 illustrates that a variety of nanotubes may be defined in terms of tube diameter, *d*, angle ϕ, and wrapping vector $\mathbf{W} = n\mathbf{a}_1 + m\mathbf{a}_2$. The "zigzag" and "armchair" nanotubes correspond to angles ϕ equal to 0° and 30° respectively. A general chiral nanotube, such as displayed in Figure 3, corresponds to a chiral angle ϕ, falling between the extremes, 0° and 30°.

A mathematical representation of a general nanotube is formed by "rolling up" the plane of Figure 4, joining the two ends of the chiral vector **W**. Variation of the tube diameter, *d*, and chiral angle, ϕ, are considered to control the properties of the various nanotubes formed.[1] That carbon nanotubes are semiconducting or metallic depending on variation of ϕ and *d* has been confirmed experimentally.[8,9] It is our expectation that if boron nanotubes could be synthesized, their electrical properties, also controlled by variation of ϕ and *d* parameters, would, like their carbon nanotube duals, be of very great interest.

Acknowledgment. We thank Professor Joe Roitberg of Hunter College for valuable topological discussions. This research was supported by an IBM Shared University Research (SUR) grant, a CUNY Research Award, a CUNY Collaborative Award, and a NASA JOVE grant.

IC980559O

(8) Wildöer, J. W. G.; Venema, L. C.; Rinzler, A. G.; Smalley, R. E.; Dekker, C. *Nature* **1998**, *391*, 59–62.
(9) Odom, T. W.; Huang, J.-L.; Kim, P.; Lieber, C. M. *Nature* **1998**, *391*, 62–64.

Chapter 17

Jianpeng Ma

Biography

Dr. Jianpeng Ma is a faculty member in the Department of Biochemistry and Molecular Biology at Baylor College of Medicine. He also holds a joint appointment in the Department of Bioengineering at Rice University. In his doctoral research, he developed algorithms for global optimization. In his postdoctoral research at Harvard University, he studied the dynamics of enzymes and supramolecular complexes such as molecular chaperonin GroEL with Professor Martin Karplus. In that period of time, Dr. Ma extensively collaborated with eminent experimental scientists such as Professor William N. Lipscomb (Harvard) and Professor Paul B. Sigler (Yale), and he was supported by postdoctoral fellowships from the Burroughs Wellcome Fund and the National Institutes of Health (NIH).

Since moving to Houston in 2000, Dr. Ma and his research group have pioneered a series of novel multiscale computational methods for simulating, refining, and modeling flexible biomolecular complexes. For example, he developed a method that reliably computes complex dynamics without atomic coordinates, which enabled functional analysis and structural refinement of biomolecules studied by experimental techniques such as cryogenic electron microscopy. His related methods also set a milestone in fiber diffraction refinement. Other major contributions include developing innovative empirical potentials and sampling algorithms for structure prediction. Most importantly, by more accurately describing large-scale anisotropic molecular motions, Dr. Ma overcame a serious algorithmic bottleneck and substantially enhanced low-resolution X-ray structural models of supramolecular complexes and membrane proteins, of which the data are becoming ever more abundant.

In 2004, in recognition of his outstanding contributions to biophysics and structural biology, the Welch Foundation awarded Dr. Ma the highly prestigious Norman Hackerman Award. In 2007, he was elected as a fellow of the American Physical Society (APS), and in 2008, as a fellow of the American Association for the Advancement of Science (AAAS).[a]

Dr. Ma lives in Houston with his wife, Dr. Qinghua Wang. They have one son, Albert Ma.

[a]In 2013, he became an Elected Fellow of the Institute for Medical Biological Engineering.

The Colonel, the Great Mentor

My association with Bill Lipscomb began in the late 1990s. At the time, I was a postdoctoral fellow at Harvard and Bill was in the Gibbs building. One of the projects his group was working on was to determine the structure of yeast chorismate mutase (YCM), a small dimeric protein that is a good candidate for studying allosteric transition. Although he was more than 70 years old, Bill went to Harvard Chemistry Library very frequently to keep himself up-to-date with the literature. One day, he noticed that there was a computational paper published on the catalytic mechanism of YCM. Since that work was based on the structure determined by his group, Bill quickly realized that the mechanism predicted in that paper had an issue. The focal point was on the protonation state of an active site glutamate. He told me his concern and I told him that this issue could be addressed by molecular dynamics simulation. At the beginning, he was skeptical about the computer simulation, but I did the simulation anyway. The result I got from my simulation was in very good agreement with what he originally speculated and, moreover, the data provided further mechanistic insights into the system. After seeing the results, Bill was very happy and he started to believe that computer simulation was a powerful tool for studying protein dynamics. This first collaboration marks the beginning of my more than ten years of apprenticeship with Bill until his final days.

Aside from being a great experimentalist, Bill was also a great theoretician. His accomplishment in quantum chemistry was enormous. Indeed, his lab has trained many outstanding theoretical chemists who are professors in many universities. That was one of the reasons that he could so quickly grasp the significance of molecular dynamics simulation in his late years and become a major advocate for such a technique. A very distinct trait of Bill Lipscomb in doing scientific research is that he had no barrier whatsoever in switching fields, and no prejudice against any techniques, be they computational or experimental. Indeed, in looking at his career, he started as a structural inorganic chemist and ended up with a Nobel Prize, then he went on to become the founding father of protein X-ray crystallography in North America, while maintaining a sizable theoretical group in his lab. When I was with him at Harvard, he said to me several times that he, in his career, barely developed any method, but he always picked an important problem and tried to solve that problem by whatever method he could use. It is sad that such a trait is so rare nowadays.

In the summer of 2000, I was getting ready to move to Texas to become an independent investigator at Baylor College of Medicine and Rice University. One day I went into his office and asked a question that many Chinese students ask their mentors — what advice or suggestions could he give me about my career? His answer was that I should pick a research subject that is "just beyond possible." What he meant was that if you pick a problem that is already "possible" at the time you pick it, then the solution of that problem will not be very exciting; if you pick a problem that is way beyond possible at the time you pick it, then you may never solve it. So, picking scientific problems that are "just beyond possible" can ensure that the scientist will deliver a breakthrough in his or her career. I valued this advice enormously and it has been one of my guiding principles in my career ever since.

Also in the summer of 2000, right before I left for Texas, I attended the 80th birthday symposium for Bill Lipscomb at Harvard. I remembered what Doug Rees said in an anecdotal session. Doug was a graduate student of Bill at Harvard and he once asked Bill which achievement in his career he was most proud of, the answer he got from Bill was "the next one!" A great scientist never rests on his past success, instead he always looks for new problems to solve.

In the year 2005, I, together with Professor Hualiang Jiang at the Shanghai Institute of Materia Medica, hosted an 85th birthday symposium for Bill Lipscomb in China. At a scientific meeting at Xi'an Jiaotong University, a student from the huge audience asked Bill a question about what it would take to get a Nobel Prize. He answered that question in many different ways, but the most significant comment I remember was that he suggested one should choose the "right mentor." That is indeed very true. Bill himself benefited from being a student of Linus Pauling who was the greatest chemist in the last century. I myself certainly benefited tremendously from my association with Bill. I learned numerous things from Bill from what he directly told me, from what I indirectly observed, or by studying his work in the past, and these all influenced me profoundly. I am sure that is how one can be consciously and subconsciously influenced by being close to a great master.

Bill Lipscomb has been a great mentor of mine. His impact on my scientific career is enormous and will continue to guide me in my future career. Bill Lipscomb, or the Colonel, will live forever in my heart.

Proc. Natl. Acad. Sci. USA
Vol. 95, pp. 14640–14645, December 1998
Biochemistry

Yeast chorismate mutase in the R state: Simulations of the active site

(allosteric enzyme/molecular dynamics/stochastic boundary condition/enzyme catalysis/protonation state)

JIANPENG MA*, XIAOFENG ZHENG*, GEORG SCHNAPPAUF†, GERHARD BRAUS†, MARTIN KARPLUS‡* AND WILLIAM N. LIPSCOMB*§

*Department of Chemistry and Chemical Biology, Harvard University, 12 Oxford Street, Cambridge, MA 02138; †Institut für Mikrobiologie und Genetik, Georg-August-Universität, Grisebachstrasse 8, D-37077 Göttingen, Germany; and ‡Laboratoire de Chimie Biophysique, Institut Le Bel, Université Louis Pasteur, 67000 Strasbourg, France

Contributed by William N. Lipscomb, September 25, 1998

ABSTRACT **The isomerization of chorismate to prephenate by chorismate mutase in the biosynthetic pathway that forms Tyr and Phe involves C5—O (ether) bond cleavage and C1—C9 bond formation in a Claisen rearrangement. Development of negative charge on the ether oxygen, stabilized by Lys-168 and Glu-246, is inferred from the structure of a complex with a transition state analogue (TSA) and from the pH-rate profile of the enzyme and the E246Q mutant. These studies imply a protonated Glu-246 well above pH 7. Here, several 500-ps molecular dynamics simulations test the stability of enzyme–TSA complexes by using a solvated system with stochastic boundary conditions. The simulated systems are (*i*) protonated Glu-246 (stable), (*ii*) deprotonated Glu-246 (unstable), (*iii*) deprotonated Glu-246 plus one H_2O between Glu-246 and the ether oxygen (unstable), (*iv*) the E246Q mutant (stable), and (*v*) addition of OH^- between protonated Glu-246 and the ether oxygen. In (*v*), a local conformational change of Lys-168 displaced the OH^- into the solvent region, suggesting a possible rate-determining step that precedes the catalytic step. In a 500-ps simulation of the enzyme complexed with the reactant chorismate or the product prephenate, no water molecule remained near the oxygen of the ligand. Calculations using the linearized Poisson–Boltzmann equation show that the effective pK_a of Glu-246 is shifted from 5.8 to 8.1 as the negative charge on the ether oxygen of the TSA is changed from −0.56 electron to −0.9 electron. Altogether, these results support retention of a proton on Glu-246 to high pH and the absence of a water molecule in the catalytic steps.**

In archaebacteria, eubacteria, fungi, and other plants, chorismate mutase (chorismate pyruvate mutase, EC 5.4.99.5) (CM) catalyzes the isomerization of chorismate to prephenate. This branch of the metabolic pathway to aromatic amino acids leads to phenylalanine and tyrosine. The other branch (the conversion of chorismate to anthranilate by anthranilate synthase) leads to tryptophan (1, 2). In *Saccharomyces cerevisiae,* CM (Fig. 1) is a homodimer consisting of two 30-kDa polypeptide chains. This yeast CM (YCM) shows homotropic activation by the substrate chorismate, approximately 10-fold allosteric activation by tryptophan, and approximately 10-fold inhibition by tyrosine (2, 3). This 100-fold modulation of YCM activity by Trp and Tyr controls the flux of chorismate into these two branches of the biosynthesis of aromatic amino acids. The structural basis for these regulatory processes has been described recently (4, 5).

The CMs catalyze the only known Claisen rearrangement in living systems (Fig. 2). It has been shown that the enzyme preferentially binds the less stable pseudodiaxial conformation of the enolpyruvate side chain (6) (Fig. 2), which is positioned thereby for the rearrangement in the chair-like transition state (7, 8). Its complex with the active site of YCM (4) is shown in Fig. 3. The mechanistic aspects of catalysis, outlined before the structures were known, include several possibilities (9), one of which was transient protonation of the ether (leaving) oxygen of the enolpyruvate side chain, and another involved hydrogen bonds to the ether oxygen to stabilize the developing negative charge on this leaving oxygen. Indeed, when an oxabicyclic diacid transition state analogue, prepared by Bartlett and coworkers (10, 11), is bound to YCM, hydrogen bonds involving the ether oxygen are inferred from its contacts with Lys-168 and Glu-246 (4). Also, the protonated form of Glu-246 is supported by pH-rate profiles of the wild-type enzyme and the Glu-246 → Gln mutant (3, 12), and by the determination of the location (13) of the active site in YCM on the basis of homology with the P protein of the CM of *Escherichia coli* (14). However, in a recent computational study (15) of docking the transition state analogue to YCM, it is argued that, around the activity maximum of about pH 6.5 when Trp is present to maximize activation, Glu-246 is unprotonated, and that its interaction with the ether (leaving) oxygen is mediated by an intervening water molecule. Because of the well known uncertainty of such modeling studies in determining specific interactions, we have performed molecular dynamics studies based on the crystal structure of YCM (4) complexed with the TSA. Simulations of the active site complexed with the TSA are performed for protonated or unprotonated Glu-246, for the mutant Gln 246, and for the structure, which has a water molecule between Glu-246 and the ether (leaving) oxygen; this last configuration was suggested as the stable structure in the modeling study (15). However, there is no evidence in the dynamics studies presented here for the presence of such an intervening water molecule in any of the structures in which the TSA is bound to YCM. A simulation is also performed for a system with an ionized water molecule modeled with its OH^- part in the active site at a position similar to that of the intervening water. Simulations indicate that this OH^- ion is transported to solvent by a local conformational change of Lys-168. No water was observed in this position during the simulations with reactant chorismate or product prephenate. Finally, the effective pK_a value of Glu-246, calculated by using the linearalized Poisson–Boltzmann method, is raised significantly so that Glu-246 is likely to retain a proton at an elevated pH.

METHODS

Initial coordinates (Protein Data Bank code 3csm) were those of the wild-type "super R" state bound to inhibitor plus Trp (ITRP

The publication costs of this article were defrayed in part by page charge payment. This article must therefore be hereby marked "*advertisement*" in accordance with 18 U.S.C. §1734 solely to indicate this fact.

Abbreviations: CM, chorismate mutase; YCM, yeast chorismate mutase; TSA, transition state analogue; RDS, rate-determining step; ECM, *Escherichia coli* chorismate mutase; BCM, *Bacillus subtilis* chorismate mutase.

§To whom reprint requests should be addressed. e-mail: lipscomb@chemistry.harvard.edu.

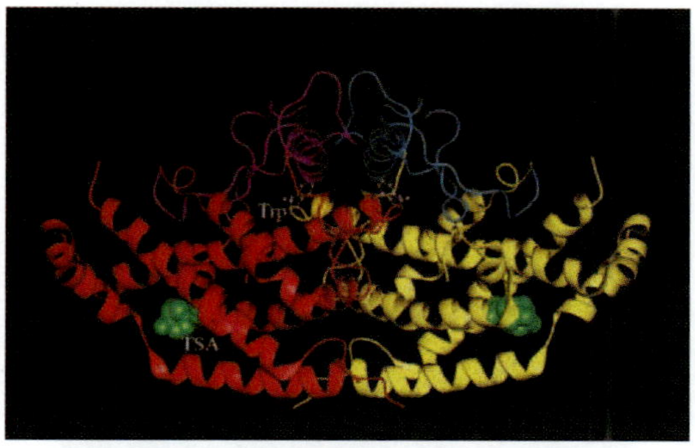

FIG. 1. Schematic representation of the dimeric structure of YCM. The allosteric domain of the left monomer (in red) is drawn in magenta and that of the right monomer (in yellow) is drawn in blue. The two effector tryptophans (in white) are in ball-and-stick form and the transition state analogues (in green) are in space-filling model. This figure, as well as Figs. 4–7 and 9, was made with QUANTA (Molecular Simulations, San Diego, CA) [(modified from (4)].

in ref. 4), in which tryptophan and the TSA are bound to the regulatory and active sites, respectively. To reduce the necessary simulation time, in accord with many other studies of active sites, the stochastic boundary molecular dynamics method (16, 17) was used; this procedure also avoided the need to model certain loops that are not well defined in the crystal structure. The CHARMM program (Molecular Simulations, San Diego, CA) (18) was used for the simulation. The all-hydrogen potential function (PARAM22) (19) was used for all protein atoms, and a modified TIP3P water model (20) was used for the solvent (21). Atomic partial charges for the transition state analogue were obtained from the Merck Molecular Force Field (Merck) (22–26) implemented for molecular dynamics in the CHARMM program. The system was separated into a reaction zone and a reservoir region, and the reaction zone was further divided into the reaction region and the buffer region (for details of stochastic boundary molecular dynamics, see ref. 17). The reference point for partitioning the system in stochastic boundary molecular dynamics was chosen as the bridge ether oxygen of the inhibitor. The reaction region around an active site was the sphere of radius r of 14 Å, the buffer region was $14 < r < 16$ Å, and the reservoir region corresponded to $r > 16$ Å; all atoms in the reservoir region were deleted. The simulation system, shown in Fig. 4, consisted of 101 protein residues (1731 atoms), a TSA, and 193 waters. Inside the reaction region, atoms were propagated by molecular dynamics (17), whereas atoms in the buffer region were propagated by Langevin

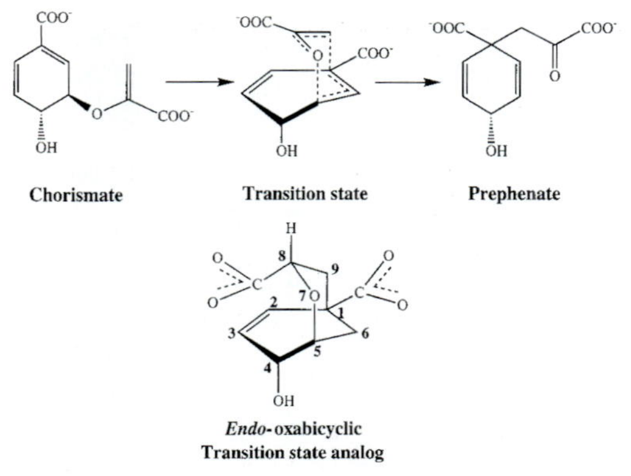

FIG. 2. Transformation of chorismate to prephenate via proposed transition state.

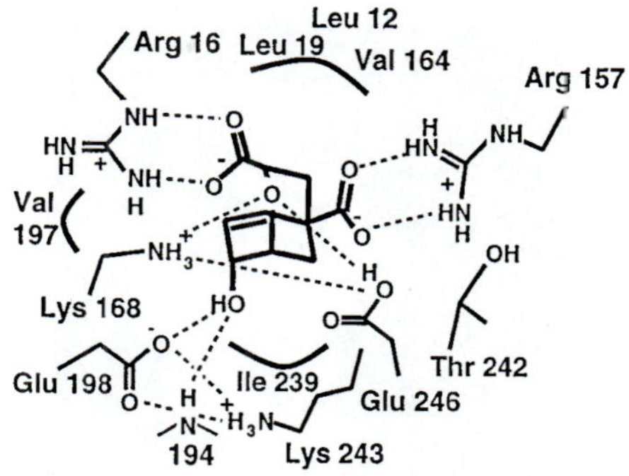

FIG. 3. Active site-binding interactions (4) [Reproduced with permission from Current Biology Ltd. (4) (Copyright 1997, Current Biology Ltd.)].

dynamics. Atoms inside the buffer region were retained by a harmonic restoring force with force constants derived from the temperature factors in the crystal structure. Water molecules were confined to the active site region by a deformable boundary (27). The friction constant in the Langevin dynamics (17) was 250 ps^{-1} for the protein atoms and 62 ps^{-1} for the water molecules. During the simulation, all of the bonds with hydrogen atoms were fixed by the SHAKE algorithm (18). A 1-fs time step was used for integration of the equations of motion during the molecular dynamic simulation. After 200 steps of minimization by using the steepest descent method and 300 steps of minimization by using the Adapted Basis Newton Raphson's method (18), the system was equilibrated for 50 ps at 300 K. This procedure was followed by a 500-ps production run. Seven simulations were performed on the enzyme-inhibitor complex with (*i*) the Glu-246 protonated, (*ii*) the Glu-246 deprotonated, (*iii*) the Glu-246 replaced by Gln-246 (the E246Q mutant), (*iv*) a water molecule between an unprotonated Glu-246 and the ether oxygen of the TSA, (*v*) an ionized water molecule modeled in a position similar to that in *iv*, such that the H^+ is attached to Glu-246 and the OH^- is placed nearby, (*vi*) the transition state analogue modeled as the reactant chorismate, and (*vii*) the transition state analogue modeled as the product prephenate.

For the evaluation of the pK_a, the linearized Poisson–Boltzmann equation was solved by using the finite difference method (for details, see refs. 28–31). For this part of the study, the whole YCM monomer structure was used. The protein coordinate was prepared by 200 steps of minimization by using the steepest descent method and 100 steps of minimization by using the Adapted Basis Newton Raphson's method. The calculation, using a modified version of the UHBD program (28) that is integrated with the CHARMM program, was carried out on a 180 × 180 × 180 grid at an initial spacing of 3 Å and converged iteratively to 0.2 Å. Dielectric constants of 80 for solvent and 20 for the interior of the protein were used. A physiological ionic strength (145 mM), and a temperature of 298 K were chosen. The Stern (ion-exclusion) layer is 2 Å. The pK_a calculation was performed by using a program developed by M. Schaefer, M. Sommer and M. Karplus (31).

RESULTS

Dynamics with Glu-246 Protonated. During the 500-ps simulation in the presence of the TSA and a protonated Glu-246, the TSA remained bound deeply buried within the four-helix bundle formed by H2, H8, H11, and H12, consistent with the crystal structure (4). The two carboxylate groups of the inhibitor form well defined salt bridges with Arg-16 and Arg-157. Except for the channel (the "allosteric pathway") between H2 and H11, the inhibitor remained inaccessible to solvent. Two arginine residues,

FIG. 4. System included in the stochastic boundary molecular dynamics simulation. The protein is in ribbon representation, the TSA is in space-filling model, and the water molecules are drawn as a stick model.

Arg-16 on H2 and Arg-204 on H11, are located near the entry of this channel. In a space-filling model (Fig. 5), one can see part of the inhibitor from outside in the direction of this channel, which may be the route for substrate binding and escape. In particular, the negatively charged inhibitor may be guided by the electrostatic field generated by the arginine residues, especially Arg-204. This channel is located furthest from the allosteric domain and facing away from the dimer interface. Inside this part of the channel, there is a hydrogen-bonding network formed by the water molecules and protein atoms (see Fig. 6).

An essential result of the simulation is the stability of the interactions between the ether bridge oxygen atom of the inhibitor and the polar side chains of Lys-168 and Glu-246. The key residue Glu-246 remains completely solvent inaccessible, as does the ether bridge oxygen; the waters nearest to the active site are separated from Glu-246 by the side chain of Lys-168. A water molecule interacts with the guanidinium group of Arg-16 (see Fig. 7). It was suggested in previous work (4) that the side chain of Glu-246 is probably protonated, based on the distance of about 3.1 Å (4) between its carboxylate group and the ether oxygen observed in the crystal structure. The present simulation clearly demonstrates that, when the proton is added to the carboxylate group of Glu-246, the structural integrity is well maintained. Also, Glu-246 remains hydrogen-bonded to Lys-168, which donates a good hydrogen bond to one of the carboxylate groups of the inhibitor (see Fig. 7). The side chain of Lys-168 is within hydrogen-bonding distance of the ether oxygen and of the main chain carbonyl group of Ile 192. Fig. 8 shows the hydrogen bond distance as a function of time for the interaction between the ether oxygen and the proton on the carboxylate oxygen of Glu-246 that is distal to Lys-168. A simulation in which the proton is added to the carboxylate oxygen atom of Glu-246 proximal to Lys-168 resulted in a flip of the side chain of Glu-246. Thus, the protonation of the oxygen distal to Lys-168 is more favorable.

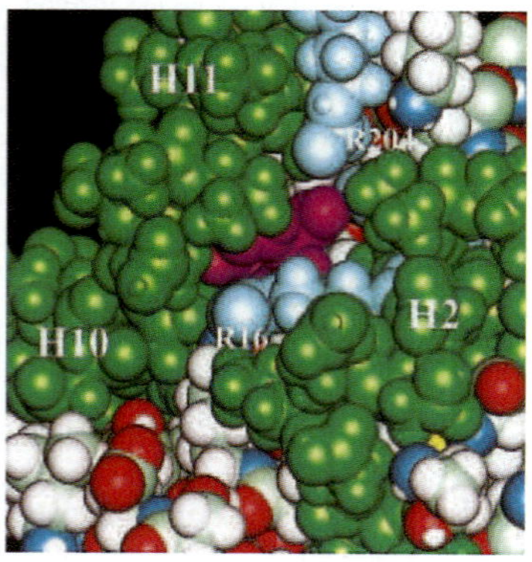

FIG. 5. Space-filling representation of the substrate channel. The helices surrounding the channel (H2, H10, and H11) are in green. The arginine side chains of R 16 and R 204 located at the entry of the channel are in light blue, and the inhibitor is in magenta.

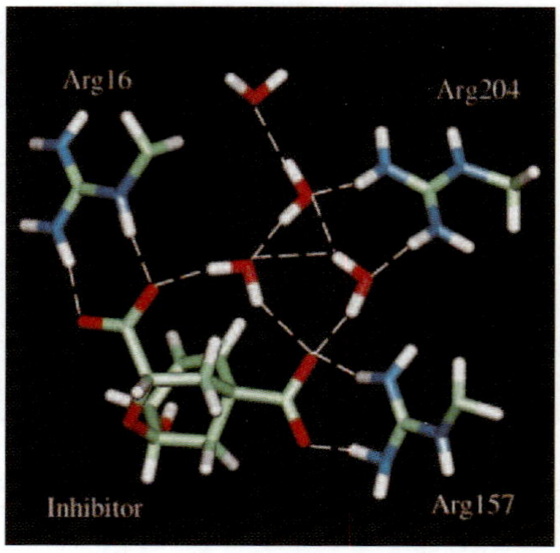

FIG. 6. Hydrogen-bonding network observed inside the substrate channel.

Dynamics with Glu-246 Unprotonated. A 500-ps dynamics simulation with the Glu-246 anion resulted in a severe distortion of the active site structure. The side chain of Glu-246, which remained hydrogen-bonded to Lys-168, was shifted to a distance of about 4.5 Å from the ether oxygen. This result is inconsistent with the x-ray crystal structure (4), in which the distance between the carboxylate of the Glu-246 side chain and the ether bridge oxygen is only about 3.1 Å. The Lys-168 side chain also moved away from its position in the crystal structure, and the hydrogen bond between Lys-168 and the carboxylate group of the inhibitor that interacts with Arg-16 was weakened (i.e., the distance increased).

Dynamics with a Gln-246 Mutant. When the E246Q mutant was simulated for 500 ps, the structure was stable. One hydrogen of the amide group of the glutamine side chain makes a hydrogen bond to the main chain carbonyl of Thr-242, whereas the other hydrogen of this amide group makes the expected hydrogen bond

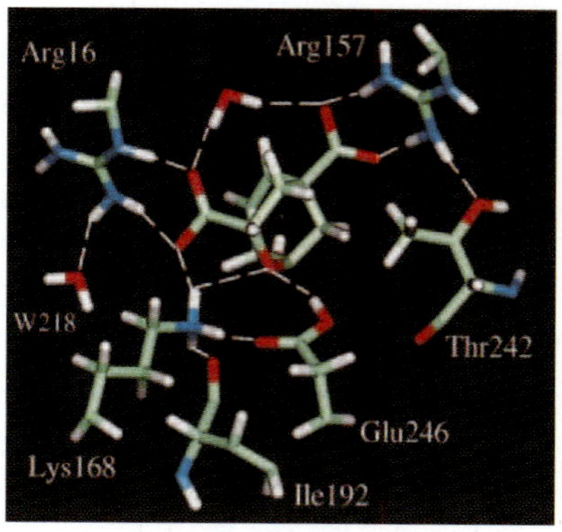

FIG. 7. Interactions between the inhibitor and the groups in the active site.

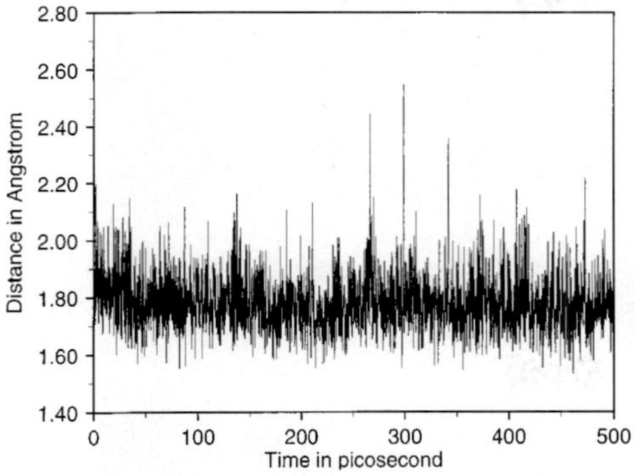

FIG. 8. Hydrogen-bonding distance as a function of time for the proton on Glu-246 and the ether oxygen of the inhibitor.

to the ether oxygen of the inhibitor. There were no significant changes in the structure.

Dynamics with a Water Molecule Between Unprotonated Glu-246 and the Ether Oxygen. To examine the possibility, suggested by the recent docking study (15), that the Glu-246 side chain interacts with the bridge oxygen through a water molecule, a 500-ps simulation was performed in which a water molecule was modeled into the space between the side chain of Glu-246 and the ether oxygen (as in Fig. 6*a* of ref. 15). The results show that the presence of a water molecule in this position has a strong effect on the binding mode of the inhibitor. Because of the size of the water molecule, the inhibitor was displaced in the early stage of simulation so that it could no longer make good hydrogen bonds to Arg-16 and Arg-157. Near the middle of the simulation (about 200 ps from the beginning of the production run), the water dissociated from the ether oxygen, though it remained bound to Glu-246. As a result, the inhibitor shifted back toward its original position and regained the hydrogen bonds to Arg-16 and Arg-157. Moreover, even though the water was modeled as hydrogen bonded to both Glu-246 and the ether oxygen, its oxygen atom is surrounded by hydrophobic groups including the methyl groups of Thr-242, Leu-19, Val-164, and a methyl group from the inhibitor. This environment, plus the lack of space, appear to make it unfavorable for the water to stay at the modeled position.

Dynamics with an Ionized Water Molecule Between Glu-246 and the Ether Oxygen. The uncatalyzed conversion of chorismate to prephenate involves no solvent H_2O/D_2O isotope effect, and a substantially asymmetric transition state: k_H/k_T is 1.15 ± 0.01 (equilibrium: 1.43) for C5—O bond breaking and 0.99 ± 0.01 (equilibrium: 0.78) for bond making at C9 (32). Catalysis by the *E. coli* CM (ECM)-T protein shows a solvent H_2O/D_2O effect of 2.2 on k_{cat} but no effect (1.0) on K_m; moreover, catalysis by ECM-T at low substrate concentrations (V_{max}/K_m conditions) is insensitive to H–T substitution at C5 or C9, thus implying a rate-determining step (RDS) before the chemical transformation (32). Catalysis by the *Bacillus subtilis* CM enzyme (BCM) shows no solvent H_2O/D_2O effect; moreover, catalysis of BCM at high substrate concentration (V_{max} conditions) indicates that k_{off} for prephenate is almost identical to k_{cat}, suggesting that under V_{max} conditions the loss of prephenate is the RDS (the "on" rate constant for prephenate is almost the same as k_{cat}/K_m for chorismate) (33).

A model for the proposed RDS that precedes the chemical transformation in YCM is examined by introducing an OH^-, presumably from an ionized water molecule that has protonated Glu-246. A 500-ps simulation was carried out, starting with an OH^- ion in a position similar to that of the intervening water in the previous simulation (see Fig. 9*a*). At the beginning of the simulation, the OH^- ion is hydrogen bonded to Lys-168. Around 50 ps, the OH^- ion moves out into the solvent as a result of a local conformational change of the side chain of Lys-168 (Fig. 9 *b–d*). Although Lys-168 acts as a gate in this process, this lysine residue is bulky enough that no water molecules enter the active site even in the most open conformation involving the complex of the enzyme with TSA, OH^-, and solvent. In fact, once the OH^- ion is displaced (Fig. 9 *d* and *e*), the side chain of Lys-168 quickly regains its original conformation in the enzyme–TSA complex (Fig. 9 *e* and *f*). Therefore, one might expect a solvent D_2O effect in YCM (Lys-168 and Glu-246) like that observed (9) for the ECM-T (Lys-37 and Glu-86), but not found (33) in BCM, which has bidentate Arg bound to the ether oxygen. In the remaining part of the active site, no large conformational change is observed. Although the transport of the OH^- ion is facilitated mainly by the motion of Lys-168, other groups also contribute. In particular, the guanidinium group of Arg-16, the main chain amide group of Asn-94, and the hydroxyl group on the TSA provide favorable interactions together with Lys-168 as the OH^- ion slides out (Fig. 9*d*). The hydroxyl group of the TSA also undergoes a displacement, but quickly recovers its original conformation (Fig. 9 *c–e*).

This striking result enlarges the possible role of the protonation of Glu-246. It reveals that the ionization of a water molecule and the displacement of the resulting hydroxide anion from the active site could provide the proton to Glu-246, and also, at least in part, could contribute to the RDS. These results provide an alternative to the proposed conformational change of the enolpyruvoyl side chain of the substrate as the RDS step (33) at low substrate concentrations.

Dynamics with Reactant Chorismate or Product Prephenate. The simulations with the reactant chorismate or the product prephenate, modeled from the transition state analogue, show that both systems are stable. The overall geometry and interactions are close to those observed in the crystal structure of the transition state analogue. Most importantly, no water molecule enters into the active site, including the vicinity of the ether oxygen. In the complexes of the enzyme fragment with either chorismate or prephenate, certain side chains, such as that of Lys-16, which makes direct contacts with the pyruvate side chain of the substrate, have larger atomic fluctuations than in the transition state analogue complex. This result is due partly to the increased flexibility of the pyruvate side chain of the substrate. The greater rigidity of the system when the TSA is present, as compared with the reactant or product, is in accord with the concept that the substrate binds tightly to the enzyme in the transition state. However, the transition state analogue is not ideal because of the lack of the stretched bonds and the developing charges of the true transition state (11). Hence, the TSA is bound more strongly than chorismate or prephenate by only a factor of about 20 to 30: the K_i is 3 μM for TSA and 70 μM for prephenate, and the K_m for chorismate is 100 μM for the BCM (33). It is expected that the real transition state would bind more tightly.

Calculation of pK_a Values of Glu-246. To complement the dynamic simulation, the pK_a values of Glu-246 have been calculated under various conditions by using the linearized Poisson–Boltzmann equation. The pK_a value of Glu-246 is mainly affected by a few important features in the active site. They are the overall low dielectric environment in the active site because of the exclusion of the bulk water, the negatively charged bridge oxygen, and the positively charged Lys-168 side chain. The first two features increase the pK_a and the third one decreases the pK_a. The effective pK_a value of Glu-246 is 5.8 (to be compared with the intrinsic pK_a of 4.4) if the bridge oxygen is modeled as an ether oxygen in TSA, and it increases to 8.1 if the partial charge on the bridge oxygen is increased from −0.56 as in TSA to −0.9 as a mimic of the actual transition state. These calculations are consistent with the results of the dynamic simulation, i.e., Glu-246 is protonated under conditions corresponding to catalysis in the active site of YCM.

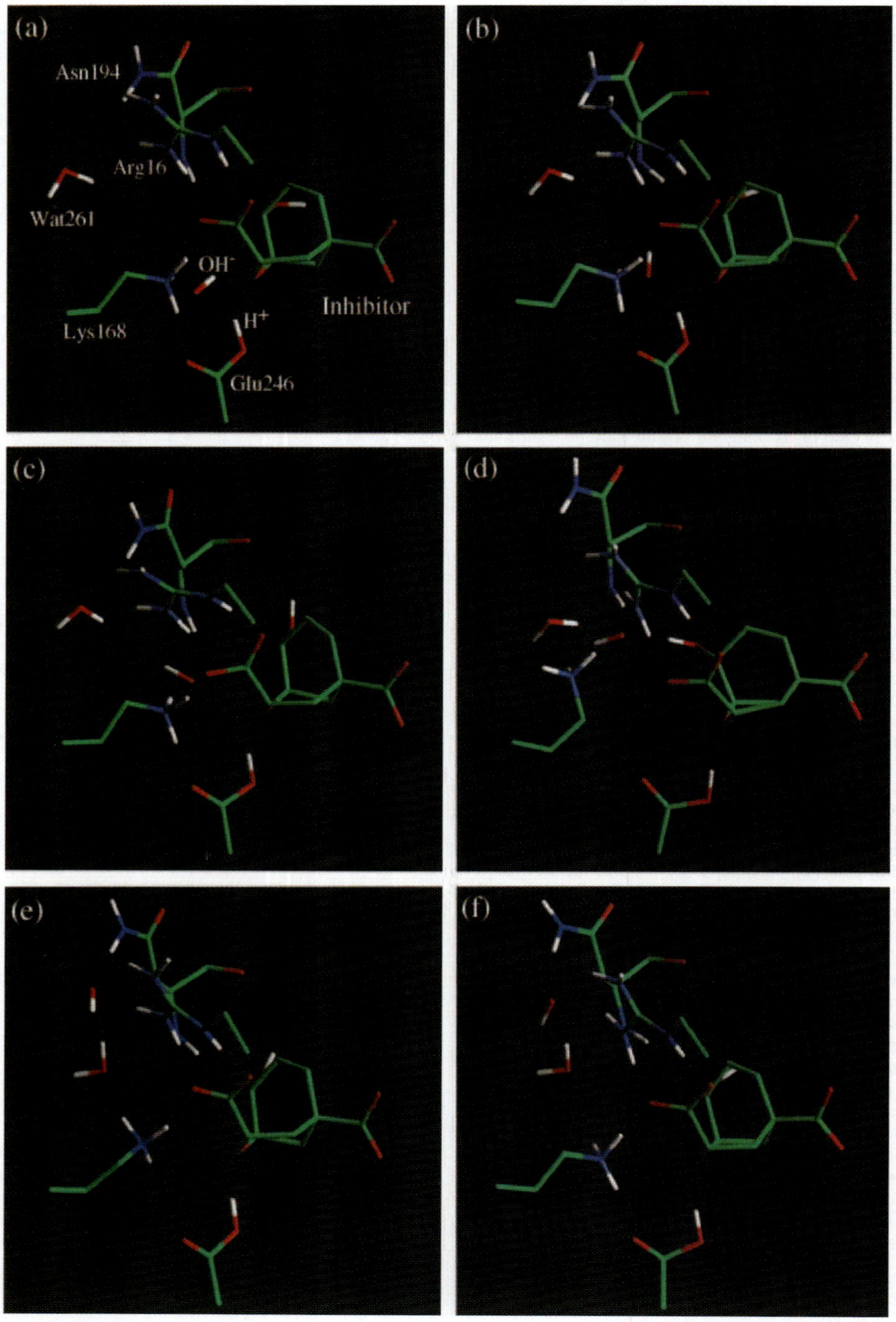

FIG. 9. Time evolution of the displacement of OH^- ion from the active site mediated by a local conformational change of Lys-168. The nearest water molecule to the active site, Wat261, is indicated explicitly.

DISCUSSION

The results of the simulations indicate that a protonated Glu-246 leads to a stable structure in which Glu-246 is hydrogen bonded to the ether oxygen, consistent with the x-ray results (4). The simulation for the complex in which Glu-246 is unprotonated and that for the complex in which there is a water molecule between Glu-246 and the ether oxygen show structural distortions that are inconsistent with the x-ray results (4). The simulation with an ionized water molecule in the active site result in a quick displacement of OH^- ion into the solvent. Also, the mutant Glu-246Gln is stable over a 500-ps simulation. No water is observed in the active site in the simulations of enzyme-reactant (chorismate) and enzyme-product (prephenate) complexes. The effective pK_a value of Glu-246 is calculated to be shifted significantly upward under catalytic conditions with a developing negative charge on the ether oxygen.

The pH-activity profile for the wild-type Glu-246 YCM (12) shows a maximum at about pH 6.5 and more than half-maximal activity between pH values of 5 and 8.5. The assignment (12) of a high pK_a for Glu-246 is strengthened by comparison of the pH-activity profile of the Glu-246 enzyme with that of the Glu-246 to Gln mutant. In this mutant (E246Q), the high end of the pH profile is extended to about 10.5, probably reflecting the pK_a of Lys-168 or some other residues. Although the simulations need to be confirmed by more quantitative studies, such as free energy simulations, the results reported here are in accord with the proposal that a protonated Glu-246, with an increased pK_a, is involved in the catalytic mechanism. The suggestion (15) of an active site water as the mediator between Glu-246 and the ether oxygen is not supported by the present results.

Although the structural and dynamic results give information about active site binding, the details of rate enhancement by the CM enzymes, typically by factors of 10^6 to 10^7 over the rate of the uncatalyzed reactions (34, 35), have yet to be analyzed qualitatively. Effects include (36) (*i*) the role of selection of the less stable pseudodiaxial conformer (about 10% of the total) of chorismate, (*ii*) the role of specific electrostatic stabilization of the ether oxygen (e.g., Arg-90 in BCM), including other hydrogen bonding and general electrostatic effects that promote the charge separation in the transition state (Arg-90 and Glu-78 in BCM), and (*iii*) the creation of a strongly hydrophobic environment. These three effects also apply to the P protein of ECM and to YCM, where Lys-39 and Lys-168, respectively, stabilize the ether oxygen, along with Gln 88 and a protonated Glu-246, respectively. Moreover, the charge separation as the transition state develops is also probably aided by Glu-52 in ECM and Glu-198 in YCM.

It is rare that the chemical transformation is the RDS in enzyme catalysis. For the BCM (36, 37), the RDS occurs before the chemical transformation at low substrate concentration (32, 33) (V_{max}/K_m conditions) and probably involves a change in the binding step or a conformational change that occurs before the chemical rearrangement. At high substrate concentrations (V_{max} conditions), the departure of the product prephenate is the rate-limiting step (33). Although no sequence homology exists between the *B. subtilis* enzyme and *E. coli* (or yeast) enzyme, the similarity of kinetic parameters suggests functional similarity (32). Both substrate binding and product leaving are expected to show large solvent compensation effects involving $\Delta H^{\ddagger}$ and $\Delta S^{\ddagger}$ (38–40), and these thermodynamic parameters may depend substantially on temperature. Thus, it may be difficult to compare these thermodynamic parameters obtained from log k versus $1/T$ plots via the Eyring equation from the different CM enzymes (41, 42) and especially to compare these parameters with those for the uncatalyzed reaction (41, 42). Molecular mechanics studies of these solvent-sensitive stages of the reaction are needed to aid in an understanding of the rate-sensitive steps in the mechanism.

This work was supported by grants GM06920 (W.N.L.) and GM30804 (M.K.) from the National Institutes of Health. J.M. was the recipient of a postdoctoral fellowship from Burroughs Wellcome Fund, Program in Mathematics and Molecular Biology, in the early part of this study. He is now a postdoctoral fellow of the National Institutes of Health.

1. Weiss, U. & Edwards, J. M. (1980) *The Biosynthesis of Aromatic Amino Compounds* (Wiley, New York).
2. Braus, G. H. (1991) *Microbiol. Rev.* **55,** 349–370.
3. Schmidheini, T., Mosch, H. U., Evans, J. N. & Braus, G. (1990) *Biochemistry* **29,** 3660–3668.
4. Sträter, N., Schnappauf, G., Braus, G. & Lipscomb, W. N. (1997) *Structure* **5,** 1437–1452.
5. Sträter, N., Hakansson, K., Schnappauf, G., Braus, G. & Lipscomb, W. N. (1996) *Proc. Natl. Acad. Sci. USA* **93,** 3330–3334.
6. Sogo, S. G., Widlanski, T. S., Hoare, J. H., Grimshaw, C. E., Berchtold, G. A. & Knowles, J. R. (1984) *J. Am. Chem. Soc.* **106,** 2701–2703.
7. Copley, S. D. & Knowles, J. R. (1985) *J. Am. Chem. Soc.* **107,** 5306–5308.
8. Copley, S. D. & Knowles, J. R. (1987) *J. Am. Chem. Soc.* **109,** 5008–5013.
9. Guilford, W. J., Copley, S. D. & Knowles, J. R. (1987) *J. Am. Chem. Soc.* **109,** 5013–5019.
10. Bartlett, P. A. & Jackson, C. R. (1985) *Biopolymers* **97,** 843–865.
11. Mader, M. M. & Bartlett, P. A. (1997) *Chem. Rev. (Washington, DC)* **97,** 1281–1301.
12. Schnappauf, G., Sträter, N., Lipscomb, W. N. & Braus, G. H. (1997) *Proc. Natl. Acad. Sci. USA* **94,** 8491–8496.
13. Xue, Y. & Lipscomb, W. N. (1995) *Proc. Natl. Acad. Sci. USA* **92,** 10595–10598.
14. Lee, A. Y., Karplus, A. P., Ganem, B. & Clardy, J. (1995) *J. Am. Chem. Soc.* **117,** 3627–3628.
15. Lin, S. L., Xu, D., Li, A., Rosen, M., Wolfson, H. J. & Nussinov, R. (1997) *J. Mol. Biol.* **271,** 838–845.
16. Brooks, C. L., III, Brunger, A. & Karplus, M. (1985) *Biopolymers* **24,** 843–865.
17. Brooks, C. L., III, & Karplus, M. (1989) *J. Mol. Biol.* **208,** 159–181.
18. Brooks, B. R., Bruccoleri, R. E., Olafson, B. D., States, D. J., Swaminathan, S. & Karplus, M. (1983) *J. Comput. Chem.* **4,** 187–217.
19. MacKerell, A. D., Jr., Bashford, D., Bellott, M., Dunbrack, R. L., Jr., Evanseck, J. D., Field, M. J., Fischer, S., Gao, J., Gao, H., Ha, S., *et al.* (1998) *J. Phys. Chem.*, Series B, 3586–3616.
20. Jorgensen, W. L. (1981) *J. Am. Chem. Soc.* **103,** 335–340.
21. Neria, E., Fisher, S. & Karplus, M. (1996) *J. Chem. Phys.* **105,** 1902–1921.
22. Halgren, T. A. (1996) *J. Comput. Chem.* **17,** 490–519.
23. Halgren, T. A. (1996) *J. Comput. Chem.* **17,** 520–552.
24. Halgren, T. A. (1996) *J. Comput. Chem.* **17,** 553–586.
25. Halgren, T. A. & Nachbar, R. B. (1996) *J. Comput. Chem.* **17,** 587–615.
26. Halgren, T. A. (1996) *J. Comput. Chem.* **17,** 616–641.
27. Brooks, C. L., III, & Karplus, M. (1983) *J. Chem. Phys.* **79,** 6312–6325.
28. Davis, M. E., Madura, J. D., Luty, B. A. & McCammon, J. A. (1991) *Comput. Phys. Commun.* **62,** 187–197.
29. Honig, B. & Nicholls, A. (1995) *Science* **268,** 1144–1149.
30. Bashford, D. & Karplus, M. (1990) *Biochemistry* **29,** 10219–10225.
31. Schaefer, M., Sommer, M. & Karplus, M. (1997) *J. Phys. Chem.* **101,** 1663–1683.
32. Addadi, L., Jaffe, E. K. & Knowles, J. R. (1983) *Biochemistry* **22,** 4494–4501.
33. Gray, J. V., Eren, D. & Knowles, J. R. (1990) *Biochemistry* **29,** 8872–8878.
34. Gorisch, H. (1978) *Biochemistry* **17,** 3700–3705.
35. Andrews, P. R., Smith, G. D. & Young, I. G. (1973) *Biochemistry* **12,** 3492–3498.
36. Chook, Y. M., Gray, J. V., Ke, H. & Lipscomb, W. N. (1994) *J. Mol. Biol.* **240,** 476–500.
37. Chook, Y. M., Ke, H. & Lipscomb, W. N. (1993) *Proc. Natl. Acad. Sci. USA* **90,** 8600–8603.
38. Hins, H. J., Shiao, D. F. & Sturtevant, J. M. (1971) *Biochemistry* **10,** 1347–1352.
39. Sturtevant, J. M. (1977) *Proc. Natl. Acad. Sci. USA* **74,** 2236–2240.
40. Naghibi, H., Tamura, A. & Sturtevant, J. M. (1995) *Proc. Natl. Acad. Sci. USA* **92,** 5597–5599.
41. Kast, P., Asif-Ullah, M. & Hilvert, D. (1996) *Tetrahedron Lett.* **37,** 2691–2694.
42. Galopin, C. C., Zhang, S., Wilson, D. B. & Ganem, B. (1996) *Tetrahedron Lett.* **37,** 8675–8678.

Structure of the full-length Shaker potassium channel Kv1.2 by normal-mode-based X-ray crystallographic refinement

Xiaorui Chen[a], Qinghua Wang[b], Fengyun Ni[c], and Jianpeng Ma[a,b,c,1]

[a]Graduate Program of Structural and Computational Biology and Molecular Biophysics, Baylor College of Medicine, One Baylor Plaza, Houston, TX 77030; [b]Verna and Marrs McLean Department of Biochemistry and Molecular Biology, Baylor College of Medicine, One Baylor Plaza, Houston, TX 77030; and [c]Department of Bioengineering, Rice University, Houston, TX 77005

Edited by William N. Lipscomb, Harvard University, Cambridge, MA, and approved April 30, 2010 (received for review January 6, 2010)

Voltage-dependent potassium channels (Kv) are homotetramers composed of four voltage sensors and one pore domain. Because of high-level structural flexibility, the first mammalian Kv structure, Kv1.2 at 2.9 Å, has about 37% molecular mass of the transmembrane portion not resolved. In this study, by applying a novel normal-mode-based X-ray crystallographic refinement method to the original diffraction data and structural model, we established the structure of full-length Kv1.2 in its native form. This structure offers mechanistic insights into voltage sensing. Particularly, it shows a hydrophobic layer of about 10 Å at the midpoint of the membrane bilayer, which is likely the molecular basis for the observed "focused electric field" of Kv1.2 between the internal and external solutions. This work also demonstrated the potential of the refinement method in bringing up large chunks of missing densities, thus beneficial to structural refinement of many difficult systems.

anisotropic *B* factors | ion channel | normal-mode refinement | voltage gating

Voltage-dependent ion channels sense change in the voltage across cell membrane and respond by allowing specific ions in/out with high selectivity and efficiency (1). Among them, voltage-dependent potassium channels (Kv) have been most extensively studied (2–6). In an action potential, the Kv channel allows intracellular potassium getting out of the cell to return the membrane potential to the polarized state, the failure of which could cause overexcitability of neuron cells, leading to neuronal diseases (7, 8). The Kv channel is a homotetramer with four voltage sensors and one central pore domain (9, 10). In each subunit, four transmembrane helices (S1, S2, S3, and S4) make up a voltage-sensor domain, and two transmembrane helices (S5 and S6) contribute to the central pore domain. There are several highly conserved positively charged residues on the S4 helix, the so-called gating charges, which respond to voltage change to open or close the central pore domain (11).

The first mammalian Kv crystal structure was of the Shaker family Kv1.2 from *Rattus norvegicus* [Protein Data Bank (PDB) ID code 2A79], determined to a resolution of 2.9 Å (10). Its structural model was refined by conventional isotropic X-ray crystallographic refinement protocol in CNS (12). Because of the high dynamic and/or static disordering, about 37% of the total molecular mass of the transmembrane portion (residues 131–421) was not resolved. For instance, all the loops connecting S1–S4 helices, and all the side chains on S1 and S3, as well as some of the side chains on S2 and S4, were missing (Fig. 1*A*). In order to obtain a clearer picture of Kv channels, the structure of a "paddle chimera" Kv was recently determined at a higher resolution of 2.4 Å (13). The improvements in the structure were achieved by crystallizing the channel in a lipid environment and by replacing the voltage-sensor paddle of Kv1.2 with the corresponding region of Kv2.1, a closely related Kv channel in the Shab family. Even though the chimera channel behaved similarly to the native Kv1.2 in electrophysiological assays, it is still desirable to have a complete structure for the native Kv1.2 channel.

A normal-mode-based X-ray crystallographic refinement method was recently developed by us and its efficiency in improving structural models was shown in a number of systems such as a supramolecular complex (14), a membrane-bound ion channel (15), a heavily glycosylated protein (16), and other systems (17). The method models anisotropic *B* factors for all atoms by using only a small set of low-frequency normal modes (14, 18, 19). Consequently, the number of parameters used in structural refinement is drastically reduced, which allows a better data-to-parameter ratio, and the structural flexibility is more accurately described than by using conventional isotropic *B*-factor refinement. Therefore, in this study, we employed this refinement method to improve the deposited Kv1.2 structure (PDB ID code 2A79) on the basis of the original diffraction data (10).

Through several iterations of normal-mode-based anisotropic *B*-factor refinement, positional refinement by REFMAC5 and manual model adjustments, we were able to lower the *R* factors and build all the missing atoms in the improved electron density maps. The normal-mode refined structure for Kv1.2 offered insights into the mechanisms of voltage sensing. In particular, a hydrophobic layer of about 10 Å was identified at the midpoint of the membrane bilayer, which is likely the molecular basis for the observed "focused electric field" of Kv1.2 between the internal and external solutions.

Results

Normal-Mode-Based Refinement of Kv1.2 Structure. To eliminate the slight differences in structural statistics because of the use of different structural refinement programs or different versions of them, we generally reminimize the structural models from PDB (14–17). In this study, the Kv1.2 structure (PDB ID code 2A79) (Fig. 1*A*) originally refined by CNS (12) was reminimized using REFMAC5 (20) in CCP4 (21). The reminimization yielded the R_{cryst} factor of 21.9% and R_{free} factor of 24.5%, slightly lower than those for the published model (22.2% for R_{cryst} factor and 25.2% for R_{free} factor) (10). This reminimized model served as the "starting model" in our normal-mode-based refinement.

To start, several parameters for normal-mode-based refinement, including the number of low-frequency modes and the value

Author contributions: J.M. designed research; X.C. and J.M. performed research; X.C., Q.W., F.N., and J.M. analyzed data; and Q.W. and J.M. wrote the paper.

The authors declare no conflict of interest.

This article is a PNAS Direct Submission.

Freely available online through the PNAS open access option.

Data deposition: The structure coordinates of the final Kv1.2 structure have been deposited in PDB with the accession code of 3LUT.

[1]To whom correspondence should be addressed. E-mail: jpma@bcm.tmc.edu.

This article contains supporting information online at www.pnas.org/lookup/suppl/doi:10.1073/pnas.1000142107/-/DCSupplemental.

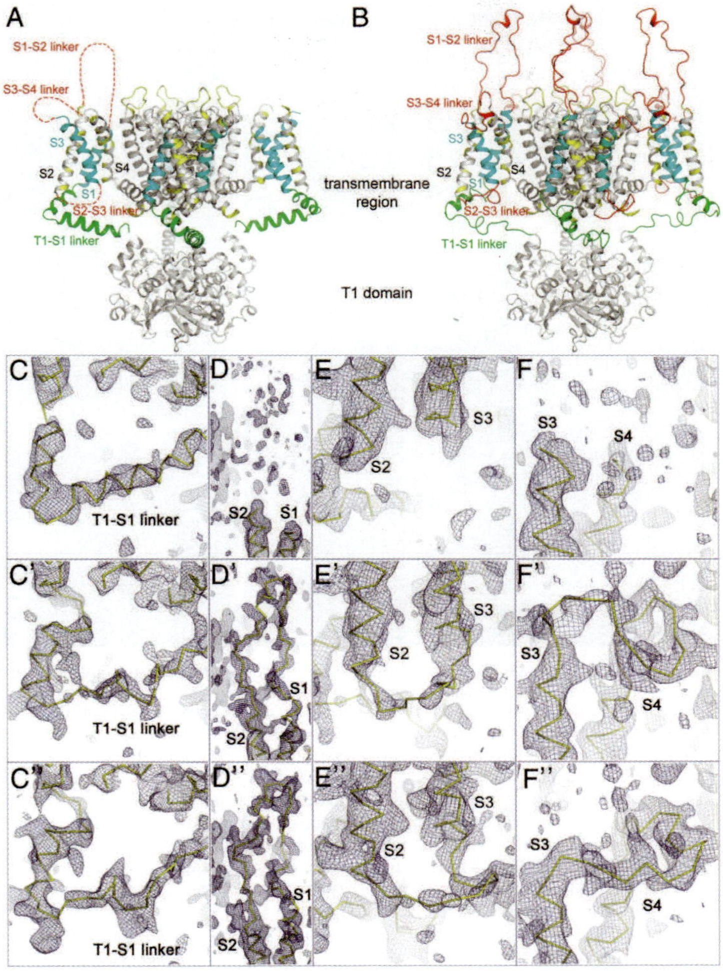

Fig. 1. Overall structural improvements. (*A*) The original Kv1.2 structure (PDB ID code 2A79). (*B*) The final Kv1.2 structure after normal-mode refinement and multicrystal averaging. In both *A* and *B*, red represents the regions newly added, including the S1-S2 linker, S2-S3 linker, and S3-S4 linker; green highlights the T1-S1 linker where large structural adjustments were made and register was determined; cyan suggests the regions (the S1 helix and S3 helix) the registers of which were undetermined in the original Kv1.2 structure but have been fixed in the normal-mode-refined structure; and yellow indicates the rebuilding of the missing side chains on S2 helix, S4 helix, and the linker between S5 and the pore helix. (*C–F''*) Examples of structural improvements in normal-mode refinement comparing the original model (*C–F*), the normal-mode model (*C'–F'*), and the final model after normal-mode refinement and multicrystal averaging (*C''–F''*), superimposed with their respective electron density maps. Shown are the T1-S1 linker region (*C, C', C''*), the S1-S2 linker (*D, D', D''*), the S2-S3 linker (*E, E', E''*), and the S3-S4 linker (*F, F', F''*). All maps were $2F_o - F_c$ maps contoured at 0.8σ.

of stiffness, were optimized by running parallel refinements (14–17). We found that 75 modes with stiffness of 30 yielded the lowest *R* factors. Thus, the number of independent thermal parameters used in the normal-mode refinement was 2,871, representing about 50% reduction in thermal parameters compared with the 4,997 independent thermal parameters for the original model (10). After replacing the isotropic *B* factors in the starting model with the normal-mode-based anisotropic *B* factors, the structure was again minimized by REFMAC5, yielding 21.8% for R_{cryst} factor and 24.2% for R_{free} factor, which were slightly lower than those of the starting model (by 0.1% and 0.3% for R_{cryst} and R_{free} factors, respectively).

The normal-mode refinement of Kv1.2 structure converged after a total of six iterative rounds of normal-mode-based anisotropic *B*-factor refinement, REFMAC5-based positional refinement, and manual adjustments. In each round, guided by the improved composite OMIT $2F_o - F_c$ maps, efforts were made to improve the completeness of the structural model step by step, including building missing main chains and side chains and determining unknown registers (Fig. 1*B*). The normal-mode Kv1.2 structure had 21.2% and 22.3% for R_{cryst} and R_{free} factors, respectively, which were 0.7% and 2.2% lower than those of the starting model (Table S1). There were a total of 851 newly added atoms that were missing in the starting model (Fig. 1*B*), accounting for ∼37% of the total mass of the transmembrane portion.

Because of the weak electron density maps for some of newly added loop regions of the transmembrane portion, in order to finalize the coordinates of atoms in these loops with more confidence, we then performed multicrystal averaging between our normal-mode model and the structure of the Kv1.2-Kv2.1 paddle chimera (see *Methods*). The improved phase from multicrystal averaging resulted in small but notable electron density shifts in

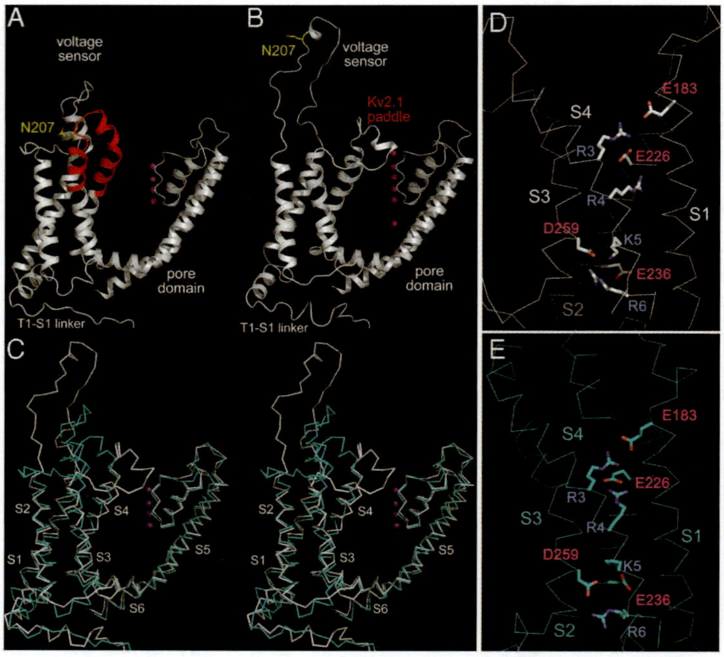

Fig. 2. Comparison of the transmembrane portions of the final Kv1.2 structure and the paddle-chimera structure. (*A*) Ribbon diagram of the paddle chimera structure with the chimera paddle region colored in red. (*B*) Ribbon diagram of the final Kv1.2 structure. The glycosylation site Asn207 (mutated from Gln) in *A* and *B* is highlighted as yellow stick. (*C*) Stereo view of the superposition of transmembrane portion of the final Kv1.2 structure (*White*) and the paddle-chimera structure (*Cyan*). Potassium ions are shown as purple spheres in *A–C*. (*D*) The voltage sensors of the final Kv1.2 structure (*White*). (*E*) The voltage sensors of the paddle-chimera structure (*Cyan*). Key residues were shown as sticks for positive gating charges (*Blue*) and their interacting negatively charged residues (*Red*).

some loop regions and the model was finalized accordingly. However, it is important to note that multicrystal averaging between the higher-resolution Kv1.2-Kv2.1 paddle chimera and the original Kv1.2 structure (PDB ID code 2A79) did not bring up the electron density for the interhelix loop regions. Thus, the normal-mode-based refinement method was critical for visualizing those missing loops. The final Kv1.2 model had R_{cryst} and R_{free} factors of 21.1% and 22.1%, which were 0.8% and 2.4% lower than those of the starting model, respectively. The modestly improved *R* factors, given the ~37% newly added atoms for the transmembrane portion, were probably because of their relatively low contribution to the diffraction data, as a result of structural disordering. The geometry parameters of the final Kv1.2 structure were similar to those of the starting model (Table S1), which made the comparison of *R* factors before and after the normal-mode-based refinement more robust.

Major Improvements of the Full-Length Kv1.2 Structure. The final full-length Kv1.2 structure had several major improvements over the starting model (Fig. 1). These improvements included: (*i*) the building of the missing main chains and side chains of the S1-S2 linker (35 residues), S2-S3 linker (10 residues), and S3-S4 linker (13 residues); (*ii*) the determination of the registers of T1-S1 linker, S1 helix, and S3 helix; (*iii*) the addition of the missing side chains on S2 helix, S4 helix, and the linker between S5 helix and the pore helix; (*iv*) the retracing of the T1-S1 linker as an extended helix instead of two helices in the original structure. All the structural adjustments were guided by the improved composite OMIT $2F_o - F_c$ electron density maps. Fig. 1 *C–F''* showed a few examples. In the original structure, the T1-S1 linker was built as two helices (Fig. 1*C*). In the final Kv1.2 structure, it was rebuilt as a random coil (Fig. 1 *C'* and *C''*). Moreover, the S1-S2 linker (35 residues) was not built because of the lack of electron density in the original structure (Fig. 1*D*). In the normal-mode refinement, electron density for this region gradually appeared and allowed the building of the full-length S1-S2 linker (Fig. 1 *D'* and *D''*). Similarly, the S2-S3 linker (Fig. 1 *E*, *E'*, and *E''*) and the S3-S4 linker (Fig. 1 *F*, *F'*, and *F''*) were also built in the newly emerged electron density.

The ability for the normal-mode-based refinement method to result in significant improvement in electron density maps and subsequent substantial model adjustments was probably related to the large phase angle shifts it caused. For instance, the final Kv1.2 differed from the starting model by 15.8° (Fig. S1*a*).

Structural Flexibility of Kv1.2. The fact that a complete model was not possible in the original structure determination suggested that the transmembrane portion of Kv1.2, in particular the voltage-sensor domains, could be highly flexible functionally and/or statically (10). The higher structural flexibility of this portion became very obvious when we compared the *B*-factor profiles of the starting model, the normal-mode model after the first round of normal-mode refinement (Fig. S1*b*), and the final Kv1.2 model (Fig. S1*c*). Although the starting model did show higher *B* factors for the transmembrane region (Fig. S1*b*, *Light Line*), the normal-mode models yielded many more features in *B*-factor profiles (Fig. S1*b*, *Dark Line*; Fig. S1*c*). For instance, the *B*-factor profile of the final Kv1.2 model suggested that all the interhelix regions of the transmembrane portion, including the T1-S1 linker (peak 1), the S1-S2 linker (peak 2), the S2-S3 linker (peak 3), the S3-S4 linker (peak 4), the loop between S5 and the pore helix (peak 5), and the loop between filter region and S6 (peak 6), be of higher mobility than the helices themselves (Fig. S1*c*). Moreover, the regions corresponding to the β-subunit and T1-domain matched very well, indicating the power of the normal-mode-based method in generating accurate *B*-factor profile with about 50% reduction in thermal parameters.

Comparison with the Paddle Chimera Structure. In the more recent paddle chimera Kv structure solved at 2.4 Å (Fig. 2*A*), the voltage sensor of Kv1.2 was replaced with the corresponding sequence of Kv2.1 (please refer to figure 1*A* in ref. 13 for details of the construct). The replaced region encompassed a total of 32 residues, 27 of which were different from those in Kv1.2 sequence. This region, often referred to as the "voltage-sensor paddle," was comprised of the C-terminal half of S3 (S3*b*), S3-S4 linker, and the N-terminal segment of S4 helix.

We compared the final Kv1.2 model (Fig. 2*B*) and the Kv1.2-Kv2.1 paddle chimera structure (Fig. 2*A*) on the basis of the least-square alignment of S5 and S6 helices, and the superposition was shown in Fig. 2*C*. Overall, the two structures were aligned very well, in particular the transmembrane helices of S1, S2, S4, S5, and S6. As expected from their intrinsic structural flexibility (Fig. S1*c*), large structural differences were noticed at the inter-helix linker regions (Fig. 2*C*). For instance, the N-terminal half of the T1-S1 linker was built as an almost straight polyglycine in the paddle chimera structure, whereas in our final Kv1.2 structure, it was traced as a loop. The 35-residue S1-S2 linker in the chimera structure contained a short helix in the middle to interact with N terminus of the S4 helix (Fig. 2*A*), whereas in our final Kv1.2 structure it was an extended loop (Fig. 2*B*). The different conformations of the S1-S2 linker were probably because of the different S4 helix in these two structures. Previous studies indicated that the middle section of the S1-S2 linker in Kv1.2, including residue 207, can be glycosylated at every position, suggesting that it be fully exposed to external solutions (22, 23). This agreed with the loop conformation of the S1-S2 linker in our final Kv1.2 model (Fig. 2*B*) but was in marked contrast to the more folded conformation of this region in the chimera structure (Fig. 2*A*).

The gating charges R3, R4, K5, and R6 on the S4 helix were all positioned at similar positions between these two structures (Fig. 2 *D* and *E*). In both structures, the gating charges were surrounded by highly conserved negatively charged residues such as E183 (S1) and E226 (S2), E236 (S2), and D259 (S3) (Fig. 2 *D* and *E*). The electrostatic interactions between the gating charges and their surrounding negatively charged partners are believed to be important for voltage gating.

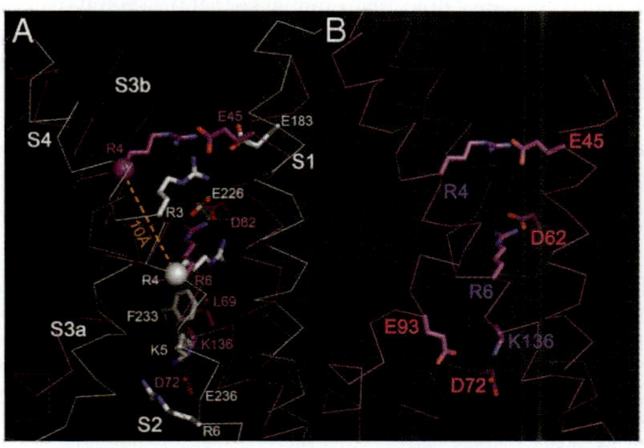

Fig. 3. Comparison of the final Kv1.2 structure and the KvAP structure (PDB ID code 1ORS). (*A*) The superposition of the voltage sensors of the final Kv1.2 structure (*White*) and the KvAP structure (*Purple*). The residues R4 in S4 helix were highlighted with a sphere at its C^α atom in two structures, and the distance between them was indicated as a dashed line in orange. (*B*) The charged residues of the voltage sensor in KvAP structure. Key residues were shown as sticks for positive gating charges (*Blue*) and their interacting negatively charged residues (*Red*).

Comparison with the Voltage Sensor of KvAP. The structure of an isolated voltage sensor of a prokaryotic Kv channel, KvAP, was previously determined (9). We compared the structure of voltage-sensor domains of KvAP and our final Kv1.2 model by aligning the S1 and S2 helices (Fig. 3*A*). It was clear that on S1–S3, all the conserved residues aligned very well, including E45 (KvAP) and E183 (Kv1.2) on S1 helix, D62 (KvAP) and E226 (Kv1.2), L69 (KvAP) and F233 (Kv1.2), D72 (KvAP) and E236 (Kv1.2) on S2 helix, E93 (KvAP) and D259 (Kv1.2) on S3 helix. However, S4 helix had rather large relative shifts between these two structures (Fig. 3*A*). Similarly, a large structural shift in S4 helix also existed between the better-quality structure of paddle chimera at 2.4 Å and KvAP. Taken together, these findings suggested that in the architecture of voltage-sensor domains, the S4 helix has the potential to translocate by at least ~10 Å (24).

Similar to those of Kv1.2 and paddle chimera, the gating charges on S4 of KvAP were surrounded by negative charges on other transmembrane helices including E45, D62, D72, and E93 (Fig. 3*B*).

A Hydrophobic Layer in Voltage-Sensor Domain in the Midpoint of the Membrane Bilayer. In the structural study of paddle chimera, a phenylalanine gap at F233 was described (13). In inspecting the voltage-sensor domain of our final Kv1.2 model, we found that near the midpoint of the membrane, there was a layer of 10 Å in thickness formed by ten completely buried residues, V172, I173, S176, I177, and F180 (S1 helix), C229, I230, and F233 (S2 helix), and A262 and I263 (S3 helix) (Fig. 4 and Table 1). Among them, there was one polar residue, S176, and nine hydrophobic residues. Sequence alignment suggested that hydrophobic residues predominate position 176 among other Kv channels (Table 1). Therefore, we collectively referred to this 10 Å-layer as "a hydrophobic layer." This hydrophobic layer was highly conserved among the Kv family (with an overall greater than 90% conservation, except for A262 that is 78% conserved) (Table 1). In the paddle chimera, the same layer of hydrophobic nature was observed (Fig. S2*a*). In KvAP, the hydrophobic layer was composed of residues A34, A35, S38, V39, and V42 (S1 helix), L65, V66, and L69 (S2 helix), A96 and L97 (S3 helix) (Fig. S2*b*). In the middle of this hydrophobic layer gating charges on S4 [R4 and K5 in Kv1.2 (Fig. 4) and paddle chimera (Fig. S2*a*), R6 and K136 in KvAP (Fig. S2*b*)], and their interacting negatively charged residues from S1–S3 helices (E226 and D259 in Kv1.2 and paddle chimera, D62 and E93 in KvAP) were buried. This thin layer of mostly hydrophobic zone was expected to separate the internal and external solutions to form a focused electric field of ~10 Å (25, 26). Consequently, the gating charges on S4 translocate a much smaller distance than the span of the entire membrane (of up to 30 Å in thickness). It was worth noting that the

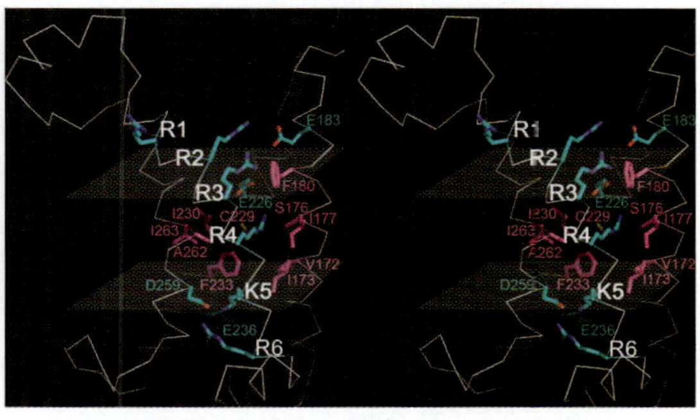

Fig. 4. A hydrophobic layer in the voltage sensor of the final Kv1.2 structure. Hydrophobic residues are colored in magenta and hydrophilic residues in cyan. Transparent planes were drawn at both the upper and lower boundaries of the hydrophobic layer to highlight its thickness.

Table 1. Comparison of solvent accessible surface area and sequence conservation of residues in Kv1.2 and KvAP that are involved in the hydrophobic layer

Residues in Kv1.2	ASA (Å^2)*	Hydrophobic/hydrophilic residues among 250 Kv sequences of the highest homology to Kv1.2	Residues in KvAP	ASA (Å^2)
V172	52	250/0	A34	4
V173	1	249/1	A35	3
S176	7	239/11	S38	5
I177	37	228/22	V39	30
F180	44	227/23	V42	11
C229	6	248/2	L65	16
I230	35	240/10	V66	9
F233	25	250/0	L69	6
A262	5	196/54	A96	8
I263	16	250/0	L97	19

*The accessible surface area (ASA) was calculated by using Areaimol in CCP4 (21).

majority of the hydrophobic core (nine out of ten residues) was from S1 and S2 helices, in agreement with their static role in the gating process to support the large conformational changes carried out by S4 helix (27, 28).

Discussion

Here we report a complete structure of the full-length Shaker family Kv1.2 in its native form based on the original diffraction data (10). This was achieved by using a normal-mode-based X-ray crystallographic refinement method that have been found to be effective in improving structural refinement in a number of biological systems (14–17). In all previous applications, the structural improvement was reflected in building missing side chains, retracing main chains, adding sugar rings, and achieving better fit between the models and the diffraction data (14–17). But in this study, the structural improvement on Kv1.2 was the addition of ~37% molecular mass of the entire transmembrane portion of the channel. Thus, this study highlights the great potential of the refinement method for improving difficult membrane protein structures.

This complete Kv1.2 structure in its native form allows a systematic comparison with previously determined structures of a Kv1.2-Kv2.1 paddle chimera (13) and a prokaryotic Kv channel, KvAP (9). Although the S1–S3 helices were aligned very well among these structures, the gating charges on S4 helix displayed large relative shifts of as much as 10 Å, which suggests that the architecture of voltage-sensor domains be designed in such a way to allow the S4 helix to translocate by at least ~10 Å in voltage gating (24).

Moreover, in the middle of the voltage-sensor domain of the final Kv1.2 structure, we found a hydrophobic layer of 10 Å in thickness formed by ten buried residues (Fig. 4 and Table 1). A further inspection of previously published structures revealed that similar hydrophobic layers also exist in the paddle chimera and in prokaryotic KvAP (Fig. S2), but the biological function of the layer was never discussed. This layer of ~10 Å separates the internal and external solutions and buries inside the gating charges and their interacting negatively charged residues. Previous studies have observed a focused electric field of ~10 Å through which the gating charges on S4 have to travel during voltage gating (25, 26). Thus, the hydrophobic layer that we identified in these three different Kv structures is most likely the molecular basis for such a focused electric field. Interestingly, MacKinnon and colleagues (29) recently described an occluded site formed by F233, E236, and D259 that is located just beneath the hydrophobic layer. Because the current Kv1.2 structure is presumably in an open state and the gating charge R1 in the closed state is probably where K5 is in the open state (29), the closed → open transition of Kv1.2 in gating is expected to involve the transpassing of the gating charges through this focused electric field, accounting for the movement of 3–4 elementary charges per voltage sensors in gating (30–33). The fact that the S1 and S2 helices contribute nine out of ten residues in this hydrophobic layer agrees well with their relative static role in the gating process to support the large conformational changes carried out by S4 helix (27, 28).

Undoubtedly, the electron density maps for Kv1.2 are weaker than those of very well-ordered systems, which is intrinsic to very flexible systems like Kv1.2 at fairly limited resolution without any noncrystallographic symmetry. Although the current normal-mode-based refinement method provided substantial density enhancement, it is unlikely to go beyond the limitation of the original data. In order to demonstrate the level of confidence we had in model building, we also included a $F_o - F_c$ map that was calculated with a region in question removed (in this case the retraced T1-S1 linker highlighted in red was removed) (Fig. S3). The $F_o - F_c$ map contoured at 1.0σ clearly justified our retracing in this region.

Although at the late stage of our refinement a multicrystal averaging was used to further improve the map, it is noteworthy that, without the normal-mode refinement, application of multicrystal averaging alone to the original Kv1.2 model was unable to bring up the large chunks of missing densities. Moreover, the fact that the normal-mode-based refinement can be used in combination with other existing methods emphasizes its versatility in assisting structural refinement of difficult systems.

The effectiveness of normal-mode-based method on refining membrane protein structures is evident in this case and in a previous study of KcsA (15). Part of such a success is probably because of the fact that all membrane proteins are very flexible and crystallized in a nonnative lipid environment so that the crystal structures may slightly deviate from their true native forms. These types of protein structural deformation in the vicinity of native form, regardless of functionally relevant or not, are most likely to follow the trajectory of low-frequency normal modes (34). By offering a better description of such structural deformation, the normal-mode-based method can substantially improve the electron density maps and consequently structural models.

Methods

The procedure of the normal-mode-based refinement was described in previous publications (14, 15). The number of the lowest-frequency modes for generating the best anisotropic *B* factors was 75. At the end of normal-mode-based refinement, multicrystal averaging was performed with DMMULTI program in CCP4 (21) to improve the map by using the known structure of Kv1.2-Kv2.1 paddle chimera channel. The detailed methods are supplied in *SI Materials*.

ACKNOWLEDGMENTS. We gratefully thank Roderick MacKinnon for his insightful comments. J.M. acknowledges support of grants from the National Institutes of Health (R01-GM067801), the National Science Foundation (MCB-0818353), and the Welch Foundation (Q-1512). Q.W. acknowledges support from the National Institutes of Health (R01-AI067839) and a Beginning-Grant-in-Aid award from the American Heart Association (0865186F).

1. Hille B (2001) *Ion Channels of Excitable Membranes* (Sinauer Associates, Sunderland), 3rd Ed.
2. Yellen G (1998) The moving parts of voltage-gated ion channels. *Q Rev Biophys* 31(3):239–295.
3. Tombola F, Pathak MM, Isacoff EY (2005) How far will you go to sense voltage?. *Neuron* 48(5):719–725.
4. Tombola F, Pathak MM, Isacoff EY (2006) How does voltage open an ion channel?. *Annu Rev Cell Dev Biol* 22:23–52.
5. MacKinnon R (2003) Potassium channels. *FEBS Lett* 555(1):62–65.
6. Mackinnon R (2004) Potassium channels and the atomic basis of selective ion conduction (Nobel Lecture). *Angew Chem Int Edit* 43:4265–4277.
7. Pongs O (1999) Voltage-gated potassium channels: From hyperexcitability to excitement. *FEBS Lett* 452(1–2):31–35.
8. Panyi G, Possani LD, Rodriguez de la Vega RC, Gaspar R, Varga Z (2006) K+ channel blockers: Novel tools to inhibit T cell activation leading to specific immunosuppression. *Curr Pharm Design* 12(18):2199–2220.
9. Jiang Y, et al. (2003) X-ray structure of a voltage-dependent K+ channel. *Nature* 423(6935):33–41.
10. Long SB, Campbell EB, Mackinnon R (2005) Crystal structure of a mammalian voltage-dependent Shaker family K+ channel. *Science* 309(5736):897–903.
11. Long SB, Campbell EB, Mackinnon R (2005) Voltage sensor of Kv1.2: Structural basis of electromechanical coupling. *Science* 309(5736):903–908.
12. Brunger AT, et al. (1998) Crystallography & NMR system: A new software suite for macromolecular structure determination. *Acta Crystallogr D* 54(Pt 5):905–921.
13. Long SB, Tao X, Campbell EB, MacKinnon R (2007) Atomic structure of a voltage-dependent K+ channel in a lipid membrane-like environment. *Nature* 450(7168): 376–382.
14. Poon BK, et al. (2007) Normal mode refinement of anisotropic thermal parameters for a supramolecular complex at 3.42-Å crystallographic resolution. *Proc Natl Acad Sci USA* 104(19):7869–7874.
15. Chen X, Poon BK, Dousis A, Wang Q, Ma J (2007) Normal-mode refinement of anisotropic thermal parameters for potassium channel KcsA at 3.2 Å crystallographic resolution. *Structure* 15(8):955–962.
16. Chen X, Lu M, Poon BK, Wang Q, Ma J (2009) Structural improvement of unliganded simian immunodeficiency virus gp120 core by normal-mode-based X-ray crystallographic refinement. *Acta Crystallogr D* 65:633–643.
17. Ni F, Poon BK, Wang Q, Ma J (2009) Application of normal-mode refinement to X-ray crystal structures at the lower resolution limit. *Acta Crystallogr D* 65(Pt 7):633–643.
18. Diamond R (1990) On the use of normal modes in thermal parameters refinement: Theory and application to the bovine pancreatic trypsin inhibitor. *Acta Crystallogr A* 46:425–435.
19. Kidera A, Go N (1990) Refinement of protein dynamic structure: Normal mode refinement. *Proc Natl Acad Sci USA* 87(10):3718–3722.
20. Murshudov GN, Vagin AA, Dodson EJ (1997) Refinement of macromolecular structures by the maximum-likelihood method. *Acta Crystallogr D* 53(Pt 3):240–255.
21. Collaborative Computational Project N (1994) The CCP4 suite: Programs for protein crystallography. *Acta Crystallogr D* 50(Pt 5):760–763.
22. Zhu J, et al. (2003) Allowed N-glycosylation sites on the Kv1.2 potassium channel S1-S2 linker: implications for linker secondary structure and the glycosylation effect on channel function. *Biochem J* 375(Pt 3):769–775.
23. Zhu J, et al. (2009) The Kv1.2 potassium channel: The position of an N-glycan on the extracellular linkers affects its protein expression and function. *Brain Res* 1251:16–29.
24. Lewis A, Jogini V, Blachowicz L, Laine M, Roux B (2008) Atomic constraints between the voltage sensor and the pore domain in a voltage-gated K+ channel of known structure. *J Gen Physiol* 131(6):549–561.
25. Ahern CA, Horn R (2004) Specificity of charge-carrying residues in the voltage sensor of potassium channels. *J Gen Physiol* 123(3):205–216.
26. Starace DM, Bezanilla F (2004) A proton pore in a potassium channel voltage sensor reveals a focused electric field. *Nature* 427(6974):548–553.
27. Posson DJ, Selvin PR (2008) Extent of voltage sensor movement during gating of shaker K+ channels. *Neuron* 59(1):98–109.
28. Pathak MM, et al. (2007) Closing in on the resting state of the Shaker K(+) channel. *Neuron* 56(1):124–140.
29. Tao X, Lee A, Limapichat W, Dougherty DA, MacKinnon R (2010) A gating charge transfer center in voltage sensors. *Science* 328(5974):67–73.
30. Zagotta WN, Hoshi T, Dittman J, Aldrich RW (1994) Shaker potassium channel gating II: Transitions in the activation pathway. *J Gen Physiol* 103(2):279–319.
31. Schoppa NE, McCormack K, Tanouye MA, Sigworth FJ (1992) The size of gating charge in wild-type and mutant Shaker potassium channels. *Science* 255(5052):1712–1715.
32. Seoh SA, Sigg D, Papazian DM, Bezanilla F (1996) Voltage-sensing residues in the S2 and S4 segments of the Shaker K+ channel. *Neuron* 16(6):1159–1167.
33. Aggarwal SK, MacKinnon R (1996) Contribution of the S4 segment to gating charge in the Shaker K+ channel. *Neuron* 16(6):1169–1177.
34. Ma J (2005) Usefulness and limitations of normal mode analysis in modeling dynamics of biomolecular complexes. *Structure* 13:373–380.

Chapter 18

Hong Guo

Biography

Dr. Hong Guo is an associate professor in the Department of Biochemistry and Cellular and Molecular Biology, University of Tennessee, Knoxville. He obtained his M.S. degree in physical chemistry from Jilin University, China, and his Ph.D. in chemical physics from Harvard University, where he first met Professor Lipscomb. Dr. Hong Guo then worked as an international postdoctoral fellow for the Natural Sciences and Engineering Research Council of Canada (NSERC) and as a research associate at the Université de Montréal and Harvard University before he joined the University of Tennessee in 2002.

His research interests are in the areas of computational biophysics and biochemistry, including the application of molecular dynamics (MD) simulations, free energy calculations, and mixed quantum mechanical/molecular mechanical (QM/MM) approaches to elucidate mechanisms of enzyme catalysis and origin of substrate specificity, to study protein–ligand and protein–protein interactions, and to understand the role of hydrogen bonding and proton transfer in protein structure and function.

Working with the Colonel

I was extremely fortunate to have the opportunity to work and interact with the Colonel, Professor Bill Lipscomb, on chorismate mutase when I was a postdoctoral fellow at Harvard University. At that time, the X-ray structures of yeast chorismate mutase complexed with a transition state analog had been determined earlier by the Colonel[1,2] which provided an opportunity for understanding the catalytic mechanism of this enzyme as well as general strategies that enzymes use to catalyze chemical reactions. The Colonel believed that computational approaches would be very useful for elucidating dynamic and functional properties of the enzyme, and we started to use the methods of molecular dynamics (MD) simulations with hybrid quantum mechanical/molecular mechanical (QM/MM) potential to study chorismate mutase based on these X-ray structures, as well as ones that had been determined earlier for *Bacillus subtilis* chorismate mutase (BsCM) by the Colonel.[3,4] I walked to the Colonel's office frequently to hear his insights on structural and biochemical properties/features of the enzyme as well as the possible explanations of the simulation data that I had obtained. Indeed, the Colonel showed me how we could combine theoretical calculations with experimental data to better understand the structure and function of chemical and biological systems. He was extremely good at it, and one of the best examples in this regard is perhaps his proposal[5] of the valence theory using three-center and two-center bonds with the support from molecular orbital calculations.[5,6] The research with the Colonel led to my first two publications on enzymes[7,8] that applied the QM/MM approach and MD simulations to understand reaction mechanisms. Our results showed that the conformational transformations from nonreactive conformers of the substrate to the reactive conformation in chorismate mutase could play an important role in the catalytic mechanism. The interaction I had with the Colonel during that period of time at Harvard has had a very important impact on my research for the last nine years as an independent researcher at University of Tennessee, Knoxville; I believe this will also be true for the rest of my scientific career. In fact, one of my main research directions now at Tennessee is the QM/MM investigation of enzymes that I started with the Colonel at Harvard on chorismate mutase.

Today, computers/computational methods and experimental techniques are unbelievably advanced compared to the time when the Colonel determined molecular structures of boranes and their relatives and performed quantum mechanical calculations based on the extended Hückel method more than a half century ago; they are unbelievably advanced even compared to ten years ago when we performed QM/MM calculations on chorismate mutases. However, we still face the same question concerning how to understand and interpret experimental data with theoretical thinking and computations that the Colonel taught us. I feel that we can always learn a great deal from the Colonel's approach and overcome some of the difficulties we may encounter in the future.

References

1. Schnappauf G., Strater N., Lipscomb W. N., and Braus G., Glutamate residue in the catalytic center of the yeast chorismate mutase restricts enzyme activity to acidic conditions. *P. Natl. Acad. Sci. USA*, **94**, 8491–8496 (1997).
2. Strater N., Schnappauf G., Braus G., and Lipscomb W. N., Mechanisms of catalysis and allosteric regulation of yeast chorismate mutase from crystal structures. *Structure*, **5**, 1437–1452 (1997).
3. Chook Y. M., Ke H., and Lipscomb W. N., Crystal structure of the monofunctional chorismate mutase from *Bacillus subtilis* and its complex with a transition state analog. *P. Natl. Acad. Sci. USA*, **90**, 8600–8603 (1993).
4. Chook Y. M., Gray J. V., Ke H., and Lipscomb W. N., The monofunctional chorismate mutase from Bacillus subtilis. *J. Mol. Biol.*, **240**, 476–500 (1994).
5. Lipscomb W. N., Nobel Lecture, December 11, 1976.
6. Hoffmann R. and Lipscomb W. N., Boron hydrides: LCAO-MO and resonance studies. *J. Chem. Phys.*, **37**, 2872–2883 (1962).
7. Guo H., Cui Q., Lipscomb W. N., and Karplus M., Substrate conformational transitions in the active site of chorismate mutase: their role in the catalytic mechanism. *P. Natl. Acad. Sci. USA*, **98**, 9032–9037 (2001).
8. Guo H., Cui Q., Lipscomb W. N., and Karplus M., Understanding the role of active-site residues in chorismate mutase catalysis from molecular dynamic simulations. *Angew. Chem. Int. Edit.*, **42**, 1508–1511 (2003).

J|A|C|S
ARTICLES
Published on Web 02/05/2005

Origin of Tight Binding of a Near-Perfect Transition-State Analogue by Cytidine Deaminase: Implications for Enzyme Catalysis

Haobo Guo, Niny Rao, Qin Xu, and Hong Guo*

Contribution from the Department of Biochemistry and Cellular and Molecular Biology, Center of Excellence in Structural Biology, University of Tennessee, Knoxville, Tennessee 37996-0840

Received October 4, 2004; E-mail: hguo1@utk.edu

Abstract: Cytidine deaminase (CDA) is a zinc metalloenzyme that catalyzes the hydrolytic deamination of cytidine to uridine. Zebularine (ZEB) binds to CDA, and the binding process leads to a near-perfect transition-state analogue (TSA) inhibitor at the active site with an estimated K_i value of 1.2×10^{-12} M. The interaction of CDA with the TSA inhibitor has become a paradigm for studying the tight TSA binding by enzymes. The formation of the TSA is catalyzed by CDA by a mechanism that is similar to the formation of the tetrahedral intermediate during the CDA-catalyzed reaction (i.e., through the nucleophilic attack of a Zn−hydroxide group on C_4). It is believed that the TSA formed at the active site is zebularine 3,4-hydrate. In this paper, it is shown from QM/MM molecular dynamics and free energy simulations that zebularine 3,4-hydrate may in fact be unstable in the enzyme and that a proton transfer from the Zn−hydroxide group to Glu-104 during the nucleophilic attack could be responsible for the very high affinity. The nucleophilic attack by the Zn−hydroxide on C_4 is found to be concerted with two proton transfers. Such concerted process allows the TSA, an alkoxide-like inhibitor, to be stabilized through a mechanism that is similar to the transition-state stabilization in the general acid−base catalysis. It is suggested that the proton transfer from the Zn−hydroxide to Glu-104, which is required to generate the general acid for protonating the leaving ammonia, may play an important role in lowering the activation barrier during the catalysis.

Introduction

Enzymes' catalytic power lies in their ability to stabilize transition state (TS) and to lower the energy barriers for enzyme-catalyzed reactions. It has been recognized that a molecule that mimics the altered structure of substrate in the transition state may capture a significant fraction of transition-state stabilization and can therefore bind to the enzyme tightly.[1−8] The precise arrangements of many transition-state analogue (TSA) inhibitors in enzyme active sites have been determined by X-ray crystallography.[9,10] These structures have led to significant insights into the interactions that might be responsible for transition-state stabilization. The transition-state analogue approach has been very useful for the design of potent inhibitors and is critical for the generation of catalytic antibodies. Nevertheless, many questions remain concerning the details of TSA binding and the energetic origin of the high binding affinities, including the role of proton transfers.

In this paper, quantum mechanical/molecular mechanical (QM/MM) molecular dynamics simulations are performed for *Escherichia coli* cytidine deaminase (CDA)[11] to study the mechanism of inhibitor binding and the energetic origin of the high binding affinity for a near-perfect TSA inhibitor. CDA catalyzes the hydrolytic deamination of cytidine to uridine (Figure 1A) and has been a subject of extensive investigations.[11−25]

(1) Wolfenden, R. *Nature* **1969**, *223*, 704−705.
(2) Wolfenden, R. *Acc. Chem. Res.* **1972**, *5*, 10−18.
(3) Wolfenden, R.; Kati, W. M. *Acc. Chem. Res.* **1991**, *24*, 209−215.
(4) Wolfenden, R. *Annu. Rev. Biophys. Bioeng.* **1976**, *5*, 271−306.
(5) Radzicka, A.; Wolfenden, R. *Methods Enzymol.* **1995**, *249*, 284.
(6) Schramm, V. L. *Annu. Rev. Biochem.* **1998**, *67*, 693−720.
(7) Morrison, J. F.; Walsh, C. T. *Adv. Enzymol. Relat. Area Mol. Biol.* **1988**, *61*, 201−301.
(8) Mader, M. M.; Bartlett, P. A. *Chem. Rev.* **1997**, *97*, 1281−1301.
(9) Lolis, E.; Petsko, G. A. *Annu. Rev. Biochem.* **1990**, *59*, 597−630.
(10) Lipscomb, W. N.; Strater, N. *Chem. Rev.* **1996**, *96*, 2375−2433.
(11) Xiang, S.; Short, S. A.; Wolfenden, R.; Carter, C. W., Jr. *Biochemistry* **1995**, *34*, 4516−4523.
(12) Betts, L.; Xiang, S.; Short, S. A.; Wolfenden, R.; Carter, C. W., Jr. *J. Mol. Biol.* **1994**, *235*, 635−656.
(13) Carlow, D. C.; Short, S. A.; Wolfenden, R. *Biochemistry* **1996**, *35*, 948−954.
(14) Carlow, D. C.; Smith, A. A.; Yang, C. C.; Short, S. A.; Wolfenden, R. *Biochemistry* **1995**, *34*, 4220−4224.
(15) Smith, A. A.; Carlow, D. C.; Wolfenden, R.; Short, S. A. *Biochemistry* **1994**, *33*, 6468−6474.
(16) Snider, M. J.; Gaunitz, S.; Ridgway, C.; Short, S. A.; Wolfenden, R. *Biochemistry* **2000**, *39*, 9746−9753.
(17) Xiang, S.; Short, S. A.; Wolfenden, R.; Carter, C. W., Jr. *Biochemistry* **1996**, *35*, 1335−1341.
(18) Xiang, S.; Short, S. A.; Wolfenden, R.; Carter, C. W., Jr. *Biochemistry* **1997**, *36*, 4768−4774.
(19) Snider, M. J.; Wolfenden, R. *Biochemistry* **2001**, *40*, 11364−11371.
(20) Lewis, J. P.; Carter, C. W., Jr.; Hermans, J.; Pan, W.; Lee, T.-S.; Yang, W. *J. Am. Chem. Soc.* **1998**, *120*, 5407−5410.
(21) Lewis, J. P.; Liu, S.; Lee, T.-S.; Yang, W. *J. Comput. Phys.* **1999**, *151*, 242−263.
(22) Frick, L.; MacNeela, J. P.; Wolfenden, R. *Bioorg. Chem.* **1987**, *15*, 100−108.
(23) Xu, Q.; Guo, H. *J. Phys. Chem. B* **2004**, *108*, 2477−2483.
(24) Frick, L.; Yang, C.; Marquez, V. E.; Wolfenden, R. *Biochemistry* **1989**, *28*, 9423−9430.
(25) Snider, M. J.; Reinhardt, L.; Wolfenden, R.; Cleland, W. W. *Biochemistry* **2002**, *41*, 415−421.

10.1021/ja0439625 CCC: $30.25

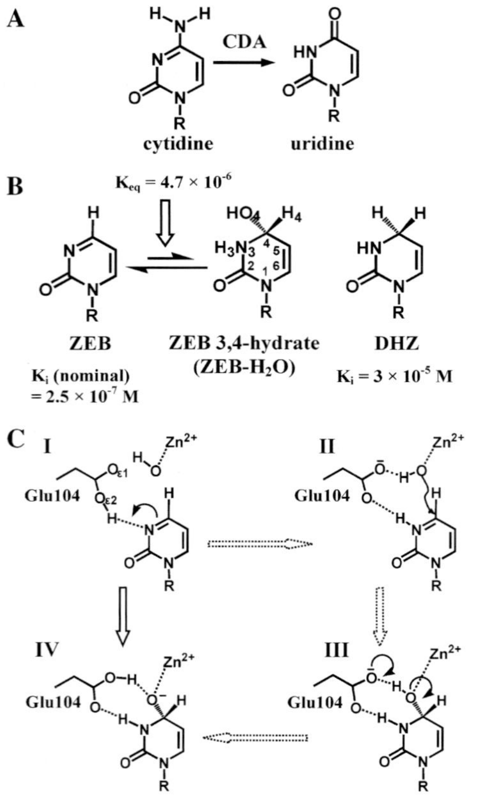

Figure 1. (A) Hydrolytic deamination of cytidine to uridine catalyzed by cytidine deaminase (CDA); R = ribose. (B) Pyrimidin-2-one ribonucleoside (zebularine or ZEB), zebularine 3,4-hydrate (ZEB−H_2O), and 3,4-dihydrozebularine (DHZ). Subscripts are used to label the atoms. ZEB−H_2O was thought to be formed through a hydration process catalyzed by CDA. However, the present QM/MM MD study suggests that ZEB−H_2O is unstable in the active site of CDA. A proton transfer from the C_4−OH to Glu-104 occurred during the MD simulations. The K_i value for ZEB−H_2O was estimated to be 1.2×10^{-12} M (ref 24) based on the equilibrium constant for hydration of ZEB in solution (4.7×10^{-6}) and the apparent K_i value observed for ZEB (2.5×10^{-7} M). (C) Possible steps during the binding (hydration) of ZEB. They include the protonation of N_3 of ZEB (from I to II), the nucleophilic attack of the Zn−hydroxide on C_4 (from II to III), and the proton transfer from the C_4−hydroxide of ZEB−H_2O to Glu-104 (from III to IV). The present free energy simulations suggest that these steps are concerted, and the actual process is I → IV, leading to an alkoxide-like TSA at the active site of CDA, as shown in IV (see text).

It has been suggested that the deamination proceeds through a tetrahedral intermediate (formed by the nucleophilic attack by a Zn−hydroxide group on C_4 of cytidine) with ammonia elimination as the major rate-determining step.[25] The solvent deuterium isotope effect indicates that there are probably two proton transfers during the formation of the tetrahedral intermediate.[25] The catalysis depends critically on the presence of Glu-104, and the mutation of Glu-104 to Ala reduces k_{cat} by 10^8-fold.[14]

Zebularine (ZEB; see Figure 1B) binds to CDA, and a near-perfect TSA is produced at the active site of the enzyme during the binding process.[11,12] The interaction of CDA with the TSA inhibitor has become a paradigm in the studies of the tight binding of TSAs by enzymes and the forces responsible for transition-state stabilization during enzyme catalysis.[2−5] The formation of this TSA is believed to be catalyzed by CDA through a mechanism that is similar to the formation of the tetrahedral intermediate during the CDA-catalyzed reaction.[10] It is thought that the TSA is the zebularine 3,4-hydrate containing a C_4−OH group (ZEB−H_2O; see Figure 1B) that is formed through the hydration of ZEB (i.e., I → II → III in Figure 1C; the I → II step may be concerted with or proceed II → III; see ref 25). The equilibrium constant for hydration of ZEB in solution is approximately 4.7×10^{-6}, and the apparent K_i value observed for ZEB is 2.5×10^{-7} M.[24] On the basis of these data, the K_i value for ZEB−H_2O was estimated to be 1.2×10^{-12} M.[24] Thus, CDA appears to be strongly inhibited by ZEB−H_2O with an affinity that exceeds the affinity for the inhibitor 3,4-dihydrozebularine (DHZ; see Figure 1B), which contains a proton in place of the C_4−OH group of ZEB−H_2O, by a factor of 10^7−10^8.[24] The origin of the large differential binding affinity (estimated to be ca. 10−13 kcal/mol) is still not clear, although it was proposed that the active site interactions involving the C_4−OH group of ZEB−H_2O could make a major contribution.[11]

In this paper, it is predicted from QM/MM molecular dynamics (MD) simulations that ZEB−H_2O is in fact unstable in the active site of CDA; a proton transfer from the C_4−OH to Glu-104 occurs during the energy minimization and MD simulations of the CDA−ZEB−H_2O complex. The QM/MM free energy simulations further suggest that the nucleophilic attack by the Zn−hydroxide group on C_4 follows I → IV in Figure 1C in a single step; that is, it is concerted with two proton transfers from $O_{\epsilon 2}$ of Glu-104 to N_3 of ZEB and from the Zn−hydroxide to $O_{\epsilon 1}$ of Glu-104. The concerted process of the nucleophilic attack and the second proton transfer to Glu-104 allows the TSA, an alkoxide-like inhibitor (see IV in Figure 1C), to be stabilized through a mechanism that is similar to the transition-state stabilization in general acid−base catalysis.[27,28] The proton-transfer process from the Zn−hydroxide to Glu-104 is also necessary to generate the general acid (Glu-104) for protonating the leaving ammonia during the enzyme catalysis. Thus, CDA seems to be able to use one step of the reaction (e.g., the proton transfer) to catalyze another step (e.g., the nucleophilic attack) to achieve the transition-state stabilization.

Methods

A fast semiempirical density-functional approach (SCC-DFTB),[29] recently implemented in the CHARMM program,[30] was used for QM/MM molecular dynamics and free energy simulations. The SCC-DFTB approach has been extensively tested against high-level quantum mechanical methods for a range of active site models containing Zn.[29] It was found that the SCC-DFTB approach reproduced structural and energetic properties rather reliably.[29] The initial coordinates for the simulations were obtained from the crystal structure of the CDA−TSA complex (1CTU),[11] which have the transition-state analogue at the active site. The TSA inhibitor was initially assumed to be ZEB−H_2O (see Figure 2A) as generally believed (see below). The MD simulations

(26) Frieden, C.; Kurtz, I. C.; Gilbert, H. R. *Biochemistry* **1980**, *19*, 5303−5309.
(27) Schowen, K. B.; Limbach, H. H.; Denisov, G. S.; Schown, R. L. *Biochim. Biophys. Acta* **2000**, *1458*, 43−62.
(28) Jencks, W. P. *Chem. Rev.* **1972**, *72*, 705−718.
(29) Elstner, M.; Cui, Q.; Munih, P.; Kaxiras, E.; Frauenheim, T.; Karplus, M. *J. Comput. Chem.* **2003**, *24*, 565−581.
(30) Brooks, B. R.; Bruccoleri, R. E.; Olafson, B. D.; States, D. J.; Swaminathan, S.; Karplus, M. *J. Comput. Chem.* **1983**, *4*, 187−217.

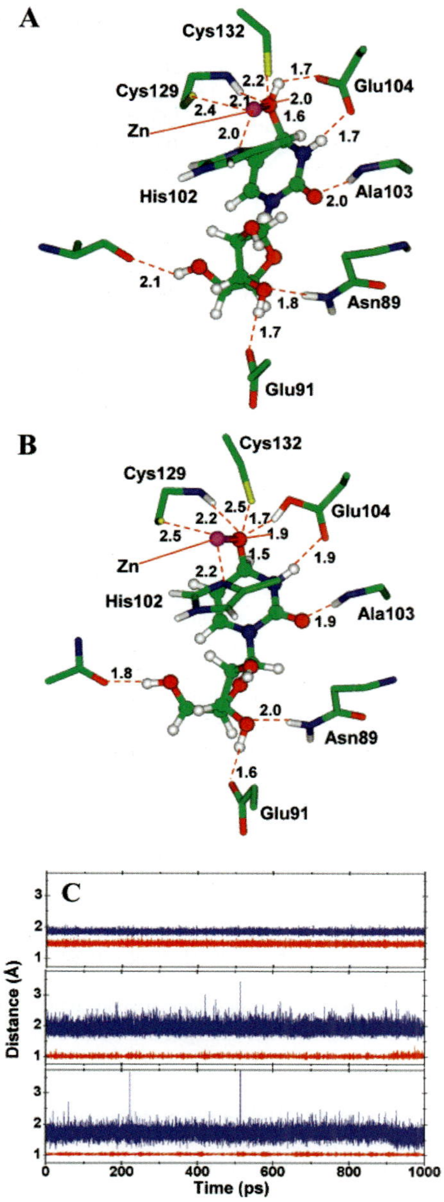

Figure 2. (A) Initial structure generated from the X-ray structure (ref 11) by addition of hydrogen atoms. The inhibitor at the active site is ZEB−H_2O. Distances are in angstroms. (B) Average active-site structure obtained from 1000 frames (50 ps). The further increase of the number of frames has little effect on the structure. A proton transfer from O_4 to Glu-104 has occurred during the energy minimization and MD simulations. (C) Internal motion of the transition-state analogue and its interactions at the active site during the MD simulations monitored by different distances. Top panel: the C_4−O_4 (red) and Zn−O_4 (blue) distances. The C_4−O_4 bond is stable with an average bond distance of 1.5 Å. Middle panel: the N_3−H_3 (red) and $O_{\epsilon 2}$ (Glu-104)···H_3 (N_3) (blue) distances. H_3 forms a covalent bond to N_3, and $O_{\epsilon 2}$ (Glu-104) accepts a hydrogen bond from N_3−H_3. Bottom panel: the $O_{\epsilon 1}$ (Glu-104)−H distance (red) and O_4···H distances. H forms a covalent bond to $O_{\epsilon 1}$(Glu-104), and O_4 of ZEB accepts a hydrogen bond from Glu-104.

were performed on the CDA−ZEB−H_2O complex. ZEB−H_2O, Zn, and Glu-104 were treated by QM and the rest of the system by MM. The all-hydrogen potential function (PARAM22)[31] was used for MM atoms. A modified TIP3P water model[32,33] was employed for the solvent. The stochastic boundary molecular dynamics method[34] was used for the QM/MM MD simulations. The system was separated into a reaction zone and a reservoir region, which was deleted; the reaction zone was further divided into the reaction region and the buffer region. The reference point for partitioning the system was chosen as the N_1 atom of ZEB−H_2O (Figure 1B). The reaction region was a sphere with radius 16 Å, and the buffer region has R equal to 16 Å $\leq R \leq$ 18 Å. Inside the reaction region, the atoms were propagated by molecular dynamics, whereas atoms in the buffer region were propagated by Langevin dynamics. The friction constants for the Langevin dynamics were 250 ps^{-1} for the protein atoms and 62 ps^{-1} for the water molecules. A 1 fs time step was used for integration of the equations of motion, and the results of the simulations were saved every 50 fs. In initiating the runs, we performed 500 steps of minimization using the steepest descent method for the protein and solvent atoms. Then minimization was performed for the entire stochastic boundary system with the Adapted Basis Newton Raphson method. The temperature of the system was gradually increased from 50 to 300 K (30 ps) and equilibrated at 300 K (40 ps). The simulations were performed on the resulting systems for several hundred picoseconds to 1 ns.

The initial model system (Figure 2A) generated from the X-ray structure containing ZEB−H_2O was found to be unstable; the proton transfer from O_4 to $O_{\epsilon 1}$ of Glu-104 occurred during the QM (SCC-DFTB)/MM energy minimization, leading to a stable alkoxide-like inhibitor at the active site (see Figure 2B and discussions below; the distances from the proton to O_4 and $O_{\epsilon 1}$ of Glu-104 are 1.66 and 1.01 Å, respectively). The resulting structure with the alkoxide-like inhibitor was found to be stable during the QM (SCC-DFTB)/MM MD simulations. To further test if this result is meaningful, several other QM/MM approaches were applied, and the initial structure for the inhibitor was also assumed to be ZEB−H_2O with the proton located on O_4 (i.e., the same initial model was used here as that in the case of the SCC-DFTB/MM calculations). The QM methods used in these additional QM/MM calculations include B3LYP/6-31G**,[35a] HF/6-31G**, B97/6-31G**,[35b] HCTH/6-31G**,[35c] and HFEX/6-31G**. Consistent with the results with the SCC-DFTB method, the proton transfer was also observed during the energy minimization with each of these QM/MM methods. The distances from the proton to O_4 ($O_{\epsilon 1}$ of Glu-104) were changed from 1 Å (1.7) in the initial structure to 1.53 Å (1.03), 1.75 Å (0.97), 1.56 Å (1.03), 1.60 Å (1.02), and 1.59 Å (0.98) from the B3LYP/6-31G**/MM, HF/6-31G**/MM, B97/6-31G**/MM, HCTH/6-31G**/MM, and HFEX/6-31G**/MM calculations, respectively. The distance between Zn and O_4 obtained from the B3LYP/6-31G**/MM, B97/6-31G**/MM, and HCTH/6-31G**/MM calculations is the same as that from the SCC-DFT/MM calculation (i.e., 1.87−1.88 Å), while the C_4−O_4 distance (1.42 Å) from these calculations is about 0.06 Å shorter. The QM/MM MD and free energy simulations were not performed with these additional QM/MM methods because they are very time-consuming. Thus, the results of the high-level QM/MM energy minimizations support the conclusion that ZEB−H_2O is unstable in the enzyme. In this paper, the alkoxide-like inhibitor in IV of Figures 1C and 2B is designated as the transition-state analogue

(31) MacKerell, A. D., Jr.; Bashford, D.; Bellott, M.; Dunbrack, R. L., Jr.; Evanseck, J. D.; Field, M. J.; Fischer, S.; Gao, J.; Guo, H; Ha, S.; et al. *J. Phys. Chem. B* **1998**, *102*, 3586−3616.
(32) Jorgensen, W. L. *J. Am. Chem. Soc.* **1981**, *103*, 335−340.
(33) Neria, E.; Fisher, S.; Karplus, M. *J. Chem. Phys.* **1996**, *105*, 1902−1921.
(34) Brooks, C. L., III; Brunger, A.; Karplus, M. *Biopolymers* **1985**, *24*, 843−865.
(35) (a) Lee, C.; Yang, W.; Parr, R. G. *Phys. Rev. B* **1988**, *37*, 785. (b) Becke, A. D. *J. Chem. Phys.* **1997**, *107*, 8554−8560. (c) Hamprecht, F. A.; Cohen, A. J.; Tozer, D. J.; Handy, N. C. *J. Chem. Phys.* **1998**, *109*, 6264−6271. (d) Zhao, Y.; Truhlar, D. G. *J. Phys. Chem. A* **2004**, *108*, 6908−6918.

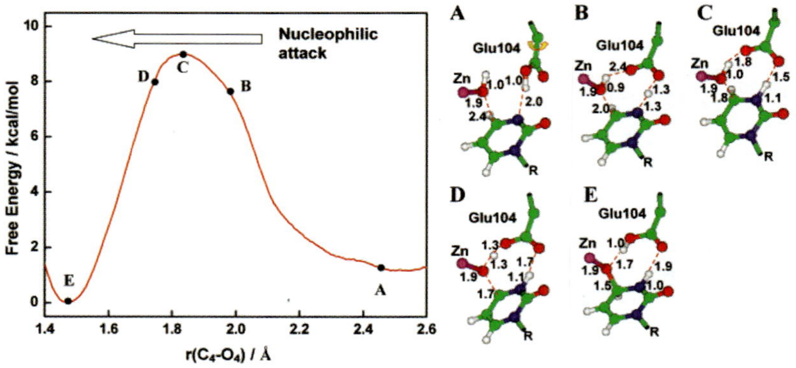

Figure 3. Free energy (potential of mean force) as a function of the C_4-O_4 distance. The average structures at different points (**A**, **B**, **C**, **D**, and **E**) during the nucleophilic attack are given. The structure at **A** is the structure for the CDA−ZEB complex (i.e., before the nucleophilic attack). The structure at **E** is the structure after the nucleophilic attack on C_4 by the Zn−hydroxide. Two proton transfers occur during the nucleophilic attack: one proton transfers from $O_{\epsilon 2}$ of Glu-104 to N_3 and the other transfers from the Zn−hydroxide to $O_{\epsilon 1}$ of Glu-104. The protonation of N_3 by Glu-104 is ahead of the deprotonation of the Zn−hydroxide. The free energy difference between **A** and **E** is ca. 1.5 kcal/mol with the structure at **E** being more stable.

(TSA) to be distinguished from the unstable ZEB−H_2O in III of Figures 1C and 2A.

The umbrella sampling method[36] implemented in the CHARMM program along with the Weighted Histogram Analysis Method (WHAM)[37] was applied to determine the changes of the free energy (potential of mean force) along three reaction coordinates starting from the stable TSA structure obtained from the MD simulations (Figure 2B). The free energy curves were obtained for the processes of the nucleophilic attack of the Zn−hydroxide on C_4 (monitored by the distance between C_4 and O_4), the proton transfer between the Zn−hydroxide and $O_{\epsilon 1}$ of Glu-104 [monitored by $r(O_{\epsilon 1}-H) - r(O_4-H)$], and the deprotonation (protonation) of N_3 by Glu-104 [monitored by $r(N_3-H_3) - r(O_{\epsilon 2}-H_3)$]. It was found that all of the three steps correspond to the same transformation between I and IV in Figure 1C; the three free energy profiles were also found to be similar. Therefore, only the free energy profile for the nucleophilic attack step is reported here. The harmonic umbrella potentials with a force constant of 100−500 kcal mol^{-1} $Å^{-2}$ were used. Twenty windows were used for the nucleophilic attack by the Zn−hydroxide, and 70 ps simulations (20 ps equilibration and 50 ps production run) were performed for each window. To estimate the effect of the proton transfer from the Zn−hydroxide to Glu-104 on the stability of the TSA, the free energy was obtained for the nucleophilic attack step with the O_4-H bond (1.00 Å) fixed by the SHAKE algorithm.[38] As a result, Glu-104 can still protonate N_3 but is unable to accept the proton from O_4.

Results

An important test for the methods of the QM/MM MD simulations is to see if the average structure generated from the simulations agrees with the one determined by X-ray crystallography. In Figure 2B, an average active-site structure based on 1000 frames in the MD trajectory is compared with the initial structure (Figure 2A) generated from the X-ray structure by addition of hydrogen atoms using the CHARMM program.[30,31] As can be seen from Figure 2A,B, the active-site structure of the CDA−TSA complex from the QM/MM simulations is generally close to the initial structure, suggesting the computational approaches used in this study are meaningful. However, there is a major difference in these two structures. For the initial structure in Figure 2A, the inhibitor was assumed to exist as ZEB−H_2O.[3−5] This is in contrast with the average structure in Figure 2B, where the proton on O_4 has already transferred to $O_{\epsilon 1}$ of Glu-104 during the energy minimization and remained on $O_{\epsilon 1}$ during the QM/MM MD simulations. Thus, the QM/MM results suggest that the structure of the CDA−ZEB−H_2O complex is likely to be unstable, and the transition-state analogue may exist as an alkoxide-like inhibitor at the active site (see Figure 2B). This suggestion is supported by the additional QM/MM calculations (see Methods).

The CDA−ZEB−H_2O complex has been the subject of a previous QM/MM MD study,[23] where Glu-104 was not treated by the QM method. As a result, Glu-104 was unable to accept a proton from O_4 or N_3, and the CDA−ZEB−H_2O complex could not change to the CDA−TSA complex or the CDA−ZEB complex. It was found that the C_4-O_4 bond of ZEB−H_2O underwent transient bond breaking and making during the MD simulations;[23] the reason for this instability was not clear at the time. In Figure 2C, the internal motion of the stable alkoxide-like transition-state analogue and its interactions with the active-site groups during the MD simulations are monitored using the C_4-O_4, Zn−O_4, N_3-H_3, $O_{\epsilon 2}$ (Glu-104)···H_3 (N_3), $O_{\epsilon 1}$ (Glu-104)−H, and O_4···H (Glu-104) distances. Consistent with the structure in Figure 2B, Glu-104 is protonated with the $O_{\epsilon 1}-H$ distance of ca. 1 Å (bottom panel, red). Moreover, the C_4-O_4 bond (top panel, red) is now stable without the transient bond breaking and making observed in the earlier study.[23] Thus, the previous observation of the transient bond breaking and making is due to the assumption that there exists a stable ZEB−H_2O at the active site. The results of the present simulations suggest that this assumption in the crystallographic least-squares refinement needs to be re-examined as well.

The change of free energy (potential of mean force) as a function of the C_4-O_4 distance is plotted in Figure 3; five average structures along the free energy curve are also given for the groups involved, representing different stages of the nucleophilic attack by the Zn−hydroxide on C_4. The average structure at **A** corresponds to the CDA−ZEB complex with a C_4-O_4 distance of about 2.4 Å. The QM/MM MD simulations

(36) Torrie, G. M.; Valleau, J. P. *Chem. Phys. Lett.* **1974**, *28*, 578−581.
(37) Kumar, M.; Bouzida, D.; Swendsen, R. H.; Kollman, P. A.; Rosenberg, J. M. *J. Comput. Chem.* **1992**, *13*, 1011−1021.
(38) Ryckaert, J. P.; Ciccotti, G.; Berendsen, H. J. C. *J. Comput. Phys.* **1977**, *23*, 327−341.

were also performed for this complex for 500 ps without the harmonic umbrella potentials. The CDA−ZEB complex was found to be stable, and the proton transfer from $O_{\epsilon 2}$ of Glu-104 to N_3 did not occur during the simulations. It appears that this proton transfer could occur only when the structure of ZEB is significantly perturbed by the nucleophilic attack with a partial sp^2 to sp^3 rehybridization of N_3.[25]

Consistent with an earlier linear-scaling quantum mechanical investigation,[20,21] $Zn-OH^-$ exists as the active species before the nucleophilic attack from the simulations. It has been suggested earlier[25] based on solvent deuterium isotopic effects that the proton from $Zn-H_2O$ does not transfer to Glu-104 before the substrate and, presumably, inhibitor binding. To examine whether the simulations could lead to the same conclusion, a model for CDA without the inhibitor was built and 500 ps QM/MM MD simulations were performed. The initial structure in Figure 2A was used as a starting structure, and $ZEB-H_2O$ was simply replaced by H_2O; this water molecule was treated by QM along with Zn and Glu-104. Consistent with the experimental suggestion, the protons on $Zn-H_2O$ were found to be stable and did not transfer to Glu-104, even though one of the protons was hydrogen-bonded to $O_{\epsilon 1}$ of Glu-104 throughout the simulations. Thus, the pK_a value of Glu-104 may be elevated due to the ZEB binding. High pK_a values (about 8−10) for the carboxyl group have been proposed for other enzymes (see refs 39 and 40).

As the nucleophilic attack proceeds to **B** with $r(C_4-O_4)$ = 2.0 Å, the average position of the proton (H_3) moves to the center between $O_{\epsilon 2}$ and N_3; the $N_3 \cdots O_{\epsilon 2}$ distance is around 2.6 Å (see the average structure at **B** in Figure 3). As it will be shown later, H_3 actually moves back and forth between $O_{\epsilon 2}$ and N_3 on a picoseconds scale. $O_{\epsilon 1}$ of Glu-104 now forms a hydrogen bond with the Zn−hydroxide. At the transition state (**C**), N_3 has already been protonated by Glu-104, while the second proton transfer from the Zn−hydroxide to $O_{\epsilon 1}$ has not started yet. The second proton transfer occurs as the nucleophilic attack on C_4 proceeds further along the reaction coordinate (see the structures at **D** and **E**). Thus, the simulations suggest that protonation of N_3 by Glu-104 is ahead of deprotonation of the Zn−hydroxide. Moreover, they indicate that the pK_a value of the Zn−hydroxide may be significantly depressed in the enzyme as a result of the C_4-O_4 bond formation. Although the locations of the two protons at **C** are similar to those in the $ZEB-H_2O$ complex shown in Figure 2A, the C_4-O_4 bond is not formed; the average distance between the C_4 and O_4 atoms at **C** is about 1.8 Å. Figure 3 shows that the free energy difference between **A** and **E** (i.e., before and after the nucleophilic attack) is ca. 1.5 kcal/mol from the simulation; the CDA−TSA complex (**E**) is more stable than the CDA−substrate analogue complex (**A**). The free energy barrier from **A** to **E** was calculated to be about 8 kcal/mol.

The dynamics of the protons at **B** and **D** are monitored in Figure 4A,B, respectively; $r(N_3-H_3)$ and $r(O_{\epsilon 2}-H_3)$ are used to monitor the motion of H_3, whereas $r(O_4-H)$ and $r(O_{\epsilon 1}-H)$ are used to monitor the motion of H. It is interesting to note from Figure 4A (top panel) that the proton (H_3) involved in the protonation of N_3 moves back and forth on a picoseconds scale

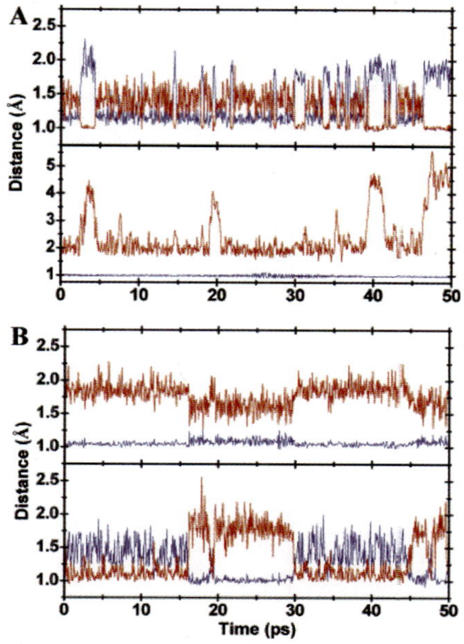

Figure 4. (A) Typical fluctuations of the covalent bond and hydrogen bond distances at **B** on the free energy curve in Figure 3. Top panel: $r(N_3-H_3)$ (blue) and $r(O_{\epsilon 2}-H_3)$ (red). The proton (H_3) moves back and forth between N_3 and $O_{\epsilon 2}$ with the average position in the middle of the two atoms (see the structure at **B** in Figure 3). Bottom panel: $r(O_4-H)$ (blue) and $r(O_{\epsilon 1}-H)$ (red). The proton is on O_4 with a bond distance of 1 Å. $O_{\epsilon 1}$ accepts a hydrogen bond from O_4-H. (B) Typical fluctuations of the covalent bond and hydrogen bond distances at **D** of the free energy curve in Figure 3. Top panel: $r(N_3-H_3)$ (blue) and $r(O_{\epsilon 2}-H_3)$ (red). The proton is on N_3 with a bond distance of 1 Å. $O_{\epsilon 2}$ accepts a hydrogen bond from N_3-H_3. Bottom panel: $r(O_4-H)$ (blue) and $r(O_{\epsilon 1}-H)$ (red). The proton moves back and forth between O_4 and $O_{\epsilon 1}$ of Glu-104 with the average position in the middle of the two atoms (see the structure at **D** in Figure 3).

between N_3 and $O_{\epsilon 2}$ at **B** (as indicated by the changes of the N_3-H_3 and $O_{\epsilon 2}-H_3$ distances), even though its average position is in the middle of the two atoms. Thus, the proton of the $N_3 \cdots H_3-O_{\epsilon 2}$ hydrogen bond spends most of the time near either N_3 or $O_{\epsilon 2}$ with a distance that is close to the one for a covalent bond (e.g., 1.0−1.15 Å). However, the corresponding bond-distance fluctuation is generally greater than those involving normal covalent bonds. This can be seen by comparing the fluctuations for the N_3-H_3 distance in Figure 4A (top panel, blue line) and 4B (top panel, blue line) where the fluctuation in the former case is more significant. This observation might be applied to some of the low-barrier hydrogen bonds widely observed in biological structures. Figure 4B (top panel) shows that the protonation of N_3 has already completed at **D** with a N_3-H_3 bond distance at about 1 Å (blue line). Figure 4B (bottom panel) shows that the second proton moves back and forth between O_4 and $O_{\epsilon 1}$ of Glu-104 when the nucleophilic attack reaches **D**; the average position of the proton is also located in the middle of the two oxygen atoms involved in the interaction (see structure **D** in Figure 3). This second proton transfer to $O_{\epsilon 1}$ of Glu-104 is completed when the nucleophilic attack proceeds further to **E** (see Figure 2B,C for the corresponding structural and dynamic information at **E**). Comparison of the fluctuations for $r(O_{\epsilon 1}-H)$ in Figure 4B (bottom panel, red line) and Figure 2C (bottom panel, red line) shows that the

(39) Deng, H.; Callender, R.; Zhu, J.; Nguyen, K. T.; Pei, D. *Biochemistry* **2002**, *41*, 10563−10569.
(40) Ma, J.; Zheng, X.; Schnappauf, G.; Braus, G.; Karplus, M.; Lipscomb, W. N. *Proc. Natl. Acad. Sci. U.S.A.* **1998**, *95*, 14640−14645.

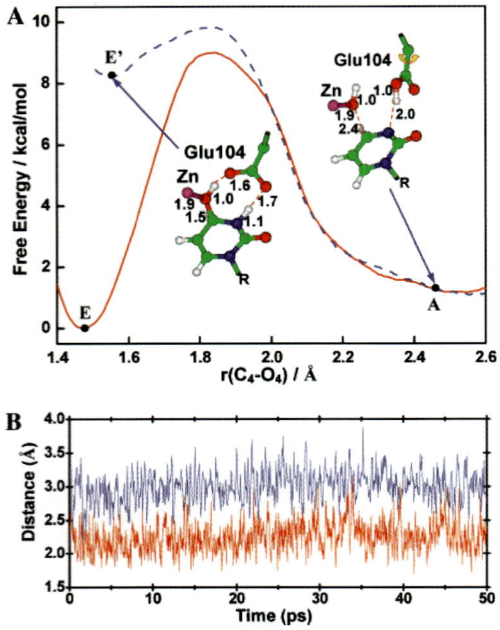

Figure 5. (A) Comparison of the free energies as functions of the C_4-O_4 distance with and without the constraint on the O_4-H bond. Red (solid line): without the constraint on the O_4-H bond (same as in Figure 3). Blue (dashed line): with the constraint on the O_4-H bond to prevent the proton transfer away from O_4. Comparison of the two curves shows that the proton transfer from O_4 to $O_{\epsilon1}$ of Glu-104 along with the possible strengthening of the interaction between the TSA and enzyme may make a very important contribution to the stability of the TSA (ca. 8 kcal/mol). (B) Fluctuations for the distance between H−N (Cys-129) and O_4 as a function of the time at **E** (red) and **E′** (blue), respectively. A hydrogen bond is formed between the backbone N−H group of Cys-129 and O_4 at **E** with an average (N)H···O_4 hydrogen bond distance of 2.2 Å. This hydrogen bond cannot be formed at **E′** containing ZEB−H_2O (i.e., obtained with the assumption that the proton H is located at O_4); the average distance is about 3 Å.

fluctuation for the $O_{\epsilon1}-H$ bond at **D** is also greater than that at the stable CDA−TSA conformation (**E**).

To estimate the effect of the second proton transfer from the Zn−hydroxide to Glu-104 on the stability of the alkoxide-like TSA at the active site, the O_4-H bond was fixed at 1.00 Å using the SHAKE algorithm.[37] As a result, Glu-104 can still protonate N_3 but is unable to accept the proton from O_4. The free energy as a function of the C_4-O_4 distance was then determined using the same procedure. The two free energy curves obtained with and without the constraint on the O_4-H bond are compared in Figure 5A. As can be seen from Figure 5A, the shapes of the two free energy curves are essentially the same in the area of $r(C_4-O_4) \geq 2$ Å. This result is expected as the O_4-H bond is intact in the early stage of the nucleophilic attack and the corresponding structures along the reaction coordinate in the two cases are similar. As is evident from Figure 5A, the free energy after the nucleophilic attack is as much as 8 kcal/mol higher for the curve with the constraint (i.e., at **E′**) than that without (i.e., at **E**). One major difference for the two free energy simulations is that the ligand in the CDA complex at **E′** is ZEB−H_2O (the structure is similar to the one in Figure 2A), while it is the alkoxide-like TSA at **E** (Figure 2B). Thus, a major stabilization of the alkoxide-like TSA complex in the active site seems to rise from the proton transfer from the Zn−hydroxide to Glu-104 in the later stage of the nucleophilic attack (i.e., after the transition state).

The alkoxide-like TSA seems to lead to a stronger interaction with the enzyme than ZEB−H_2O. Indeed, as can be seen from Figure 5B, the hydrogen bond between the amide group of Cys-129 and O_4 cannot be formed in the CDA−ZEB−H_2O complex (**E′**) during the MD simulations; the average distance between H(N) and O_4 is about 3 Å. A similar observation was made in the earlier study.[23] By contrast, a stable hydrogen bond between the amide group of Cys-129 and O_4 can be formed in the CDA−TSA complex with an average hydrogen bond distance of 2.2 Å. The formation of this hydrogen bond is likely to make additional contribution to stabilize the alkoxide-like TSA in the active site. Figure 5A shows that ZEB−H_2O at **E′** is more stable by ~2 kcal/mol than the structure at the transition state for the case with the constraint, presumably due to the stronger interactions at the active site as a result of the C_4-O_4 bond formation.[23] Nevertheless, this effect of hydrogen bonding is much smaller than the effect of the proton transfer involving the O_4-H group and Glu-104.

Discussion

Although the binding of ZEB results in a near-perfect TSA at the active site of CDA,[11,12] the origin of the high binding affinity for this inhibitor is still not clear. It is believed that the TSA is zebularine 3,4-hydrate containing a C_4-OH group (ZEB−H_2O; see Figure 1B) that is formed through the hydration of ZEB at the active site. Comparison of the structures of CDA complexed with different inhibitors has led to the suggestion that the active-site interactions involving the C_4-OH group of ZEB−H_2O may make a major contribution to the high binding affinity.[11] However, the proton on the C_4-OH group is invisible in the X-ray structure, and the possibility that it may be located on Glu-104 cannot be simply ruled out. Indeed, the present QM/MM MD simulations showed that ZEB−H_2O is in fact unstable in the active site of CDA. A proton transfer from the C_4-OH to Glu-104 occurred during the energy minimizations, and the resulting structure was stable during the QM/MM MD simulations. This proton-transfer process leads to the formation of a stable alkoxide-like TSA (or simply TSA) at the active site (see IV in Figures 1C and 2B) which may not be distinguishable from ZEB−H_2O in the X-ray structure. The average structure of the CDA−alkoxide-like TSA complex obtained from the simulations is generally closer to the X-ray structure than the average structure of the CDA−ZEB−H_2O complex at **E′** in Figure 5A (except for the location of H), supporting the existence of the TSA, rather than the unstable ZEB−H_2O, at the active site. For instance, the X-ray structure showed that there is a hydrogen bond between the backbone N−H group of Cys-129 and O_4.[11] The simulations, however, suggested that whereas this hydrogen bond is stable in the TSA complex, it could not be formed for the ZEB−H_2O complex (Figure 5B), due to the poor geometries for the formation of the two hydrogen bonds between the O_4-H group and $O_{\epsilon1}$ and between the N−H group of Cys-129 and the electron lone-pairs of O_4 simultaneously.

The QM/MM free energy simulations were used to understand the mechanism of the inhibitor binding and the energetic origin of the high binding affinity of the TSA. It was shown that the nucleophilic attack by the Zn−hydroxide group on C_4 was concerted with two proton-transfer processes from $O_{\epsilon 2}$ of Glu-104 to N_3 of ZEB and from the Zn−hydroxide to $O_{\epsilon 1}$ of Glu-104, leading to the alkoxide-like TSA (i.e., I → IV, Figure 1C). The protonation of N_3 by Glu-104 was found to be completed before the nucleophilic attack reaches the transition state, while deprotonation of the Zn−hydroxide started mainly after the transition state was reached (see Figure 3). No stable ZEB−H_2O was observed during the free energy simulations, supporting the conclusion from the MD simulations that this species is unstable at the active site (see above). The results reported here seem to be consistent with the previous study of ^{15}N isotope effects on CDA-catalyzed deamination of cytidine,[25] which proposed that two or more protons were transferred during the formation of the tetrahedral intermediate. However, the existence of the alkoxide-like TSA in the active site and the absence of a stable ZEB−H_2O from the simulations indicate that the tetrahedral intermediate formed during the enzyme catalysis might be alkoxide-like as well, resembling a second tetrahedral intermediate proposed earlier,[12] although a detailed study is still necessary.

The equilibrium constant for hydration of ZEB to generate ZEB−H_2O in solution was estimated to be approximately 4.7 × 10^{-6},[24] and the equilibrium constant for generating the corresponding alkoxide-like compound is probably even lower. Figure 3 shows that the CDA−TSA complex (**E**) is about 1.5 kcal/mol more stable than the CDA−ZEB complex (**A**) based on the free energy simulations. Thus, the enzyme is able to reduce the instability (or increase the stability) of the TSA relative to ZEB by at least 9 kcal/mol compared to the case in solution. This effect is close to the previous estimate (about 10 kcal/mol) based on experimental data.[3,5,11,24] However, the nucleophilic attack of the Zn−hydroxide on C_4 did not lead to ZEB−H_2O from our study, and the stabilization of the TSA cannot be simply attributed to the interactions involving the C_4−OH group, as proposed previously. Indeed, when the O_4−H covalent bond was fixed by SHAKE in the free energy simulations (see Figure 5A, blue dashed line), the resulting CDA−ZEB−H_2O complex (i.e., at **E′** in Figure 5A) was found to be as much as 6−7 kcal/mol less stable than the CDA−ZEB complex (i.e., at **A** in Figure 5A). Thus, the proton transfer from Zn−hydroxide to Glu-104 along with the possible strengthening of the interaction involving the TSA (see above) may provide a stabilizing effect of about 8 kcal/mol for the CDA−TSA complex. This effect is much greater than the effect of the hypothetical hydrogen-bonding interaction involving the C_4 hydroxyl group of ZEB−H_2O, which was estimated to be only about 2 kcal/mol (Figure 5A).

Conclusions

The QM/MM MD and free energy simulations have been performed on CDA to understand the inhibitor binding and the origin of the high binding affinity for TSA. Zebularine 3,4-hydrate was found to be unstable at the active site of CDA from the simulations. It was suggested that the nucleophilic attack by the Zn−hydroxide group on C_4 is concerted with two proton-transfer processes, leading to the formation of a stable alkoxide-like TSA at the active site. Such concerted process allows the TSA to be stabilized through a mechanism that is similar to the transition-state stabilization in general acid−base catalysis.[27,28] The proton transfer from the Zn−hydroxide to Glu-104 is necessary to generate the general acid (Glu-104) for protonating the leaving ammonia during the enzyme catalysis. Thus, CDA seems to be able to use this proton transfer to stabilize the alkoxide-like tetrahedral intermediate[26] and therefore achieve transition-state stabilization. The results may have important implications for some other zinc enzymes as well, where a proton transfer involving a Glu or Asp residue interacting with a Zn-activated hydroxide nucleophile could play an important role during the catalysis.[10,41]

Acknowledgment. This work was supported by the Center of Excellence in Structural Biology, University of Tennessee. We are grateful for the computational resources from the Center for Computational Sciences, Oak Ridge National Laboratory, the National Center for Supercomputing Applications (NCSA), and the Scientific Computing Facilities of Boston University. We thank Professor Martin Karplus for a gift of the CHARMM program.

JA0439625

(41) Christianson, D. W.; Cox, J. D. *Annu. Rev. Biochem.* **1999**, *68*, 33−57.

Chapter 19

Hualiang Jiang

Biography

Dr. Hualiang Jiang was born in Wujin county, Jiangsu province, China, on January 10, 1965. He obtained his bachelor's degree from the Department of Chemistry, Nanjing University in 1987. In September 1989, he entered the East China Normal University, and received his Masters degree in physical chemistry (quantum chemistry) in 1992. From September 1992 to July 1995 he studied in the Shanghai Institute of Materia Medica (SIMM), Chinese Academy of Sciences and in 1995 he received his Ph.D. in organic chemistry.

Dr. Jiang is currently a professor at SIMM and is also the deputy director of the Institute and the director of the Drug Discovery and Design Center. He is the chief scientist of a 973 Program (a National Basic Research Program in China) and acts as a member for the scientific committee of several major research programs in China, such as a 863 Program (a National High-Tech Research and Development Program in China) in biology and medical technology, a National Basic Research Program in protein science, and a major research project of the National Natural Science Foundation.

He also serves on the editorial advisory boards for several journals including *Journal of Medicinal Chemistry* (senior editor) and *ChemMedChem*. He has been awarded the Natural Sciences Award of China, 5th prize in the Young Scientist Awards of China, the Natural Sciences Award of Shanghai, and was named a Top 10 Outstanding Scientist of Shanghai (2001–2003).

A Great Man

In 2010, Professor Jianpeng Ma at the Baylor College of Medicine invited me to write something for the book celebrating the 90th birthday of Professor William N. Lipscomb. I did not expect that Professor Lipscomb would leave us forever on April 14, 2011. Different people have different lives, some people's lives are heavier than a mountain, while others are lighter than a feather. Professor Lipscomb was not only a great scientist but also a great man. I write this article to celebrate his life and to mourn his loss.

In September 2004, I organized a course for the European Molecular Biology Organization called Methods for Protein Simulations and Drug Design. Through Jianpeng, I invited Professor Lipscomb to give a keynote lecture at the workshop. It was my first time meeting him. He arrived in Shanghai several days early, before the workshop, thus I had time to chat with him. This discourse has deeply influenced and will continue to influence me, particularly in my scientific career.

We discussed a variety of topics, including sciences, life, and his career in research and teaching. One day, Jianpeng and I accompanied him to visit Tongli, a small town close to Shanghai where there are still many Ming and Qing dynasty buildings. We had lunch in a small restaurant. At the table, when we were discussing the field of protein folding, Professor Lipscomb told us a historic story regarding the early days of the field of protein folding. In 1929, a Chinese protein scientist named Hsien Wu announced his research results at the 13th International Congress of Physiology in Boston. Wu proposed that protein denaturation was a purely conformational change, i.e. protein unfolding, without chemical alterations of the protein. In 1931, Wu published these important results in the *Chinese Journal of Physiology* in a paper entitled "Studies on denaturation of proteins. XIII. A theory of denaturation."[1] According to Professor Lipscomb, Hsien Wu was not only the first to propose the crucial theory of protein denaturation, but also the first to propose the prototype of helix structure of peptides, though Hsien Wu used the term "spiro." Both ideas of Wu's were later popularized by Professor Linus Pauling, Professor Lipscomb's mentor. According to Professor Lipscomb, "In fact, Linus Pauling got several ideas and inspirations from Wu's paper. Unfortunately, when he and Alfred Mirsky published their first paper about protein denaturation in *PNAS*,[2] they did not cite Wu's paper. I guess Pauling's idea about peptide's α-helix structure was also from Wu's paper, because he mentioned Wu's result in a footnote of one of his *PNAS* papers, although he did not formally cite Wu's paper." After Professor Lipscomb went back to the US, he mailed me copies of Hsien Wu's papers and several of Pauling's papers, and highlighted the places he mentioned at Tongli. I greatly admire Professor Lipscomb for his fairness in crediting Hsien Wu for his original theory of protein denaturation.

The second thing I should mention is that Professor Lipscomb was an excellent teacher. His skill in teaching not only influenced his students, among them three went on to win Nobel Prizes, but also influenced me. During the same visit in 2004, we chatted for about two hours in a small meeting room at our institute, the Shanghai Institute of Materia Medica (SIMM), Chinese Academy of Sciences. I asked him, "How can one become a great scientist, just like you?" "Like me?" he replied with a benevolent smile, "One thing you should do is teaching. I have taught 25 courses at Harvard. Teaching can reconstruct your knowledge structure, which is beneficial to your research." At exactly that time, Professor Xuehong Qian, the president of the East China University of Science and Technology (ECUST), invited me to establish the School of Pharmacy and to be its first dean. I was actually hesitating. The advice from Professor Lipscomb prompted me to accept Professor Qian's invitation. Now, the School of Pharmacology at ECUST is developing very well and has been one of the top ten pharmacology schools in China. Since then, I have also started to teach courses at ECUST. Indeed, my research has benefited enormously from teaching and interacting with students.

In August 2005, the Chinese Chemical Society invited Professor Lipscomb to give a plenary lecture at the 40th IUPAC congress held in Beijing. Thus, Professor Lipscomb visited China again. This time, Jianpeng and I hosted an 85th birthday symposium for him at SIMM. He was very happy to see so many of his previous associates attending his birthday symposium. This was the last time I saw him.

It is difficult to differentiate life and death: some people live long but contribute little; some people continue to contribute even after their death. Professor Lipscomb has passed away; however, his spirit will live in my heart forever, encouraging me to continue to climb the scientific mount.

References

1. Wu H., Studies on denaturation of proteins. XIII. A theory of denaturation. *Chinese J. Physiol.*, **5**, 321–344 (1931).
2. Mirsky, A. E. and Pauling, L., On the structure of native, denatured, and coagulated proteins. *Proc. Natl. Acad. Sci. USA*, **22**, 439–447 (1936).

Conformational transition of amyloid β-peptide

Yechun Xu*†, Jianhua Shen*†, Xiaomin Luo*, Weiliang Zhu*, Kaixian Chen*, Jianpeng Ma*‡§¶, and Hualiang Jiang*¶‖

*Center for Drug Discovery and Design, State Key Laboratory of Drug Research, Shanghai Institute of Materia Medica, Shanghai Institutes for Biological Sciences, Chinese Academy of Sciences, and Graduate School of Chinese Academy of Sciences, 555 Zuchongzhi Road, Shanghai 201203, China; ‖School of Pharmacy, East China University of Science and Technology, Shanghai 200237, China; ‡Department of Biochemistry and Molecular Biology, Baylor College of Medicine, One Baylor Plaza, BCM-125, Houston, TX 77030; and §Department of Bioengineering, Rice University, Houston, TX 77005

Communicated by William N. Lipscomb, Harvard University, Cambridge, MA, February 17, 2005 (received for review December 21, 2004)

The amyloid β-peptides (Aβs), containing 39–43 residues, are the key protein components of amyloid deposits in Alzheimer's disease. To structurally characterize the dynamic behavior of $A\beta_{40}$, 12 independent long-time molecular dynamics (MD) simulations for a total of 850 ns were performed on both the wide-type peptide and its mutant in both aqueous solution and a biomembrane environment. In aqueous solution, an α-helix to β-sheet conformational transition for $A\beta_{40}$ was observed, and an entire unfolding process from helix to coil was traced by MD simulation. Structures with β-sheet components were observed as intermediates in the unfolding pathway of $A\beta_{40}$. Four glycines (G_{25}, G_{29}, G_{33}, and G_{37}) are important for $A\beta_{40}$ to form β-sheet in aqueous solution; mutations of these glycines to alanines almost abolished the β-sheet formation and increased the content of the helix component. In the dipalmitoyl phosphatidylcholine (DPPC) bilayer, the major secondary structure of $A\beta_{40}$ is a helix; however, the peptide tends to exit the membrane environment and lie down on the surface of the bilayer. The dynamic feature revealed by our MD simulations rationalized several experimental observations for $A\beta_{40}$ aggregation and amyloid fibril formation. The results of MD simulations are beneficial to understanding the mechanism of amyloid formation and designing the compounds for inhibiting the aggregation of Aβ and amyloid fibril formation.

molecular dynamics simulation

Alzheimer's disease (AD), a neurodegenerative disorder, is pathologically characterized by the presence of extracellular senile plaques and intracellular neurofibrillary tangles in the brain (1). The major components of the plaques are amyloid β-peptides (Aβs) consisting of 39–43 residues that are proteolytically derived from the widely distributed transmembrane amyloid precursor glycoprotein (APP) (2–4). The amyloid hypothesis suggests that misfolding of Aβ leads to its dysfunctions and fibrillization; the latter is associated with a cascade of neuropathogenetic events to produce the cognitive and behavioral decline hallmarks of AD (4–6). Recently, however, compelling evidence has emerged that not the fibrillar aggregates but the dominant β-sheet structure of oligomers and fibril intermediates (protofibrils) are neurotoxic and might be the determinant pathogenic factor in AD (7–9). For example, Walsh *et al.* (8) demonstrated that cerebral microinjection of cell medium containing abundant Aβ monomers and oligomers but no amyloid fibrils markedly inhibited hippocampal long-term potentiation in rats *in vivo*. The conflict between the amyloid hypothesis and these new emerging experimental observations has stimulated more and more researchers to study the earliest phases of Aβ assembly and to explore the intrinsic properties of Aβs and the conformational dynamic behaviors of the Aβ monomer (9–15).

In addition, it has been established experimentally that the major secondary structure adopted by Aβ depends on the environment. The Aβ monomer favors an α-helix structure in a membrane or membrane-mimicking environment such as ionic detergents (15–19). In contrast, Aβ exists mainly as a random coil with few components of α-helix and/or β-sheet conformations in aqueous solution (14, 18, 20–22). However, x-ray diffraction measurements on oriented fibril bundles have indicated an extended β-sheet structure for Alzheimer's β-amyloid fibrils, with peptide chains in β-strand conformations running approximately perpendicular to the fibril axis and hydrogen bonds between peptide chains in each layer occurring roughly parallel to the fiber axis (23, 24). The dominant structure of Aβ in membrane or membrane-like mediums, aqueous solution, and fibrils leads to a postulation that a conversion from an α-helix or random coil to a β-sheet conformation occurs during (or before) the aggregation of Aβ (25–28). Similar conformational conversions have also been found in proteins of other diseases (29–31). Thus, a rigorous mechanistic study of such conformational transition is critical to understanding and development of therapeutic approaches for the conformational disorders in general, and AD in particular.

Unfortunately, because of the fibril's extreme insolubility and the monomer's high propensity to aggregate, investigation of the above conformational conversions of Aβ in the atomic detail by means of experimental methods is still intractable (32). Computational simulation with its extremely high time resolution and atomic level representation has been increasingly used in understanding the complex conformational features of polypeptides and in predicting structural preferences (33–38). Simulations have been performed by other groups with a view to elucidating the conformational transition and assembly mechanism of Aβ (10, 12, 13, 32, 39–42) However, these studies are based on the modeled structures of Aβ, i.e., fragments or synthesized analogs of Aβ, rather than the physiologically produced Aβ in the simulations. In addition, the environment was not considered explicitly, and the simulation time was not long enough to discover the significant conformations.

Here, we report on the result of conformational transitions of monomer $A\beta_{40}$ addressed by a series of long time molecular dynamics (MD) simulations. We examined the conformational transitions of $A\beta_{40}$ in both aqueous solution and lipid bilayers. By running six MD simulations in aqueous solution, we show that conformational transition from α-helix to coil occurs through helix/β-sheet mixed conformations, which is consistent with the recent experimental finding by Kirkitadze *et al.* (11) that helix/β-sheet mixed conformations are possible intermediates for Aβ oligomerization. Remarkably, these MD simulations addressed that G_{25}, G_{29}, G_{33}, and G_{37} play essential roles for $A\beta_{40}$ β-sheet formation, and a segment consisting of residues 24–40 may be the core for Aβ oligomerization. This finding was confirmed by running another four MD simulations on the mutant of these glycines. By running an additional two MD simulations, we demonstrate that $A\beta_{40}$ adopts helix as its major secondary structure in the lipid bilayer. However, $A\beta_{40}$ has a tendency to move out of the lipid environment toward the surface lipid

---

Abbreviations: AD, Alzheimer's disease; Aβ, amyloid β-peptide; $A\beta_{40}$, residues 1–40 of wild-type amyloid β-peptide; $MA\beta_{40}$, residues 1–40 of mutant amyloid β-peptide; MD, molecular dynamics; DSSP, defined secondary structure of proteins; APP, amyloid precursor glycoprotein; DPPC, dipalmitoyl phosphatidylcholine.

†Y.X. and J.S. contributed equally to this work.

¶To whom correspondence may be addressed. E-mail: hljiang@mail.shcnc.ac.cn or jpma@bcm.tmc.edu.

bilayer. These observations are beneficial to understanding the mechanism of amyloid formation and designing compounds for inhibiting the aggregation of Aβ.

Methods

Simulation Systems. Initial coordinates for the wild-type Aβ_{40} with a sequence of DAEFRHDSGYEVHHQKLVFFA-EDVG$_{25}$SNKG$_{29}$AIIG$_{33}$LMVG$_{37}$GVV was taken from the NMR structures determined in aqueous SDS micelles at pH 5.1 (15) (PDB entry 1BA4). For the simulations of Aβ_{40} (or mutated Aβ_{40}) in aqueous solution, the peptide was first put into a suitably sized box, of which the minimal distance from the peptide to the box wall was 1.5 nm. Then the box was solvated with the simple point charge (SPC) water model (43). The peptide/water system was submitted to energy minimization. Afterward, counterions were added to the system to provide a neutral simulation system. The whole system was subsequently minimized again. To set up the Aβ_{40}/dipalmitoyl phosphatidylcholine (DPPC) bilayer, one lipid molecule was first removed from a preequilibrated DPPC bilayer to fit in the C-terminal fragment of the peptide. The peptide/lipids system was subsequently solvated with the SPC water model. The resulting system was submitted to energy minimization to remove unfavorable contacts and equilibrated for 500 ps with positional restraints on the peptide atoms. Counterions were subsequently added to compensate for the net positive charge of the system. Finally, the whole system was minimized again.

MD Simulations. MD simulations were carried out by using the GROMACS package (44, 45) with constant number, pressure, and temperature (NPT) and periodic boundary conditions. The GROMACS force field (46) was applied for the peptide, and the lipid parameters adopted were those used in previous MD studies of lipid bilayers (47–50). The linear constraint solver (LINCS) method (51) was used to constrain bond lengths, allowing an integration step of 2 fs. Electrostatic interactions were calculated with the Particle-Mesh Ewald algorithm (52, 53). A constant pressure of 1 bar was applied with a coupling constant of 1.0 ps (54). For the peptide/DPPC bilayer system the pressure was performed independently in the *x*, *y*, and *z* directions of the whole system. Water, lipids, and peptide were coupled separately to a temperature bath, 300 K for peptide/water systems and 323 K for peptide/DPPC bilayer systems, by using a coupling time of 0.1 ps (54).

MD simulations were run on a 128-CPU Silicon Graphics (Mountainview, CA) Origin 3800 server. Analyses were performed by using facilities within the GROMACS package. Secondary structure analyses were carried out employing the defined secondary structure of proteins (DSSP) method (55). Structural diagrams were prepared by using MOLSCRIPT (56) and then rendered with RASTER3D (57).

Results

To probe how the environment and the sequence specificity affect the conformational transitions of Aβ, three models for MD simulations were designed. To make certain the conformational transition is the intrinsic character of Aβ rather than a stochastic output of simulations, several MD simulations were carried out for each system under different initial conditions by assigning different initial velocities on each atom of the simulation systems. In model I, the wild-type Aβ_{40} was solvated with water molecules. Six simulations were conducted for this model: five simulations (A1–A5) adopted different initial velocities assigned on each atom; one simulation (A6) was performed under a higher temperature (400 K). In model II, four glycines (G$_{25}$, G$_{29}$, G$_{33}$, and G$_{37}$) of Aβ_{40} were mutated by alanines, and then the Aβ_{40} mutant (MAβ_{40}) was solvated with water molecules. Four simulations were performed on this model: three simulations (B1–B3) were conducted with different initial velocities assigned on each atom; one simulation (B4) was performed at a higher temperature (400 K). In model III, Aβ_{40} was embedded in the environment of the DPPC bilayer, and two simulations (C1 and C2) were performed in two different sizes of the DPPC bilayers (127 and 63 DPPC lipids). In total, 12 MD simulations were performed on Aβ_{40} and MAβ_{40}. More detailed information for these MD simulations is listed in Table 1.

Table 1. The detailed information for 12 MD simulations

Models	Trajectory	Temperature, K	Time, ns	Mutation	Solvent
I	A1	300	150		Water
	A2	300	50		
	A3	300	50		
	A4	300	50		
	A5	300	50		
	A6	400	100		
II	B1	300	50	$G_{25}G_{29}G_{33}G_{37}$ mutated into $A_{25}A_{29}A_{33}A_{37}$	Water
	B2	300	50		
	B3	300	50		
	B4	400	50		
III	C1	323	100		DPPC
	C2	323	100		

Conformational Transition of Aβ_{40} in Aqueous Solution. Six simulations for model I were performed to explore the conformational dynamics of wild-type Aβ_{40} in aqueous solution. The profile of secondary structures of Aβ_{40} along trajectory A1 is shown in Fig. 1*a*. Obviously, the initial long α-helical structure of the peptide disappeared after several nanoseconds; in particular, the α-helix of residues 24–37 was converted into several short β-strands connected with turns during the first 10 ns. Afterward, three to four β-strands, each comprising two residues, existed throughout the remainder of the simulation. However, the central hydrophobic region (residues 16–21) always kept the helical structure throughout the whole 150-ns simulation, with occasional local deviations from α-helix to 5-helix or 3-helix. The remainder of

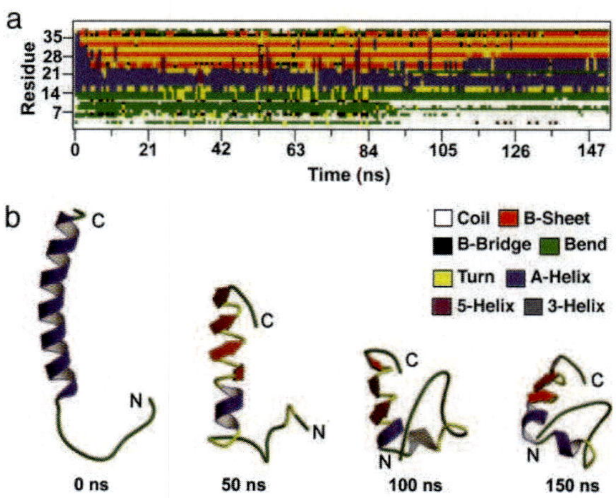

Fig. 1. Structural analyses of Aβ_{40} in trajectory A1. (*a*) Secondary structures as function of time for Aβ_{40} in trajectory A1 as calculated by DSSP. The structures were analyzed every 750 ps. (*b*) The snapshot structures of Aβ_{40} extracted from trajectory A1, in which the secondary structure is according to DSSP analysis. C and N denote C-terminal and N-terminal, respectively.

Table 2. The statistical helix and β-sheet components of $A\beta_{40}$ from 10 MD trajectories

Models	Trajectory	Helix,* %	β-sheet, %
I	A1	20	14
	A2	25	9
	A3	22	14
	A4	35	11
	A5	28	10
		Avg 26†	Avg 12†
II	B1	53	0
	B2	53	0
	B3	54	0
		Avg 53†	Avg 0†
III	C1	38	0
	C2	44	0
		Avg 41†	Avg 0†

*The sum of α-helix, 3-helix, and 5-helix.
†Averaged value of each model.

the peptide, residues 1–14, mostly adopts a random coil structure. The snapshot structures of $A\beta_{40}$ extracted from trajectory A1 (shown in Fig. 1*b*) visualize the secondary structural transformation of $A\beta_{40}$ described above (Fig. 1*a*).

The transformations of secondary structures of $A\beta_{40}$ along the trajectories of simulations A2–A5 are displayed in Fig. 6, which is published as supporting information on the PNAS web site. The profiles of the secondary structure of simulations A2–A5 is similar to that of simulation A1 (Figs. 1 and 6). These five simulations revealed that, in aqueous solution and room temperature (300 K), $A\beta_{40}$ is statistically composed by ≈26% helix, ≈12% β-sheet, and ≈62% coil components (Table 2). This finding is in good agreement with the NMR observations (14, 18, 22), which indicated that, in aqueous solution, the monomer of $A\beta_{40}$ favors random coil structure, but α-helix and β-sheet conformations still exist. Interestingly, the MD simulations indicate that helix/β-strand mixed conformations exist in the trajectory of the $A\beta_{40}$ conformational transition. This result is consistent with the experimental results by Soto *et al.* (16) and Kirkitadze *et al.* (11). Notably, all of the five trajectories pointed out that $V_{24}G_{25}$, $K_{28}G_{29}$, $I_{32}G_{33}$, and $V_{36}G_{37}$ form short β-strands, indicating that four glycines (G_{25}, G_{29}, G_{33}, and G_{37}) are important for $A\beta_{40}$ to form β-sheet in aqueous solution.

In another 100-ns MD simulation (simulation A6), the temperature was increased to 400 K, which accelerated the process of $A\beta_{40}$ structure transition. The result is shown in Fig. 2. Similar to simulations A1–A5, β-sheet consisting of three or four β-strands near the C terminus of $A\beta_{40}$ (residues 24–37) was observed in this simulation (Fig. 2*a*). The β-sheet lasted to ≈38 ns, and then a short α-helix was formed instead of the β-sheet in the following 10 ns. After ≈50 ns, the secondary structure of residues 24–37, which adopts a β-sheet structure in trajectories A1–A5, was destroyed, and the entire $A\beta_{40}$ peptide mainly adopted random coil structure throughout the rest of the simulation time. However, there are a few short-lived β-strands that appeared in the different positions of the peptide from 50 ns to 100 ns. Different from simulations A1–A5, the helical structure of residues 16–21 lasted only ≈13 ns (Fig. 2*a*). Remarkably, this simulation captured the whole process of $A\beta_{40}$ conformational transition from α-helix to random coil through helix/β-sheet mixed conformations (Fig. 2). These explored conformational changes provide a significant interpretation for the recent result of the kinetic study of $A\beta$ fibrillogenesis by Kirkitadze *et al.* (11), suggesting that helix/β-sheet mixed conformations are possible intermediates for $A\beta$ oligomerization.

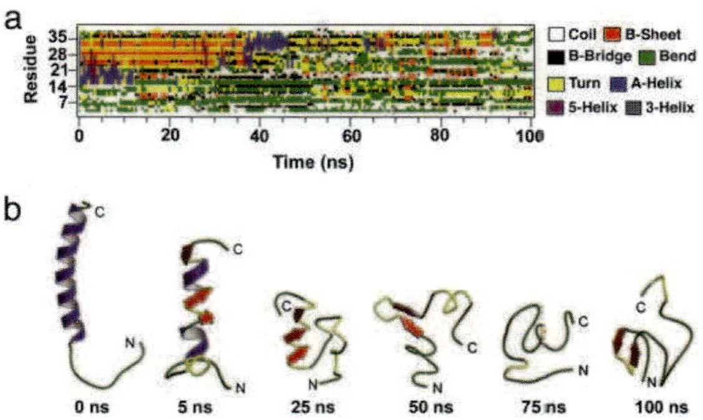

Fig. 2. Structural analyses of $A\beta_{40}$ in trajectory A6. (*a*) Secondary structures as function of time for $A\beta_{40}$ in trajectory A6 as calculated by DSSP. The structures were analyzed every 500 ps. (*b*) The snapshot structures of $A\beta_{40}$ extracted from trajectory A6, in which the secondary structure is according to DSSP analysis. C and N denote C-terminal and N-terminal, respectively.

Mutation of Four Glycines. To illustrate the importance of G_{25}, G_{29}, G_{33}, and G_{37} in the β-sheet formation, four MD simulations (B1–B4) were performed on a mutant of $A\beta_{40}$ ($MA\beta_{40}$, model II), in which the four glycines were mutated to alanines, for alanine is an appropriate helix former in aqueous solution. The simulation results are shown in Figs. 3 and 4. Fig. 3*a* exhibits the time dependence of secondary structure altered for $MA\beta_{40}$ in trajectory B1, and those in trajectories B2 and B3 are shown in Fig. 6. Unlike the wild-type $A\beta_{40}$, ≈53% residues of $MA\beta_{40}$ adopt helical structures in aqueous solution (Table 2); the main components of the helical structures are α-helix and 5-helix (Fig. 3). It is noteworthy that, throughout the three 50-ns simulations, no β-sheet structure was observed for residues 24–37 of the $MA\beta_{40}$. MD simulation (B4) at 400 K for $MA\beta_{40}$ revealed a similar result to that of simulations B1–B3 (Fig. 4): the major component of secondary structure is helix; residues 15–20 and 24–34 adopted helix structures throughout the simulation; long lime durable β-sheet was not observed although a few β-stands were formed sporadically near the N terminus (Fig. 4*a*). We even continued the simulation for $MA\beta_{40}$ at 400 K to 100 ns, where

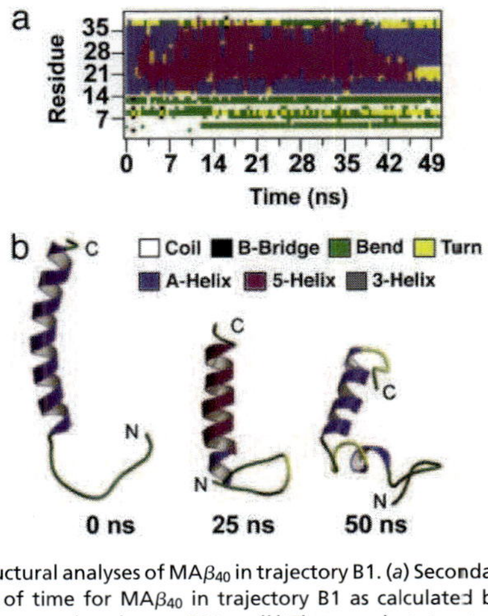

Fig. 3. Structural analyses of $MA\beta_{40}$ in trajectory B1. (*a*) Secondary structures as function of time for $MA\beta_{40}$ in trajectory B1 as calculated by DSSP. The structures were analyzed every 500 ps. (*b*) The snapshot structures of $MA\beta_{40}$ extracted from trajectory B1, in which the secondary structure is according to DSSP analysis. C and N denote C-terminal and N-terminal, respectively.

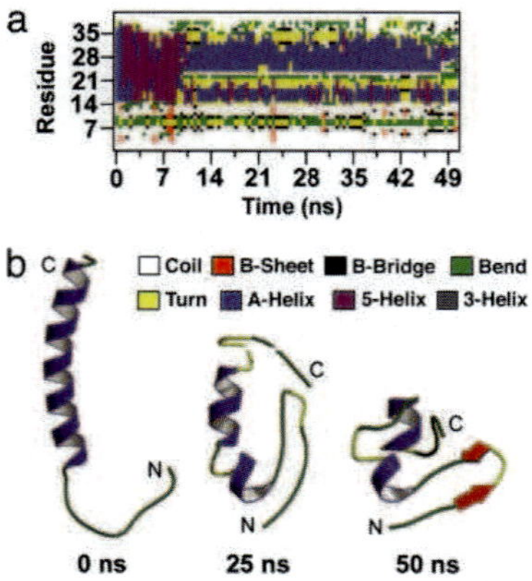

Fig. 4. Structural analyses of $MA\beta_{40}$ in trajectory B4. (*a*) Secondary structures as function of time for $MA\beta_{40}$ in trajectory B4 as calculated by DSSP. The structures were analyzed every 500 ps. (*b*) The snapshot structures of $MA\beta_{40}$ extracted from trajectory B4, in which the secondary structure is according to DSSP analysis. C and N denote C-terminal and N-terminal, respectively.

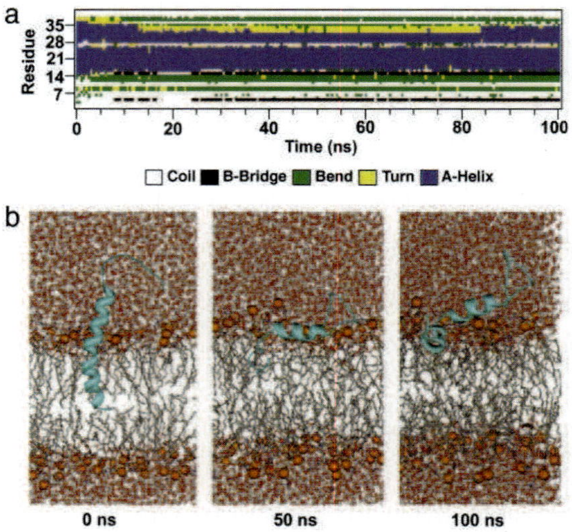

Fig. 5. Structural analyses of $A\beta_{40}$ in trajectory C1. (*a*) Secondary structures as function of time for $A\beta_{40}$ in trajectory C1 as calculated by DSSP. The structures were analyzed every 500 ps. (*b*) The snapshot structures of $A\beta_{40}$ extracted from trajectory C1, in which the peptide is drawn as ribbon, phosphate atoms of lipid are shown as brown balls, the other atoms of the lipid are labeled as sticks, water molecules are displayed as balls and sticks, and the front half of the bilayer is not shown for the sake of clarity.

the characterized β-sheets formed in the wild-type peptide were not observed (Fig. 6*B4*). All these results indicated that the mutations stabilized the helix structure of $A\beta_{40}$ and illustrated that the four glycines exerted significant control over the process in the formation of β-sheet.

Inserting $A\beta_{40}$ into DPPC Bilayers. $A\beta$ is a cleavage product of the single transmembrane protein APP. β-secretase cleaves the extracellular domain of APP to generate the N terminus of $A\beta$, and then γ-secretase performs an unusual proteolysis in the middle of the transmembrane domain of APP to produce the C terminus of $A\beta$ (58). Residues 29–40 of $A\beta_{40}$ are from the transmembrane domain of APP. This special secreted pathway of $A\beta$ raises several questions. What happens to $A\beta$ at the moment when it is completely cut from APP? How does $A\beta_{40}$ undergo conformational transition in the membrane environment? To answer these questions, model III was designed to examine the conformational dynamics of $A\beta_{40}$ in the membrane through inserting residues 29–40 of $A\beta_{40}$ into a DPPC lipid bilayer.

Two simulations (C1 and C2) were performed on model III at 300 K. Different sizes of the DPPC bilayer were applied for these two simulations, respectively: 127 DPPC lipids for simulation C1 and 63 DPPC lipids for simulation C2. The profile of secondary structure change of $A\beta_{40}$ (DSSP plot) and several snapshot structures extracted from trajectory C1 are shown in Fig. 5. The DSSP plot illustrates that residues 15–35 kept α-helical structure during the 100-ns simulation except that residues 32–35 were once switched to turn structure from 14 to 85 ns but recover to helical structure after 85 ns. The rest of the residues of $A\beta_{40}$ other than residues 15–35 mainly adopted coil or bend structures. During the 100-ns MD simulation in the DPPC bilayer, β-sheet structures were not observed, helical structure dominated the secondary structure, and $\approx$41% of the residues adopted helical conformation (Table 2). In addition, snapshot structures demonstrate that, during the simulation, $A\beta_{40}$ moves to the interface between DPPC lipids and water molecules, and, after $\approx$10 ns, part of $A\beta_{40}$ lies down on the surface of the lipid bilayer. As the simulation goes on, $A\beta_{40}$ tends to totally exit the lipid bilayer (Fig. 5). Similar results were obtained from simulation C2 (Fig. 6).

Discussion

The conformational transitions of $A\beta_{40}$ associated with $A\beta$ aggregation and neurotoxicity were studied by using MD simulation method. Detailed analyses were performed for the conformational transition of $A\beta_{40}$ based on 12 MD trajectories. The results demonstrate that $A\beta_{40}$ adopts different conformations in aqueous solution and in the DPPC bilayer, which are in agreement with the experimental results (14–22). Six MD simulations (A1–A6) with different initial velocities assigned to the atoms and different temperatures consistently produced the same result that conformational transition of $A\beta_{40}$ from α-helix to β-sheet occurs in aqueous solution (Figs. 1 and 2). Our MD simulations captured the whole process of $A\beta_{40}$ conformational transition from α-helix to coil structure through an intermediate β-sheet structure (Fig. 2). In addition, the MD simulations revealed that a segment of residues 24–37 is the core of $A\beta_{40}$ to form β-sheet. In particular, four glycines (G_{25}, G_{29}, G_{33}, and G_{37}), which normally destabilize helical structure, are essential to the β-sheet formation. Substitution of these four glycines of $A\beta_{40}$ by alanines increases the stability of the helical structure. β-sheets were not observed for residues 24–37 in another three MD simulations (B1–B3) on the $A\beta_{40}$ mutant ($MA\beta_{40}$), and the dominant secondary structure of $MA\beta_{40}$ remained as helix (Figs. 3 and 4). This finding indicates that the sequence of $A\beta_{40}$ is specific to forming β-sheet in aqueous solution. This conclusion is in agreement with the opinion of de la Paz and Serrano (59), who found that amyloid fibril formation depends on the peptide sequences based on positional scanning mutagenesis on a designed amyloid peptide.

In the DPPC bilayer, helix dominates the secondary structure of $A\beta_{40}$, and no β-strand was observed in the two 100-ns MD simulations (C1 and C2) (Fig. 5*a* and Table 2). This result agrees well with the NMR determination of $A\beta_{40}$ in membrane-mimicking environment (15, 17, 19). In addition, the segment of $A\beta_{40}$ outside the membrane (residues 1–28) bends toward the surface of the DPPC bilayer, forming electrostatic and van der

Waals interactions to the polar heads of the lipids (Fig. 7, which is published as supporting information on the PNAS web site). These interactions facilitate $A\beta_{40}$ to completely leave the lipids, moving toward the surface of the DPPC bilayer (Fig. 5*b*). Even at the interface between aqueous solution and the lipid bilayer, $A\beta_{40}$ still keeps helix as its dominant secondary structure.

Abundant experimental data indicated that the major component of the plaques in AD patient's brain is β-peptides, which aggregate into amyloid fibrils with cross-β-structure (2, 23, 24). An essential question concerning the plaques is the origin of the amyloid fibrils formation. Taken together, 12 long-time MD simulations provide a possible mechanism of amyloid fibrils aggregation. Cleaved by β- and γ-secretases, $A\beta$ peptides were released from APP (58). Firstly, residues 29–40 of $A\beta_{40}$ are embedded into the cellular membrane; this environment is beneficial to stabilizing the helical structure of $A\beta$ (Fig. 5*a*). Our MD simulations (C1 and C2) indicate that $A\beta$ tends to leave the pure DPPC lipid bilayer (Fig. 5*b*). In fact, the concentration of cholesterol in the cytofacial leaflet of brain synaptic plasma membrane is lower in aged people (60), and $A\beta$ cannot insert into the membrane containing less cholesterol (61). Once $A\beta$ exits the membrane and enters into the extracellular solution, the segment of residues 29–40 has a high tendency to form short β-sheets (Figs. 1 and 2). Accordingly, we suggest that the short β-sheet formation of residues 29–40 is a potential driving force for $A\beta$s aggregating into amyloid fibrils, implying that small molecules or peptides that can bind tightly with this segment may be developed as inhibitory therapeutic agents.

Although it is clear from our MD simulations that amyloid fibrils formation of $A\beta_{40}$ is associated with its dynamic properties, a complete description of the mechanism of how $A\beta_{40}$ monomers aggregate together requires further analysis using both experimental and theoretical methods.

This work was supported by State Key Program of Basic Research of China Grant 2002CB512802; the Key Project for New Drug Research from the Chinese Academy of Sciences; 863 Hi-Tech Program Grants 2002AA233061, 2001AA235051, 2001AA235041, 2002AA104270, 2002AA233011, and 2003AA235030; National Natural Science Foundation of China Grants 20372069, 29725203, and 20072042; and Shanghai Science and Technology Commission Grant 02DJ14006 J.M. was partially supported by National Science Foundation CAREER Award MCB-0237796 (U.S.A.); he is also a recipient of the Award for Distinguished Young Scholars Abroad from the National Natural Science Foundation of China.

1. Selkoe, D. J. (1991) *Neuron* **6,** 487–498.
2. Miller, D. L., Papayannopoulos, I. A., Styles, J., Bobin, S. A., Lin, Y. Y., Biemann, K. & Iqbal, K. (1993) *Arch. Biochem. Biophys.* **301,** 41–52.
3. Kang, J., Lemaire, H. G., Unterbeck, A., Salbaum, J. M., Masters, C. L., Grzeschik, K. H., Multhaup, G., Beyreuthe, K. & Muller-Hill, B. (1987) *Nature* **325,** 733–736.
4. Mattson, M. P. (1997) *Physiol. Rev.* **77,** 1081–1132.
5. Selkoe, D. J. (2001) *Physiol. Rev.* **81,** 741–766.
6. Hardy, J. & Selkoe, D. J. (2002) *Science* **297,** 353–356.
7. McLean, C. A., Cherny, R. A., Fraser, F. W., Fuller, S. J., Smith, M. J., Beyreuther, K., Bush, A. I. & Masters, C. L. (1999) *Ann. Neurol.* **46,** 860–866.
8. Walsh, D. M., Klyubin, I., Fadeeva, J. V., Cullen, W. K., Anwyl, R., Wolfe, M. S., Rowan, M. J. & Selkoe, D. J. (2002) *Nature* **416,** 535–539.
9. Rosenblum, W. I. (2002) *Neurobiol. Aging* **23,** 225–230.
10. Hwang, W., Zhang, S. G., Kamm, R. D. & Karplus, M. (2004) *Proc. Natl. Acad. Sci. USA* **101,** 12916–12921.
11. Kirkitadze, M. D., Condron, M. M. & Teplow, D. B. (2001) *J. Mol. Biol.* **312,** 1103–1119.
12. Straub, J. E., Guevara, J., Huo, S. H. & Lee, J. P. (2002) *Acc. Chem. Res.* **35,** 473–481.
13. MA, B. Y. & Nussinov, R. (2002) *Proc. Natl. Acad. Sci. USA* **99,** 14126–14131.
14. Zhang, S., Iwata, K., Lachenmann, M. J., Peng, J. W., Li, S., Stimson, E. R., Lu, Y. A., Felix, A. M., Maggio, J. E. & Lee, J. P. (2000) *J. Struct. Biol.* **130,** 130–141.
15. Coles, M., Bicknell, W., Watson, A. A., Fairlie, D. P. & Craik, D. J. (1998) *Biochemistry* **37,** 11064–11077.
16. Soto, C., Castano, E. M., Frangione, B. & Inestrosa, N. C. (1995) *J. Biol. Chem.* **270,** 3063–3067.
17. Sticht, H., Bayer, P., Willbold, D., Dames, S., Hilbich, C., Beyreuthe, K., Frank, R. W. & Rosch, P. (1995) *Eur. J. Biochem.* **233,** 293–298.
18. Barrow, C. J., Yasuda, A., Kenny, P. T. M. & Zagorski, M. G. (1992) *J. Mol. Biol.* **225,** 1075–1093.
19. Shao, H. Y., Jao, S. C., Ma, K. & Zagorski, M. G. (1999) *J. Mol. Biol.* **285,** 755–773.
20. Serpell, L. (2000) *Biochim. Biophys. Acta* **1502,** 16–30.
21. Good, T. A. & Murphy, R. M. (1995) *Biochem. Biophys. Res. Commun.* **207,** 209–215.
22. Barrow, C. J. & Zagorski, M. G. (1991) *Science* **253,** 179–182.
23. Blake, C. & Serpell, L. (1996) *Structure* **4,** 989–998.
24. Sunde, M., Serpell, L. C., Bartlam, M., Fraser, P. E., Pepys, M. B. & Blake, C. C. F. (1997) *J. Mol. Biol.* **273,** 729–739.
25. Teplow, D. B. (1998) *Amyloid* **5,** 121–142.
26. Pike, C. J., Walencewicz-Wasserman, A. J., Kosmoski, J., Cribbs, D. H., Glabe, C. G. & Cotman, C. W. (1995) *J. Neurochem.* **64,** 253–265.
27. Walter, M. F., Mason, P. E. & Mason, R. P. (1997) *Biochem. Biophys. Res. Commun.* **233,** 760–764.
28. Lee, J. P., Stimson, E. R., Ghilardi, J. R., Mantyh, P. W., Lu, Y. A., Felix, A. M., Llanos, W., Behbin, A., Cummings, M., Van Criekinge, M., *et al.* (1995) *Biochemistry* **34,** 5191–5200.
29. Pan, K., Baldwin, M., Nguyen, J., Gasset, M., Serban, A., Groth, D., Mehlhorn, I., Huang, Z., Fletterick, R. J., Cohen, F. E. & Prusiner, S. B. (1993) *Proc. Natl. Acad. Sci. USA* **90,** 10962–10966.
30. Weinreb, P. H., Zhen, W., Poon, A. W., Conway, K. A. & Lansbury, P. T., Jr. (1996) *Biochemistry* **35,** 13709–13715.
31. Conway, K. A., Harper, J. D. & Lansbury, P. T., Jr. (1998) *Nat. Med.* **4,** 1318–1320.
32. Mager, P. P., Penke, B., Walter, R., Harkany, T. & Hartig, W. (2002) *Curr. Med. Chem.* **9,** 1763–1780.
33. Daura, X., Jaun, B., Seebach, D., van Gunsteren, W. F. & Mark, A. E. (1998) *J. Mol. Biol.* **280,** 925–932.
34. Duan, Y. & Kollman, P. A. (1998) *Science* **282,** 740–744.
35. Simmerling, C., Strockbine, B. & Roitberg, A. E. (2002) *J. Am. Chem. Soc.* **124,** 11258–11259.
36. Snow, C. D., Nguyen, H., Pande, V. S. & Gruebele, M. (2002) *Nature* **420,** 102–106.
37. Mayor, U., Guydosh, N. R., Johnson, C. M., Grossmann, J. G., Sato, S., Jas, G. S., Freund, S. M. V., Alonso, D. O. V., Daggett, V. & Fersht, A. R. (2003) *Nature* **421,** 863–867.
38. Zangi, R., de Vocht, M. L., Robillard, G. T. & Mark, A. E. (2002) *Biophys. J.* **83,** 112–124.
39. Massi, F. & Straub, J. E. (2001) *Biophys. J.* **81,** 697–709.
40. Mager, P. P. (1998) *Med. Res. Rev.* **18,** 403–430.
41. Klimov, D. K. & Thirumalai, D. (2003) *Structure* **11,** 295–307
42. Zanuy, D., Ma, B. Y. & Nussinov, R. (2003) *Biophys. J.* **84,** 1884–1894.
43. Berendsen, H. J. C., Postma, J. P. M., van Gunsteren, W. F. & Hermans, J. (1981) in *Intermolecular Forces*, ed. Pullman, B. (Reidel, Dordrecht, The Netherlands), pp. 331–342.
44. Berendsen, H. J. C., van der Spoel, D. & van Drunen, R. (1995) *Comput. Phys. Commun.* **91,** 43–56.
45. Lindahl, E., Hess, B. & van der Spoel, D. (2001) *J. Mol. Model.* **7,** 306–317.
46. van Gunsteren, W. F. & Berendsen, H. J. C. (1987) *Gromos-87 Manual* (Biomos, Groningen, The Netherlands).
47. Tieleman, D. P., Sansom, M. S. P. & Berendsen, H. J. C. (1999) *Biophys. J.* **76,** 40–49.
48. Berger, O., Edholm, O. & Jahnig, F. (1997) *Biophys. J.* **72,** 2002–2013.
49. Marrink, S. J., Berger, O., Tieleman, P. & Jahnig, F. (1998) *Biophys. J.* **74,** 931–943.
50. Tieleman, D. P., Berendsen, H. J. C. & Sansom, M. S. P. (1999) *Biophys. J.* **76,** 1757–1769.
51. Hess, B., Bekker, H., Berendsen, H. J. C. & Fraaije, J. G. E. M. (1997) *J. Comp. Chem.* **18,** 1463–1472.
52. Darden, T., York, D. & Pedersen, L. (1993) *J. Chem. Phys.* **98,** 10089–10092.
53. Essmann, U., Perera, L., Berkowitz, M. L., Darden, T., Lee, H. & Pedersen, L. G. (1995) *J. Chem. Phys.* **103,** 8577–8592.
54. Berendsen, H. J. C., Postma, J. P. M., van Gunsteren, W. F., DiNola, A. & Haak, J. R. (1984) *J. Chem. Phys.* **81,** 3684–3690.
55. Kabsch, W. & Sander, C. (1983) *Biopolymers* **22,** 2577–2637.
56. Kraulis, P. J. (1991) *J. Appl. Crystallogr.* **24,** 946–950.
57. Merrit, E. A. & Murphy, M. E. P. (1994) *Acta Crystallogr.* **D50,** 869–873.
58. Esler, W. P. & Wolfe, M. S. (2001) *Science* **293,** 1449–1454.
59. de la Paz, M. L. & Serrano, L. (2004) *Proc. Natl. Acad. Sci. USA* **101,** 87–92.
60. Igbavboa, U., Avdulov, N. A., Schroeder, F. & Wood, W. G. (1996) *J. Neurochem.* **66,** 1717–1725.
61. Ji, S. R., Wu, Y. & Sui, S. F. (2002) *J. Biol. Chem.* **277,** 6273–6279.

Chapter 20

Qinghua Wang

Biography

Dr. Qinghua Wang is a faculty member in the Department of Biochemistry and Molecular Biology at Baylor College of Medicine. In her doctoral research with Professor Alan R. Fersht at Cambridge University, she utilized a novel homology-based approach to improve the thermostability of the apical domain of molecular chaperonin GroEL, commonly known as the "minichaperone." Her structural studies on minichaperones and its more stable mutants have shed important new light on the nature of GroEL-substrate interactions and on the molecular forces that can substantially stabilize a protein.

In her postdoctoral research at Harvard University, Dr. Wang demonstrated that the fusion domain of influenza virus hemagglutinin is itself a necessary and sufficient membrane-fusion machinery. This discovery provided support for the existence of ancestral membrane-fusion proteins common to a wide spectrum of viral membrane fusion proteins including HIV envelope protein gp120/gp41 and Ebola virus glycoprotein. Her study at Harvard was supported by a postdoctoral fellowship from the Irvington Institute for Immunological Research.

Since joining the faculty at Baylor College of Medicine, Dr. Wang and her research group have made significant contributions to a better understanding of the structures and functions of influenza virus hemagglutinin. They were the first to report a set of crystal structures of influenza B virus hemagglutinin in different functional states. These long-sought structures provided the critical framework for further investigation of influenza B virus. More importantly, these new structures of the distant cousins of influenza A virus hemagglutinin allowed the identification of important regions conserved in both influenza A and B virus hemagglutinin for their common functions of membrane fusion and receptor binding.

Dr. Wang lives in Houston with her husband, Dr. Jianpeng Ma, and their son, Albert Ma.

William Lipscomb: We All Miss You!

I first met Bill Lipscomb, the Colonel, shortly after I joined Dr. Don Wiley's group at Harvard University as a postdoctoral fellow, in the summer of 2000. It was on the occasion of the Colonel's 80th birthday symposium. It was such a wonderful event that I enjoyed every moment of it ranging from the scientific talks given by the Colonel's past associates to the evening dinner during which the Colonel gave a scientific presentation reporting the intriguing results from the fruitful collaboration between him and Jianpeng. It was also at this symposium that I first met Jean and their daughter, Jenna.

In November 2001, Don tragically passed away in an accident. Everyone in the lab was so sad and lost. In March 2002, a memorial meeting was held to celebrate the life of Don with his family. It was reassuring for me to see the Colonel at the meeting. The Colonel, as Don's mentor, was exceptionally strong. He was able to talk about Don's early days with him with humor and laughter, making the statement that, "losing Don is like losing a son of mine." The Colonel's presence and speech at the memorial meeting had meant so much to my fellow labmates in Don's group that we were able to regroup and move forward in our scientific careers as Don would have wanted us to. Thanks, Colonel, for your extraordinary strength!

The last time that I saw the Colonel was at his residence on Garden Street in Cambridge, MA, on the occasion of his 90th birthday celebration in the winter of 2009. It was shortly after the announcement that two of his former associates, Drs. Thomas Steitz and Ada Yonath, were awarded the 2009 Nobel Prize in Chemistry. Both Tom and Ada were invited to this small informal gathering prior to the birthday celebration. It was so wonderful to witness the conversation between the Colonel and two of his Nobel associates and so lovely to see Ada again, who is a dear friend of mine. We also talked about the possibility that the Colonel may see more of his past affiliates becoming Nobel Laureates. Even our then four-year-old son loved the gathering, and has kept asking us to go back to Boston. Just as Jianpeng and I were planning a trip to visit the Colonel in the summer of 2011, the sad news of the Colonel's passing away arrived. While we mourn his death, we all know that the spirit of the Colonel will live on forever through so many of his past associates and his spiritual students! Colonel, we all miss you!

Structural basis for receptor specificity of influenza B virus hemagglutinin

Qinghua Wang*†, Xia Tian*, Xiaorui Chen‡, and Jianpeng Ma*‡§

*Department of Biochemistry and Molecular Biology, Baylor College of Medicine, One Baylor Plaza, BCM-125, Houston, TX 77030; ‡Graduate Program in Structural Computational Biology and Molecular Biophysics, Baylor College of Medicine, One Baylor Plaza, Houston, TX 77030; and §Department of Bioengineering, Rice University, 6100 Main Street, Houston, TX 77005

Communicated by William N. Lipscomb, Harvard University, Cambridge, MA, September 4, 2007 (received for review July 3, 2007)

Receptor-binding specificity of HA, the major surface glycoprotein of influenza virus, primarily determines the host ranges that the virus can infect. Influenza type B virus almost exclusively infects humans and contributes to the annual "flu" sickness. Here we report the structures of influenza B virus HA in complex with human and avian receptor analogs, respectively. These structures provide a structural basis for the different receptor-binding properties of influenza A and B virus HA molecules and for the ability of influenza B virus HA to distinguish human and avian receptors. The structure of influenza B virus HA with avian receptor analog also reveals how mutations in the region of residues 194 to 196, which are frequently observed in egg-adapted and naturally occurring variants, directly affect the receptor binding of the resultant virus strains. Furthermore, these structures of influenza B virus HA are compared with known structures of influenza A virus HAs, which suggests the role of the residue at 222 as a key and likely a universal determinant for the different binding modes of human receptor analogs by different HA molecules.

egg adaptation | sialic acid | cross-species transmission

Influenza virus, including type A and type B, remains a major cause of morbidity and mortality in humans. In a typical influenza infection, the first close contact with the host cell is the binding of the virus to its cellular receptors, terminal sialic acids of glycoproteins and glycolipids (1, 2). This binding is mediated by HA, a major surface glycoprotein of influenza A or B virus.

Influenza A virus infects a wide variety of hosts including humans, avians, equines, and swines. The HA molecules of influenza A viruses isolated from different hosts differ in their ability to recognize different cellular receptors in which the linkage between sialic acid and galactose of the carbohydrate chain is either $\alpha(2,3)$ or $\alpha(2,6)$. HA molecules of human viruses preferentially bind to $\alpha(2,6)$-linked receptors, those of avian and equine viruses bind to $\alpha(2,3)$-linked, and those of swine viruses bind to both (1–9). Thus, $\alpha(2,3)$-linked and $\alpha(2,6)$-linked receptors are also referred to as avian and human receptors, respectively. X-ray crystallographic studies of influenza A virus HAs have defined the receptor-binding site as a shallow depression at the top of the molecule, formed by a set of conserved residues (10–19). We now have a detailed understanding of the molecular determinants for the different receptor specificity of different subtypes of influenza A virus HAs, in particular for H1, H2, and H3 subtypes that infect humans (1–3, 10–28). For H2 and H3 HAs, it is predominantly determined by residues at 226 and 228 (Leu-226 and Ser-228 in human HAs and Gln-226 and Gly-228 in avian HAs, following the numbering convention for H3 HA) (5, 7, 11, 21, 29–31). However, H1 HAs are different in this regard: although human and avian H1 HAs recognize $\alpha(2,6)$-linked or $\alpha(2,3)$-linked receptors, respectively, they both have Gln-226 and Gly-228. Instead, residues at other sites appear to determine the receptor specificity (8, 16, 28, 32). For instance, mutations at residues 190 and 225 in human 1918 H1 HA completely changed the receptor specificities (28, 32). The different receptor specificity of HAs appears to be a major barrier for cross-species transmission (2). This justifies the ongoing global efforts to identify mutations on avian H5N1 HA that would render the virus the ability to rapidly infect humans (18, 25–27). Such knowledge will allow prompt actions to stop H5N1 pandemics as soon as a pandemic strain emerges.

Despite the obvious significance of influenza B virus in epidemiology (33), in sharp contrast to our extensive knowledge on the receptor-binding specificity of influenza A virus HAs, very little is known about influenza B virus HA from a structural standpoint (34). For example, it was unknown how the binding modes of influenza B virus HA with its cellular receptors differ from those of influenza A virus HAs, particularly in the light of the rather low sequence identity (≈25%) between them (35, 36). Moreover, it was unclear whether the HA of influenza B virus, which circulates exclusively in humans, preferentially binds to $\alpha(2,6)$-linked receptors as human influenza A virus HAs or binds equally well to $\alpha(2,3)$-linked and $\alpha(2,6)$-linked receptors (6, 37–41). This was further complicated by the probability of altered receptor specificity due to egg adaptation of cultured viruses (6, 40). For instance, influenza B virus HA tends to lose its glycosylation at residue 194 when grown in embryonated chicken eggs as a result of the mutations at residues 194 or 196 (42). Those mutant HAs apparently have an augmented affinity for $\alpha(2,3)$-linked receptors that are abundant in chicken eggs (40). However, it remained to be determined whether the mutations themselves, or the loss of glycosylation, is responsible for such changes in affinity.

To address these unresolved issues, we have determined the crystal structures of HA of the influenza B/HongKong/8/73 strain (hereafter referred to as influenza B/HK HA) in complex with $\alpha(2,3)$-linked and $\alpha(2,6)$-linked receptor analogs, respectively. These structures provide a detailed understanding of the interactions by which influenza B virus HA distinguishes human and avian receptors. To elucidate the structural basis for the differences in receptor specificity between influenza A and B virus HAs, the structures of influenza B/HK HA are compared with those of human influenza A virus H3 HA (X31), the most extensively studied HA molecule (10–14). Furthermore, by comparing the structure of influenza B/HK HA with a number of known structures of influenza A virus HAs complexed with $\alpha(2,6)$-linked human receptor analogs, we propose an important role of the residue at 222 as likely a universal determinant for the different binding modes of human receptor analogs by different types and subtypes of HA molecules.

Results

Receptor-Binding Site of Influenza B/HK HA. The complex structures of influenza B/HK HA with $\alpha(2,3)$-linked and $\alpha(2,6)$-linked

Author contributions: Q.W. designed research; Q.W., X.T., and X.C. performed research; Q.W. and J.M. analyzed data; and Q.W. wrote the paper.

The authors declare no conflict of interest.

Data deposition: The atomic coordinates have been deposited in the Protein Data Bank, www.pdb.org (PDB ID codes 2RFT and 2RFU).

†To whom correspondence should be addressed. E-mail: qinghuaw@bcm.tmc.edu.

Table 1. Crystallographic statistics of B/HK HA–receptor complex structures

	LSTa	LSTc
Data collection statistics		
Space group	*P*321	*P*321
Unit cell, Å	$a = b = 98.33$, $c = 135.99$	$a = b = 98.37$, $c = 135.90$
No. of unique reflections	19,145	19,705
Resolution, Å	13–2.7	20–2.7
Completeness, %	89.9 (87.6)	92.6 (89.9)
$I/\sigma I$	16.6 (2.1)	20.9 (2.6)
R_{merge}, %	4.7 (29.1)	5.3 (29.7)
Refinement statistics		
Resolution, Å	10–2.8	20–2.8
R/R_{free}, %	27.99/29.83	29.48/31.07
rms bond length, Å	0.008	0.008
rms bond angle, °	1.48	1.52

Numbers in parentheses are for the highest-resolution shell.

receptor analogs were determined to 2.8-Å resolution with one subunit per asymmetric unit (Table 1). The final model for each subunit contains HA1 1 to 342 and HA2 1 to 169. Despite the relatively low sequence identity between them, ≈20% for HA1 and 30% for HA2, influenza B/HK HA shares an overall similar fold to influenza A virus HAs (1, 2, 10–19). It has an elongated membrane-proximal domain and a globular membrane-distal domain (Fig. 1*a*). The biological unit of HA is a homotrimer of HA1/HA2 subunits (Fig. 1*a*). The receptor-binding site is located at the top of the membrane-distal domain (Fig. 1*a*) formed by 190-helix (HA1 193 to 202) at the top, 240-loop (HA1 237 to 242) as the left edge, and 140-loop (HA1 136 to 143) as the right edge (Fig. 1*b*). Four residues from the HA1 chain, Phe-95, Trp-158, His-191, and Tyr-202, constitute the base of the receptor-binding site, all of which are absolutely conserved among all known sequences of influenza B virus HAs (35, 36, 43–50). In addition to π-stacking interactions among these four aromatic residues, the receptor-binding site is further stabilized by a hydrogen bond between His-191 and Tyr-202 and four more hydrogen bonds between Asp-193 and Ser-240 (Fig. 1*b*). The latter interactions make the left entrance of the binding site narrower by 0.3 Å (Fig. 1*b*) than that of human H3 HA (between Glu-190 and Ser-228).

One unique feature of the receptor-binding site of influenza B virus HA is a phenylalanine residue at position 95, which is in sharp contrast to a tyrosine residue for all influenza A virus HAs at equivalent position (Fig. 1*c*). This difference could be at least partially responsible for the different shape of the receptor-binding site and the different orientation of Trp-158 of influenza B virus HAs (Fig. 1*c*). Moreover, this subsite of B/HK HA binding site is further reshaped by the one-residue deletion after His-191 and a substitution of a proline residue by Ser-192 (41) (Fig. 1*c*). The structural difference of this subsite can be appreciated by a shift of 3.6 Å between the Cα atoms of Asp-193 in B/HK HA and its equivalent residue, Ser-186, in H3 HA (Fig. 1*c*).

In the region of 190-helix, the receptor-binding sites of B/HK HA and H3 HA are very similar. As previously suggested based on primary sequence alignment (41), the right edge of B/HK HA binding site is also very similar to that of human H3 HA and provides the conserved main-chain and side-chain atoms for interactions with the sialic acid moiety (Sia-1) of the receptors. These atoms include the main-chain carbonyl of residue Thr-139, the side-chain hydroxyl of Ser-140, and the main-chain amide and carbonyl of Gly-141 (equivalent to Gly-135, Ser-136, and Asn-137 in H3 HA, respectively). However, despite the conserved interactions, the residue Ser-140 of B/HK HA is located

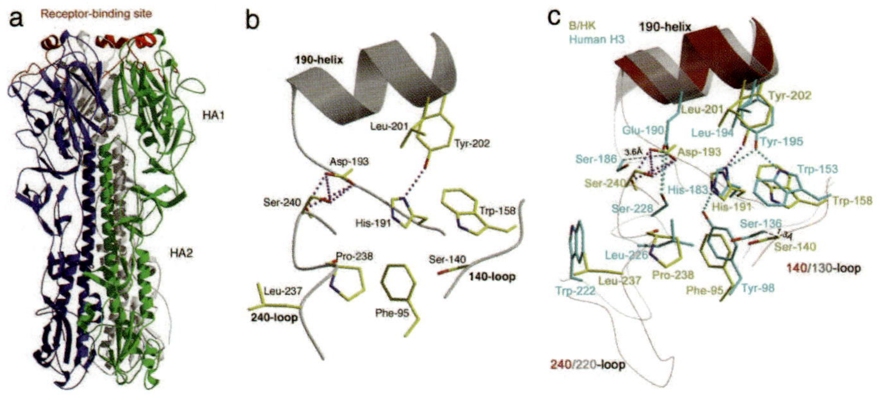

Fig. 1. Structure and receptor-binding site of influenza B/HK HA. (*a*) The overall structure of B/HK HA with three subunits differently colored. There is one receptor-binding site in each subunit at the distal end of molecule (away from the membrane), highlighted in red for clarity. (*b*) Close-up view of the receptor-binding site of B/HK HA. Important residues forming the receptor-binding site are explicitly drawn, with hydrogen-bonding interactions among them shown as dashed lines. (*c*) Comparison of the receptor-binding sites of B/HK (red) and human H3 (gray) HAs. Their corresponding residues are shown in yellow and cyan, respectively. Two large structural shifts between them, at residues Ser-140 and Asp-193, respectively, are indicated by black dashed lines with the distances labeled. The hydrogen-bonding interactions are shown as dashed lines.

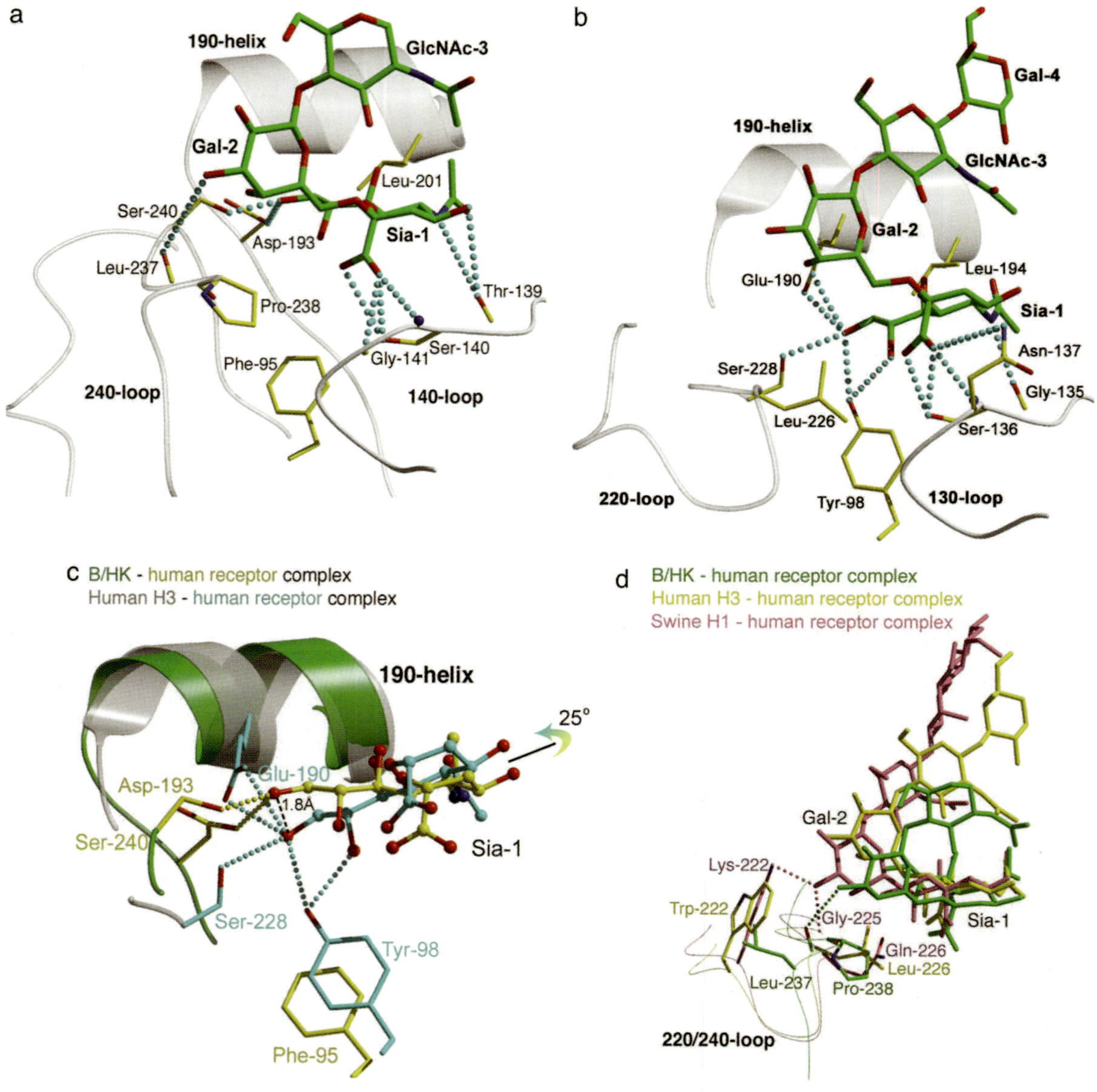

Fig. 2. Human receptor complexes. (*a*) The interaction of human receptor analogs in the receptor-binding site of B/HK HA. The hydrogen-bonding interactions are shown as dashed lines in cyan. The asialo portion of the receptors refers to the sugar rings other than Sia-1. (*b*) The interaction of human receptor analogs in the receptor-binding site of human H3 HA. The hydrogen-bonding interactions are shown as dashed lines in cyan. It is clear that the receptor-binding site of human H3 HA has an overall stronger interaction network with the bound human receptor analog than B/HK HA does. (*c*) Comparison of interactions of B/HK and human H3 HAs with the Sia-1 moiety. The ≈25° tilting of the Sia-1 ring around the C5–O6 axis is indicated by an arrow. The up-shift of the glycerol moiety of the Sia-1 in B/HK HA is also labeled. (*d*) The superposition of B/HK (green), human H3 (yellow), and swine H1 (purple) HAs in complex with human receptor analogs. It is clear that the interactions of Gal-2 ring with the residue at 222 affect the conformations of bound human analogs.

by ≈1.3 Å further away from the 240-loop than that of human H3 HA (Fig. 1*c*).

The left edge of the B/HK HA receptor-binding site harbors residues Pro-238 and Ser-240, which are at equivalent positions to residues 226 and 228 in H3 HA, respectively. Both residues 226 and 228 of H3 HA have been implicated as major determinants for receptor specificity (1, 2). Because of a four-residue insertion (residues 233 to 236) proceeding Leu-237 in influenza B/HK HA, this part of the receptor-binding site is markedly different from that of human H3 HA (Fig. 1*c*). The four-residue expansion results in much looser packing between the 240-loop and its proceeding loop (residues 230 to 236). Thus, to compensate for the interactions provided by Trp-222 in human H3 HA (see later for more discussion of its functional role), the side chain of Leu-237 in B/HK HA is flipped over to be located similarly to Trp-222 (Fig. 1*c*). The highly conserved residue Ser-240 is located by ≈3.6 Å higher than its equivalent residue, Ser-228, in human H3 HA. Although the side chain of Pro-238

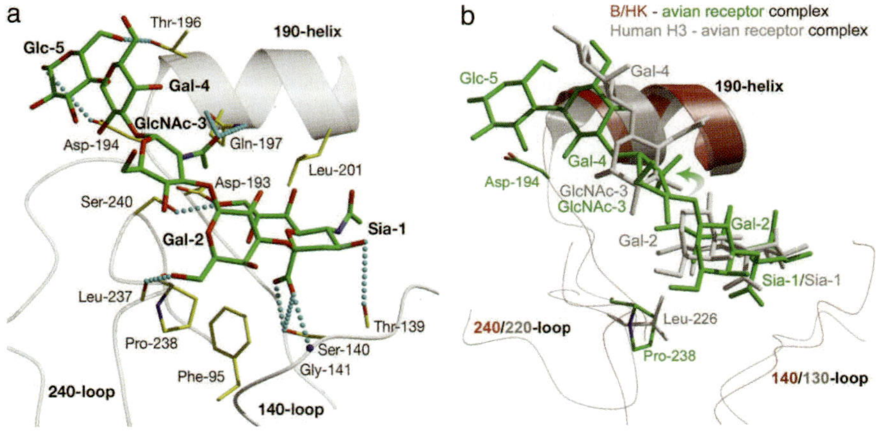

Fig. 3. Avian receptor complexes. (*a*) The interaction of avian receptor analogs in the receptor-binding site of B/HK HA. The hydrogen-bonding interactions are shown as dashed lines in cyan. (*b*) Comparison of interactions of B/HK and human H3 HAs with avian receptor analogs. The ≈60° rotation of the GlcNAc-3 ring of the avian receptor analog in the receptor-binding site of B/HK HA is indicated by an arrow.

is shorter than that of Leu-226 in H3 HA, the main chains of the 240-loop in B/HK HA are shifted toward the 140-loop so that the side-chain atoms responsible for interacting with the receptors are similarly located. Together with the ≈1.3-Å shift of the 140-loop in B/HK HA, the lower opening of the B/HK HA binding site (between Ser-140 and Pro-238) is ≈1.4 Å wider than that of human H3 HA (between Ser-136 and Leu-226). These structural differences between the receptor-binding sites of influenza B virus and human H3 HAs may account for the observed difference in receptor affinity and specificity between influenza A and B viruses (41).

Binding of Human Receptor Analog by Influenza B Virus HA. $\alpha(2,6)$-linked sialopentasaccharide LSTc, lactoseries tetrasaccharide c, was used as a human receptor analog (14) to study the interactions with influenza B/HK HA (Fig. 2*a*). Between Sia-1 and B/HK HA, there are two hydrogen bonds made by the main-chain amide and carbonyl of Gly-141, two hydrogen bonds by the side-chain hydroxyl of Ser-140, and two hydrogen bonds by the main-chain carbonyl of Thr-139 (Fig. 2*a*). On the other side of Sia-1, each of the side chains of residues Asp-193 and Ser-240 contributes one hydrogen bond with the 9-hydroxyl group of Sia-1. Except in only one case for Gly-141 (44, 45), all of those residues that interact with the Sia-1 moiety are absolutely conserved among different field isolates of influenza B viruses (35, 36, 43–50), highlighting their importance in receptor binding for viral infection.

To compare, we show the human H3 HA–human receptor complex in Fig. 2*b* (11). In contrast to the total of six hydrogen bonds between Sia-1 of LSTc and 140-loop of B/HK HA, five hydrogen bonds are found between Sia-1 and 130-loop of human H3 HA, among which the new hydrogen bond between the side chain of Asn-137 and Sia-1 of LSTc is at a distance of 2.7 Å. Moreover, the glycerol moiety of Sia-1 makes more extensive interactions with the binding site of H3 HA. There are two hydrogen bonds with the glycerol moiety of Sia-1 by the hydroxyl of the tryosine residue (Tyr-98) and three additional hydrogen bonds by the side chains of Glu-190 and Ser-228 (Fig. 2*b*). The overall less extensive interactions between the Sia-1 moiety and B/HK HA provide an explanation for the experimental observation that influenza B viruses in general have lower binding affinity for Neu5Ac (representing only the Sia-1 moiety) than influenza A H1 and H3 viruses (41). It appears that the interactions with the receptor-binding site, in particular those made by Tyr-98, pull the ring of Sia-1 down deeper into the pocket in the case of influenza A HAs (Fig. 2*c*). In contrast, in B/HK HA where a Phe-95 is present, the glycerol moiety of Sia-1 exclusively hydrogen-bonds with the side chains of Asp-193 and Ser-240. The higher position of residue Ser-240 in the structure, in the absence of any hydrogen-bonding interaction from the base of the binding site, pulls up the glycerol side of Sia-1. Consequently, the glycerol moiety of Sia-1 in B/HK HA is ≈1.8 Å higher than that in human H3 HA, with an ≈25° tilting of the Sia-1 ring around the C5–O6 axis (Fig. 2*c*).

In addition to the Sia-1 moiety, Gal-2 and GlcNAc-3 of LSTc are clearly visible in the B/HK HA–LSTc complex (Fig. 2*a*). However, only fragmented density is observed for the terminal Gal-4 and Glc-5, which are presumably disordered. The majority of the interactions between LSTc and HA is made by the Sia-1 moiety, whereas the asialo portion of LSTc does not make significant contacts with B/HK HA, except for a relatively weak hydrogen bond between Gal-2 and the main-chain carbonyl of Leu-237 (at a distance of 3.9 Å) (Fig. 2*a*). The overall conformation of bound LSTc in B/HK HA is the most similar to that of LSTc in the complex with human H3 HA. Thus, if it follows the path of LSTc in human H3 HA complex (Fig. 2*b*), LSTc in B/HK HA would exit the receptor-binding site from the right-hand side of 190-helix (Fig. 2 *a* and *b*).

Binding of Avian Receptor Analog by Influenza B Virus HA. $\alpha(2,3)$-linked sialopentasaccharide LSTa, lactoseries tetrasaccharide a, was used as an avian receptor analog (14) to study the interactions with influenza B/HK HA (Fig. 3*a*). The full-length LSTa is clearly visible (Fig. 3*a*) in the complex structure. Because it is

Table 2. Comparison of residues interacting with human receptor analogs in different HA proteins

Strain	Residue at 226	Residue at 222	Conformation of LSTc
Swine 1930 H1	Gln	Lys	Extended
Human 1934 H1	Gln	Lys	Extended
Avian H3	Gln	Trp	Folded
Human H3	Leu	Trp	Folded
Swine H9	Leu	Leu	Folded
Human B/HK	Pro-238	Leu-237	Folded

sitting 0.4–0.8 Å higher in the binding site than that of LSTc, the Sia-1 moiety of LSTa makes two fewer hydrogen bonds with B/HK HA that are contributed by the main-chain carbonyl of Thr-139 and the main-chain amide of Gly-141, respectively (Fig. 2*a*). Besides the Sia-1 moiety, Gal-2, GlcNAc-3, and Glc-5 all interact with the B/HK HA receptor-binding site. Gal-2 forms a hydrogen bond with the main-chain carbonyl of Leu-237; GlcNAc-3 forms two hydrogen bonds with the side chain of Gln-197, and Glc-5 interacts with Asp-194 and Thr-196. In contrast to LSTc, LSTa exits the receptor-binding site from the N terminus of 190-helix (Fig. 3*a*).

The bound LSTa in B/HK HA is in a cis-conformation, most similar to that of LSTa in human H3 (Fig. 3*b*) and swine H9 HAs, and in sharp contrast to the trans-conformation of LSTa in avian H3 (17), avian H5 (15), and human 1934 H1 HAs (16). In the cis-conformation, the hydrophobic side chain of residue Pro-238 and the hydrogen bond between Gal-2 and main-chain carboxyl of Leu-237 play an important role, just as Leu-226 in human H3 and swine H9 HAs (11, 15). In the B/HK HA–LSTa complex, the GlcNAc-3 rotates around the Gal-2–GlcNAc-3 bond by ≈60° in relation to GlaNAc-3 of LSTa in human H3 HA. It is this rotation that directs the rest of LSTa toward the 190-helix and allows for the interaction of Glc-5 with Asp-194 and Thr-196 (Fig. 3*b*). As a result, LSTa fits more snugly in the receptor-binding site of B/HK HA. In sharp contrast, LSTa in human H3 HA moves away from the 190-helix at as early as GlcNAc-3 (Fig. 3*b*).

The significant contributions of the asialo portion of avian receptor to the interaction with influenza B virus HA are consistent with its substantially higher affinity for larger receptors than for Neu5Ac (41). In sharp contrast, the H1 and H3 subtype HAs have comparable affinity for larger receptors to that of free Neu5Ac (41), which agrees well with the less extensive interactions made by the asialo portion of LSTa with those HAs (11, 14, 16).

Discussion

One unique feature of the B/HK HA receptor-binding site is the presence of Phe-95 in place of tyrosine as ordinarily found in all known influenza A virus HAs. It is interesting to note that a Tyr-98→Phe mutation of human H3 HA abolished erythrocyte binding, and the mutant virus could not infect mutant MDCK cells with a reduced level of cell-surface sialic acid, underscoring the importance of the hydrogen bonds that Tyr-98 makes with His-183 and with receptors inside the receptor-binding site of H3 HA (24). However, in the case of influenza B/HK HA, although these hydrogen bonds are absent, the capability of the virus to bind to cell-surface receptors is not abolished. Given the close evolutionary relationship between influenza A and B virus HAs, it remains an open question as to whether Phe-95 in influenza B virus HA is a direct result of selective pressure to optimize the receptor-binding site or it is simply a random mutation that has survived in evolution. In other words, it needs to be determined whether the receptor-binding site of influenza B virus HA is still able to accommodate the extra hydroxyl group of a tyrosine residue at position 95. Further investigation is needed to help address this question.

Naturally influenza B virus has been detected only in humans. When adapted to growth in chicken eggs, changes in the properties of influenza B virus had been noticed (51). Egg adaptation frequently caused the loss of a glycosylation site at residues 194 to 196 (42) and affects both antigenicity (52–54) and receptor-binding properties of HA (40, 53). The egg-adapted influenza B virus variants, with the absence of the glycans at residue 194, tend to grow better in chicken eggs than mammalian cell propagated ones of the same origin (54). The B/HK HA–LSTa complex structure provides a highly plausible explanation for why the loss of the glycosylation site at residues 194 to 196, as in the case of influenza B/HK HA, is advantageous for egg-adapted influenza B virus (54). It is because the glycans attached to Asn-194 in wild-type influenza B virus HA severely interfere with the binding of α(2,3)-linked sialic acids that are abundant in chicken egg allantone (40), thereby hindering the propagation of influenza B virus. Thus, the growth environment of chicken eggs imposes a strong selective pressure on influenza B virus toward losing the glycans at residue 194, with concomitant enhanced affinity for α(2,3)-linked sialic acids (40). Interestingly, the gain or loss of a glycosylation site at residue 194 seems to be an effective tool used by naturally occurring influenza B virus to modify its antigenicity (55, 56). Similar strategy was also found by egg-adapted influenza A virus (2, 40).

Superposition of all known structures of HA complexed with human receptor analog LSTc on the Sia-1 moiety reveals two different conformations of the bound analog (Fig. 2*d*): a slightly more extended conformation observed in 1930 H1 and 1934 H1 HAs and a more folded conformation as seen in avian H3, human H3, swine H9, and B/HK HAs. In the more extended conformation, the Gal-2 moiety inserts deeply into the receptor-binding site and interacts with the main-chain carbonyl of residue 225 as well as with the side chain of residue Lys-222. In sharp contrast, the Gal-2 moiety in the more folded conformation is positioned by ≈1–2 Å higher in the binding site (16). To understand the structural basis for the observed conformational differences of LSTc, we compared the residues at position 226 and 222 that interact with LSTc in the complex structures (Table 2). Residue 226 has been implicated in determining the preferential binding of H3 HA with α(2,6)-linked or α(2,3)-linked receptor analogs: Leu-226 prefers α(2,6)-linked receptors, whereas Gln-226 prefers α(2,3)-linked receptors (2, 20, 21). However, despite the difference at residue 226, the α(2,6)-linked receptor LSTc binds to both avian and human H3 HAs in a folded conformation, pointing to an important role of Trp-222 that is common for both H3 HAs (Table 2). In agreement with this finding, swine 1930 H1 and human 1934 H1 HAs, both having the same residue as avian H3 HA at 226 (Gln-226) but different residues at 222 (Lys-222 for both H1 HAs), bind to human receptor analogs in a different, more extended conformation. Thus, it appears that residue 222 is likely a major determinant for the conformations of bound α(2,6)-linked receptor analogs. In particular, a hydrophilic residue such as lysine at 222 offers favorable hydrogen bonding interaction with Gal-2 to pull it to a lower position. In contrast, a hydrophobic residue at position 222, tryptophan in avian and human H3 HAs, or leucine in swine H9 and human B/HK HAs, tends to push Gal-2 up to a higher position, possibly to avoid unfavorable interactions between hydrophilic groups on Gal-2 and hydrophobic groups on residue 222 (Fig. 2*d*). In the case of B/HK HA, a weak hydrogen bond between the main-chain carbonyl of Leu-237 and Gal-2 (Fig. 2*a*) brings Gal-2 to a slightly lower position than its equivalence in human H3 and swine H9 HAs.

Materials and Methods

Influenza B/HK HA protein was digested with bromelain to remove the C-terminal transmembrane sequence of HA2 (57). Pure B/HK HA protein at 15 mg/ml was used for crystallization trials. Crystals of diamond shape were grown in 100 mM Pipes (pH 6.5) and 2.5 M ammonium sulfate at 18°C within a week to $100 \times 100 \times 50$ μm. Crystals were soaked in mother liquor containing 16 mM LSTa or LSTc (Sigma) for 30 min and then transferred to mother liquor containing 9.0% glycerol before they were flash-frozen in liquid nitrogen. The diffraction data were collected at Advanced Photon Source beamline BM14C. All data sets were processed by using the HKL2000 package (58). The structure of unbound B/HK HA was used as a phasing model (Q.W., F. Cheng, M. Lu, X.T., and J.M., unpublished work). Model building was carried out by using the program O (59), and structural refinement was carried out by using CNS (60) and REFMAC5 in CCP4 (61). Ten percent reflections of the native data set were set aside for calculating R_{free} factor. The quality of the model was analyzed by using PROCHECK in CCP4 (61). Structural alignment of B/HK HA with influenza A virus HA molecules was performed by using LSQKAB in CCP4 (61) superimposed at the 140-loop, 240-loop, 190-helix, and the regions that harbor the key residues His-191 and Trp-158 of the receptor-binding site. Figures were made by using MOLSCRIPT software (62) and rendered by using RASTER3D (63).

We are indebted to D. J. Stevens (MRC National Institute for Medical Research, London, U.K.) for providing protein samples and to D. C. Wiley, J. J. Skehel, and S. C. Harrison for guidance and stimulating discussions. We gratefully thank the staff members at Advanced Photon Source beamline BM14C for their wonderful technical assistance. J.M. received support from National Institutes of Health Grant GM067801 and Welch Foundation Grant Q-1512.

1. Wiley DC, Skehel JJ (1987) *Annu Rev Biochem* 56:365–394.
2. Skehel JJ, Wiley DC (2000) *Annu Rev Biochem* 69:531–569.
3. Rogers GN, Paulson JC (1983) *Virology* 127:361–373.
4. Rogers GN, D'Souza BL (1989) *Virology* 173:317–322.
5. Connor RJ, Kawaoka Y, Webster RG, Paulson JC (1994) *Virology* 205:17–23.
6. Gambaryan AS, Tuzikov AB, Piskarev VE, Yamnikova SS, Lvov DK, Robertson JS, Bovin NV, Matrosovich MN (1997) *Virology* 232:345–350.
7. Matrosovich M, Zhou N, Kawaoka Y, Webster R (1999) *J Virol* 73:1146–1155.
8. Matrosovich MN, Gambaryan AS, Teneberg S, Piskarev VE, Yamnikova SS, Lvov DK, Robertson JS, Karlsson KA (1997) *Virology* 233:224–234.
9. Ito T, Suzuki Y, Takada A, Kawamoto A, Otsuki K, Masuda H, Yamada M, Suzuki T, Kida H, Kawaoka Y (1997) *J Virol* 71:3357–3362.
10. Wilson IA, Skehel JJ, Wiley DC (1981) *Nature* 289:366–373.
11. Weis W, Brown JH, Cusack S, Paulson JC, Skehel JJ, Wiley DC (1988) *Nature* 333:426–431.
12. Watowich SJ, Skehel JJ, Wiley DC (1994) *Structure (London)* 2:719–731.
13. Sauter NK, Hanson JE, Glick GD, Brown JH, Crowther RL, Park SJ, Skehel JJ, Wiley DC (1992) *Biochemistry* 31:9609–9621.
14. Eisen MB, Sabesan S, Skehel JJ, Wiley DC (1997) *Virology* 232:19–31.
15. Ha Y, Stevens DJ, Skehel JJ, Wiley DC (2001) *Proc Natl Acad Sci USA* 98:11181–11186.
16. Gamblin SJ, Haire LF, Russell RJ, Stevens DJ, Xiao B, Ha Y, Vasisht N, Steinhauer DA, Daniels RS, Elliot A, *et al.* (2004) *Science* 303:1838–1842.
17. Ha Y, Stevens DJ, Skehel JJ, Wiley DC (2003) *Virology* 309:209–218.
18. Stevens J, Blixt O, Tumpey TM, Taubenberger JK, Paulson JC, Wilson IA (2006) *Science* 312:404–410.
19. Stevens J, Corper AL, Basler CF, Taubenberger JK, Palese P, Wilson IA (2004) *Science* 303:1866–1870.
20. Rogers GN, Daniels RS, Skehel JJ, Wiley DC, Wang XF, Higa HH, Paulson JC (1985) *J Biol Chem* 260:7362–7367.
21. Rogers GN, Paulson JC, Daniels RS, Skehel JJ, Wilson IA, Wiley DC (1983) *Nature* 304:76–78.
22. Stevens J, Blixt O, Glaser L, Taubenberger JK, Palese P, Paulson JC, Wilson IA (2006) *J Mol Biol* 355:1143–1155.
23. Kelm S, Paulson JC, Rose U, Brossmer R, Schmid W, Bandgar BP, Schreiner E, Hartmann M, Zbiral E (1992) *Eur J Biochem* 205:147–153.
24. Martin J, Wharton SA, Lin YP, Takemoto DK, Skehel JJ, Wiley DC, Steinhauer DA (1998) *Virology* 241:101–111.
25. Yamada S, Suzuki Y, Suzuki T, Le MQ, Nidom CA, Sakai-Tagawa Y, Muramoto Y, Ito M, Kiso M, Horimoto T, *et al.* (2006) *Nature* 444:378–382.
26. Auewarakul P, Suptawiwat O, Kongchanagul A, Sangma C, Suzuki Y, Ungchusak K, Louisirirotchanakul S, Lerdsamran H, Pooruk P, Thitithanyanont A, *et al.* (2007) *J Virol* 81:9950–9955.
27. Yang ZY, Wei CJ, Kong WP, Wu L, Xu L, Smith DF, Nabel GJ (2007) *Science* 317:825–828.
28. Glaser L, Stevens J, Zamarin D, Wilson IA, Garcia-Sastre A, Tumpey TM, Basler CF, Taubenberger JK, Palese P (2005) *J Virol* 79:11533–11536.
29. Nobusawa E, Aoyama T, Kato H, Suzuki Y, Tateno Y, Nakajima K (1991) *Virology* 182:475–485.
30. Naeve CW, Hinshaw VS, Webster RG (1984) *J Virol* 51:567–569.
31. Vines A, Wells K, Matrosovich M, Castrucci MR, Ito T, Kawaoka Y (1998) *J Virol* 72:7626–7631.
32. Tumpey TM, Maines TR, Van Hoeven N, Glaser L, Solorzano A, Pappas C, Cox NJ, Swayne DE, Palese P, Katz JM, Garcia-Sastre A (2007) *Science* 315:655–659.
33. Air GM, Compans RW (1983) in *Genetics of Influenza Viruses*, eds Palese P, Kingsbury DW (Springer, Vienna), pp 21–69.
34. Tung CS, Goodman JL, Lu H, Macken CA (2004) *J Gen Virol* 85:3249–3259.
35. Krystal M, Elliott RM, Benz EW, Jr, Young JF, Palese P (1982) *Proc Natl Acad Sci USA* 79:4800–4804.
36. Krystal M, Young JF, Palese P, Wilson IA, Skehel JJ, Wiley DC (1983) *Proc Natl Acad Sci USA* 80:4527–4531.
37. Suzuki Y, Nakao T, Ito T, Watanabe N, Toda Y, Xu G, Suzuki T, Kobayashi T, Kimura Y, Yamada A, *et al.* (1992) *Virology* 189:121–131.
38. Gambaryan AS, Matrosovich MN (1992) *J Virol Methods* 39:111–123.
39. Gambaryan AS, Piskarev VE, Yamskov IA, Sakharov AM, Tuzikov AB, Bovin NV, Nifant'ev NE, Matrosovich MN (1995) *FEBS Lett* 366:57–60.
40. Gambaryan AS, Robertson JS, Matrosovich MN (1999) *Virology* 258:232–239.
41. Matrosovich MN, Gambaryan AS, Tuzikov AB, Byramova NE, Mochalova LV, Golbraikh AA, Shenderovich MD, Finne J, Bovin NV (1993) *Virology* 196:111–121.
42. Robertson JS, Naeve CW, Webster RG, Bootman JS, Newman R, Schild GC (1985) *Virology* 143:166–174.
43. Air GM, Gibbs AJ, Laver WG, Webster RG (1990) *Proc Natl Acad Sci USA* 87:3884–3888.
44. Xu G, Suzuki T, Tahara H, Kiso M, Hasegawa A, Suzuki Y (1994) *J Biochem (Tokyo)* 115:202–207.
45. Xu G, Horiike G, Suzuki T, Miyamoto D, Kumihashi H, Suzuki Y (1996) *Biochem Biophys Res Commun* 224:815–818.
46. Rota PA, Hemphill ML, Whistler T, Regnery HL, Kendal AP (1992) *J Gen Virol* 73:2737–2742.
47. Nakagawa N, Nukuzuma S, Haratome S, Go S, Nakagawa T, Hayashi K (2002) *J Clin Microbiol* 40:3068–3070.
48. Muyanga J, Matsuzaki Y, Sugawara K, Kimura K, Mizuta K, Ndumba I, Muraki Y, Tsuchiya E, Hongo S, Kasolo FC, *et al.* (2001) *Arch Virol* 146:1667–1679.
49. Lindstrom SE, Hiromoto Y, Nishimura H, Saito T, Nerome R, Nerome K (1999) *J Virol* 73:4413–4426.
50. Shaw MW, Xu X, Li Y, Normand S, Ueki RT, Kunimoto GY, Hall H, Klimov A, Cox NJ, Subbarao K (2002) *Virology* 303:1–8.
51. Burnet FM, Bull DR (1943) *Aust J Exp Biol Med Sci* 21:55–69.
52. Schild GC, Oxford JS, de Jong JC, Webster RG (1983) *Nature* 303:706–709.
53. Robertson JS, Bootman JS, Newman R, Oxford JS, Daniels RS, Webster RG, Schild GC (1987) *Virology* 160:31–37.
54. Saito T, Nakaya Y, Suzuki T, Ito R, Saito H, Takao S, Sahara K, Odagiri T, Murata T, Usui T, *et al.* (2004) *J Med Virol* 74:336–343.
55. Nakagawa N, Kubota R, Maeda A, Nakagawa T, Okuno Y (2000) *J Clin Microbiol* 38:3467–3469.
56. Nakagawa N, Kubota R, Maeda A, Okuno Y (2004) *J Clin Microbiol* 42:3295–3297.
57. Dopheide TA, Ward CW (1981) *J Gen Virol* 52:367–370.
58. Otwinowski Z, Minor W (1997) *Methods Enzymol* 276:307–326.
59. Jones TA, Zou JY, Cowan SW, Kjeldgaard (1991) *Acta Crystallogr A* 47:110–119.
60. Brünger AT, Adams PD, Clore GM, DeLano WL, Gros P, Grosse-Kunstleve RW, Jiang JS, Kuszewski J, Nilges M, Pannu NS, *et al.* (1998) *Acta Crystallogr D* 54:905–921.
61. Collaborative Computational Project N (1994) *Acta Crystallogr D* 50:760–763.
62. Kraulis PJ (1991) *J Appl Crystallogr* 24:946–950.
63. Merritt EA, Murphy ME (1994) *Acta Crystallogr D* 50:869–873.

Portraits in Science
Professor William Nunn Lipscomb Jr.

WOLCOTT GIBBS
RVMFORD PROFESSOR
HARVARD VNIVERSITY
PATHFINDER IN AMERICAN CHEMISTRY

Structure and Function of
Chemistry and Biology

HTLV-1
gp21

There is no law that says
Things must get better
Structure and Function of Chemistry and Biology
Structure and Function of Chemistry and Biology

福如東海長流水

WOLCOTT GIBBS MEMORIAL LABORATORY

WOLCOTT GIBBS MEMORIAL LABORATORY